# Lecture Notes in Computer Science     664
Edited by G. Goos and J. Hartmanis

Advisory Board: W. Brauer   D. Gries   J. Stoer

M. Bezem  J.F. Groote (Eds.)

# Typed Lambda Calculi and Applications

International Conference
on Typed Lamda Calculi and Applications
TLCA '93
March, 16-18, 1993, Utrecht, The Netherlands
Proceedings

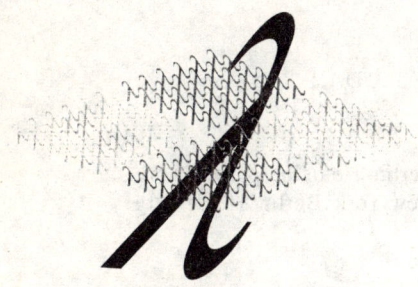

Springer-Verlag
Berlin Heidelberg New York
London Paris Tokyo
Hong Kong Barcelona
Budapest

Series Editors

Gerhard Goos
Universität Karlsruhe
Postfach 69 80
Vincenz-Priessnitz-Straße 1
W-7500 Karlsruhe, FRG

Juris Hartmanis
Cornell University
Department of Computer Science
4130 Upson Hall
Ithaca, NY 14853, USA

Volume Editors

Marc Bezem
Jan Friso Groote
Department of Philosophy, Utrecht University
Heidelberglaan 8, 3584 CS Utrecht, The Netherlands

CR Subject Classification (1991): F.4.1, F.3.0, D.1.1

ISBN 3-540-56517-5 Springer-Verlag Berlin Heidelberg New York
ISBN 0-387-56517-5 Springer-Verlag New York Berlin Heidelberg

This work is subject to copyright. All rights are reserved, whether the whole or part of the material is concerned, specifically the rights of translation, reprinting, re-use of illustrations, recitation, broadcasting, reproduction on microfilms or in any other way, and storage in data banks. Duplication of this publication or parts thereof is permitted only under the provisions of the German Copyright Law of September 9, 1965, in its current version, and permission for use must always be obtained from Springer-Verlag. Violations are liable for prosecution under the German Copyright Law.

© Springer-Verlag Berlin Heidelberg 1993
Printed in Germany

Typesetting: Camera ready by author/editor
Printing and binding: Druckhaus Beltz, Hemsbach/Bergstr.
45/3140-543210 - Printed on acid-free paper

# Preface

TLCA'93 is the first international conference on Typed Lambda Calculi and Applications. It has been organised by the Department of Philosophy of Utrecht University, with kind assistance of the Centre for Mathematics and Computer Science (CWI) in Amsterdam, both in The Netherlands. Thanks go to these organisations for providing a financial backup that enabled us to undertake the organisation.

The 51 papers submitted were mostly of very good quality. From these, 29 papers were selected for presentation at the conference. We are glad, and even a little proud, to be able to say that almost all leading researchers in the area of Typed Lambda Calculi contribute to these proceedings. We thank the program committee for their careful selection of the best among the submitted papers.

Given the current developments in the area of typed lambda calculi and their applications, we think and hope that this conference will mark the beginning of a tradition of conferences covering this field.

Utrecht, January 1993                                                    The Editors

## Program Committee

H. Barendregt, Nijmegen (chair)
T. Coquand, Chalmers
M. Dezani, Torino
G. Huet, Rocquencourt
M. Hyland, Cambridge

J.W. Klop, Amsterdam
J.C. Mitchell, Stanford
R. Nederpelt, Eindhoven
H. Schwichtenberg, München
R. Statman, Pittsburg

## Referees

M. Abadi
R.C. Backhouse
S.J. van Bakel
F. Barbanera
E. Barendsen
L.S. van Benthem Jutting
S. Berardi
U. Berger
I. Bethke
M. Bopp
V.A.J. Borghuis
W. Buchholz
F. Cardone
M. Coppo

W. Dekkers
G. Dowek
J.-P. van Draanen
P. Dybjer
H.M.M. ten Eikelder
H. Geuvers
P. Giannini
G. Gonthier
Ph. de Groote
C. Hemerik
C. Murthy
V. van Oostrom
Chr. Paulin-Mohring

B.C. Pierce
E. Poll
F. van Raamsdonk
D. Rémy
S. Ronchi
F. Rouaix
H. Schellinx
A. Setzer
W. Sieg
J. Smith
B. Venneri
R. de Vrijer
P. Weis

## Organising Committee

M. Bezem, Utrecht University
J.F. Groote, Utrecht University
F. Snijders, CWI, Amsterdam

# Table of Contents

Y. Akama
On Mints' reduction for ccc-calculus   1

Th. Altenkirch
A formalization of the strong normalization proof for system F in LEGO   13

S. van Bakel
Partial intersection type assignment in applicative
term rewriting systems   29

F. Barbanera, S. Berardi
Extracting constructive content from classical logic via
control-like reductions   45

F. Barbanera, M. Fernández
Combining first and higher order rewrite systems with
type assignment systems   60

N. Benton, G. Bierman, V. de Paiva, M. Hyland
A term calculus for intuitionistic linear logic   75

U. Berger
Program extraction from normalization proofs   91

G. Castagna, G. Ghelli, G. Longo
A semantics for $\lambda\&$-early: a calculus with overloading and early binding   107

P. Di Gianantonio, F. Honsell
An abstract notion of application   124

G. Dowek
The undecidability of typability in the lambda-pi-calculus   139

G. Ghelli
Recursive types are not conservative over $F_\leq$   146

Ph. de Groote
The conservation theorem revisited   163

J.M.E. Hyland, C.-H.L. Ong
Modified realizability toposes and strong normalization proofs   179

B. Jacobs
Semantics of lambda-I and of other substructure lambda calculi   195

B. Jacobs, T. Melham
Translating dependent type theory into higher order logic — 209

A. Jung, A. Stoughton
Studying the fully abstract model of PCF within its
continuous function model — 230

A. Jung, J. Tiuryn
A new characterization of lambda definability — 245

H. Leiß
Combining recursive and dynamic types — 258

D. Leivant, J.-Y. Marion
Lambda calculus characterizations of poly-time — 274

J. McKinna, R. Pollack
Pure type systems formalized — 289

T. Nipkow
Orthogonal higher-order rewrite systems are confluent — 306

D.F. Otth
Monotonic versus antimonotonic exponentiation — 318

Chr. Paulin-Mohring
Inductive definitions in the system Coq; rules and properties — 328

B.C. Pierce
Intersection types and bounded polymorphism — 346

G. Plotkin, M. Abadi
A logic for parametric polymorphism — 361

K. Sieber
Call-by-value and nondeterminism — 376

J. Springintveld
Lower and upper bounds for reductions of types in $\lambda\underline{\omega}$ and $\lambda P$ — 391

M. Takahashi
$\lambda$-calculi with conditional rules — 406

P. Urzyczyn
Type reconstruction in $F_\omega$ is undecidable — 418

Author Index — 433

# On Mints' Reduction for ccc-Calculus

Yohji Akama*

Department of Information Science, Tokyo Institute of Technology
Tokyo, 152, Japan (e-mail: akama@is.titech.ac.jp)

**Abstract.** In this paper, we present a divide-and-conquer lemma to infer the SN+CR (Strongly Normalization and Church-Rosser) property of a reduction system from that property of its subsystems. Then we apply the lemma to show the property of Mints' reduction for ccc-calculus with restricted $\eta$-expansion and restricted $\pi$-expansion. In the course of the proof, we obtain some relations of the two restricted expansions against traditional reductions. Among others, we get a simple characterization of the restricted $\eta$-expansion in terms of traditional $\beta$- and $\eta$-reductions, and a similar characterization for the restricted $\pi$-expansion.

## 1 Introduction

By ccc-calculus, we mean a deduction system for equations of typed $\lambda$-terms, essentially due to Lambek-Scott[9]. The types are generated from the type constant T and other atomic types by means of *implication* ($\varphi \supset \psi$) and *product* ($\varphi \times \psi$). We use $\varphi, \psi, \sigma, \tau, \ldots$ as meta-variables for types.

The terms of ccc-calculus are generated from the constant $*^T$ of type T and denumerable variables $x^\varphi, y^\varphi, z^\varphi, \ldots$ of each type $\varphi$ by means of *application* $(u^{\varphi \supset \psi} v^\varphi)^\psi$, *abstraction* $(\lambda x^\varphi. u^\psi)^{\varphi \supset \psi}$, *left-projection* $(\mathrm{l} u^{\varphi \times \psi})^\varphi$, *right-projection* $(\mathrm{r} u^{\varphi \times \psi})^\psi$, and *pairing* $\langle u^\varphi, v^\psi \rangle^{\varphi \times \psi}$. We use $s^\varphi, t^\varphi, u^\varphi, v^\varphi, \ldots$ as meta-variables for terms of each type $\varphi$.

This system is to deduce equations of the form $u^\varphi = v^\varphi$ and it consists of the usual equational axioms and rules(i.e., reflexive, symmetric, and transitive laws and substitution rules) and the following proper axioms[2]:

($\beta$) $(\lambda x. u)v = u[x := v]$, where $[x := v]$ is a substitution;
(l) $\mathrm{l} \langle u_\mathrm{l}, u_\mathrm{r} \rangle = u_\mathrm{l}$;
(r) $\mathrm{r} \langle u_\mathrm{l}, u_\mathrm{r} \rangle = u_\mathrm{r}$;
($\eta$) $\lambda x. ux = u$ if variable $x$ does not occur free in $u$;
($\pi$) $\langle \mathrm{l} u, \mathrm{r} u \rangle = u$;
(T) $u^\mathrm{T} = *^\mathrm{T}$.

---

* Supported by JSPS Fellowship for Japanese Junior Scientists
[2] In this paper, the type forming operator $\supset$ is right associative. And the type-superscripts of terms are often omitted. The term forming operators l and r have higher precedences than the application operator. As usual, we denote $\alpha$-congruence by $\equiv$. We refer to [1] as the standard text.

The proper axioms reflect the properties of cartesian closed category(CCC, for short), e.g. the axiom (T) reflects the property of the terminal object.

In relation to its decision problems, coherence problem and the like, it is desirable to have a reduction system to generate the calculus which is both Church-Rosser and (weakly or strongly) normalizable.

A naive idea to get such a system is to read each proper axiom $(X)$ as a rewriting rule $\to_X$(to rewrite the left-hand side to the right-hand side where $X = \beta, \mathrm{l}, \mathrm{r}, \eta, \pi$) and $\to_\mathrm{T}$(to rewrite $u^\mathrm{T}$ to $*^\mathrm{T}$ when $u \not\equiv *^\mathrm{T}$). But, unfortunately, the system so obtained is not Church-Rosser[9], although it is strongly normalizable.

So, Mints[10] introduced a new reduction system by replacing the rules $\to_\eta$ and $\to_\pi$ with a certain restriction of $\eta$-expansion(denoted by $\to_{\eta^*}$) and that of $\pi$-expansion(denoted by $\to_{\pi^*}$), respectively. Thus his reduction system consists of the basic reduction $\to_\mathrm{B} := \to_\beta \cup \to_\mathrm{l} \cup \to_\mathrm{r}$, the restricted expansion $\to_\mathrm{E} := \to_{\eta^*} \cup \to_{\pi^*}$, and the terminal reduction $\to_\mathrm{T}$. Then it is proved by Čubrić[3] that the reduction $\to_\mathrm{BET} := \to_\mathrm{B} \cup \to_\mathrm{E} \cup \to_\mathrm{T}$ is weakly normalizable and Church-Rosser. We call the reduction $\to_\mathrm{BET}$ Mints' reduction.

In literature, the restricted $\eta$-expansion $\to_{\eta^*}$ also arose from the study of program transformation. Hagiya[6] introduced the same notion $\to_{\eta^*}$ independently in order to study $\omega$-order unification modulo $\beta\eta$-equality[7] in simply-typed $\lambda$-calculus. A weaker version of $\to_{\eta^*}$ also appeared in Prawitz[11] in connection with proof theory.

The feature of $\to_{\eta^*}$, in addition to the strong normalization property, is that a term in $\to_{\eta^*}$-normal form explicitly reflects the structure of its type. For example, if a term $t^{\varphi_1 \supset \varphi_2 \supset \cdots \supset \varphi_n \supset \tau}$ is in $\to_{\eta^*}$-normal form then $t \equiv \lambda x_1 x_2 \cdots x_n . s$. The $\to_\mathrm{BE}$-normal form is known as expanded normal form[11] in proof theory.

Thus, the reductions $\to_{\eta^*}$ and $\to_{\pi^*}$ are useful reductions in category theory, proof theory, and computer science. Nevertheless these reductions have not been fully investigated, because of their context-sensitiveness and non-substitutivity.

The main theorem of this paper is:

- Mints' reduction $\to_\mathrm{BET}$ satisfies SN+CR property.

Based on the result, we also show that

- The reduction $\to_{\eta^*}$ is exactly an $\eta$-expansion which is not $\beta$-expansion.
- The reduction $\to_{\pi^*}$ is exactly a $\pi$-expansion which is not a finite series of $(\to_\mathrm{l} \cup \to_\mathrm{r})$-expansion.

Mints' reduction $\to_\mathrm{BET}$ inherits the above mentioned annoying properties of the reduction $\to_\mathrm{E} := \to_{\eta^*} \cup \to_{\pi^*}$. In proving that $\to_\mathrm{BET}$ satisfies SN+CR property, to overcome the annoyance, we will use the divide-and-conquer technique: We separate the annoying part $\to_\mathrm{E}$ from the rest $\to_\mathrm{BT} := \to_\mathrm{B} \cup \to_\mathrm{T}$ in the reduction $\to_\mathrm{BET}$, and apply the following lemma with $\to_R := \to_\mathrm{E}$ and $\to_S := \to_\mathrm{BT}$:

**Lemma.** If two binary relations $\to_R$ and $\to_S$ on a set $U (\neq \emptyset)$ have SN+CR property, then so does $\to_{SR} := \to_S \cup \to_R$, provided that we have

$$\forall u, v \in U \left( u \to_S v \implies u^R \xrightarrow{+}_S v^R \right),$$

where $u^R$ and $v^R$ are the $\to_R$-normal forms of $u$ and $v$ respectively, and $\stackrel{+}{\to}_S$ is the transitive closure of $\to_S$.

Thus our proof of the SN+CR property of $\to_{BET}$ amounts to verifying the same property of $\to_E$ and $\to_{BT}$ separately and verifying the following:

**Claim.** For any terms $u$ and $v$, if $u \to_{BT} v$ then $u^E \stackrel{+}{\to}_{BT} v^E$.

The SN+CR property of $\to_{BET}$ has been already proved by Jay[8] and independently by the author, by means of a modified version of Girard's reducibility method. Comparing the divide-and-conquer method with the reducibility method, the advantages of the former are summarized as follows:

- The claim above clarifies the relation of $\to_E$-normalization against the reduction $\to_{BT}$.
- The method is less tedious in that we need not prove the weak Church-Rosser property of $\to_{BET}$.

Besides, of independent interest might be the lemma above.

The basic idea of the lemma is found implicitly in Hagiya's proof [5] of the strong normalization property of $\to_\beta \cup \to_{\eta^*}$ for simply-typed $\lambda$-calculus.

The outline of this paper is as follows: In Sect. 2, the precise definition of Mints' reduction is given. In Sect. 3, the lemma is proved. In Sect. 4, $\to_E$ is proved to satisfy SN+CR property. In Sect. 5, at first the $\to_E$-normal form of an arbitrary term $t^\varphi$ is described by induction on the structure of $t^\varphi$ and $\varphi$. Then based on the description, we verify above mentioned claim. It then yields the property of Mints' reduction $\to_{BET}$ as above, since $\to_{BT}$ is known to satisfy the property by the standard argument. As a corollary to the main theorem, we give simple characterizations of $\to_{\eta^*}$ and $\to_{\pi^*}$. In Sect. 6, we discuss another proof method of our main theorem and relating topics.

## 2  Mints' Reduction

**Definition 1.** The subterm occurrence $u$ in a term $C[uv]$ is said to be *functional*, while the occurrence $u$ in $C[lu]$ or in $C[ru]$ is *projective*($C[\ ]$ stands for an arbitrary one-hole context).

We define binary relations $\to_{\eta^*}$ and $\to_{\pi^*}$ on terms.

**Definition 2 ($\eta^*$-reduction).** $C[u] \to_{\eta^*} C[\lambda z.\, uz] \stackrel{\text{def}}{\Longleftrightarrow}$

(0) $z$ does not occur free in $u$,
(1) The occurrence of $u$ in $C[u]$ is non-functional, and
(2) $u$ is not an abstraction.

In this case the occurrence of $u$ in $C[u]$ is called a $\eta^*$-*redex*.

**Definition 3** ($\pi^*$-reduction). $C[u] \to_{\pi^*} C[\langle lu, ru \rangle] \overset{\text{def}}{\iff}$

(1) The occurrence of $u$ in $C[u]$ is non-projective, and
(2) $u$ is not a pairing.

In this case the occurrence of $u$ in $C[u]$ is called a $\pi^*$-redex.

The union of reductions $\to_{\eta^*}$ and $\to_{\pi^*}$ is denoted by $\to_E$. We define *Mints'* reduction $\to_{\text{BET}}$ to be $\to_B \cup \to_E \cup \to_T$, where the reduction $\to_B$ is $\to_\beta \cup \to_l \cup \to_r$.

*Example 1.*
(1) When $y \to_{\eta^*} v \equiv \lambda z. yz$, we have $yu \not\to_{\eta^*} vu$ and $(\lambda x. w) \not\to_{\eta^*} \lambda z.(\lambda x. w)z$. Thus the reduction $\to_{\eta^*}$ is context-sensitive and non-substitutive.
(2) When $y$ is of type $(\varphi \supset \psi) \supset \gamma$ and $xy$ of type $\sigma \times \tau$, we have

$$
\begin{array}{ccc}
xy & \to_{\pi^*} & \langle l(xy), r(xy) \rangle \\
\downarrow_{\eta^*} & & \downarrow_{\eta^*} \\
& & \langle l(x(\lambda z. yz)), r(xy) \rangle \\
& & \downarrow_{\eta^*} \\
x(\lambda z. yz) & \to_{\pi^*} & \langle l(x(\lambda z. yz)), l(x(\lambda z. yz)) \rangle
\end{array}
$$

## 3 The Lemma

**Lemma 4.** If $\to_R$ and $\to_S$ are binary relations on a set $U(\neq \emptyset)$ and they have SN+CR property, then so does $\to_{SR} := \to_S \cup \to_R$, provided that we have

$$\forall a, b \in U \left( a \to_S b \implies a^R \overset{*}{\to}_S b^R \right),$$

where $a^R$ and $b^R$ are the $\to_R$-normal forms of $a$ and $b$ respectively, and $\overset{+}{\to}_S$ is the transitive closure of the relation $\to_S$.

*Proof.* We first prove the strong normalization property of the relation $\to_{SR}$ by contradiction.

Assume there exists an infinite $\to_{SR}$-reduction sequence:

(1) $\qquad a_0 \to_{SR} a_1 \to_{SR} a_2 \to_{SR} \cdots$ .

Then, (1) contains infinitely many $\to_S$-steps. Because otherwise there exists a terminal part of the reduction sequence (1) that is an infinite $\to_R$-reduction sequence. This contradicts the strong normalization property of $\to_R$.

In the sequence (1), if $a_i \to_S a_{i+1}$ then $a_i^R \overset{+}{\to}_S a_{i+1}^R$ because of the condition. If $a_i \to_R a_{i+1}$ then $a_i^R \equiv a_{i+1}^R$ by the Church-Rosser property of $\to_R$. Therefore, $a_0^R \overset{*}{\to}_S a_1^R \overset{*}{\to}_S a_2^R \overset{*}{\to}_S \cdots$ and it contains infinitely many $\overset{+}{\to}_S$-steps. This contradicts the strong normalization property of $\to_S$.

Next, we check the Church-Rosser property of $\to_{SR}$. If $b \overset{*}{_{SR}\leftarrow} a \overset{*}{\to}_{SR} c$ for $a, b, c \in U$, then we infer $b^R \overset{*}{_S\leftarrow} a^R \overset{*}{\to}_S c^R$ by the same manner as above, so there exists $e \in U$ such that $b^R \overset{*}{\to}_S e \overset{*}{_S\leftarrow} c^R$ by virtue of the Church-Rosser property of $\to_S$. Hence $b \overset{*}{\to}_R b^R \overset{*}{\to}_S e \overset{*}{_S\leftarrow} c^R \overset{*}{_R\leftarrow} c$. □

# 4 The SN+CR Property of $\to_E$

**Definition 5.** Let $u \to_E v$ by contracting an $X$-redex $U$ in $u$ ($X$ is either $\eta^*$, or $\pi^*$). We say $u \to_X v$ is *inner* if $U \not\equiv u$ (in this case we write $u \stackrel{\text{in}}{\to}_E v$).

**Lemma 6.** The reduction $\to_E$ is strongly normalizable. Hence so is $\stackrel{\text{in}}{\to}_E$.

*Proof.* The proof due to [10] is carried by assigning to each term $t$ a natural number $\#t$ so that

$$t \to_E t' \text{ implies } \#t > \#t' \text{ for any terms } t \text{ and } t'.$$

In order to define $\#t$, for each E-redex $R$ in $t$ let

$$d(R) = 2^{\sum |\varphi_i|}$$

where $\varphi_1, \ldots, \varphi_n$ are the types of E-redexes in $t$ which contain $R$, and each $|\varphi_i|$ represents the length[3] of the type $\varphi_i$. Then define

$$\#t = \sum d(R_j)$$

where $R_j$ are all (occurrences of) redexes in $t$. For details we refer to Čubrić [3]. □

In fact, we can prove that among $\to_E$-reduction sequences beginning with a term $t$, the maximal[4] leftmost one (denoted by $\mathcal{L}t$) is the longest. This is verified by giving an inductive description of $\mathcal{L}t$ and that of the length of $\mathcal{L}t$.

**Lemma 7.** The reductions $\to_E$ and $\stackrel{\text{in}}{\to}_E$ are weakly Church-Rosser.

*Proof.* In the usual manner. □

**Theorem 8.** The reductions $\to_E$ and $\stackrel{\text{in}}{\to}_E$ have SN+CR property.

*Proof.* By virtue of Newman's lemma combined with Lemmas 6 and 7. □

# 5 The SN+CR Property of $\to_{BET}$

## 5.1 Describing the $\to_E$-Normal Forms

We will describe the $\to_E$-normal form of a term $t^\varphi$ by induction on the structure of $t^\varphi$ and $\varphi$.

**Definition 9.** A term is called *designating* if it is of the form $\lambda x.\, u$, $\langle u, v \rangle$, or $*$. A non-designating term (i.e., that of the form $x, uv, \text{l}u$, or $\text{r}u$) is called *neutral*.

---
[3] the number of occurrences of atomic types and type forming operators $\supset$ and $\times$.
[4] i.e. infinite or ending with a normal form.

**Fact 1.** Any neutral term is of the form:

$$\overline{f_k}\left(\cdots \left(\overline{f_1}\,s\,\overline{u_1}\right)\cdots\right)\overline{u_k}$$

where $k \geq 0$; $\overline{f_i}(1 \leq i \leq k)$ is a finite sequence of l or r ; $\overline{u_i}(1 \leq i \leq k)$ is a finite sequence of terms; and $s$ is either an atom, a $\beta$-redex or a $f$-redex($f =$ l or r). Recall that the term forming operator l and r have a higher precedences than the application operator.

In other words, any neutral term is of the form:

$$\left(\overline{f_k}\left(\cdots\left(\left(\overline{f_1}\,s\right)\overline{u_1}\right)\cdots\right)\right)\overline{u_k} ,$$

where

$$s \equiv \begin{cases} a & \text{for some atom } a \text{ ; or} \\ (\lambda x.\,s_1)s_2 & \text{for some variable } x \text{ and terms } s_1 \text{ and } s_2 \text{ ; or} \\ g\langle s_1, s_2\rangle & \text{for some terms } s_1 \text{ and } s_2, \text{ and } g = \text{l or r .} \end{cases}$$

*Proof.* By induction on the structure of terms. □

**Lemma 10.**
(1) Every term $w^\varphi$ has a unique $\to_\mathrm{E}$-normal form and a unique $\xrightarrow{\text{in}}_\mathrm{E}$-normal form, which is denoted by $(w^\varphi)^\mathrm{E}$ and $(w^\varphi)^\mathrm{e}$, respectively.
(2) For each term $w^\varphi$, the terms $(w^\varphi)^\mathrm{E}$ and $(w^\varphi)^\mathrm{e}$ satisfy the relation:

$$(w^\varphi)^\mathrm{E} \equiv \begin{cases} A_\varphi[(w^\varphi)^\mathrm{e}] & \text{if } w^\varphi \text{ is neutral;} \\ (w^\varphi)^\mathrm{e} & \text{if } w^\varphi \text{ is designating,} \end{cases}$$

where, for each term $s^\varphi$ of type $\varphi$, a term $A_\varphi[s^\varphi]$ of the same type is defined as follows:

$$A_\varphi[s^\varphi] \equiv \begin{cases} s^\varphi & \text{if } \varphi \text{ is an atomic type;} \\ \langle A_\sigma[\mathrm{l}s^\varphi],\, A_\tau[\mathrm{r}s^\varphi]\rangle & \text{if } \varphi = \sigma \times \tau \text{ ;} \\ \lambda z^\sigma.\,A_\tau[s^\varphi A_\sigma[z^\sigma]] & \text{if } \varphi = \sigma \supset \tau \text{ and } z^\sigma \text{ is a fresh variable.} \end{cases}$$

*Proof.*
(1) By Theorem 8.
(2) By induction on the structure of $\varphi$ using the following facts:

$$(\lambda z.\,u)^\mathrm{E} \equiv \lambda z.\,u^\mathrm{E} \equiv (\lambda z.\,u)^\mathrm{e}, \quad \langle u, v\rangle^\mathrm{E} \equiv \langle u^\mathrm{E}, v^\mathrm{E}\rangle \equiv \langle u, v\rangle^\mathrm{e},$$
$$a^\mathrm{e} \equiv a, \qquad (fu)^\mathrm{e} \equiv f(u^\mathrm{e}),$$
$$(tu)^\mathrm{e} \equiv (t^\mathrm{e})(u^\mathrm{E}), \qquad (t^\psi)^\mathrm{E} \equiv (t^\psi)^\mathrm{e}$$

where $a$ is an atom, $f =$ l or r, and $\psi$ is an atomic type. □

**Lemma 11.**

(1) If $w^\varphi$ is a neutral term, say $w^\varphi \equiv \overline{f_k}\left(\cdots\left(\overline{f_1}\,s\,\overline{u_1}\right)\cdots\right)\overline{u_k}$ for some $k \geq 0$, $\overline{f_1}, \cdots \overline{f_k}, \overline{u_1}, \cdots, \overline{u_k}$, and

$$s \equiv \begin{cases} a & \text{for some atom } a\,;\text{ or} \\ (\lambda x.\,s_1)s_2 & \text{for some variable } x \text{ and terms } s_1 \text{ and } s_2\,;\text{ or} \\ g\langle s_1,\,s_2\rangle & \text{for some terms } s_1 \text{ and } s_2, \text{ and } g = \mathrm{l} \text{ or } \mathrm{r}\,, \end{cases}$$

then

$$(w^\varphi)^{\mathrm{e}} \equiv \begin{cases} \overline{f_k}\left(\cdots\left(\overline{f_1}\,a\,\overline{u_1^{\mathrm{E}}}\right)\cdots\right)\overline{u_k^{\mathrm{E}}} & \text{if } s \equiv a\,; \\ \overline{f_k}\left(\cdots\left(\overline{f_1}((\lambda x.\,s_1{}^{\mathrm{E}})s_2{}^{\mathrm{E}})\overline{u_1^{\mathrm{E}}}\right)\cdots\right)\overline{u_k^{\mathrm{E}}} & \text{if } s \equiv (\lambda x.\,s_1)s_2\,; \\ \overline{f_k}\left(\cdots\left(\overline{f_1}\,g\langle s_1{}^{\mathrm{E}},\,s_2{}^{\mathrm{E}}\rangle\overline{u_1^{\mathrm{E}}}\right)\cdots\right)\overline{u_k^{\mathrm{E}}} & \text{if } s \equiv g\langle s_1, s_2\rangle\,. \end{cases}$$

Here $\overline{u^{\mathrm{E}}} \equiv (u)^{\mathrm{E}}\,(u')^{\mathrm{E}}\,(u'')^{\mathrm{E}}\cdots$ if $\overline{u} \equiv u\,u'\,u''\cdots$ .

(2) If $w^\varphi$ is a designating term, then

$$(w^\varphi)^{\mathrm{e}} \equiv \begin{cases} \lambda x.\,s_1{}^{\mathrm{E}} & \text{if } w^\varphi \equiv \lambda x.\,s_1; \\ \langle s_1{}^{\mathrm{E}},\,s_2{}^{\mathrm{E}}\rangle & \text{if } w^\varphi \equiv \langle s_1, s_2\rangle\,; \\ * & \text{if } w^\varphi \equiv *. \end{cases}$$

*Proof.* By using the facts in the proof of Lemma 10. □

### 5.2 The Relation of $\to_{\mathrm{E}}$-Normalization against $\to_{\mathrm{BT}}$

Here, the claim in Sect. 1 will be proved.

**Lemma 12.**
(1) For any terms $w^\varphi$ and $u^\psi$, and a variable $z^\psi$, we have

$$A_\varphi[w^\varphi]\left[z^\psi := u^\psi\right] \equiv A_\varphi\left[w^\varphi\left[z^\psi := u^\psi\right]\right]\,.$$

(2) Suppose $X$ stand for B or T. For any type $\varphi$ and terms $s^\varphi$ and $t^\varphi$ ,

$$\text{if } s^\varphi \to_X t^\varphi \text{ then } A_\varphi[s^\varphi] \xrightarrow{+}_X A_\varphi[t^\varphi]\,.$$

*Proof.*
(1) By induction on the structure of type $\varphi$.
(2) Similarly for (1). □

From now on, Lemma 12(2) is often used without mentioning.

**Lemma 13.** For any terms $s^\psi, t^\varphi$ and variable $x^\varphi$, we have

$$\left(s^\psi\right)^{\mathrm{E}}\left[x^\varphi := (t^\varphi)^{\mathrm{E}}\right] \xrightarrow{*}_{\mathrm{B}} \left(s^\psi\left[x^\varphi := t^\varphi\right]\right)^{\mathrm{E}}\,.$$

*Proof.* We will prove the statement by the transfinite induction on $\omega \cdot |\varphi| +$ (the length of $s$).

The key case is where $t^\varphi$ is neutral and $s^\psi$ is non-atomic of the form

$$s^\psi \equiv \overline{f_k}\left(\cdots \left(\overline{f_1}\, x\, \overline{u_1}\right)\cdots\right) \overline{u_k}$$

for some $k \geq 0$, $\overline{f_1}, \cdots \overline{f_k}$, $\overline{u_1}, \cdots, \overline{u_k}$. Without loss of generality we assume that $\overline{f_1}$ and $\overline{u_1}$ are not simultaneously empty.

We claim that:

(*) $\quad \overline{f_1}\, t^{\mathrm{E}}\, \overline{(u_1\theta)}^{\mathrm{E}} \xrightarrow{*}_{\mathrm{B}} \left(\overline{f_1}\, t\, \overline{(u_1\theta)}\right)^{\mathrm{E}},$

where $\theta$ and $\theta^{\mathrm{E}}$ stand for substitutions $[x := t]$ and $[x := t^{\mathrm{E}}]$, respectively, and $\overline{u\theta} \equiv u\theta\ u'\theta\ u''\theta \cdots$ if $\overline{u} \equiv u\ u'\ u'' \cdots$ .

Once this is established, the statement $s^{\mathrm{E}}\theta^{\mathrm{E}} \xrightarrow{*}_{\mathrm{B}} (s\theta)^{\mathrm{E}}$ of the lemma is derived as follows:

$s^{\mathrm{E}}\theta^{\mathrm{E}} \equiv\ A_\psi\left[\overline{f_k}(\cdots \left(\overline{f_1}\, t^{\mathrm{E}}\overline{(u_1^{\mathrm{E}}\theta^{\mathrm{E}})}\right)\cdots)\overline{(u_k^{\mathrm{E}}\theta^{\mathrm{E}})}\right]$
by Lemmas 10 and 11;

$\xrightarrow{*}_{\mathrm{B}} A_\psi\left[\overline{f_k}(\cdots \left(\overline{f_1}\, t^{\mathrm{E}}\overline{(u_1\theta)^{\mathrm{E}}}\right)\cdots)\overline{(u_k\theta)^{\mathrm{E}}}\right]$
by induction hypothesis
(each $\overline{u_i}$ is a sequence of proper subterms of $s$);

$\xrightarrow{*}_{\mathrm{B}} A_\psi\left[\overline{f_k}(\cdots \left(\overline{f_2}\left(\overline{f_1}\, t\, \overline{(u_1\theta)}\right)^{\mathrm{E}}\overline{(u_2\theta)}^{\mathrm{E}}\right)\cdots)\overline{(u_k\theta)^{\mathrm{E}}}\right]$ \quad by (*);

$\equiv\ A_\psi\left[\overline{f_k}(\cdots \left(\overline{f_2}\, z\, \overline{(u_2\theta)}^{\mathrm{E}}\right)\cdots)\overline{(u_k\theta)^{\mathrm{E}}}\right]\left[z := \left(\overline{f_1}\, t\, \overline{(u_1\theta)}\right)^{\mathrm{E}}\right]$
for some fresh variable $z$, by Lemma 12;

$\equiv\ \left(\overline{f_k}(\cdots \left(\overline{f_2}\, z\, \overline{(u_2\theta)}\right)\cdots)\overline{(u_k\theta)}\right)^{\mathrm{E}}\left[z := \left(\overline{f_1}\, t\, \overline{(u_1\theta)}\right)^{\mathrm{E}}\right]$
by Lemmas 10 and 11.

Now we can apply the induction hypothesis to last line, because the type of the term $\overline{f_1}\, t\, \overline{(u_1\theta)}$ is strictly shorter than that of $t$ (recall that $\overline{f_1}$ and $\overline{(u_1\theta)}$ are not simultaneously empty). Then we get,

$$s^{\mathrm{E}}\theta^{\mathrm{E}} \xrightarrow{*}_{\mathrm{B}} \left(\overline{f_k}(\cdots \left(\overline{f_1}\, t\, \overline{(u_1\theta)}\right)\cdots)\overline{(u_k\theta)}\right)^{\mathrm{E}} \equiv (s\theta)^{\mathrm{E}}.$$

The claim (*) can be verified as follows:

$\overline{f_1}\, t^{\mathrm{E}}\overline{(u_1\theta)}^{\mathrm{E}} \xrightarrow{*}_{\mathrm{B}} \left(\overline{f_1}t\right)^{\mathrm{E}}\overline{(u_1\theta)}^{\mathrm{E}}$
by simple calculation in view of Lemma 10
and the facts in its proof;

$\xrightarrow{*}_{\mathrm{B}} \left(\overline{f_1}\, t\, \overline{(u_1\theta)}\right)^{\mathrm{E}}$ \quad as explained below:

In case $\overline{u_1}$ is empty, it is trivial. So assume otherwise and $\overline{u_1} \equiv u_1^{(1)} \cdots u_1^{(m_1)}$ ($m_1 > 0$). Then $\overline{f_1} t$ is of implication type. Since we assumed that $t$ is neutral, so is $\overline{f_1} t$. By virtue of Church-Rosser property of the reduction $\to_E$,

$$\left(\overline{f_1} t\right)^E \equiv \left(\lambda z_1. \overline{f_1} t z_1\right)^E \equiv \lambda z_1. \left(\overline{f_1} t z_1\right)^E, \quad \text{for some fresh variable } z_1 .$$

Therefore,

$$\left(\overline{f_1} t\right)^E \overline{(u_1 \theta)}^E \to_B \left(\left(\overline{f_1} t z_1\right)^E \left[z_1 := \left(u_1^{(1)} \theta\right)^E\right]\right) \left(u_1^{(2)} \theta\right)^E \cdots \left(u_1^{(m_1)} \theta\right)^E .$$

We can apply the induction hypothesis to last line, since the type of the term $u_1^{(1)} \theta$ is strictly shorter than that of $t$. Hence,

$$\left(\overline{f_1} t\right)^E \overline{(u_1 \theta)}^E \xrightarrow{+}_B \left(\overline{f_1} t \left(u_1^{(1)} \theta\right)\right)^E \left(u_1^{(2)} \theta\right)^E \cdots \left(u_1^{(m_1)} \theta\right)^E .$$

By iterating this argument, we finally get

$$\left(\overline{f_1} t\right)^E \overline{(u_1 \theta)}^E \xrightarrow{*}_B \left(\overline{f_1} t \overline{(u_1 \theta)}\right)^E .$$

This completes the proof of (*), thus, that of the key case. The other cases are proved similarly. □

Now we can prove our claim in Sect. 1, as follows.

**Lemma 14.** Suppose $X$ stand for B or T. For any terms $u$ and $v$,

$$\text{if } u \to_X v \text{ then } u^E \xrightarrow{+}_X v^E .$$

*Proof.* By induction on the structure of $u$. The key case is where $u$ is neutral and $u \to_X v$ is a "head reduction", that is:

$$u \equiv \overline{f_k} \left(\cdots \left(\overline{f_1} s \overline{u_1}\right) \cdots\right) \overline{u_k} \to_X \overline{f_k} \left(\cdots \left(\overline{f_1} t \overline{u_1}\right) \cdots\right) \overline{u_k} \equiv v$$

for some $k \geq 0$; $\overline{f_1}, \cdots \overline{f_k}$; $\overline{u_1}, \cdots, \overline{u_k}$, $s$, and $t$. Here $s$, $t$, and $X$ are given in the following table: where $x$ is a variable; $s_1$ and $s_2$ are terms; and $g = $ l or r.

**Table 1.** The cases for $s$, $t$, and $X$

| Subcase | $s$ | $t$ | $X$ |
|---|---|---|---|
| Subcase 1 | $(\lambda x. s_1) s_2$ | $s_1[x := s_2]$ | B |
| Subcase 2 | $g \langle s_1, s_2 \rangle$ | $s_g$ | B |
| Subcase 3 | a term of type T | $*^T$ | T |

In subcase 3, $u \equiv s \to_T * \equiv v$. So, $u^E \equiv s^E \to_T * \equiv v^E$.
In the other subcases, when

$$(\ddagger) \qquad s^e \xrightarrow{+}_B t^E$$

is established, the statement is derived as follows:

$$u^E \equiv A\left[\overline{f_k}\left(\cdots\left(\overline{f_1}\, s^e\, \overline{u_1{}^E}\right)\cdots\right)\overline{u_k{}^E}\right]$$
by Lemmas 10 and 11;

$$\xrightarrow{+}_B A\left[\overline{f_k}\left(\cdots\left(\overline{f_1}\, t^E\overline{u_1{}^E}\right)\cdots\right)\overline{u_k{}^E}\right] \qquad \text{by } (\ddagger);$$

$$\equiv A\left[\overline{f_k}\left(\cdots\left(\overline{f_1}\, z\, \overline{u_1{}^E}\right)\cdots\right)\overline{u_k{}^E}\right]\left[z := t^E\right]$$
for some fresh variable $z$, by Lemma 12;

$$\equiv \left(\overline{f_k}\left(\cdots\left(\overline{f_1}\, z\, \overline{u_1}\right)\cdots\right)\overline{u_k}\right)^E\left[z := t^E\right]$$

$$\xrightarrow{*}_B \left(\overline{f_k}\left(\cdots\left(\overline{f_1}\, t\, \overline{u_1}\right)\cdots\right)\overline{u_k}\right)^E \qquad \text{by Lemma 13;}$$

$$\equiv v^E\ .$$

We will prove the claim ($\ddagger$). In subcase 1:

$$\begin{aligned} s^e &\equiv \left(\lambda x.\, s_1{}^E\right) s_2{}^E \rightarrow_B s_1{}^E\left[x := s_2{}^E\right] \\ &\xrightarrow{*}_B \left(s_1[x := s_2]\right)^E \qquad \text{by Lemma 13;} \\ &\equiv t^E\ . \end{aligned}$$

In subcase 2, similarly derived.

This completes the proof of the key case. The other cases are proved by the straightforward application of the induction hypothesis on the structure of $s$. □

**Theorem 15.** Mints' reduction $\rightarrow_{BET}$ has SN+CR property.

*Proof.* The reduction $\rightarrow_{BT}$ is proved to be strongly normalizable by the reducibility method. Besides, it is checked to be weakly Church-Rosser. Thus it is also Church-Rosser by virtue of Newman's lemma.

On the other hand, the SN+CR property of the reduction $\rightarrow_E$ is established by Theorem 8.

Hence, we get the property of $\rightarrow_{BET}$ by Lemma 4. □

**Corollary 16 (characterization of $\rightarrow_{\eta\bullet}$ and $\rightarrow_{\pi\bullet}$).**

(1) $u \rightarrow_{\eta\bullet} v \iff v \rightarrow_\eta u$ and $v \not\rightarrow_\beta u$ .

(2) $u \rightarrow_{\pi\bullet} v \iff v \rightarrow_\pi u$ and not $v \xrightarrow{*}_{lr} u$ where $\rightarrow_{lr} := \rightarrow_l \cup \rightarrow_r$ .

*Proof.*

(1) If-part is readily derived from the definition of $\to_{\eta^*}$. Only-if-part is proved as follows: Let $u \to_{\eta^*} v$. By the definition of $\to_{\eta^*}$, we have $v \to_\eta u$. If $v \to_\beta u$, then we will get an infinite $\to_{BE}$-reduction sequence

$$v \to_\beta u \to_{\eta^*} v \to_\beta u \to_{\eta^*} \cdots ,$$

which contradicts the strong normalization property of $\to_{BET}$.
(2) Similar. □

## 6 Discussion

By this paper's method, we can verify that the SN+CR property holds for various subsystems of Mints reduction $\to_{BET}$. We can prove not only the claim in Sect. 1 but also the following:

For any terms $u$ and $v$, if $u \to_S v$ then $u^R \overset{\pm}{\to}_S v^R$
provided
(1) $\to_R = \to_{\eta^*}$ or $\to_{\pi^*}$; and
(2) $\to_S = \bigcup \Sigma$, for $\Sigma \subseteq \{\to_\beta, \to_l, \to_r, \to_{\eta^*}, \to_{\pi^*}, \to_T\}$ where $\to_S \notin \Sigma$.

So, the reduction $\to_R \cup \to_S$ satisfies the property.

Curien and Di Cosmo[2] proved the weak normalization property and Church-Rosser property of the conservative extension of ccc-calculus, which is second-order.

Mints' reduction can be extended with adding the second-order version of the reduction $\to_\beta$ and $\to_{\eta^*}$ (denoted by $\to_{2\beta}$ and $\to_{2\eta^*}$). In this system, we can define the pairing operator and the reductions $\to_l$ and $\to_r$ in the standard fashion [4] by way of the reductions $\to_\beta$ and $\to_{2\beta}$, but neither $\to_\pi$ nor $\to_{\pi^*}$ in spite of the existence of the reductions $\to_{2\eta^*}$ and $\to_{\eta^*}$.

Another extension of Mints' reduction is to add natural numbers and the iterator with the following rewriting rules:

$$\left(\mathrm{It}_\varphi\left(u^\varphi, v^{\varphi \supset \varphi}, 0^N\right)\right)^\varphi \to_B u^\varphi ,$$
$$\left(\mathrm{It}_\varphi\left(u^\varphi, v^{\varphi \supset \varphi}, (Sw^N)^N\right)\right)^\varphi \to_B \left(v^{\varphi \supset \varphi}\left(\mathrm{It}_\varphi\left(u^\varphi, v^{\varphi \supset \varphi}, w^N\right)\right)^\varphi\right)^\varphi .$$

Here, N is a distinguished atomic type, $0^N$ is for zero, and S for the successor. $\mathrm{It}_\varphi$ is a term forming operator for each type $\varphi$, such that $(\mathrm{It}_\varphi(u^\varphi, v^{\varphi \supset \varphi}, w^N))^\varphi$ is a term if so are $u^\varphi, v^{\varphi \supset \varphi}$, and $w^N$.

This extended system is proved to satisfy SN+CR property by a modified version of the reducibility method[8]. But, unfortunately, when we try to prove this fact by this paper's method, we have difficulty to show

$$\left(\left(\mathrm{It}_\varphi\left(u^\varphi, v^{\varphi \supset \varphi}, (Sw^N)^N\right)\right)^\varphi\right)^E \overset{\pm}{\to}_B \left(v^{\varphi \supset \varphi}\left(\mathrm{It}_\varphi\left(u^\varphi, v^{\varphi \supset \varphi}, (Sw^N)^N\right)\right)^\varphi\right)^E,$$

which is a part of the claim in Sect. 1.

Finally we note that, our main claim can be derived from the results in Čubrić[3]: in particular, $\to_{BT}$ and $\to_E$ commute(his Proposition 4.1). Indeed, suppose $u \to_{BT} v$. Then applying his Proposition 4.1 to $u^E \;_E{\overset{*}{\leftarrow}}\; u \to_{BT} v$, we have $u^E \overset{*}{\to}_{BT} w \;_E{\overset{*}{\leftarrow}}\; v$ for some $w$. As he pointed out(in the proof of his Proposition 4.2), $\to_{BT}$ preserves $\to_E$-normality, hence the term $w$ is in $\to_E$ normal. By the Church-Rosser property of $\to_E$, we have $w \equiv v^E$. Since $\to_E$ never deletes redexes but only copies or creates redexes, we get $u^E \overset{+}{\to}_{BT} w^E \equiv v^E$.

Čubrić also shows that for any term $u$, $u^{(BET)} \equiv (u^E)^{(BT)}$ (Proposition 4.2 [3]). In order to obtain these properties of $\to_{BET}$, he exploits minimal development of $\to_E$, the notion of which is defined in terms of residuals. Our approach is free from residuals, and in fact independently obtained from his.

## Acknowledgement

The author would like to thank to Masami Hagiya and Masako Takahashi for fruitful discussions. I also thank Djordje Čubrić who pointed out the alternative proof of our main claim by his method.

## References

1. H.P. Barendregt. *The Lambda Calculus, Its Syntax and Semantics*. North Holland, second edition, 1984.
2. P.-L. Curien and R. Di Cosmo. A confluent reduction for the $\lambda$-calculus with surjective pairing and terminal object. In *Proceedings ICALP '91, Madrid*, pages 291–302. Springer-Verlag, 1991.
3. D.Čubrić. On free CCC. manuscript, March 12, 1992.
4. J.-Y. Girard, Y. Lafont, and P. Taylor. *Proofs and Types*. Cambridge Theoretical Computer Science. Cambridge University Press, 1989.
5. M. Hagiya. Personal communication, Aug. 1991.
6. M. Hagiya. From programming-by-example to proving-by-example. In T. Ito and A.R. Meyer, editors, *Theoretical Aspects of Computer Software. Proceedings*, pages 387–419. Springer-Verlag, 1991. Lecture Notes in Computer Science, 526.
7. G.P. Huet. A unification algorithm for typed lambda-calculus. *Theoretical Computer Science*, 1:27–57, 1975.
8. C. Barry Jay. Long $\beta\eta$ normal forms and confluence(revised). LFCS, Department of Computer Science, University of Edinburgh, February 1992.
9. J. Lambek and P.J. Scott. *Introduction to higher order categorical logic*, volume 7 of *Cambridge studies in advanced mathematics*. Cambridge University Press, 1986.
10. G.E. Mints. Theory of categories and theory of proofs. i. In *Urgent Questions of Logic and the Methodology of Science*, 1979. [in Russian], Kiev.
11. D. Prawitz. Ideas and results in proof theory. In J.E. Fenstad, editor, *Proceedings of the Second Scandinavian Logic Symposium*, pages 235–307. North-Holland, 1971.

This article was processed using the LaTeX macro package with LLNCS style

# A Formalization of the Strong Normalization Proof for System F in LEGO [*][†]

Thorsten Altenkirch
Laboratory for Foundations of Computer Science
Department of Computer Science
University of Edinburgh
Email address : alti@dcs.ed.ac.uk

December 8, 1992

### Abstract

We describe a complete formalization of a strong normalization proof for the Curry style presentation of System F in LEGO. The underlying type theory is the Calculus of Constructions enriched by inductive types. The proof follows Girard et al [GLT89], i.e. we use the notion of *candidates of reducibility*, but we make essential use of general inductive types to simplify the presentation. We discuss extensions and variations of the proof: the extraction of a normalization function, the use of *saturated sets* instead of candidates, and the extension to a Church Style presentation. We conclude with some general observations about Computer Aided Formal Reasoning.

## 1 Introduction

I am going to describe a complete formalization of a strong normalization proof for System F in LEGO [LP92]. The proof[1] uses the tactics provided by the LEGO system. However, in the end we can extract a typed $\lambda$-term which represents the proof. The proof is complete, i.e. there are no non-logical axioms used.[2] This implies in particular that the proof is intuitionistic. As a consequence of

---

[*]When doing this research I have been supported by a SIEMENS studentship. This research was also partially supported by the ESPRIT BRA on Logical Frameworks and a SERC grant.

[†]This is a revision of the LFCS report ECS-LFCS-92-230 "Brewing Strong Normalization Proofs with LEGO".

[1]The complete proof text can be obtained by anonymous ftp from ftp.dcs.ed.uk or by email request. The directory is pub/alti; the file is snorm.tar.Z.

[2]To show the fourth Peano axiom $\forall_{n \in \mathcal{N}} n+1 \neq 0$ and its variants we need *large eliminations*. They have been investigated by Benjamin Werner recently [Wer92].

this we can extract a normalization function together with its verification from the proof.

Normalization proofs are quite a fashionable subject for formal proofs. Stefano Berardi worked on a strong normalization proof for System F in the Pure Constructions also using LEGO [Ber91]. Catarina Coquand [Coq92a] did a normalization proof for simply typed $\lambda$ calculus using ALF [Mag92].

One reason why strong normalization proofs are interesting candidates for formalization is that they are fairly intricate and that they require a complete formalization and understanding of the calculus involved. Another reason is that everybody who works in the area of Type Theory or formal proofs knows them and studies them anyway.

## 2 Using LEGO

LEGO is a proof development system based on Type Theory which has been implemented by Randy Pollack. A good introduction to the use of LEGO can be found in [Hof92], where LEGO is used for program verification.

### 2.1 The Type Theory

The *standard* Type Theory used in LEGO is the Extended Calculus of Constructions (ECC) [Luo90]. However, here we do not exploit the extensions introduced by Luo: universes and strong $\Sigma$-types — we only require one predicative universe over the Pure Calculus of Constructions. We also diverge from Luo's proposal to use the predicative universes (**Type**) for computations, instead using the impredicative universe **Prop** for both: logic and computations . This is expressed by the definition **Set=Prop**. This means that philosophically we follow Martin-Löf's identification of propositions and types but we differ in that we accept impredicative quantification.

Many formal proofs in the Calculus of Constructions use the so called *impredicative encodings* of inductive types (e.g. see [Alt90]). These encodings have a number of disadvantages, in particular we have to assume an induction axiom to prove anything about them. This induction axiom is not interpreted computationally and destroys the computational interpretation of propositions because we now have non-canonical elements in every type. To overcome this problem we introduce inductive types explicitly and introduce elimination constants *à la* Martin-Löf together with typed reduction rules (e.g. [NPS90]). Our approach to inductive types is similar to the one chosen in Coq [D[+]91] and is based on [CP89].

| Usual notation | LEGO notation | Remarks |
|---|---|---|
| $\lambda x : A.M$ | [x:A]M  *or*<br>[x\|A]M | If [x\|A]M is used the argument is inferred when the function is applied. |
| $\Pi_{x:A} B$ | {x:A}B  *or*<br>{x\|A}M | {x\|A}B is used for the typing of [x\|A]M. |
| $MN$ | M N  *or*<br>M\|N | M\|N is used to supply an implicit argument. |

Figure 1: Syntax of terms in LEGO

## 2.2 The Logic

The basic logical connectives (/\,\/,not and Ex) are defined using an impredicative encoding. However, instead of Leibniz Equality we define propositional equality EQ as an inductive type in the same way as in Martin-Löf Type Theory.

Theoretically, it would have been better to use inductive types for all logical connectives, because they come with stronger elimination rules and it seems a bit of a waste to introduce impredicativity just to encode basic logical connectives. However, the current Refine tactic of LEGO is tuned for the impredicative encodings.

## 2.3 Inductive types

So far inductive types have not been implemented in the LEGO system but we can use typed rewriting rules to realize them. In the proof I use my own syntax for mu types and definitions by primitive recursion which is put as a comment into the proof. This is then expanded into LEGO code — I will explain this by a simple example:

```
mu[Nat:Set](zero:Nat,succ:Nat->Nat)
```

is the type of natural numbers. We translate this into LEGO by *introducing* the type and the constructor as constants. We also add an elimination constant:

```
$[RecNat : {P:Nat->Type}
           (P zero)->({n:Nat}(P n)->(P (succ n)))
        -> {n:Nat}P n];
```

We declare typed rewriting rules which correspond to primitive recursion on natural numbers. These definitions extend LEGO's definitional equality and we can perform computations with them. They also have the effect that no non-canonical elements are introduced because every occurrence of RecNat is eventually eliminated.

If we want to define a function over natural numbers, we declare it first in an ML-like fashion:

```
add : Nat->Nat->Nat
rec add zero n = n
  | add (succ m) n = succ (add m n)
```

which can be (mechanically) translated into the following LEGO code using the recursor (here we use a derived non-dependent recursor RecNatN[3] to simplify the typing):

```
[RecNatN[C|Type] = RecNat ([_:Nat]C)];

[add = RecNatN ([n:Nat]n)
               ([m:Nat][add_m:Nat->Nat][n:Nat]succ (add_m n))];
```

We can only define functions by primitive recursion in this way, but note that we get more than the usual primitive recursive functions because we have higher order functions.

It is interesting to consider inductive types with dependent constructors like the type of vectors:

```
[A:Set]mu[Vec:Nat->Set](v_nil:Vec zero,
                        v_cons:A->{n|Nat}(Vec n)->(Vec succ n))
```

or the family of finite sets:

```
mu[Fin:Nat->Set](f_zero:{n:Nat}Fin (succ n),
                 f_succ:{n|Nat}(Fin n)->(Fin (succ n)))
```

Vectors resemble lists but differ in that the length of the sequence is part of its type. Therefore we have Vec: Set->Nat->Set in contrast to List : Set->Set, i.e. Vec A $3$[4] is the type of sequences of type A of length 3. Finite sets are a representation of subsets of natural numbers less than a certain number, i.e. Fin $n$ corresponds to $\{i \mid i < n\}$.

When we defined the elimination constant for Nat we allowed eliminations over an arbitrary universe — we call these *large eliminations*. We can use them to prove the fourth Peano axiom as in Martin-Löf Type Theory with universes. However, it is interesting to observe that there is a purely computational use of *large eliminations*: we can apply them to realize the following *run-time-error-free* lookup function for vectors:[5]

---

[3] We adopt the convention that $R\mathbf{N}$ stands for the non-dependent version of recursor $R$.

[4] The official LEGO syntax for this is Vec A (succ (succ (succ zero))).

[5] I.e. v_nth (f_succ (f_zero 3)) : (Vec A 5)->A extracts the second element out of a sequence of five.

```
v_nth : {n|Nat}(Fin n)->(Vec A n)->A

rec v_nth|(succ n) (f_zero n) (v_cons a _) = a
  | v_nth|(succ n) (f_succ i) (v_cons _ l) = v_nth|n i l
```

Using these error-free functions not only simplifies the verification of functions using vectors; it also allows, in principle a more efficient compilation of code involving dependent types.[6]

Another use of dependent inductive types is the definition of predicates as the initial semantics of a set of Horn clauses. An example is the definition of the predicate $\leq$ (LE) for natural numbers:

```
mu[LE:Nat->Nat->Set](
   le0:{n:Nat}LE zero n,
   le1:{m,n|Nat}(LE m n)->(LE (succ m) (succ n)))
```

# 3  A guided tour through the formal proof

In the following I am going to explain the formalized proof. For more detailed information it may be worthwhile to obtain the complete LEGO code.

## 3.1  The untyped $\lambda$-calculus

We define untyped $\lambda$-terms (Tm) using de Bruijn indices [dB72] as the following inductive type:

```
mu[Tm:Set](var : Nat->Tm,
           app : Tm->Tm->Tm,
           lam : Tm->Tm)
```

We define the operations weakening weak : Nat->Tm->Tm[7] (introduction of dummy variables) and substitution subst : Nat->Tm->Tm->Tm by primitive recursion over the structure of terms.

The first parameter indicates the number of bound variables — weak0 and subst0 are defined as abbreviations, i.e. subst0 M N substitutes the free variable with index 0 in M by N.[8]

---

[6] The idea to use dependent types to avoid run-time-errors was first proposed by Healfdene Goguen to me.

[7] Huet [Hue92] calls the operation *lift*. I prefer the name *weak* because it corresponds to the notion of weakening in the typed case.

[8] Although we only use weak0 and subst0 in the following definition we really have to *export* the general versions because we have to use them whenever we want to prove anything about substitution or weakening in general (i.e. for terms containing $\lambda$-abstractions).

In the course of the proof we need a number of facts about weakening and substitution. An example is the following proposition stating that under certain conditions we can interchange substitution and weakening:

```
{l',l:Nat}{M,N:Tm}(LE l' l)
->(EQ (subst (succ l) (weak l' M) N) (weak l' (subst l M N)))
```

Stating and proving this sort of lemma takes up a lot of time when doing the formalization whereas in the informal proof one would just appeal to some intuition about substitution and bound variables.

We define the one-step reduction relation[9] by the following inductive type:

```
mu[Step:Tm->Tm->Set](
    beta : {M,N:Tm}Step (app (lam M) N) (subst0 M N),
    app_l : {M,M',N:Tm}(Step M M')->(Step (app M N) (app M' N)),
    app_r : {M,M',N:Tm}(Step M M')->(Step (app N M) (app N M')),
    xi : {M,N:Tm}(Step M M')->(Step (lam M) (lam M')) )
```

This amounts to translating the usual Horn clauses defining the reduction relation into the constructors for an inductive type.

## 3.2 Strong Normalization

One of the main technical contributions which simplify the formalization of the proof is the definition of the predicate *strongly normalizing* by the following inductive type:[10]

```
mu[SN:Tm->Set](
    SNi : {M:Tm}({N:Tm}(Step M N)->(SN N))->(SN M))
```

In other words: we define SN as the set of elements for which Step is well founded.

More intuitively: SN holds for all normal forms because for them the premise of SNi is vacuously true. Now we can also show that all terms which reduce in one step to a normal form are SN and so on for an arbitrary number of steps. On the other hand these are all the terms for which SN holds because SN is defined inductively.

---

[9] We are going to define Red (the reflexive, transitive closure of Step) later (section 4.1). Note, however, that we never need it for the strong normalization proof.

[10] It is interesting to note that this inductive type is not algebraic or equivalently does not correspond to a specification by a set of Horn formulas. However, the requirement of (strict) positivity as stated in [CP89] is fulfilled.

We will use the non-dependent version of the recursor [11]

```
RecSNN : {P:Tm->Type}
    ({M:Tm}({N:Tm}(Step M N)->SN N)->({N:Tm}(Step M N)->P N)->P M)
    ->{M|Tm}(SN M)->P M];
```

to simulate induction over the length of the longest reduction of a strongly normalizing term — in terms of [GLT89] this is induction over $\nu(M)$. Observe that we never have to formalize the concept of the length of a reduction or to define the partial function $\nu$. [12] It is also interesting that the important property that SN is closed under reduction shows up as the destructor for this type (SNd).

## 3.3 System F

The type expressions of System F have essentially the same structure as untyped $\lambda$-terms. However, in contrast to the definition of Tm we will use a dependent type here, which makes the number of free variables explicit. This turns out to be useful when we define the semantic interpretation of types later.[13]

```
mu[Ty:Nat->Set](t_var : {n|Nat}(Fin n)->(Ty n),
                arr   : {n|Nat}(Ty n)->(Ty n)->(Ty n),
                pi    : {n|Nat}(Ty (succ n))->(Ty n) )
```

Ty $i$ represents type expressions with $i$ free variables.

When defining weakening and substitution for Ty we observe that the types actually tell us how these operations behave on free variables:

```
t_weak  : {l:Nat}(Ty (add l n))->(Ty (succ (add l n)))
t_subst : {l:Nat}(Ty (add (succ l) n))->(Ty n)->(Ty (add l n))
```

Although these functions are operationally equivalent to weak and subst we have to put in more effort to implement them. We do this by deriving some special recursors from the standard recursor.[14]

We now define contexts and derivations as:

```
[Con[m:Nat] = Vec (Ty m)];

mu[Der:{m,n|Nat}(Con m n)->Tm->(Ty m)->Set](
```

---

[11] This corresponds to the principle of Noetherian Induction [Hue80].
[12] Note that *bounded* and *noetherian* coincide for $\beta$-reduction because it is finitely branching (König's lemma).
[13] We could have used a dependent type for Tm as well, but we never need to reason about the number of free variables of an untyped term.
[14] It seems that we could save a lot of effort here by using Thierry Coquand's idea of considering definitions by pattern matching as primitive [Coq92b].

```
Var : {m,n|Nat}{G:Con m n}{i:Fin n}
        Der G (var (Fin2Nat i)) (v_nth i G)
App : {m,n|Nat}{G|Con m n}{s,t|Ty m}{M,N|Tm}
        (Der G M (arr s t))
        -> (Der G N s)
        -> (Der G (app M N) t)
Lam : {m,n|Nat}{G|Con m n}{s,t|Ty m}{M|Tm}
        (Der (v_cons s G) M t)
        -> (Der G (lam M) (arr s t))
Pi_e: {m,n|Nat}{G|Con m n}{s:Ty (succ m)}{t:Ty m}{M|Tm}
        (Der G M (pi s))
        -> (Der G M (t_subst0 s t))
Pi_i: {m,n|Nat}{G|Con m n}{s:Ty (succ m)}{M|Tm}
        (Der (v_map t_weak0 G) M s)
        -> (Der G M (pi s)) )
```

In the rule `Var` we use `Fin` because this rule is only applicable to integers smaller than the length of the context. Here we have to coerce it to a natural number first (`Fin2Nat`) because `var` requires `Nat` as an argument.

`v_map t_weak0 G` means that all the types in `G` are weakened — this is equivalent to the usual side condition in the standard definition of Π-introduction. It is nice to observe how well the types of `t_subst0` and `t_weak0` fit for the definition of the rules.

## 3.4 Candidates

One of the essential insights about strong normalization proofs is that they require another form of induction than proofs of other properties of typed λ-calculi like the *subject reduction property* or the *Church-Rosser property*. We cannot show strong normalization just by induction over type derivations or by induction over the length of a reduction. We have to apply another principle which may best be described as an *induction over the meaning of types*.

We have to find a family of sets of terms,[15] here called the *Candidates of Reducibility*, such that we can show the following things:

1. Every Candidate only contains strongly normalizing terms.

2. For every operation on types we can define a semantic operation on sets of terms such which is closed under candidates. Another way to express this is to say that the Candidates constitute a logical predicate.

3. Candidates are sound, i.e. every term which has a type is also in the semantic interpretation of the type.

---
[15] The term *set* becomes a bit overloaded. Here we mean a set in the classical sense, i.e. the extension of a predicate and not the universe `Set`. A set of terms corresponds to the type `Tm->Set`.

Putting these things together we will obtain that every typable term is strongly normalizing.

In the definition of *Candidates of Reducibility* CR:(Tm->Set)->Set we follow [GLT89]:[16]

```
[neutr[M:Tm] = {M':Tm}not (EQ (lam M') M)];

[P:Tm->Set]
[CR1 = {M|Tm}(P M)->(SN M)]
[CR2 = {M|Tm}(P M)->{N:Tm}(Step M N)->(P N)]
[CR3 = {M|Tm}(neutr M)->
        ({N:Tm}(Step M N)->(P N))->(P M)]
[CR = CR1 /\ CR2 /\ CR3];
Discharge P;
```

We define **neutr** as the set of terms which are not generated by the constructor for the arrow type — **lam**.[17] CR1 places an upper bound on candidates: they may only contain strongly normalizing terms. CR2 says that candidates have to be closed under reduction and CR3 is essentially SNi restricted to neutral terms.

The essence of this definition lies in the possibility of proving the following lemmas:

CR_var *Candidates contain all variables*

{P:Tm->Set}(CR P)->{i:Nat}P (var i);

We need this not only for the following lemmas, but also for the final corollary when we want to deduce strong normalization from soundness for non-empty contexts.

This is a trivial consequence of CR3 because variables are neutral terms in normal form.

CR_SN *There is a candidate set*

CR SN

The choice is arbitrary but the simplest seems to be SN. The proof is trivial: just apply SNd for CR2 and SNi for CR3.

CR_ARR *Candidates are closed under the semantic interpretation of arrow types.*

{P,R:Tm->Set}(CR P)->(CR R)->(CR (ARR P R))

---

[16] [P:Tm->Set] ... Discharge P; means that P is $\lambda$-abstracted from all definitions in between.
[17] If we generalize this to systems with inductive types we have to include their *constructors* as well.

where

```
[ARR[P,R:Tm->Set] = [M:Tm]{N:Tm}(P N)->(R (app M N))];
```

The proof of `CR3` for `ARR P R` is actually quite hard and requires an induction using `RecSNN` which corresponds to the reasoning using $\nu(N)$ in [GLT89].

**CR_PI** *Candidates are closed under the interpretation of $\Pi$-types*

```
{F:(Tm->Set)->(Tm->Set)}
    ({P:Tm->Set}(CR P)->(CR (F P)))
    -> (CR (PI F));
```

where

```
[PI[F:(Tm->Set)->(Tm->Set)] =
    [M:Tm]{P:Tm->Set}(CR P)->(F P M)];
```

At this point we really need impredicativity for the proof. However, it is interesting to observe how simple this lemma is technically: we do not apply any induction — we just have to show that `CR` is closed under arbitrary non-empty intersections.

**Lam_Sound** *The rule of arrow introduction (Lam) is semantically sound for candidate sets.*

```
{P,R:Tm->Set}(CR P)->(CR R)->
    {M:Tm}({N:Tm}(P N)->(R (subst0 M N)))
    ->(ARR P R (lam M));
```

Observe that we could not have proved this lemma for arbitrary subsets of `SN`. The proof requires a nested induction using `RecSNN` which corresponds to an induction over $\nu(M) + \nu(N)$.

## 3.5 Proving strong normalization

We now have all the ingredients for the proof, we just have to put them together.

We proceed by defining an interpretation function. Types are interpreted by functions from sequences of sets of terms to sets of terms, the length of the sequence depending on the number of free type variables:[18]

---

[18] We have to use another type of vectors (large vectors) `VEC:Nat->Type(0)->Type(0)` instead of `Vec:Nat->Set->Set`. Unfortunately, this sort of polymorphism cannot be expressed in the current implementation of LEGO — i.e. we have to duplicate the definitions.

```
Int : {m|Nat}(Ty m)->(VEC (Tm->Set) m)->(Tm->Set)

rec Int|m (t_var i) = [v:VEC (Tm->Set) m]V_nth i v
  | Int|m (arr s t) = [v:VEC (Tm->Set) m]ARR (Int s v) (Int t v)
  | Int|m (pi t)    = [v:VEC (Tm->Set) m]
              PI ([P:Tm->Set]Int t (V_cons P v))
```

We can show by a simple induction that every interpretation of a type preserves candidates (`CR_Int`) by exploiting `CR_ARR` and `CR_PI`.

We extend this to an interpretation of judgements, i.e. pairs of types and contexts (`Mod`). The idea is that `Mod G M T v` holds iff by substituting all the variables in `M` by terms of the corresponding interpretation of the types in `G` we end up with an element of `Int T v`:[19]

```
Mod : {m,n|Nat}(Con m n)->Tm->(Ty m)->(VEC (Tm->Set) m)->Set

rec Mod m zero      empty        M T v = Int T v M
  | Mod m (succ n) (v_cons S G) M T v =
         {N:Tm}(Int S v N)->(Mod G (subst0 M (rep_weak0 N n)) T v)
```

We use `Mod` to state soundness (`Int_Sound`), i.e. that `Der G M T` implies `Mod G M T v` if all free type variables are interpreted by candidates:

```
{m,n|Nat}{G|Con m n}{M|Tm}{T|Ty m}(Der G M T)->
        {v:VEC (Tm->Set) m}({i:Fin m}CR (V_nth i v))
        -> (Mod G M T v);
```

The proof of soundness proceeds by induction over derivations. Essentially we only have to apply `Lam_sound` to show that the rule `Lam` is sound. The soundness of application `App` follows directly from the definition of `ARR`. To verify soundness for the rules which are particular to System F we do not need additional properties of `CR` but we have to verify that `t_weak` and `t_subst` are interpreted correctly with respect to `Int`. Again these intuitively simple lemmas are quite hard to show formally.

To conclude strong normalization:

```
{m,n|Nat}{G|Con m n}{M|Tm}{T|Ty m}(Der G M T)->(SN M)
```

we have to put `Int_sound` and `CR_Int` together to show that every term is in the interpretation of a candidate; and by definition candidates only contain strongly normalizing terms. There are two technical complications: to show the theorem for terms with free term variables we exploit `CR_var`; to show it for a derivation with free type variables we have to supply a candidate — here we use `CR_SN`. Note that the choice is arbitrary but that it is essential that `CR` is not empty.

---

[19] `rep_weak0` is the iterated application of `weak0`. It is necessary to apply weakening here because we do not get parallel substitution by a repeated application of `subst0`.

# 4 Alternatives and extensions

Apart from the first section I have not yet formalized the following ideas, but I have checked them on paper.

## 4.1 Extracting a normalization function

The proof not only tells us that every typable term is strongly normalizing but it is also possible to derive a function which computes the normal form. This seems to be a case where it is actually more straightforward to give a proof that every strongly normalizable term has a normal form than to program it directly.

To specify normalization we need a notion of reduction (`Red`) — which is just the transitive reflexive closure of `step`:

```
mu[Red:Tm->Tm->Set](
    r_refl : {M:Tm}(Red M M),
    step  : {M1,M2,M3|Tm}(Step M1 M2)->(Red M2 M3)->(Red M1 M3) )
```

and we define the predicate *normal form*:

```
[nf[M:Tm] = {M':Tm}not (Step M M')]
```

Now we want to show `norm_lem`:

```
{M:Tm}(SN M)->Ex[M':Tm](Red M M')/\(nf M')
```

which can be done using `RecSNN` — it turns out that we need decidability of normal form as a lemma:

```
{M:Tm}(nf M)\/(Ex[M':Tm]Step M M')
```

Actually, this is even stronger, because it also gives us a choice of a reduct for terms not in normal form (this is the point where we specify the strategy of reduction).

If we use the strong sum to implement `Ex`, instead of the weak impredicative encoding, we can use `norm_lem` to derive:

```
norm    : {M:Tm}(SN M)->Tm
norm_ok : {M:Tm}{p:SN M}(Red M (norm M p)) /\ (nf (norm M p))
```

## 4.2 Saturated Sets

In many strong normalization proofs the notion of *Saturated Sets* is used instead of *Candidates of Reducibility* (e.g. [Luo90], [B+91]). It is relatively easy to change the proof to use saturated sets: all we have to do is to replace CR by SAT and prove that it has the same properties as CR.

The definition of SAT used in the literature seems to be quite hard to formalize in Type Theory. Therefore, we use an equivalent formulation, exploiting the concept of weak head reduction:

```
mu[W_Hd_Step:Tm->Tm->Set](
        wh_beta : {M,N:Tm}W_Hd_Step (app (lam M) N) (subst0 M N),
        wh_app_1 : {M,M',N:Tm}(W_Hd_Step M M')
                        ->(W_Hd_Step (app M N) (app M' N)) )
```

Now we can define SAT analogously to CR:

```
[P:Tm->Set]
[SAT1 = {M|Tm}(P M)->(SN M)]
[SAT2 = {M|Tm}(neutr M)->(SN M)
                ->({N:Tm}(W_Hd_Step M N)->(P N))->(P M)]
[SAT = SAT1 /\ SAT2];
Discharge P;
```

Luo shows that CR P implies SAT P ([Luo90], page 95) and remarks that the converse does not hold because saturated sets do not have to be closed under reduction. An example is the set of all strongly normalizing terms whose weak head normal form is neutral or equal to $\lambda x.II$, which is saturated but not closed under reduction.

If we want to show CR_ARR and Lam_sound formally we have to use RecSNN in a manner similar to the original proof. Therefore using saturated sets does not seem to simplify the proof.

## 4.3 Reduction for Church terms

We have only done the proofs for the Curry style systems — so one obvious question is how hard it would be to extend this proof to the Church style presentation, i.e. to terms with explicit type information. In the case of simply typed $\lambda$-calculus this is straightforward because every reduction on a typed term corresponds to one on its untyped counterpart and vice versa. However, this reasoning does not generalize to System F because here we have additional (second order) reductions on typed terms.

This problem is usually solved by arguing that the second order reductions are terminating anyway. Another way, maybe more amenable to formalization, would be to extend the notion of untyped terms and reduction:

```
mu[Tm:Set](...,
            T_Lam : Tm->Tm,
```

```
            T_App : Tm->Tm)

mu[Step:Tm->Tm->Set](...,
        Beta : {M:Tm}(Step (T_App (T_Lam M)) M) )
```

Note that `T_Lam` does not actually bind any term-variables but corresponds to second order abstraction for typed terms; analogously `T_App` is used as a dummy type application where the type is omitted.[20]

It does not seem hard to extend the proof to this notion of untyped terms. We have to extend the notion of neutrality, and the soundness of `Pi_i`, which was trivial so far, has to be proved as an additional property of `CR`. The result for Church terms now follows by observing that for the extended notion of untyped terms reductions coincide with the typed terms.

## 5  Future work

My original motivation for doing this formalization was actually to understand better how to extend the standard proofs to systems with inductive types, i.e. I was interested in checking whether a proof on paper was correct.

In [GLT89] strong normalization for System T is proved as an extension of simply typed $\lambda$-calculus. Here the type `Nat` is interpreted by `SN`. This seems to be a rather arbitrary choice and one consequence is that one has to do an additional induction to show that the elimination rule for `Nat` is sound.[21] This technique does not seem to generalize to non-algebraic types (at least one would need induction over an ordinal greater than $\omega$). We avoid this problem by using an impredicative trick on the level of semantics and apply this technique to the System T enriched by a type of notations for countable ordinals (we call this system $T^\Omega$). Here the usual structure of the proof is preserved, and the soundness of the elimination rules is trivial. I hope that I can now use the insights gained by doing this experiment to give strong normalisation proof for a system using a general notion of inductive types like the one we used to formalize this proof.[22]

## 6  Conclusions

The formalization of the proof in LEGO was done in a fairly short period of time — two weeks from typing in the first definition until proving the last lemma.

---

[20] It may just be a curiosity, but this version of untyped terms corresponds to (a special case of) partial terms. In [Pfe92] it is shown that type inference for partial terms is undecidable, which is still open for the usual notion of untyped terms.

[21] See [GLT89], p.49, case 4. of the proof.

[22] This should be compared with [Men88].

This does not account for the time needed to understand the proof or how to use LEGO properly.

When doing the formalization, I discovered that the core part of the proof (here proving the lemmas about CR) is fairly straightforward and only requires a good understanding of the paper version. However, in completing the proof I observed that in certain places I had to invest much more work than expected, e.g. proving lemmas about substitution and weakening. Although some may consider this as a point against formal proofs, I believe it is actually useful to check that basic definitions really reflect our intuitions about them. Many subtle errors could have been avoided this way.

However, the fact that formalizing the proof after understanding it was not so much of an additional effort seems to justify the believe that *Computer Aided Formal Reasoning* may serve as a useful tool in mathematical research in future. Using formal proofs simplifies the validation of results: we do not have do understand a subtle proof to know that the result is true, we only have to check that the formalization of the statement is correct. The process of checking whether the proof term really validates the statement can be completely mechanized. This sort of validation is not only relevant for mathematics but may prove to have a role in the area of program verification.

# Acknowledgements

Thanks to Randy Pollack for implementing LEGO and for his encouragement to pursue this project. I would also like to thank the following people for helpful discussions, help in proof reading, etc: Matt Fairtlough, Healfdene Goguen, Martin Hofmann, Zhaohui Luo, Benjamin Pierce and Benjamin Werner. Thanks to the referees for their comments.

# References

[Alt90]   Thorsten Altenkirch. Impredicative representations of categorical datatypes, thesis proposal, October 1990.

[B$^+$91]  Henk Barendregt et al. Summer school on $\lambda$ calculus, 1991.

[Ber91]   Stefano Berardi. Girard's normalisation proof in LEGO. unpublished draft, 1991.

[Coq92a]  Catarina Coquand. A proof of normalization for simply typed lambda calculus written in ALF. In *Workshop on Logical Frameworks*, 1992. Preliminary Proceedings.

[Coq92b]  Thierry Coquand. Pattern matching with dependent types. In *Workshop on Logical Frameworks*, 1992. Preliminary Proceedings.

[CP89]   Thierry Coquand and Christine Paulin. Inductively defined types. In Peter Dybjer et al., editors, *Proceedings of the Workshop on Programming Logic*, 1989.

[D+91]   Gilles Dowek et al. *The Coq Proof Assistant User's Guide*. INRIA-Rocquencourt — CNRS-ENS Lyon, 1991. Version 5.6.

[dB72]   N.G. de Bruijn. Lambda calculus notation with nameless dummies, a tool for automatic formula manipulation, with application to the church-rosser theorem. *Indag. Math.*, 34, 1972.

[Gal90]  J.H. Gallier. On Girard's "Candidats de Reducibilité". In Piergiogio Oddifreddi, editor, *Logic and Computer Science*. Academic Press, 1990.

[GLT89]  J.-Y. Girard, Y. Lafont, and P. Taylor. *Proofs and Types*. Cambridge University Press, 1989.

[Hof92]  Martin Hofmann. Formal development of functional programs in type theory — a case study. LFCS Report ECS-LFCS-92-228, University of Edinburgh, 1992.

[Hue80]  Gérard Huet. Confluent reductions. *JACM*, 27(4):797–821, 1980.

[Hue92]  Gérard Huet. Initiation au lambda calcul. Lecture notes, 1992.

[LP92]   Zhaohui Luo and Robert Pollack. The LEGO proof development system: A user's manual. LFCS report ECS-LFCS-92-211, University of Edinburgh, 1992.

[Luo90]  Zhaohui Luo. *An Extended Calculus of Constructions*. PhD thesis, University of Edinburgh, 1990.

[Mag92]  Lena Magnusson. The new implementation of ALF. In *Workshop on Logical Frameworks*, 1992. Preliminary Proceedings.

[Men88]  N.P. Mendler. *Inductive Definition in Type Theory*. PhD thesis, Cornell University, 1988.

[NPS90]  Bengt Nordström, Kent Petersson, and Jan Smith. *Programming in Martin-Löf's Type Theory*. Oxford University Press, 1990.

[Pfe92]  Frank Pfenning. On the undecidability of partial polymorphic type reconstruction. Technical Report CMU-CS-92-105, Carnegie Mellon University, January 1992.

[Wer92]  Benjamin Werner. A normalization proof for an impredicative type system with large eliminations over integers. In *Workshop on Logical Frameworks*, 1992. Preliminary Proceedings.

# Partial Intersection Type Assignment in Applicative Term Rewriting Systems

Steffen van Bakel*

Department of Informatics, Faculty of Mathematics and Informatics,
University of Nijmegen, Toernooiveld 1, 6525 ED Nijmegen, The Netherlands.
steffen@cs.kun.nl

## Abstract

This paper introduces a notion of partial type assignment on applicative term rewriting systems that is based on a combination of an essential intersection type assignment system, and the type assignment system as defined for ML [16], both extensions of Curry's type assignment system [11]. Terms and rewrite rules will be written as trees, and type assignment will consists of assigning intersection types function symbols, and specifying the way in which types can be assigned to nodes and edges between nodes. The only constraints on this system are local: they are imposed by the relation between the type assigned to a node and those assigned to its incoming and out-going edges. In general, given an arbitrary typeable applicative term rewriting system, the subject reduction property does not hold. We will formulate a sufficient but undecidable condition typeable rewrite rules should satisfy in order to obtain this property.

## Introduction

In the recent years several paradigms have been investigated for the implementation of functional programming languages. Not only the lambda calculus [5], but also term rewriting systems [15] and term graph rewriting systems [7] are topics of research. Lambda calculus (or rather combinator systems) forms the underlying model for the functional programming language Miranda [22], term rewriting systems are used in the underlying model for the language OBJ [14], and term graph rewriting systems is the model for the language Clean [8, 17].

The lambda calculus, term rewriting systems and graph rewriting systems themselves are type free, whereas in programming the notion of types plays an important role. Type assignment to programs and objects is in fact a way of performing abstract interpretation that provides necessary information for both compilers and programmers. Since the lambda calculus is a fundamental basis for many functional programming languages, a type assignment system for the pure untyped lambda calculus, capable of deducing meaningful and expressive types, has been a topic of research for many years.

There exists a well understood and well defined notion of type assignment on lambda terms, known as the Curry type assignment system [11] which expresses abstraction and

---
*Partially supported by the Esprit Basic Research Action 3074 "Semagraph" and the Netherlands Organisation for the advancement of pure research (N.W.O.).

application. Many of the now existing type assignment systems for functional programming languages are based on (extensions of) the Curry type assignment system.

In [9, 6] the intersection type discipline for the lambda calculus is presented, an extension of Curry's type assignment system. The extension consists of allowing more than one type for term-variables and adding a type constant '$\omega$' and, next to the type constructor '$\rightarrow$', the type constructor '$\cap$'. This yields a type assignment system that is very powerful: it is closed under $\beta$-equality. Because of this power, type assignment in this system (and even in the system that does not contain $\omega$, see [2]) is undecidable. The essential type assignment system as presented in this paper is a restriction of the intersection type discipline presented in [6] that satisfies all properties of that system, and is also an extension of the Curry type assignment system. The main advantage of the essential system over the intersection system is that the set of types assignable to a term is significantly smaller.

Most functional programming languages, like Miranda for instance, allow programmers to specify an algorithm (function) as a set of rewrite rules. The type assignment systems incorporated in most term rewriting languages are in fact extensions of type assignment systems for a(n extended) lambda calculus, and although it seems straightforward to generalize those systems to the (significantly larger) world of term rewriting systems, it is at first look not evident that those borrowed systems have still all the properties they possessed in the world of lambda calculus. For example, type assignment in term rewriting systems in general does not satisfy the subject reduction property: i.e. types are not preserved under rewriting, as illustrated in [4]. In order to be able to study the details of type assignment for term rewriting systems, a formal notion of type assignment on term rewriting systems is needed, that is more close to the approach of type assignment in lambda calculus than the algebraic one [12].

The aim of this paper is to present a formal notion of type assignment on term rewriting systems that is close to those defined for the lambda calculus and use intersection types.

The notion of type assignment presented here for term rewriting systems is based on both an essential type assignment system for the lambda calculus and the polymorphic type assignment system for the functional programming language ML [16]. The polymorphic aspect of type assignment can be found in the use of an environment, that provides a type for every function symbol $F$; for every occurrence of $F$ the way in which its type can be obtained from the one provided by the environment is specified.

Intersection types are studied because they are a good means to perform abstract interpretation, better than Curry types, also even better than the kind of types used in languages like ML. Also, the notion of type assignment presented in this paper could be extended to the world of term graph rewriting systems, and in that world intersection types are the natural tool to type nodes that are shared. Moreover, intersection types seem to be promising for use in functional programming languages, since they seem to provide a good formalism to express overloading (see also [20]).

In this paper we define applicative term rewriting systems (ATRS), a slight extension of the term rewriting systems as defined in [15], as the term rewriting systems that contain a special binary operator $Ap$. The applicative term rewriting systems defined in this paper are extensions to those suggested by most functional programming languages in that they do not discriminate against the varieties of function symbols that can be used in patterns.

In [4] and [3] partial type assignment systems for (left linear) applicative term rewriting systems are presented. The system presented here can be seen as a variant of those systems; the main difference between those two systems and the one presented here are in the set of types that can be assigned to nodes and edges: Curry types in [4], intersection types of Rank 2 in [3], and strict intersection types in this one. Also, type assignment in those systems is decidable, but in the one presented in this paper it is not.

Throughout this paper, the symbol $\varphi$ (often indexed, like in $\varphi_i$) will be a type-variable; when writing a type-variable $\varphi_i$, sometimes only the index $i$ is used, so as to obtain more readable types. Greek symbols like $\alpha$, $\beta$, $\gamma$, $\mu$, $\nu$, $\eta$, $\rho$, $\sigma$ and $\tau$ (often indexed) will range over types. To avoid parentheses in the notation of types, '$\rightarrow$' is assumed to associate to the right – so right-most, outer-most brackets will be omitted – and, as in logic, '$\cap$' binds stronger than '$\rightarrow$'. The symbol $B$ is used for bases.

Because of the restricted length of this paper, all results are presented without proofs.

## 1 Essential type assignment for the lambda calculus

In this section we present the essential type assignment system, a restricted version of the system presented in [6], together with some of its properties. The major feature of this restricted system is, compared to that system, a restricted version of the derivation rules and it is based on a set of strict types.

Strict types are the types that are strictly needed to assign a type to a term in the system of [6]. We will assume that $\omega$ is the same as an intersection over zero elements: if $n = 0$, then $\sigma_1 \cap \cdots \cap \sigma_n = \omega$, so $\omega$ does not occur in an intersection subtype. Moreover, intersection type schemes (so also $\omega$) occur in strict types only as subtypes at the left hand side of an arrow type scheme. We could have omitted the type constant $\omega$ completely from the presentation of the system, because we can always assume that $n = 0$ in $\sigma_1 \cap \cdots \cap \sigma_n$, but some of the definitions and the results we obtain are more clear when $\omega$ is dealt with explicitly.

### 1.1 Essential type assignment

**Definition 1.1.1** (cf. [2]) i) $\mathcal{T}_s$, the set of *strict types*, is inductively defined by:
    a) All type-variables $\varphi_0, \varphi_1, \ldots \in \mathcal{T}_s$.
    b) If $\tau, \sigma_1, \ldots, \sigma_n \in \mathcal{T}_s$ ($n \geq 0$), then $\sigma_1 \cap \cdots \cap \sigma_n \rightarrow \tau \in \mathcal{T}_s$.
ii) $\mathcal{T}_S$, the set of *strict intersection types*, is defined by: If $\sigma_1, \ldots, \sigma_n \in \mathcal{T}_s$ ($n \geq 0$), then $\sigma_1 \cap \cdots \cap \sigma_n \in \mathcal{T}_S$.
iii) On $\mathcal{T}_S$, the relation $\leq_S$ is defined by:
    a) $\forall\, 1 \leq i \leq n\ (n \geq 1)\, [\, \sigma_1 \cap \cdots \cap \sigma_n \leq_S \sigma_i\, ]$.
    b) $\forall\, 1 \leq i \leq n\ (n \geq 0)\, [\, \sigma \leq_S \sigma_i\, ] \Rightarrow \sigma \leq_S \sigma_1 \cap \cdots \cap \sigma_n$.
    c) $\sigma \leq_S \tau \leq_S \rho \Rightarrow \sigma \leq_S \rho$.
iv) We define the relation $\leq_E$ on $\mathcal{T}_S$ like the relation $\leq_S$, that is only defined for strict intersection types, but we add an extra alternative.
    d) $\rho \leq_E \sigma\ \&\ \tau \leq_E \mu \Rightarrow \sigma \rightarrow \tau \leq_E \rho \rightarrow \mu$.
v) On $\mathcal{T}_S$, the relation $\sim_E$ is defined by: $\sigma \sim_E \tau \Leftrightarrow \sigma \leq_E \tau \leq_E \sigma$.
vi) A *statement* is an expression of the form $M:\sigma$, where $M \in \Lambda$ and $\sigma \in \mathcal{T}_S$. $M$ is the *subject* and $\sigma$ the *predicate* of $M:\sigma$.
vii) A *basis* is a set of statements with only distinct variables as subjects.
    If $\sigma_1 \cap \cdots \cap \sigma_n$ is a predicate in a basis, then $n \geq 1$.

$\mathcal{T}_S$ may be considered modulo $\sim_E$. Then $\leq_E$ becomes a partial order, and in this paper we consider types modulo $\sim_E$.

Unless stated otherwise, if $\sigma_1 \cap \cdots \cap \sigma_n$ is used to denote a type, then by convention all $\sigma_1, \ldots, \sigma_n$ are assumed to be strict. Notice that $\mathcal{T}_s$ is a proper subset of $\mathcal{T}_S$.

**Definition 1.1.2** i) *Essential type assignment* and *essential derivations* are defined by the following natural deduction system (where all types displayed are strict, except $\sigma$ in the derivation rules ($\rightarrow$I) and ($\leq_E$)):

$$(\rightarrow I): \quad \frac{\begin{array}{c}[x:\sigma]\\ \vdots\\ M:\tau\end{array}}{\lambda x.M:\sigma\rightarrow\tau}\ (a) \qquad (\leq_E): \quad \frac{x:\sigma \quad \sigma \leq_E \tau}{x:\tau}$$

$$(\rightarrow E): \quad \frac{M:\sigma_1\cap\cdots\cap\sigma_n\rightarrow\tau \quad N:\sigma_1 \ldots N:\sigma_n}{MN:\tau}\ (n \geq 0)$$

(a) If $x:\sigma$ is the only statement about $x$ on which $M:\tau$ depends.

If $M:\sigma$ is derivable from $B$ using an essential derivation, we write $B \vdash_e M:\sigma$.

ii) We define $\vdash_E$ by: $B \vdash_E M:\sigma$ if and only if: there are $\sigma_1, \ldots, \sigma_n$ ($n \geq 0$) such that $\sigma = \sigma_1 \cap \cdots \cap \sigma_n$ and for every $1 \leq i \leq n$ $B \vdash_e M:\sigma_i$.

Although the derivation rule ($\leq_E$) is not allowed on all terms, we can prove that if $B \vdash_E M:\sigma$, and $\sigma \leq_E \tau$, then $B \vdash_E M:\tau$. It is then easy to prove that type assignment in this system is closed under $\eta$-reduction. It is also possible to prove that the essential type assignment system satisfies the main properties of the BCD-system:

**Property 1.1.3** i) $B \vdash_E M:\sigma$ & $M =_\beta N \Rightarrow B \vdash_E N:\sigma$.
ii) $\exists B, \sigma [ B \vdash_E M:\sigma$ & $B, \sigma$ $\omega$-free $] \Leftrightarrow M$ has a normal form.
iii) $\exists B, \sigma [ B \vdash_E M:\sigma$ & $\sigma \neq \omega ] \Leftrightarrow M$ has a head normal form. ∎

## 1.2 Operations on pairs

In this subsection we present three different operations on pairs of <*basis, type*>, namely substitution, expansion, and lifting as defined in [1]. The operation of substitution deals with the replacement of type-variables by types and is a slight modification of the one normally used; this modification is needed to make sure that substitution is closed on strict types. The operation of expansion replaces types by the intersection of a number of copies of that type and coincides with the one given in [10, 21]. The operation of lifting deals with the introduction of extra (types to) statements in the basis of a derivation, or introduces extra types to term-variables that are bound.

Substitution is normally defined on types as the operation that replaces type-variables by types. For strict types this definition would not be correct. For example, the replacement of $\varphi$ by $\omega$ would transform $\sigma \rightarrow \varphi$ (or $\sigma \cap \varphi$) into $\sigma \rightarrow \omega$ ($\sigma \cap \omega$), which is not a strict type. Therefore, for strict types substitution is not defined as an operation that replaces type-variables by types, but as a mapping from types to types.

**Definition 1.2.1** [1] i) The *substitution* $(\varphi := \alpha) : T_S \to T_S$, where $\varphi$ is a type-variable and $\alpha \in T_s \cup \{\omega\}$, is defined by:
  a) $(\varphi := \alpha)(\varphi) = \alpha$.
  b) $(\varphi := \alpha)(\varphi') = \varphi'$, if $\varphi \neq \varphi'$.
  c) $(\varphi := \alpha)(\sigma \to \tau) = \omega$, if $(\varphi := \alpha)(\tau) = \omega$.
  d) $(\varphi := \alpha)(\sigma \to \tau) = (\varphi := \alpha)(\sigma) \to (\varphi := \alpha)(\tau)$, if $(\varphi := \alpha)(\tau) \neq \omega$.
  e) $(\varphi := \alpha)(\sigma_1 \cap \cdots \cap \sigma_n) = (\varphi := \alpha)(\sigma_1') \cap \cdots \cap (\varphi := \alpha)(\sigma_m')$,
  where $\{\sigma_1', \ldots, \sigma_m'\} = \{\sigma_i \in \{\sigma_1, \ldots, \sigma_n\} \mid (\varphi := \alpha)(\sigma_i) \neq \omega\}$.
ii) If $S_1$ and $S_2$ are substitutions, then so is $S_1 \circ S_2$, where $S_1 \circ S_2(\sigma) = S_1(S_2(\sigma))$.
iii) $S(B) = \{x{:}S(\alpha) \mid x{:}\alpha \in B \ \& \ S(\alpha) \neq \omega\}$.
iv) $S(<B, \sigma>) = <S(B), S(\sigma)>$.

Notice that in part (i.e), if for $1 \leq i \leq n$ $(\varphi := \alpha)(\sigma_i) = \omega$, then $(\varphi := \alpha)(\sigma_1 \cap \cdots \cap \sigma_n) = \omega$.

The operation of expansion is an operation on types that deals with the replacement of (sub)types by an intersection of a number of copies of that type. In this process it can be that also other types need to be copied. An expansion indicates not only the type to be expanded, but also the number of copies that has to be generated.

**Definition 1.2.2** [1] i) The *last variable* of a strict type is defined by:
  a) The last variable of $\varphi$ is $\varphi$.
  b) The last variable of $\sigma_1 \cap \cdots \cap \sigma_n \to \tau$ $(n \geq 0)$ is the last variable of $\tau$.
ii) A strict type $\sigma$ is said to *end with* $\varphi$, if $\varphi$ is the last variable of $\sigma$.

**Definition 1.2.3** [1] For every $\mu \in T_s$, $n \geq 2$, basis $B$ and $\sigma \in T_S$, the quadruple $<\mu, n, B, \sigma>$ determines an *expansion* $E_{<\mu,n,B,\sigma>} : T_S \to T_S$, that is constructed as follows.
  i) The set of type-variables $\mathcal{V}_\mu(<B, \sigma>)$ is constructed by:
    a) If $\varphi$ occurs in $\mu$, then $\varphi \in \mathcal{V}_\mu(<B, \sigma>)$.
    b) Let $\tau$ be a strict (sub)type occurring in $<B, \sigma>$, with last variable $\varphi_0$. If $\varphi_0 \in \mathcal{V}_\mu(<B, \sigma>)$, then for all type-variables $\varphi$ that occur in $\tau$: $\varphi \in \mathcal{V}_\mu(<B, \sigma>)$.
  ii) Suppose $\mathcal{V}_\mu(<B, \sigma>) = \{\varphi_1, \ldots, \varphi_m\}$. Choose $m \times n$ different type-variables $\varphi_1^1, \ldots, \varphi_1^n, \ldots, \varphi_m^1, \ldots, \varphi_m^n$, such that each $\varphi_j^i$ does not occur in $<B, \sigma>$, for $1 \leq i \leq n$ and $1 \leq j \leq m$. Let $S_i$ be the substitution that replaces every $\varphi_j$ by $\varphi_j^i$.
  iii) $E_{<\mu,n,B,\sigma>}(\alpha)$ is obtained by traversing $\alpha$ top-down and replacing, in $\alpha$, a subtype $\beta$ that ends with an element of $\mathcal{V}_\mu(<B, \sigma>)$ by $S_1(\beta) \cap \cdots \cap S_n(\beta)$.
  iv) $E_{<\mu,n,B,\sigma>}(B') = \{x{:}E_{<\mu,n,B,\sigma>}(\rho) \mid x{:}\rho \in B'\}$.
  v) $E_{<\mu,n,B,\sigma>}(<B', \sigma'>) = <E_{<\mu,n,B,\sigma>}(B'), E_{<\mu,n,B,\sigma>}(\sigma')>$.

The last operation on pairs defined in this subsection is the operation of lifting.

**Definition 1.2.4** [1] A *lifting* $L$ is denoted by a pair $<<B_0, \tau_0>, <B_1, \tau_1>>$ such that $\tau_0 \leq_E \tau_1$ and $B_1 \leq_E B_0$, and is defined by:
  i) $L(\sigma) = \tau_1$ if $\sigma = \tau_0$; $L(\sigma) = \sigma$ otherwise.
  ii) $L(B) = B_1$ if $B = B_0$; $L(B) = B$ otherwise.
  iii) $L(<B, \sigma>) = <L(B), L(\sigma)>$.

**Definition 1.2.5** [1] A *chain* is an object $<O_1,\ldots,O_n>$, with each $O_i$ an operation of substitution, expansion or lifting; $<O_1,\ldots,O_n>(<B,\sigma>) = O_n(\cdots(O_1(<B,\sigma>))\cdots)$.

It is possible to prove the principal type property for the essential type assignment system, in the same way as done in [21] for the BCD-system. The operations needed for this proof are substitution, expansion, and lifting, and it is possible to show that all pairs for a term can be generated by chains that exist of expansions, and substitutions (in that order) and that end with one lifting. Moreover, all three operations can be proven to be sound on all pairs.

## 2  Type assignment in Applicative Term Rewriting Systems

In this paper we study type assignment on applicative term rewriting systems, which is a slight extension of the term rewriting systems as defined in [15]. Applicative term rewriting systems are defined as term rewriting systems that (can) contain a special binary operator *Ap*, this in contrast to the *pure* applicative term rewriting systems, that contain *only* the binary operator *Ap*.

### 2.1  Applicative Term Rewriting Systems

The motivation for the use of applicative term rewriting systems instead of the general term rewriting systems can be illustrated by the following: There is a clear translation (embedding) of combinator systems into term rewriting systems, in which the implicit application of the world of combinators is made explicit. The kind of term rewriting system that is needed for such a translation contains only one function symbol, called *Ap*, and is therefore often called an applicative term rewriting system. A translation of for example Combinatory Logic (CL)

$$S\,x\,y\,z = x\,z\,(y\,z)$$
$$K\,x\,y\ \ = x$$
$$I\,x\ \ \ \ = x$$

into such a term rewriting system then looks like:

$$Ap(Ap(Ap(S,x),y),z) \Rightarrow Ap(Ap(x,z),Ap(y,z))$$
$$Ap(Ap(K,x),y) \Rightarrow x$$
$$Ap(I,x) \Rightarrow x$$

The definition of applicative systems we present in this paper is, however, more general: in the systems we consider, *Ap* is a *special* function symbol; in particular it is *one of the function symbols*, not the only one. To distinguish between the term rewriting systems that contain *only* the function symbol *Ap* and those that contain *Ap* next to other function symbols, we call the former the *pure* applicative term rewriting systems.

We prefer to see the symbols $S$, $K$ and $I$ as functions, with 3, 2 and 1 operands respectively. This means that we have to introduce extra rewrite rules to express the Curried versions of these symbols. Moreover, to get some computational power, some rewrite rule starting with *Ap* should be added. Such an extended CL system could look like:

$$S(x,y,z) \Rightarrow Ap(Ap(x,z),Ap(y,z))$$
$$Ap(S_2(x,y),z) \Rightarrow S(x,y,z)$$
$$Ap(S_1(x),y) \Rightarrow S_2(x,y)$$
$$Ap(S_0,x) \Rightarrow S_1(x)$$

$$
\begin{aligned}
K(x,y) &\Rightarrow x \\
Ap(K_1(x),y) &\Rightarrow K(x,y) \\
Ap(K_0,x) &\Rightarrow K_1(x) \\
I(x) &\Rightarrow x \\
Ap(I_0,x) &\Rightarrow I(x)
\end{aligned}
$$

We consider the applicative rewriting systems, because they are far more general than the subclass of systems in which there exists only the function symbol $Ap$. Moreover, they are a natural extension of those rewrite systems considerd in papers on type assignment on term rewriting systems that follow the 'algebraic' approach [12], and are also the kind of rewrite systems an effecient implementation of a functional language would be based upon [18]. Since the pure applicative term rewriting systems are a subclass of the applicative term rewriting systems, all results obtained in this paper are also valid for that subclass.

We take the view that in a rewrite rule a certain symbol is defined; it is this symbol to which the structure of the rewrite rule gives a type. We treat $Ap$ as a predefined symbol; the symbol $Ap$ is neglected when we are looking for the symbol that is defined in a rewrite rule.

The type assignment system we present in this paper is a partial system in the sense of [19]: we not only define how terms and rewrite rules can be typed, but assume that every function symbol already has a type, which structure is usually motivated by a rewrite rule. There are several reasons to do so.

First of all a term rewriting system can contain symbols that is not the defined symbol of a rewrite rule (such a symbol is called a constant). A constant can appear in a rewrite rule more or less as a symbol that 'has to be there', but for which it is impossible to determine any functional characterisation, apart from what is demanded by the immediate context. If we provide a type for every constant, then we can formulate some consistency requirement, by saying that the types used for a constant must be related to the provided type.

Moreover, even for every defined symbol there must be some way of determining what type can be used for an occurrence. Normally the rewrite rules that define such a symbol are investigated, and from analyzing the structure of those rules the 'most general type' for that symbol can be constructed. Instead of for defined symbols investigating all their defining rules every time the symbol is encountered, we can store the type of the symbol in a mapping from symbols to types, and use this mapping instead. Of course it makes no difference to assume the existence from the start of such a mapping from symbols (both defined and constant) to types, and to define type assignment using that mapping (in the following such a mapping is called an 'environment').

**Definition 2.1.1** (cf. [15, 4]) An *Applicative Term Rewriting System* (ATRS) is a pair $(\Sigma, \mathbf{R})$ of an *alphabet* or *signature* $\Sigma$ and a set of *rewrite rules* $\mathbf{R}$.
  i) The alphabet $\Sigma$ consists of:
    a) A countable infinite set of variables $x_1, x_2, x_3, \ldots$ (or $x, y, z, x', y', \ldots$).
    b) A non empty set $\Sigma_0$ of *function symbols* $F, G, \ldots$, each equipped with an 'arity'.
    c) A special binary operator, called *application* ($Ap$).
  ii) The set of *terms* (or *expressions*) 'over' $\Sigma$ is $Ter(\Sigma)$ and is defined inductively:
    a) $x, y, z, \ldots \in Ter(\Sigma)$.
    b) If $F \in \Sigma_0 \cup \{Ap\}$ is an $n$-ary symbol, and $T_1, \ldots, T_n \in Ter(\Sigma)$, then $F(T_1, \ldots, T_n) \in Ter(\Sigma)$.

**Definition 2.1.2** (cf. [15, 4]) Let $(\Sigma, \mathbf{R})$ be an ATRS.
i) A *replacement* is a map $R : Ter(\Sigma) \to Ter(\Sigma)$ satisfying $R(F(T_1, \ldots, T_n)) = F(R(T_1), \ldots, R(T_n))$ for every $n$-ary function symbol $F \in \Sigma_0 \cup \{Ap\}$ (here $n \geq 0$). We also write $T^R$ instead of $R(T)$.
ii) a) A *rewrite rule* $\in \mathbf{R}$ is a pair $(Lhs, Rhs)$ of terms $\in Ter(\Sigma)$. Often a rewrite rule will get a name, e.g. $\mathbf{r}$, and we write $\mathbf{r} : Lhs \Rightarrow Rhs$. Three conditions will be imposed:
   1) $Lhs$ is not a variable.
   2) The variables occurring in $Rhs$ are contained in $Lhs$.
   3) For every $Ap$ in $Lhs$, the left hand argument is not a variable.
   b) A rewrite rule $\mathbf{r} : Lhs \Rightarrow Rhs$ determines a set of *rewrites* $Lhs^R \Rightarrow Rhs^R$ for all replacements $R$. The left hand side $Lhs^R$ is called a *redex*; it may be replaced by its *'contractum'* $Rhs^R$ inside a context $C[\ ]$; this gives rise to *rewrite steps*: $C[Lhs^R] \Rightarrow_\mathbf{r} C[Rhs^R]$.
iii) In a rewrite rule, the leftmost, outermost symbol in the left hand side that is not an $Ap$, is called the *defined symbol* of that rule. If the symbol $F$ is the defined symbol of a rewrite rule $\mathbf{r}$, then $\mathbf{r}$ *defines* $F$. $F$ is a *defined symbol*, if there is a rewrite rule that defines $F$. $Q \in \Sigma_0$ is called a *constant symbol* if $Q$ is not a defined symbol.
iv) For every defined symbol $F$ with arity $n \geq 1$, there are $n$ additional rewrite rules that define the function symbols $F_0$ upto $F_{n-1}$ as follows:
$$Ap(F_{n-1}(x_1, \ldots, x_{n-1}), x_n) \Rightarrow F(x_1, \ldots, x_n)$$
$$Ap(F_{n-2}(x_1, \ldots, x_{n-2}), x_{n-1}) \Rightarrow F_{n-1}(x_1, \ldots, x_{n-1})$$
$$\vdots$$
$$Ap(F_0, x_1) \Rightarrow F_1(x_1)$$
The added rules with $F_{n-1}, \ldots, F_1, F_0$, etc. give in fact the 'Curried'-versions of $F$.

Part (ii.a.3) of definition 2.1.2 is added in order to avoid rewrite rules with left hand sides like $Ap(x, y)$, because such a rule would not have a defined symbol.

We will, for the sake of simplicity, assume that rewrite rules are not mutually recursive; for those rules the definition of defined symbol, and also that of type assignment, would be more complicated, and this would unnecessarily obscure this paper.

In the following definition we give a special applicative term rewriting system.

**Definition 2.1.3** *Applicative Combinatory Logic* (ACL) is the ATRS $(\Sigma, \mathbf{R})$, where $\Sigma_0 = \{S, S_2, S_1, S_0, K, K_1, K_0, I, I_0\}$, and $\mathbf{R}$ contains the rewrite rules
$$S(x, y, z) \Rightarrow Ap(Ap(x, z), Ap(y, z))$$
$$K(x, y) \Rightarrow x$$
$$I(x) \Rightarrow x.$$

For ACL we have for example the following rewriting sequence:
$$S(K_0, S_0, I_0) \Rightarrow Ap(Ap(K_0, I_0), Ap(S_0, I_0)) \Rightarrow Ap(K_1(I_0), Ap(S_0, I_0)) \Rightarrow K(I_0, Ap(S_0, I_0)) \Rightarrow I_0.$$

Notice that a term like $K_1(I_0)$ itself cannot be rewritten. This corresponds to the fact that in CL the term $KI$ is not a redex. Because ACL is Curry-closed, it is in fact combinatory complete: every lambda term can be translated into a term in ACL; for details of such a translation, see [5, 13].

**Example 2.1.4** If the left hand side of a rewrite rule is $F(T_1, \ldots, T_n)$, then the $T_i$ need not be simple variables, but can be terms as well, as for example in the rewrite rule $H(S_2(x,y)) \Rightarrow S_2(I_0, y)$.

It is also possible that for a certain symbol $F$, there are more than one rewrite rule that define $F$. For example the rewrite rules $F(x) \Rightarrow x$, $F(x) \Rightarrow Ap(x,x)$ are legal.

## 2.2 Essential type assignment in ATRS's

Partial intersection type assignment on an ATRS $(\Sigma, R)$ is defined as the labelling of nodes and edges in the tree-representation of terms and rewrite rules with types in $T_S$. In this labelling, we use that there is a mapping that provides a type in $T_S$ for every $F \in \Sigma_0 \cup \{Ap\}$. Such a mapping is called an environment.

**Definition 2.2.1** Let $(\Sigma, R)$ be an ATRS.
i) A mapping $\mathcal{E} : \Sigma_0 \cup \{Ap\} \to T_S$ is called an *environment* if $\mathcal{E}(Ap) = (1 \to 2) \to 1 \to 2$, and for every $F \in \Sigma_0$ with arity $n$, $\mathcal{E}(F) = \mathcal{E}(F_{n-1}) = \cdots = \mathcal{E}(F_0)$.
ii) For $F \in \Sigma_0$ with arity $n \geq 0$, $\sigma \in T_S$, and $\mathcal{E}$ an environment, the environment $\mathcal{E}[F := \sigma]$ is defined by:
   a) $\mathcal{E}[F := \sigma](G) = \sigma$, if $G \in \{F, F_{n-1}, \ldots, F_0\}$.
   b) $\mathcal{E}[F := \sigma](G) = \mathcal{E}(G)$, otherwise.

Since $\mathcal{E}$ maps all $F \in \Sigma_0$ to types in $T_S$, no function symbol is mapped to $\omega$.

Type assignment on applicative term rewriting systems is defined in two stages. In the next definition we define type assignment on terms, in definition 2.2.6 we define type assignment on term rewrite rules.

**Definition 2.2.2** Let $(\Sigma, R)$ be an ATRS, and $\mathcal{E}$ an environment.
i) We say that $T \in Ter(\Sigma)$ *is typeable by* $\sigma \in T_S$ *with respect to* $\mathcal{E}$, if there exists an assignment of types to edges and nodes that satisfies the following constraints:
   a) The root edge of $T$ is typed with $\sigma$; if $\sigma = \omega$, then the root edge is the only thing in the term-tree that is typed.
   b) The type assigned to a function node containing $F \in \Sigma_0 \cup \{Ap\}$ (where $F$ has arity $n \geq 0$) is $\tau_1 \cap \cdots \cap \tau_m$, if and only if for every $1 \leq i \leq m$ there are $\sigma_1^i, \ldots, \sigma_n^i \in T_S$, and $\sigma_i \in T_S$, such that $\tau_i = \sigma_1^i \to \cdots \to \sigma_n^i \to \sigma_i$, the type assigned to the $j$-th ($1 \leq j \leq n$) out-going edge is $\sigma_j^1 \cap \cdots \cap \sigma_j^m$, and the type assigned to the incoming edge is $\sigma_1 \cap \cdots \cap \sigma_m$.

$$\begin{array}{c} \Big\downarrow \sigma_1 \cap \cdots \cap \sigma_m \\ F:(\sigma_1^1 \to \cdots \to \sigma_n^1 \to \sigma_1) \cap \cdots \cap (\sigma_1^m \to \cdots \to \sigma_n^m \to \sigma_m) \\ \sigma_1^1 \cap \cdots \cap \sigma_1^m \swarrow \quad \cdots \quad \searrow \sigma_n^1 \cap \cdots \cap \sigma_n^m \\ \sigma_2^1 \cap \cdots \cap \sigma_2^m \qquad\qquad \sigma_{n-1}^1 \cap \cdots \cap \sigma_{n-1}^m \end{array}$$

   c) If the type assigned to a function node containing $F \in \Sigma_0 \cup \{Ap\}$ is $\tau$, then there is a chain $C$, such that $C(\mathcal{E}(F)) = \tau$.

ii) Let $T \in \mathit{Ter}(\Sigma)$ be typeable by $\sigma$ with respect to $\mathcal{E}$. If $B$ is a basis such that for every statement $x{:}\tau$ occurring in the typed term-tree there is a $x{:}\tau' \in B$ such that $\tau' \leq_E \tau$, we write $B \vdash_{\mathcal{E}} T{:}\sigma$.

Notice that if $B \vdash_{\mathcal{E}} T{:}\sigma$, then $B$ can contain more statements than needed to obtain $T{:}\sigma$. Notice also that parts (i.a) and (ii) are not in conflict so for every $B$ and $T$: $B \vdash_{\mathcal{E}} T{:}\omega$.

A typical example for part (i.b) of definition 2.2.2 is the symbol $Ap$; for every occurrence of $Ap$ in a term-tree, there are $\sigma$ and $\tau$ such that the following is part of the term-tree.

$$\begin{array}{c} \downarrow \tau \\ Ap{:}(\sigma{\to}\tau){\to}\sigma{\to}\tau \\ \sigma{\to}\tau \swarrow \qquad \searrow \sigma \end{array}$$

Notice that the type the environment provides for $Ap$ is crucial; it is the type suggested by the ($\to$E) derivation rule, and gives structure to the type assignment.

**Example 2.2.3** The term $S(K_0, S_0, I_0)$ can be typed with the type $7{\to}7$, under the assumption that: $\mathcal{E}(S) = (1{\to}2{\to}3){\to}(4{\to}2){\to}1{\cap}4{\to}3$, $\mathcal{E}(K) = 5{\to}\omega{\to}5$, $\mathcal{E}(I) = 6{\to}6$.

$$\begin{array}{c} \downarrow 7{\to}7 \\ S{:}((7{\to}7){\to}\omega{\to}7{\to}7) \to \omega \to (7{\to}7) \to 7{\to}7 \\ \swarrow \qquad \downarrow \qquad \searrow \\ K_0{:}(7{\to}7){\to}\omega{\to}7{\to}7 \qquad S_0 \qquad I_0{:}7{\to}7 \end{array}$$

Notice that to obtain the type $((7{\to}7){\to}\omega{\to}7{\to}7){\to}\omega{\to}(7{\to}7){\to}7{\to}7$ for $S$, we have used the chain $<(1 := 7{\to}7), (2 := \omega), (3 := 7{\to}7), (4 := \omega)>$, and that the node containing $S_0$ is not typed since the incoming edge is typed with $\omega$. If we define $D(x) \Rightarrow Ap(x, x)$, then we can even check that for example $D(S(K_0, S_0, I_0))$ and $D(I_0)$ are both typeable by $8{\to}8$.

The following definition introduces some terminology and notations for bases.

**Definition 2.2.4** i) The relation $\leq_E$ is extended to bases by: $B \leq_E B'$ if and only if for every $x{:}\sigma' \in B'$ there is an $x{:}\sigma \in B$ such that $\sigma \leq_E \sigma'$.
ii) If $B_1, \ldots, B_n$ are bases, then $\Pi\{B_1, \ldots, B_n\}$ is the basis defined as follows: $x{:}\sigma_1 \cap \cdots \cap \sigma_m \in \Pi\{B_1, \ldots, B_n\}$ if and only if $\{x{:}\sigma_1, \ldots, x{:}\sigma_m\}$ is the set of all statements whose subject is $x$ that occur in $B_1 \cup \ldots \cup B_n$.

Notice that if $n = 0$, then $\Pi\{B_1, \ldots, B_n\} = \emptyset$.

In the next definition we introduce the notion of used bases. The idea is to collect all types assigned to term-variables that are actually used for the typed term-tree, but the collected types need not occur in the original bases themselves.

**Definition 2.2.5** i) The *used bases of* $B \vdash_{\mathcal{E}} T{:}\sigma$ are inductively defined by:
   a) $\sigma \in \mathcal{T}_s$.
   1) $T \equiv x$. Take $\{x{:}\sigma\}$.
   2) $T \equiv F(T_1, \ldots, T_n)$. There are $\sigma_1, \ldots, \sigma_n$ such that for every $1 \leq i \leq n$ $B \vdash_{\mathcal{E}} T_i{:}\sigma_i$. Let for $1 \leq i \leq n$, $B_i$ be a used basis of $B \vdash_{\mathcal{E}} T_i{:}\sigma_i$.

Take $\Pi\{B_1, \ldots, B_n\}$.

b) If $\sigma = \sigma_1 \cap \cdots \cap \sigma_n$ ($n \geq 0$), then for every $1 \leq i \leq n$ $B \vdash_{\mathcal{E}} T:\sigma_i$. Let for every $1 \leq i \leq n$, $B_i$ be a used basis of $B \vdash_{\mathcal{E}} T:\sigma_i$. Take $\Pi\{B_1, \ldots, B_n\}$.

ii) A basis $B$ *is used for* $T:\sigma$ *with respect to* $\mathcal{E}$ if and only if there is a basis $B'$ such that $B' \vdash_{\mathcal{E}} T:\sigma$ and $B$ is a used basis of $B' \vdash_{\mathcal{E}} T:\sigma$.

Notice that in part (i.b), if $n = 0$, then $\sigma = \omega$, and $\Pi\{B_1, \ldots, B_n\} = \emptyset$.

We will say '$B$ is used for $T:\sigma$' instead of '$B$ is used for $T:\sigma$ with respect to $\mathcal{E}$'. A used basis for a statement $T:\sigma$ is not unique, but the results of this paper do not depend on the actual structure of such a basis, only on its existence. Thanks to the notion of used basis, we can give a clear definition of a typeable rewrite rule and a typeable rewrite system. The condition '$B$ is used for $Lhs:\sigma$' in definition 2.2.6 (i.a) is crucial. Just saying:

We say that $Lhs \Rightarrow Rhs \in \mathbf{R}$ with defined symbol $F$ *is typeable with respect to* $\mathcal{E}$, if there are basis $B$, and type $\sigma \in \mathcal{T}_s$ such that: $B \vdash_{\mathcal{E}} Lhs:\sigma$ and $B \vdash_{\mathcal{E}} Rhs:\sigma$,

would give a notion of type assignment that is not closed under rewriting (i.e. does not satisfy the subject reduction property), and is not a natural extension of the essential intersection type assignment system for the $\lambda$-calculus. For an example of the first, take the rewrite system $I(x) \Rightarrow x$, $K(x,y) \Rightarrow x$, $F(I_0) \Rightarrow I_0$, $G(x) \Rightarrow F(x)$.

Take the environment $\mathcal{E}(G) = (5 \rightarrow \omega \rightarrow 5) \rightarrow 6 \rightarrow 6$, $\mathcal{E}(F) = (3 \rightarrow 3) \rightarrow 4 \rightarrow 4$, $\mathcal{E}(K) = 2 \rightarrow \omega \rightarrow 2$, $\mathcal{E}(I) = 1 \rightarrow 1$. Take $B = \{x:(7 \rightarrow 7) \cap (5 \rightarrow \omega \rightarrow 5)\}$, then $B \vdash_{\mathcal{E}} G(x):6 \rightarrow 6$, and $B \vdash_{\mathcal{E}} F(x):6 \rightarrow 6$. Notice that $\vdash_{\mathcal{E}} G(K_0):7 \rightarrow 7$, but not $\vdash_{\mathcal{E}} F(K_0):7 \rightarrow 7$.

Therefore, a minimal requirement for subject reduction is to demand that all types assigned to term-variables in the typed term-tree for the right hand side of a rewrite rule already occurred in the typed term-tree for the left hand side. This is accomplished by restricting the possible bases to those that contain nothing but the types actually used for the left hand side.

**Definition 2.2.6** Let $(\Sigma, \mathbf{R})$ be an ATRS, and $\mathcal{E}$ an environment.

i) We say that $Lhs \Rightarrow Rhs \in \mathbf{R}$ with defined symbol $F$ *is typeable with respect to* $\mathcal{E}$, if there are basis $B$, type $\sigma \in \mathcal{T}_s$, and an assignment of types to nodes and edges such that:

  a) $B$ is used for $Lhs:\sigma$ and $B \vdash_{\mathcal{E}} Rhs:\sigma$.
  b) In $B \vdash_{\mathcal{E}} Lhs:\sigma$ and $B \vdash_{\mathcal{E}} Rhs:\sigma$, all nodes containing $F$ are typed with $\mathcal{E}(F)$.

ii) We say that $(\Sigma, \mathbf{R})$ *is typeable with respect to* $\mathcal{E}$, if every $\mathbf{r} \in \mathbf{R}$ is typeable with respect to $\mathcal{E}$.

Condition (i.b) of definition 2.2.6 is in fact added to make sure that the type provided by the environment for a function symbol $F$ is not in conflict with the rewrite rules that define $F$. By restricting the type that can be assigned to the defined symbol to the type provided by the environment, we are sure that the rewrite rule is typed using that type, and not using some other type. Since by part (i.b) of definition 2.2.6 all occurrences of the defined symbol in a rewrite rule are typed with the same type, type assignment of rewrite rules is actually defined using Milner's way of dealing with recursion.

It is easy to check that if $F$ is a function symbol with arity $n$, and all rewrite rules that define $F$ are typeable, then there are $\gamma_1, \ldots, \gamma_n, \gamma$ such that $\mathcal{E}(F) = \gamma_1 \rightarrow \cdots \rightarrow \gamma_n \rightarrow \gamma$.

The use of an environment and part (i.c) of definition 2.2.2 introduces a notion of polymorphism into our type assignment system. The environment returns the 'principal type' for a function symbol; this symbol can be used with types that are 'instances' of its principal type.

**Example 2.2.7** Typed versions of some of the rewrite rules given in definition 2.1.3, under the assumption that: $\mathcal{E}(S) = (1{\to}2{\to}3){\to}(4{\to}2){\to}1\cap 4{\to}3$, $\mathcal{E}(K) = 5{\to}\omega{\to}5$, $\mathcal{E}(I) = 6{\to}6$.

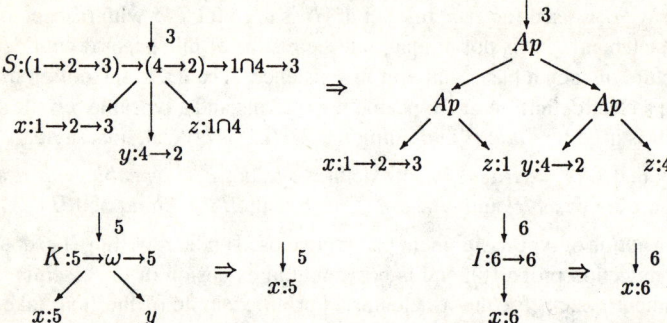

Notice that the node containing $y$ in the rule for $K$ is not typed; its incoming edge is typed with $\omega$.

**Example 2.2.8** Typed versions of the the rewrite rules given in example 2.1.4, using: $\mathcal{E}(H) = (1{\to}2){\to}(3{\to}4)\cap 1{\to}4$, $\mathcal{E}(S) = (7{\to}8{\to}9){\to}(10{\to}8){\to}7\cap 10{\to}9$, $\mathcal{E}(I) = 11{\to}11$, $\mathcal{E}(F) = 6\cap(5{\to}6)\cap 5{\to}6$.

$$\begin{array}{c}
\downarrow (3{\to}4)\cap 1{\to}4 \\
H:(1{\to}2){\to}(3{\to}4)\cap 1{\to}4 \\
\downarrow \\
S_2:(1{\to}3{\to}2){\to}(1{\to}3){\to}1{\to}2 \\
\swarrow \qquad \searrow \\
x:1{\to}3{\to}2 \qquad y:1{\to}3
\end{array}
\Rightarrow
\begin{array}{c}
\downarrow (3{\to}4)\cap 1{\to}4 \\
S_2:((3{\to}4){\to}3{\to}4){\to}(1{\to}3){\to}(3{\to}4)\cap 1{\to}4 \\
\swarrow \qquad \searrow \\
I_0:(3{\to}4){\to}3{\to}4 \qquad y:1{\to}3
\end{array}$$

$$\begin{array}{c}
\downarrow 6 \\
F:6\cap(5{\to}6)\cap 5{\to}6 \\
\downarrow \\
x:6\cap(5{\to}6)\cap 5
\end{array}
\Rightarrow
\begin{array}{c}
\downarrow 6 \\
x:6
\end{array}
\qquad
\begin{array}{c}
\downarrow 6 \\
F:6\cap(5{\to}6)\cap 5{\to}6 \\
\downarrow \\
x:6\cap(5{\to}6)\cap 5
\end{array}
\Rightarrow
\begin{array}{c}
\downarrow 6 \\
Ap \\
\swarrow \searrow \\
x:5{\to}6 \quad x:5
\end{array}$$

## 2.3 Soundness of operations

It is possible to show that the three operations on pairs (substitution, expansion, and lifting) are sound on typed term-trees. For the operations of substitution and expansion it is also possible to show that part (i.c) of definition 2.2.2 is sound in the following sense: if there is an operation $O$ (either a substitution or an expansion) such that $O(\mathcal{E}(F)) = \sigma$, then for every type $\tau \in \mathcal{T}_s$ such that $\sigma \leq_S \tau$, the rewrite rules that define $F$ are typeable with respect to the changed environment $\mathcal{E}[F := \tau]$. It is not possible to prove such a property for the operation of lifting.

**Theorem 2.3.1** *Soundness of substitution.* Let $S$ be a substitution.
  i) If $B \vdash_{\mathcal{E}} T{:}\sigma$, then $S(B) \vdash_{\mathcal{E}} T{:}S(\sigma)$.
  ii) If $B$ is used for $T{:}\sigma$, then $S(B)$ is used for $T{:}S(\sigma)$.
  iii) Let $\mathbf{r}: Lhs \Rightarrow Rhs$ be a rewrite rule typeable with respect to the environment $\mathcal{E}$, and let $F$ be the defined symbol of $\mathbf{r}$. Then $\mathbf{r}$ is typeable with respect to $\mathcal{E}[F := S(\mathcal{E}(F))]$. ∎

**Theorem 2.3.2** *Soundness of expansion.* Let $E$ be an expansion such that $E(<B, \sigma>) = <B', \sigma'>$.
  i) If $B \vdash_{\mathcal{E}} T{:}\sigma$, then $B' \vdash_{\mathcal{E}} T{:}\sigma'$.
  ii) If $B$ is used for $T{:}\sigma$, then $B'$ is used for $T{:}\sigma'$.
  iii) Let $\mathbf{r}: Lhs \Rightarrow Rhs$ be a rewrite rule typeable with respect to the environment $\mathcal{E}$, and let $F$ be the defined symbol of $\mathbf{r}$. If $E(\mathcal{E}(F)) = \tau \in \mathcal{T}_S$, then for every $\mu \in \mathcal{T}_s$ such that $\tau \leq_S \mu$, $\mathbf{r}$ is typeable with respect to $\mathcal{E}[F := \mu]$. ∎

Notice that in part (iii) the relation $\leq_S$ is used, not $\leq_E$.

**Theorem 2.3.3** *Soundness of lifting.* If $B \vdash_{\mathcal{E}} T{:}\sigma$ and $L$ is a lifting, then $L(B) \vdash_{\mathcal{E}} T{:}L(\sigma)$. ∎

Obviously not every lifting performed on a pair $<B, \sigma>$ such that $B$ is used for $T{:}\sigma$ produces a pair with this same property. Since type assignment of rewrite rules is defined using the notion of used bases, it is clear that lifting cannot be a sound operation on rewrite rules. This can also be illustrated by the rewrite system $I(x) \Rightarrow x$, $F(I_0) \Rightarrow I_0$, that is typeable with respect to the environment $\mathcal{E}_1(I) = 1 \rightarrow 1$, $\mathcal{E}_1(F) = (2 \rightarrow 2) \rightarrow 3 \rightarrow 3$. Notice that $(2 \rightarrow 2) \rightarrow 3 \rightarrow 3 \leq_E (2 \rightarrow 2) \cap 4 \rightarrow 3 \rightarrow 3$, so $<<\emptyset, (2 \rightarrow 2) \rightarrow 3 \rightarrow 3>, <\emptyset, (2 \rightarrow 2) \cap 4 \rightarrow 3 \rightarrow 3>>$ is a lifting. It is not possible to show that the rewrite rule that defines $F$ is typeable with respect to $\mathcal{E}[F := (2 \rightarrow 2) \cap 4 \rightarrow 3 \rightarrow 3]$, since all types in $(2 \rightarrow 2) \cap 4$ should be types for $I$.

Combining the above results for the different operations, we have:

**Theorem 2.3.4** i) If $B \vdash_{\mathcal{E}} T{:}\sigma$ then for every chain $C$ such that $C(<B, \sigma>) = <B', \sigma'>$: $B' \vdash_{\mathcal{E}} T{:}\sigma'$.
  ii) If $B$ is used for $T{:}\sigma$, and $C$ is a chain that contains no lifting, then:
    if $C(<B, \sigma>) = <B', \sigma'>$, then $B'$ is used for $T{:}\sigma'$.
  iii) Let $\mathbf{r}: Lhs \Rightarrow Rhs$ be a rewrite rule typeable with respect to the environment $\mathcal{E}$, and let $F$ be the defined symbol of $\mathbf{r}$. If $C$ is a chain that contains no lifting, then: if $C(\mathcal{E}(F)) = \tau \in \mathcal{T}_S$, then for every $\mu \in \mathcal{T}_s$ such that $\tau \leq_S \mu$, $\mathbf{r}$ is typeable with respect to $\mathcal{E}[F := \mu]$. ∎

# 3  The loss of the subject reduction property

By definition 2.1.2(i), if a term $T$ is rewritten to the term $T'$ using the rewrite rule $Lhs \Rightarrow Rhs$, there is a subterm $T_0$ of $T$, and a replacement R, such that $Lhs^R = T_0$, and $T'$ is obtained by replacing $T_0$ by $Rhs^R$. The subject reduction property for this notion of reduction is: If $B \vdash_{\mathcal{E}} T{:}\sigma$, and $T$ can be rewritten to $T'$, then $B \vdash_{\mathcal{E}} T'{:}\sigma$.

This is of course an important property of reduction systems. To guarantee the subject reduction property, we should accept only those rewrite rules $Lhs \Rightarrow Rhs$, that satisfy:

For all replacements R, bases $B$ and types $\sigma$: if $B \vdash_{\mathcal{E}} Lhs^R : \sigma$, then $B \vdash_{\mathcal{E}} Rhs^R : \sigma$.
because then we are sure that all possible rewrites are safe.

Definitions 2.2.1, 2.2.2 and 2.2.6 define what a type assignment should be, just using the strategy as used in languages like for example Miranda. Unfortunately, it is not sufficient to guarantee the subject reduction property. Take for example the definition of $H$ as in example 2.1.4, and the following environment $\mathcal{E}_0(H) = ((1{\rightarrow}2){\rightarrow}3){\rightarrow}(1{\rightarrow}2){\rightarrow}2$, $\mathcal{E}_0(S) = (1{\rightarrow}2{\rightarrow}3){\rightarrow}(1{\rightarrow}2){\rightarrow}1{\rightarrow}3$, $\mathcal{E}_0(I) = 1{\rightarrow}1$. The rule that defines $H$ is typeable with respect to $\mathcal{E}_0$:

$$\begin{array}{c}
\downarrow (1{\rightarrow}2){\rightarrow}2 \\
H{:}((1{\rightarrow}2){\rightarrow}3){\rightarrow}(1{\rightarrow}2){\rightarrow}2 \\
\downarrow \\
S_2{:}((1{\rightarrow}2){\rightarrow}1{\rightarrow}3){\rightarrow}((1{\rightarrow}2){\rightarrow}1){\rightarrow}(1{\rightarrow}2){\rightarrow}3 \\
\swarrow \qquad \searrow \\
x{:}(1{\rightarrow}2){\rightarrow}1{\rightarrow}3 \qquad y{:}(1{\rightarrow}2){\rightarrow}1
\end{array}
\quad \Rightarrow \quad
\begin{array}{c}
\downarrow (1{\rightarrow}2){\rightarrow}2 \\
S_2{:}((1{\rightarrow}2){\rightarrow}1{\rightarrow}2){\rightarrow}((1{\rightarrow}2){\rightarrow}1){\rightarrow}(1{\rightarrow}2){\rightarrow}2 \\
\swarrow \qquad \searrow \\
I_0{:}(1{\rightarrow}2){\rightarrow}1{\rightarrow}2 \qquad y{:}(1{\rightarrow}2){\rightarrow}1
\end{array}$$

If we take the term $H(S_2(K_0, I_0))$ then it is easy to see that the rewrite is allowed, and that this term will be rewritten to $S_2(I_0, I_0)$. Although the first term is typeable with respect to $\mathcal{E}_0$:

$$\begin{array}{c}
\downarrow (4{\rightarrow}5){\rightarrow}5 \\
H{:}((4{\rightarrow}5){\rightarrow}4{\rightarrow}5){\rightarrow}(4{\rightarrow}5){\rightarrow}5 \\
\downarrow \\
S_2{:}((4{\rightarrow}5){\rightarrow}(4{\rightarrow}5){\rightarrow}4{\rightarrow}5){\rightarrow}((4{\rightarrow}5){\rightarrow}4{\rightarrow}5){\rightarrow}(4{\rightarrow}5){\rightarrow}4{\rightarrow}5 \\
\swarrow \qquad \searrow \\
K_0{:}(4{\rightarrow}5){\rightarrow}(4{\rightarrow}5){\rightarrow}4{\rightarrow}5 \qquad I_0{:}(4{\rightarrow}5){\rightarrow}4{\rightarrow}5
\end{array}$$

the term $S_2(I_0, I_0)$ is not typeable with respect to $\mathcal{E}_0$ with the type $(4{\rightarrow}5){\rightarrow}5$.

We should emphasize that the loss of the subject reduction property is not caused by the fact that we use intersection types. The environment $\mathcal{E}_0$ maps function symbols to Curry-types, so even for a notion of type assignment based on Curry-types types are not preserved under rewriting.

In [4] and [3] two restrictions (variants) of the notion of type assignment of this paper are discussed for which a decidable and sufficient condition is formulated that rewrite rules should satisfy in order to reach the subject reduction property.

The construction of those conditions is made using a notion of principal pairs; the condition a rewrite rule should satisfy is that the principal pair for the left hand side term is also a pair for the right hand side term. For the notion of type assignment defined in this section we are not able to formulate this condition in a constructive way, since it is not clear how we should define *the* principal pair for a term. This problem is overcome in [4] and [3] by defining a most general unification algorithm for types and defining principal pairs using that algorithm. At this moment there is no general unification algorithm for types in $T_S$ that works well on all types, so we cannot take this approach.

For the notion of type assignment as defined in this paper the only result we can obtain is to show that *if* a left hand side of a rewrite rule has *a* principal pair and using that pair the rewrite rule can be typed, then rewriting using this rule is safe with respect to subject reduction.

**Definition 3.1** i) Let $T \in \text{Ter}(\Sigma)$. A pair $<P, \pi>$ is called *a principal pair for* $T$ *with respect to* $\mathcal{E}$, if $P \vdash_\mathcal{E} T{:}\pi$ and for every $B, \sigma$ such that $B \vdash_\mathcal{E} T{:}\sigma$ there is a chain $C$ such that $C(<P, \pi>) = <B, \sigma>$.

ii) The definition of a *safe type assignment with respect to* $\mathcal{E}$ is the same as the one for a type assignment as defined in definition 2.2.6, by replacing condition (i.a) by:

$<B, \sigma>$ is a principal pair for $Lhs$ with respect to $\mathcal{E}$, and $B \vdash_\mathcal{E} Rhs{:}\sigma$.

Then rewrite rule $Lhs \Rightarrow Rhs$ is called a *safe rewrite rule*.

Notice that we do not show that every typeable term *has* a principal pair with respect to $\mathcal{E}$; at the moment we cannot give a construction of such a pair for every term. But even with this non-constructive approach we can show that the condition is sufficient.

**Theorem 3.2** *The condition is sufficient.* Let $\mathbf{r} : Lhs \Rightarrow Rhs$ be a safe rewrite rule. Then for every replacement R, basis $B$ and a type $\mu$: if $B \vdash_\mathcal{E} Lhs^R{:}\mu$, then $B \vdash_\mathcal{E} Rhs^R{:}\mu$. ∎

# Future work

We intend to formulate a decidable condition rewrite rules should satisfy in order to obtain the subject reduction property. In the near future characterizations of typeable rewrite systems will be looked for, like for example normalizability of non-recursive typeable systems.

# References

[1] S. van Bakel. Principal type schemes for the Strict Type Assignment System. Technical Report 91-6, Department of Computer Science, University of Nijmegen, 1991.

[2] S. van Bakel. Complete restrictions of the Intersection Type Discipline. *Theoretical Computer Science*, 102:135–163, 1992.

[3] S. van Bakel. Partial Intersection Type Assignment of Rank 2 in Applicative Term Rewriting Systems. Technical Report 92-03, Department of Computer Science, University of Nijmegen, 1992.

[4] S. van Bakel, S. Smetsers, and S. Brock. Partial Type Assignment in Left Linear Applicative Term Rewriting Systems. In J.-C. Raoult, editor, *Proceedings of CAAP '92. 17th Colloquim on Trees in Algebra and Programming, Rennes, France*, volume 581 of *Lecture Notes in Computer Science*, pages 300–321. Springer-Verlag, 1992.

[5] H. Barendregt. *The lambda calculus: its syntax and semantics*. North-Holland, Amsterdam, revised edition, 1984.

[6] H. Barendregt, M. Coppo, and M. Dezani-Ciancaglini. A filter lambda model and the completeness of type assignment. *The Journal of Symbolic Logic*, 48(4):931–940, 1983.

[7] H.P. Barendregt, M.C.J.D. van Eekelen, J.R.W. Glauert, J.R. Kennaway, M.J. Plasmeijer, and M.R. Sleep. Term graph rewriting. In *Proceedings of PARLE, Parallel Architectures and Languages Europe, Eindhoven, The Netherlands*, volume 259-II of *Lecture Notes in Computer Science*, pages 141–158. Springer-Verlag, 1987.

[8] T. Brus, M.C.J.D. van Eekelen, M.O. van Leer, and M.J. Plasmeijer. Clean - A Language for Functional Graph Rewriting. In *Proceedings of the Third International Conference on*

*Functional Programming Languages and Computer Architecture, Portland, Oregon, USA*, volume 274 of *Lecture Notes in Computer Science*, pages 364–368. Springer-Verlag, 1987.

[9] M. Coppo and M. Dezani-Ciancaglini. An Extension of the Basic Functionality Theory for the λ-Calculus. *Notre Dame, Journal of Formal Logic*, 21(4):685–693, 1980.

[10] M. Coppo, M. Dezani-Ciancaglini, and B. Venneri. Principal type schemes and λ-calculus semantics. In J. R. Hindley and J. P. Seldin, editors, *To H. B. Curry, Essays in combinatory logic, lambda-calculus and formalism*, pages 535–560. Academic press, New York, 1980.

[11] H.B. Curry and R. Feys. *Combinatory Logic*, volume 1. North-Holland, Amsterdam, 1958.

[12] N. Dershowitz and J.P. Jouannaud. Rewrite systems. In J. van Leeuwen, editor, *Handbook of Theoretical Computer Science*, volume B, chapter 6, pages 245–320. North-Holland, 1990.

[13] M. Dezani-Ciancaglini and J.R. Hindley. Intersection types for combinatory logic. *Theoretical Computer Science*, 100:303–324, 1992.

[14] K. Futatsugi, J. Goguen, J.P. Jouannaud, and J. Meseguer. Principles of OBJ2. In *Proceedings $12^{th}$ ACM Symposium on Principles of Programming Languages*, pages 52–66, 1985.

[15] J.W. Klop. Term Rewriting Systems. Report CS-R9073, Centre for Mathematics and Computer Science, Amsterdam, 1990.

[16] R. Milner. A theory of type polymorphism in programming. *Journal of Computer and System Sciences*, 17:348–375, 1978.

[17] E.G.J.M.H. Nöcker, J.E.W. Smetsers, M.C.J.D. van Eekelen, and M.J. Plasmeijer. Concurrent Clean. In *Proceedings of PARLE '91, Parallel Architectures and Languages Europe, Eindhoven, The Netherlands*, volume 506-II of *Lecture Notes in Computer Science*, pages 202–219. Springer-Verlag, 1991.

[18] S. Peyton Jones and J. Salkild. The spineless tagless G-machine. In *Functional Programming Languages and Computer Architecture*, pages 184–201. ACM press, 1989.

[19] F. Pfenning. Partial Polymorphic Type Inference and Higher-Order Unification. In *Proceedings of the 1988 conference on LISP and Functional Programming Languages*, volume 201 of *Lecture Notes in Computer Science*, pages 153–163. Springer-Verlag, 1988.

[20] B.C. Pierce. *Programming with Intersection Types and Bounded Polymorphism*. PhD thesis, Carnegie Mellon University, School of Computer Science, Pittsburgh, 1991. CMU-CS-91-205.

[21] S. Ronchi della Rocca and B. Venneri. Principal type schemes for an extended type theory. *Theoretical Computer Science*, 28:151–169, 1984.

[22] D.A. Turner. Miranda: A non-strict functional language with polymorphic types. In *Proceedings of the conference on Functional Programming Languages and Computer Architecture*, volume 201 of *Lecture Notes in Computer Science*, pages 1–16. Springer-Verlag, 1985.

# Extracting Constructive Content from Classical Logic via Control-like Reductions

Franco Barbanera *   Stefano Berardi

Universita' di Torino
Dipartimento di Informatica
Corso Svizzera, 185
10149 Torino (Italy)
e-mail: { barba,stefano} @di.unito.it

**Abstract.** Recently there has been much interest in the problem of finding the computational content of classical reasoning. One of the most appealing directions for the computer scientist to tackle such a problem is the relation which has been established between classical logic and lambda calculi with control operators, like Felleisen's control operator $\mathcal{C}$. In this paper we introduce a typed lambda calculus with the $\mathcal{C}$ operator corresponding to Peano Arithmetic, and a set of reduction rules related to the ones of the usual control calculi with $\mathcal{C}$. We show how these rules, which are proved to be strongly normalizing, can be used to extract witnesses from proofs of $\Sigma_1^0$ sentences in Peano Arithmetic.

## 1 Introduction

There has been in the last years in the computer science community a wider and wider interest in the problem of finding the computational content of classical proofs. This interest has been motivated by the ambitious purpose of extending to classical logic the well known proofs-as-programs paradigm of constructive logic. The basis of all the presents investigations are quite old results by Kreisel [9] and Friedman [6], showing that classical and intuitionistic provability coincide for $\Sigma_1^0$ formulas. Now, since from the programming point of view constructivity is needed just for this class of formulas, the problems that the research has to cope with is that of finding techniques to extract *directly* constructive content from classical proofs without having to pass through translations to intuitionistic logic, and to understand the algorithmic meaning of classical constructions. In the direction of finding new logical systems in which it could be better understood and extracted the constructive content of classical proofs we can mention the works of Girard [7] and Parigot [11]. Game theory has been instead investigated as an environment in which it is possible to interpret the constructive content of classical connectives and proofs (see, among others, Coquand [3]). In this last direction an interpretation of classical logic in terms of a valuation semantic has been given by the authors in [1].
However, one of the most appealing directions for the computer scientist, not only the theoretical one, is that which has begun being developed with works by Griffin

---
* To Roberta Gottardi

[8] and Murthy [10] (indeed it was these works that made raise the interest for the topic we are speaking about). In these works a strong relation has been established between classical logic and the control operators usually added in functional programming languages in order to have in them some of the good features of imperative languages. To isolate and study the properties of control operators in functional languages people like Felleisen investigated pure calculi like the *pure* control language consisting in the pure lambda calculus enriched with a particular operator $\mathcal{C}$ ($\lambda_\mathcal{C}$) [4],[5]. $\mathcal{C}$ can be rightly considered as an abstraction of the actual control operators. For this language it is possible to give a machine-like operational semantics or to define a calculus extending the notion of $\beta$-reduction of the $\lambda$-calculus. Griffin noted that classical proofs are connected, in a "Proofs-as-Programs" relation, to the terms of such a calculus, which allows precise mathematical reasoning about programs equivalences under various possible evaluators. Later on Murthy showed that from a term of $\lambda_\mathcal{C}$ corresponding to a proof of a $\Sigma_1^0$ formula it is possible to extract the witness of the formula by reducing the term using the reductions simulating the evaluation of the term via the machine-like operational semantics. Unfortunately the extraction of the witness was proved to be possible only using particular reduction strategies, like call-by-name and call-by-value, leaving open the problem of arbitrary reduction strategies. To face such a problem the authors studied the problem of strong normalization for the calculus $\lambda_\mathcal{C}$, a study that led in [2] to a proof of strong normalization for a system ($\lambda_{\mathcal{C}\tau^-}$) similar to $\lambda_\mathcal{C}$, where the reduction rules present in $\lambda_\mathcal{C}$ are somewhat restricted and where other reduction rules are added. In the present paper we show that this set of reduction rules, even if not the same as that of $\lambda_\mathcal{C}$, can be used for the purpose of extracting witnesses from classical proofs of $\Sigma_1^0$ sentences. In particular we define in Section 2, $\mathbf{PA}_\mathcal{C}$, a richer system than $\lambda_{\mathcal{C}\tau^-}$. $\mathbf{PA}_\mathcal{C}$ as a logic corresponds to Peano Arithmetic. The reduction rules of $\mathbf{PA}_\mathcal{C}$ are now the version for Peano Arithmetic of the rules of $\lambda_{\mathcal{C}\tau^-}$. We prove in Section 4 a result of strong normalization for $\mathbf{PA}_\mathcal{C}$ by means of a reduction preserving translation to $\lambda_{\mathcal{C}\tau^-}$ (indeed to a system slightly stronger). This result, together with a theorem on the shape of normal forms (Section 5), enable one to define an algorithm for the extraction of witnesses from $\Sigma_1^0$ sentences of Peano Arithmetic (Section 3). Then, by means of the strong normalization result, it is possible to capture the multi-valued nature of classical proofs.

## 2 The system $\mathbf{PA}_\mathcal{C}$

In this section we introduce system $\mathbf{PA}_\mathcal{C}$, which is a version of first order Peano-Arithmetic. In $\mathbf{PA}_\mathcal{C}$ formulas of Peano-Arithmetic are viewed as types of a type-system where terms denote proofs.

We begin by defining the terms of Peano arithmetic (PA-terms) in our context. They are built out of numeral variables ($n, m, n_1, \ldots$), the constant 0, the successor function s and recursion. In order to define recursion in a simple way we shall use also abstraction and application.

**Definition 1 PA-terms.** (i) PA-terms are all the terms which it is possible to build using the following *term-formation rules*.

Let $n$ be a numeral variable, $A, B$ arrow types built out of the type constant $Int$.

$$n : Int \qquad\qquad 0 : Int$$

$$\frac{\begin{array}{c}[n:A]\\ \vdots\\ u:B\end{array}}{\lambda n.u : A \to B} \qquad \frac{u_1 : A \to B \quad u_2 : A}{(u_1 u_2) : B}$$

$$\frac{u : Int}{su : Int} \qquad \frac{u : Int \quad p : A \quad f : Int \to A \to A}{\text{Rec}(u, p, f) : A}$$

(ii) On PA-terms the following and well known notions of reductions are defined:

$$\beta) \quad (\lambda x.p)q \quad \to_1 p[q/x]$$

$$\text{Rec}_0) \ \text{Rec}(0, p, f) \ \to_1 p$$

$$\text{Rec}_S) \ \text{Rec}(su, p, f) \to_1 f(u)\text{Rec}(u, p, f)$$

We denote by $\to_{1PA}$ the union of all the above notions of reductions, and by $\to_{PA}$ its reflexive and transitive closure.

(iii) We denote by $\simeq$ the least congruence obtained out of $\to_{PA}$

In what follows, a generic PA-terms will be denoted by $u, v, t, \ldots$.

**Lemma 2.** *PA-terms strongly normalizes.*

The formulas (types) of $\mathbf{PA}_\mathcal{C}$ are a subset of the formulas built out of atomic formulas using the connectives $\to$ and $\forall$. The atomic formulas are the "falsehood" ($\bot$) and the formulas built out of the equality predicate between natural numbers, i.e. formulas of the form $t = u$ where $t$ and $u$ are PA-terms of type $Int$.
Negation, conjunction, disjunction and existential quantification can be defined as usual:

$$\neg A =_{Def} A \to \bot$$

$$A \vee B =_{Def} \neg A \to B$$

$$A \wedge B =_{Def} \neg(\neg A \vee \neg B)$$

$$\exists n.A =_{Def} \neg \forall n.\neg A.$$

We restrict the set of well-formed formulas by forbidding formulas to have *strict* subformulas of the form $\neg\neg A$. We also forbid $\bot$ to occur on the lefthand side of $\to$ (like in $\bot \to A$) and to be universally quantified (i.e. we forbid $\forall n.\bot$).
Hence $\neg\neg A$ and $\neg(B \to \neg A)$ $(A, B \neq \bot)$ are formulas of our system, while $A \to \neg\neg B$ and $A \to (\bot \to B)$ are not.
The motivation of such restrictions is that we shall get strong normalization for terms of $\mathbf{PA}_\mathcal{C}$ by means of a reduction preserving translation to a system ($\lambda_{\mathcal{C}^T-}$) where similar restrictions are strictly necessary for its proof of strong normalization.

However, the given restrictions on formulas of **PA**$_C$ are not a serious problem for our purpose, i.e. that of extracting witnesses from proofs of formulas of the form $\exists n.A$, since, as we shall see, it is always possible to transform the proof of a theorem in such a way that the given conditions on negations are respected.

The formal definition of well-formed formula (type) of **PA**$_C$ runs as follows:

**Definition 3.** The sets of *Positive formulas* (P), *Negative formulas* (N), *PosNeg formulas* (PN) and *Double negated formulas* (NN) are defined by the following grammars:

$$P ::= u = t \mid P \to P \mid P \to \neg P \mid \neg P \to P \mid \neg P \to \neg P \mid \forall n.P \mid \forall n.\neg P$$
$$N := \neg P$$
$$PN := P \mid N$$
$$NN := \neg\neg P.$$

The set of *Formulas* (F) of **PA**$_C$ is the union of the sets defined above with $\perp$, i.e. it is defined by

$$F := \perp \mid P \mid N \mid NN$$

We define now a set of "pseudoterms" on which **PA**$_C$-terms are based and a set of inference rules.

**Definition 4.** The set of *Pseudoterms* of **PA**$_C$ is defined by the following grammar:

$$p ::= x \mid PA_{id} \mid PA_{sym}(p) \mid PA_{trans}(p,p) \mid PA_{bot}(p) \mid$$
$$\lambda x.p \mid \lambda n.p \mid (pp) \mid (pu) \mid (\mathcal{C}p) \mid \text{IND}(u,p,p)$$

where x is the category of **PA**$_C$-term variables, n that of PA-term variables (numeral variables) and u that of PA-terms.

A *term* of **PA**$_C$ will be a pseudoterm representing a correct proof for a well-formed formula. Correctness for a pseudoterm is derived using the following rules which we divide into two groups: atomic and logical rules.

**PA**$_C$-rules.

- *Atomic rules*

$$PA_{id}) \ \overline{PA_{id} : (n = n)} \qquad PA_{sym}) \ \frac{p : (u = t)}{PA_{sym}(p) : (t = u)}$$

$$PA_{trans}) \ \frac{p : (u = v) \quad q : (v = t)}{PA_{trans}(p,q) : (u = t)} \qquad PA_{bot}) \ \frac{p : (s0 = 0)}{PA_{bot}(p) : \perp}$$

- *Logical rules*

Let $A$ be a PosNeg formula, $B$ a PosNeg formula or $\bot$, $P$ a positive formula and $T$ a formula $\neq \bot$. [2]

$$var) \; \frac{}{x^A : A}$$

$$\rightarrow I) \; \frac{\begin{array}{c} x:A \\ \vdots \\ p:B \end{array}}{\lambda x^A.p : A \rightarrow B} \qquad \rightarrow E) \; \frac{p_1 : A \rightarrow B \quad p_2 : A}{(p_1 p_2) : B}$$

$$\forall I) \; \frac{p:A}{\lambda n.p : \forall n.A} \; (*) \qquad \forall E) \; \frac{p : \forall n.A(n) \quad u : Int}{pu : A(u)}$$

$$C) \; \frac{p : \neg\neg P}{Cp : P} \qquad \text{IND)} \; \frac{u : Int \quad p : T(0) \quad f : \forall n.T(n) \rightarrow T(sn)}{\mathsf{IND}(u,p,f) : T(u)}$$

$$Conv) \; \frac{p : T(u)}{p : T(u')} \; \text{if} \; u \simeq u'$$

(*) for all $x^B \in FV(p)$ we assume that $n \notin FV(B)$.

We shall say that a term $p$ represents the proof of a formula $A$ (is a term of type $A$) if it is possible to derive $p : A$.

The set of all terms of $\mathbf{PA}_C$ will be denoted by $Terms(\mathbf{PA}_C)$.

We shall denote by $\Gamma \vdash_{\mathbf{PA}_C} p : A$ the fact that $p : A$ is derivable in $\mathbf{PA}_C$ from the set of assumptions $\Gamma$.

Rules $(\mathbf{PA}_{id}), (\mathbf{PA}_{sym}), (\mathbf{PA}_{trans})$ and $(\mathbf{PA}_{bot})$ will be often referred to generically as *atomic rules*.

A term (proof) is said to be *closed* if it does not contain free $\mathbf{PA}_C$-term variables. It is said to be *PA-closed* if it does not contain free PA-term variables.

We introduce now a set of notions of reduction on $\mathbf{PA}_C$-terms, which will be used as a means to extract witnesses from closed proofs of existential formulas.

## $\mathbf{PA}_C$-reduction rules

- *Functional reductions:*

$\beta) \quad (\lambda x.p)q \quad \rightarrow_1 p[q/x]$

$\beta_\forall) \quad (\lambda n.p)u \quad \rightarrow_1 p[u/n] \quad (*)$

$\mathsf{IND}_0) \; \mathsf{IND}(0,p,f) \rightarrow_1 p$

$\mathsf{IND}_s) \; \mathsf{IND}(su,p,f) \rightarrow_1 f(u)(\mathsf{IND}(u,p,f))$

$Comp) \; u \rightarrow_1 u' \quad$ if $u$ and $u'$ are PA-terms and $u \rightarrow_{PA} u' \quad (^*_*)$

---

[2] We have to assume types of this sort in order to get correct types in the conclusions of the rules

($*$) The substitution applies to types as well
($^*_*$) Note that this rule does not imply that PA-terms are indeed $PA_C$-terms, but only that we can reduce PA-terms using PA-reduction when they are subterm of $PA_C$-terms.
- *Control reductions:*

$$C_L)\quad (Cp)q \to_1 C\lambda k.p(\lambda f.k(fq))\ (^1)$$

$$C_{L\forall})\ (Cp)u \to_1 \lambda k.p(\lambda f.k(fu))\ (^2)$$

$$C_R)\quad p(Cq) \to_1 C\lambda k.q(\lambda a.k(pa))\ (^3)$$

- *Prawitz-style reductions for classical logic:*

$$C'_L)\quad (Cp)q \to_1 \lambda x.p(\lambda f.(fq)x)\ (^4)$$

$$C'_{L\forall})\ (Cp)u \to_1 \lambda x.p(\lambda f.(fu)x)\ (^5)$$

$$C'_R)\quad p(Cq) \to_1 \lambda x.q(\lambda a.(pa)x)\ (^6)$$

$$C''_R)\quad p(Cq) \to_1 q(\lambda a.(pa))\quad\quad (^7)$$

$$C_{At})\ C\lambda x.xq \to_1 q \quad\quad\quad\quad\quad (^8)$$

$$Triv)\ E[p]\quad \to_1 p \quad\quad\quad\quad\quad (^9)$$

Provisos:
($^1$) $p$ has to have type of the form $\neg\neg(A \to P)$
($^2$) $p$ has to have type of the form $\neg\neg(\forall n.P)$
($^3$) $p$ has to have type of the form $A \to P$
($^4$) $p$ has to have type of the form $\neg\neg(A \to \neg P)$
($^5$) $p$ has to have type of the form $\neg\neg(\forall n.\neg P)$
($^6$) $p$ has to have type of the form $A \to \neg P$
($^7$) $p$ has to have type of the form $\neg P$
($^8$) $q$ has to have *atomic* type and to be formed only by atomic rules
($^9$) $E[-]$ is a context $\neq [-]$ with type $\bot$, $p$ has type $\bot$ and $FV(p) \subseteq FV(E[p])$

$\to$ will denote the reflexive and transitive closure of $\to_1$.

In the reductions defined above we have not put the type decorations for sake of simplicity. We give below the reduction rules with all the type decorations.

Functional reductions :

$\beta)\quad ((\lambda x^A.p^B)^{A\to B}q^A)^B \quad\quad\quad\quad \to_1 (p[q/x])^B$
$IND_0)\ IND(0, p^{T(0)}, f^{\forall n.T(n)\to T(Sn)})^{T(0)} \quad \to_1 p^{T(0)}$
$IND_S)\ IND((su)^{Int}, p^{T(0)}, f^{Int \to T(n) \to T(Sn)})^{T(u)} \to_1 (f(u)(IND(u,p,f)))^{T(u)}$

Control reductions :

$\mathcal{C}_L)$ $((\mathcal{C}(p)^{\neg\neg(A\to P)})^{A\to P}q^A)^P \to_1$
$\quad(\mathcal{C}(\lambda k^{\neg P}.(p(\lambda f^{A\to P}.(k(fq))^\bot)^{\neg(A\to P)})^\bot)^{\neg\neg P})^P$

$\mathcal{C}_{L\forall})$ $((\mathcal{C}(p)^{\neg\neg(\forall n.P)})^{\forall n.P}u^{Int})^P \to_1$
$\quad(\mathcal{C}(\lambda k^{\neg P}.(p(\lambda f^{\forall n.P}.(k(fu))^\bot)^{\neg(\forall n.P)})^\bot)^{\neg\neg P})^P$

$\mathcal{C}_R)$ $(p^{A\to P}(\mathcal{C}(q)^{\neg\neg A})^A)^P \to_1$
$\quad(\mathcal{C}(\lambda k^{\neg P}.(q(\lambda a^A.(k(pa)^P)^\bot)^{\neg A})^\bot)^{\neg\neg P})^P$

Prawitz-style reductions for classical logic :

$\mathcal{C}'_L)$ $((\mathcal{C}(p)^{\neg\neg(A\to\neg P)})^{A\to\neg P}q^A)^{\neg P} \to_1$
$\quad(\lambda x^P.(p(\lambda f^{A\to\neg P}.((fq)^{\neg P}x)^\bot)^{\neg(A\to\neg P)})^\bot)^{\neg P}$

$\mathcal{C}'_{L\forall})$ $((\mathcal{C}(p)^{\neg\neg(\forall n.\neg P)})^{\forall n.\neg P}u^{Int})^{\neg P} \to_1$
$\quad(\lambda x^P.(p(\lambda f^{\forall n.\neg P}.((fu)^{\neg P}x)^\bot)^{\neg(\forall n.\neg P)})^\bot)^{\neg P}$

$\mathcal{C}'_R)$ $(p^{A\to\neg P}(\mathcal{C}(q)^{\neg\neg A})^A)^{\neg P} \to_1$
$\quad(\lambda x^P.(q(\lambda a^A.((pa)^{\neg P}x)^\bot)^{\neg A})^\bot)^{\neg P}$

$\mathcal{C}''_R)$ $(p^{\neg A}(\mathcal{C}(q)^{\neg\neg A})^A)^\bot \to_1 (q(\lambda a^A.(pa)^\bot)^{\neg A})^\bot$

$Triv)$ $(E[p^{At}])^{At} \to_1 p^{At}$

$\mathcal{C}_{At})$ $(\mathcal{C}(\lambda x^{\neg At}.(xq^{At})^\bot)^{\neg\neg At})^{At} \to_1 q^{At}$

It is now easy to see the motivation of the name "Control-like" reductions given in the title of the paper. Reductions $\mathcal{C}_R$, $\mathcal{C}_L$ and $\mathcal{C}_{L\forall}$ are exactly the typed version of the reductions of Felleisen's calculus [4][5] which allows reasoning about equivalences of functional programs enriched with the control operator $\mathcal{C}$. However, as it is possible to see from the provisos of our rules, the applicability of the above rules in our calculus is somewhat restricted. To deal with the case when the terms $(\mathcal{C}M)N$ or $M(\mathcal{C}N)$ have a negative type we have introduced rules $\mathcal{C}'_R$, $\mathcal{C}'_L$ and $\mathcal{C}'_{L\forall}$. This because, otherwise, the use of the other rules above would have led to terms having non well-formed formulas as types. Rules $\mathcal{C}'_R$, $\mathcal{C}'_L$ and $\mathcal{C}'_{L\forall}$, instead of moving the continuation outside, make it disappear. This is possible because a triple negation is intuitionistically equivalent to a negation. The name Prawitz-style reductions comes out of the fact that the rules of the group we dubbed with such a name are quite similar to the reductions defined by Prawitz to show the consistency of classical logic [12]. Rule $\mathcal{C}''_R$ is an instance of the general rule $E[\mathcal{C}M] \to_1 M(\lambda x.E[x])$ for continuations and has been introduced in order to deal with the case of the elimination of negation.

It is easy to see that the system we have defined does not have the Church-Rosser property. This means that we will be able to get, out of a proof, different answers (witnesses).

*Remark.* Note that we do not have in our calculus the operator $\mathcal{A}$ (abort) of usual control calculi. In a typed context the operator $\mathcal{A}$ should have a rule of introduction of the form

$$\frac{p:\bot}{\mathcal{A}_T p:T}$$

where $\mathcal{A}_T$ is a type labelled version of the operator $\mathcal{A}$. For the abort operator its related reduction rule should be something like

$$E[(\mathcal{A}_T p^\bot)^T] \to_1 p^\bot,$$

where $E[-] \neq [-]$, $T \neq \bot$ and $FV(p) \subseteq FV(E[p])$.

However, like in most control calculi, the $\mathcal{A}$ operator can be defined out of the operators we have, and the reduction rule derived. It not difficult to see that by defining, for $P$ positive (and using "_" as notation for a new fresh variable),

$$\mathcal{A}_P p^\bot =_{Def} \mathcal{C}\lambda\_^{\neg P}.p$$

$$\mathcal{A}_{\neg P} p^\bot =_{Def} \lambda\_^{P}.p$$

$$\mathcal{A}_{\neg\neg P} p^\bot =_{Def} \lambda\_^{\neg P}.p,$$

the reduction rule for $\mathcal{A}$ shown above can be derived from the reduction $(Triv)$.

**Definition 5.** Let $k$ be an integer and $p$ a term.
i) $k$ is a bound for $p$ if the reduction tree of $p$ has a finite height $\leq k$.
ii) $p$ strongly normalizes if it has a bound.

Then a term strongly normalizes iff its reduction tree is finite.
This definition has to be preferred to the usual one, i.e. "each reduction sequence out of $p$ is finite", because the latter is intuitionistically weaker than the former (classically, they are equivalent through König's Lemma).

The main property enjoyed by system $\mathbf{PA}_\mathcal{C}$, and that will be essential for the extraction of witnesses is that of strong normalization.

**Theorem 6 Strong Normalization for $\mathbf{PA}_\mathcal{C}$.** *Terms of $\mathbf{PA}_\mathcal{C}$ strongly normalize.*

The proof of the theorem above will be the argument of Section 4

## 3 Witness Extraction through $\mathbf{PA}_\mathcal{C}$-reductions

In this section we shall show how to use the strong normalization result for $\mathbf{PA}_\mathcal{C}$-reductions in order to extract witnesses from closed and PA-closed proofs of formulas of the form

$$\exists n. f(n) = 0$$

where $f$ is the term denoting a primitive recursive function.
Note that the use of the proposition $f(n) = 0$ is not restrictive, since in Peano Arithmetic it is always possible to put any decidable proposition in this form.

One could argue that because of our restrictions, the proofs of existential formulas which it is possible to give in $\mathbf{PA}_\mathcal{C}$ are not all the possible proofs of Peano-Arithmetic. This however is not true. The main restriction of our system is that of preventing two or more consecutive negations inside a formula. This indeed causes no problem at all since, given a proof in Peano-Arithmetic (in which we assume not to have formulas of the form $\bot \to A^3$), it is always possible to simply process it in order to get a proof where no two or more consecutive negations are present inside a formula. More formally we can state the following

---

[3] Of course it is possible to process the proof in order to get rid also of this sort of formulas.

**Lemma 7.**

$$x : A \vdash_{\mathbf{PA}} p : B \;\Rightarrow\; x : \phi_1(A) \vdash_{\mathbf{PA}_\mathcal{C}} \chi(p) : \phi(B)$$

where $\vdash_{\mathbf{PA}}$ denotes derivability in Peano-Arithmetic, $\chi$ is a function from proofs of Peano-Arithmetic to proofs of $\mathbf{PA}_\mathcal{C}$ which it is not difficult to define by induction on the structure of the proofs and $\phi$ and $\phi_1$ are mapping from arbitrary formulas to $\mathbf{PA}_\mathcal{C}$-formulas inductively defined as follows:

1. $\phi_1(P^{At}) = P$    2. $\phi_1(A \to B) = \phi_1(A) \to \phi_1(B)$ if $B \neq \bot$
3. $\phi_1(\neg^{2k} A) = \phi_1(A)$    4. $\phi_1(\neg^{2k+1} A) = \neg \phi_1(A)$
5. $\phi_1(\forall n.A) = \forall n.\phi_1(A)$

1. $\phi(P^{At}) = P$    3. $\phi(A \to B) = \phi_1(A \to B)$ if $B \neq \bot$
2. $\phi(\neg A) = \neg \phi_1(A)$   4. $\phi(\forall n.A) = \forall n.\phi_1(A)$

Witness extraction is possible since we can rely, beside the strong normalization theorem, on the following

**Theorem 8.** *Let $p$ be a normal proof, closed and PA-closed, such that*

$$\vdash_{\mathbf{PA}_\mathcal{C}} p : \neg \forall n. \neg f(n) = 0$$

*Then $p$ is of the form $\lambda x^{\forall n. \neg f(n) = 0}.(xu) q^{f(n)=0}$ where $q^{f(n)=0}$ is formed only by atomic rules.*

The proof of this theorem will be given in Section 5.

Now, given a term $p$, closed and PA-closed, and such that

$$\vdash_{\mathbf{PA}_\mathcal{C}} p : \neg \forall n. \neg f(n) = 0$$

to get a witness of its we simply put it in normal form. From this normal form we can take the witness since, by the theorem above, we have that $p$ has to be of the shape

$$\lambda x^{\forall n. \neg f(n)=0}.(xu) q^{f(u)=0}$$

where $u$ is a witness.

Note that a proof can contain more than one witnesses; which one we take depends on the reduction strategy used.

## 4 Strong Normalization for $\mathbf{PA}_\mathcal{C}$

In this section we provide the proof of strong normalization for system $\mathbf{PA}_\mathcal{C}$. The proof consists in a reduction preserving translation to the type system $\lambda_{\mathcal{C}_T^-}$, which it is already known to be strongly normalizable. System $\lambda_{\mathcal{C}_T^-}$ is similar to $\mathbf{PA}_\mathcal{C}$. In a sense it could be considered as a propositional version of it. It was introduced in [2], where a proof of strong normalization for it is given using a non trivial version of Girard's candidates of reducibility method. Indeed the system $\lambda_{\mathcal{C}_T^-}$ we present here contains slight extensions with respect to its original version in [2]. These extensions consist in the introduction of a recursion operator and of a successor function, together with the introduction of the types $Int$ and $Id$.

## 4.1 The System $\lambda_{\mathcal{C}T^-}$

System $\lambda_{\mathcal{C}T^-}$ is a typed lambda calculus enriched with the control operator $\mathcal{C}$.
The types of this system are a subset of the simple types à la Curry, i.e. of the types built out of atomic types $Int$ and $Id$ and using the connectives $\to$ ("implication") and $\bot$ ("falsehood"). The negation is defined as usual. The restriction we give to the types à la Curry are the same as in $\mathbf{PA}_\mathcal{C}$. Indeed the restrictions on formulas of $\mathbf{PA}_\mathcal{C}$ are due to the restrictions in $\lambda_{\mathcal{C}T^-}$, in turn due to the proof method used in [2] to prove strong normalization for $\lambda_{\mathcal{C}T^-}$. We briefly recall such restrictions: we forbid types to have *strict* subtypes of the form $\neg\neg A$. We also forbid $\bot$ to occur on the lefthand side of $\to$ (like in $\bot \to A$).

We define now a set of "pseudoterms" and a set of typing rules. The set of terms of system $\lambda_{\mathcal{C}T^-}$ will be the pseudoterms having a correct type.

**Definition 9.** (i) The set of *Pseudoterms* of $\lambda_{\mathcal{C}T^-}$ is defined by the following grammar:
$$M ::= 0 \mid x^T \mid \mathsf{s}M \mid \lambda x.M \mid (MM) \mid \mathcal{C}M \mid \mathsf{Ind}(M, M, M)$$
(ii) *Typing Rules*:
Let $A, B$ be PosNeg types, $P$ a positive type, $T$ a type $\neq \bot$ and $M, N$ pseudoterms.

$$var)\ \frac{}{x^A : A} \qquad (0 : Int)$$

$$\to I)\ \frac{M : B}{\lambda x^A.M : A \to B} \qquad \to E)\ \frac{M : A \to B \quad N : A}{MN : B}$$

$$\neg I)\ \frac{M : \bot}{\lambda x^A.M : \neg A} \qquad \neg E)\ \frac{M : \neg A \quad N : A}{MN : \bot}$$

$$\mathsf{s}I)\ \frac{M : Int}{\mathsf{s}M : Int} \qquad \mathsf{Ind}I)\ \frac{U : Int \quad N : T \quad F : Int \to T \to T}{\mathsf{Ind}(U, N, F) : T}$$

$$\neg\neg E)\ \frac{M : \neg\neg P}{\mathcal{C}M : P}$$

We call then *term* a pseudo term having a correct type.

(iii) *Reduction Rules*:
The reduction rules are the corresponding in $\lambda_{\mathcal{C}T^-}$ of the following reductions of $\mathbf{PA}_\mathcal{C}$: $\beta, \mathsf{Ind}_0, \mathsf{Ind}_s, \mathcal{C}_L, \mathcal{C}_R, \mathcal{C}'_L, \mathcal{C}'_R, \mathcal{C}''_R, \mathcal{C}_{At}, Triv$ with the same provisos, but for rule $\mathcal{C}_{At}$ for which the provisos is the following

$p$ has to have type $Int$ or $Id$ and to be *algebraic*. A term of type $Int$ is algebraic if it is built out only of variables and $0$ by means of rules $(\mathsf{s}I)$ and $(\to E)$; A term of type $Id$ is algebraic if it is formed only by variables.

**Theorem 10.** *Terms of system $\lambda_{\mathcal{C}T^-}$ are strongly normalizable*

The theorem above is provable by a slight modification of the proof method of [2].

## 4.2 The reduction Preserving Translation

We define now a translation from formulas of $\mathbf{PA}_\mathcal{C}$ to types of $\lambda_{\mathcal{C}\mathcal{T}^-}$ and then one from terms of the former system to the latter which will be proved to be reduction-preserving.

**Definition 11.** (i) The translation $\underline{\ }: Formulas(\mathbf{PA}_\mathcal{C}) \to Types(\lambda_{\mathcal{C}\mathcal{T}^-})$ is inductively defined as follows:

1. $\underline{\bot} = \bot$
2. $\underline{t=u} = Id$
3. $\underline{A \to B} = \underline{A} \to \underline{B}$
4. $\underline{\forall n.A} = Int \to \underline{A}$

(i) The translation $\underline{\ }: Terms(\mathbf{PA}_\mathcal{C}) \to Terms(\lambda_{\mathcal{C}\mathcal{T}^-})$ is inductively defined as follows:

1. $\underline{x^A} = x^{\underline{A}}$
2. $\underline{\mathsf{PA}_{id}} = r_1$
3. $\underline{\mathsf{PA}_{sym(bot)}(p)} = r_{2(4)}\underline{p}$
4. $\underline{\mathsf{PA}_{trans}(p,q)} = r_3 \underline{p}\,\underline{q}$
5. $\underline{\lambda x.p} = \lambda \underline{x}.\underline{p}$
6. $\underline{p_1 p_2} = \underline{p_1}\,\underline{p_2}$
7. $\underline{\lambda n.p} = \lambda \underline{n}.\underline{p}$
8. $\underline{pu} = \underline{p}\underline{u}$
9. $\underline{\mathcal{C}p} = \mathcal{C}\underline{p}$
10. $\underline{\mathsf{IND}(u,p,f)} = \mathsf{Ind}(u^*, \underline{p}, \underline{f})$

where, for $u$ PA-terms, $u^*$ is obtained out of $u$ by simply replacing Ind for Rec and $r_i$, for $i = 1, 2, 3, 4$, are $\lambda_{\mathcal{C}\mathcal{T}^-}$-term variables with the following types: $r_1 : Id$, $r_{2(4)} : Id \to Id$ and $r_3 : Id \to Id \to Id$.

If $\Gamma = \{x_1 : A_1, \ldots, x_n : A_n\}$ we define $\underline{\Gamma} = \{\underline{x_1} : \underline{A_1}, \ldots, \underline{x_n} : \underline{A_n}\}$.

It is possible to show now that the above translation from terms to terms is well defined.

**Lemma 12.**
$$\Gamma \vdash_{\mathbf{PA}_\mathcal{C}} p : A \Rightarrow \underline{\Gamma} \vdash_{\lambda_{\mathcal{C}\mathcal{T}^-}} \underline{p} : \underline{A}.$$

**Proof.** Easy induction. □

**Lemma 13 Reduction Preservation.**
$$p \to_1 p' \Rightarrow \underline{p} \to_1 \underline{p'}$$

**Proof.** Easy checking of all the possible reductions. □

We are now able to prove Theorem 6, i.e. that terms of system $\mathbf{PA}_\mathcal{C}$ are strongly normalizing.

**Proof of Theorem 6**
Easy from Theorem 10 and Lemma 13, from which it is possible to show that an infinite reduction sequence out of a term $p$ in $\mathbf{PA}_\mathcal{C}$ would induce an infinite reduction sequence out of the term $\underline{p}$ of $\lambda_{\mathcal{C}\mathcal{T}^-}$, which is impossible. □

## 5 The Shape of Normal Forms

This section will be devoted to prove Theorem 8.

We define now the usual notion of proof theory of *main branch* and define the set of *branches* for the system $\mathbf{PA}_\mathcal{C}$. The notion of main branch we define is the usual one used in natural deduction in which we take into account also a possible main branch of the PA-term of the first argument of a term $\mathsf{Rec}(u, p, f)$.

We just use a particular notation in order to denote a main branch, i.e. we shall denote a main branch by a sequence of symbols, each of them denoting a particular rule of the branch.

The following symbols are the introduction symbols: $0$, $\lambda_{Int}$, $\mathsf{s}$, $\lambda$, $\lambda_\forall$, $\mathcal{C}$ and correspond to the rules for introducing 0, for the abstraction and for introducing s of Definition 1 and to the $\mathbf{PA}_\mathcal{C}$-rules $(\forall I)$ and $(\mathcal{C})$, respectively. The elimination symbols are $Ap_{Int}$, $\mathsf{Rec}$, $Ap$, $Ap_\forall$, $\mathsf{IND}$, and correspond to the rules for the application of PA-terms and the introduction of the symbol $\mathsf{Rec}$ of Definition 1 and to the $\mathbf{PA}_\mathcal{C}$-rules $(\to E)$, $(\forall E)$ and $(\mathsf{IND})$, respectively.

**Definition 14.** *Let p be a term.*

(i) *We define the* main branch *of $p$ ($\mathbf{MB}(p)$) as its leftmost branch. For instance, if $p = \lambda n.\mathcal{C}(\mathsf{IND}(\mathsf{s}0, P, Q)(y\mathcal{C}\lambda x.zR))$ then $\mathbf{MB}(p) = <\lambda_\forall, \mathcal{C}, Ap, \mathsf{IND}, \mathsf{s}, 0>$.*

(ii) *We define the set $\mathbf{B}(p)$ of the* branches *of $p$ as the set of all the maximal main branches of all its subterms but subterms which are PA-terms. For $p$ of the example above we then get $\mathbf{B}(p) =$*
*$\{<\lambda_\forall, \mathcal{C}, Ap, \mathsf{IND}, \mathsf{s}, 0>, \mathbf{MB}(P), \mathbf{MB}(Q), <Ap, y>, <\mathcal{C}, \lambda, Ap, z>, \mathbf{MB}(R)\}$*

In the following we shall denote by $At$ a generic atomic formula. For sake of semplicity when we say that a sequence is formed only by eliminations, the intended meaning will be "eliminations followed by a variable".

We shall denote by $\overline{PA}$ a generic sequence of $PA_x$'s.

**Lemma 15.** *Let $u$ be a PA-term in normal form. Then either $\mathbf{MB}(u)$ begins with an introduction or it is formed only by eliminations followed by a variable.*

**Proof.** By induction un $u$. □

**Lemma 16.** *Let $p^A$ be a proof in $\mathbf{PA}_\mathcal{C}$ in normal form.*

(i) *If $A$ is not atomic then for $\mathbf{MB}(p)$ there is one of the following three possibilities:*
  1. *it is a variable*
  2. *it begins with an introduction symbol*
  3. *it is formed by elimination symbols only.*

(ii) *If $A$ is atomic then for $\mathbf{MB}(p)$ there is one of the following five possibilities:*
  1. *it is a variable*
  2. *it begins with the symbol of an atomic rule*
  3. *it is $<\mathcal{C}, x>$ (i.e. is has the shape $\mathcal{C}x^{\neg\neg At}$)*
  4. *it begins with $<\mathcal{C}, \lambda>$ (i.e. is has the shape $\mathcal{C}\lambda x.M^\perp$)*

5. it is formed by elimination symbols only.

**Proof.** For both (i) and (ii) we proceed by induction on the structure of $p$. □

**Lemma 17.** *Let $p^A$ be a proof in $PA_C$ in normal form. Then $MB(p)$ is formed by three subsequences, $MB(p) = <s_1, s_2, s_3>$, where $s_1$ is formed only by introduction symbols, $s_2$ is the concatenation of subsequences of the following form in whatever order: $<\overline{PA}>, <C, x>, <C, \lambda>$ and $s_3$ is formed by elimination symbols preceeded by a variable.*

**Proof.** If $A$ is atomic then the thesis follows by induction on $p$ using Lemma 16(ii). Otherwise, by induction on $p$.

- $p$ is a variable.
  Immediate.
- $p$ ends with an introduction.
  If the immediate subterm of $p$ is of atomic type we have finished since we have already considered the case with atomic type, otherwise the thesis follows by the induction hypothesis.
- $p$ ends with an elimination.
  The thesis follows by Lemma 16(i). □

**Lemma 18.** *Let $p$ be a normal proof, PA-closed and such that*

$$y : \neg At, x : \forall n. \neg P(n) \vdash_{PA_C} p : \neg At.$$

*Besides, let $MB(p)$ be formed only by eliminations. Then either $p \equiv y$ or $p \equiv xu$ for some (closed) PA-term $u$.*

**Proof.** Since $MB(p)$ is formed only by eliminations, there is no discharging of variables in the main branch and hence $MB(p)$ ends with $x$ (it cannot end with a free variable of type $Int$ since $p$ is PA-closed) or $y$. Thus, the main branch contains no IND, and $p \equiv y$ or $p \equiv xu$. □

**Lemma 19.** *Let $p$ be a normal proof different from a variable, PA-closed and such that*

$$y : \neg At, x : \forall n. \neg P(n) \vdash_{PA_C} p : A.$$

*Besides, let $MB(p)$ be formed only by eliminations. Then either $A \equiv \neg At$ or $A \equiv \bot$.*

**Proof.** Since $MB(p)$ is formed only by eliminations, there is no discharging of variables and hence $MB(p)$ ends with $x$ or $y$ (it cannot end with a free variable of type $Int$ since $p$ is PA-closed). Thus the main branch contains no IND and we may have : $p \equiv xu : \neg P(n)$, $xua : \bot$, $y : \neg At$, $ya : \bot$. □

**Lemma 20.** *Let $p$ be a normal proof, PA-closed and such that*

$$y : \neg At, x : \forall n. \neg P(n) \vdash_{PA_C} p : At$$

*where $At \not\equiv \bot$.*
*Then each variable occurrence in $p$ is in some subterm of the form $z^{\neg At} q^{At}$ or $(xu)q$.*

**Lemma 21.** *Let $C[q^{At}]^\perp$ be a proof in normal form, PA-closed, containing a subproof $q$ of atomic type $At \not\equiv \perp$, such that*

$$x : \forall n. \neg P(n) \vdash_{\mathbf{PA}_c} C[q^{At}] : \perp$$

*where $C[\,]$ is a context which does not bind any free variable of $q$. Then $q$ is formed only by atomic rules (and hence closed).*

**Lemma 22.** *Let $p$ be a normal proof, closed and PA-closed, such that*

$$\vdash_{\mathbf{PA}_c} p : \neg \forall n. \neg P(n)$$

*with $P$ atomic.*
*Then $p$ is of the form $\lambda x^{\forall n. \neg P(n)}.(xu)q^{P(u)}$ where $q^{P(u)}$ is formed only by At-rules. In natural deduction form:*

$$\frac{\dfrac{\forall n.\neg P(n) \quad u : Int}{\neg P(u)} \quad \begin{array}{c}\Pi\\P(u)\end{array}}{\dfrac{\perp}{\neg \forall n.\neg P(n)}}$$

**Proof.** By Lemma 17(i) $p$ has necessarily to begin with an introduction rule, otherwise it would be formed only by eliminations and hence it would not be closed. Then we have

$$p \equiv \lambda x^{\forall n. \neg P(n)}.p'$$

and

$$z : \forall n. \neg P(n) \vdash_{\mathbf{PA}_c} p' : \perp.$$

It is easy to see that, by the restriction on the form of formulas in $\mathbf{PA}_c$, $p'$ has to end either with a $(\rightarrow E)$-rule or with an atomic rule. Let us deal separately with the two cases.

- $p'$ ends with $(\rightarrow E)$.
  Then $p' \equiv p_1 q$ and $x : \forall n.\neg P(n) \vdash_{\mathbf{PA}_c} p_1 : \neg B$. Since $p'$ ends with an elimination, we have that, by Lemma 17, $p_1$ is formed only by eliminations. Then we can apply Lemma 19, getting $\neg B \equiv \neg At (\equiv \neg P(u))$. It now follows from Lemma 18 that $p$ is of the form $\lambda x^{\forall n. \neg P(n)}.(xu)q^{P(u)}$. Since $x : \forall n.\neg P(n) \vdash_{\mathbf{PA}_c} ((xu)q)^\perp$ we can apply Lemma 21, obtaining that $q$ is formed only by atomic rules.

- $p'$ ends with an atomic rule.
  We prove that such a case is not possible. If $p'$ ends with an atomic rule this has necessarily to be $\mathsf{PA}_{bot}$ and hence $p' \equiv \mathsf{PA}_{bot}(q^{At})$. Since we have that $x : \forall n. \neg P(n) \vdash_{\mathbf{PA}_c} (\mathsf{PA}_{bot}(q^{At}))^\perp$ we can apply Lemma 21 obtaining that $q$ is formed only by atomic rules and therefore $\mathsf{PA}_{bot}(q)$ would be a closed proof of $\perp$, which is impossible. □

**Acknowledgements.** We wish to thank Mariangiola Dezani and Mario Coppo for their gentle guidance through the perilous waters of the scientific life.

# References

1. Barbanera F., Berardi S. A constructive valuation interpretation for classical logic and its use in witness extraction. Proceedings of Colloquium on Trees in Algebra and Programming (CAAP), LNCS 581, Springer Verlag, 1992.
2. Barbanera F., Berardi S. Continuations and simple types: a strong normalization result. Proceedings of the ACM SIGPLAN Workshop on Continuations. Report No STAN-CS-92-1426 Stanford University. San Francisco, June 1992.
3. Coquand T. A Game-theoric semantic of Classical Logic, 1992, submitted to JSL
4. M. Felleisen, R. Hieb, The revised report on the syntactic theories of sequential control. Technical report 100, University of Rice, Houston, 1989. To appear in *Theoretical Computer Science*.
5. M. Felleisen, D. Friedman, E. Kohlbecker, and B. Duba, Reasoning with continuations. In "Proceedings of the First Annual Symposium on Logic in Computer Science", pages 131-141,1986.
6. Friedman H. Classically and intuitionistically provably recursive functions. In Scott D.S. and Muller G.H. editors, *Higher Set Theory*, vol.699 of *Lecture Notes in Mathematics*, 21-28. Springer Verlag, 1978.
7. Girard Y.J. A new constructive logic : Classical Logic, in MSCS, n. 1, 1990.
8. Timothy G. Griffin. A formulas-as-types notion of control. In "Conference Record of the Seventeenth Annual ACM Symposium on Principles of Programming Languages", 1990.
9. Kreisel G. Mathematical significance of consistency proofs. *Journal of Symbolic Logic*, 23:155-182, 1958.
10. Murthy C. Extracting constructive content from classical proofs. Ph.d. thesis, Department of Computer science, Cornell University, 1990.
11. Parigot M. lambda-mu-calculus: an algorithmic interpretation of classical natural deduction. Manuscript.
12. Prawitz D. Validity and normalizability of proofs in 1-st and 2-nd order classical and intuitionistic logic. In *Atti del I Congresso Italiano di Logica*, Napoli, Bibliopolis, 11-36, 1981.

# Combining First and Higher Order Rewrite Systems with Type Assignment Systems

Franco Barbanera[1][*] and Maribel Fernández[2]

[1] Dipartimento di Informatica, Corso Svizzera 185, 10149 Torino, Italy.
E-mail: barba@di.unito.it
[2] LRI Bât. 490, CNRS/Université de Paris-Sud, 91405 Orsay Cedex, France.
E-mail: maribel@lri.lri.fr

**Abstract.** General computational models obtained by integrating different specific models have recently become a stimulating and promising research subject for the computer science community. In particular, combinations of algebraic term rewriting systems and various $\lambda$-calculi, either typed or untyped, have been deeply investigated. In the present paper this subject is addressed from the point of view of *type assignment*. A powerful type assignment system, the one based on intersection types, is combined with term rewriting systems, first and higher order, and relevant properties of the resulting system, like strong normalization and confluence, are proved.

## 1 Introduction

The interactions between term rewriting systems and $\lambda$-calculus have been widely studied in the last years ([10], [16], [6], [7], [20], [3], [2], [15]). An important motivation of these works is the need of establishing a formal setting integrating two closely-related models of computation: the one based on algebraic reduction of algebraic terms and the one based on $\beta$-reduction of $\lambda$-terms.

Since both models are based on term reduction, the most natural combined model is given by the combination of the two kinds of reductions on mixed terms, i.e., on $\lambda$-algebraic-terms. An interesting question to investigate is under which conditions the properties of each reduction relation transfer to their combination. *Confluence* and *strong normalization* (or *termination*) are the most important properties of reductions: strong normalization ensures that all reduction sequences are finite, whereas confluence ensures determinacy, because if two different terms $u$ and $v$ can be obtained by reducing some term $t$, there must exist a common reduct of $u$ and $v$. The question above is whether the combined reduction relation inherits these properties when they are true of each reduction relation alone.

If we consider untyped languages, the attempt to combine both sorts of computations raises serious problems. In [16] it is proved that in order to add a confluent term rewriting system to the untyped $\lambda$-calculus preserving confluence,

[*] To Roberta Gottardi

restrictions have to be imposed on the term rewriting system. If such restrictions are omitted, it is proved in [10] that confluence and strong normalization of the combination hold only if we impose severe restrictions on the set of mixed terms. On the other hand, if we consider typed languages (typed versions of λ-calculus and typed term rewriting systems) things work out nicely: confluence and termination are *modular* properties in this case, that is, the system resulting from the combination of a confluent and terminating first-order many-sorted rewrite system and typed λ-calculus is again confluent and terminating. This result is shown in [7] and [20] for polymorphic (or second order) λ-calculus, by using the computability predicate method developed by Girard and Tait (see [11]).

In [3], types are considered from another perspective: the author investigated a version of λ-calculus where types are assigned to λ-terms by means of an inference system, showing that the same good properties that are preserved by combining typed λ-calculi with first order term rewriting systems hold even when dealing with type assignment systems. He considered in particular the intersection type discipline described in [4]. The strength of the intersection type assignment system lies in the fact that it yields a characterization of strongly normalizable terms as the class of terms having a type according to the rules of this type assignment system [21]. Besides, the strong normalization property of the combination of the intersection type assignment system and first order rewrite systems is used in [2] to extend Gallier and Breazu-Tannen's result about modularity of strong normalization, to more powerful type disciplines, such as Girard's system $F_w$ and the Calculus of Constructions.

Briefly, at this point it is well-known that strong typing (either in typed languages or in type assignment systems) is the key for the modularity of confluence and termination in combinations of first order rewrite systems and λ-calculus.

Up to now we have only mentioned theoretical motivations for the combination of two computational models. But the combination of term rewriting systems and λ-calculus was also studied as an interesting alternative in the design of new programming languages. J.-P. Jouannaud and M. Okada were the first to investigate from this point of view the combination of λ-calculus and term rewriting systems, defining a new programming paradigm that allows first order definitions of abstract data types and operators, as in equational languages like OBJ, and the full power of λ-calculus for defining functions of higher types, as in functional languages like ML.

The following example, taken from [15], shows a simple program that can not be written in one of the above-mentioned traditional languages alone, but can be written in the unified language.

*append nil l = l*
*append (cons x l) l' = cons x (append l l')*
*append (append l l') l'' = append l (append l' l'')*
*map X nil = nil*
*map X (cons x l) = cons (X x) (map X l)*

Here, the first order function *append* is defined algebraically (the third rule establishes the associativity of append on lists, and makes the definition non-recursive), while the higher order function *map* is defined recursively on the structure of lists.

Adding λ-expressions to this kind of higher order rewrite systems (i.e., allowing to substitute a λ-term for higher order variables) we obtain the usual paradigm of functional programming languages extended with equational (algebraic) definitions of operators and types. This is what Jouannaud and Okada called "algebraic functional paradigm", i.e., the combination of first order algebraic equations and λ-calculus enriched with higher order functional constants defined by higher order equations.

In [15] it is shown that for algebraic functional languages built upon typed versions of λ-calculus (simply typed λ-calculus or polymorphic λ-calculus) the combination with confluent and terminating rewrite rules is modular provided that the set of higher order rules satisfies certain conditions (the so-called *general schema*) and the first order rules are *non-duplicating* (that is, the number of occurrences of any variable in the right-hand side of a rule is less or equal than the number of occurrences in the left-hand side).[3] But the authors left open the question of whether other type disciplines for λ-calculus interact as nicely with "well-behaved" term rewriting systems. In this paper we address the problem of the modularity of confluence and strong normalization properties for algebraic functional languages with type inference, a way of dealing with types that has several advantages in practice and is used in powerful functional programming languages, notably ML [12] and MIRANDA [23].

More precisely, we consider the combination of first and higher order rewriting systems with type assignment systems for λ-calculus, and we prove that the same good properties that are preserved when combining simple and polymorphic λ-calculus with rewrite systems in algebraic functional languages hold even when dealing with types in this modified perspective. We consider in particular the intersection type assignment system, since it is a very expressive and powerful system, as already mentioned.

Our proof of strong normalization is based on the computability predicate method. We use a predicate that results from the generalization to our type assignment system of the one defined in [15]. For the proof of confluence however, we didn't apply the same method (although a proof based on computability predicates can be done as suggested in [15]). Instead, we give a simple proof using the strong normalization result proved previously.

The results we present here can be seen on one hand as a generalization of [15] because of the introduction of type assignment systems, and on the other hand as a generalization of [3] by taking into account higher order rewrite systems. It is also interesting to note that our modularity results extend Dougherty's results for untyped systems [10], since there are sets of mixed terms that do not satisfy Dougherty's conditions but are typeable in our system.

---

[3] Although not explicitly required in [15], the restriction to non-duplicating first order rules is necessary for the modularity of termination.

## 2 Type assignment systems

Type assignment systems (also called type inference systems) are formal systems for assigning types to untyped terms. These systems are defined by specifying a set of terms, a set of types one assigns to terms and a set of type-inference rules. We shall refer to [5] and [14] for all the notions about $\lambda$-calculus we do not define explicitly, and to [9] for definitions and notations for term rewriting systems.

### 2.1 The intersection type assignment system

In this section we present a variant of the intersection type assignment system ($\vdash^\wedge$) defined in [4], [21]. See also [8] and [13] for motivations and applications of intersection systems.

The set $T_\wedge$ of *intersection types* is inductively defined by:

1. $\alpha_1, \alpha_2, \ldots \in T_\wedge$ (*type variables*, a denumerable set we shall denote by $V_T$)
2. $\sigma, \tau \in T_\wedge \Rightarrow \sigma \to \tau \in T_\wedge$ (*arrow types*).
3. $\sigma, \tau \in T_\wedge \Rightarrow \sigma \wedge \tau \in T_\wedge$ (*intersection types*).

The elements of $T_\wedge$ are considered modulo associativity, commutativity and idempotency of $\wedge$. In the following $\bigwedge_{i \in I} \sigma_i$, where $I = \{i_1, \ldots, i_n\}$, will be a shorthand for $\sigma_{i_1} \wedge \ldots \wedge \sigma_{i_n}$.

The *terms* we assign types to are the terms of the pure $\lambda$-calculus ($\Lambda$). A *statement* is an expression of the form $M : \sigma$ where $\sigma \in T_\wedge$ and $M \in \Lambda$. $M$ is the *subject* of the statement. A *basis* (the set of assumptions a statement depends on) is a set of statements with only variables as subjects. Without loss of generality we assume that in a basis there are no two statements with the same subject. If $x$ does not occur in the basis $B$ then $B, x : \sigma$ denotes the basis $B \cup \{x : \sigma\}$.

The axioms and type inference rules of the intersection type assignment system $\vdash^\wedge$ are the following:

$$(Ax) \quad B, x : \sigma \vdash^\wedge x : \sigma$$

$$(\to I) \quad \frac{B, x : \sigma \vdash^\wedge M : \tau}{B \vdash^\wedge \lambda x.M : \sigma \to \tau} \qquad (\to E) \quad \frac{B \vdash^\wedge M : \sigma \to \tau \quad B \vdash^\wedge N : \sigma}{B \vdash^\wedge (MN) : \tau}$$

$$(\wedge I) \quad \frac{B \vdash^\wedge M : \sigma \quad B \vdash^\wedge M : \tau}{B \vdash^\wedge M : \sigma \wedge \tau} \qquad (\wedge E) \quad \frac{B \vdash^\wedge M : \sigma \wedge \tau}{B \vdash^\wedge M : \sigma}$$

We shall express the fact that it is possible to derive $B \vdash^\wedge M : \sigma$ in system $\vdash^\wedge$ by simply stating $B \vdash^\wedge M : \sigma$. A term will be called *typeable* in $\vdash^\wedge$ if there exists a basis $B$ and a type $\sigma$ such that $B \vdash^\wedge M : \sigma$.

Using system $\vdash^\wedge$ it is possible to characterize $\lambda$-terms having a relevant functional property, namely $\beta$-strong normalization.

**Theorem 1 [21], [17].** *A term of the pure $\lambda$-calculus is $\beta$-strongly normalizable if and only if it is typeable in the intersection type assignment system.*

## 2.2 The $\wedge TRH$ type assignment system

As stated in the introduction we now extend the system $\vdash^\wedge$ with algebraic rewriting capabilities. First of all, we consider a denumerable set $S$ of *sorts* and we define by induction the set $T_S$ of *arrow ground types based on $S$*:

1. If $s \in S$ then $s \in T_S$.
2. If $\sigma_1, \ldots, \sigma_n, \sigma \in T_S$ then $\sigma_1 \ldots \sigma_n \to \sigma \in T_S$.

An *extended signature* $\mathcal{F}$ contains *first* and *higher order function symbols*: $\mathcal{F} = \bigcup_{\tau \in T_S} \mathcal{F}_\tau$, where $\mathcal{F}_\tau$ denotes the set of function symbols of type $\tau$ ($\tau \neq \tau'$ implies $\mathcal{F}_\tau \cap \mathcal{F}_{\tau'} = \emptyset$). In case $\tau \equiv s_1 \ldots s_n \to s$ where $s_1, \ldots, s_n, s \in S$, $\mathcal{F}_\tau$ is a set of first order function symbols. We denote by $\Sigma$ the set of all first order function symbols and by $f, g$ its elements. We use $F, G$ to denote higher order function symbols. When it is clear from the context, we use $f$ to denote a generic (first or higher order) element of $\mathcal{F}$.

Let $\mathcal{X} = \bigcup_{\tau \in T_S} \mathcal{X}_\tau$, where $\mathcal{X}_\tau$ denotes a denumerable set of variables of type $\tau$ ($\tau \neq \tau'$ implies $\mathcal{X}_\tau \cap \mathcal{X}_{\tau'} = \emptyset$). $\mathcal{X}$ contains first order variables, i.e. variables whose type is a sort, and higher order variables. We use $x, y, \ldots$ to denote variables and $X, Y, \ldots$ when we want to emphasize that they are higher order variables.

The set $T(\mathcal{F}, \mathcal{X})_\sigma$ of *higher order algebraic terms of type $\sigma$* is inductively defined by:

1. If $f \in \mathcal{F}_{\sigma_1 \ldots \sigma_n \to \sigma}$, $t_1 \in T(\mathcal{F}, \mathcal{X})_{\sigma_1}, \ldots, t_n \in T(\mathcal{F}, \mathcal{X})_{\sigma_n}$ ($n \geq 0$) then $ft_1 \ldots t_n \in T(\mathcal{F}, \mathcal{X})_\sigma$.
2. If $X \in \mathcal{X}_{\sigma_1 \ldots \sigma_n \to \sigma}$, $t_1 \in T(\mathcal{F}, \mathcal{X})_{\sigma_1}, \ldots, t_n \in T(\mathcal{F}, \mathcal{X})_{\sigma_n}$ ($n \geq 0$) then $Xt_1 \ldots t_n \in T(\mathcal{F}, \mathcal{X})_\sigma$.

The set of higher order algebraic terms is $T(\mathcal{F}, \mathcal{X}) = \bigcup_\tau T(\mathcal{F}, \mathcal{X})_\tau$. It includes the set $T(\Sigma, \mathcal{X}) = \bigcup_{s \in S} T(\Sigma, \mathcal{X})_s$ of first order algebraic terms.

The rewriting systems we consider are composed of:

1. a set $R$ of *first order rewrite rules* whose elements are denoted by $r : t \to t'$ ($t, t' \in T(\Sigma, \mathcal{X})$). Obviously, in order to have a reasonable system, for each rewriting rule $r : t \to t'$ there exists $s \in S$ such that $t, t' \in T(\Sigma, \mathcal{X})_s$, i.e., the rewriting rules of $R$ are assumed sound with respect to types. Besides, in $r : t \to t'$, $t$ cannot be a variable and the set of variables in $t'$ must be a subset of the variables in $t$.
2. a set $HOR$ of *higher order rewrite rules* satisfying the *general schema* [15]:

$$F\vec{l}[\vec{X}, \vec{x}]\vec{Y}\vec{y} \to v[(F\vec{r_1}[\vec{X}, \vec{x}]\vec{Y}\vec{y}), \ldots, (F\vec{r_m}[\vec{X}, \vec{x}]\vec{Y}\vec{y}), \vec{X}, \vec{x}, \vec{Y}, \vec{y}]$$

where $\vec{X}, \vec{Y}$ are sequences of higher order variables, $\vec{x}, \vec{y}$ are sequences of first order variables, $F$ is a higher-order function symbol that can not appear in $\vec{l}, v, \vec{r_1}, \ldots, \vec{r_m}$, both sides of the rule have the same type (i.e. they belong to $T(\mathcal{F}, \mathcal{X})_\tau$ for some $\tau$), $\vec{l}, \vec{r_1}, \ldots, \vec{r_m}$ are all terms of sort type, $\vec{X} \subseteq \vec{Y}$ and $\forall i \in [1..m]$, $\vec{l} \triangleright_{mul} \vec{r_i}$ (where $\triangleleft$ is the strict subterm ordering and $mul$ denotes multiset extension).

The reduction rules above can be considered as the definitions of new functionals of the (programming) language. There is a restriction we have to impose on the set $HOR$: mutually recursive definitions are forbidden. This restriction, however, together with the one that the higher order variables in $\vec{l}$ must also appear in the left hand side definition of $F$ could be easily removed (the first by introducing product types and packing mutually recursive definitions in a same product and the second by reasoning on a transformed version of $F$). The restriction that terms in $\vec{r_i}$ are subterms of terms in $\vec{l}$ instead is essential in the proof of strong normalization of the combined language, as we shall see later.

Although restricted, this schema is interesting from a practical point of view: it allows, for example, the introduction of functional constants of higher order types by primitive recursion (or structural induction) on a first order data structure, as the function $map$ showed in the introduction (see [15] for more examples and applications). Moreover, for higher order systems satisfying the general schema all the properties we are interested in are valid also when simply typed $\lambda$-terms are allowed in right-hand sides. So, from now on we consider an extended version of higher order systems whose rules satisfy the general schema above, but where there may be simply typed $\lambda$-terms built from $\mathcal{X}$ and $\mathcal{F}$ in $v$. See [14] for missing definitions on simply typed $\lambda$-calculus.

Let us describe now the *extended type assignment system* $\wedge TRH$ ($\wedge TRH$ stands for intersection type assignment system with an algebraic rewrite system $R \cup HOR$). Let $S$ be the set of sorts and $\mathcal{F}$ the $S$-sorted algebraic signature of $R \cup HOR$. $\wedge TRH$ is the type assignment system obtained by extending the intersection type assignment system as follows:

1. The set $T_{\wedge RH}$ of types is defined by adding to case 1 of the inductive definition of $T_\wedge$ a constant type $s$ for each sort $s \in S$.
2. The terms we consider are pure $\lambda$-terms built out of a denumerable set of variables and a set of constants. The set of constants has exactly the function symbols of $\mathcal{F}$ as elements. Note that we do not consider the types of function symbols and variables here. This means that even if $S = \{int, bool\}$ and $f \in \mathcal{F}_{int \to int}$, a term such as $\lambda x.f\, True$ is well-formed, although it can not be given a type.
3. For each function symbol $f \in \mathcal{F}_{\sigma_1 \ldots \sigma_n \to \sigma}$ we add to the axioms of the intersection type assignment system the following axiom:
    $(Ax) \quad B \vdash f : \sigma_1 \to \ldots \to \sigma_n \to \sigma \quad$ for all bases $B$.
   This axiom states that $f$ can be given only its right type. Indeed what we assign to $f$ is the curryfication of its type since in $\lambda$-calculus all functions are thought of as unary higher order functions.
4. The inference rules and the definition of basis are the same as in the intersection type assignment system.

We shall use "⊢" as deduction symbol for $\wedge TRH$. Note that $\wedge TRH$ does not treat function symbols as variables. Instead, it assigns to each function symbol the very ground type it has in the typed signature we consider, avoiding in this way the problems dealt with in [1] for the preservation of types along rewriting. In the following section we prove that our rewriting relation preserves types.

We say that a term $M$ is *typeable* if there exist a basis $B$ and a type $\sigma$ such that $B \vdash M : \sigma$. The *set of typeable terms* is denoted by $\Lambda_{\wedge RH}$.

The $\wedge$-*algebraic terms* (a subset of $\Lambda_{\wedge RH}$) are inductively defined by:

1. If $f \in \mathcal{F}_{\sigma_1 \ldots \sigma_n \to \sigma}$ $(n \geq 0)$, $M_1, \ldots, M_n$ are $\wedge$-algebraic terms and $fM_1 \ldots M_n$ is typeable then $fM_1 \ldots M_n$ is a $\wedge$-algebraic term.
2. If $X \in \mathcal{X}$ and $M_1, \ldots, M_n$ $(n \geq 0)$ are $\wedge$-algebraic terms, then $XM_1 \ldots M_n$ is a $\wedge$-algebraic term.

We define in the obvious way the set of first order $\wedge$-algebraic terms.

Note that in the definition of $\wedge$-algebraic terms the variables are considered untyped. It is not difficult to check that each (first order) $\wedge$-algebraic term is typeable (with a sort type) by means of a deduction in which rule $(\wedge I)$ is not used, and rule $(\wedge E)$ is applied only as first rule. Rule $(\wedge E)$ is necessary since in a $\wedge$-algebraic term the same variable can occur whereas in the corresponding term of $T(\mathcal{F}, \mathcal{X})$ different variables are required. For example, if $f \in \mathcal{F}_{int,bool \to int}$ then $fxx$ is a $\wedge$-algebraic term (it can be assigned the type $int$ in a basis where $x$ has type $int \wedge bool$) but it is not an algebraic term.

Terms in $T(\mathcal{F}, \mathcal{X})$ can be seen as $\wedge$-algebraic if types are not taken into consideration (to emphasize this change of perspective, the $\wedge$-algebraic term corresponding to $t \in T(\mathcal{F}, \mathcal{X})$ will be denoted by **t**), but for the converse a renaming of variables may be necessary in order to avoid type conflicts. Let us formally define the mapping $\varsigma : \wedge\text{-}algebraic\ terms \to T(\mathcal{F}, \mathcal{X})$. Let $P$ be a $\wedge$-algebraic term and let $\rho_P$ be a renaming of occurrences of variables such that if $x$ occurs in $P$ in positions where arguments of different types $\tau_1, \ldots, \tau_q$ are required (to get a term in $T(\mathcal{F}, \mathcal{X})$), then, for $i = 1 \ldots q$, $\rho$ replaces $x$ by the new variable $x_i$ of type $\tau_i$ in all the positions where an argument of type $\tau_i$ is required. We define $\varsigma(P) = P\rho_P$. For example, if $f \in \mathcal{F}_{int,bool \to int}$ then $\varsigma(fxx) = fxy$.

Of course, there are many ways of defining $\rho_P$ and each possible $\rho_P$ gives a different $\varsigma(P)$, but it is sufficient to choose a canonical $\rho_P$ for each $P$ to have $\varsigma$ uniquely defined on $\wedge$-algebraic terms.

On terms in $\Lambda_{\wedge RH}$ the *$\beta$-reduction relation*, denoted by $\to_\beta$, is defined as usual. Moreover for terms in $\Lambda_{\wedge RH}$ we can define new reduction relations, each of them induced by an algebraic or higher order rewriting rule.

In the following we denote substitutions by $\{x_1 \mapsto N_1, \ldots, x_n \mapsto N_n\}$. Postfix notation is used for their application. When applying a substitution to a $\lambda$-term we must take care of bound variables, as usual.

We will denote by $\to_R$ and $\to_{HOR}$ the reduction relations induced by $R$ and $HOR$ respectively in $T(\mathcal{F}, \mathcal{X})$. $R$ and $HOR$ also induce reductions relation in $\Lambda_{\wedge RH}$. Since it will be always clear from the context, we shall use the same notation for reductions on algebraic terms and on terms in $\Lambda_{\wedge RH}$. Let us define the $\to_R$ and $\to_{HOR}$ *reduction relations* on $\Lambda_{\wedge RH}$:

Let $M, M' \in \Lambda_{\wedge RH}$ and $r : t \to t' \in R \cup HOR$. $M \to^r M'$ iff there exists a subterm $P$ of $M$ such that $P \equiv \mathbf{t}\{x_1 \mapsto N_1, \ldots, x_n \mapsto N_n\}$ (where obviously $N_1, \ldots, N_n \in \Lambda_{\wedge RH}$) and $M'$ is obtained out of $M$ by replacing the term $\mathbf{t}'\{x_1 \mapsto N_1, \ldots, x_n \mapsto N_n\}$ for $P$. Note that since $N_1, \ldots, N_n \in \Lambda_{\wedge RH}$ they can contain $\lambda$-terms. $M \to_R M'$ (resp. $M \to_{HOR} M'$) iff $M \to^r M'$ for some $r : t \to t' \in R$ (resp. $r : t \to t' \in HOR$). By $\to^*_R, \to^*_{HOR}$ we denote the reflexive and transitive closure of $\to_R, \to_{HOR}$ respectively.[4]

**Property 1** *For each reduction $M \to_R N$ (resp. $M \to_{HOR} N$) where $M, N$ are $\wedge$-algebraic terms, there exists a reduction $\varsigma(M) \to_R \varsigma(N)$ (resp. $\varsigma(M) \to_{HOR} \varsigma(N)$).*

**Theorem 2 (Termination of $\to_R$ and $\to_{HOR}$ on $\wedge$-algebraic terms).** *If $R$ (resp. HOR) is a terminating rewrite system, i.e. strongly normalizing on algebraic terms, then $\to_R$ (resp. $\to_{HOR}$) is strongly normalizing on $\wedge$-algebraic terms.*

Theorem 2 is a direct consequence of Property 1. Note that the definition of the reduction relations only for typeable terms is crucial for strong normalization to hold. If we defined algebraic reductions without care of the typability of terms we could have terms in $\Lambda_{\wedge RH}$ not strongly normalizable even if $R$ and $HOR$ are strongly normalizing systems: it is in fact easy to find strongly normalizing many-sorted term rewrite systems which fail to be strongly normalizing if we do not take types into consideration i.e. if all sorts are identified. See [22] for examples.

Now, to prove a similar result about confluence we need one more property.

**Property 2** *Let $M$ be a $\wedge$-algebraic term and $N$ an algebraic term such that $\varsigma(M) \to_R N$ (resp. $\varsigma(M) \to_{HOR} N$), then $M \to_R N\rho_M^{-1}$ (resp. $M \to_{HOR} N\rho_M^{-1}$).*

As a consequence of Properties 1, 2 and the fact that for $\wedge$-algebraic terms $M, N$ such that $M \to_{R \cup HOR} N$ we can use either $\rho_M$ or $\rho_N$ in $\varsigma(N)$, we have

**Theorem 3 (Confluence of $\to_R$ and $\to_{HOR}$ on $\wedge$-algebraic terms).** *If $R$ (resp. HOR) is confluent then $\to_R$ (resp. $\to_{HOR}$) is confluent on $\wedge$-algebraic terms.*

In the following $\to_{mix}$ will denote the reduction relation on terms of $\Lambda_{\wedge RH}$ obtained by the union of $\to_\beta, \to_R$ and $\to_{HOR}$. $\to^*_{mix}$ will denote its reflexive and transitive closure.

Now we are going to show an important property of $\to_{mix}$, the so-called *Subject Reduction* property, which states that $\to_{mix}$ preserves the types of terms.

---

[4] We have chosen to define first algebraic terms and reductions on them, and later to include this sort of reductions on terms of our type assignments by using a translation from algebraic terms to $\wedge$-algebraic terms. A different approach, more frequent in the literature, would have been that of first defining the set $\Lambda_{\wedge RH}$ of terms and then the reduction relations on $\wedge$-algebraic terms. We have refrained from this latter choice since, by being in an intersection type assignment context it would have led, in our opinion, to a more cumbersome definition.

**Lemma 4 (Rewriting preserves types).** *If $B \vdash M : \tau$ and $M \to_{R \cup HOR} M'$ then $B \vdash M' : \tau$.*

The assumption of ground types for the symbols in $\mathcal{F}$ is crucial for rewriting to be type preserving. If we accept non-ground types with the possibility of instantiating the type variables, then rewriting is not type preserving in general, as shown in [1].

**Theorem 5 (Subject Reduction).** *If $B \vdash M : \tau$ and $M \to^*_{mix} N$ then $B \vdash N : \tau$.*

## 3 Strong Normalization and Confluence of $\to_{mix}$

In [3] it is shown that $\to_R \cup \to_\beta$ is confluent and strongly normalizing in $\Lambda_{\wedge RH}$ whenever $R$ is confluent and strongly normalizing on algebraic terms. But unfortunately, to extend this result to $\to_R \cup \to_{HOR} \cup \to_\beta$ we have to require one extra condition on the first order rewrite system $R$: for each rule $r : t \to t' \in R$, each variable must "appear less" in $t'$ than in $t$, more precisely, the number of occurrences of each variable in $t'$ must be less or equal than its number of occurrences in $t$. This is because otherwise we could easily code Toyama's example of non-termination [22] by using a first order rule where this condition is not satisfied: Consider the system

$f01x \to fxxx$
$FX \to 1$
$FX \to 0$

where $F$ is a higher order function symbol, $f$ is a first order function symbol, $0, 1$ are first order constants and $X, x$ are variables. There is an infinite reduction sequence out of the term $f(FX)(FX)(FX)$:

$$f(FX)(FX)(FX) \to f0(FX)(FX) \to f01(FX) \to f(FX)(FX)(FX) \ldots$$

Confluence of $HOR$ does not suffice to restore the strong normalization property. This can be seen by coding the example of Barendregt and Klop (see [22]).

We will say that $R$ is *non-duplicating* (or *conservative*) when it satisfies the condition above. What we shall prove in the following is that if $R$ is non-duplicating, confluent and strongly normalizing, and $HOR$ is confluent (and satisfies the general schema), $\to_{mix}$ is confluent and strongly normalizing. More precisely, we shall prove that all terms in $\Lambda_{\wedge RH}$ are $\to_{mix}$-strongly normalizable and that $\to_{mix}$ is a confluent reduction relation for terms in $\Lambda_{\wedge RH}$ provided that there are no critical pairs between rules in $HOR$ and between rules in $R$ and $HOR$ (that is, only $R$ can have critical pairs).

Remark that these properties extend [10]: D.J. Dougherty showed that strong normalization and confluence are modular properties for sets of "stable terms". Intuitively, stability means that algebraic rewriting can not create new abstractions. Formally, if a set $T$ of terms is $R$-stable then for all $M \in T$ such that $M$ is in

$\beta$-normal form and $M \to_R N$, $N$ is in $\beta$-normal form. But $\Lambda_{\wedge RH}$ is not $R \cup HOR$-stable in general: Consider the rewriting system $HOR$ containing only one rule $FX0 \to X$, where $F \in \mathcal{F}_{(int \to int), int \to (int \to int)}$ and $0 \in \mathcal{F}_{int}$. This is a higher order system satisfying the general schema. Let $M \equiv (F(\lambda x.x)0)0 \in \Lambda_{\wedge RH}$. $M$ is in $\beta$-normal form, but $M \to_{HOR} (\lambda x.x)0 \to_\beta 0$.

## 3.1 Strong Normalization of $\to_{mix}$

Our proof is a generalization of the proof given in [15]. It is done in two parts: in the first one we give the definition of a predicate *Red* on types and terms, usually called *computability predicate*, and prove some properties of *Red*, the most important one stating that if *Red* holds for a term then the term is strongly normalizable. In the second part *Red* is shown to hold for each term in $\Lambda_{\wedge RH}$.

We shall call *base type* a type in $S \cup V_T$. Let $t$ be a term and $\tau$ a type. $Red(t, \tau)$ holds if there exists a basis $B$ such that $B \vdash t : \tau$ and

1. $\tau = \tau_1 \wedge \tau_2$
   $Red(t, \tau_1)$ and $Red(t, \tau_2)$.
2. $\tau \neq \tau_1 \wedge \tau_2$
   (a) $t = \lambda x.v$ (then $\tau = \sigma \to \rho$)
       for all $w$ such that $Red(w, \sigma)$ we have $Red(v\{x \mapsto w\}, \rho)$.
   (b) $t \neq \lambda x.v$ (such a $t$ will be called *neutral*)
       either $\tau$ is a base type and $t$ is strongly normalizable
       or
       for all $v$ such that $t \to_{mix} v$, $Red(v, \tau)$.

Note that by the Knaster-Tarski Theorem, this recursive definition has a least fixpoint since the underlying functional is monotonic.

We say that a term $t$ is *computable* if $\exists \tau$ such that $Red(t, \tau)$.

**Property 3** *Red satisfies the standard properties of computability predicates:*

- **(Fact 1)** *A term of a base type is computable if and only if it is strongly normalizable.*
- **(C1)** $Red(u, \tau)$ *implies $u$ is strongly normalizable.*
- **(C2)** *If $Red(u, \tau)$ and $u \to_{mix}^* v$ then $Red(v, \tau)$.*
- **(C3)** *If $u$ is neutral, $B \vdash u : \sigma$ for some $B, \sigma$ (i.e., $u$ is typeable) and for all $v$ such that $u \to_{mix} v$, $Red(v, \sigma)$ holds, then also $Red(u, \sigma)$ holds.*

The proofs are by induction on the definition of *Red*.

We need in the following the notions of alien subterm and cap, which we define as in [15]: Let $u$ be a $\Lambda_{\wedge RH}$-term. An *alien subterm of $u$* is a maximal subterm of $u$ which is not of the form $ft_1 \ldots t_n$ for some $f \in \Sigma_{s_1 \ldots s_n \to s}$. We denote by $alien(t)$ the multiset of alien subterms of $t$. The *cap* of $u$ is the $\wedge$-algebraic term obtained by replacing alien subterms by variables (the same variable for equal alien subterms). In other words, the cap of $u$ is the $\wedge$-algebraic term $u[x_1]_{p_1} \ldots [x_n]_{p_n}$ where $x_1, \ldots, x_n$ are new variables, $alien(u) = [u|_{p_1}, \ldots, u|_{p_n}]$

where square brackets denote multiset, and $x_i = x_j$ if $u|_{p_i} \equiv u|_{p_j}$. For example, the cap of $(\lambda x.M)N$ is $z$ and the cap of $f(\lambda y.y)(\lambda y.y)$ is $fzz$ if $f \in \Sigma$.

The following lemma will be used in the proof of the Strong Normalization Theorem.

**Lemma 6 (Principal Case).** *If $R$ is conservative and strongly normalizing, any $\Lambda_{\wedge RH}$-term of the form $ft_1 \ldots t_n$, where $f \in \Sigma$, is $\to_{mix}$-strongly normalizable whenever its alien subterms are $\to_{mix}$-strongly normalizable.*

*Proof.* We interpret a term $t$ by the pair $(alien(t), cap(t))$ of the multiset of its alien subterms and its cap. Multisets of alien subterms are compared in the multiset extension of $\to_{mix} \cup \triangleright$ (recall that $\triangleleft$ is the strict subterm relation), and caps are compared in the algebraic reduction ordering $\to_R$. This is a well-founded ordering since $\to_{mix}$ is strongly normalizing on alien subterms by hypothesis and $\to_R$ is strongly normalizing on $\wedge$-algebraic terms.

Now, assume $t \to_{mix} t'$. We have two cases:

1. If $t \to_{mix} t'$ in a position inside an alien subterm then $alien(t) \, (\to_{mix} \cup \triangleright)_{mul} \, alien(t')$. Note that we need to consider $\triangleright$ since the result of reducing an alien subterm of $t$ can be a term whose root belongs to $\Sigma$ and so only its strict subterms can be alien subterms of $t'$.

2. If $t \to_{mix} t'$ in a cap position (it is an algebraic reduction) then $alien(t') \subseteq alien(t)$ since $R$ is conservative, and $cap(t) \to_R cap(t')$.

In both cases the interpretation decreases (strictly). Hence there is no infinite reduction sequence. $\square$

**Theorem 7 (Strong Normalization of $\to_{mix}$ in $\Lambda_{\wedge RH}$).** *If $R$ is conservative and strongly normalizing and $HOR$ satisfies the general schema then $\to_{mix}$ is strongly normalizing on $\Lambda_{\wedge RH}$.*

*Proof.* We say that a substitution $\{x_1 \mapsto u_1, \ldots, x_n \mapsto u_n\}$ is computable in a basis $B = x_1 : \sigma_1, \ldots, x_n : \sigma_n$ if $Red(u_1, \sigma_1), \ldots, Red(u_n, \sigma_n)$ hold. In order to prove the theorem we shall prove a stronger property: *For any term $u$ with free variables $x_1, \ldots, x_n$ and such that $x_1 : \sigma_1, \ldots, x_n : \sigma_n \vdash u : \sigma$, if $\gamma = \{x_1 \mapsto u_1, \ldots, x_n \mapsto u_n\}$ is computable in $B = x_1 : \sigma_1, \ldots, x_n : \sigma_n$ then $Red(w, \sigma)$ holds for $w = u\gamma$.* The theorem follows by C1, putting $u_i = x_i$.

To prove the property we apply noetherian induction. Let $F_1, \ldots, F_n$ be the higher order function symbols of $\mathcal{F}$. We interpret the term $w = u\gamma$ by the triple $< i, u, \{\gamma\} >$, where $i$ is the maximal index of the functional constants belonging to $u$ and $\{\gamma\}$ is the multiset $\{x\gamma \mid x \in Var(u)\}$. These triples are compared in the ordering $(>_N, \diamond, (\to_{mix} \cup \triangleright)_{mul})_{lex}$, denoted by $\gg$ in the following and by $\gg_n$ when we want to indicate that the $n$th element of the triple has decreased but not the $(n-1)$ first ones. Here, $>_N$ denotes the standard ordering on natural numbers, $\diamond$ stands for the well-founded encompassment ordering, that is, $u \diamond v$ if $u \neq v$ and $u|_p = v\delta$ for some position $p \in u$ and substitution $\delta$, and $lex$, $mul$ denote respectively the lexicographic and multiset extension of an ordering. Note that encompassment contains strict subterm. Besides, since $\gamma$ is computable, $\to_{mix}$ is well-founded on $\gamma$ by property (C1), and because the union of the strict subterm relationship ($\triangleright$) with a terminating rewrite relation is well-founded [9], the relation $(\rightarrow_{mix} \cup \triangleright)_{mul}$ is well-founded on $\gamma$. Hence, $\gg$ is a well-founded ordering on the triples.

The induction hypothesis is: For all $v \equiv t\gamma'$ such that $w \gg v$, $y_1, \ldots, y_m$ are the free variables of $t$, $y_1 : \tau_1, \ldots, y_m : \tau_m \vdash t : \tau$ and $\gamma'$ is computable in $y_1 : \tau_1, \ldots, y_m : \tau_m$, $Red(v, \tau)$ holds.

There are ten cases in the proof, depending upon the properties of $w = u\gamma$.

1. Let $u = \lambda x.v$, hence $w = \lambda x.v\gamma$. We know that $B \vdash \lambda x.v : \sigma$, then there exist $\tau_i, \rho_i$ such that $B, x : \tau_i \vdash v : \rho_i$ and $\sigma = \bigwedge_{i \in I} \tau_i \rightarrow \rho_i$. Assume that $t$ is a computable term of type $\tau_i$, and define $\gamma' = \gamma \cup \{x \mapsto t\}$, hence $\gamma'$ is computable in $B, x : \tau_i$. Now, $u\gamma \gg_2 v\gamma'$, hence $Red(v\gamma', \rho_i)$ holds by the induction hypothesis. Therefore, by definition of $Red$, for all $i \in I$ $Red(w, \tau_i \rightarrow \rho_i)$ holds, and then $Red(w, \sigma)$ holds.

2. Let $u \neq \lambda x.v$. If $w$ is irreducible, then $Red(w, \sigma)$ holds by definition of $Red$. Otherwise, let $w \rightarrow_{mix} w'$ at position $p$. We must show $Red(w, \sigma)$ itself, or (by property (C3)) $Red(w', \sigma)$.

   (a) Let $p \notin \mathcal{F}Dom(u)$, where $\mathcal{F}Dom(u)$ is the set of positions of function symbols in $u$ union the set of non-leaf variable positions (i.e. positions of higher order variables applied to some arguments). This means that the reduction $w \rightarrow_{mix} w'$ takes place inside the substitution $\gamma$, in this case $p$ is a position of a leaf variable in $u$ or it is not a position in $u$. Note that this case takes care of the case where $u$ is a variable.
   
   Let $p = qp'$, $u|_q = x_i \in \mathcal{X}$, $u' = u[z]_q$ where $z$ is a new variable. If $x_i$ occurs only once in $u$ then we consider $\gamma' = \gamma|_{dom(\gamma) - \{x_i\}} \cup \{z \mapsto w'|_q\}$, otherwise $\gamma' = \gamma \cup \{z \mapsto w'|_q\}$. Note that $w|_q \rightarrow_{mix} w'|_q$ at position $p'$. Since $w|_q \in \{\gamma\}$, $Red(w|_q, \sigma_i)$ holds by assumption, then $Red(w'|_q, \sigma_i)$ by property (C2), hence $\gamma'$ is computable in $B, z_i : \sigma_i$. Now, $u = u'$ modulo renaming of variables if the variable $u|_q$ has exactly one occurrence in $u$, otherwise $u \diamond u'$. In the first case $u\gamma \gg_3 u'\gamma'$, and $u\gamma \gg_2 u'\gamma'$ in the second case. In both cases $Red(u'\gamma', \sigma)$ holds, and $u'\gamma' \equiv w'$.

   (b) Let $p \in \mathcal{F}Dom(u)$, with $p \neq \varepsilon$ ($p$ is not the root position). Then $u \diamond u|_p$, hence $u\gamma \gg_2 (u|_p)\gamma = (u\gamma)|_p$. Let $\tau_i$ be the types assigned to $u|_p$ in the derivation of $B \vdash u : \sigma$, then $Red((u\gamma)|_p, \tau_i)$ holds by the induction hypothesis and hence by definition of $Red$, $Red((u\gamma)|_p, \bigwedge \tau_i)$. Let $u' = u[z]_p$ for some new variable $z$, and $\gamma' = \gamma \cup \{z \mapsto (u\gamma)|_p\}$, then $\gamma'$ is computable in $B, z : \bigwedge \tau_i$, and $B, z : \bigwedge \tau_i \vdash u' : \sigma$. Now $u \diamond u'$, hence $u\gamma \gg_2 u'\gamma' = w$, hence $Red(w, \sigma)$.

   (c) Let $p \in \mathcal{F}Dom(u)$ and $p = \varepsilon$. We distinguish now the cases:

      i. $u = (\lambda x.v)t$ and $w = (\lambda x.v\gamma)t\gamma$. Hence $u \diamond t$, hence $u\gamma \gg_2 t\gamma$, and by the induction hypothesis we get $Red(t\gamma, \tau_i)$ where $\tau_i$ are the types assigned to $t$ in $B \vdash u : \sigma$, and hence $Red(t\gamma, \bigwedge \tau_i)$. Let now $\gamma' = \gamma \cup \{x \mapsto t\gamma\}$, hence $\gamma'$ is computable in $B, x : \bigwedge \tau_i$. Now $u \diamond v$, hence $u\gamma \gg_2 v\gamma' = w'$ and $Red(w', \sigma)$ holds.

      ii. $u = yt$ and $y\gamma = \lambda x.v$. Hence $w = (\lambda x.v)(t\gamma)$ and $w' = v\gamma'$ where $x\gamma' = t\gamma$. Since $u \diamond t$, $u\gamma \gg_2 t\gamma$, hence $\gamma'$ is computable in $B, x : \bigwedge \tau_i$

where $\tau_i$ are the types assigned to $t$ in the derivation of $B \vdash u : \sigma$. Now, $\lambda x.v$ is computable by assumption ($Red(\lambda x.v, \sigma_i)$ holds), hence by definition of $Red$, we have $Red(v\gamma', \sigma)$. Hence, $Red(w', \sigma)$.

iii. $w = u\gamma \rightarrow^r w'$, where $r \in R \cup HOR$. Assume that $u(\varepsilon) = V$ is an arbitrary algebraic symbol or a higher-order variable, and that $u \diamond V z_1 \ldots z_q$ but $u$ is not equal to $V z_1 \ldots z_q$ up to renaming of variables (cases 2(c)iv, 2(c)v and 2(c)vi below deal with $u = V z_1 \ldots z_q$). Let $u'_i = u|_i$ and $z_i \gamma' = u'_i \gamma$. $u \diamond u'_i$, hence $u\gamma \gg_2 u'_i \gamma$. Then, if $B \vdash u'_i : \sigma'_i$, $Red(u'_i \gamma, \sigma'_i)$ holds. Hence $\gamma'$ is computable in $B, z_1 : \sigma'_1, \ldots, z_q : \sigma'_q$. But $u\gamma \gg_2 (V z_1 \ldots z_q)\gamma' = w$ and $B, z_1 : \sigma'_1, \ldots, z_q : \sigma'_q \vdash V z_1 \ldots z_q : \sigma$. Hence $Red(w, \sigma)$.

iv. $u = F_k z_1 \ldots z_q$ and $w = u\gamma \rightarrow^r w'$ at the root. Then $i$ (the first component of the interpretation) is $k$. Hence $w = (F_k z_1 \ldots z_q)\gamma = F_k \vec{l}[\vec{M}, \vec{s}]\vec{N}\vec{t} \rightarrow^r v[(F_k \vec{r_1}[\vec{M}, \vec{s}]\vec{N}\vec{t}), \ldots, (F_k \vec{r_m}[\vec{M}, \vec{s}]\vec{N}\vec{t}), \vec{M}, \vec{s}, \vec{N}, \vec{t}]$. $\vec{l}[\vec{M}, \vec{s}], \vec{M} \subseteq \vec{N}, \vec{N}, \vec{t}$ are all terms in $\{\gamma\}$, hence are computable by hypothesis. Then terms in $\vec{l}[\vec{M}, \vec{s}]$ are strongly normalizable by *Fact 1* since they are of base type by definition of $HOR$. Terms in $\vec{s}$ are of base type and strongly normalizable as subterms of strongly normalizable terms, hence they are computable by *Fact 1*. Let $\vec{X}\gamma' = \vec{M}$ and $\vec{x}\gamma' = \vec{s}$. Since $\vec{M}$ and $\vec{s}$ are computable, $\gamma'$ is computable. Now $\forall j \in [1..m]$, $F_k \notin \vec{r_j}$ (by definition of $HOR$), hence $(F_k z_1 \ldots z_q)\gamma \gg_1 r_j \gamma'$, hence $r_j \gamma'$ is computable. Let now $\gamma''$ be the computable substitution such that $u\gamma'' = (F_k \vec{r_j}[\vec{X}, \vec{x}]\vec{Y}\vec{y})\gamma'$. Since $\vec{l} \rhd_{mul} \vec{r_j}$, and $\rhd$ is closed under substitution, then $\vec{l}\gamma' \rhd_{mul} \vec{r_j}\gamma'$, hence $(F_k z_1 \ldots z_q)\gamma \gg_3 (F_k z_1 \ldots z_q)\gamma''$, hence $(F_k z_1 \ldots z_q)\gamma''$ is computable. Let now $\gamma'''$ be the substitution such that $u\gamma \rightarrow^r v\gamma'''$ and that we have just proved to be computable. Since $F_j \in v$ implies $j < i$ (by definition of $HOR$) then $u\gamma \gg_1 v\gamma'''$, hence $w' = v\gamma'''$ is computable and we get $Red(w', \sigma)$ ($\sigma$ is the only type that can be assigned to $w'$).

v. Let $u = f z_1 \ldots z_q$ and $w = u\gamma$. Since $u\gamma$ is of base type, it is computable iff it is strongly normalizable. The latter property follows from Lemma 6 (Principal Case).

vi. Finally, let $u = X x_1 \ldots x_q$ and $w = u\gamma \rightarrow w'$ at the root. Then $w$ is either $(f x_1 \ldots x_q)\gamma$ or $(F_k x_1 \ldots x_q)\gamma$ or (without loss of generality $\gamma$ can be assumed idempotent) $(f t_1 \ldots t_j x_1 \ldots x_q)\gamma$ or $(F_k t_1 \ldots t_j x_1 \ldots x_q)\gamma$, which have already been shown computable. This finishes the proof.

## 3.2 Confluence of $\rightarrow_{mix}$

We recall first the definition of confluence. A reduction relation $\rightarrow$ is *confluent* if for all $t$ such that $t \rightarrow^* v_1$ and $t \rightarrow^* v_2$ for some $v_1, v_2$, there exists $v_3$ such that $v_1 \rightarrow^* v_3$ and $v_2 \rightarrow^* v_3$. Local confluence is a closely related (weaker) property: $\rightarrow$ is *locally confluent* if for all $t$ such that $t \rightarrow v_1$ and $t \rightarrow v_2$ for some $v_1, v_2$, there exists $v_3$ such that $v_1 \rightarrow^* v_3$ and $v_2 \rightarrow^* v_3$. For strongly normalizing relations, local confluence is equivalent to confluence (Newman's Lemma [18]). So, since we have already proved that $\to_{mix}$ is strongly normalizing on $\Lambda_{\wedge RH}$, it is sufficient to prove local confluence.

First we have to prove that the confluence of $\to_R$ for algebraic terms transfers to terms in $\Lambda_{\wedge RH}$, and that $HOR$ is confluent when there are no critical pairs.

**Lemma 8.** *If $R$ is confluent and strongly normalizing on the set of algebraic terms then $\to_R$ is locally confluent on $\Lambda_{\wedge RH}$.*

For terminating first order rewrite systems it is well-known that the absence of critical pairs implies confluence. The same property holds for the class of higher order rewrite systems we consider (Lemma 9 below). However, it is not true for arbitrary higher order rewrite systems, as shown in [19]. There the same lemma is proved for another class of higher order rewrite systems that differ from ours in that rules are required to be of base type and $\lambda$-terms are allowed in left-hand sides of rules. We do not need to impose the first restriction because we are assuming that all left-hand sides are algebraic.

**Lemma 9.** *Let $HOR$ be a higher order rewrite system satisfying the general schema. If there is no critical pair then $\to_{HOR}$ is confluent on $\Lambda_{\wedge RH}$.*

**Theorem 10 (Confluence of $\to_{mix}$ in $\Lambda_{\wedge RH}$).** *If $R$ is confluent over the set of algebraic terms, and there are no critical pairs between rules in $R$ and $HOR$ and between rules in $HOR$, then $\to_{mix}$ is locally confluent on $\Lambda_{\wedge RH}$.*

*Proof.* Since on $\Lambda_{\wedge RH}$ $\to_\beta$ is confluent, $\to_{HOR}$ is confluent (by Lemma 9) and $\to_R$ is confluent (by Lemma 8), it is sufficient to prove that for all $t$ such that $t \to_{mix} v_1$ at position $p$ using one of the reduction relations, and $t \to_{mix} v_2$ at position $p.q$ using a different reduction relation, there exists $v_3$ such that $v_1 \to^*_{mix} v_3$ and $v_2 \to^*_{mix} v_3$.

Let $t'$ be the term obtained out of $t$ by replacing the subterm at position $p.q$ by a new variable $x$. Since there are no critical pairs, $t'$ is still reducible at position $p$: $t' \to_{mix} v'$. If $x$ appears in $v'$ at positions $m_1, \ldots, m_n$ then $t\,|_{p.q}$ appears in $v_1$ at the same positions. Let $t''$ be the term obtained after reducing $v_1$ at positions $m_1, \ldots, m_n$. Then $v_2 \to_{mix} t''$ at position $p$. □

For example, the class of higher order rewriting systems defining higher order functions by primitive recursion (structured recursion) on first order data structures, satisfies the hypothesis and then $\to_{mix}$ is confluent in this case.

**Acknowledgments:** We would like to thank Val Breazu-Tannen for his comments on the proof of confluence. Special thanks to Jean-Pierre Jouannaud and Mariangiola Dezani for many fruitful discussions and for reading a previous version of this paper. Thanks also to Giuseppe Castagna for his help in proving strong normalization, and to Delia Kesner and Walid Sadfi for their support.

# References

1. S. van Bakel, S. Smetsers, and S. Brock. Type assignment in left linear applicative term rewriting systems. In *Proc. of CAAP'92. Colloquium on Trees in Algebra and Programming, Rennes, France, 1992*, 1992.

2. F. Barbanera. Adding algebraic rewriting to the calculus of constructions: Strong normalization preserved. In *Proc. of the 2nd Int. Workshop on Conditional and Typed Rewriting*, 1990.
3. F. Barbanera. Combining term rewriting and type assignment systems. *International Journal of Foundations of Computer Science*, 1:165–184, 1990.
4. H. Barendregt, M. Coppo, and M. Dezani-Ciancaglini. A filter $\lambda$-model and the completeness of type assignment. *Journal of Symbolic Logic*, 48(4):931–940, 1983.
5. H. P. Barendregt. *The Lambda Calculus, its Syntax and Semantics*. North Holland, Amsterdam, 2nd ed., 1984.
6. Val Breazu-Tannen. Combining algebra and higher-order types. In *Proc. 3rd IEEE Symp. Logic in Computer Science, Edinburgh*, July 1988.
7. Val Breazu-Tannen and Jean Gallier. Polymorphic rewriting conserves algebraic strong normalization. *Theoretical Computer Science*, 1990. to appear.
8. F. Cardone and M. Coppo. Two extensions of Curry's type inference system. In P. Odifreddi, editor, *Logic and Computer Sciennce*. Academic Press, 1990.
9. Nachum Dershowitz and Jean-Pierre Jouannaud. Rewrite systems. In J. van Leeuwen, editor, *Handbook of Theoretical Computer Science*, volume B, pages 243–309. North-Holland, 1990.
10. Daniel J. Dougherty. Adding algebraic rewriting to the untyped lambda calculus. In *Proc. 4th Rewriting Techniques and Applications, Como, LNCS 488*, 1991.
11. J.-Y. Girard, Y. Lafont, and P. Taylor. *Proofs and Types*. Cambridge Tracts in Theoretical Computer Science. Cambridge University Press, 1989.
12. M. Gordon, R. Milner, and C. Wadsworth. Edinburgh LCF. *Lecture Notes in Computer Science*, 78, 1979.
13. R. Hindley. Types with intersection, an introduction. *Formal aspects of Computing*, 1990.
14. R. Hindley and J. Seldin. *Introduction to Combinators and $\lambda$-calculus*. Cambridge University Press, 1986.
15. Jean-Pierre Jouannaud and Mitsuhiro Okada. Executable higher-order algebraic specification languages. In *Proc. 6th IEEE Symp. Logic in Computer Science, Amsterdam*, pages 350–361, 1991.
16. J. W. Klop. Term rewriting systems: a tutorial. *EATCS Bulletin*, 32:143–182, June 1987.
17. D. Leivant. Typing and computational properties of lambda expressions. *Theoretical Computer Science*, 44:51–68, 1986.
18. M. H. A. Newman. On theories with a combinatorial definition of 'equivalence'. *Ann. Math.*, 43(2):223–243, 1942.
19. T. Nipkow. Higher order critical pairs. In *Proc. IEEE Symp. on Logic in Comp. Science*, Amsterdam, 1991.
20. Mitsuhiro Okada. Strong normalizability for the combined system of the types lambda calculus and an arbitrary convergent term rewrite system. In *Proc. ISSAC 89, Portland, Oregon*, 1989.
21. G. Pottinger. A type assignment for the strongly normalizable $\lambda$-terms. In *To H.B. Curry: Essays on Combinatory Logic, Lambda-Calculus and Formalism*, pages 561–578. Academic Press, 1980.
22. Y. Toyama. Counterexamples to termination for the direct sum of term rewriting systems. *Information Processing Letters*, 25:141–143, April 1987.
23. D. A. Turner. Miranda: A non-strict functional language with polymorphic types. In *Functional Programming Languages and Computer Architecture, Nancy, LNCS 201*. Springer-Verlag, September 1985.

# A Term Calculus for Intuitionistic Linear Logic

Nick Benton[1], Gavin Bierman[1], Valeria de Paiva[1] and Martin Hyland[2]

[1] Computer Laboratory, University of Cambridge, UK
[2] Department of Pure Mathematics and Mathematical Statistics, University of Cambridge, UK

**Abstract.** In this paper we consider the problem of deriving a term assignment system for Girard's Intuitionistic Linear Logic for both the sequent calculus and natural deduction proof systems. Our system differs from previous calculi (e.g. that of Abramsky [1]) and has two important properties which they lack. These are the *substitution property* (the set of valid deductions is closed under substitution) and *subject reduction* (reduction on terms is well-typed). We also consider term reduction arising from cut-elimination in the sequent calculus and normalisation in natural deduction. We explore the relationship between these and consider their computational content.

## 1 Intuitionistic Linear Logic

Girard's Intuitionistic Linear Logic [3] is a refinement of Intuitionistic Logic where formulae must be used exactly once. Given this restriction the familiar logical connectives become divided into *multiplicative* and *additive* versions. Within this paper, we shall only consider the multiplicatives.

Intuitionistic Linear Logic can be most easily presented within the sequent calculus. The linearity constraint is achieved by removing the *Weakening* and *Contraction* rules. To regain the expressive power of Intuitionistic Logic, we introduce a new logical operator, !, which allows a formula to be used as many times as required (including zero). The fragment we shall consider is given in Fig. 1.

We use capital Greek letters $\Gamma, \Delta$ for sequences of formulae and $A, B$ for single formulae. The system has multiplicative conjunction or tensor, $\otimes$, linear implication, $\multimap$, and a logical operator, !. The *Exchange* rule simply allows the permutation of assumptions. In what follows we shall consider this rule to be implicit, whence the convention that $\Gamma, \Delta$ denote multisets (and not sequences).

The '! rules' have been given names by other authors. $!_{\mathcal{L}-1}$ is called *Weakening*, $!_{\mathcal{L}-2}$ *Contraction*, $!_{\mathcal{L}-3}$ *Dereliction* and $(!_{\mathcal{R}})$ *Promotion*[3]. We shall use these terms throughout this paper. In the *Promotion* rule, $!\Gamma$ means that every formula in the set $\Gamma$ is modal, in other words, if $\Gamma$ is the set $\{A_1, \ldots, A_n\}$, then $!\Gamma$ denotes the set $\{!A_1, \ldots, !A_n\}$. We shall defer the question of a term assignment system until Section 3.

## 2 Linear Natural Deduction

In the natural deduction system, originally due to Gentzen [11], but expounded by Prawitz [9], a deduction is a derivation of a proposition from a finite set of assumption

---
[3] Girard, Scedrov and Scott [5] prefer to call this rule *Storage*.

$$\frac{}{A \vdash A} \text{ Identity}$$

$$\frac{\Gamma, A, B, \Delta \vdash C}{\Gamma, B, A, \Delta \vdash C} \text{ Exchange}$$

$$\frac{\Gamma \vdash B \quad B, \Delta \vdash C}{\Gamma, \Delta \vdash C} \text{ Cut}$$

$$\frac{\Gamma \vdash A}{\Gamma, I \vdash A} (I_\mathcal{L}) \qquad \frac{}{\vdash I} (I_\mathcal{R})$$

$$\frac{\Gamma, A, B \vdash C}{\Gamma, A \otimes B \vdash C} (\otimes_\mathcal{L}) \qquad \frac{\Gamma \vdash A \quad \Delta \vdash B}{\Gamma, \Delta \vdash A \otimes B} (\otimes_\mathcal{R})$$

$$\frac{\Gamma \vdash A \quad \Delta, B \vdash C}{\Gamma, \Delta, A \multimap B \vdash C} (\multimap_\mathcal{L}) \qquad \frac{\Gamma, A \vdash B}{\Gamma \vdash A \multimap B} (\multimap_\mathcal{R})$$

$$\frac{\Gamma \vdash B}{\Gamma, !A \vdash B} !_{\mathcal{L}-1} \qquad \frac{\Gamma, !A, !A \vdash B}{\Gamma, !A \vdash B} !_{\mathcal{L}-2}$$

$$\frac{\Gamma, A \vdash B}{\Gamma, !A \vdash B} !_{\mathcal{L}-3} \qquad \frac{!\Gamma \vdash A}{!\Gamma \vdash !A} (!_\mathcal{R})$$

**Fig. 1.** (Multiplicative) Intuitionistic Linear Logic

packets, using some predefined set of inference rules. More specifically, these packets consist of a multiset of propositions, which may be empty. This flexibility is the equivalent of the Weakening and Contraction rules in the sequent calculus. Within a deduction, we may "discharge" any number of assumption packets. Assumption packets can be given natural number labels and applications of inference rules can be annotated with the labels of those packets which it discharges.

We might then ask what restrictions need we make to natural deduction to make it linear? Clearly, we need to withdraw the concept of packets of assumptions. A packet must contain exactly one proposition, i.e. a packet is now equivalent to a proposition. A rule which used to be able to discharge many packets of the same proposition, can now only discharge the one. Thus we can label every proposition with a *unique* natural number. We derive the inference rules given in Fig. 2.

The ($\multimap_\mathcal{I}$) rule says that we can discharge exactly one assumption from a deduction to form a linear implication. The ($\multimap_\mathcal{E}$) rule looks similar to the ($\supset_\mathcal{E}$) rule of Intuitionistic Logic. However it is implicit in the linear rule that the assumptions of the two upper deductions are disjoint, i.e. their set of labels are disjoint. This upholds the fundamental feature of linear natural deduction; that all assumptions must have *unique* labels.

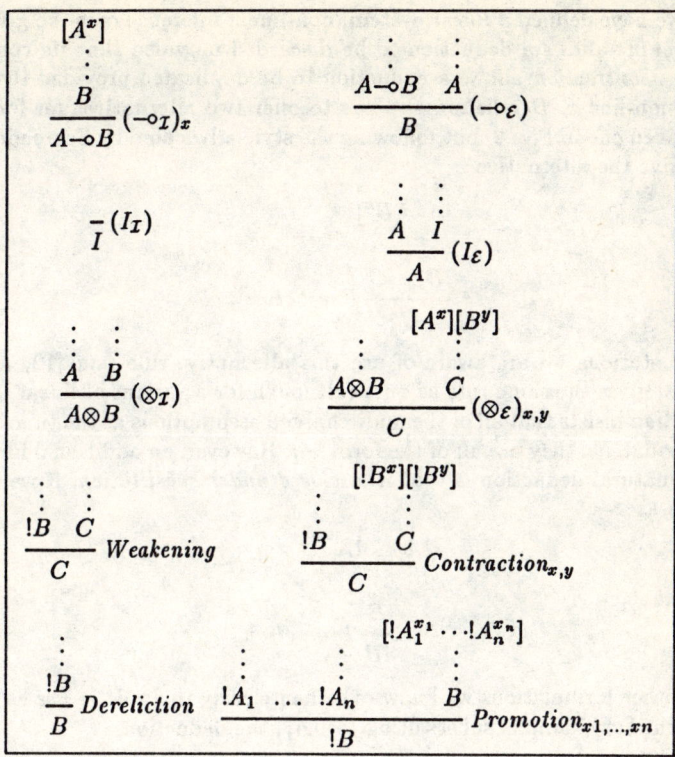

**Fig. 2.** Inference Rules in Linear Natural Deduction

The $(\otimes_\mathcal{I})$ rule is similar to the $(\wedge_\mathcal{I})$ rule of Intuitionistic Logic. It has the same restriction of disjointness of upper deduction assumptions as $(-\circ_\mathcal{E})$. In Linear Logic this makes $\otimes$ a *multiplicative* connective. The $(\otimes_\mathcal{E})$ rule is slightly surprising. Traditionally in Intuitionistic Logic we provide two projection rules for $(\wedge_\mathcal{E})$, namely

$$\frac{A \wedge B}{A} \qquad \frac{A \wedge B}{B}$$

But Intuitionistic Linear Logic decrees that a multiplicative conjunction can *not* be projected over; but rather both components must be used[4]. In the $(\otimes_\mathcal{E})$ rule, both components of the pair $A \otimes B$ are used in the deduction of $C$.

Rules that are of a similar form to $(\otimes_\mathcal{E})$ have been considered in detail by Schroeder-Heister [10]. The astute reader will have noticed the similarity between our $(\otimes_\mathcal{E})$ rule and the $(\vee_\mathcal{E})$ rule of Intuitionistic Logic. This is interesting as we know that $(\vee_\mathcal{E})$ is not very well behaved as a logical rule [4, Chapter 10].

---

[4] Projections are only defined for *additive* conjunction.

Since we have defined a *linear* system, non-linear inference must be given explicitly. *Weakening* allows a deduction to be discarded provided that its conclusion is non-linear. *Contraction* allows a deduction to be duplicated provided that its conclusion is non-linear. *Dereliction* appears to offer two alternatives for formulation. We have given one in Fig. 2, but following the style advocated by Schroeder-Heister, we could give the alternative

$$\frac{!B \quad \begin{array}{c} [B^x] \\ \vdots \\ C \end{array}}{C} \textit{Dereliction}'_x$$

Most presentations we are aware of use this alternative rule (e.g. [12, 7, 6]); only O'Hearn [8] gives the same rule as ours (although for a variant of linear logic).

*Promotion* insists that all of the undischarged assumptions at the time of application are modal, i.e. they are all of the form $!A_i$. However, an additional fundamental feature of natural deduction is that it is *closed under substitution*. If we had taken *Promotion* as

$$\frac{\begin{array}{c} !A_1 \cdots !A_n \\ \vdots \\ B \end{array}}{!B} \textit{Promotion}$$

(as in all other formulations we know of), then clearly this rule is *not* closed under substitution. For example, substituting for $!A_1$, the deduction

$$\frac{C \multimap !A_1 \quad C}{!A_1} (\multimap_\varepsilon)$$

we get the following deduction

$$\frac{\begin{array}{cc} \dfrac{C \multimap !A_1 \quad C}{!A_1} (\multimap_\varepsilon) & \cdots !A_n \\ \vdots \\ B \end{array}}{!B} \textit{Promotion}$$

which is no longer a valid deduction (the assumptions are not all modal.) We conclude that *Promotion* must be formulated as in Fig. 2, where the substitutions are given explicitly[5].

It is possible to present natural deduction rules in a 'sequent-style', where given a sequent $\Gamma \vdash A$, the multiset $\Gamma$ represents all the undischarged propositions so far in the deduction, and $A$ represents conclusion of the deduction. We can still label the undischarged assumptions with a unique natural number, but we refrain from doing

---

[5] Prawitz [9, p.79] encountered similar problems when defining the rule for introduction of necessitation. He defined a notion of essentially modal formulae and needed to keep track of dependencies in the derivation.

so. This formulation should not be confused with the sequent calculus formulation, which differs by having operations which act on the left and right of the turnstile, rather than rules for the introduction and elimination of logical constants.

We now apply the Curry-Howard Correspondence to derive a term assignment system for this natural deduction formulation of Intuitionistic Linear Logic. The Curry-Howard Correspondence essentially annotates each stage of the deduction with a "term", which is an encoding of the construction of the deduction so far. This means that a logic can be viewed as a type system for a term assignment system. The Correspondence also links proof normalisation to term reduction. We shall use this feature in Section 4.

The term assignment system obtained is given in a 'sequent-style' in Fig. 3. We should point out that the unique natural number labels used above, are replaced by (the more familiar) unique variable names.

$$x:A \vdash x:A$$

$$\frac{\Gamma, x:A \vdash e:B}{\Gamma \vdash \lambda x^A.e : A \multimap B}\ (\multimap_\mathcal{I}) \qquad \frac{\Gamma \vdash e : A \multimap B \quad \Delta \vdash f : A}{\Gamma, \Delta \vdash ef : B}\ (\multimap_\mathcal{E})$$

$$\vdash * : I \qquad \frac{\Gamma \vdash e : A \quad \Delta \vdash f : I}{\Gamma, \Delta \vdash \text{let } f \text{ be } * \text{ in } e : A}\ (I_\mathcal{E})$$

$$\frac{\Gamma \vdash e : A \quad \Delta \vdash f : B}{\Gamma, \Delta \vdash e \otimes f : A \otimes B}\ (\otimes_\mathcal{I}) \quad \frac{\Gamma \vdash e : A \otimes B \quad \Delta, x:A, y:B \vdash f:C}{\Gamma, \Delta \vdash \text{let } e \text{ be } x \otimes y \text{ in } f : C}\ (\otimes_\mathcal{E})$$

$$\frac{\Delta_1 \vdash e_1 :! A_1 \ \cdots \ \Delta_n \vdash e_n :! A_n \quad x_1 :! A_1, \ldots, x_n :! A_n \vdash f : B}{\Delta_1, \ldots, \Delta_n \vdash \text{promote } e_1, \ldots, e_n \text{ for } x_1, \ldots, x_n \text{ in } f :! B}\ Promotion$$

$$\frac{\Gamma \vdash e :! A \quad \Delta \vdash f : B}{\Gamma, \Delta \vdash \text{discard } e \text{ in } f : B}\ Weakening \quad \frac{\Gamma \vdash e :! A \quad \Delta, x :! A, y :! A \vdash f : B}{\Gamma, \Delta \vdash \text{copy } e \text{ as } x, y \text{ in } f : B}\ Contraction$$

$$\frac{\Gamma \vdash e :! A}{\Gamma \vdash \text{derelict}(e) : A}\ Dereliction$$

**Fig. 3.** Term Assignment System for Linear Natural Deduction

We note at once a significant property of the term assignment system for linear natural deduction. Essentially the terms code the derivation trees so that any valid term assignment has a *unique* derivation.

**Theorem 1 (Unique Derivation).** *For any term t and proposition A, if there is a valid derivation of the form $\Gamma \vdash t : A$, then there is a unique derivation of $\Gamma \vdash t : A$.*

*Proof.* By induction on the structure of $t$. □

We are now in a position to consider the question of substitution. In previous work [12], it was shown that substitution does not hold for the term assignment systems considered hitherto. Some thought that this represented a mismatch between the semantics and syntax of linear logic. We can now see that this is not the case. For our system, the substitution property holds.

**Theorem 2 Substitution.** *If $\Gamma \vdash a : A$ and $\Delta, x : A \vdash b : B$ then $\Gamma, \Delta \vdash b[a/x] : B$*

*Proof.* By induction on the derivation $\Delta, x : A \vdash b : B$. □

Before we continue, a quick word concerning the *Promotion* rule. At first sight this seems to imply an ordering of the $e_i$ and $x_i$ subterms. However, the *Exchange* rule (which does not introduce any additional syntax) tells us that any such order is really just the effect of writing terms in a sequential manner on the page. This paper is not really the place to discuss such syntactical questions. Perhaps proof nets (or a variant of them) are the answer.

## 3 Sequent Calculus

Here we shall consider briefly term assignment for the sequent calculus. For the sequent calculus there is no Curry-Howard Correspondence because we can encode the same proof in many ways. We have to have some further insight into the logic before we can produce the terms. There are (at least) two ways of doing this: semantically or proof-theoretically. More specifically we can either use a model to suggest the term system or use the well-known relationship between sequent calculus and natural deduction directly. Of course, both methods should converge to a single solution! In our case both these methods lead to the term assignment system given in Fig. 4. Our term system is essentially the same as Abramsky's [1] except for the *Dereliction* and *Promotion* rules.

## 4 Proof Normalisation

Within natural deduction we can produce so-called "detours" in a deduction, which arise where we introduce a logical constant and then eliminate it immediately afterwards. We can define a procedure called *normalisation* which systematically eliminates such detours from a deduction. A deduction which has no such detours is said to be in *normal form*.

We can define the normalisation procedure by considering each pair of introduction and elimination rules in turn, of which there are six. Here we shall just give an example:

$$x:A \vdash x:A$$

$$\frac{\Gamma \vdash e:A \quad \Delta, x:A \vdash f:B}{\Gamma, \Delta \vdash f[e/x]:B} Cut$$

$$\frac{\Gamma \vdash e:A \quad \Delta, x:B \vdash f:C}{\Gamma, g:A \multimap B, \Delta \vdash f[(ge)/x]:C} (\multimap_{\mathcal{L}}) \qquad \frac{\Gamma, x:A \vdash e:B}{\Gamma \vdash \lambda x^A.e:A \multimap B} (\multimap_{\mathcal{R}})$$

$$\frac{\Gamma \vdash e:A}{\Gamma, x:I \vdash \text{let } x \text{ be } * \text{ in } e:A} (I_{\mathcal{L}}) \qquad \frac{}{\vdash *:I} (I_{\mathcal{R}})$$

$$\frac{\Delta, x:A, y:B \vdash f:C}{\Delta, z:A \otimes B \vdash \text{let } z \text{ be } x \otimes y \text{ in } f:C} (\otimes_{\mathcal{L}}) \qquad \frac{\Gamma \vdash e:A \quad \Delta \vdash f:B}{\Gamma, \Delta \vdash e \otimes f:A \otimes B} (\otimes_{\mathcal{R}})$$

$$\frac{\Gamma \vdash e:B}{\Gamma, z:!A \vdash \text{discard } z \text{ in } e:B} Weakening \qquad \frac{\Gamma, x:!A, y:!A \vdash e:B}{\Gamma, z:!A \vdash \text{copy } z \text{ as } x,y \text{ in } e:B} Contraction$$

$$\frac{\Gamma, x:A \vdash e:B}{\Gamma, z:!A \vdash e[\text{derelict}(z)/x]:B} Dereliction \qquad \frac{\overline{x}:!\Gamma \vdash e:A}{\overline{y}:!\Gamma \vdash \text{promote } \overline{y} \text{ for } \overline{x} \text{ in } e:!A} Promotion$$

Fig. 4. Term Assignment System for Sequent Calculus

- *Promotion* with *Contraction*

$$\frac{\frac{\begin{array}{c}\vdots \\ !A_1 \ \ldots \ !A_n\end{array} \quad \begin{array}{c}[!A_1]\ldots[!A_n] \\ \vdots \\ B\end{array}}{!B} Prom. \quad \begin{array}{c}[!B][!B] \\ \vdots \\ C\end{array}}{C} Cont.$$

normalises to

$$\frac{\dfrac{[!A_1]\ldots[!A_n] \quad \begin{array}{c}[!A_1]\ldots[!A_n]\\ \vdots \\ B\end{array}}{!B} Prom. \quad \dfrac{[!A_1]\ldots[!A_n] \quad \begin{array}{c}[!A_1]\ldots[!A_n]\\ \vdots \\ B\end{array}}{!B} Prom.}{\begin{array}{c}\vdots \\ C\end{array} \quad \begin{array}{c}\vdots \\ C\end{array} \quad !A_1 \ \ldots \ !A_n} Cont.*$$

As mentioned earlier, the Curry-Howard Correspondence tells us that we can relate proof normalisation to term reduction. Hence we can annotate the proof tree transformations to produce the (one-step) term reduction rules, which are given in full in Fig. 5. The astute reader will have noticed our use of shorthand in the last two rules. Hopefully, our notation is clear; for example, the term discard $e_i$ in $u$ represents the term discard $e_1$ in ... discard $e_n$ in $u$.

| | |
|---|---|
| $(\lambda x^A.t)e$ | $\to_\beta t[e/x]$ |
| let $*$ be $*$ in $e$ | $\to_\beta e$ |
| let $e \otimes t$ be $x \otimes y$ in $u$ | $\to_\beta u[e/x, t/y]$ |
| derelict(promote $e_i$ for $x_i$ in $t$) | $\to_\beta t[e_i/x_i]$ |
| discard (promote $e_i$ for $x_i$ in $t$) in $u$ | $\to_\beta$ discard $e_i$ in $u$ |
| copy (promote $e_i$ for $x_i$ in $t$) as $y, z$ in $u$ | $\to_\beta$ copy $e_i$ as $x'_i, x''_i$ in $u$[promote $x'_i$ for $x_i$ in $t/y$, promote $x''_i$ for $x_i$ in $t/z$] |

**Fig. 5.** One-step $\beta$-reduction rules

### Commuting Conversions

We follow a similar presentation to that of Girard [4, Chapter 10]. We use the shorthand notation

$$\frac{C \quad \vdots}{D} r$$

to denote an elimination of the premise $C$, where the conclusion is $D$ and the ellipses represent possible other premises. This notation covers the six elimination rules: $(-\circ_\mathcal{E})$, $(I_\mathcal{E})$, $(\otimes_\mathcal{E})$, *Contraction*, *Weakening* and *Dereliction*. We shall follow Girard and commute the $r$ rule upwards. Here we shall just give an example.

– Commutation of *Contraction*

$$\cfrac{\cfrac{!B \quad \cfrac{[!B][!B]}{\vdots} \quad C}{C} \text{Contraction} \quad \vdots}{D} r$$

which commutes to

$$\cfrac{!B \qquad \cfrac{[!B][!B] \quad \vdots \atop \vdots \; C \; \vdots \atop D}{D}r}{D} \text{Contraction}$$

Again we can use the Curry-Howard Correspondence to get the term conversions. We give (all) the term conversions in Fig. 6. We use the symbol $\to_c$ to denote a commuting conversion.

These commuting conversions, although traditionally dismissed, appear to have some computational significance—they appear to reveal further $\beta$-redexes which exist in a term. Let us consider an example; the term

$$(\text{copy } e \text{ as } x, y \text{ in } \lambda z^{!A}.\text{discard } z \text{ in } x \otimes y)g$$

is in normal form. We can apply a commuting conversion to get the term

$$\text{copy } e \text{ as } x, y \text{ in } (\lambda z^{!A}.\text{discard } z \text{ in } x \otimes y)g$$

which has an (inner) $\beta$-redex. From an implementation perspective, such conversions would ideally be performed at compile-time (although almost certainly not at runtime). Again, as mentioned earlier, a better (i.e. less sequential) syntax might make such conversions unnecessary.

We can now prove subject reduction; namely that ($\beta$ and commuting) reduction ($\to_{\beta,c}$) is well-typed. Again this property was thought not to hold [6, 8].

**Theorem 3 (Subject Reduction).** *If $\Gamma \vdash e : A$ and $e \to_{\beta,c} f$ then $\Gamma \vdash f : A$.*

*Proof.* By induction on the derivation of $e \to_{\beta,c} f$. □

It is evident that the above theorem also holds for $\to_{\beta,c}^*$ the reflexive and transitive closure of $\to_{\beta,c}$.

## 5 Cut Elimination for Sequent Calculus

In this section we consider cut elimination for the sequent calculus formulation of Intuitionistic Linear Logic. Suppose that a derivation in the term assignment system of Fig. 4 contains a cut:

$$\cfrac{\overline{\Gamma \vdash e : A}^{D_1} \qquad \overline{\Delta, x : A \vdash f : B}^{D_2}}{\Gamma, \Delta \vdash f[e/x] : B} \text{Cut}$$

If $\Gamma \vdash e : A$ is the direct result of a rule $D_1$ and $\Delta, x : A \vdash f : B$ the result of a rule $D_2$, we say that the cut is a $(D_1, D_2)$-cut. A step in the process of eliminating cuts in the derivation tree will replace the subtree with root $\Gamma, \Delta \vdash f[e/x] : B$ with a tree with root of the form

|  |  |
|---|---|
| (let $e$ be $x \otimes y$ in $f)g$ | $\to_c$ let $e$ be $x \otimes y$ in $(fg)$ |
| let (let $e$ be $x \otimes y$ in $f$) be $p \otimes q$ in $g$ | $\to_c$ let $e$ be $x \otimes y$ in (let $f$ be $p \otimes q$ in $g$) |
| discard (let $e$ be $x \otimes y$ in $f$) in $g$ | $\to_c$ let $e$ be $x \otimes y$ in (discard $f$ in $g$) |
| copy (let $e$ be $x \otimes y$ in $f$) as $p, q$ in $g$ | $\to_c$ let $e$ be $x \otimes y$ in (copy $f$ as $p, q$ in $g$) |
| let (let $e$ be $x \otimes y$ in $f$) be $*$ in $g$ | $\to_c$ let $e$ be $x \otimes y$ in (let $f$ be $*$ in $g$) |
| derelict(let $e$ be $x \otimes y$ in $f$) | $\to_c$ let $e$ be $x \otimes y$ in (derelict($f$)) |
| | |
| (let $e$ be $*$ in $f)g$ | $\to_c$ let $e$ be $*$ in $(fg)$ |
| let (let $e$ be $*$ in $f$) be $p \otimes q$ in $g$ | $\to_c$ let $e$ be $*$ in (let $f$ be $p \otimes q$ in $g$) |
| discard (let $e$ be $*$ in $f$) in $g$ | $\to_c$ let $e$ be $*$ in (discard $f$ in $g$) |
| copy (let $e$ be $*$ in $f$) as $p, q$ in $g$ | $\to_c$ let $e$ be $*$ in (copy $f$ as $p, q$ in $g$) |
| let (let $e$ be $*$ in $f$) be $*$ in $g$ | $\to_c$ let $e$ be $*$ in (let $f$ be $*$ in $g$) |
| derelict(let $e$ be $*$ in $f$) | $\to_c$ let $e$ be $*$ in (derelict($f$)) |
| | |
| (discard $e$ in $f)g$ | $\to_c$ discard $e$ in $(fg)$ |
| let (discard $e$ in $f$) be $p \otimes q$ in $g$ | $\to_c$ discard $e$ in (let $f$ be $p \otimes q$ in $g$) |
| discard (discard $e$ in $f$) in $g$ | $\to_c$ discard $e$ in (discard $f$ in $g$) |
| copy (discard $e$ in $f$) as $p, q$ in $g$ | $\to_c$ discard $e$ in (copy $f$ as $p, q$ in $g$) |
| let (discard $e$ in $f$ be $*$ in $g$ | $\to_c$ discard $e$ in (let $f$ be $*$ in $g$) |
| derelict(discard $e$ in $f$) | $\to_c$ discard $e$ in (derelict($f$)) |
| | |
| (copy $e$ as $x, y$ in $f)g$ | $\to_c$ copy $e$ as $x, y$ in $(fg)$ |
| let (copy $e$ as $x, y$ in $f$) be $p \otimes q$ in $g$ | $\to_c$ copy $e$ as $x, y$ in (let $f$ be $p \otimes q$ in $g$) |
| discard (copy $e$ as $x, y$ in $f$) in $g$ | $\to_c$ copy $e$ as $x, y$ in (discard $f$ in $g$) |
| copy (copy $e$ as $x, y$ in $f$) as $p, q$ in $g$ | $\to_c$ copy $e$ as $x, y$ in (copy $f$ as $p, q$ in $g$) |
| let (copy $e$ as $x, y$ in $f$) be $*$ in $g$ | $\to_c$ copy $e$ as $x, y$ in (let $f$ be $*$ in $g$) |
| derelict(copy $e$ as $x, y$ in $f$) | $\to_c$ copy $e$ as $x, y$ in (derelict($f$)) |

Fig. 6. Commuting Conversions

$$\Gamma, \Delta \vdash t : B$$

The terms in the remainder of the tree may be affected as a result.

Thus to ensure that the cut elimination process extends to derivations in the term assignment system, we must insist on an equality $f[e/x] = t$, which we can read from left to right as a term reduction. In fact we must insist on arbitrary substitution instances of the equality, as the formulae in $\Gamma$ and $\Delta$ may be subject to cuts in the derivation tree below the cut in question.

In this section we are mainly concerned to describe the equalities/reductions which result from the considerations just described. Note however that we cannot be entirely blithe about the process of eliminating cuts at the level of the propositional logic. As we shall see, not every apparent possibility for eliminating cuts should be realized in practice. This is already implicit in our discussion of natural deduction.

As things stand there are 11 rules of the sequent calculus aside from *Cut* (and *Exchange*) and hence 121 a priori possibilities for $(D_1, D_2)$-cuts. Fortunately most of these possibilities are not computationally meaningful in the sense that they have no effect on the terms. We say that a cut is *insignificant* if the equality $f[e/x] = t$ we derive from it as above is actually an identity (up to $\alpha$-equivalence) on terms (so in executing the cut the term at the root of the tree does not change). Let us begin by considering the insignificant cuts.

First note that any cut involving an axiom rule

$$\frac{}{x : A \vdash x : A} Identity$$

is insignificant; and the cut just disappears (hence instead of 121 we must now account for 100 cases). These 100 cases of cuts we will consider as follows: 40 cases of cuts the form $(R, D)$ as we have 4 right rules and 10 others; 24 cases of cuts of the form $(L, R)$ as we have 6 left-rules and 4 right ones and finally 36 cases of cuts of the form $(L, L)$. Let us consider these three groups in turn.

Firstly we observe that there is a large class of insignificant cuts of the form $(R, D)$ where $R$ is a right rule: $(\otimes_\mathcal{R})$, $(I_\mathcal{R})$, $(-\circ_\mathcal{R})$, *Promotion*. Indeed all such cuts are insignificant with the following exceptions:

– *Principal cuts.* These are the cuts of the form $((\otimes_\mathcal{R}),(\otimes_\mathcal{L}))$, $((I_\mathcal{R}),(I_\mathcal{L}))$, $((-\circ_\mathcal{R}),(-\circ_\mathcal{L}))$, (*Promotion*, *Dereliction*), (*Promotion*, *Weakening*), (*Promotion*, *Contraction*) where the cut formula is introduced on the right and left of the two rules.
– Cases of the form $(R, Promotion)$ where $R$ is a right rule. Here we note that cuts of the form $((\otimes_\mathcal{R}), Promotion)$, $((I_\mathcal{R}), Promotion)$ and $((-\circ_\mathcal{R}), Promotion)$ cannot occur; so the only possibility is (*Promotion, Promotion*).

Next any cut of the form $(L, R)$ where $L$ is one of the left rules $(\otimes_\mathcal{L})$, $(I_\mathcal{L})$, $(-\circ_\mathcal{L})$, *Weakening*, *Contraction*, *Dereliction* and $R$ is one of the simple right rules $(\otimes_\mathcal{R})$, $(I_\mathcal{R})$, $(-\circ_\mathcal{R})$ is insignificant (18 cases). Also cuts of the form $((-\circ_\mathcal{L}), Promotion)$ and $(Dereliction, Promotion)$ are insignificant (2 cases). This is one of the things we gain by having actual substitutions in the $(-\circ_\mathcal{L})$ and *Dereliction* rules. Thus there remains four further cases of cuts of the form $(L, Promotion)$ where $L$ is a left rule.

Lastly the 36 cuts of the form $(L_1, L_2)$, where the $L_i$ are both left rules. Again we derive some benefit from our rules for $(-\circ_\mathcal{L})$ and *Dereliction*: cuts of the form $((-\circ_\mathcal{L}), L)$ and $(Dereliction, L)$ are insignificant. There are hence 24 remaining cuts of interest.

We now summarize the cuts of which we need to take some note. They are:

– Principal cuts. There are six of these.
– Secondary Cuts. The single (strange) form of cut: (*Promotion, Promotion*) and the four remaining cuts of form $(L, Promotion)$ where $L$ is a left rule other than $(-\circ_\mathcal{L})$ or *Dereliction*.
– Commutative Cuts. The twenty-four remaining cuts of the form $(L_1, L_2)$ just described.

## 5.1 Principal Cuts

The first three cases are entirely familiar and we simply state the resulting $\beta$-rules.

$$\text{let } f \otimes g \text{ be } x \otimes y \text{ in } h \; \triangleright \; h[f/x, g/y] \tag{1}$$

$$\text{let } * \text{ be } * \text{ in } h \triangleright h \qquad (2)$$

$$h[(\lambda x^A.f)g/y] \triangleright h[f[g/x]/y] \qquad (3)$$

We shall consider in detail the principal cuts involving the *Promotion* rule.

- (*Promotion, Dereliction*)-cut. The derivation

$$\cfrac{\cfrac{!\Gamma \vdash B}{!\Gamma \vdash !B} \text{ Promotion} \qquad \cfrac{B, \Delta \vdash C}{!B, \Delta \vdash C} \text{ Dereliction}}{!\Gamma, \Delta \vdash C} \text{ Cut}$$

is reduced to

$$\cfrac{!\Gamma \vdash B \qquad B, \Delta \vdash C}{!\Gamma, \Delta \vdash C} \text{ Cut}$$

This reduction yields the following term reduction.

$$(f[\text{derelict}(q)/p])[\text{promote } y_i \text{ for } x_i \text{ in } e/q] \triangleright f[e/p] \qquad (4)$$

- (*Promotion, Weakening*)-cut. The derivation

$$\cfrac{\cfrac{!\Gamma \vdash B}{!\Gamma \vdash !B} \text{ Promotion} \qquad \cfrac{\Delta \vdash C}{!B, \Delta \vdash C} \text{ Weakening}}{!\Gamma, \Delta \vdash C} \text{ Cut}$$

is reduced to

$$\cfrac{\Delta \vdash C}{!\Gamma, \Delta \vdash C} \text{ Weakening*}$$

where *Weakening\** corresponds to many applications of the *Weakening* rule.

This gives the term reduction

$$\text{discard (promote } e_i \text{ for } x_i \text{ in } f) \text{ in } g \triangleright \text{discard } e_i \text{ in } g \qquad (5)$$

- (*Promotion, Contraction*)-cut. The derivation

$$\cfrac{\cfrac{!\Gamma \vdash B}{!\Gamma \vdash !B} \text{ Promotion} \qquad \cfrac{!B, !B, \Delta \vdash C}{!B, \Delta \vdash C} \text{ Contraction}}{!\Gamma, \Delta \vdash C} \text{ Cut}$$

is reduced to

$$\cfrac{\cfrac{!\Gamma \vdash B}{!\Gamma \vdash !B} \text{ Promotion} \qquad \cfrac{\cfrac{!\Gamma \vdash B}{!\Gamma \vdash !B} \text{ Promotion} \qquad !B, !B, \Delta \vdash C}{!\Gamma, !B, \Delta \vdash C} \text{ Cut}}{\cfrac{!\Gamma, !\Gamma, \Delta \vdash C}{!\Gamma, \Delta \vdash C} \text{ Contraction*}} \text{ Cut}$$

or to the symmetric one where we cut against the other $B$ first. This gives the term reduction

$$\text{copy (promote } e_i \text{ for } x_i \text{ in } f) \text{ as } y, y' \text{ in } g \;\triangleright$$
$$\text{copy } e_i \text{ as } z_i, z'_i \text{ in } g[\text{promote } z_i \text{ for } x_i \text{ in } f/y, \text{promote } z'_i \text{ for } x_i \text{ in } f/y'] \tag{6}$$

Note that the three cases of cut elimination above involving *Promotion* are only considered by Girard, Scedrov and Scott [5] when the context ($!\Gamma$) is empty. If the context is non-empty these are called *irreducible cuts*.

The principal cuts correspond to the $\beta$-reductions in natural deduction. Hence the reductions that we have just given are almost the same as those given in Fig. 5. The differences arise because in the sequent calculus some 'reductions in context' are effected directly by the process of moving cuts upwards. Hence some of the rules just given appear more general.

## 5.2 Secondary Cuts

We now consider the cases where the *Promotion* rule is on the right of a cut rule. The first case is the 'strange' case of cutting *Promotion* against *Promotion*, then we have the four cases $(\otimes_\mathcal{L})$, $(I_\mathcal{L})$, *Weakening* and *Contraction* against the rule *Promotion*.

- (*Promotion*, *Promotion*)-cut. The derivation

$$\cfrac{\cfrac{!\Gamma \vdash B}{!\Gamma \vdash !B}\text{ Promotion} \quad \cfrac{!B, !\Delta \vdash C}{!B, !\Delta \vdash !C}\text{ Promotion}}{!\Gamma, !\Delta \vdash !C}\text{ Cut}$$

reduces to

$$\cfrac{\cfrac{\cfrac{!\Gamma \vdash B}{!\Gamma \vdash !B}\text{ Promotion} \quad !B, !\Delta \vdash C}{!\Gamma, !\Delta \vdash C}\text{ Cut}}{!\Gamma, !\Delta \vdash !C}\text{ Promotion}$$

Note that it is always possible to permute the cut upwards, as all the formulae in the antecedent are nonlinear.

This gives the term reduction

$$\text{promote (promote } z \text{ for } x \text{ in } f) \text{ for } y \text{ in } g \;\triangleright$$
$$\text{promote } w \text{ for } z \text{ in } (g[\text{promote } z \text{ for } x \text{ in } f/y]) \tag{7}$$

- $((\otimes_\mathcal{L}), \text{\textit{Promotion}})$-cut. The derivation

$$\cfrac{\cfrac{A, E, \Gamma \vdash !B}{A \otimes E, \Gamma \vdash !B}(\otimes_\mathcal{L}) \quad \cfrac{!\Delta, !B \vdash C}{!B, !\Delta \vdash !C}\text{ Promotion}}{A \otimes E, \Gamma, !\Delta \vdash !C}\text{ Cut}$$

reduces to

$$\dfrac{\dfrac{A,E,\Gamma \vdash !B \qquad \dfrac{!B,!\Delta \vdash C}{!B,!\Delta \vdash !C}\,Promotion}{A,E,\Gamma,!\Delta \vdash !C}\,Cut}{A\otimes E,\Gamma,!\Delta \vdash !C}\,(\otimes_{\mathcal{L}})$$

This gives the term reduction

$$\mathsf{promote}\,(\mathsf{let}\,z\,\mathsf{be}\,x,y\,\mathsf{in}\,f)\,\mathsf{for}\,w\,\mathsf{in}\,g\ \triangleright\ \mathsf{let}\,z\,\mathsf{be}\,x,y\,\mathsf{in}\,(\mathsf{promote}\,f\,\mathsf{for}\,w\,\mathsf{in}\,g) \qquad (8)$$

- $((I_{\mathcal{L}}), Promotion)$-**cut.** The derivation

$$\dfrac{\dfrac{\Gamma \vdash !B}{I,\Gamma \vdash !B}\,(I_{\mathcal{L}}) \qquad \dfrac{!\Delta,!B \vdash C}{!B,!\Delta \vdash !C}\,Promotion}{I,\Gamma,!\Delta \vdash !C}\,Cut$$

reduces to

$$\dfrac{\dfrac{\Gamma \vdash !B \qquad \dfrac{!B,!\Delta \vdash C}{!B,!\Delta \vdash !C}\,Promotion}{\Gamma,!\Delta \vdash !C}\,Cut}{I,\Gamma,!\Delta \vdash !C}\,(I_{\mathcal{L}})$$

This gives the term reduction

$$\mathsf{promote}\,(\mathsf{let}\,z\,\mathsf{be}\,*\,\mathsf{in}\,f)\,\mathsf{for}\,w\,\mathsf{in}\,g\ \triangleright\ \mathsf{let}\,z\,\mathsf{be}\,*\,\mathsf{in}\,(\mathsf{promote}\,f\,\mathsf{for}\,w\,\mathsf{in}\,g) \qquad (9)$$

- $(Weakening, Promotion)$-**cut.** The derivation

$$\dfrac{\dfrac{\Gamma \vdash !B}{!A,\Gamma \vdash !B}\,Weakening \qquad \dfrac{!\Delta,!B \vdash C}{!B,!\Delta \vdash !C}\,Promotion}{!A,\Gamma,!\Delta \vdash !C}\,Cut$$

reduces to

$$\dfrac{\dfrac{\Gamma \vdash !B \qquad \dfrac{!B,!\Delta \vdash C}{!B,!\Delta \vdash !C}\,Promotion}{\Gamma,!\Delta \vdash !C}\,Cut}{!A,\Gamma,!\Delta \vdash !C}\,Weakening$$

This gives the term reduction

$$\mathsf{promote}\,(\mathsf{discard}\,x\,\mathsf{in}\,f)\,\mathsf{for}\,y\,\mathsf{in}\,g\ \triangleright\ \mathsf{discard}\,x\,\mathsf{in}\,(\mathsf{promote}\,f\,\mathsf{for}\,y\,\mathsf{in}\,g) \qquad (10)$$

- (*Contraction, Promotion*)-cut. The derivation

$$\cfrac{\cfrac{!A,!A,\Gamma \vdash !B}{!A,\Gamma \vdash !B}\text{Contraction} \qquad \cfrac{!\Delta,!B \vdash C}{!B,!\Delta \vdash !C}\text{Promotion}}{!A,\Gamma,!\Delta \vdash !C}\text{Cut}$$

reduces to

$$\cfrac{!A,!A,\Gamma \vdash !B \qquad \cfrac{!B,!\Delta \vdash C}{!B,!\Delta \vdash !C}\text{Promotion}}{\cfrac{!A,!A,\Gamma,!\Delta \vdash !C}{!A,\Gamma,!\Delta \vdash !C}\text{Contraction}}\text{Cut}$$

This gives the term reduction

promote (copy $x$ as $y, z$ in $f$) for $y$ in $g$ ▷ copy $x$ as $y, z$ in (promote $f$ for $y$ in $g$)  (11)

One is tempted to suggest that perhaps the reason why the rule *Promotion* gives us reductions with some sort of computational meaning is because this rule is not clearly either a left or a right rule. It introduces the connective on the right (so it is mainly a right rule), but it imposes conditions on the context on the left. Indeed there does not appear to be any analogous reductions in natural deduction.

## 5.3  Commutative Cuts

Next we consider briefly the 24 significant cuts of the form $(L_1, L_2)$ where the $L_i$ are both left rules. These correspond case by case to the commutative conversions for natural deduction considered in Section 4. For the most part the reduction rules we obtain from cut elimination are identical with those in Fig. 6. The exceptions are the cases where $(-\circ_\mathcal{L})$ is the (second) rule above the cut. In these cases we obtain in place of the first rules in the four groups of six in Fig. 6, the following stronger rules:

$$v[(\text{let } z \text{ be } x \otimes y \text{ in } t)u/w] \longrightarrow \text{let } z \text{ be } x \otimes y \text{ in } v[tu/w]$$

$$v[(\text{let } z \text{ be } * \text{ in } t)u/w] \longrightarrow \text{let } z \text{ be } * \text{ in } v[tu/w]$$

$$v[(\text{discard } z \text{ in } t)u/w] \longrightarrow \text{discard } z \text{ in } v[tu/z]$$

$$v[(\text{copy } z \text{ as } x, y \text{ in } t)u/w] \longrightarrow \text{copy } z \text{ as } x, y \text{ in } v[tu/w]$$

## 6  Conclusions and Further Work

In this paper we have considered the question of a term assignment system for both the natural deduction and sequent calculus system. We have also considered the question of reduction in both systems; in natural deduction resulting from normalisation of proofs and in the sequent calculus from the process of eliminating cuts. Our systems have important properties which previous proposals lack, namely closure under substitution and subject reduction.

In the full version of this paper [2] we give a general categorical model for our calculus. We also show how the term assignment system can be alternatively suggested by our semantics.

This paper represents preliminary work; much more remains. Given our term calculus it is our hope that this refined setting should shed new light on various properties of the $\lambda$-calculus such as Church-Rosser, strong normalisation and optimal reductions.

## References

1. Samson Abramsky. Computational interpretations of linear logic. Technical Report 90/20, Department of Computing, Imperial College, London, October 1990.
2. Nick Benton, Gavin Bierman, Valeria de Paiva, and Martin Hyland. Term assignment for intuitionistic linear logic. Technical Report 262, Computer Laboratory, University of Cambridge, August 1992.
3. Jean-Yves Girard. Linear logic. *Theoretical Computer Science*, 50:1–101, 1987.
4. Jean-Yves Girard, Yves Lafont, and Paul Taylor. *Proofs and Types*, volume 7 of *Cambridge Tracts in Theoretical Computer Science*. Cambridge University Press, 1989.
5. Jean-Yves Girard, Andre Scedrov, and Philip Scott. Bounded linear logic: A modular approach to polynomial time computability. *Theoretical Computer Science*, 97:1–66, 1992.
6. Patrick Lincoln and John Mitchell. Operational aspects of linear lambda calculus. In *Proceedings of Symposium on Logic in Computer Science*, pages 235–246, June 1992.
7. Ian Mackie. Lilac: A functional programming language based on linear logic. Master's thesis, Department of Computing, Imperial College, London, September 1991.
8. P.W. O'Hearn. Linear logic and interference control (preliminary report). In *Proceedings of Conference on Category Theory and Computer Science*, volume 530 of *Lecture Notes in Computer Science*, pages 74–93, September 1991.
9. Dag Prawitz. *Natural Deduction*, volume 3 of *Stockholm Studies in Philosophy*. Almqvist and Wiksell, 1965.
10. Peter Schroeder-Heister. A natural extension of natural deduction. *The Journal of Symbolic Logic*, 49(4):1284–1300, December 1984.
11. M.E. Szabo, editor. *The Collected Papers of Gerhard Gentzen*. North-Holland, 1969.
12. Philip Wadler. There's no substitute for linear logic. Draft Paper, December 1991.

# Program Extraction from Normalization Proofs

Ulrich Berger

Berger@Mathematik.Uni-Muenchen.de
Mathematisches Institut
der Ludwig–Maximilians–Universität München
Theresienstraße 39, 8000 München 2, Germany

**Abstract.** From a constructive proof of strong normalization plus uniqueness of the long beta-normal form for the simply typed lambda-calculus a normalization program is extracted using Kreisel's modified realizability interpretation for intuitionistic logic. The proof – which uses Tait's computability predicates – is not completely formalized: Induction is done on the meta level, and therefore not a single program but a family of program, one for each term is obtained. Nevertheless this may be used to write a short and efficient normalization program in a type free programming functional programming language (e.g. LISP) which has the interesting feature that it first evaluates a term $r$ of type $\rho$ to a functional $|r|$ of type $\rho$ and then collapses $|r|$ to the normal form of $r$.

## 1 Introduction

In this paper we extract a normalization algorithm from a constructive proof of strong normalization for the simply typed $\lambda$–calculus.

The extraction method we will use is the *modified realizability interpretation* for intuitionistic logic introduced by Kreisel [KR59]. Variants of it can be found in most constructive type theories and their implementations. We will confine ourselves to a very restricted system sufficient for our examples and will introduce refinements allowing us to avoid redundant parameters in the extracted programs. In Section 1 a complete description of this interpretation is given and its correctness is proved.

In Section 2 we will prove strong normalization using Taits computability predicates. Tait introduced computability in [TA67] (he called it "reducibility") for proving weak normalization and Troelstra in [TR73] proved strong normalization with this method. Here we will prove strong normalization plus a kind of unicity of normal form: Call $s$ the *long normal form* of $r$, written $\mathbf{N}(r, s)$, if every $\beta$–reduction sequence starting with $r$ terminates with a $\beta$–normal term having $s$ as its $\eta$–expansion. We will prove that every simple typed term has a long normal form, which by definition is unique.

We will partly formalize this proof and extract a nice program computing the long normal form. The formalization will be incomplete in two respects: Firstly, structural induction on types and terms will be done on the meta level. This is just to keep the formal system simple. Secondly, only the relatively small "logical" part of the proof will be formalized whereas the rather complicated "arithmetical" part analyzing $\beta$-reduction is packed into axioms which do not affect the extracted program as long as they are *Harrop-formulas*.

Because induction is not formalized we do *not* formally prove $\forall r \exists s.\mathbf{N}(r,s)$ but just $\exists s.\mathbf{N}(\mathbf{r},s)$ for each term $\mathbf{r}$. Consequently we do not formally extract a *single* normalization program $\Omega$: terms $\to$ terms but a family $\Omega_{\mathbf{r}}$ of closed expressions denoting the normal form of $\mathbf{r}$. To actually get the normal form we have to evaluate, i.e. normalize $\Omega_{\mathbf{r}}$. Thus we just replaced the problem of normalizing $\mathbf{r}$ by the problem of normalizing $\Omega_{\mathbf{r}}$. But look: We can normalize $\Omega_{\mathbf{r}}$ by simply pressing Return (or calling eval in LISP) since its a closed expression of ground type. Thus $\beta$-reduction is done by the compiler (emboddied by eval) and hence in the program code no procedures performing $\beta$-reduction appear.

A more sophisticated view of this program is offered by the fact that $\Omega_{\mathbf{r}}$ is of the form $\Phi r$ for some closed expression of type $\rho \to$ **terms** (if $\rho$ is the type of $\mathbf{r}$). Thus, denotationally, the normal form of $r$ is $|\Phi||r|$, i.e. our program first evaluates $r$ to a functional $|r|$ of type $\rho$ and then, via $\Phi$, collapses $|r|$ to the normal form of $r$ (an algorithm with similar properties is defined in [Co92]). Consequently if two terms are semantically equal then they have the same normal form, i.e. they are $\beta\eta$-provably equal. Using this fact a very strong completeness theorem for the $\beta\eta$-calculus can be proved. This has been worked out in [BS91].

With the same method we may also compute the *short* i.e. $\beta\eta$-normal form. Semantically, we just have to replace '$\eta$-expansion' by '$\eta$-normal form' in the definition of $\mathbf{N}(r,s)$, and one function symbol has to be interpreted differently. The proof and the program code remain unchanged.

We hope that this case study will show that the technique of program extraction is not only theoretically interesting but also of practical value, in particular that it may yield elegant, short and efficient algorithms, is manageable by hand even for complicated proofs (because often only a small part of the proof has to be formalized), helps understanding programs and supports controlled modification of programs and may help discovering new algorithms.

## 2  Modified realizability for first order minimal logic

In this section we review the essentials of modified realizability. A detailed proof theoretic analysis of various realizability notions can be found in [TR73].

The main idea of a realizability interpretation is to make explicit the constructive meaning of a statement explicit or more formally to define what it means that an object *realizes* a formula. Two typical clauses of this definition are:

($\exists$)  A pair $(a,b)$ realizes $\exists x.\varphi(x)$ iff $b$ realizes $\varphi(a)$.
($\to$)  A function $f$ realizes $\varphi \to \psi$ iff for every $a$ realizing $\varphi$, $f(a)$ realizes $\psi$.

Clearly by ($\exists$) the formulas '$a$ realizes $\varphi$' will not contain existential quantifiers any more, and by ($\rightarrow$) nested implications will give rise to realizing functionals of higher type. Therefore logic with $\exists$ for first order objects is interpreted in a logic without $\exists$ for higher type functionals. Kreisel [KR59] has observed that from an intuitionistic proof $d$ of a formula $\varphi$ one can easily extract a program $\text{ep}(d)$ and a proof that $\text{ep}(d)$ realizes $\varphi$.

## 2.1 Minimal logic

**Definition** First order minimal logic $\mathbf{M_1}$ for first order terms is given by

*Sorts* $\iota, \iota_1, \ldots$.

*Terms* Variables $x^\iota$, constants $\mathbf{c}^\iota$, $\mathbf{f}^{\iota_1 \rightarrow \cdots \rightarrow \iota_k \rightarrow \iota} r_1^{\iota_1} \ldots r_k^{\iota_k}$, short $\mathbf{f}^{\vec{\iota} \rightarrow \iota} \vec{t}^{\vec{\iota}}$ or $\mathbf{f}\vec{t}$ (**f** a function symbol, $\vec{t}$ terms).

*Formulas* Atomic formulas $\mathbf{P}\vec{t}^{\vec{\iota}}$ (**P** a predicate symbol), $\varphi \rightarrow \psi$, $\forall x^\iota \varphi$, $\exists x^\iota.\varphi$.

*Derivations* We define $\lambda$–terms $d^\varphi$ for natural deduction proofs of formulas $\varphi$ together with the set $\text{FA}(d)$ of free assumptions in $d$:

$(ass)$    $u^\varphi$,   $\text{FA}(u) = \{u\}$
$(\rightarrow^+)$    $(\lambda u^\varphi . d^\psi)^{\varphi \rightarrow \psi}$,   $\text{FA}(\lambda u.d) = \text{FA}(d) \backslash \{u\}$
$(\rightarrow^-)$    $(d^{\varphi \rightarrow \psi} e^\varphi)^\psi$,   $\text{FA}(de) = \text{FA}(d) \cup \text{FA}(e)$
$(\forall^+)$    $(\lambda x^\iota . d^\varphi)^{\forall x^\iota . \varphi}$,   $\text{FA}(\lambda x.d) = \text{FA}(d)$
     provided $x^\iota \notin \text{FV}(\varphi)$ for any $u^\varphi \in \text{FA}(d)$,
$(\forall^-)$    $(d^{\forall x^\iota . \varphi} t^\iota)^{\varphi_x[t]}$,   $\text{FA}(dt) = \text{FA}(d)$
$(\exists^+)$    $\langle t, d^{\varphi_x[t]} \rangle^{\exists x.\varphi}$,   $\text{FA}(\langle t,d \rangle) = \text{FA}(d)$
$(\exists^-)$    $[d^\psi, u^\varphi, e^{\exists x.\varphi}]^\psi$,   $\text{FA}([d,u,e]) = (\text{FA}(d)\backslash\{u\}) \cup \text{FA}(e)$,
     provided $x \notin \text{FV}(\psi)$ and $x \notin \text{FV}(\chi)$ for any $v^\chi \in \text{FA}(d)\backslash\{u\}$

Usually we will omit type and formula indices in derivations if they are uniquely determined by the context or if they are not relevant, and sometimes we will write $d : \varphi$ instead of $d^\varphi$.

Now we specify the logic in which $\mathbf{M_1}$ will be interpreted.

**Definition** The negative fragment $\mathbf{M_1^-}(\lambda)$ of first order minimal logic for simply typed $\lambda$–terms is given by

*Types* All sorts $\iota$ of $\mathbf{M_1}$, $\rho \rightarrow \sigma$.

*Terms* Variables $x^\rho$ of all types $\rho$, constants $\mathbf{c}^\rho$ including all constants and function symbols of $\mathbf{M_1}$, $(\lambda x^\rho r^\sigma)^{\rho \rightarrow \sigma}$, $(r^{\rho \rightarrow \sigma} s^\rho)^\sigma$.

*Formulas* $\mathbf{P}\vec{r}^{\vec{\rho}}$ (all predicate symbols of $\mathbf{M_1}$ included), $\varphi \rightarrow \psi$, $\forall x^\rho.\varphi$.

*Derivations* As for $\mathbf{M_1}$ without the $\exists$ rules. The $\forall$ rules are extended to variables of all types.

$\mathbf{M}_1^-(\lambda)$ is to be understood modulo identification of terms interconvertible by the $\alpha\beta\eta$-rules.

To save parenthesis we adopt the following conventions:
$\rho_1 \to \ldots \to \rho_k \to \rho$ (short $\vec{\rho} \to \rho$) means $\rho_1 \to (\rho_2 \to \ldots \to (\rho_k \to \rho))$, same for implications.
$\lambda\vec{x}.r\vec{s}$ means $\lambda x_1 \ldots \lambda x_k(\ldots(rs_1)\ldots s_l)$,
$\forall\vec{x}.\vec{\varphi} \to \psi$ means $\forall x_1 \ldots \forall x_k(\varphi_1 \to \ldots \to (\varphi_k \to \psi)\ldots)$.

Since in the next subsection we will be concerned with finite sequences of types and terms we also introduce the abbreviations
$\vec{\rho} \to \vec{\sigma} := \vec{\rho} \to \sigma_1, \ldots, \vec{\rho} \to \sigma_l$, $\vec{rs} := r_1\vec{s}, \ldots, r_k\vec{s}$, $\lambda\vec{x}.\vec{r} := \lambda\vec{x}.r_1, \ldots, \lambda\vec{x}.r_k$,
$\vec{r}^{\vec{\rho}} = r^{\vec{\rho}} := r_1^{\rho_1}, \ldots, r_k^{\rho_k}$.

## 2.2 Modified realizability

First we define for every $\mathbf{M}_1$-formula $\varphi$ the type $\tau(\varphi)$ of its realizing objects. It is obtained by simply ignoring all dependencies of formulas from objects. In general $\varphi$ will be realized by a finite sequence of terms, hence the $\tau(\varphi)$ will be a finite sequence of types.

**Definition**

$\tau(\mathbf{P}\vec{t}) := \epsilon$ (the empty sequence), $\tau(\varphi \to \psi) := \tau(\varphi) \to \tau(\psi)$, $\tau(\forall x^\iota \varphi) := \iota \to \tau(\phi)$, $\tau(\exists x^\iota \varphi) := \iota, \tau(\phi)$.

Next we formally define the meaning of 'a modified realizes $\varphi$'.

**Definition** To every $\mathbf{M}_1$ formula $\varphi$ and every sequence of terms $\vec{r}^{\tau(\varphi)}$ a $\mathbf{M}_1^-(\lambda)$-formula $\vec{r}\operatorname{mr}\varphi$ is defined by

$\epsilon \operatorname{mr} \mathbf{P}\vec{t} := \mathbf{P}\vec{t}$

$\vec{r} \operatorname{mr} (\varphi \to \psi) := \forall\vec{x}.\vec{x} \operatorname{mr} \varphi \to \vec{r}\vec{x} \operatorname{mr} \psi$ ($\vec{x}$ fresh variables of type $\tau(\varphi)$)

$\vec{r} \operatorname{mr} \forall x^\iota \varphi := \forall x^\iota.\vec{r}x \operatorname{mr} \varphi$

$r, \vec{r} \operatorname{mr} \exists x^\iota \varphi := \vec{r} \operatorname{mr} \varphi_x[r]$

**Example 1** Consider the formula $\forall x^\iota \exists y^\iota \varphi(x,y)$ where $\varphi$ is *negative*, i.e. contains no existential quantifier. Clearly $\tau(\varphi) = \epsilon$ and $\epsilon \operatorname{mr} \varphi = \varphi$. Hence $\tau(\forall x^\iota \exists y^\iota \varphi(x,y)) = \iota \to \iota$ and $r \operatorname{mr} \forall x^\iota \exists y^\iota \varphi(x,y) = \forall x \varphi(x, rx)$.

Now we extract programs from proofs.

**Definition** Relative to a fixed one-to-one mapping from $\mathbf{M}_1$-assumption variables $u^\psi$ to finite sequences of variables $\vec{x}_u^{\tau(\psi)}$ for every $\mathbf{M}_1$-derivation $d^\varphi$ we define a finite sequence of $\lambda$-terms $\operatorname{ep}(d)^{\tau(\varphi)}$ called the *extracted program* of $d$.

$$\text{ep}(u^\varphi) := \vec{x}_u^{\tau(\varphi)}$$

$$\text{ep}(\lambda u^\varphi . d^\psi) := \lambda \vec{x}_u^{\tau(\varphi)} . \text{ep}(d)$$

$$\text{ep}(d^{\varphi \to \psi} e^\varphi) := \text{ep}(d)\text{ep}(e)$$

$$\text{ep}(\lambda x^\iota d^\varphi) := \lambda x^\iota \text{ep}(d)$$

$$\text{ep}(d^{\forall x^\iota . \varphi} t^\iota) := \text{ep}(d)t$$

$$\text{ep}(\langle t, d^{\varphi_x[t]}\rangle) := t, \text{ep}(d)$$

$$\text{ep}([d^\psi, u^\varphi, e^{\exists y . \varphi}]) := \text{ep}(d)_y[\text{first}(\text{ep}(e))]_{x_u}[\text{rest}(\text{ep}(e))]$$

**Example 2** Consider the following $\mathbf{M_1}$-derivation (containing the free assumption variable $v$)

$$\lambda z^\iota \lambda u^{\exists y^\iota \mathbf{P} z y}[\langle \mathbf{f} y, v^{\forall z \forall y . \mathbf{P} z y \to \mathbf{P}(\mathbf{g} z)(\mathbf{f} y)} z y w\rangle, w^{\mathbf{P} z y}, u] : \forall z . \exists y \mathbf{P} z y \to \exists y \mathbf{P}(\mathbf{g} z) y$$

The extracted program is $\lambda z \lambda x_u \mathbf{f} x_u : \iota \to \iota \to \iota$.

The axioms used in a derivation $d$ of a specification appear as free assumption variables $u^\psi$, and the extracted program $\text{ep}(d)$ possibly will contain the corresponding free variables or constants $\vec{x}_u^{\tau(\psi)}$. To execute such programs these constants must be given a computational meaning. For example constants corresponding to induction axioms have to be interpreted by recursors. However there are also axioms which need not be interpreted. Namely those axioms $u^\psi$ with $\tau(\psi) = \epsilon$. Because then $\vec{u}_x$ is $\epsilon$. For instance the assumption $v^{\forall z \forall y . \mathbf{P} z y \to \mathbf{P}(\mathbf{g} z)(\mathbf{f} y)}$ in Example 2, although necessary for the proof, does not appear in the extracted program. Formulas $\psi$ such that $\tau(\psi) = \epsilon$ are called *Harrop formulas*. The fact that (true) Harrop formulas may be used freely as axioms without effect on the extracted program is very important in practice. It means that large parts of complex proofs need not be formalized for the purpose of program extraction. In the formal proof of strong normalization in the next section we will extremely benefit from this possibility.

Finally we have to prove that extracted programs realize the formulas derived.

**Soundness Theorem**

Fix a one-to-one mapping from $\mathbf{M_1}$-ass. variables $u : \psi$ to $\mathbf{M_1^-}(\lambda)$-ass. variables $\tilde{u}^{\vec{x}_u \text{ mr } \psi}$. Then to every $\mathbf{M_1}$-derivation $d : \varphi$ one can construct a $\mathbf{M_1^-}(\lambda)$-derivation $\mu(d)^{\text{ep}(d) \text{ mr } \varphi}$ with $\text{FA}(\mu(d)) \subseteq \{\tilde{u} | u \in \text{FA}(d)\}$.

**Proof:** We define $\mu(d)$ by induction on $d$.

(i) $\mu(u) := \tilde{u}$

(ii) $\mu(\lambda u^\varphi d^\psi) := \lambda \vec{x}_u^{\tau(\varphi)} \lambda \tilde{u}^{\vec{x}_u \text{ mr } \varphi} . \mu(d)$

(iii) $\mu(d^{\varphi \to \psi} e^{\varphi}) := \mu(d)\mathrm{ep}(e)\mu(e)$

(iv) $\mu(\lambda x^{\iota} d^{\varphi}) := \lambda x^{\iota} \mu(d)$

(v) $\mu(d^{\forall x \varphi} t) := \mu(d) t$

(vi) $\mu(\langle t, d^{\varphi_y[t]} \rangle) := \mu(d)$

(vii) $\mu([d^{\psi}, u^{\varphi}, e^{\exists y \varphi}]) := \mu(d)_y[\mathrm{first}(\mathrm{ep}(e))]_{x_u}[\mathrm{rest}(\mathrm{ep}(e))]_{\tilde{u}}[\mu(e)]$

Remember that $\beta$–equal terms are identified. Therefore we defined correct derivations of the right formulas. QED

## 2.3 Refinements

During the program extraction process many parts of a derivation are erased. Thus it may happen that a variable $x$ occuring free in $d$ does not occur in $\mathrm{ep}(d)$ and therefore the program $\lambda x.\mathrm{ep}(d)$ extracted from $\lambda x.d$ has the redundant input parameter $x$. For instance this happened with the variable $z$ in Example 2.

Often the situation is even worse. In programs $\mathrm{ep}(d)$ there may occur variables which disappear when normalizing $\mathrm{ep}(d)$. To detect these redundancies one would have to normalize the whole program which for big programs is very undesirable if not impossible (practically). Therefore it is important to remove redundant variables already *during* and not only *after* the extraction. We will do this by labelling certain occurences of quantifiers and quantifier rules to indicate that the quantified variable is computationally redundant. For the labelled $(\forall)^+$ rules stronger variable conditions have to be required ensuring that the quantified variable will indeed not be used int the extracted program. In the realizability interpretation then all labeled parts are simply ignored.

It should be emphasized that this is just an ad hoc construction which might well be subsumed by more elaborate optimizations existing in Coq [PW92] or MX [HA87].

**Definition** We add to $\mathbf{M_1}$ the quantification $\{\forall x\}\varphi$ with the proof rules

$(\{\forall\}^+)$   $(\{\lambda x\} d^{\varphi})^{\{\forall x\}\varphi}$, provided $x \notin \mathrm{FV}(\varphi)$ for any $u^{\varphi} \in \mathrm{FA}(d)$ and $x \notin \mathrm{CV}(d)$.

$(\{\forall\}^-)$   $(d^{\{\forall x\}\varphi}\{t\})^{\varphi_x[t]}$

$\mathrm{FA}(d)$ is defined as before. Therefore if all curly brackets are removed we get derivations in the old sense.

The set $\mathrm{CV}(d)$ of *computationally relevant variables* of a derivation $d$ is defined as follows:

$CV(u) := \emptyset$, $CV(\lambda ud) := CV(d)$, $CV(de) := CV(d) \cup CV(e)$,
$CV(\lambda xd) := CV(d) \setminus \{x\}$, $CV(dt) := CV(d) \cup FV(t)$, $CV(\langle t, d \rangle) := FV(t) \cup CV(d)$,
$CV([d, u, e^{\exists y \varphi}]) := (CV(d) \setminus \{y\}) \cup CV(e))$, $CV(\{\lambda x\}d) := CV(d)$, $CV(d\{t\}) := CV(d)$.

To the definitions of $\tau(\varphi)$, $r$ mr $\varphi$ and ep$(d)$ we add the following clauses:
$\tau(\{\forall x\}\varphi) := \tau(\varphi)$.
$r$ mr $\{\forall x\}\varphi := r$ mr $\varphi$.
ep$(\{\lambda x\}d) :=$ ep$(d)$, ep$(d\{t\}) :=$ ep$(d)$.

**Example 3** Let us refine the derivation in Example 2 by introducing some curly brackets.

$$\{\lambda z\}\lambda u^{\exists y}{}^{\iota}\mathbf{P}^{zy}[\langle \mathbf{f}y, v^{\{\forall z\}\{\forall y\}}.\mathbf{P}zy \to \mathbf{P}(\mathbf{g}z)(\mathbf{f}y)\{z\}\{y\}w\rangle, w^{\mathbf{P}zy}, u]$$

with end formula $type\{\forall z\}.\exists y\mathbf{P}zy \to \exists y\mathbf{P}(\mathbf{g}z)y$. Now the extracted program is $\lambda x_u \mathbf{f} x_u : \iota \to \iota$.

The following may be proved easily by induction on derivations using the stronger variable condition for the new quantifier.

**CV–Lemma**   $FV(\text{ep}(d)) \subseteq CV(d) \cup \bigcup \{x_u | u^\varphi \in FA(d)\}$

It remains to complete the proof of the Soundness Theorem. We add the following clauses to the definition of $\mu(d)$:

(viii) $\mu(\{\lambda x\}d^\varphi) := \mu(d)$,

(ix) $\mu(d^{\{\forall x\}\varphi}\{t\}) := \mu(d)_x[t]$,

To show for (ix) that $\mu(d)$ is a correct derivation of the right formula the CV-lemma is needed.

## 3  Strong normalization

An interesting feature of modified realizability is the fact that even if we start with a first order specification $\varphi$ the program extracted from a proof of $\varphi$ may contain higher type constructs. An excellent example is provided by the Tait/Troelstra proof of strong normalization for the simply typed $\lambda$–calculus [TR73]. Part of this proof is a lemma stating that $SC_\rho(r^\rho)$ implies that $r$ is strongly normalizable. The property $SC_\rho(r^\rho)$ ('$r^\rho$ is strongly computable') is defined using $n$ implications nested to the left, where $n$ is the type level of $\rho$. Consequently the type of the program extracted from the proof of this lemma will have at least type level $n + 1$.

The lemma mentioned above is proved by induction on $\rho$. Formalizing it would require a definition of $SC_\rho(r)$ as a *single* proposition with parametrs $\rho$ and $r$. For this we would need a rather strong system allowing primitive recursive definitions of propositions (such a system is used for a normalization proof in [GS92]). To keep the system simple, we therefore will do this induction on the meta level.

## 3.1 Chosing the language and the axioms

To formulate the normalization problem we have to fix a formal language. Clearly the function constants we use will be the basic constructors of the algebra of typed λ–terms. Here we have no big choice because constants may appear in the extracted program and therefore we must know how to compute them. For the predicate symbols the situation is different. We may use arbitrarily complex predicates as primitive. The only restriction is that the axioms used to describe them must be expressible by Harrop formulas, since then they will not appear in the program. Finding a good system of predicates and axioms is in fact the most difficult and interesting part of the whole work. Usually this can be done only after a careful anlysis of an informal proof of the specification.

In the following we will informally introduce and motivate some predicates on λ–terms. Those named by a bold face letter will be treated as primitive in our formal language. All facts proved on them will be used as axioms in the formal proof of our normalization theorem.

**Definition**

1. Let $r$ be a simply typed λ–term in β–normal form. Assume

$$r = \lambda x_1^{\rho_1} \ldots \lambda x_k^{\rho_k}.y^{\sigma_1 \to \ldots \to \sigma_l \to \tau_1 \to \ldots \to \rho_m \to \iota} s^{\sigma_1} \ldots s^{\sigma_l}$$

which is the general form of a term in β–normal form. Then the *long normal form* or *η-expanded form* $r^*$ is defined by recursion on the maximum of the level of the type of $r$ and the levels of the $\sigma$ with $x^\sigma \in \mathrm{FV}(r)$

$$r^* := \lambda x_1 \ldots \lambda x_k \lambda z_1 \ldots \lambda z_m . y s_1^* \ldots s_l^* z_1^* \ldots z_m^*$$

where $z_1, \ldots z_m$ are fresh variables of the right type. Clearly $(\lambda x r)^* = \lambda x r^*$, and if $s^{\rho \to \sigma}$ is not an abstraction then $s^* = \lambda x (sx)^*$ with a fresh variable $x^\rho$.

2. Let $r$ and $s$ be terms of type $\rho$.
$\mathbf{N}_\rho(r, s) :\Leftrightarrow$  Every maximal sequence of β–reductions beginning with $r$ is finite and ends with a term $t$ such that $t^* = s$.

If $\mathbf{N}(r, s)$ we again say that $s$ is the *long normal form* or simply the *normal form* of $r$.

We will show that every term has a long normal form, which implies that every term is strongly normalizing in the usual sense. The key to the result is the following

**Fact 1** *If* $x^\rho \notin \mathrm{FV}(r^{\rho \to \sigma})$ *and* $\mathbf{N}(rx, s)$ *then* $\mathbf{N}(r, \lambda x s)$.

**Proof:** Assume $x^\rho \notin \mathrm{FV}(r^{\rho \to \sigma})$ and $\mathbf{N}(rx, s)$. Consider a β–reduction sequence $r = r_1 \to_\beta \ldots$. This sequence must be finite since otherwise we would obtain an infinite reduction sequence $rx = r_1 x \to_\beta \ldots$ contradicting $\mathbf{N}(rx, s)$. Assume that $r = r_1 \to_\beta \ldots$ ends with $r_n$. We have to show that $r_n^* = \lambda x s$.

*Case 1* $r_n = \lambda yt$ for some $t$. Then $rx \to_\beta \ldots \to_\beta (\lambda yt)x \to_\beta t_y[x]$ where $t_y[x]$ is in $\beta$–normal form. Because $\mathbf{N}(rx, s)$ this implies $t_y[x]^* = s$ and therefore

$$r_n^* =_\alpha (\lambda x t_y[x])^* = \lambda x t_y[x]^* = \lambda x s$$

*Case 2* $r_n$ is not an abstraction. Then $rx \to_\beta \ldots \to_\beta r_n x$ where $r_n x$ is in $\beta$–normal form. Hence $(r_n x)^* = s$ and therefore $r_n^* = \lambda x (r_n x)^* = \lambda x s$   QED

To use Fact 1, one certainly has to compute a variable not free in a given term. This computation becomes very easy if one provides extra information on the Free variables of a term.

**Definition** For a term $r$ of type $\rho$ and a number $k$ let $\mathbf{F}_\rho(r, k) :\Leftrightarrow \forall l . x_l \in FV(r) \to l < k$.

Clearly $\mathbf{F}(r, k)$ implies $x_k \notin FV(r)$ and $\mathbf{F}(rx_k, k+1)$. Therefore we may restate Fact 1

**Fact 1'**   If $\mathbf{F}(r, k)$ and $\mathbf{N}(rx_k, s)$ then $\mathbf{N}(r, \lambda x_k s)$ and $\mathbf{F}(rx_k, k+1)$.

**Definition**   $SN_\rho(r) := \forall k . \mathbf{F}_\rho(r, k) \to \exists s \mathbf{N}_\rho(r, s)$

We will prove $SN(r)$ for every term $r$. Note that $SN(r)$ is just an abbreviation and not a primitive predicate.

We will use two further predicates, describing normalization in the **Arguments** of an Applicative term and a kind of **Head** reduction.

**Definition**

1.   $\mathbf{A}_\rho(r^\rho, s^\rho) :\Leftrightarrow r$ has the form $xr_1 \ldots r_k$, $s$ has the form $xs_1 \ldots s_k$ and $\mathbf{N}(r_1, s_1), \ldots, \mathbf{N}(r_k, s_k)$.

2.   $\mathbf{H}_\rho(r\rho, r'^\rho) :\Leftrightarrow r$ is of the form $(\lambda xt)s\vec{s}$ with $SN(s)$ and $r' = t_x[s]\vec{s}$.

**Fact 2**   $\mathbf{H}(r, r') \to \mathbf{N}(r', s) \to \mathbf{N}(r, s)$

**Proof:** This is proved by a standard argument.

Facts on **A** are fairly trivial and may be found in the axiom list below.

Now we may specify the language for the normalization problem and the axiom system on which our formal proof will be based.

*Sorts*   nat, $\lambda_\rho$ (terms of type $\rho$) and $V_\rho$ (variables of type $\rho$) for all types $\rho$.

*Constants*   $0^{\text{nat}}$, $(+1)^{\text{nat}\to\text{nat}}$, $(\mathbf{x}^\rho)^{V_\rho}$ for every variable $x^\rho$, $\mathbf{mvar}_\rho^{\text{nat}\to V_\rho}$, $\mathbf{abst}_{\rho,\sigma}^{V_\rho \to \lambda_\sigma \to \lambda_{\rho\to\sigma}}$, $\mathbf{app}_{\rho,\sigma}^{\lambda_{\rho\to\sigma} \to \lambda_\rho \to \lambda_\sigma}$ and $\mathbf{mterm}_\rho^{V_\rho \to \lambda_\rho}$ for all types $\rho, \sigma$.

We abbreviate $\mathbf{mvar}_\rho(k)$ by $x_k^\rho$ and $\mathbf{mterm}_\rho(x)$ by $x$. Furthermore for every $\lambda$–term $r^\rho$ we will denote by $\mathbf{r}$ the closed term of sort $\lambda_\rho$ with value $r$ in the intended model.

*Predicate symbols* $\mathbf{N}_\rho, \mathbf{A}_\rho, \mathbf{H}_\rho\colon(\lambda_\rho,\lambda_\rho)$ and $\mathbf{F}_\rho\colon(\lambda_\rho,\mathrm{nat})$ for all types $\rho$.

The following Harrop formulas will be used as axioms, i.e. free assumptions. They consist of Fact 1' and Fact 2 and other trivial facts. By $\{\forall r_1,\ldots,r_k\}\varphi$ we mean $\{\forall r_1\}\ldots\{\forall r_k\}\varphi$ where $\{\forall r\}$ is the new quantifier introduced in the previous section.

$$
\begin{aligned}
&(\mathrm{Ax}1_{\rho,\sigma}) && \{\forall k, r^{\lambda_{\rho\to\sigma}}, s^{\lambda_\rho}\}.\mathbf{F}(r,k) \to\to \mathbf{N}(\mathrm{app}(r,x_k),s) \\
&&& \hspace{6em} \to \mathbf{N}(r,\mathrm{abst}(x_k,s)) \\
&(\mathrm{Ax}2_\iota) && \{\forall r^{\lambda_\iota}, s^{\lambda_\iota}\}.\mathbf{A}_\iota(r,s) \to \mathbf{N}_\iota(r,s) \\
&(\mathrm{Ax}3_\rho) && \{\forall x^{V_\rho}\}\mathbf{A}(x,x) \\
&(\mathrm{Ax}4_{\rho,\sigma}) && \{\forall r_1^{\lambda_{\rho\to\sigma}}, r_2^{\lambda_{\rho\to\sigma}}, s_1^{\lambda_\rho}, s_2^{\lambda_\rho}\}.\mathbf{A}(r_1,r_2) \to \mathbf{N}(s_1,s_2) \to \\
&&& \hspace{6em} \to \mathbf{A}(\mathrm{app}(r_1,s_1),\mathrm{app}(r_2,s_2)) \\
&(\mathrm{Ax}5_\rho) && \{\forall r^{\lambda_\rho}, s^{\lambda_\rho}, t^{\lambda_\rho}\}.\mathbf{H}(r,s) \to \mathbf{N}(s,t) \to \mathbf{N}(r,t) \\
&(\mathrm{Ax}6_{x^\rho,x^\beta,r^\sigma}) && \{\forall s^{\lambda_\rho}, \vec{s}^{\lambda_\beta}\}.\mathrm{SN}_\rho(s) \to \mathbf{H}_\sigma(\mathrm{app}(\mathrm{abst}(\mathbf{x},\mathbf{r}_{\vec{\mathbf{x}}}[\vec{s}]),s),\mathbf{r}_{\mathbf{x}\vec{\mathbf{x}}}[s\vec{s}]) \\
&(\mathrm{Ax}7_{\rho,\sigma}) && \{\forall r^{\lambda_{\rho\to\sigma}}, s^{\lambda_{\rho\to\sigma}}, t^{\lambda_\rho}\}.\mathbf{H}(r,s) \to \mathbf{H}(\mathrm{app}(r,t),\mathrm{app}(s,t)) \\
&(\mathrm{Ax}8_{\rho,\sigma}) && \{\forall k, r^{\lambda_{\rho\to\sigma}}\}.\mathbf{F}(r,k) \to \mathbf{F}(\mathrm{app}(r,x_k^\rho), k+1) \\
&(\mathrm{Ax}9_{\rho,\sigma}) && \{\forall k, r^{\lambda_{\rho\to\sigma}}, s^{\lambda_\rho}\}.\mathbf{F}(\mathrm{app}(r,s),k) \to \mathbf{F}(s,k) \\
&(\mathrm{Ax}10_{\rho,\sigma}) && \{\forall k, r^{\lambda_{\rho\to\sigma}}, s^{\lambda_\rho}\}.\mathbf{F}(\mathrm{app}(r,s),k) \to \mathbf{F}(r,k) \\
&(\mathrm{Ax}11_\rho) && \{\forall k, r^{\lambda_\rho}, s^{\lambda_\rho}\}.\mathbf{F}(r,k) \to \mathbf{H}(r,s) \to \mathbf{F}(s,k) \\
&(\mathrm{Ax}12_{r^\rho,k}) && \mathbf{F}(\mathbf{r},\mathbf{k})
\end{aligned}
$$

The last axiom for all $r$ and $k$ such that $\mathbf{F}(\mathbf{r},\mathbf{k})$ is true.

## 3.2 Proof and program extraction

Now we will prove $\mathrm{SN}(r)$ for every term $r$ on the axiomatic basis set up above. The proof will be organized in three lemmas and a concluding normalization theorem. For every lemma first an informal argument is given, then the corresponding proof terms are presented and (sub)programs are extracted. As already mentioned induction will be done on the meta level. Hence we will define (primitive recursive) families of derivations indexed by types or $\lambda$-terms.

For better readability we will write $\mathrm{abst}(x,s)$ instead of $\mathrm{abst}xs$ and $\mathrm{app}(r,s)$ instead of $\mathrm{app}rs$.

The crucial trick (due to Tait) for proving $\mathrm{SN}(r)$ is to show something stronger, namely that every term is *strongly computable*.

**Definition** We define formulas $\mathrm{SC}_\rho(r^{\lambda_\rho})$ by recursion on $\rho$.

$$\mathrm{SC}_\iota(r^{\lambda_\iota}) := \mathrm{SN}_\iota(r) = \forall k.\mathbf{F}(r,k) \to \exists s \mathbf{N}_\iota(r,s)$$

$$\mathrm{SC}_{\rho\to\sigma}(r^{\lambda_{\rho\to\sigma}}) := \{\forall s^{\lambda_\rho}\}.\mathrm{SC}_\rho(s) \to \mathrm{SC}_\sigma(rs)$$

**Lemma 1**

(a) $\text{SC}_\rho(r) \to \text{SN}(r)$

(b) $\text{SA}_\rho(r) \to \text{SC}_\rho(r)$

**Proof:** Induction on $\rho$.

For a ground type $\iota$ part (a) is trivial, and part (b) holds by Ax4.
Now assume we have a type $\rho \to \sigma$.
To prove (a) assume $\text{SC}_{\rho\to\sigma}(r)$ and $\mathbf{F}(r,k)$. Show $\mathbf{N}(r,s)$ for some $s$. Since $\mathbf{A}(x_k^\rho, x_k^\rho)$ by Ax3, clearly $\text{SA}_\rho(x_k)$. Hence $\text{SC}_\rho(x_k)$ by I.H.(b). Together with the assumption $\text{SC}_{\rho\to\sigma}(r)$ this implies $\text{SC}_\sigma(rx_k)$. Hence by I.H.(a) $\text{SN}(rx_k)$. Since $\mathbf{F}(rx_k, k+1)$ by Ax8, we get $\mathbf{N}(rx_k, t)$ for some $t$. Now $\mathbf{N}(r, \lambda x_k t)$ by Ax1.

To prove (b) assume $\text{SA}_{\rho\to\sigma}(r)$ and $\text{SC}_\rho(s)$. Show $\text{SC}_\sigma(rs)$. By I.H.(b) it will suffice to show $\text{SA}_\sigma(rs)$: Assume $\mathbf{F}(rs, k)$. Then clearly $\mathbf{F}(r,k)$ by Ax10. Hence $\mathbf{A}(r, r')$ for some $r'$ since we assumed $\text{SA}_{\rho\to\sigma}(r)$. Furthermore $\text{SN}(s)$ by I.H.(a) and hence $\mathbf{N}(s, s')$ for some $s'$ since $\mathbf{F}(s, k)$ by Ax9. Now by Ax4, $\mathbf{A}(r, r')$ and $\mathbf{N}(s, s')$ imply $\mathbf{A}(rs, r's')$. Hence we have proved $\text{SA}_\sigma(rs)$.  QED

**Proof terms**  We define closed derivations

$$\mathrm{d}_\rho^{1(a)} \colon \{\forall r^{\lambda_\rho}\}.\text{SC}_\rho(r) \to \text{SN}_\rho(r)$$

$$\mathrm{d}_\rho^{1(b)} \colon \{\forall r^{\lambda_\rho}\}.\text{SA}_\rho(r) \to \text{SC}_\rho(r)$$

where 'closed' of course means that all free assumptions are axioms.

$\mathrm{d}_\iota^{1(a)} := \{\lambda r^{\lambda_\iota}\}\lambda u^{\text{SC}_\iota(r)}.u$

$\mathrm{d}_\iota^{1(b)} := \{\lambda r^{\lambda_\iota}\}\lambda g^{\text{SA}_\iota(r)}\lambda k \lambda v^{\mathbf{F}(r,k)}.[\langle s, (\text{Ax2})\{r\}\{s\}w\rangle, w^{\mathbf{A}(r,s)}, gkv]$

$\mathrm{d}_{\rho\to\sigma}^{1(a)} := \{\lambda r^{\lambda_{\rho\to\sigma}}\}\lambda u^{\text{SC}_{\rho\to\sigma}(r)}\lambda k \lambda v^{\mathbf{F}(r,k)}.[\langle \mathbf{abst}(x_k, s), \tilde{d}_1\rangle, w,]$

$\qquad \mathrm{d}_\sigma^{1(a)}\{\mathbf{app}(r, x_k)\}(u\{r\}(\mathrm{d}_\rho^{1(b)}\{x_k\}\lambda l. \lambda p^{\mathbf{F}(x_k, l)}\langle x_k, \tilde{d}_2\rangle))(k+1)\tilde{d}_3$

where

$w : \mathbf{N}(\mathbf{app}(r, x_k), s)$,
$\tilde{d}_1 := (\text{Ax1})\{k\}\{r\}\{s\}vw : \mathbf{N}(r, \mathbf{abst}(x_k, s))$,
$\tilde{d}_2 := (\text{Ax3})\{x_k\} : \mathbf{A}(x_k, x_k)$,
$\tilde{d}_3 := (\text{Ax8})\{k\}\{r\}v : \mathbf{F}(\mathbf{app}(r, x_k^\rho), k+1)$.

$\mathrm{d}_{\rho\to\sigma}^{1(b)} := \{\lambda r^{\lambda_{\rho\to\sigma}}\}\lambda g^{\text{SA}_\rho(r)}\{\lambda s^{\lambda_\rho}\}\lambda v^{\text{SC}_\rho(s)}.\mathrm{d}_\sigma^{1(b)}\{\mathbf{app}(r,s)\}\lambda k\lambda w^{\mathbf{F}(\mathbf{app}(r,s),k))}$

$\qquad [[\langle \mathbf{app}(r', s'), \tilde{d}_1\rangle, w_2^{\mathbf{N}(s,s')}, \mathrm{d}_\rho^{1(a)}\{s\}vk\tilde{d}_2], w_1^{\mathbf{A}(r,r')}, gk\tilde{d}_3]$

where

$\tilde{d}_1 := (\text{Ax4})\{r\}\{r'\}\{s\}\{s'\}w_1w_2 : \mathbf{A}(\mathbf{app}(r,s), \mathbf{app}(r',s'))$,
$\tilde{d}_2 := (\text{Ax9})\{k\}\{r\}\{s\} : \mathbf{F}(s,k)$,
$\tilde{d}_3 := (\text{Ax10})\{k\}\{r\}\{s\} : \mathbf{F}(r,k)$.

**Programs extracted** Define $(\rho) := \tau(\text{SC}_\rho(r))$. We have $(\iota) = \text{nat} \to \iota$ and $(\rho \to \sigma) = (\rho) \to (\sigma)$. Let

$$\Phi_\rho := \text{ep}(d_\rho^{1(a)}) : (\rho) \to (\text{nat} \to \lambda_\rho)$$
$$\Psi_\rho := \text{ep}(d_\rho^{1(b)}) : (\text{nat} \to \lambda_\rho) \to (\rho)$$

We have

$$\begin{aligned}
\Phi_\iota &= \lambda u^{(\iota)} u \\
\Psi_\iota &= \lambda g^{(\iota)} \lambda k. gk =_\eta \lambda g^{(\iota)} g \\
\Phi_{\rho \to \sigma} &= \lambda u^{(\rho \to \sigma)} \lambda k. \mathbf{abst}(x_k, \Phi_\sigma(u(\psi_\rho(\lambda l.x_k)))(k+1)) \\
\Psi_{\rho \to \sigma} &= \lambda g^{\text{nat} \to \lambda_\rho} \lambda v^{(\rho)}. \Psi_\sigma \lambda k. \mathbf{app}(gk, \Phi_\rho vk)
\end{aligned}$$

**Lemma 2** $\quad\quad\quad \text{SC}_\rho(r') \to \mathbf{H}(r,r') \to \text{SC}_\rho(r)$

**Proof:** Induction on $\rho$.
For the base case assume $\text{SC}_\iota(r')$, i.e. $\text{SN}(r')$, $\mathbf{H}(r,r')$ and $\mathbf{F}(r,k)$. Show $\mathbf{N}(r,s)$ for some $s$. By Ax11, $\mathbf{F}(r',k)$ and hence $\mathbf{N}(r',s)$ for some $s$. Therefore $\mathbf{N}(r,s)$ by Ax5.
For the induction step assume $\text{SC}_{\rho \to \sigma}(r')$, $\mathbf{H}(r,r')$ and $\text{SC}_\rho(s)$. Show $\text{SC}_\sigma(rs)$. We have $\text{SC}_\sigma(r's)$ by assumption and $\mathbf{H}(rs, r's)$ by Ax7. Hence $\text{SC}_\sigma(rs)$. QED

**Proof terms** Define $d_\rho^2 : \{\forall r, r'\}.\text{SC}_\rho(r') \to \mathbf{H}(r,r') \to \text{SC}_\rho(r)$

$$d_\iota^2 := \{\lambda r\}\{\lambda r'\}\lambda u^{\text{SN}(r')} \lambda v_1^{\mathbf{H}(r,r')} \lambda k \lambda v_2^{\mathbf{F}(r,k)}.[\langle s, \tilde{d}_1 \rangle, w^{\mathbf{N}(r',s)}, uk\tilde{d}_2]$$

where

$\tilde{d}_1 := (\text{Ax5})\{r\}\{r'\}\{s\}v_1w : \mathbf{N}(r,s)$,
$\tilde{d}_2 := (\text{Ax11})\{k\}\{r\}\{r'\}v_2v_1 : \mathbf{F}(r',k)$.

$d_{\rho \to \sigma}^2 := \{\lambda r\}\{\lambda r'\}\lambda u^{\text{SC}_{\rho \to \sigma}(r')}\lambda v_1^{\mathbf{H}(r,r')}\{\lambda s\}\lambda v^{\text{SC}_\rho(s)}.$

$$d_\sigma^2\{\mathbf{app}(r,s)\}\{\mathbf{app}(r',s)\}(uw)\tilde{d}$$

where $\tilde{d} := (\text{Ax7})\{r\}\{r'\}\{s\}v : \mathbf{H}(\mathbf{app}(r,s), \mathbf{app}(r',s))$.

**Programs extracted** Let $\Gamma_\rho := \text{ep}(d_\rho^2) : (\rho) \to (\rho)$. We have

$$\begin{aligned}
\Gamma_\iota &= \lambda u^{(\iota)} \lambda k.uk =_\eta \lambda uu \\
\Gamma_{\rho \to \sigma} &= \lambda u^{(\rho \to \sigma)} \lambda w^{(\rho)}.\Gamma_\sigma(uw)
\end{aligned}$$

Clearly $\Gamma_\rho =_{\beta\eta} \lambda u^{(\rho)} u$ by induction on $\rho$.

**Lemma 3**

Let $r$ be a term of type $\rho$ with free variables among $x_1^{\sigma_1}, \ldots, x_k^{\sigma_k}$. Then for all $s_1^{\sigma_1}, \ldots, s_k^{\sigma_k}$ s.t. $\mathrm{SC}_{\sigma_1}(s_1), \ldots, \mathrm{SC}_{\sigma_k}(s_k)$ we have $\mathrm{SC}_\rho(r_{x_1,\ldots,x_k}[s_1, \ldots, s_k])$.

**Proof:** Induction on $\rho$. We write $r[\vec{s}]$ for $r_{x_1,\ldots,x_k}[s_1, \ldots, s_k]$ etc.

$\underline{x_i}$  Assume $\mathrm{SC}_{\vec{\sigma}}(\vec{s})$. Since $x_i[\vec{s}] = s_i$ we may use the assumption $\mathrm{SC}_{\sigma_i}(s_i)$.

$\underline{r^{\rho \to \sigma} s^\rho}$  Assume $\mathrm{SC}_{\vec{\sigma}}(\vec{s})$. By I.H. $\mathrm{SC}_{\rho\to\sigma}(r[\vec{s}])$ and $\mathrm{SC}_\rho(s[\vec{s}])$. Hence $\mathrm{SC}_\sigma((rs)[\vec{s}])$.

$\underline{\lambda x^\rho r^\sigma}$  Assume $\mathrm{SC}_{\vec{\sigma}}(\vec{s})$. Show $\mathrm{SC}_{\rho\to\sigma}(\lambda x r[\vec{s}])$. Therefore we assume $\mathrm{SC}_\rho(s)$ and show $\mathrm{SC}_{\rho\to\sigma}((\lambda x r[\vec{s}])s)$. By I.H. $\mathrm{SC}_\sigma(r[\vec{s}]_x[s])$. Furthermore $\mathrm{SN}(s)$ by Lemma 1(a). Hence $\mathbf{H}((\lambda x r[\vec{s}])s, r[\vec{s}]_x[s])$ by Ax6. Therefore $\mathrm{SC}_{\rho\to\sigma}((\lambda x r[\vec{s}])s)$ by Lemma 2.  QED

**Proof terms**  For every $\lambda$–term term $r^\rho$ and variables $\vec{x}^{\vec{\sigma}} \supseteq \mathrm{FV}(r)$ we define

$$\mathrm{d}^3_{r,\vec{x}} : \{\forall \vec{s}\}.\mathrm{SC}_{\vec{\sigma}}(\vec{s}) \to \mathrm{SC}_\rho(\mathbf{r}_{\vec{x}}[\vec{s}])$$

$$\begin{aligned}
\mathrm{d}^3_{x_i,\vec{x}} &:= \{\lambda \vec{s}\}\lambda \vec{u}.u_i \\
\mathrm{d}^3_{rs,\vec{x}} &:= \{\lambda \vec{s}\}\lambda \vec{u}.\mathrm{d}^3_{r,\vec{x}}\{\vec{s}\}\vec{u}(\mathrm{d}^3_{s,\vec{x}}\{\vec{s}\}\vec{u}) \\
\mathrm{d}^3_{\lambda x^\rho r^\sigma,\vec{x}} &:= \{\lambda \vec{s}\}\lambda \vec{u}\{\lambda s\}\lambda u^{\mathrm{SC}_\rho(s)}.\mathrm{d}^2_\sigma\{\tilde{t}\}\{\tilde{t}'\}(\mathrm{d}^3_{r,\sigma\vec{\sigma}}\{s\}\{\vec{s}\}u\vec{u})\tilde{d}
\end{aligned}$$

where

$\tilde{t} := \mathbf{app}(\mathbf{abst}(\mathbf{x}, \mathbf{r}_{\vec{x}}[\vec{s}]), s)$,
$\tilde{t}' := \mathbf{r}_{\vec{x}}[\vec{s}]\mathbf{x}[s]$,
$\tilde{d} := (\mathrm{Ax}6_{x,\vec{x},r})(\mathrm{d}^{1(\mathrm{a})}_\rho\{s\}u) : \mathbf{H}(\tilde{t}, \tilde{t}')$.

**Programs extracted**  Let $\Delta_{r^\rho, \vec{x}^{\vec{\sigma}}} := \mathrm{ep}(\mathrm{d}^3_{r^\rho, \vec{x}^{\vec{\sigma}}}) : (\vec{\sigma}) \to (\rho)$. We have

$$\begin{aligned}
\Delta_{x_i,\vec{x}} &= \lambda \vec{u}^{(\vec{\sigma})}.u_i \\
\Delta_{rs,\vec{x}} &= \lambda \vec{u}.\Delta_{r,\vec{x}}\vec{u}(\Delta_{s,\vec{x}}\vec{u}) \\
\Delta_{\lambda x^\rho r^\sigma,\vec{x}} &= \lambda \vec{u}\lambda u.\Gamma_\sigma(\Delta_{r,\sigma\vec{\sigma}}u\vec{u}) =_{\beta\eta} \lambda \vec{u}\lambda u.(\Delta_{r,\sigma\vec{\sigma}}u\vec{u})
\end{aligned}$$

An easy induction on $r$ shows that

$$\Delta_{r,\vec{x}} =_{\beta\eta} \lambda \vec{x} r[(\iota)/\iota | \iota \in r]$$

In particular $\Delta_{r,\vec{x}}$ and $\lambda \vec{x} r$ have the same underlying type free terms modulo $\beta\eta$. This will be important when implementing the normalization program.

**Normalization Theorem**

For every $r^\rho$ there is an $s$ such that $\mathbf{N}(r, s)$.

**Proof:** Let $\vec{x}^{\vec{\sigma}} = \mathrm{FV}(r)$. Since $\mathrm{SA}_{\vec{\sigma}}(\vec{x})$ by Ax3, we get $\mathrm{SA}_{\vec{\sigma}}(\vec{x})$ by Lemma 1(b). Now $\mathrm{SC}_\rho(r)$ by Lemma 3 and therefore $\mathrm{SN}(r)$ by Lemma 1(a). Using Ax12 we get $\mathbf{N}(r, s)$ for some $s$.  QED

**Proof terms** For every term $r^\rho$ we define $d_r^N : \exists s N(r,s)$ Let $\vec{x} = x_1^{\sigma_1},\ldots,x_n^{\sigma_n}$
$:= FV(r)$ and let $k$ be any number such that $\mathbf{F}(\mathbf{r},\mathbf{k})$ holds. Then

$$d_{r,\vec{x},k}^N := d_\rho^{1(a)}\{\mathbf{r}\}(d_{r,\vec{x}}^3\{\vec{\mathbf{x}}\}\tilde{d}_1\ldots\tilde{d}_n)\mathbf{k}(Ax12_{r,k})$$

where for $i \in \{1,\ldots,n\}$ $\tilde{d}_i := d_{\sigma_i}^{1(b)}\{\mathbf{x_i}\}\lambda l \lambda p^{\mathbf{F(x_i},l)}\langle \mathbf{x_i},(Ax3)\{\mathbf{x_i}\}\rangle : SC_{\sigma_i}(\mathbf{x_i})$

**Programs extracted** Let $\Omega_{r^\rho} := \mathrm{ep}(d_{r^\rho}^N) : \lambda_\rho$. We have

$$\Omega_{r^\rho} = \Phi_\rho(\Delta_{r^\rho,\vec{x}}(\Psi_{\sigma_1}\lambda l\mathbf{x_1})\ldots(\Psi_{\sigma_n}\lambda l\mathbf{x_n}))\mathbf{k}$$

By the soundness theorem we also have a proof of $\mathbf{N}(r^\rho,\Omega_{r^\rho})$. Therefore this formula is true in the intended model, i.e. the value $|\Omega_{r^\rho}|$ of $\Omega_{r^\rho}$ is the normal form of $r$.

The same proof may also be viewed as showing that every term has a $\beta\eta$-normal form. We just have to replace '$\eta$-expansion' by '$\eta$-normal form' in the definition of $N(r,s)$, and the function symbol **abst** has to be interpreted differently according to the following fact for the new $\mathbf{N}$

If $x^\rho \notin FV(r^{\rho\to\sigma})$ and $\mathbf{N}(rx,s)$ then $\mathbf{N}(r,\mathrm{cabst}(x,s))$.

Where cabst (conditional **abstraction**) is defined by **Definition** $\mathrm{cabst}(x,r) := s$ if $r = sx$ and $x \notin FV(s)$ and $\mathrm{cabst}(x,r) := \lambda x r$ otherwise.

The proof and the program code remain unchanged.

## 3.3 Implementation

In a functional programming language we may compute $|\Omega_{r^\rho}|$ in two steps:
1. Compute $\Omega_{r^\rho}$ from $r$ and $\rho$. 2. Evaluate the term $\Omega_{r^\rho}$, i.e. press **Return**.

If the language accepts type free $\lambda$-terms then we can do even better. We may replace the subterm $\Delta_{r^\rho,\vec{x}}$ by $\lambda\vec{x}r$ because as already remarked $\Delta_{r,\vec{x}}$ and $\lambda\vec{x}r$ have the same underlying type free terms modulo $\beta\eta$. Furthermore we may compute $|\Phi_\rho|$ and $|\Psi_\rho|$ separately as (primitive) recursive functionals of $\rho$. Hence

$$\begin{aligned}|\Omega_{r^\rho}| &= |\Phi_\rho(\Delta_{r^\rho,\vec{x}}(\Psi_{\sigma_1}\lambda l\mathbf{x_1})\ldots(\Psi_{\sigma_n}\lambda l\mathbf{x_n}))\mathbf{k}| \\ &= |\Phi_\rho((\lambda\vec{x}r)(\Psi_{\sigma_1}\lambda l\mathbf{x_1})\ldots(\Psi_{\sigma_n}\lambda l\mathbf{x_n}))\mathbf{k}| \\ &= |\Phi_\rho|(|\lambda\vec{x}r|(|\Psi_{\sigma_1}||\lambda l\mathbf{x_1}|)\ldots(|\Psi_{\sigma_n}||\lambda l\mathbf{x_n}|))|\mathbf{k}| \\ &= |\Phi_\rho|(|r|^\eta)k\end{aligned}$$

where $\eta$ is the environment binding every variable $x^\sigma$ free in $r$ to the functional $|\Psi_\sigma||\lambda l x|$.

If the programming language in addition provides an explicit procedure **eval** for evaluation, like LISP does for instance, then we may use this for writing a clever program taking a term $r$ and its type $\rho$ as input and computing the normal form of $r$. Below we present this program in the LISP dialect SCHEME. For simplicity we write it only for normalizing *closed* terms. Therefore we can forget about the environment $\eta$ and may take 0 for $k$.

```
(define (norm r rho)
   (((phi rho) (eval r)) 0))
```

The procedure phi and with it the procedure psi are defined by simultaneous recursion.

```
(define (phi type)
  (if (ground-type? type)
      (lambda (u) u)
      (let ((psi-rho (psi (arg-type type)))
    (phi-sigma (phi (val-type type))))
(lambda (u)
  (lambda (k)
    (let ((xk (mvar k)))
      (abst xk ((phi-sigma (u (psi-rho (lambda (l) xk))))
(+ k 1))))))))))

(define (psi type)
  (if (ground-type? type)
      (lambda (g) g)
      (let ((phi-rho (phi (arg-type type)))
    (psi-sigma (psi (val-type type))))
(lambda (g)
  (lambda (u)
    (psi_sigma (lambda (k) (app (g k) ((phi-rho u) k)))))))))

(define (mvar k)
  (string->symbol (string-append "x" (number->string) k '(int))))
(define (abst x r) (list 'lambda (list x) r))
(define app list)

(define ground-type? symbol?)
(define arg-type car)
(define val-type cadr)
```

Note that the type of the value of (phi rho) depends on rho. Therefore this program is not typable (e.g. in ML).

An extension of this algorithm to $\beta$–conversion plus term rewriting is implemented in a proof system developed by Schwichtenberg [SCH90].

## 4 Conclusion

In this paper program extraction via modified realizability is used for deriving a good normalization program from a formal proof of strong normalization for the simply typed $\lambda$-calculus.

It was not the main objective to provide a machine checked correctness proof but rather to show that this method can be a valuable tool for developing and understanding programs. From this point of view the fact that many difficult parts of the proof were hidden behind axioms should not be regarded as cheating. It rather demonstrates the main advantage of the realizability method, namely the possibility of dividing a proof into a computationally relevant part and into a part responsible for correctness.

## References

[BS91]   U. Berger, H. Schwichtenberg, An Inverse of the Evaluation Functional for Typed $\lambda$-calculus, Proceedings of the Sixth Annual IEEE Symposium on Logic in Computer Science, Amsterdam, (1991) 203–211.

[Co92]   C. Coquand, A proof of normalization for simply typed lambda calculus written in ALF, Proceedings of the 1992 Workshop on Types for Proofs and Programs, Båstad, Sweden (1992) 80–87.

[Ha87]   S. Hayashi, PX: a system extracting programs from proofs, Kyoto University, Japan (1987).

[GS92]   V. Gaspes, J. S. Smith, Machine Checked Normalization Proofs for Typed Combinator Calculi, Proceedings of the 1992 Workshop on Types for Proofs and Programs, Båstad, Sweden (1992) 168–192.

[Kr59]   G. Kreisel, Interpretation of analysis by means of functionals of finite type, in Constructivity in Mathematics (1959) 101–128.

[PW92]   C. Paulin-Mohring, B. Werner, Synthesis of ML programs in the system Coq, Submitted to the Journal of Symbolic Computations (1992).

[Sch90]  H. Schwichtenberg, Proofs as Programs, Leeds: Proof Theory '90 (P. Aczel, H. Simmons, Editors 1990).

[Ta67]   W. W. Tait, Intensional Interpretation of Functionals of Finite Type I, Journal of Symbolic Logic 32(2) (1967) 198–212.

[Tr73]   A. S. Troelstra, Metamathematical Investigation of Intuitionistic Arithmetic and Analysis, SLNM 344 (1973).

# A semantics for λ&-*early*: a calculus with overloading and early binding
### (Extended Abstract)

Giuseppe Castagna[1]   Giorgio Ghelli[2]   Giuseppe Longo[3]

[1] DISI (Univ. Genova) and LIENS-DMI (Paris): castagna@dmi.ens.fr
[2] Dipartimento d'Informatica, Corso Italia 40, Pisa, ITALY: ghelli@di.unipi.it
[3] LIENS(CNRS)-DMI, Ecole Normale Sup., 45 rue d'Ulm, Paris, FRANCE: longo@dmi.ens.fr

**Abstract.** The role of λ−calculus as core functional language is due to its nature as "pure" theory of functions. In the present approach we use the functional expressiveness of typed λ-calculus and extend it with our understanding of some relevant features of a broadly used programming style: Object Oriented Programming (OOP). The core notion we focus on, yields a form of "dependency on input types" (or "on the types of the inputs") and formalizes "overloading" as implicitly used in OOP.
The basis of this work has been laid in [CGL92a], where the main syntactic properties of this extension have been shown. In this paper, we investigate an elementary approach to its mathematical meaning. The approach is elementary, as we tried to follow the most immediate semantic intuition which underlies our system, and yet provide a rigorous mathematical model. Indeed, our semantics provides an understanding of a slightly modified version of the system in [CGL92a]: we had, so far, to restrict our attention only to "early binding".
In order to motivate our extended λ-calculus, we first survey the key features in OOP which inspired our work. Then we summarize the system presented in [CGL92a] and introduce the variant with "early binding". Finally we present the model.

## 1 Introduction

The role of λ-calculus as core functional language is due to its nature as "pure" theory of functions: just application, $MN$, and functional abstraction, $\lambda x.M$, define it. In spite of the "minimality" of these notions, full computational expressiveness is reached, in the type-free case. In the typed case, expressiveness is replaced by the safety of type-checking. Yet, the powerful feature of implicit and explicit polymorphism may be added. With polymorphism, one may have type variables, which apparently behave like term variables: they are meant to vary over the intended domain of types, they appear in applications and one may λ-abstract w.r.t. them. The functions depending on type variables, though, have a very limited behavior. A clear understanding of this is provided by a simple remark in [Gir72], were second order λ-calculus was first proposed: no term taking types as inputs can "discriminate" between different types. More precisely, if one extends system F by a term $M$ such that on input type $V$ returns 0 and 1 on $U$, then normalization is lost. Second order terms, then, are "essentially" constant, or "parametric". Indeed, the notion of parametricity has been the object of a deep investigation, since [Rey84] (see also [ACC93] and [LMS92] for recent investigations). In the present approach we use the functional expressiveness of λ-calculus and extend it with some relevant features of a broadly used programming style: Object Oriented Programming. Indeed, the core notion we want to focus on, yields a form of "dependency on input types" (or "on the types of the inputs").

The basis of this work has been laid in [CGL92a], were the main syntactical properties of the system below have been shown. In this paper, we investigate an *elementary* approach to its mathematical meaning. The approach is elementary, as we tried to follow the most immediate semantic intuition which underlies our system, and yet provide a rigorous mathematical model. A more general (categorical) understanding of what we mean by "dependency on input types" should be a matter of further investigation, possibly on the grounds of the concrete construction below. Indeed, our model provides an understanding of a slightly modified version of the system in [CGL92a]: we had, so far, to restrict our attention only to "early binding" (see the discussion below).

In order to motivate our extended $\lambda$-calculus, we first survey the key features in Object Oriented programming which inspired our work, then, in section 3, we summarize the system presented in [CGL92a] and develop some further syntactic properties, instrumental to our semantic approach. Section 4, introduces the variant with "early binding". Section 5 presents the model.

## 2 Abstract view of object-oriented paradigms

Object oriented programs are built around *objects*. An object is a programming unit which associates data with the operations that can use or affect these data. These operations are called *methods*; the data they affect are the *instance variables* of the object. In short an object is a programming unit formed by a data structure and a group of procedures which affect it. The instance variables of an object are private to the object itself; they can be accessed only through the methods of the object. An object can only respond to *messages* which are *sent* or *passed* to it. A *message* is simply the name of a method that was designed for that object.

Message passing is a key feature of object-oriented programming. As a matter of facts, every object oriented program consists in a set of objects which interact by exchanging messages. Since objects are units that associate data and procedures, they are very similar to instances of abstract data types and, thus, message passing may be viewed as the application of a function defined on the abstract data type. However, message passing is a peculiar feature, as we show in the next section.

### 2.1 Message passing

Every language has its own syntax for messages. We use the following one:

$$message \bullet receiver$$

The *receiver* is an object (or more generally an expression returning an object); when it receives a message, the run-time system selects among the methods defined for that object the one whose name corresponds to passed message; the existence of such a method should be statically checked (i.e. verified at compile time) by a type checking algorithm.

There are two ways to understand message passing. The first way roughly consists in considering an object as a record whose fields contain the methods defined for the object, and whose labels are the messages of the corresponding methods [4]. This implementation has suggested the modeling of the objects as proposed in [Car88] and known as the "objects as records" analogy. This modeling has been adopted by a large number of authors (see [Bru91, BL90, CL91, CW85, CCH+89, CHC90, CMMS91, Ghe91a, Mey88, Pie92, PT93, Rém89, Wan87, Wan89]).

The second way to implement message passing is the one used in the language CLOS (see [Kee89]), introduced in the context of typed functional languages in [Ghe91b], where message passing is functional application in which the message is (the identifier of) the function and

---

[4] Indeed the real organization is a bit more complicate since methods are not searched directly in the object but in the *class* of the object

the receiver is its argument. We have based our modeling of object-oriented programming on this intuition, and for this reason our work is alternative and somewhat detached from the type-driven research done so far in object-oriented programming. However, in order to formalize this approach, ordinary functions (e.g. lambda-abstraction and application) do not suffice. Indeed the fact that a method *belongs* to a specific object implies that the implementation of *message passing* is different from the one of the custom function application (i.e., the $\beta$-reduction). The main characteristics that distinguish methods from functions are the following:

- *Overloading*: Two objects can respond differently to the same message. For instance, the code executed when sending a message **inverse** to an object representing a matrix will be different from the one executed when the same message is sent to an object representing a real number. Though the same message behaves uniformly on objects of the same kind (e.g. on all objects of *class* matrix). This feature is known as *overloading* since we overload the same operator (in this case **inverse**) by different operations; the actual operation depends on the type of the operands. Thus *messages are identifiers of overloaded functions* and in message passing the *receiver* is the first argument of an overloaded function, and the one on whose type is based the selection of the code to be executed. Each method constitutes a branch of the overloaded function referred by the message it is associated to.
- *Late Binding*: The second crucial difference between function applications and message passing is that a function is bound to its meaning at compile time while the meaning of a method can be decided only at run-time when the receiving object is known. This feature, called *late-binding*, is one of the most powerful characteristics of object-oriented programming (see the excursus below). In our approach, it will show up in the the combination between overloading and subtyping. Indeed we define on types a partial order which concerns the utilization of values: a value of a certain type can be used where a value of a supertype is required. In this case the exact type of the receiver cannot be decided at compile time since it may change (notably decrease) during computation. For example, suppose that a graphic editor is coded using an object-oriented style, defining *Line* and *Square* types as subtypes of *Picture* with a method *draw* defined on all of them, and suppose that $x$ is a formal parameter of a function, with type *Picture*. If the compile time type of the argument is used for branch selection (early binding) an overloaded function application like the following one

$$\lambda x^{Picture}.(\ldots(draw \bullet x)\ldots)$$

is always executed using the *draw* code for *pictures*. Using late binding, each time the whole function is applied, the code for *draw* is chosen only when the $x$ parameter has been bound and evaluated, on the basis of the run-time type of $x$, i.e. according to whether $x$ is bound to a line or to a square.

Therefore in our model overloading with late binding is the basic mechanism.

**Excursus** (late vs. dynamic binding)   *Overloaded operators can be associated to a specific operation using either "early binding" or "late binding". This distinction applies to languages where the type which is associated at compile time to an expression can be different (less informative) from the type of the corresponding value, at run time. The example above with Line and Picture should be clarifying enough. Note though that what here we call late binding, in object-oriented languages is usually referred as dynamic binding (see for example [Mey88, NeX91]). Late and dynamic binding (or "dynamic scoping") are yet two distinct notions. Early vs. late binding has to do with overloading resolution, while static vs. dynamic binding means that a name is connected to its meaning using a static or a dynamic scope. However this mismatch is only apparent, and it is due to the change of perspective between our approach and the one of* *the languages cited above: in [Mey88] and [NeX91], for example, the suggested understanding is that a message identifies a method, and the method (i.e. the meaning of the message) is dynamically connected to the message; in our approach a message identifies an overloaded function (thus a set of methods) and it will always identify this function (thus it is statically bounded[5]) but the selection of the branch is performed by late binding.*

The use of overloading with late binding automatically introduces a new distinction between message passing and ordinary functions. As a matter of fact, overloading with late binding requires a restriction in the evaluation technique of arguments: while ordinary function application can be dealt with by either call-by-value or call-by-name, overloaded application with late binding can be evaluated only when the run-time type of the argument is known, i.e. when the argument is fully evaluated (closed and in normal form). In view of our analogy "messages as overloaded functions" this corresponds to say that message passing (i.e. overloaded application) acts by call-by-value or, more generally, only closed and normal terms respond to messages.

Thus to start a formal study on the base of this intuition, we have defined in [CGL92a, CGL92b] an extension of the typed lambda calculus that could model these features. We will not give here a detailed description of the calculus, since the reader can find it in the papers cited above. We just recall the underlying intuition and the formal definitions of this calculus.

## 3   The λ&-calculus

An overloaded function is constituted by a set of ordinary functions (i.e. lambda-abstractions), each one forming a different branch. To glue together these functions in an overloaded one we have chosen the symbol &; thus we have added to the simply typed lambda calculus the term

$$(M\&N)$$

which intuitively denotes an overloaded function of two branches, $M$ and $N$, that will be selected according to the type of the argument. We must distinguish ordinary application from the application of an overloaded function since, as we tried to explain in the previous section, they constitute different mechanism. Thus we use "•" to denote the overloaded application and · for the usual one. For technical reasons we determined to build overloaded functions as it is customary with lists, starting by an *empty* overloaded function that we denote by $\varepsilon$, and concatenating new branches by means of &. Thus in the term above $M$ is an overloaded function while $N$ is a regular function, which we call a "branch" of the resulting overloaded function. Therefore an overloaded function with $n$ branches $M_1, M_2, \ldots M_n$ can be written as

$$((\ldots((\varepsilon\&M_1)\&M_2)\ldots)\&M_n)$$

The type of an overloaded function is the ordered set of the types of its branches.[6] Thus if $M_i: U_i \to V_i$ then the overloaded function above has type

$$\{U_1 \to V_1, U_2 \to V_2, \ldots, U_n \to V_n\}$$

---

[5] As a matter of fact, messages obey to an intermediate scoping rule: they have a "dynamically extensible" meaning. If the type *Picture* is defined with the method *draw*, then the meaning of the *draw* method is fixed for any object of type picture, like happens with static binding. However, if later a new type *Circle* is added to the graphic editor, the set of possible meanings for the *draw* message is dynamically extended with the method for *Circle* and the function in the previous example will use the correct method for circles, even if circles did not exist when the function was defined. This combination of late binding and dynamic extensibility is one of the keys of the high reusability of object-oriented code.

[6] This is just a first approximation; see later for the exact meaning of overloaded types.

and if we pass to this function an argument $N$ of type $U_j$ then the selected branch will be $M_j$. That is:
$$(\varepsilon\&M_1\&\ldots\&M_n)\bullet N \;\triangleright^*\; M_j\bullet N \tag{*}$$
We have also a subtyping relation on types. Its intuitive meaning is that $U \leq V$ iff any expression of $U$ can be safely used in the place of an expression of $V$. An overloaded function can be used in the place of another when for each branch of the latter there is one branch in the former that can substitute it; thus, an overloaded type $U$ is smaller than another overloaded type $V$ iff for any arrow type in $V$ there is at least one smaller arrow type in $U$.

Due to subtyping, the type of $N$ in the expression above may not match any of the $U_i$ but it may be a subtype of one of them. In this case we choose the branch whose $U_i$ "best approximates" the type, say, $U$ of $N$; i.e. we select the branch $z$ s.t. $U_z = \min\{U_i | U \leq U_i\}$.

It is important to notice that, because of subtyping, in this system types evolve during computation. This reflects the fact that, in languages with subtypes, the run-time types of the values of an expression are not necessarily equal to its compile-time type, but are always subtypes of that compile-time type. In the same way, in this system, the types of all the reducts of an expression are always smaller than the type of the expression itself.

In our system, not every set of arrow types can be considered an overloaded type. A set of arrow types is an overloaded type iff it satisfies these two conditions:

$$U_i \leq U_j \Rightarrow V_i \leq V_j \tag{1}$$
$$U_i \Downarrow U_j \Rightarrow \text{there exists a unique } z \in I \text{ such that } U_z = \inf\{U_i, U_j\} \tag{2}$$

where $U_i \Downarrow U_j$ means that $U_i$ and $U_j$ are downward compatible, i.e. they have a common lower bound.

Condition (1) is a consistency condition, which assures that during computation the type of a term may only decrease. In a sense, this takes care of the common need for some sort of covariance of the arrow in the practice of programming. More specifically if we have a two-branched overloaded function $M$ of type $\{U_1 \to V_1, U_2 \to V_2\}$ with $U_2 < U_1$ and we pass it a term $N$ which at compile-time has type $U_1$ then the compile-time type of $M\bullet N$ will be $V_1$; but if the normal form of $N$ has type $U_2$ then the run-time type of $M\bullet N$ will be $V_2$ and therefore $V_2 < V_1$ must hold. The second condition concerns the selection of the correct branch: we said before that if we apply an overloaded function of type $\{U_i \to V_i\}_{i \in I}$ to a term of type $U$ then the selected branch has type $U_j \to V_j$ such that $U_j = \min_{i \in I}\{U_i | U \leq U_i\}$; condition (2) assures the existence and uniqueness of this branch.[7] While condition (1) is essential to avoid run-time type errors, condition (2) is just the simplest way to deal with the multiple inheritance problem, that is the problem of selecting one branch when many possible choices have been defined; adopting a different solution for this problem would not change the essence of the approach (see [Ghe91b]).

Finally we have to include *call-by-value* and *late-binding*. This can simply be done by requiring that a reduction as (*) can be performed only if $N$ is a closed normal form, and that the chosen branch depends on the type of the reduced term. This is late-binding since the branch choice cannot be performed before evaluating the argument, and this choice does not depend on the compile-time type of the expression which generated the value, but on the run-time type of the value itself. The formal description of the calculus is given by the following definitions:

**Pretypes**
$$V ::= A \mid V \to V \mid \{V_1' \to V_1'', \ldots, V_n' \to V_n''\}$$

---

[7] By the way note how these conditions are very related to the regularity condition discussed in [GM89], in the quite different framework of order-sorted algebras and order-sorted rewriting systems

## Subtyping

The subtyping preorder relation is predefined as a partial lattice[8] on atomic types, and it is extended to higher pretypes in the following way:

$$\frac{U_2 \leq U_1 \quad V_1 \leq V_2}{U_1 \to V_1 \leq U_2 \to V_2} \qquad \frac{\forall i \in I, \exists j \in J \quad U'_j \to V'_j \leq U''_i \to V''_i}{\{U'_j \to V'_j\}_{j \in J} \leq \{U''_i \to V''_i\}_{i \in I}}$$

Since the set $(A, \leq)$ of atomic types is a partial lattice, the set of all types can be easily shown to be a partial lattice too

## Types

1. $A \in$ **Types**
2. if $V_1, V_2 \in$ **Types** then $V_1 \to V_2 \in$ **Types**
3. if for all $i, j \in I$
   (a) $(U_i, V_i \in$ **Types**$)$ and
   (b) $(U_i \leq U_j \Rightarrow V_i \leq V_j)$ and
   (c) $(U_i \Downarrow U_j \Rightarrow \exists! h \in I \quad U_h = \inf\{U_i, U_j\})$
   then $\{U_i \to V_i\}_{i \in I} \in$ **Types**

## Terms (where $V$ is a type)

$$M ::= x^V \mid \lambda x^V M \mid M \cdot M \mid \varepsilon \mid M \&^V M \mid M \bullet M$$

## Type-checking Rules

[TAUT] $\qquad\qquad\qquad\qquad x^V : V$

[$\to$ INTRO] $\qquad\qquad\qquad \dfrac{M : V}{\lambda x^U . M : U \to V}$

[$\to$ ELIM$_{(\leq)}$] $\qquad\qquad \dfrac{M : U \to V \quad N : W \leq U}{M \cdot N : V}$

[TAUT$_\varepsilon$] $\qquad\qquad\qquad \varepsilon : \{\}$

[$\{\}$INTRO] $\qquad \dfrac{M : W_1 \leq \{U_i \to V_i\}_{i \leq (n-1)} \quad N : W_2 \leq U_n \to V_n}{(M \&^{\{U_i \to V_i\}_{i \leq n}} N) : \{U_i \to V_i\}_{i \leq n}}$

[$\{\}$ELIM] $\qquad \dfrac{M : \{U_i \to V_i\}_{i \in I} \quad N : U \quad U_j = \min_{i \in I}\{U_i | U \leq U_i\}}{M \bullet N : V_j}$

## Reduction

The reduction $\triangleright$ is the compatible closure of the following notion of reduction:

$\beta$) $(\lambda x^S . M) N \triangleright M[x^S := N]$

$\beta_\&$) If $N : U$ is closed and in normal form, $U_j = \min_{i=1..n}\{U_i | U \leq U_i\}$ and $(M_1 \& M_2) : \{U_i \to V_i\}_{i=1..n}$ then

$$(M_1 \& M_2) \bullet N \triangleright \begin{cases} M_1 \bullet N \text{ for } j < n \\ M_2 \cdot N \text{ for } j = n \end{cases}$$

---

[8] i.e. it must satisfy the following constraints: $A \Downarrow A' \Rightarrow \exists \inf\{A, A'\}$ and $A \Uparrow A' \Rightarrow \exists \sup\{A, A'\}$

**Main Theorems**
For this calculus we have proved the following fundamental theorems (see [CGL92b]):
- *Type Uniqueness*: Every well-typed term possesses a unique type.
- *Generalized Subject Reduction*: Let $M:U$. If $M \triangleright^* N$ then $N:U'$, where $U' \leq U$.
- *Strong Normalization*: There is no infinite sequence of reductions starting from a well-typed term
- *Church-Rosser*: Every term possesses a unique normal form.

## 3.1 The completion of overloaded types

This section presents some general, syntactic, properties of (overloaded) types, which may be viewed as some sort of "preprocessing" on the syntactic structures and which provide by this an interface towards our semantic constructions. The intuitive semantics of overloaded types is rather different from the usual meaning of types. It will require an extension of the usual model constructions of typed $\lambda$-calculus. Notice first that types directly affect the computation: the output value of a $\beta_{\&}$ reduction (in the previous section) explicitly depends on the type of the $(M_1 \& M_2)$ term, and on the type of the argument $N$.[9] Moreover as already mentioned, we understand overloaded types also according to the prevailing role of types in typed languages: they are "type-checkers". To put it otherwise, (overloaded) types are used to check the "dimension" of programs, similarly as in Physics where, by a "dimensional analysis", one checks that in an equation, say, a force faces a force etc.. Thus, in particular, an overloaded type is the collection of programs that "type-check", or that, when fed with "acceptable" inputs, give outputs of the due type.

These will be the crucial semantic differences between arrow types and overloaded types, since arrow types will keep their usual, more restrictive, meaning as "collection of functions or morphisms identified by the input and output types". For example, let $M:U \to V$, then $M$ will be interpreted as a function from the meaning of $U$ (or any of its subtypes) to the meaning of $V$, as sets (or objects in a category of sets) and $U \to V$ will be interpreted as the collection of such functions. Robust structural properties of the model we propose will allow to understand easily how a function in $U \to V$ can be applied to elements of a subtype of $U$, as if it were in $U$.

The situation is more complex as for overloaded types and terms. Take for instance $M: \{U \to V, U' \to V\}$; in this case, $M$ may be applied to any $N:W \leq U$ and to any $N':W' \leq U'$, by rule [{} ELIM], similarly to what happens with regular function, but now the behavior of the function can be different *depending on the type of the argument*. Thus two crucial properties need to be described explicitly in the semantics of overloaded terms. First, output values depend on types; second, as a type may have infinitely many subtypes and the choice of the branch depends on "$\leq$", overloaded semantic functions explicitly depend on infinitely many types. By this, we will consider $M$ as an element of a larger set of functions, roughly the indexed product (see later)

$$\prod_{W \leq U, U'} (W \to V)$$

In other words, the interpretation of $U \to V$ will be given by the usual set of functions from the interpretation of $U$ to the interpretation of $V$ (in a suitable categorical environment), while the meaning of $\{U \to V, U' \to V\}$ will directly take care of the possibility of applying overloaded functions to all the subtypes of the argument types.

---

[9] The fact that terms depend on types should not be confused with the rather different situation of "dependent types" where types depend on terms

This interpretation of overloaded types and terms, is the consequence of our definition of the preorder on types, an issue which requires some more discussion. Consider, say, $\{U \to V\}$ and $\{U \to V, U' \to V'\}$. Clearly, $\{U \to V, U' \to V'\} \leq \{U \to V\}$. The opposite inclusion is also possible here, in contrast to record types; namely, if $U \to V \leq U' \to V'$, say, one also has $\{U \to V, U' \to V'\} \geq \{U \to V\}$: indeed, "$\leq$" is not a partial order, as it is not antisymmetric.

This preorder relation makes perfectly sense w.r.t. type-checking. Suppose, say, that $M: \{U_i \to V_i\}_{i\in J} \leq \{U'_i \to V'_i\}_{i\in I}$. Then the intended meaning is that $M$ can be fed with any input $N$ which would be acceptable for a term $M'$ in $\{U'_i \to V'_i\}_{i\in I}$, and that the output can be used in any context where $M' \cdot N$ would be accepted. Indeed, let $N: U'_i$ and $C[\ ]$ be a context where a value of type $V'_i$ can be put. Then, for some $j \in J$, $U_j \to V_j \leq U'_i \to V'_i$, so that $U'_i \leq U_j$ and $V_j \leq V'_i$, hence the application $M \cdot N$ type-checks and can be used in the context $C[M \cdot N]$. Our "$\leq$" is the least (or less fine) preorder that one can define with this property.

The rest of this section is devoted to the understanding of some general (syntactic) properties of $\leq$ and of the equivalence relation "$\sim$" it induces.

**Definition 1.** *Given types $U$ and $V$, set $U \sim V$ if $U \leq V$ and $V \leq U$.*

**Remark 2** *If $\{U \to V\} \sim \{U \to V, U' \to V'\}$ then $U' \leq U$ and $V' \sim V$. Indeed, one must have $U \to V \leq U' \to V'$, so that $U' \leq U$ and $V \leq V'$, but also $V' \leq V$ by the formation rules of overloaded types (the monotonicity of "$\to$"). Similarly if the second case applies. This gives the intuitive meaning of the equivalence: a type $U' \to V'$ can be freely added or removed from an overloaded type if there is another type $U \to V$ which "subsumes" it, i.e. such that $U \geq U'$ but $V \sim V'$.*

Our next step is now the definition of the "completion" of overloaded types; intuitively, the completion of an overloaded type is formed by adding all the subsumed types, so that two equivalent overloaded types should be transformed, by completion, in *essentially* the same completed type. For this purpose and for the purpose of their semantics, we now adopt a different notation for overloaded types. We preferred to use the current notation, so far, in order to match the programmer's intuition, but the entire paper could be reworked with the notation below. Write $\Downarrow H$, if the collection H of types has a lower bound.

**Notation 3 (g.o.t.)** *A general overloaded type (g.o.t.) is a pair $(K, out)$ where $K$ is a set of types, i.e. a subset of Type, and out is a function from $K$ to Type such that:*
1. $H \subseteq K$ and $\Downarrow H$ imply that there is $V \in K$, $V = \inf H$.[10]
2. *out is monotone w.r.t. the subtype preorder.*

*We may write $\{U \to out(U)\}_{U \in K}$ for a general overloaded type.*

Clearly, any g.o.t. $(K, out)$, with a finite $K$, defines a (well formed) overloaded type in the usual sense and conversely.[11]

The preorder on g.o.t is the extension of the one given in the previous section.

**Definition 4 (completion).** *Let $\{U \to out(U)\}_{U \in K}$ be a g.o.t.. Then its completion $\{U \to \widehat{out}(U)\}_{U \in \widehat{K}}$ is the g.o.t. given by: $\widehat{K} = \{U' | \exists U \in K \ \ U' \leq U\}$ and $\widehat{out}(U') = out(\min\{U \in K | U' \leq U\})$.*

**Fact 5** *The completion of a g.o.t. $\{U \to out(U)\}_{U \in K}$ is a well defined g.o.t..*

---
[10] Notice that V is not required to be the unique inf.
[11] Notice, however, that many different overloaded types correspond to a unique g.o.t., since firstly, the domain $K$ of a g.o.t. is not ordered and, secondly, $K$ could contain two equivalent types $U$ and $V$; in this case just one of the two types appears in the domain of the corresponding overloaded type

Clearly, the completion is an idempotent operation. Note also that, even for a singleton $K = \{U\}$, $\widehat{K}$ may be infinite.

**Fact 6** *By completion, one obtains an equivalent type, that is:*
$$\{U \to out(U)\}_{U \in K} \sim \{U \to \widehat{out}(U)\}_{U \in \widehat{K}}$$

The idea is to interpret overloaded types by using their completions, in the model. However, as several preliminary facts may be stated at the syntactic level, we preferred to define syntactic completions and work out their properties. The following, 7, is the most important one, which guarantees the monotonicity of completion. Note that subtyping between overloaded types is contravariant w.r.t. the collections $\widehat{K}$ and $\widehat{H}$. The reader familiar with the semantics of records as indexed products (see [BL90]) may observe analogy with that contravariant understanding of records. Indeed in [CGL92a] we showed that record types may be coded as particular overloaded types.

**Theorem 7.** *Let $(K, out)$ and $(H, out')$ be g.o.t.. Then*
$$\{U \to out'(U)\}_{U \in H} \leq \{U \to out(U)\}_{U \in K} \Leftrightarrow \widehat{K} \subseteq \widehat{H} \text{ and } \forall U \in \widehat{K} \ U \to \widehat{out'}(U) \leq U \to \widehat{out}(U)$$

**Corollary 8.** *Let $(K, out)$ and $(H, out')$ be g.o.t.; then:*
$$\{U \to out'(U)\}_{U \in H} \sim \{U \to out(U)\}_{U \in K} \Leftrightarrow \widehat{K} = \widehat{H} \text{ and } \forall U \in \widehat{K}. \widehat{out'}(U) \sim \widehat{out}(U)$$

In conclusion, completions are not exactly canonical representatives of equivalence classes, but at least push the differences between two overloaded types one level inside the types. In this way in the interpretation of types we will be able to get rid of the differences between equivalent types by iterating completion at all the levels inside the type structure. The fact that a type is equivalent to its completion makes it clear that an overloaded type does not describe the structure of the corresponding functions (e.g. how many different branches they have) but just which are the contexts where they can be inserted.

## 4 Early Binding

We have presented in our introduction a limited overview of object-oriented languages. These languages are characterized by an interplay of many features, namely encapsulation, overloading, late-binding, dynamic-binding, subtyping and inheritance. We have selected three of them — overloading, late binding and subtyping— that, in our opinion, suffice to model the relevant features of a class-based object-oriented language. Not that these features are exclusive to this approach: overloading existed long before object-oriented languages (FORTRAN already used it) while subtyping, even if it was first suggest by object-oriented paradigms, has been included in other different paradigms (e.g. EQLOG [GM85], LIFE [AKP91] or Quest [CL91]). But their combination is peculiar to object-oriented programming. And exactly the interplay of all these features makes the object-oriented approach so useful in the large-scale software production.

In our system we do not deal with encapsulation and dynamic-binding, but we think they can be obtained using existential and reference types (even if, admittedly, there is still much to be done about the combination of existential types and object oriented programming), while our calculus accounts for the other aspects although we had not the occasion to show it in this paper.

At the semantic level, our system presents three main technical challenges. The first is the true dependence of overloaded functions from types. The second is the fact that subtyping is not an order relation. The third is the distinction between run-time types and compile-time types. In the present paper we just concentrate on the first two aspects, which already requires some technical efforts, while we will avoid the third problem by taking into consideration only a subsystem of full λ& where the type of the arguments of overloaded functions is "frozen", i.e. is the same at compile-time and at run-time. This is just a first step in the direction of defining a semantics for the full system.

The resulting system is somehow intermediate between late-binding and early-binding overloading. It features early-binding, since for any application of an overloaded function the type which will be used to perform branch selection is already known at compile-time, as happens for example with arithmetic operators in imperative languages. It has still a form of late-binding since, as overloaded functions are first class values which can be the result of expression evaluation, it is not possible to get rid of branch selection at compile time. For example, if a function applies a formal parameter $x$, which has the type of an overloaded function, to an argument whose type is $U$, even if we know that branch selection will be based on $U$, branch selection cannot be statically performed since the function associated to $x$ is unknown. However this restriction is meaningful for its own sake since when overloaded functions are not first-class (i.e. when no variable possess an overloaded type) this exactly corresponds to to the "classical" (i.e. with early binding) implementation of overloading in imperative languages: the standard example is the operator + which is defined both on *reals* and *integers*, though a different code is used according to the type of the argument[12]. What happens is that a *preprocessor* scans *at compile time* the text of a program looking for all occurrences of + and it substitutes them by a call to the appropriated code, depending on whether they are applied to reals or integers. The same is true for paradigms which have a cleverer use of overloading: for example, when the programmer can define its own overloaded operators. This is possible in Haskell; in this language the implementation of overloading is based on strong theoretical grounds as shown in [WB89]; in that paper it is also shown how the selection in overloading can be solved at compile time by the use of a preprocessor. In fact, as far as we know, languages that use in an explicit way overloading (and not implicitly as it is done in object-oriented programming via the method definitions) base the selection of the code on the type possessed by the arguments at compile time.

We obtain this "half-way early-binding" restriction of our system simply by adding explicit coercions and imposing that every argument of an overloaded function is coerced. A coercion $c_V$ is just an function which, informally, does nothing, but which cannot be reduced, so that the type of all the residuals of a term $c_V(M)$ is always $V$. Thus to model overloading with early binding we require that a coercion freezes the type of the argument of an overloaded function up to branch selection. Thus we have to change the system in the following way: Pretypes, Types and Subtyping rules are as before. Terms are now:

$$M::= x^V \mid \lambda x^V.M \mid M \cdot M \mid c_V(M) \mid \varepsilon \mid M \& M \mid M \bullet c_V(M)$$

We have to add to the rules of type-checking the one for coercions:

[COERCION] $$\frac{M:U \leq V}{c_V(M):V}$$

Finally we have to define the reduction on the coercion; the minimal modification required is the addition of this rule to our notion of reduction:

(*coerce*) $\qquad c_V(M) \circ N \triangleright M \circ N$

---

[12] This example is sometimes misleading because of the fact that the codes for the to branches must give the same results when applied to integers (being integer numbers a subset of real numbers); this extra property is proper to this example and it has nothing to do with overloading: in general the branches do not need to be related on the values they return

where o denotes either · or •. This rule is needed since otherwise coercions could prevent some $\beta$ or $\beta\&$ reduction. This rule does not interfere with our use of coercions, since it only allows us to reduce the left hand side, but not the right hand side, of an application. Another rule that could be added to the calculus is

$$c_V(c_U(M)) \triangleright c_V(M)$$

However it does not bring any interesting modification to the system, so that we prefer not to include it. We are now able to give the denotational semantics of this system.

## 5 Semantics

### 5.1 PER as a model

In this section we give the basic structural ideas which will allow us to interpret the syntax of $\lambda\&$-*early*. Namely, we state which geometric or algebraic structures may interpret arrow and overloaded types; terms will be their elements and will be interpreted in full details in section 5.3.

A general model theory of $\lambda\&$-*early* may be worth pursuing as an interesting development on the grounds of the concrete model below. Indeed, by some general categorical tools, one may even avoid to start with a model of type-free lambda calculus, but this may require some technicalities from Category Theory (see [AL91]). Thus we use here a classic model of type-free lambda calculus and a fundamental type structure out of it. We refer to the construction of $\mathcal{P}\omega$ in [Sco76] and [AL91, LM91] for the construction of the CCC of Partial Equivalence Relations (**PER**) on and arbitrary model of type-free $\lambda$-calculus $\mathcal{D}$.

**Notation 9** *Let $A$ be a symmetric and transitive relation on $\mathcal{D}$, we set, for $n, m \in \mathcal{D}$:*

$nAm \Leftrightarrow n$ *is related to $m$ by $A$*

$\text{dom}(A) = \{n | nAn\}$

$\lceil n \rceil_A = \{m | mAn\}$ *(the equivalence class of $n$ with respect to $A$)*

$Q(A) = \{\lceil n \rceil_A | n \in \text{dom}(A)\}$ *(the quotient set of $A$).*

The semantics of ordinary types is given as described for **PER** models in the references. Just recall that $[U \to V] = [U] \to [V]$, where $m(A \to B)n$ iff $\forall p, q(pAq \Rightarrow (m \cdot p)B(n \cdot q))$

Before going into the semantics of the other types, we briefly introduce the meaning of "subtypes", in view of the relevance this notion has in our language. The semantics of subtypes over **PER** is given in terms of "subrelations", (see [BL90]).

**Definition 10 (subtypes).** Let $A, B \in \textbf{PER}$. Define: $A \leq B$ iff $\forall n, m.(nAm \Rightarrow nBm)$

Note that, given $n \in \text{dom}(A)$ and $A \leq B$, we may view the passage from $\lceil n \rceil_A$ to $\lceil n \rceil_B$ as an obvious coercion.

**Definition 11 (semantic coercions).** Let $A, B \in \textbf{PER}$ with $A \leq B$. Define $c_{AB} \in \textbf{PER}[A, B]$ by $\forall n \in \text{dom}(A)\ c_{AB}(\lceil n \rceil_A) = \lceil n \rceil_B$

**Remark 12** *By this definition, for any $a \in Q(A)$, $c_{AB}(a) \supseteq a$*

For the sake of conciseness if $U$ and $V$ are syntactic types, we denote by $c_{UV}$ the semantic coercion $c_{[U][V]}$

Note that semantic coercions don't do any work, as type-free computations, but, indeed, change the "type" of the argument, i.e. its equivalence class and the equivalence relation where it lives. Thus they are realized by the indexes of the type free identity map, among others, and they are meaningful maps in the typed structure.

Since terms will be interpreted as equivalence classes in (the meaning as p.e.r.'s of) their types, we need to explain what the application of an equivalence class to another equivalence class may mean, as, so far, we only understand the application "·" between elements of the underlying type-free structure $(\mathcal{D}, \cdot)$.

**Definition 13 (application).** Let $A, A'$ and $B$ be p.e.r.'s, with $A' \leq A$. Define then, for $n(A \to B)n$ and $mA'm$, $\lceil n \rceil_{A \to B} \cdot \lceil m \rceil_{A'} = \lceil n \cdot m \rceil_B$.

Note that this is well defined, since $mA'm'$ implies $mAm'$ and, thus, $n \cdot mBn' \cdot m'$, when $n(A \to B)n'$. This is clearly crucial for the interpretation of our "arrow elimination rule". We end this section on subtyping by two technical lemmas that will be heavily used in the next sections.

**Lemma 14 (monotonicity of application).** *Let $a, b, a', b'$ be equivalence classes such that the applications $a \cdot b$ and $a' \cdot b'$ are well defined (i.e. $a \in Q(A_1 \to A_2)$ and $b \in Q(B)$ with $B \leq A_1$, and similarly for $a'$ and $b'$). If $a \subseteq a'$ and $b \subseteq b'$ then $a \cdot b \subseteq a' \cdot b'$*

**Lemma 15 (irrelevance of coercions).** *Let $A, A'$ and $B$ be p.e.r.'s, with $A' \leq A$. Assume that $n(A \to B)n$ and $mA'm$. Then*
$\lceil n \rceil_{A \to B} \cdot c_{A'A}(\lceil m \rceil_{A'}) = \lceil n \rceil_{A \to B} \cdot \lceil m \rceil_A = \lceil n \cdot m \rceil_B = \lceil n \rceil_{A \to B} \cdot \lceil m \rceil_{A'} = c_{A \to BA' \to B}(\lceil n \rceil_{A \to B}) \cdot \lceil m \rceil_{A'}$

## 5.2 Overloaded types as Products

Overloaded functions are similar, in a sense, to records; in the first case the basic operation is selection of a function depending on a type, while in the second case it is selection of a field depending on a label. Consequently, subtyping is strictly related too: theorem 7 shows that, working with the completion of types, subtyping is the same in the two cases.[13] So we will interpret overloaded types as indexed product types, similarly as it was done for record types in [BL90]. We will need to use, though, completions and infinite products.

Recall that, in set theory, given a set $A$ and a function $G : A \to \mathbf{Set}$ (**Set** is the category of sets and set-theoretical maps), one defines the indexed product:

$$\bigotimes_{a \in A} G(a) = \{f \mid f : A \to \cup_{a \in A} G(a) \text{ and } f(a) \in G(a)\}$$

If $A$ happens to be a subset of $\mathcal{D}$ and $G : A \to \mathbf{PER}$, then the resulting product may be viewed as a p.e.r. on $\mathcal{D}$, as follows.

**Definition 16.** Let $A \subseteq \mathcal{D}$ and $G : A \to \mathbf{PER}$. Define the p.e.r. $\prod_{a \in A} G(a)$ by
$$n(\prod_{a \in A} G(a))m \text{ iff } \forall a \in A \; n \cdot aG(a)m \cdot a$$

**Remark 17 (empty product)** *Notice that, by the definition above, for any $G$:*
$$\prod_{a \in \emptyset} G(a) = \mathcal{D} \times \mathcal{D}$$

---

[13] However we cannot get rid of overloaded type in λ&-early by encoding them as product types, using the technique developed in [CL91] for record types, since the completion of an overloaded type is an infinite structure, and also since we want to lay foundations which can be used to study the whole late-binding version of λ&.

Clearly, $\prod_{a \in A} G(a)$ is a well defined p.e.r. and may be viewed as a collection of computable functions, relatively to $\mathcal{D}$: the maps computed by the elements of equivalence classes in $\prod_{a \in A} G(a)$, a map for each equivalence class. That is, by the usual abuse of language, we may identify functions and equivalence classes and write:

$$f \in \prod_{a \in A} G(a) \text{ iff } f \in \bigotimes_{a \in A} G(a) \text{ and } \exists n \in \mathcal{D}.\forall a \in A. f(a) = \lceil n \cdot a \rceil_{G(a)}. \tag{3}$$

We then say that $n$ **realizes** $f$.

The idea now is to consider the type symbols as a particular subset of $\mathcal{D}$. Namely, assume that each type symbol $U$ is associated, in an injective fashion, to an element $n$ in $\mathcal{D}$, the semantic code of $U$ in $\mathcal{D}$. Call [**Type**]$\subseteq \mathcal{D}$ the collection of semantic codes of types. The choice of the set of codes is irrelevant, provided that
- it is in a bijection with **Type**;
- the induced topology on [**Type**] is the discrete topology.

These assumptions clearly pose no problem, in view of the cardinality and the topological structure of the model $\mathcal{D}$ we chose. For example, enumerate the set of type symbols and fix [**Type**] to be the collection of singletons $\{\{i\}|i \in \omega\}$ of $\mathcal{P}\omega$ (**Type** is countable as each type has a finite representation). We then write $T_n$ for the type-symbol associated to code $n$[14] and, given $K \subseteq$ **Type**, we set $[K] = \{n | T_n \in K\}$.

We can now interpret as a p.e.r. any product indexed over a subset $[K]$ of [**Type**]. Indeed, this will be the semantic tool required to understand the formalization of overloading we proposed: in $\lambda\&$, the value of terms or procedures may depend on types. This is the actual meaning of overloaded terms: they apply a procedure, out a finite set of possible ones, according to the type of the argument. As terms will be functions in the intended types (or equivalence classes of their realizers), our choice functions will go from codes of types to (equivalence classes in) the semantic types.

**Remark 18** *The reader may observe that there is an implicit higher order construction in this: terms may depend on types. However:*
- *in view of the countable (indeed finite) branching of overloaded terms and types, we do not need higher order models to interpret this dependency;*
- *note though that the intended meaning of an overloaded term is a function which depends on a possibly infinite set of input types, as it accepts terms in any subtype of the $U_i$ types in the $\{U_i \to V_i\}$ types. Whence the use of g.o.t.'s and completions.*
- *known higher order systems (System F, Calculus of Constructions...) would not express our "true" type dependency, where different types of the argument may lead to essentially different computations. This was mentioned in the introduction and it is understood in the* **PER** *model of these calculi by a deep fact: the product indexed over (uncountable) collections of types is isomorphic to an intersection (see [LM91]). A recent syntactic understanding of this phenomenon may be found in [LMS92]*

Now we are ready to define the semantics of overloaded types as products. In view of the fact that we want to interpret subtyping, which is a preorder, by an order relation in the model, we will use completion to get rid of "irrelevant differences" between overloaded types.

**Definition 19.** The semantics of overloaded types is given by

$$[\{U \to out(U)\}_{U \in K}] = \prod_{n \in [\widehat{K}]} [T_n \to \widehat{out}(T_n)]$$

where $[\widehat{K}] = \left\{ n \Big| T_n \in \widehat{K} \right\}$

---

[14] Remember that, despite the letter $n$, $n$ is a singleton, not just an integer.

This is a well defined meaning over **PER**, by definition 16, where $A = [\widehat{K}]$ and $G: [\widehat{K}] \to$ **PER** is given by $G(n) = [T_n \to \widehat{out}(T_n)]$. It clearly extends to g.o.t.'s, as we only need that $K$ is countable, here. Now we are finally in the position to check that the preorder on types is interpreted as the partial order "$\leq$" on **PER**.

**Theorem 20.** *If $U \leq V$ is derivable, then $[U] \leq [V]$ in* **PER**.

It should be clear that we presented here a core extension of the typed $\lambda$-calculus with overloading and subtyping. Note though that it suffices to represent records, as shown in [CGL92a]. As for recursively defined procedures, the natural extension of our language by a fixed point operator over terms may represent them. A sound model, as a substructure of our semantics, may be derived from the work in [Ama90].

## 5.3 The semantics of terms

We can now give meaning to terms of the $\lambda$&-calculus, with the use of coercions introduced in the previous section.

Syntactic coercions were denoted by $c_V$ where $V$ was the type the argument was coerced to; the type-checker assured that this type was greater than the type of the argument of $c_V$. In the semantics we need also to know the type of the argument since the semantic coercions are "typed functions", from a p.e.r. to another: thus, we denote semantic coercions between p.e.r.'s $[U]$ and $[V]$ by $c_{UV}$; the double indexation distinguishes them from the syntactic symbol. Also, in the following we index a term by type as a shorthand to indicate that the term possesses that type.

An environment $e$ for typed variables is a map $e: Var \to \bigcup_{A \in \mathbf{PER}} A$ such that $e(x^U) \in [U]$. Thus each typed variable is interpreted as an equivalence class in its semantic type. This will be now extended to the interpretation of terms by an inductive definition, as usual.

In spite of the heavy notation, required by the blend of subtyping and overloading, the intuition in the next definition should be clear. The crucial point 6 gives meaning to an overloaded term by a function which lives in an indexed product (as it will be shown formally below): the product is indexed over (indexes for) types and the output of the function is the (meaning of the) term or computation that one has to apply. Of course, this is presented inductively. Some coercions are required as $M_1$ and $M_2$ may live in smaller types than the ones in $\&^{\{V_i \to W_i\}_{i \leq n}}$. Then, in point 7, this term is actually applied to the term argument of the overloaded term.

**Definition 21 (semantics of terms).** Let $e: Var \longrightarrow \bigcup_{A \in PER} A$ be an environment. Set then:

1. $[\varepsilon]_e = \mathcal{D}$, the only equivalence class in the p.e.r. $\mathcal{D} \times \mathcal{D}$ (see remark 17)
2. $[x^U]_e = e(x^U)$
3. $[\lambda x^U.M^V]_e = \lceil n \rceil_{[U \to V]}$ where n is a realizer of $f$ such that $\forall u \in [U].f(u) = [M^V]_{e[x/u]}$
4. $[M^{U \to V} N^W]_e = [M^{U \to V}]_e [N^W]_e$
5. $[c^V(M^U)]_e = c_{UV}([M^U]_e)$ (the semantic coercion)
6. Let $(M_1 \&^{\{V_i \to W_i\}_{i \leq n}} M_2): \{U \to out(U)\}_{U \in \{V_i\}_{i \leq n}}$ with $M_1: T_1 \leq \{V_i \to W_i\}_{i < n}$ and $M_2: T_2 \leq V_n \to W_n$.
   Set then $[M_1 \& M_2]_e = f$ such that, given $i \in [\widehat{\{V_i\}_{i \leq n}}]$ and $Z = \min\{U \in \{V_i\}_{i \leq n} | T_i \leq U\}$, one has
   $$f(i) = \begin{cases} c_{T_2(T_i \to \widehat{out}(T_i))}([M_2]_e) & \text{if } Z = V_n \\ (c_{T_1(\{V_i \to W_i\}_{i < n})}[M_1]_e)(i) & \text{else} \end{cases}$$
7. $[M^{\{U \to out(U)\}_{U \in K}} \bullet c_V(N^W)]_e = [M^{\{U \to out(U)\}_{U \in K}}]_e(i)[c_V(N^W)]_e$ where $T_i = V$.

**Remark 22** *Notice that this semantics does not interpret reduction as equality. Indeed:*

$$[(\lambda x^V.Q^U)P^W]_e = [\lambda x^V.Q^U]_e[P^W]_e$$
$$= \lceil n \rceil_{[V \to U]}[P^W]_e \quad \text{with } n \text{ as in point 2 of definition 21}$$
$$= \lceil n \rceil_{[V \to U]}(c_{WV}[P^W]_e) \quad \text{by lemma 15}$$
$$= [Q^U]_{e[x/c_{WV}[P^W]_e]} \quad \text{by point 3 of definition 21}$$

*In general,* $[Q]_{e[x/c_{WV}[P^W]_e]}$ *is different from* $[Q[x/P^W]]_e$. *For example, if* $Q = x$, *the two expressions evaluate to* $c_{WV}[P^W]_e$ *and to* $[P^W]_e$ *respectively. This will be more generally understood in 25.*

The soundness of this definition is split into two theorems and is proved right below. We recall first, in a lemma, that $\mathcal{P}\omega$ is an "injective" topological space.

**Lemma 23** (injectivity). *Let $Y$ be a topological space and $X \subseteq Y$, a subspace with the induced topology. Then any continuous $h : X \to \mathcal{P}\omega$ can be extended to a continuous $\hbar : Y \to \mathcal{P}\omega$. Indeed, $\hbar$ is given by $\hbar(y) = \sqcup\{\sqcap\{h(x)|x \in X \cap U\}|y \in U\}$. (The proof is easy; see [Sco76] for this and more properties of $\mathcal{P}\omega$).*

**Theorem 24** (soundness w.r.t. type-checking). *If $N:U$ then, for any environment $e$, $[N]_e$ is well defined and $[N]_e \in [U]$.*

**Lemma 25** (substitution). $[Q[x/P]]_e \subseteq [Q]_{e[x/p]}$ *where* $p = c_{T'T}[P]_e$, $x{:}T$, $P{:}T' \leq T$

The immediate consequence of the work done so far is an interesting fact. It says that we have obtained a simple model of "reduction" and not, as customary in denotational semantics, of "conversion". This is precisely stated by the following theorem.

**Theorem 26** (soundness wrt reductions). $M^U \triangleright N^V \Rightarrow [M^U]_e \supseteq [N^V]_e$

**Remark 27** *Clearly, theorem 26 specializes to the implicative fragment of our calculus, which is simply typed $\lambda$-calculus with subtyping. Thus, by a simple observation of the properties of PER, we spotted a mathematical model of the reduction predicate "$\triangleright$" between terms of $\lambda$-calculi, instead of conversion "=". The non-syntactic models so far constructed could only give mathematical meaning to the theory of "=" between $\lambda$-terms and $\beta$-reduction was interpreted as the "=".*

*It is important to notice, however, that the decrease of the size of the equivalence class which is the interpretation of a term is not directly related to the reduction process, but to the fact that types decrease during computation. In fact, if you consider two terms $M$ and $c_V(M)$ and apply the same reduction steps to both of them, while the semantics of $M$ can decrease, any time its type changes, the semantics of $c_V(M)$ remain fixed, even if the same reduction steps are executed.*

## 6 Summary

As already mentioned in the introduction, there is a general understanding that polymorphism, as intended in $\lambda$-calculus, is not compatible with "procedures depending on input types". As pointed out in [Gir72], one cannot extend II order $\lambda$-calculus with a term giving different output terms according to different input types. Indeed, in [LMS92], it is shown that terms depending on types use types as "generic", i.e. the value on just one type determines the value everywhere. This is why, in order to express an explicit type dependency, it was not sufficient to extend simply typed $\lambda$-calculus by type variables, and we proposed an entirely new feature, based on "finite branching of terms", in order to formalize the dependency we wanted. Moreover, the use of late-binding and subtyping added expressiveness to the system. Indeed, the expressive power of the syntax poses some problems.

First, the use of a pre-order between types is handled, at a syntactic level in section 2, by the notion of completion. But then our finite branching immediately becomes an infinite one: this is indeed what is actually meant in the syntax, by the rules, as we allow terms to work also on inputs inhabiting types *smaller* than the intended one. Thus, the intended function depending on the type of the input, may depend on an infinity of input types, implicitly. This must be made explicit in the semantics.

Finally, we considered types as "coded" in the semantics, by using their indexes also as meaning in the model. Then the implicit polymorphism of our approach shows up in the semantics by the interpretation of overloaded functions as elements of an (infinite) indexed product.

To sum up, in this phase of the work we focused on the problems related to type dependency and to the lack of antireflexivity of the subtyping relation. The result obtained will be the basis for the definition of a denotational semantics for the full language, dealing also with the problem of late-binding.

# References

[ACC93]   M. Abadi, L. Cardelli, and P.-L. Curien. Formal parametric polymorphism. In *Ann. ACM Symp. on Principles of Programming Languages*. ACM, ACM-press, 1993. Extended abstract.

[AKP91]   H Aït-Kaci and A. Podelski. Towards a meaning of LIFE. Technical Report 11, Digital, Paris Research Laboratory, June 1991.

[AL91]    A. Asperti and G. Longo. *Categories, Types and Structures: An Introduction to Category Theory for the Working Computer Scientist*. MIT-Press, 1991.

[Ama90]   R. Amadio. Domains in a realizability framework. Technical Report 19, Laboratoire d'Informatique, Ecole Normale Supérieure - Paris, 1990.

[BL90]    K.B. Bruce and G. Longo. A modest model of records, inheritance and bounded quantification. *Information and Computation*, 87(1/2):196–240, 1990. A preliminary version can be found in *3rd Ann. Symp. on Logic in Computer Science*, 1988.

[Bru91]   K.B. Bruce. The equivalence of two semantic definitions of inheritance in object-oriented languages. In *Proceedings of the 6th International Conference on Mathematical Foundation of Programming Semantics*, 1991. to appear.

[Car88]   Luca Cardelli. A semantics of multiple inheritance. *Information and Computation*, 76:138–164, 1988. A first version can be found in Semantics of Data Types, LNCS 173, 51-67, Springer-Verlag, 1984.

[CCH+89]  P.S. Canning, W.R. Cook, W.L. Hill, J. Mitchell, and W.G. Orthoff. F-bounded quantification for object-oriented programming. In *ACM Conference on Functional Programming and Computer Architecture*, September 1989.

[CGL92a]  G. Castagna, G. Ghelli, and G. Longo. A calculus for overloaded functions with subtyping. In *ACM Conference on LISP and Functional Programming*, pages 182–192, San Francisco, July 1992. ACM Press. Extended abstract.

[CGL92b]  G. Castagna, G. Ghelli, and G. Longo. A calculus for overloaded functions with subtyping. Technical Report 92-4, Laboratoire d'Informatique, Ecole Normale Supérieure - Paris, February 1992.

[CHC90]   W.R. Cook, W.L. Hill, and P.S. Canning. Inheritance is not subtyping. *17th Ann. ACM Symp. on Principles of Programming Languages*, January 1990.

[CL91]    L. Cardelli and G. Longo. A semantic basis for Quest. *Journal of Functional Programming*, 1(4):417–458, 1991.

[CMMS91]  L. Cardelli, S. Martini, J.C. Mitchell, and A. Scedrov. An extension of system F with subtyping. In T. Ito and A.R. Meyer, editors, *Theoretical Aspects of Computer Software*, pages

750–771. Springer-Verlag, September 1991. LNCS 526 (preliminary version). To appear in *Information and Computation*.

[CW85] L. Cardelli and P. Wegner. On understanding types, data abstraction and polymorphism. *Computing Surveys*, 17(4):471–522, December 1985.

[Ghe91a] G. Ghelli. Modelling features of object-oriented languages in second order functional languages with subtypes. In J.W. de Bakker, W.P. de Roever, and G. Rozenberg, editors, *Foundations of Object-Oriented Languages*, number 489 in LNCS, pages 311–340, Berlin, 1991. Springer-Verlag.

[Ghe91b] G. Ghelli. A static type system for message passing. In *Proc. of OOPSLA '91*, 1991.

[Gir72] J-Y. Girard. Interprétation fonctionelle et élimination des coupures dans l'arithmetique d'ordre supérieur. Thèse de doctorat d'état, 1972. Université Paris VII.

[GM85] Joseph Goguen and José Meseguer. EQLOG: equality, types and generic modules for logic programming. In deGroot and Lindstrom, editors, *Functional and Logic Programming*. Prentice-Hall, 1985.

[GM89] J.A. Goguen and J. Meseguer. Order-sorted algebra I: Equational deduction for multiple inheritance, overloading, exceptions and partial operations. Technical Report SRI-CSL-89-10, Computer Science Laboratory, SRI International, July 1989.

[Kee89] S.K. Keene. *Object-Oriented Programming in COMMON LISP: A Programming Guide to CLOS*. Addison-Wesley, 1989.

[LM91] G. Longo and E. Moggi. Constructive natural deduction and its $\omega$-set interpretation. *Mathematical Structures in Computer Science*, 1(2):215–253, 1991.

[LMS92] G Longo, K. Milsted, and S. Soloviev. The genericity theorem and parametricity in functional languages. Technical report, Digital, Paris Reserch Lab., 1992.

[Mey88] Bertrand Meyer. *Object-Oriented Software Construction*. Prentice-Hall International Series, 1988.

[NeX91] NeXT Computer Inc. *NeXTstep-concepts. Chapter 3: Object-Oriented Programming and Objective-C*, 2.0 edition, 1991.

[Pie92] B.C. Pierce. Type-theoretic foundations for object-oriented programming. Technical report, LFCS, Departement of Computer Science, University of Edinburgh, May 1992. Lecture note for a short course.

[PT93] B.C. Pierce and D.N. Turner. Object-oriented programming without recursive types. In *10th Ann. ACM Symp. on Principles of Programming Languages*. ACM-Press, 1993.

[Rém89] D. Rémy. Typechecking records and variants in a natural extension of ML. In *Ann. ACM Symp. on Principles of Programming Languages*, 1989.

[Rey84] J.C. Reynolds. Polymorphism is not set-theoretic. *LNCS*, 173, 1984.

[Sco76] D. Scott. Data-types as lattices. *S. I. A. M. J. Comp.*, 5:522–587, 1976.

[Wan87] Mitchell Wand. Complete type inference for simple objects. In *2nd Ann. Symp. on Logic in Computer Science*, 1987.

[Wan89] Mitchell Wand. Type inference for record concatenation and multiple inheritance. In *4th Ann. Symp. on Logic in Computer Science*, 1989.

[WB89] Philip Wadler and Stephen Blott. How to make "ad-hoc" polymorphism less "ad-hoc". In *16th Ann. ACM Symp. on Principles of Programming Languages*, pages 60–76, 1989.

# An Abstract Notion of Application

Pietro Di Gianantonio, Furio Honsell

Dipartimento di Matematica e Informatica, Università di Udine,
via Zanon 6, I-33100 Udine, ITALY
e-mail: pietro@udmi5400.cineca.it, honsell@uduniv.cineca.it

Many concrete notions of function application, suitable for interpreting typed lambda calculi with recursive types, have been introduced in the literature. These arise in different fields such as set theory, multiset theory, type theory and functor theory and are apparently unrelated. In this paper we introduce the general concept of *applicative exponential structure* and show that it subsumes all these notions. Our approach is based on a generalization of the notion of *intersection type*. We construe all these structures in a finitary way, so as to be able to utilize uniformly a general form of type assignment system for defining the interpretation function. Applicative exponential structures are just combinatory algebras, in general. Our approach suggests a wide variety of entirely new concrete notions of function application; e.g. in connection with boolean sets. Applicative exponential structures can be used for modeling various forms of non-deterministic operators.

## 1 Introduction

Various natural concrete models of the notion of function application arising in λ-calculus have been discovered since the early seventies. G.Plotkin [13], inspired by earlier work of Scott, was the first to define a set-theoretical notion of application. By means of it he built a set theoretical model for untyped λ-calculus. Since then, various other natural notions of application were discovered by Scott [15] and Engeler [7] in set theory, by Coppo, Dezani and Venneri [5, 3] in type theory and by Girard [8], Ore [12] and Lamarche [11] in functor theory, in the theory of multisets and in analytic function theory. All these concrete notions of application give rise to concrete structures, albeit not always categories, which can be used as domains for denotational semantics. More precisely these structures are rich enough to model the behavior of application in typed λ-calculi with recursive types and appropriate constructors, destructors and fixed point operators.

These notions of application, although apparently different, seem to share a common pattern. In this paper we try to capture this pattern by introducing a notion of algebraic structure, termed *applicative exponential structure*, which we show to be general enough subsumes all the concrete notions mentioned earlier. In particular we define a general framework in which one can easily and uniformly express all classical constructions. One of the key features of our approach is the use of a generalized notion of type, inspired by that of "intersection type" [3], for providing a finitary description of the structures under consideration. This analysis allows for the use of a uniform kind of type assignment system for defining the interpretation of the λ-calculus language. Intuitively types are understood as finite elements of the domain, possibly having some coefficients; and a term has a given type if its denotation is approximated by a given type. Special care has to be taken in order to deal with the coefficients. This is a particularly interesting way of presenting the interpretation function since, besides being finitary in nature, it constitutes an endogenous logic in the sense of Abramsky [1]. Moreover it can provide a proof theoretic analysis of the fine structure of the models. This technique was initially introduced for the study of filter models [4] but was later applied to Girard's qualitative domains in coherent semantics [10] and quantitative domains [6].

We think that our approach is successful and fruitful since, besides illuminating on the idea underlying so many apparently unrelated notions of function application, it suggests also a wide variety of new concrete alternatives. Particularly appealing and potentially interesting is the notion of application which arises in connection with boolean sets. This notion yields a sort of "boolean valued" model of the λ-calculus. It is closely related to that which arises if we carry out Plotkin's original construction in a Boolean-valued model of set theory. This kind of construction can prove to be quite interesting for modeling programming languages which feature non-deterministic operators. Moreover, it seems to open a new area of applications of model theoretic concepts to the semantics of programming languages.

For simplicity we shall not deal in this paper with the whole language of typed λ-calculus with reflexive types as in [14] or [12]. We will discuss only the case of the untyped λ-calculus language, as an interesting and important example of a reflexive type. All the results can be extended with little difficulty to the more general case.

Somehow unexpectedly, the abstract structure introduced in the paper, and many of the concrete examples given, do not model λ-calculus in the strongest possible way. In general the, so called, ξ-rule fails and these are only combinatory algebras and not lambda models nor lambda algebras. Surprisingly enough the general construction does not seem to be amenable to a simple categorical presentation, unless further equivalence relations are superimposed. These will be discussed in a forthcoming paper.

A final remark is in order. We could have presented these results following more closely the approach of Girard[8] and Lamarche [11]. This would amount to use as main source of inspiration the notion of analytic function in complex analysis. No substantial difference would arise. The approach via analytic functions is in fact "dual" to the one used here. In this paper we will only describe very briefly this alternative approach in Appendix A.

The paper is organized as follows. In section 2 we define the structures normally utilized for modeling λ-calculus using a style which focuses on the properties of the interpretation function. In section 3 we present some classical and new constructions of concrete models of the lambda calculus and gradually introduce our general framework. In section 4 we give the definition of applicative exponential structure and prove the main theorem of this paper, i.e.: applicative exponential structures are combinatory algebras. Finally in section 5 we give more examples of concrete applicative exponential structures yet uninvestigated and we outline a possible use of applicative exponential structures for modeling non-deterministic operators.

Finally the authors would like to gratefully acknowledge Fabio Alessi and Simona Ronchi della Rocca for helpful discussions in the early stages of this work.

## 2 Combinatory Structures

Throughout the paper we assume the reader familiar with standard notions and notations in Lambda Calculus and Combinatory Logic as in [2]. Several different applicative structures, i.e. structures with a binary operation defined on them, have been introduced in the literature for interpreting the language of λ-calculus: combinatory algebras, lambda algebras and lambda models. These differ by the strength of the equalities which they enforce on interpretations of λ-terms. Usually combinatory algebras are defined without any reference to the interpretation function using the standard combinators S, K and I. Contrary to this tradition will define uniformly all these structures in the style of [10]. By so doing the connection with type assignment systems in the sequel will be clearer.

The language $\Lambda$ of *λ-calculus* is defined as usual by: M::= x | MN | λx.M. Terms which do not have abstracted subterms will be called *applicative terms*.

**Definition 1** (à la Hindley Longo)
1) An applicative structure $A$, is an algebra $\langle A : \text{Set}, \circ : A \times A \to A \rangle$;
2) An environment is a function $\rho : \text{Var} \to A$. The set of environments is denoted by Env.
3) An interpretation of $\Lambda$ is a function $[\![\ ]\!]: \text{Env} \to (\Lambda \to A)$. As usual $\xi[x/a]$ denotes the environment defined as $\xi[x/a](x)=a$ and $\xi[x/a](y)=\xi(y)$ if $x \neq y$.
4) A *combinatory structure*, c.s. for short, is a pair $\langle A, [\![\ ]\!]: \text{Env} \to (\Lambda \to A) \rangle$ consisting of an applicative structure and an interpretation function defined over it;
5) A *lambda model* is a c.s. where the interpretation function satisfies the following

properties:
i) $[\![ x ]\!]_\rho = \rho(x)$
ii) $[\![ MN ]\!]_\rho = [\![ M ]\!]_\rho \circ [\![ N ]\!]_\rho$
iii) $[\![ \lambda x.M ]\!]_\rho \circ a = [\![ M ]\!]_{\rho[x/a]}$
iv) $[\![ \lambda x.M ]\!]_\rho = [\![ \lambda y.([x/y]M) ]\!]_\rho$ provided $y \notin FV(M)$
v) $[\![ M ]\!]_\rho = [\![ M ]\!]_\xi$ provided $\rho(x) = \xi(x)$ for $x \in FV(M)$
vi) $(\forall a . [\![ M ]\!]_{\rho[x/a]} = [\![ N ]\!]_{\rho[x/a]}) \Rightarrow [\![ \lambda x.M ]\!]_\rho = [\![ \lambda x.N ]\!]_\rho$

6) A *lambda algebra* is a c.s. where the interpretation function satisfies conditions i) - v) above and also the following rule:
($\xi$) $\vdash_\lambda M = N \Rightarrow [\![ \lambda x.M ]\!]_\rho = [\![ \lambda x.N ]\!]_\rho$
where $\vdash_\lambda$ denotes derivability in the theory of $\lambda$-calculus.

7) A *combinatory algebra* is a c.s. where the interpretation function satisfies only conditions i)-v) above.  △

It is well known that lambda models are lambda algebras and lambda algebras are combinatory algebras. The following proposition illustrates the importance of combinatory algebras and establishes the equivalence of our definition with the usual ones.

**Proposition 1.** Let $\langle A, [\![\ ]\!] \rangle$ be a combinatory structure. The following properties are equivalent:
a) $A$ can be extended to a combinatory algebra $\langle A, [\![\ ]\!]' \rangle$;
b) There exist distinguished constants K and S in A such that for all constants $a, b, c \in A$:
$((K \circ a) \circ b) = a$ and $(((S \circ a) \circ b) \circ) c = ((a \circ c) \circ (b \circ c))$.
c) The interpretation function $[\![\ ]\!]$: Env $\to (\Lambda \to A)$ satisfies conditions i) and ii) in Definition 1, and moreover for all applicative term $M \in \Lambda$ such that $FV(M) \subseteq \{x_1,...,x_n\}$, there is a constant c, such that $[\![M]\!]_\xi = [\![cx_1...x_n]\!]_\xi$, for $x \in$ Env. This property is usually called *functional completeness*.

**Proof.** Standard  △

# 3 Concrete Models of Application

In this section we present some classical constructions of concrete models of the untyped $\lambda$-calculus language, some alternative presentations of these and some entirely new models. As remarked in the introduction any of these could be turned into a full-fledged domain structure for denotational semantics, but for lack of space we shall not do it here.

Perhaps the best known example of a natural concrete model of lambda calculus is the Plotkin-Engeler set theoretical model see [13, 7]. This model construction is closely related to the Filter Model constructions in [3, 5, 4], where arbitrary sets are replaced by particular ones called filters. Interestingly enough, Plotkin-Engeler Model and the Filter Model are indeed lambda models, the model introduced in [5], on the other hand, is only a lambda algebra. Nonetheless this latter structure, which we call Intersection Algebra, is quite remarkable since it is the first example of a lambda algebra which has not been defined by purely syntactical means.

In the literature the applicative structure underlying Plotkin-Engeler set theoretical model is defined as follows:

**Definition 2.** The applicative structure $\langle \mathscr{B}, \circ \rangle$ is defined inductively by:
$B_0$ is an arbitrary set of atoms,
$B_{n+1} \equiv B_0 \cup \{ (\beta, b) \mid \beta \subseteq B_n, \beta \text{ finite}, b \in B_n \}$
$\mathscr{B} \equiv \wp(\cup_n B_n)$
given $U, V \in \mathscr{B}$ we put $U \circ V \equiv \{ b \mid (\beta, b) \in U, \beta \subseteq V \}$.  △

We now give an alternative presentation of the above structure. This will be the first example of the standard format which will be used throughout the paper for presenting concrete applicative structures. The definition of the general notion of applicative exponential structure in Section 4 will build upon the shape of this format. The original presentation of Engeler's model, i.e.Definition 2, was given just for introductory purposes.

We need first some notation. Given a set A and a set B we denote by $[A \to B]$ the set of functions from A to B. If B contains a distinguished point 0, we denote by $[A \xrightarrow{o} B]$ the set of functions from A to B with value almost everywhere 0. As will become clear in the sequel these particular functions are introduced essentially as a useful "trick" for encoding functions with a finite domain, without having to bother about issues of definedness. According to this intended meaning, given $r \in [A \xrightarrow{o} B]$, dom(r) will denote the set $\{a \mid a \in A, r(a) \neq 0\}$; and the term $\{a_1:x_1,...,a_n:x_n\}$ will denote the function $s:A \to B$ defined by

$$s(a) \equiv \begin{cases} x_i \text{ if } a=a_i \text{ for some i } 1 \leq i \leq n \\ 0 \text{ otherwise} \end{cases}$$

Finally **2** will denote the boolean algebra of truth-values where "false" is taken to be 0 and "true" to be 1.

**Definition 3.** Let $J: \mathbb{N} \to$ Set be inductively defined by:
$J(0) \equiv J_0$, an arbitrary set of atoms, $J(n+1) \equiv J(0) \cup ([J(n) \xrightarrow{o} 2] \times J(n))$.
Now put $\mathbf{J} \equiv \cup_n J(n)$ and $\mathscr{J} \equiv [\mathbf{J} \to 2]$.
Let $\langle \mathscr{J}, \circ \rangle$ be the applicative structure where application is defined by:

$$(f \circ g)(j) \equiv \sum_{r \in [\mathbf{J} \xrightarrow{o} 2]} f((r,j)) \times \prod_{k \in \text{dom}(r)} (r(k) \Rightarrow g(k))$$

where $f, g \in \mathscr{J}$, $j \in \mathbf{J}$, $\Rightarrow$ is logical implication, + is disjunction and × is conjunction. ∆

Notice that any function $r \in [A \xrightarrow{o} 2]$ is indeed the characteristic function of a finite subset of A, just as any function $f: A \to 2$ is the characteristic function of an arbitrary subset of A.

The expression $\prod_{k \in \text{dom}(r)} (r(k) \Rightarrow g(k))$, used above, is therefore true, i.e. equal to 1 if and only if the finite set represented by r is a subset of the set represented by g.

What we have done in this new presentation amounts to substituting subsets with their characteristic functions. It is now easy to show that $\langle \mathscr{B}, \circ \rangle$ and $\langle \mathscr{J}, \circ \rangle$ are isomorphic.

Filter Models and Intersection Algebras can be accounted for similarly as follows. The structure $\langle \mathscr{J}, \circ \rangle$ is precisely the applicative structure underlying the Intersection Algebra in [5]. The applicative structure underlying the filter model in [3], instead, is obtained by taking $\mathscr{J}$ in Definition 3 to be the set of only those subsets of **J** which are *filters*. Elements of **J** behave in fact like intersection types. A filter f is a subset of **J** which is upwards closed under the order relation ≤ induced by the following rule:

$$\frac{j \leq j' \quad \forall k \in \text{dom}(r). \exists k' \in \text{dom}(r'). \ k' \leq k}{(r,j) \leq (r',j')}$$

We are now ready to turn these applicative structures into combinatory structures. As remarked in the introduction, we will utilize throughout the paper,*type assignment systems* in the sense of [3,4] to define the interpretation function. This is made possible because elements of **J** behave as a generalized "intersection types" [3,4]. In general we construe type assignment systems as formal systems for establishing assignment judgements of the form $\beta \vdash M:j$ where $M \in$ Term, $j \in \mathbf{J}$ and $\beta$ is a (multi)set of assumptions of the shape x:j'. Apart from the particular choice of the set **J**, the type assignment systems in the paper will vary greatly in the structural rules assumed in the formal system. The intended meaning of the judgement $\beta \vdash M:j$ will be made formally precise case by case, but it will always mean something of the kind "under the assumptions recorded in $\beta$ the interpretation of M depends on the type j". This is both an "endogenous logic" and a "logical" presentation or "finitary" presentation of the structure, in the sense of Abramsky, [1].

In order to define the interpretation function in the Plotkin-Engeler set theoretical model, in the Intersection Algebra and in the Filter Model we shall utilize judgements of the form $\beta \vdash M:j$ where the bases $\beta$ are sets i.e. functions $\beta \in [(\text{Var} \times \mathbf{J}) \xrightarrow{o} 2]$.

**Definition 4.**
1) Consider the following set of rules:

(axiom)     $\dfrac{}{\{(x,j):1\} \vdash x:j}$     provided $j \in \mathbf{J}$

(abstraction)     $\dfrac{\beta + \{(x,j_1):1,\ldots,(x,j_k):1\} \vdash M:j}{\beta \vdash \lambda x.M:(\{j_1:1,\ldots,j_k:1\},j)}$     provided $\forall j\, \beta((x,j)) = 0$

(application)     $\dfrac{\beta \vdash M:(\{j_1:1,\ldots,j_k:1\},j) \quad \beta_1 \vdash N:j_1 \ldots \beta_k \vdash N:j_k}{\beta + \sum_{1 \leq i \leq k} \beta_i \vdash MN:j}$

(weakening)     $\dfrac{\beta \vdash M:j}{\beta + \beta' \vdash M:j}$

($\eta$)     $\dfrac{\beta \vdash \lambda x.Mx:j}{\beta \vdash M:j}$     provided $x \notin FV(M)$

boolean operations are extended pointwise to bases.
2) Let $S_1$ be the type assignment system consisting of the rules{(axiom), (abstraction), (application)}, $S_2$ be the type assignment system consisting of the rules {(axiom), (abstraction), (application), (weakening)} and $S_3$ be the type assignment system consisting of the rules{(axiom), (abstraction), (application), (weakening), ($\eta$)}.
3) Let the interpretation functions $[\![\ ]\!]^i$: Env $\to (\Lambda \to \mathcal{F}^i)$, ($i \in \{1,2,3\}$), be defined as

$$[\![ M ]\!]^i_\rho(j) \equiv \sum_{\beta \vdash_i M:j} \left( \prod_{k \in \text{dom}(\beta)} (\beta(k) \Rightarrow \rho(k)) \right)$$

where we write $\beta \vdash_i M:j$ to indicate that the judgment $\beta \vdash M:j$ can be derived in the system $S_i$ ($i \in \{1,2,3\}$); boolean operations are extended pointwise to environments which are taken to be functions $\rho : (\text{Var} \times \mathbf{J}) \to 2$; the expression $\sum_{\beta \vdash_i M:j}$ denotes the disjunction over provable judgments in $S_i$; and finally $\mathcal{F}^1$ and $\mathcal{F}^2$ are $\mathcal{F}$ of Definition 3 above, while $\mathcal{F}^3$ is the set of filters on $\mathbf{J}$.     $\Delta$

The definition above illustrates how type assignment systems can be used to define in a finitary way interpretation functions. The proposition below illustrates what we have achieved so far.

**Proposition 2.** The combinatory structure $\langle \mathcal{F}^1, [\![\ ]\!]^1 \rangle$ is the intersection algebra of [5], the combinatory structure $\langle \mathcal{F}^2, [\![\ ]\!]^2 \rangle$ is Plotkin-Engeler's lambda model while $\langle \mathcal{F}^3, [\![\ ]\!]^3 \rangle$ is the Filter Model [3].

**Proof.** A tedious but routine verification that conditions in Definition 1 are satisfied.     $\Delta$

Once we have presented Plotkin-Engeler's model in the format $\langle \mathcal{F}^2, [\ ]^2 \rangle$ it is natural to generalize the role of the Boolean Algebra **2** in the construction of Definiton 3 to an arbitrary boolean algebra B. What we obtain is then an entirely new class of combinatory structures. This kind of construction is very closely related to that which would arise if we carried out the construction of $\langle \mathcal{F}, \circ \rangle$ in a Boolean-valued universe of Set Theory: i.e. using boolean sets instead of ordinary sets. We cannot follow up this here. We give just appropriate generalizations of Definitions 3,4 and Proposition 2 to the case of an arbitrary complete boolean algebra.

**Definition 5.** Let B be a complete boolean algebra and $A_o$ an arbitrary sets of constants
1) the combinatory structure $\langle \mathcal{F}_B, \circ \rangle$ is defined as follows:
let $J_B: \mathbb{N} \to \text{Set}$ be inductively defined by

$J_B(0) \equiv A_o$ and $J_B(n+1) \equiv J_B(0) \cup ([J_B(n) \xrightarrow{o} B] \times J_B(n))$

now put $J_B \equiv \cup_n J_B(n)$ and let $\mathcal{F}_B \equiv [J_B \to B]$, application over $\mathcal{F}_B$ is defined by

$$(f \circ g)(j) \equiv \sum_{r \in [J_B \xrightarrow{o} B]} f((r,j)) \times \prod_{k \in \text{dom}(r)} (r(k) \Rightarrow g(k))$$

where $f, g \in \mathcal{F}_B$ and $j \in J_B$.
2) Let $S_{B1}$ be the type assignment system consisting of the following rules and:

(axiom) $$\frac{}{\{(x,j):1\} \vdash x:j}$$

(abstraction) $$\frac{\beta + \{(x,j_1):b_1,\ldots,(x,j_k):b_k\} \vdash M:j}{\beta \vdash \lambda x.M:(\{j_1:b_1,\ldots,j_k:b_k\},j)} \quad \text{provided } \forall j\ \beta((x,j)) = 0$$

(application) $$\frac{\beta \vdash M:(\{j_1:b_1,\ldots,j_k:b_k\},j) \quad \beta_1 \vdash N:j_1 \ldots \beta_k \vdash N:j_k}{\beta + \sum_{1 \le i \le k}(b_i \times \beta_i) \vdash MN:j}$$

(weakening) $$\frac{\beta \vdash M:j}{\beta + \beta' \vdash M:j}$$

where $j \in J_B$ and $b \in B$;
3) let $S_{B2}$ be the subsystem obtained from $S_{B1}$ by omitting (weakening);
4) The interpretation functions $[\![\ ]\!]^{Bi}$: Env $\to (\Lambda \to \mathcal{F}_B)$ ($i \in \{1,2\}$), are defined by:

$$[\![ M\sigma ]\!]^{Bi}_\rho(j) \equiv \sum_{\beta \vdash_{Bi} M:j} \left( \prod_{k \in \text{dom}(\beta)} (\beta(k) \Rightarrow \rho(k)) \right). \qquad \Delta$$

**Proposition 3**
i) The combinatory structures $\langle\langle \mathcal{F}_B, \circ \rangle, [\![\ ]\!]^i \rangle$ ($i \in \{1,2\}$) are combinatory algebras;
ii) If B is not trivial then $\langle\langle \mathcal{F}_B, \circ \rangle, [\![\ ]\!]^i \rangle$ ($i \in \{1,2\}$) are not lambda algebra

**Proof.** i) Omitted, since it is a special case of the proof of Proposition 5 below.
ii) Let $B \equiv \lambda xyz.x(yz)$ denote the usual composition combinator, one has immediately that $\vdash_\lambda \lambda yz.Bx(Byz) = \lambda yz.B(Bxy)z$ i.e. compostion of functions is associative. A tedious computation shows that $[\![ \lambda xyz.Bx(Byz) ]\!]^i \ne [\![ \lambda xyz.B(Bxy)z ]\!]^i$. $\qquad \Delta$

It is interesting to notice that the role played by **J** in the above Proposition is again akin to an "intersection type" to whom a "boolean weight" has been attached. The corresponding type assignment system is then built so as to take into account also this non-standard "weight". Boolean coefficients appear in the hypotheses and consequently, the structural rule of *contraction* is replaced by a sort of *boolean contraction*.. This "boolean set" construction however, does not even give rise to a category. We conjecture that in order to turn the structures $\langle\langle \mathscr{F}_B, \circ\rangle, [\![\ ]\!]^i\rangle$ ($i \in \{1,2,\}$) into $\lambda$-algebras we need to define a suitable quotient.

So far we have only considered examples where sets, be these ordinary or boolean, were used in defining types, i.e. elements of **J**. Yet more examples of combinatory structures can be obtained by "repeating" Plotkin-Engeler's construction using multisets in the definition of types. Surprisingly enough this construction amounts to the construction carried out by Ore [12]. This in turn was introduced as a simplified version of the notion due to Girard of *quantitative domain* [8]. Loosely speaking Girard's construction corresponds to Plotkin-Engeler's construction using arbitrary set-valued functions in place of multisets. Here we will analyze in detail only the multiset case.

Ore's notion of domain can be put into our framework by replacing the boolean algebra B in Definition 5 with $\mathbb{N}$, the set of natural numbers. Of course, the boolean operations which appear in the definiton of application and interpretation, which in turn generalized the simple set theoretic concepts of Plotkin-Engeler's algebra, have to be replaced here with arithmetic operation on natural number, i.e. disjunction with addition, conjunction with multiplication and logical implication with exponentiation. It comes almost as a surprise that under this twist of perspective, Girard-Ore's construction can be naturally related to Plotkin-Engeler's. Notice the close similarity between the following definition and Definition 5.

**Definition 6.** Let $A_o$ an arbitrary sets of atoms, the combinatory structure $\langle \mathscr{F}_N, \circ \rangle$ is defined as follows: let $J_N: \mathbb{N} \to$ Set be inductively defined by

$J_N(0) \equiv A_o$ and $J_N(n+1) \equiv J_N(0) \cup ([J_N(n) \xrightarrow{o} \mathbb{N}] \times J_N(n))$

put $J_N \equiv \cup_n J_N(n)$ and let $\mathscr{F}_N \equiv [J_N \to (\mathbb{N} \cup \{\infty\})]$, application over $\mathscr{F}_N$ is defined by

$$(f \circ g)(j) \equiv \sum_{r \in [J_N \xrightarrow{o} \mathbb{N}]} f((r, j)) \times \prod_{k \in \text{dom}(r)} (r(k) \Rightarrow g(k))$$

where $f, g \in \mathscr{F}_N$, $j \in J_N$, arithmetic operations are extended to $\mathbb{N} \cup \{\infty\}$ in the obvious way and $\Rightarrow$ is the usual operation of exponentiation. $\triangle$

The above definition can be easily modified to encompass the case of a notion of domain intermediate between that of Girard's quantitative domain and Ore's domain. This notion of domain is obtained using the set *Card* of cardinals in place of the set ($\mathbb{N} \cup \{\infty\}$). In order to avoid the use of proper classes we can always think of Card as the set of cardinals smaller than a given inaccessible cardinal. The structure $\langle \mathscr{F}_N^C, \circ \rangle$ is then obtained taking $\mathscr{F}_N^C$ to be $[J_N \to Card]$, the definition of application remaining unchanged. As remarked in the introduction, presentations in [12],[8] and[11] rely heavily on the notion of analytic function. In fact both in $\langle \mathscr{F}_N^C, \circ \rangle$ and $\langle \mathscr{F}_N, \circ \rangle$ we can interpret the formula defining the application as the evaluation of an analytic function. See Appendix A for a brief illustration of this alternative viewpoint.

Going back to the structure in Definition 6, the interpretation of a $\lambda$-term M with respect to $\langle \mathscr{F}_N, \circ \rangle$ can be given again following the familiar pattern using a type assignment system. Again, in fact , elements of $J_N$ can play the role of generalized "intersection types". The coefficients being now integers. In this case however it is slightly more complex. In order to define the interpretation function it is necessary to introduce an equivalence relation on proofs of typing judgements to take care of multiplicities.

**Definition 7.**
1) The type assignment system N consists of the following rules:

(axiom)
$$\frac{}{\{(x,j):1\} \vdash x:j}$$

(abstraction)
$$\frac{\beta + \{(x,j_1):n_1,\ldots,(x,j_k):n_k\} \vdash M:j}{\beta \vdash \lambda x.M:(\{j_1:n_1,\ldots,j_k:n_k\},j)} \quad \text{provided } \forall j \, \beta((x,j)) = 0$$

(application)
$$\frac{\beta \vdash M:(\{j_1:n_1,\ldots,j_k:n_k\},j) \quad \beta_{i;t_i} \vdash N:j_i \quad 1 \leq i \leq k \quad 1 \leq t_i \leq n_i}{\beta + \sum_{\substack{1 \leq i \leq k \\ 1 \leq t_i \leq n_i}} \beta_{i;t_i} \vdash MN:j}$$

Here a basis $\beta$ can be seen as a finite multiset of hypothesis of the form x:j, accordingly arithmetic operations are extended to bases in the natural way.

2) The equivalence relation $\equiv$ on proofs of typing judgements is the finest equivalence relation satisfying the following two conditions:
a) $\equiv$ is a congruence relation on the structure of the proof, i.e.:

$$\Delta_i \equiv \Delta'_i \Rightarrow \frac{\Delta_1\ldots\Delta_i\ldots\Delta_n}{\Delta} \equiv \frac{\Delta_1\ldots\Delta'_i\ldots\Delta_n}{\Delta}$$

b) For every permutation $\sigma$ of the set $\{1,\ldots,k\}$

$$\frac{\beta \vdash M:(\{j_1:n_1,\ldots,j_k:n_k\},j) \quad \beta_{i;t_i} \vdash N:j_i \quad 1 \leq i \leq k \quad 1 \leq t_i \leq n_i}{\beta + \sum_{\substack{1 \leq i \leq k \\ 1 \leq t_i \leq n_i}} \beta_{i;t_i} \vdash MN:j} \equiv \frac{\beta \vdash M:(\{j_1:n_1,\ldots,j_k:n_k\},j) \quad \beta_{\sigma(i);t_{\sigma(i)}} \vdash N:j_{\sigma(i)} \quad 1 \leq i \leq k \quad 1 \leq t_i \leq n_i}{\beta + \sum_{\substack{1 \leq i \leq k \\ 1 \leq t_i \leq n_i}} \beta_{i;t_i} \vdash MN:j}$$

3) The interpretation of a $\lambda$-term M in the applicative structure $\langle \mathscr{F}_N, \circ \rangle$ is defined by:

$$[\![ M^\sigma ]\!]^N \rho(j) = \sum_\beta \rho^\beta \times \sum_{\Delta \in \{[\beta \vdash M:j]\}} 1 = \sum_\beta \sum_{\Delta \in \{[\beta \vdash M:j]\}} \rho^\beta$$

where we use the abbreviation $g^\beta$ for $\prod_{k \in \text{dom}(\beta)} (\beta(k) \Rightarrow g(k))$ ; $[\Delta]$ denotes the equivalence class modulo $\equiv$ of $\Delta$ and $\{[G]\}$ denotes the set of equivalence classes of proofs having the judgement G as conclusion. $\triangle$

The definition of the equivalence $\equiv$ between proofs can be motivated as follows. When we apply the rule (application) we must fix an order on the domain of the function $\{(j_1:n_1)\ldots(j_k:n_k)\}$. This order is completely arbitrary. The condition b) in Definition 7.2. sets two proofs to be equivalent if they differ just up to the order chosen on the domain of the function $\{(j_1:n_1)\ldots(j_k:n_k)\}$.
We are now ready to establish the properties of $\langle \mathscr{F}_N, \circ \rangle$.

**Proposition 4.** The combinatory structure $\langle \langle \mathscr{F}_N, \circ \rangle, [\![ \; ]\!]^N \rangle$ is a lambda algebra but not a lambda model.
**Proof.** The proof that $\langle \langle \mathscr{F}_N, \circ \rangle, [\![ \; ]\!]^N \rangle$ is a combinatory algebra is given in Appendix B.

For lack of space we omit the routine proof that it is a λ-algebra, see[12]. The following counterexample shows that the ξ-rule fails in $\langle\langle \mathscr{F}_N, \circ\rangle, [\![\ ]\!]^N\rangle$: let $\rho(x)=((\{j_1:1\},j):\infty)$ and $\rho(y)=((\{j_1:2\},j):\infty)$, now $\forall a.[\![xz]\!]_{\rho[z/a]}=[\![yz]\!]_{\rho[z/a]}$ but $[\![\lambda z.xz]\!]_\rho=[\![\lambda z.yz]\!]_\rho$.   △

## 4 The General Case

After having gone through various different examples in the previous section, we are now ready to introduce the general notion of *applicative exponential structure*. This notion subsumes all the concrete notions of application presented up to now. It denotes a general kind of structure where function application can be adequately defined via a type-assignment system. It arises from the abstract characterizations of the structure of the coefficients which are applied either to the "types" or to the "points" of the domains in the previous examples. It turns out infact, that two different kinds of coefficients are actually involved in the construction; a fact this, which was never apparent in the previous examples.

**Definition 8.** A *preexponential structure* consists of a triple $\langle A, E, \Rightarrow \rangle$ where:
1) $A \equiv \langle A, +_A, \times_A, 0_A, 1_A \rangle$ is an infinitary commutative semiring, i.e. an algebraic structure where $+_A$ is an infinitary commutative, associative operation over A, with identity $0_A$, and $\times_A$ is an associative, commutative binary operation over A which distributes over $+_A$, with identity $1_A$.
2) $E \equiv \langle E, +_E, \times_E, 0_E, 1_E \rangle$ is a commutative semiring, i.e. $+_E$ is a binary associative and commutative operation over E with identity $0_E$, and $\times_E$ is an associative, commutative binary operation over E which distributes over $+_E$, with identity $1_E$.
3) there is a binary operation $\Rightarrow : E \times A \to A$ satisfying the following axioms:
   a) $(e_1 +_E e_2) \Rightarrow a = (e_1 \Rightarrow a) \times_A (e_2 \Rightarrow a)$
   b) $e \Rightarrow (a_1 \times_A a_2) = (e \Rightarrow a_1) \times_A (e \Rightarrow a_2)$
   c) $(e_1 \Rightarrow (e_2 \Rightarrow a)) = e_1 \times_E e_2 \Rightarrow a$
   d) $0_E \Rightarrow a = 1_A$  and  $1_E \Rightarrow a = a$   △

As will become clear in the following preexponential structures will be the abstract coefficients of our genral notion of applicative structure. Elements of E will be the "weights" of the "types" while elements of A will be the coefficients of the points.

Condition 3) above shows that the function $\Rightarrow$ satisfies essentially the properties of an exponential function. In the proof of Proposition 4 an essential property of exponentiation over natural numbers is crucial: Newton's binomial expansion. The corresponding identity over booleans, necessary for proving Proposition 3, on the other hand is trivial. The notion of *exponential structure* below, is the appropriate abstract setting for carrying out the analogue of the "binomial expansion" over a preexponential structure, where elements of E play the role of exponents (whence the name) and elements of A play the role of bases. Subscripts are omitted.

**Definition 9.** Let $\langle A, E, \Rightarrow \rangle$ be a *preexponential structure*.
1) An element e of E is called *unitary* if $e \Rightarrow \sum_{j\in J} a_j = \sum_{j\in J} (e \Rightarrow a_j)$ holds for all $\sum_{j\in J} a_j$. The set of unitary elements of E is denoted with $U_E$.
2) Given $e \in E$, a function $f \in \mathbb{N} \xrightarrow{o} U_E \cup \{0\}$ is a *unitary decomposition* of e if $e = \sum_{j\in dom(f)} f(j)$. Given $e \in E$ the set of all unitary decompositions of e is denoted with $U(e)$
3) An *exponential structure* is a preexponential structure $\langle A, E, \Rightarrow \rangle$ together with a function
$H : E \to ([\mathbb{N} \xrightarrow{o} U_E \cup \{0\}])$ which satisfies the following two axioms:
   Ax1) For each $e \in E$ there is a unitary decomposition of E;
   Ax2) The function H chooses a unitary decomposition for each element of E    △

This is the least intuitive definition among the ones given so far. But its complexity is rewarding. We can now safely use exponential structures as possible coefficients in the machinery that we put to work in the previous section for defining concrete combinatory algebras.

**Definition 10.** Given an exponential structure $\langle A, E, \Rightarrow, H \rangle$ and an arbitrary set of constants $C_o$, an *applicative exponential structure*, a.e.s. for short, over $\langle A, E, \Rightarrow, H \rangle$ is the c.s $\langle\langle \mathscr{F}_E^A, \circ \rangle, [\![\ ]\!]\rangle$ defined as follows:

define inductively $J_E: \mathbb{N} \to \text{Set}$ by $J_E(0) \equiv C_o$ and
$J_E(n+1) \equiv J_E(0) \cup (([J_E(n) \xrightarrow{o} E] \times J_E(n));$
put $J_E \equiv \cup_n J_E(n)$ and $\mathscr{F}_E^A = [J_E \to A]$, and, for f, $g \in \mathscr{F}_E^A$ and $j \in J_E$, let application over $\mathscr{F}_E^A$ be defined by $(f \circ g)(j) \equiv \sum_{r \in [J_E \xrightarrow{o} \mathbb{N}]} f((r,j)) \times g^r$.

The interpretation function $[\![\ ]\!] : \text{Env} \to (\Lambda \to \mathscr{F}_E^A)$ is defined via the type assignment system $S_E^A$ by: $[\![ M ]\!]_\rho(j) = \sum_\beta \rho^\beta \times \sum_{\Delta \in \{[\beta \vdash M:j]\}} 1 = \sum_\beta \sum_{\Delta \in \{[\beta \vdash M:j]\}} \rho^\beta$

for $j \in J_E$ and $M \in \Lambda$.

The system $S_E^A$ consists of the following three rules:

(axiom) $\dfrac{}{\{(x, j): 1\} \vdash x:j}$ provided $j \in J_E$

(abstraction) $\dfrac{\beta + \{(x,j_1):e_1,\ldots,(x,j_k):e_k\} \vdash M:j}{\beta \vdash \lambda x.M:(\{j_1:e_1,\ldots,j_k:e_k\},j)}$ provided $\forall j\ \beta((x,j)) = 0$

(application) $\dfrac{\beta \vdash M:(\{j_1:e_1,\ldots,j_k:e_k\},j) \quad \beta_{i:t_i} \vdash N:j_i \quad 1 \leq i \leq k \quad t_i \in \text{dom}(H(e_i))}{\beta + \sum_{\substack{1 \leq i \leq k \\ t_i \in \text{dom}(H(e_i))}} H(e_i)(t_i) \times \beta_{i:t_i} \vdash MN:j}$

The equivalence relation on proofs used in the definition of $[\![\ ]\!]$ is that of Definition 7 and the abbreviations $g^r$, $\rho^\beta$ are those of Definition 7. $\Delta$

The following Proposition is the main result of the paper.

**Proposition 5.** Applicative exponential structures are combinatory algebras.

**Proof.** The proof is very similar to the one given in Appendix B. $\Delta$

One can easily check that all the constructions in Section 3, fall under the above general definition. The only non-trivial issue is the choice of the function H in the definition of the exponential structure. Whenever E is instantiated by $\mathbb{N}$ there is only one choice possible. Whenever E is a boolean algebra the choice is immaterial. Moreover in the latter case the summation over equivalence classes of proofs, in the definition of the interpretation function, is irrelevant.

## 5  Applications and Directions for Future Works

The general notion of applicative exponential structure suggests other notions of function application and hence other interesting constructions of concrete combinatory algebras. Here are few examples.

1) Lamarche [11] discusses at length the case of an a.e.s. $\langle\langle \mathcal{F}_N^A, \circ \rangle, [\![\ ]\!] \rangle$ where $\mathbf{A}$ is a complete Heyting algebra. This is a slight variation of Ore's construction. In this case we have to define the exponential $(\Rightarrow) : \mathbb{N} \times \mathbf{A} \to \mathbf{A}$ by:

$$n \Rightarrow a := \begin{cases} a \times \ldots \times a & \text{if } n \geq 1 \\ 1 & \text{if } n = 0 \end{cases}$$

2) Using the set of positive rational numbers $\mathbf{Q}^+$, or the set of positive real numbers $\mathbb{R}^+$, we can form the exponential algebras $\langle \mathbf{Q}^+ \cup \{\infty\}, \mathbb{N}, \Rightarrow, U \rangle$ and $\langle \mathbb{R}^+ \cup \{\infty\}, \mathbb{N}, \Rightarrow, U \rangle$ where $\mathbb{N}$ is the set of natural numbers, addition and multiplication are the standard ones and $\Rightarrow$ is the usual exponential, the function U decomposes each natural number n in a sum of 1's. In the corresponding λ-algebras $\langle\langle \mathcal{F}_N^{Q^+\cup\{\infty\}}, \circ \rangle, [\![\ ]\!] \rangle$ and $\langle\langle \mathcal{F}_N^{R^+\cup\{\infty\}}, \circ \rangle, [\![\ ]\!] \rangle$ the points can be interpreted as formal power series, i.e. the description of analytic functions either on positive rational or on positive real numbers. More examples of this kind can be defined by restricting the range of the relevant operators to suitable intervals.

3) Given any complete Heyting algebra G and indicating with $\text{In}_G$ the set of invertible elements in G, $\text{In}_G = \{a \in G \mid \exists \bar{a}.\ a+\bar{a}=1, a\times\bar{a}=0\}$, $\langle G, \text{In}_G, \Rightarrow \rangle$ is an exponential algebra. In this case the function I decomposes every element $a$ of $\text{In}_G$ in the a sum containing the single element $a$. Combinatory algebras $\langle\langle \mathcal{F}_{\text{In}_G}^G, \circ \rangle, [\![\ ]\!] \rangle$ generated as in Definition 10 by this kind of exponential algebras generalize the boolean c.s. of Definition 6.

4) Let G be a finite Heyting algebra then $\langle G, G, \Rightarrow, H \rangle$ is an exponential algebra for an appropriate choice of H, which always exists. The c.s. $\langle\langle \mathcal{F}_G^G, \circ \rangle, [\![\ ]\!] \rangle$ generated as in Definition 10 is yet another example of a combinatory algebra.

The ideas outlined in this paper need to be investigated further. First of all one can try to strengthen the conditions in the definition of a.e.s. so as to obtain always lambda algebras. In another direction one can try to define a coherence predicate on the elements of $\mathbf{J}$ so as to be able to subsume notions of domain which involve stable functions. Finally one should explore the relation between the notions of domain arising in this setting, which for instance, are not necessarily ω-algebraic, and those which are normally used in connection with Scott Domains. An abstract notion of implication between elements of $\mathbf{J}$ can be possibly introduced, which could be used to introduce a general notion of filter.

The structures introduced in this paper can turn out to be quite useful from the point of view of programming language semantics. For example, one can easily get a plethora of different denotational semantics for a simple functional language featuring a non-detrministic **or** operator. For any particular applicative exponential structure based on the exponential structure $\langle A, E, \Rightarrow \rangle$, one can give a denotation to the non-deterministic **or** in terms of the operators $+_A$ and $\times_A$. One can take it to be, for instance, an operation $[\![\mathbf{or}]\!]:[\mathbf{J} \to \mathbf{A}]^2 \to [\mathbf{J} \to \mathbf{A}]$ defined by applying pointwise on $\mathbf{J}$ a suitable weighted average. The intuition behind this is that the meaning of **or** is that of evaluating either the left hand with a suitable weight or the right hand with another weight. According to the particular choice made one gets different flavours of non-deterministic operators. Some of these are interesting in themselves and can illuminate our intuition of non-determinism. For example, applicative exponential structures based on boolean sets, where weights are thought of as sets of favorable events, yield a kind of non-determinsm which is settled once and forall before the computation is started; the resulting coefficient being the set of favorable events for a given result of a computation. Semantics based on multisets are more directly related to the frequency with which a given result is obtained following different computations, see [8,12]. Semantics based on real valued sets are finally closer to real probabilities.

## References

1. S.Abramsky: Domain Theory in Logical Form. Annals of Pure and Applied Logic, (1991)
2. H.Barendregt: Lambda Calculus: its Syntax and Semantics revised version. Studies in Logic. Amsterdam: North Holland 1984
3. H.Barendregt, M.Coppo, M.Dezani Ciancaglini: A Filter Lambda Model and the Completeness of Type Assignment. Journal fo Symbolic Logic, 48, 4 (1983)
4. M.Coppo, M.Dezani Ciancaglini, F.Honsell, G.Longo: Extended Type Structures and Filter Lambda Models. In: G.Longo et al. (eds.): Logic Colloquium '82. Amsterdam: North Holland 1983
5. M.Coppo, M.Dezani Ciancaglini, B.Venneri: Principal Type Schemes and Lambda Calculus Semantics. In: J.Seldin et al. (eds): To H.B.Curry: Essays. Academic Press 1980
6. P.Di Gianantonio, F.Honsell: A General Type Assignment System for an Abstract Notion of Domain. Talks given at the 4th and 6th Meetings of the Jumelage Typed Lambda Calculus. Edinburgh, October 1989 and Paris, January 1991
7. E.Engeler: Algebras and Combinators. Berichte des Instituts f. Informatik 32, ETH, Zurich 1979
8. J.Y.Girard: Normal Functors Power Series and Lambda Calculus. Annals of Pure and Applied Logic, 37, 2 (1988)
9. R.Hindley, G.Longo: Lambda Calculus Models and Extensionality. Zeit. f. Math. Logik u. Grund. d. Math., 26 (1980)
10. F.Honsell, S.Ronchi della Rocca: Reasoning about interpretations in qualitative Lambda Models. In: M.Broy et al. (eds.) Programming Concepts and Methods. 1990
11. F.Lamarche: Quantitative Domains and Infinitary Algebras. Unpublished manuscript, 1990
12. Ch.-E.Ore: Introducing Girard's quantitative domains. PhD Thesis, Research Report 113. University of Oslo 1988
13. G.Plotkin: A set-theoretical definition of application. Memorandum MIP-R-95, School of Artificial Intelligence, University of Edinburgh, 1972
14. G.Plotkin: Domains for Denotational Semantics, course notes, Stanford 1985
15. D.Scott: Some philosophical issues concerning theories of combinators, lambda calculus and computer science theory. In LNCS 37, Springer Verlag, 1975
16. D.Scott: Data Types as Lattices. SIAM Journal of computing, 5 (1976)

## Appendix A

Using the notation introduced in Section 3, we outline, in the 7 points below, the alternative viewpoint under which applicative exponential structure can be considered. This viewpoint focuses on the notion of formal power series and analytic function as a justification for the definition of application in $\langle \mathscr{F}_N^C, \circ \rangle$ and $\langle \mathscr{F}_N, \circ \rangle$.

1) $J_N$ is taken to be a set of variables. ( In the following $J_N$ will be denoted simply by $J$.)

2) any $r \in [J \xrightarrow{o} N]$ is viewed as a monomial in the variables $J$:
i.e. r corresponds to the monomial $\prod_{k \in \text{dom}(r)} k^{r(k)}$

3) an element g∈ $\mathscr{F}_N \equiv [J \to (\mathbb{N} \cup \{\infty\})]$ is viewed as a vector consisting of **J** components;
the value $\prod_{k \in \text{dom}(r)} (r(k) \Rightarrow g(k))$ is then the result of the evaluation of the monomial r with respect to the vector g;

4) a function $h: (J \overset{o}{\to} \mathbb{N}) \to (\mathbb{N} \cup \{\infty\})$ assigns a coefficient to every monomial and therefore determines a formal power series in the variables **J** : i.e.

$$\sum_{r \in [J_N \overset{o}{\to} \mathbb{N}]} h(r) \times \prod_{k \in \text{dom}(r)} k^{r(k)} ;$$

5) a function $h' : J \to ([J \overset{o}{\to} \mathbb{N}] \to (\mathbb{N} \cup \{\infty\}))$ associates a formal power series to each variable in **J** and is therefore a vector valued formal power series. These functions are usually called analytic;

6) the following isomorphisms hold:
$\mathscr{F}_N \equiv J \to (\mathbb{N} \cup \{\infty\}) \equiv (J(0) \cup ((J \overset{o}{\to} \mathbb{N}) \times J)) \to (\mathbb{N} \cup \{\infty\}) \equiv$
$\equiv (J(0) \to (\mathbb{N} \cup \{\infty\})) \times (((J \overset{o}{\to} \mathbb{N}) \times J) \to (\mathbb{N} \cup \{\infty\})) \equiv$
$\equiv (J(0) \to (\mathbb{N} \cup \{\infty\})) \times (J \to ((J \overset{o}{\to} \mathbb{N}) \to (\mathbb{N} \cup \{\infty\})))$

therefore an element $f \in \mathscr{F}_N$ is a vector of **J** components and containes the representation of a vector valued formal power series;

7) f ∘ g is the vector obtained applying the analytic functions described by f to the vector g.

## Appendix B

Proof of Proposition 4.
First we need a lemma.

**Lemma.** The set $\{[\beta \vdash MN : j]\}$ can be decomposed in disjoint singletons in the following way:

$$\{[\beta \vdash MN:j]\} = \bigcup_{\substack{\alpha = \{(j_1:n_1)...(j_k:n_k)\}}} \bigcup_{\substack{\beta' \\ \beta_{1;1}...\beta_{1;n_1} \\ \vdots \\ \beta_{k;1}...\beta_{k;n_k} \\ \beta = \beta' + \beta_{1;1} + ... + \beta_{k;nk}}} \bigcup_{\substack{\Delta' \in \{[\beta \vdash M:j]\} \\ \Delta_{1;1}...\Delta_{1;n_1} \\ \vdots \\ \Delta_{k;1}...\Delta_{k;nk} \\ \Delta_{i;h} \in \{[\beta_{i;h} \vdash N:j_i]\}}} \left\{ \left[ \frac{\Delta' \Delta_{1;1}...\Delta_{k;nk}}{\beta \vdash MN:j} \right] \right\}$$

**Proof** (lemma). The above formula just enumerates all possible equivalence classes of proofs of judgments $\beta \vdash MN:j$. It is not difficult to see that the formula is correct observing that:
1) in all the proofs of $\beta \vdash MN:j$ the last rule applied is the application rule.
2) in the formula we consider just one of the possible orderings of the domain of the function $\alpha = \{(j_1:n_1)...(j_k:n_k)\}$. △

**Proof** (Proposition)
*i* ) We must prove $[\![ x ]\!]_\rho = \rho(x)$, by definition we have:

$[\![ x ]\!]_\rho(j) = \sum_\beta \sum_{\Delta \in \{[\beta \vdash x:j]\}} \rho^\beta$

the only proofs of judgments of the form: $\beta \vdash x:j$ are given by an application of the (axiom)-rule $\dfrac{}{\{(x,j):1\} \vdash x:j}$ , so $[\![ x ]\!]_\rho(j) = \rho^{\{(x,j),\,1\}} = (\rho(x,j))^1 = \rho(x)(j)$.

*ii*) We must prove $[\![ MN ]\!]_\rho(j) = ([\![M]\!]_\rho \circ [\![N]\!]_\rho)(j)$; expanding the term on the left of this equality we have:

$[\![ MN ]\!]_\rho(j) = \sum_\beta \sum_{\Delta \in \{[\beta \vdash MN:j]\}} \rho^\beta = $ (by the previous lemma)

$= \sum_\beta \sum_{\substack{\alpha = \{(j_1:n_1)\ldots(j_k:n_k)\}}} \sum_{\substack{\beta' \\ \beta_{1;1}\ldots\beta_{1;n1} \\ \vdots \\ \beta_{k;1}\ldots\beta_{k;nk} \\ \beta=\beta'+\beta_{1;1}+\ldots+\beta_{k;nk}}} \sum_{\substack{\Delta' \in \{[\beta \vdash M:j]\} \\ \Delta_{1;1}\ldots\Delta_{1;n1}\ldots\Delta_{k;1}\ldots\Delta_{k;nk} \\ \Delta_{i;h} \in \{[\beta_{i;h} \vdash N:j_i]\}}} \rho^\beta =$

$= \sum_{\substack{\alpha = \{(j_1:n_1)\ldots(j_k:n_k)\}}} \sum_{\substack{\beta' \\ \beta_{1;1}\ldots\beta_{1;n1}\ldots\beta_{k;1}\ldots\beta_{k;nk}}} \sum_{\substack{\Delta' \in \{[\beta \vdash M:j]\} \\ \Delta_{1;1}\ldots\Delta_{1;n1} \\ \vdots \\ \Delta_{k;1}\ldots\Delta_{k;nk} \\ \Delta_{i;h} \in \{[\beta_{i;h} \vdash N:j_i]\}}} \rho^{\beta' + \beta_{1;1}+\ldots+\beta_{k,nk}}$

Expanding the term on the right we obtain:

$([\![M]\!]_\rho \circ [\![N]\!]_\rho)(j) = \sum_{\alpha = \{(j_1:n_1)\ldots(j_k:n_k)\}} [\![M]\!]_\rho(\alpha,j) \times ([\![N]\!]_\rho(j_1))^{n_1} \times \ldots \times ([\![N]\!]_\rho(j_k))^{n_k} =$

$= \sum_{\alpha = \{(j_1:n_1)\ldots(j_k:n_k)\}} \left( \sum_\beta \sum_{\Delta \in \{[\beta \vdash M:(\alpha;j)]\}} \rho^\beta \right) \times \left( \sum_\beta \sum_{\Delta \in \{[\beta \vdash N:(j_1)]\}} \rho^\beta \right)^{n_1} \times \ldots$

$\ldots \times \left( \sum_\beta \sum_{\Delta \in \{[\beta \vdash N:(j_k)]\}} \rho^\beta \right)^{n_k} =$

$= \sum_{\substack{\alpha = \{(j_1:n_1)\ldots(j_k:n_k)\}}} \sum_{\substack{\beta' \\ \beta_{1,1}\ldots\beta_{1,n1}\ldots\beta_{k,1}\ldots\beta_{k,nk}}} \sum_{\substack{\Delta' \in \{[\beta \vdash M:j]\} \\ \Delta_{1;1}\ldots\Delta_{1;n1} \\ \vdots \\ \Delta_{k;1}\ldots\Delta_{k;nk} \\ \Delta_{i;h} \in \{[\beta_{i;h} \vdash N:j_i]\}}} \rho^{\beta'} \times \rho^{\beta_{1;1}} \times \ldots \times \rho^{\beta_{k,nk}}$

and this expression is equal to what we obtained before expanding the other term.

*iii*) We must prove that $([\![ \lambda x.M ]\!]_\rho \circ a)(j) = [\![M]\!]_{\rho[a/x]}(j)$; by definition we have the following equalities,

$([\![ \lambda x.M ]\!]_\rho \circ a)(j) = \sum_\alpha ([\![ \lambda x.M ]\!]_\rho(\alpha;j)) a^\alpha =$

$$= \sum_\alpha \left( \sum_\beta \sum_{\Delta \in \{[\beta \vdash \lambda x.M:(\alpha,j)]\}} \rho^\beta \right) \times a^\alpha =$$

$$= \sum_\alpha \left( \sum_\beta \sum_{\Delta \in \{[\beta+x:\alpha \vdash M:j] \mid x \notin \beta\}} \rho^\beta \right) \times a^\alpha = \quad \text{(using distributivity)}$$

$$= \sum_\alpha \sum_\beta \sum_{\Delta \in \{[\beta+x:\alpha \vdash M:j] \mid x \notin \beta\}} \rho^\beta \times a^\alpha =$$

$$= \sum_\alpha \sum_\beta \sum_{\Delta \in \{[\beta+x:\alpha \vdash M:j] \mid x \notin \beta\}} \rho[a/x]^{\beta+\{x:\alpha\}} =$$

$$= \sum_{\beta'} \sum_{\Delta \in \{[\beta' \vdash M:j]\}} \rho[a/x]^{\beta'} = [\![ M ]\!]_{\rho[a/x]}(j) \quad \text{(by definition)}.$$

*iv*) We must prove that $\forall j. [\![ \lambda x.M ]\!]_\rho(j) = [\![ \lambda y.([x/y]M) ]\!]_\rho(j)$ provided $y \notin FV(M)$. By definition :

$$[\![ \lambda x.M ]\!]_\rho(j) = \sum_\beta \sum_{\Delta \in \{[\beta \vdash \lambda x.M:j]\}} \rho^\beta =$$

if $y^\sigma \notin FV(M)$ then there is a bijection between proofs of $\beta \vdash \lambda x.M:j$ and proofs of $\beta \vdash \lambda y.([x/y]M):j$ and these correspondence preserves the equivalence relation on proofs, thus

$$\sum_\beta \sum_{\Delta \in \{[\beta \vdash \lambda x.M:j]\}} \rho^\beta = \sum_\beta \sum_{\Delta \in \{[\beta \vdash \lambda y.([x/y]M):j]\}} \rho^\beta = [\![ \lambda x([x/y]M) ]\!]_\rho(j)$$

*v*) We must prove that $\forall j. [\![ M ]\!]_\rho(j) = [\![ \lambda x.M ]\!]_\xi(j)$ provided $\rho(x) = \xi(x)$ for $x \in FV(M)$
By definition $[\![ M ]\!]_\rho(j) = \sum_\beta \rho^\beta \times \sum_{\Delta \in \{[\beta \vdash M:j]\}} 1$

it is easy to show by induction that in $\beta \vdash M:j$ the domain of $\beta$ contains only the free variables of M. Moreover $\rho^\beta$ is equal to $\xi^\beta$ if $\rho$ and $\xi$ are equal on the domain of $\beta$. $\Delta$

# The Undecidability of Typability in the Lambda-Pi-Calculus

Gilles Dowek

School of Computer Science, Carnegie Mellon University
Pittsburgh, PA 15213-3890, U.S.A.

**Abstract.** The set of pure terms which are typable in the $\lambda\Pi$-calculus in a given context is not recursive. So there is no general type inference algorithm for the programming language Elf and, in some cases, some type information has to be mentioned by the programmer.

## Introduction

The programming language Elf [13] is an extension of $\lambda$-Prolog in which the clauses are expressed in a $\lambda$-calculus with dependent types ($\lambda\Pi$-calculus [8]). Since this calculus verifies the propositions-as-types principle, a proof of a proposition is merely a term of the calculus. Using this property of the $\lambda\Pi$-calculus, the programmer can either express a proposition and let the machine search for a proof of this proposition (as in usual logic programming) or express both a proposition and its proof and let the machine check that this proof is correct (as in proof-verification systems). Thus Elf can be used both to express logic programs and to reason about of their properties.

A *type inference algorithm* for a given language is an algorithm which assigns a type to each variable of a program. Thus, when such an algorithm exists, the types of the variables do not need to be mentioned by the programmer. As an example, a type inference algorithm for the language ML is given in [3].

We show here that the set of pure terms which are typable in the $\lambda\Pi$-calculus in a given context is not recursive. So there is no general type inference algorithm for the language Elf and, in some cases, some type information has to be mentioned by the programmer.

As already remarked in [3], typing a term requires the solution of a unification problem. Typing a term in the simply typed $\lambda$-calculus (and in ML) requires the solution of a first order unification problem and thus typability is decidable in these languages.

Typing a term in the $\lambda\Pi$-calculus requires the solution of a unification problem which is also formulated in the $\lambda\Pi$-calculus. Unification in the $\lambda\Pi$-calculus has been shown to be undecidable (third order unification in [9], then second order unification in [7] and third order pattern matching in [4]), i.e. there is no algorithm that decides if such a unification problem has a solution. But in order to prove the undecidability of typability in the $\lambda\Pi$-calculus we need to prove that there is no algorithm that decides if a unification problem *produced by a typing problem* has a solution. Unification problems produced by typing problems are very restricted and the undecidability proofs of unification have to be adapted to this class of problems. We show here that the proof of [9] can easily be adapted.

## 1 The Lambda-Pi-Calculus

We follow [1] for a presentation of the $\lambda\Pi$-calculus. The set of *terms* is inductively defined by

$$T ::= Type \mid Kind \mid x \mid (T\ T) \mid \lambda x : T.T \mid \Pi x : T.T$$

In this note, we ignore variable renaming problems. A rigorous presentation would use de Bruijn indices. The terms $Type$ and $Kind$ are called *sorts*, the terms $x$ *variables*, the terms $(t\ t')$ *applications*, the terms $\lambda x : t.t'$ *abstractions* and the terms $\Pi x : t.t'$ *products*. The notation $t \to t'$ is used for $\Pi x : t.t'$ when $x$ has no free occurrence in $t'$.

Let $t$ and $t'$ be terms and $x$ a variable. We write $t[x \leftarrow t']$ for the term obtained by substituting $t'$ for $x$ in $t$. We write $t \cong t'$ when the terms $t$ and $t'$ are $\beta$-equivalent ($\beta\eta$-equivalence can also be considered without affecting the proof given here).

A *context* is a list of pairs $< x, T >$ (written $x : T$) where $x$ is a variable and $T$ a term.

We define inductively two judgements: $\Gamma$ *is well-formed* and $t$ *has type* $T$ *in* $\Gamma$ ($\Gamma \vdash t : T$) where $\Gamma$ is a context and $t$ and $T$ are terms.

$$\overline{[\ ]\ \text{well-formed}}$$

$$\frac{\Gamma \vdash T : s}{\Gamma[x : T]\ \text{well-formed}} s \in \{Type, Kind\}$$

$$\frac{\Gamma\ \text{well-formed}}{\Gamma \vdash Type : Kind}$$

$$\frac{\Gamma\ \text{well-formed}\quad x : T \in \Gamma}{\Gamma \vdash x : T}$$

$$\frac{\Gamma \vdash T : Type \quad \Gamma[x : T] \vdash T' : s}{\Gamma \vdash \Pi x : T.T' : s} s \in \{Type, Kind\}$$

$$\frac{\Gamma \vdash \Pi x : T.T' : s \quad \Gamma[x : T] \vdash t : T'}{\Gamma \vdash \lambda x : T.t : \Pi x : T.T'} s \in \{Type, Kind\}$$

$$\frac{\Gamma \vdash t : \Pi x : T.T' \quad \Gamma \vdash t' : T}{\Gamma \vdash (t\ t') : T'[x \leftarrow t']}$$

$$\frac{\Gamma \vdash T : s \quad \Gamma \vdash T' : s \quad \Gamma \vdash t : T \quad T \cong T'}{\Gamma \vdash t : T'} s \in \{Type, Kind\}$$

A term $t$ is said to be *well-typed* in a context $\Gamma$ if there exists a term $T$ such that $\Gamma \vdash t : T$.

The reduction relation on well-typed terms is strongly normalizable and confluent. Thus each well-typed term has a unique normal form and two terms are equivalent if they have the same normal form [8] ([6] [15] [2] if $\beta\eta$-equivalence is considered).

A term $t$ well-typed in a context $\Gamma$ has a unique type modulo equivalence.

A normal term $t$ well-typed in a context $\Gamma$ has either the form

$$t = \lambda x_1 : T_1....\lambda x_n : T_n.(x \; c_1 \; ... \; c_n)$$

where $x$ is a variable or a sort or

$$t = \lambda x_1 : T_1....\lambda x_n : T_n.\Pi x : P.Q$$

The *head symbol* of $t$ is $x$ in the first case and, by convention, the symbol $\Pi$ in the second. The *top variables* of $t$ are the variables $x_1, ..., x_n$.

## 2 Typability in the Lambda-Pi-Calculus

**Definition 1.** A term $t$ of type $T$ in a context $\Gamma$ is said to be an *object* in $\Gamma$ if $\Gamma \vdash T : Type$.

**Proposition 2.** *If a term $t$ is an object in a context $\Gamma$ then it is either a variable, an application or an abstraction. If it is an application $t = (u \; v)$ then both terms $u$ and $v$ are objects in $\Gamma$, if it is an abstraction $t = \lambda x : U.u$ then the term $u$ is an object in the context $\Gamma[x : U]$.*

**Definition 3.** The set of *pure terms* is inductively defined by

$$T ::= x \mid (T \; T) \mid \lambda x.T$$

**Definition 4.** Let $t$ be an object in a context $\Gamma$, the *content* of $t$ ($|t|$) is the pure term defined by induction over the structure of $t$ by
- $|x| = x$,
- $|(t \; t')| = (|t| \; |t'|)$,
- $|\lambda x : U.t| = \lambda x.|t|$.

A pure term $t$ is said to be *typable* in a context $\Gamma$ if there exists a term $t'$ well-typed in an extension $\Gamma\Delta$ of $\Gamma$ such that $t'$ is an object in $\Gamma\Delta$ and $t = |t'|$.

*Remark.* Typing a pure term in a context $\Gamma$ is assigning a type to bound variables and to some of the free variables, while the type of the other free variables is given in the context $\Gamma$. When the context $\Gamma$ is empty, then typing a term in $\Gamma$ is assigning a type to both bound and free variables.

**Proposition 5.** *Typability in the empty context is decidable in the $\lambda\Pi$-calculus.*

*Proof.* Pure terms typable in the empty context in the $\lambda\Pi$-calculus and in the simply typed $\lambda$-calculus are the same [8] and typability is decidable in simply typed $\lambda$-calculus [3].

## 3 Post Correspondence Problem

**Definition 6.** Post Correspondence Problem
A *Post correspondence problem* is a finite set of pairs of words over the two letters alphabet $\{A, B\} : \{<\varphi_1, \psi_1>, ..., <\varphi_n, \psi_n>\}$. A *solution* to such a problem is a non empty sequence of integers $i_1, ..., i_p$ such that

$$\varphi_{i_1}...\varphi_{i_p} = \psi_{i_1}...\psi_{i_p}$$

**Theorem 7.** *(Post [14]) It is undecidable whether or not a Post problem has a solution.*

## 4 Undecidability of Typability in the Lambda-Pi-Calculus

Let us consider the context
$\Gamma = [T : Type; a : T \to T; b : T \to T; c : T; d : T; P : T \to Type;$
$$F : \Pi x : T.((P\ x) \to T)]$$

**Definition 8.** (Huet [9]) Let $\varphi$ be a word in the two letters alphabet $\{A, B\}$, we define by induction on the length of $\varphi$ the term $\hat{\varphi}$ well-typed in $\Gamma$ and the pure term $\tilde{\varphi}$ as follows

$$\hat{\varepsilon} = \lambda y : T.y$$
$$\hat{A\varphi} = \lambda y : T.(a\ (\hat{\varphi}\ y))$$
$$\hat{B\varphi} = \lambda y : T.(b\ (\hat{\varphi}\ y))$$

$$\tilde{\varphi} = |\hat{\varphi}|$$

**Proposition 9.** *Let $\{<\varphi_1, \psi_1>, ..., <\varphi_n, \psi_n>\}$ be a Post problem, the non empty sequence $i_1, ..., i_p$ is a solution to this problem if and only if*

$$(\hat{\varphi_{i_1}}\ (...(\hat{\varphi_{i_p}}\ c)...)) \cong (\hat{\psi_{i_1}}\ (...(\hat{\psi_{i_p}}\ c)...))$$

**Proposition 10.** *If $g$ is a term such that the term $(g\ a\ ...\ a)$ ($n$ symbols $a$) is well-typed and is an object in an extension $\Gamma\Delta$ of $\Gamma$ then the term $g$ is well-typed in the context $\Gamma\Delta$ and its type is equivalent to the term*

$$\Pi x_1 : T \to T....\Pi x_n : T \to T.(\beta\ x_1\ ...\ x_n)$$

*for some term $\beta$ of type $(T \to T) \to ... \to (T \to T) \to Type$ in the context $\Gamma\Delta$.*

*Proof.* By induction on $n$.

**Proposition 11.** *Let $t, u_1, ..., u_n, v$ be normal terms such that $(t\ u_1\ ...\ u_n)$ is a well-typed term and its normal form is $v$. The head symbol of the $t$ is either the head symbol of $v$ or a top variable of $t$.*

*Proof.* Let $x$ be the head symbol of $t$. If $x$ is not a top variable of $t$ then the head symbol of the normal form of $(t\ u_1\ ...\ u_n)$ is also $x$, so $x$ is the head symbol of $v$.

**Proposition 12.** *Let $t$ be a normal term of type $(T \to T) \to ... \to (T \to T) \to T$ in the context $\Gamma$ such that the normal form of $(t\ \lambda y : T.y\ ...\ \lambda y : T.y)$ is equal to $c$. Then the term $t$ has the form*

$$t = \lambda x_1 : T \to T....\lambda x_n : T \to T.(x_{i_1}\ (...(x_{i_p}\ c)...))$$

*for some sequence $i_1, ..., i_p$.*

*Proof.* By induction on the number of variable occurrences in $t$.

**Theorem 13.** *It is undecidable whether or not a pure term is typable in a given context.*

*Proof.* Consider a Post problem $\{<\varphi_1, \psi_1>, ..., <\varphi_n, \psi_n>\}$. We construct the pure term $t$ such that $t$ is typable in $\Gamma$ if and only if the Post problem has a solution.

$$\begin{aligned}
t = \lambda f.\lambda g.\lambda h.(&f\ (g\ a\ ...\ a) \\
&(h\ (g\ \tilde{\varphi}_1\ ...\ \tilde{\varphi}_n)) \\
&(h\ (g\ \tilde{\psi}_1\ ...\ \tilde{\psi}_n)) \\
&(F\ c\ (g\ \lambda y.y\ ...\ \lambda y.y)) \\
&(F\ d\ (g\ \lambda y.d\ ...\ \lambda y.d)))
\end{aligned}$$

Assume this term is typable and call $\alpha$ the type of $g$. The term $(g\ a\ ...\ a)$ is well-typed and is an object in $\Gamma\Delta$ so

$$\alpha \cong \Pi x_1 : T \to T....\Pi x_n : T \to T.(\beta\ x_1\ ...\ x_n)$$

where $\beta$ is a term of type $(T \to T) \to ... \to (T \to T) \to Type$ in $\Gamma\Delta$.

Then all the variables $y$ bound in the terms $\tilde{\varphi}_i$, $\tilde{\psi}_i$, $\lambda y.y$ and $\lambda y.d$ have type $T$. The term $(g\ \hat{\varphi}_1\ ...\ \hat{\varphi}_n)$ has the type $(\beta\ \hat{\varphi}_1\ ...\ \hat{\varphi}_n)$, so from the well-typedness of the term $(h\ (g\ \hat{\varphi}_1\ ...\ \hat{\varphi}_n))$ we get that the type of the variable $h$ has the form $\Pi x : \gamma.\gamma'$ and

$$\gamma \cong (\beta\ \hat{\varphi}_1\ ...\ \hat{\varphi}_n)$$

in the same way, from the well-typedness of the term $(h\ (g\ \hat{\psi}_1\ ...\ \hat{\psi}_n))$ we get

$$\gamma \cong (\beta\ \hat{\psi}_1\ ...\ \hat{\psi}_n)$$

so

$$(\beta\ \hat{\varphi}_1\ ...\ \hat{\varphi}_n) \cong (\beta\ \hat{\psi}_1\ ...\ \hat{\psi}_n)$$

From the well-typedness of the term $(F\ c\ (g\ \lambda y : T.y\ ...\ \lambda y : T.y))$ we get

$$(\beta\ \lambda y : T.y\ ...\ \lambda y : T.y) \cong (P\ c)$$

At last from the the well-typedness of the term $(F\ d\ (g\ \lambda y : T.d\ ...\ \lambda y : T.d))$ we get
$$(\beta\ \lambda y : T.d\ ...\ \lambda y : T.d) \cong (P\ d)$$
Since the term $\beta$ has type $(T \to T) \to ... \to (T \to T) \to Type$, the head symbol of the normal form of the term $\beta$ cannot be a top variable of $\beta$, so it is the variable $P$ and we have
$$\beta \cong \lambda x_1 : T \to T....\lambda x_n : T \to T.(P\ (\delta\ x_1\ ...\ x_n))$$
For some term $\delta$ of type $(T \to T) \to ... \to (T \to T) \to T$. We get
$$(\delta\ \hat{\varphi}_1\ ...\ \hat{\varphi}_n) \cong (\delta\ \hat{\psi}_1\ ...\ \hat{\psi}_n)$$
$$(\delta\ \lambda y : T.y\ ...\ \lambda y : T.y) \cong c$$
$$(\delta\ \lambda y : T.d\ ...\ \lambda y : T.d) \cong d$$
The second equality shows that the normal form of the term $\delta$ has the form
$$\lambda x_1 : T \to T....\lambda x_n : T \to T.(x_{i_1}\ (...(x_{i_p}\ c)...))$$
for some sequence $i_1, ..., i_p$. The third equality shows that $p > 0$ and the first one that
$$(\hat{\varphi}_{i_1}\ (...(\hat{\varphi}_{i_p}\ c)...)) \cong (\hat{\psi}_{i_1}\ (...(\hat{\psi}_{i_p}\ c)...))$$
so the sequence $i_1, ..., i_p$ is a solution to the Post problem.

Conversely assume that the Post problem has a solution $i_1, ..., i_p$, then by giving the following types to the variables $f$, $g$ and $h$
$$f : (P\ (a\ (...(a\ c)...))) \to T \to T \to T \to T \to T$$
$$g : \Pi x_1 : T \to T....\Pi x_n : T \to T.(P\ (x_{i_1}\ (...(x_{i_p}\ c)...)))$$
$$h : (P\ (\hat{\varphi}_{i_1}\ (...(\hat{\varphi}_{i_p}\ c)...))) \to T$$
and the type $T$ to all the other variables of the term $t$, we get a term $t'$ well-typed in $\Gamma$, which is an object and such that $t = |t'|$.

*Remark.* Along the way, we have proved that in the simply typed $\lambda$-calculus, the unification problems of the form
$$(f\ t_1\ ...\ t_n) = (f\ t'_1\ ...\ t'_n)$$
$$(f\ u_1\ ...\ u_n) = u'$$
$$(f\ v_1\ ...\ v_n) = v'$$
where $t_i, t'_i, u_i, u', v_i, v'$ are closed terms and $f$ a third order variable are undecidable.

It is decidable if each of these equations has a solution or not (since the first one is flexible-flexible [10] [11] and the others third order matching problems [5]), but it is undecidable whether or not they have a solution *in common*. If the variable $f$ is second order the problems of this form are decidable since the second order matching algorithm [11] [12] produces a finite complete set of closed solutions.

# Acknowledgements

The author thanks Frank Pfenning for many stimulating and helpful discussions on this problem and Pawel Urzyczyn for his careful reading of a previous draft of this paper.

# References

1. H. Barendregt, Introduction to Generalized Type Systems, *Journal of Functional Programming* **1, 2** (1991) 125–154.
2. Th. Coquand, An Algorithm for Testing Conversion in Type Theory, *Logical Frameworks*, G. Huet and G. Plotkin (Eds.), Cambridge University Press (1991).
3. L. Damas, R. Milner, Principal Type-Scheme for Functional Programs, *Proceedings of Principles of Programming Languages* (1982).
4. G. Dowek, L'Indécidabilité du Filtrage du Troisième Ordre dans les Calculs avec Types Dépendants ou Constructeurs de Types (The Undecidability of Third Order Pattern Matching in Calculi with Dependent Types or Type Constructors), *Comptes Rendus à l'Académie des Sciences* **I, 312, 12** (1991) 951–956.
5. G. Dowek, Third Order Matching is Decidable, *Proceedings of Logic in Computer Science* (1992) 2–10.
6. H. Geuvers, The Church-Rosser Property for $\beta\eta$-reduction in Typed Lambda Calculi, *Proceedings of Logic in Computer Science* (1992) 453–460.
7. W.D. Goldfarb, The Undecidability of the Second-Order Unification Problem, *Theoretical Computer Science* **13** (1981) 225–230.
8. R. Harper, F. Honsell, G. Plotkin, A Framework for Defining Logics, *Proceedings of Logic in Computer Science* (1987) 194–204.
9. G. Huet, The Undecidability of Unification in Third Order Logic, *Information and Control* **22** (1973) 257–267.
10. G. Huet, A Unification Algorithm for Typed $\lambda$-calculus, *Theoretical Computer Science* **1** (1975) 27–57.
11. G. Huet, Résolution d'Équations dans les Langages d'Ordre 1, 2, ..., $\omega$, Thèse de Doctorat d'État, Université de Paris VII (1976).
12. G. Huet, B. Lang, Proving and Applying Program Transformations Expressed with Second Order Patterns, *Acta Informatica* **11** (1978) 31–55.
13. F. Pfenning, Logic Programming in the LF Logical Framework, *Logical Frameworks*, G. Huet and G. Plotkin (Eds.), Cambridge University Press (1991).
14. E. L. Post, A Variant of a Recursively Unsolvable Problem, *Bulletin of American Mathematical Society* **52** (1946) 264–268.
15. A. Salvesen, The Church-Rosser Theorem for Pure Type Systems with $\beta\eta$-reduction, Manuscript, University of Edinburgh (1991).

# Recursive Types Are not Conservative over $F_\leq$
## (*extended abstract*)

## Giorgio Ghelli[1]

**Abstract.** $F_\leq$ is a type system used to study the integration of inclusion and parametric polymorphism. $F_\leq$ does not include a notion of recursive types, but extensions of $F_\leq$ with recursive types are widely used as a basis for foundational studies about the type systems of functional and object-oriented languages. In this paper we show that adding recursive types results in a non conservative extension of the system. This means that the algorithm for $F_\leq$ subtyping (the kernel of the algorithm for $F_\leq$ typing) is no longer complete for the extended system, even when it is applied only to judgements where no recursive type appears, and that most of the proofs of known properties of $F_\leq$ do not hold for the extended system; this is the case, for example, for Pierce's proof of undecidability of $F_\leq$. However, we prove that this non conservativity is limited to a very special class of subtyping judgements, the "diverging judgements" introduced in [Ghe]. This last result implies that the extension of $F_\leq$ with recursive types could be still useful for practical purposes.

## 1 Introduction

$F_\leq$ is a minimal language integrating subtyping and bounded parametric polymorphism. It is a simplification of the language Fun introduced in [CaWe]; the essential difference is that Fun has recursive types and values, which are missing in $F_\leq$. $F_\leq$ was introduced in [CuGhe] (see also [BrLo], [CaMaMiSce] and [CaLo]). Extensions of $F_\leq$ are being exploited as a foundational tool in the research on the extension of the type system of functional languages with modules or with object-oriented features. We refer to [CaWe], [GheTh], [CuGhe] and [CaMaMiSce] for more motivations and details about $F_\leq$.

In all the models of object-oriented type systems, as in many other contexts, recursive types play a central role. It has been argued (see, e.g., [AmCa]) that recursion can be added to $F_\leq$ quite painlessly, as a feature which is in some way "orthogonal" to the rest of the system. In this paper we show that this is not the case: the mere "existence" of recursive types changes the subtyping relations between non recursive types.

This result shows that "transitivity elimination" for $F_\leq$, in the form proved in [CuGhe], is not valid in $\mu F_\leq$. Transitivity elimination means, essentially, that every provable subtyping can be proved without using the transitivity rule below:

$$\frac{\Gamma \vdash T \leq U \quad \Gamma \vdash U \leq V}{\Gamma \vdash T \leq V}$$

Transitivity elimination is essential, from the point of view of type-checking. A subtype checker uses deduction rules backward to generate subproblems from a subtype checking problem; this approach is not possible with the transitivity rule, which transforms the problem $\Gamma \vdash T \leq V$ in the infinite disjunction of all the ($\Gamma \vdash T \leq U$, $\Gamma \vdash U \leq V$) problems which

---

[1] Dipartimento di Informatica, Università di Pisa, Corso Italia 40, I-56125, Pisa, Italy, ghelli@di.unipi.it. This work was carried out with the partial support of E.E.C., Esprit Basic Research Action 6309 FIDE2 and of "Progetto Finalizzato Sistemi Informatici e Calcolo Parallelo" of the Italian National Research Council under grant No.91.00877.PF69.

can be generated by choosing any type U. From the point of view of theoretical studies, the transitivity rule makes it impossible to reason about subtyping judgements via induction on the types involved in their proof, since not only are T and V in the premises not proper subterms of T and V in the result, but, moreover, there is no relation, a priori, between the size of U and that of T and V.

The fact that "transitivity elimination" does not hold has at least three important consequences:

- Implementation of type checking: the standard type-checking algorithm for $F_\leq$, which is complete for that system, is no longer complete when applied to the recursion-free fragment of $\mu F_\leq$. In other words, extending that algorithm to the recursive case does not mean just adding code to deal with recursive types, but means altering its behavior when dealing with non recursive types too. At present it is not clear how this can be accomplished, thought our results might give some suggestions.

- Language design: the only known proof of the existence of a minimum type for $F_\leq$, which is based on transitivity elimination, is no longer valid for $\mu F_\leq$. The existence of a minimum type is essential for the usability of the system as a basis for a programming language and is important for writing efficient type-checkers. We conjecture that $\mu F_\leq$ still enjoys that property, but we do not know how to prove it.

- Semantics: an interpretation of derivable $F_\leq$ judgements which is defined by induction on their derivation proof, is coherent when it depends only on the judgement and not on the proof. Coherence is a crucial property of interpretations. In [CuGhe] a set of "coherence equations" was defined, which allow us to reduce the problem of checking the coherence of any coercion-based model of $F_\leq$ to the validity of those equations in that model. With non-conservativity, satisfaction of these equations could be not sufficient to ensure even the coherence of the interpretation of $\mu F_\leq$ judgements not involving recursive types.

$F_\leq$ subtyping is known to be undecidable [Pie]. When an undecidable proof system is conservatively extended, its undecidability is inherited by the extended system, since otherwise the decision procedure for the "bigger" system could be used to decide provability in the "smaller" one. Hence, our result implies that undecidability of $F_\leq$ is not immediately inherited by $\mu F_\leq$, but has to be proved independently. Moreover in this paper we show that Pierce's proof of undecidability cannot be trivially rephrased in the context of $\mu F_\leq$.

However, as a partial conservativity result, we show that the set of subtyping judgements which can be expressed in $F_\leq$ but proved only in $\mu F_\leq$ is contained in a very peculiar class of judgements, the "diverging" judgements, i.e. those which make the standard subtype checking algorithm diverge. This result is important since it shows that a strict relation exists between the undecidability of $F_\leq$ and our non-conservativity result, and suggests a way to prove that recursive types are conservative over the decidable variants of $F_\leq$, like Kernel Fun (see [Ghe] and [GhePie]) and $F_\leq$ without Top (see [Ka]). Furthermore, it has been argued that the behavior of a type checking algorithm on diverging judgements is of no practical interest, since there is, in practice, no possibility that a real program contains such a judgement. This point of view is supported by the results in [Ghe] which describe the structure of the diverging judgements, which seems far more complex than what is actually used in practice. Adopting this point of view, our partial conservativity result means that, restricting the attention to "practically relevant" judgements, $\mu F_\leq$ is conservative over $F_\leq$.

The paper is structured as follows. In Section 2 we give the preliminary definitions about $F_\leq$ and $\mu F_\leq$. In Section 3 we prove non conservativity of $\mu F_\leq$ over $F_\leq$. In Section 4 we characterize the set of provable or diverging judgements as the maximal relation enjoying some closure properties. This result is used in Section 5 to prove the partial conservativity result. All the sketched proofs are fully developed in the full paper.

## 2 Definitions

### 2.1 $F_\leq$

The syntax of $F_\leq$ subtyping judgements is defined as follows ($F_\leq$ terms and typing judgements have no relevance in this paper):

| | | |
|---|---|---|
| **Types** | T | ::= t \| Top \| T→T \| ∀t≤T. T |
| **Environments** | Γ | ::= () \| Γ, t≤T |
| **Judgements** | J | ::= Γ ⊢ T ≤ T |

$F_\leq$ types are either variables, Top, function types (T→T) or universally quantified types, used to give a type to polymorphic functions, where the type variable can be bounded. $F_\leq$ subtype rules are listed in Appendix A. We will often ignore function types in the following since, from the point of view of subtyping, they behave exactly like ∀ types where the quantified variable does not appear in the codomain. Variables are actually just names for their De Bruijn indexes, so that judgements and types are always considered modulo α conversion and all substitutions must be performed in a capture free way.

Due to transitivity, $F_\leq$ subtype rules are not a good tool for reasoning about non-derivability of judgements. This problem was solved in [CuGhe], by proving that any derivable judgement admits a normal form proof, i.e. a proof where transitivity is used only to prove judgements like Γ ⊢ t ≤ U as a consequence of Γ ⊢ t ≤ Γ(t) and Γ ⊢ Γ(t) ≤ U where Γ(t) is the bound of t in Γ (i.e., if Γ = (Γ', t≤T, Γ'') then Γ(t)$=_{def}$ T). The existence of such a normal form proof for any subtyping judgement is what we called transitivity elimination in the introduction.

It is easy to show that an algorithm exists which, given any derivable subtyping judgement, rebuilds a normal form proof tree for that judgement. That algorithm can be described by the set of rewrite rules below. These rules transform a derivable judgement into the premises of the last subtyping rule applied in the normal form proof of that judgement.

| | | | | |
|---|---|---|---|---|
| (top) | | Γ ⊢ T ≤ Top | ▷ | **true** |
| (varId) | | Γ ⊢ u ≤ u | ▷ | **true** |
| (varTrans) | T≠u, T≠Top ⇒ | Γ ⊢ u ≤ T | ▷ | Γ ⊢ Γ(u)[2] ≤ T |
| (∀dom) | | Γ ⊢ ∀t≤T.U' ≤ ∀t≤T'.U | ▷ | Γ ⊢ T' ≤ T |
| (∀cod) | | Γ ⊢ ∀t≤T.U' ≤ ∀t≤T'.U | ▷ | Γ, t≤T' ⊢ U' ≤ U[3] |

The algorithm tries to build a normal-form proof tree starting from the conclusion and building (backward) all the branches by following all the possible rewriting chains generated by the conclusion. If some chain generates an irreducible judgement, different from **true**, then the algorithm fails. When all chains terminate with **true**, the set of these

---

[2] Variables in Γ(u) must be renamed to avoid capture; e.g., we could rename all variables in Γ⊢u≤T so that no variable name is used twice.

[3] We could write Γ ⊢ ∀t≤T.U' ≤ ∀t'≤T'.U ▷ Γ, t'≤T' ⊢ [t'/t]U' ≤ U to emphasize that t and t' can be different.

chains represents a proof tree for the original judgement[4], as depicted in Figure 1, and the algorithm terminates with success.

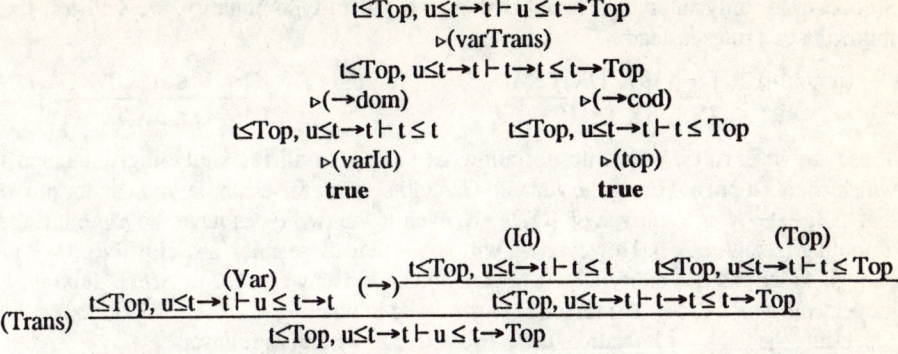

Fig. 1: Correspondence between the chains generated by a judgement and its proof.

*Proposition 2.1*: The algorithm described by the five rules above is correct and complete with respect to $F_\leq$ subtyping, i.e. it terminates with success on all and only the derivable judgements, while it may either fail or loop forever with non derivable judgements [CuGhe,Ghe]. The looping behavior cannot be avoided, since $F_\leq$ subtyping is undecidable [Pie].

## 2.2 $\mu F_\leq$

$\mu F_\leq$ extends $F_\leq$ by adding recursive types:

**Types**      $T ::= t \mid X \mid Top \mid T \rightarrow T \mid \forall t \leq T. T \mid \mu X.T \rightarrow T \mid \mu X.\forall t \leq T. T$

For clarity, we use two different families $t$ and $X$ for, respectively, bounded and recursive type variables. $\forall$ and $\mu$ bind the corresponding variables. We forbid recursive types whose body is either $t, X, Top$ or $\mu X.T$. This is a minor restriction which simplifies some proofs without restricting the expressive power of the system. In fact a recursive type denotes the regular tree which constitutes its infinite unfolding. So, $\mu X.Top$, $\mu X.t$ and $\mu X.Y$ ($X \neq Y$) can be forbidden, since they would denote just Top, t and Y. $\mu X.X$ is forbidden, since it does not mean anything, and this is the only essential restriction. Finally, $\mu X.\mu Y.T$ can be forbidden since its unfolding is the same as $\mu X.[X/Y]T$ [AmCa].

Different notions of type equality and subtyping can be defined for recursive types (see [AmCa] for an excellent discussion). The essential requirement is that a recursive type must be equal, or at least isomorphic, to any of its unfoldings. This requirement can be expressed by the following rule[5] (note the use of $\vdash_\mu$ for $\mu F_\leq$ judgements):

---

[4] Actually the rebuilt tree is slightly different from the normal form proof, since a) the (exp) rule does not transform $\Gamma \vdash t \leq T$ into the axiom $\Gamma \vdash t \leq \Gamma(t)$ plus $\Gamma \vdash \Gamma(t) \leq T$, but takes for granted the proof of the axiom and generates only $\Gamma \vdash \Gamma(t) \leq T$; b) the rewrite rules prove $\Gamma \vdash T \leq T$ by analyzing step by step the structure of T, applying identity only on atoms; these are just algorithmic optimizations which do not affect the correspondence between the two approaches.

[5] A less practical kind of recursive types can be obtained by not requiring that $\mu X.T = [\mu X.T/X]T$, but simply that for each recursive type there exist two functions, $fold_{(\mu X.T)}:[\mu X.T/X]T \rightarrow \mu X.T$ and $unfold_{(\mu X.T)}:\mu X.T \rightarrow [\mu X.T/X]T$, such that $fold(unfold(a)) \triangleright a$ and $unfold(fold(a)) \triangleright a$; this approach is adopted, for example, in Cardelli's Fsub system [Ca]. Our result does not apply to this approach.

(unfold) $\dfrac{\Gamma \vdash_\mu \mu X.T \text{ type}}{\Gamma \vdash_\mu \mu X.T = [\mu X.T/X]T}$

Since we are only interested in subtyping, and not in type equality, we will use the following two rules instead:

(unfold-l) $\dfrac{\Gamma \vdash_\mu [\mu X.T/X]T \leq U}{\Gamma \vdash_\mu \mu X.T \leq U}$     (unfold-r) $\dfrac{\Gamma \vdash_\mu U \leq [\mu X.T/X]T}{\Gamma \vdash_\mu U \leq \mu X.T}$

These are weak rules, which do not allow us to deduce all the subtyping judgements which could be proved using the Amadio-Cardelli system; for example you cannot prove that $\mu X.t \to t \to X$ is a subtype of $\mu X.t \to X$, even if the two types have the same infinite unfolding (see [AmCa]). However we will show that these rules are already powerful enough to extend $F_\leq$ subtyping in a non-conservative way, i.e. to prove subtyping judgements which can be expressed but not proved in plain $F_\leq$.

Using the backward notation, these two rules are written as follows:

$\Gamma \vdash_\mu \mu X.T \leq U \; \triangleright \; \Gamma \vdash_\mu [\mu X.T/X]T \leq U$     (unfold-l)
$\Gamma \vdash_\mu U \leq \mu X.T \; \triangleright \; \Gamma \vdash_\mu U \leq [\mu X.T/X]T$     (unfold-r)

*Proposition 2.2*: Adding the two rules above to the five rules for $F_\leq$ subtyping we obtain a rule system for $\mu F_\leq$ judgements which is sound but not complete.

The fact that the seven-rule system is sound can be easily verified, by showing that they actually build, in a backward fashion, a valid $\mu F_\leq$ proof. The fact that they are not complete, even with respect to the weak recursive subtyping rules adopted, can be verified by observing that the judgement, provable in $\mu F_\leq$ but not in $F_\leq$, which we will exhibit in the next section is not proved by the seven-rule system. Non completeness of this set of rules is what we call "loss of transitivity elimination", since the only difference between the backward rules and the complete system is that the backward rules only use the limited form of transitivity formalized by the (varTrans) rule.

Soundness and completeness of the five rules means that we can use them to prove both derivability and non-derivability in $F_\leq$, while the seven-rule set can be used only as an alternative way to prove derivability in $\mu F_\leq$.

## 3 The Counterexample

In this section we will present a counterexample to the conservativity conjecture, i.e. an $F_\leq$ judgement which cannot be derived in $F_\leq$ but can be derived in $\mu F_\leq$.

*Notation*: To improve readability, we will exploit the following conventions:

$\forall t.T$     stands for     $\forall t \leq \text{Top}.T$
$-T$     stands for     $T \to \text{Top}$

And we will exploit the following derived rules:

($\forall$d)  $\Gamma \vdash \forall t.T \leq \forall u \leq U.T$  $\triangleright$  $\Gamma, u \leq U \vdash [u/t]T \leq T$
(-)  $\Gamma \vdash -T \leq -U$  $\triangleright$  $\Gamma \vdash U \leq T$
(var)  $\Gamma \vdash t \leq \text{Rename}(\Gamma(t))$  $\triangleright$  **true**

(in the last rule Rename($\Gamma(t)$) stands for any $\alpha$-variant of the bound of $t$ in $\Gamma$).

Let $B = \forall u.-\forall v \leq u.-u$. Then the following judgement cannot be derived in $F_\leq$:

$t_0 \leq B \quad \vdash \quad t_0 \quad \leq \quad \forall t_1 \leq t_0.-t_0$

This can be proved by showing that the complete $F_\le$ subtyping algorithm described in the previous section diverges as follows:

let $B = \forall u.-\forall v \le u.-u$:

| | | | | |
|---|---|---|---|---|
| $t_0 \le B$ | $\vdash$ | $t_0$ | $\le \forall t_1 \le t_0.-t_0$ | ▷(varTrans) |
| $t_0 \le B$ | $\vdash$ | $\forall t_1.-\forall t_2 \le t_1.-t_1$ | $\le \forall t_1 \le t_0.-t_0$ | ▷($\forall$) |
| $t_0 \le B, t_1 \le t_0$ | $\vdash$ | $-\forall t_2 \le t_1.-t_1$ | $\le -t_0$ | ▷(-) |
| $t_0 \le B, t_1 \le t_0$ | $\vdash$ | $t_0$ | $\le \forall t_2 \le t_1.-t_1$ | |

The judgement above is a special case (i=1) of the family below:

$$J_i = t_0 \le B, \ldots, t_{i+1} \le t_i \vdash t_0 \le \forall t_{i+2} \le t_{i+1}.-t_{i+1}$$

For any $i$ that judgement is transformed in the judgement $J_{i+1}$:

| | | | | |
|---|---|---|---|---|
| $t_0 \le B, \ldots, t_{i+1} \le t_i$ | $\vdash$ | $t_0$ | $\le \forall t_{i+2} \le t_{i+1}.-t_{i+1}$ | ▷(varTrans) |
| $t_0 \le B, \ldots, t_{i+1} \le t_i$ | $\vdash$ | $\forall t_{i+2}.-\forall t_{i+3} \le t_{i+2}.-t_{i+2}$ | $\le \forall t_{i+2} \le t_{i+1}.-t_{i+1}$ | ▷($\forall$) |
| $t_0 \le B, \ldots, t_{i+1} \le t_i, t_{i+2} \le t_{i+1}$ | $\vdash$ | $-\forall t_{i+3} \le t_{i+2}.-t_{i+2}$ | $\le -t_{i+1}$ | ▷(-) |
| $t_0 \le B, \ldots, t_{i+1} \le t_i, t_{i+2} \le t_{i+1}$ | $\vdash$ | $t_{i+1}$ | $\le \forall t_{i+3} \le t_{i+2}.-t_{i+2}$ | ▷(varTrans) |
| $t_0 \le B, \ldots, t_{i+1} \le t_i, t_{i+2} \le t_{i+1}$ | $\vdash$ | $t_i$ | $\le \forall t_{i+3} \le t_{i+2}.-t_{i+2}$ | ▷(varTrans) |
| ... | | | | ▷(varTrans) |
| $t_0 \le B, \ldots, t_{i+1} \le t_i, t_{i+2} \le t_{i+1}$ | $\vdash$ | $t_0$ | $\le \forall t_{i+3} \le t_{i+2}.-t_{i+2}$ | |

This implies that the algorithm never terminates on the original judgement, hence, by completeness, that judgement cannot be proved in $F_\le$[6].

We show now that in $\mu F_\le$ there exists a (recursive) type T, such that:

(a) $t_0 \le B \vdash t_0 \le T$
(b) $t_0 \le B \vdash T \le \forall t_1 \le t_0.-t_0$

So that, by transitivity, the judgement above is derivable in $\mu F_\le$; T is the following type:

$$T = \mu X.\forall t \le X.-X.$$

We give the backward representation of a proof of the two judgements (a) and (b) above. When a $\forall$ judgement can be rewritten in two different ways we indicate with (1) the beginning of the rewriting chain of the first premise and with (2) the beginning of the rewriting chain of the second premise. Note that, while to prove non-derivability of the judgement above in $F_\le$ we *needed* the backward representation, in this case, where we are only interested in the derivability of the judgement, we could write its proof in the usual tree form. We still resort to the backward representation to make it easier for the reader to compare the behavior of these two judgements with the behavior of the judgement above; the tree form of a proof of the judgement can be easily recovered.

| | | | | | | |
|---|---|---|---|---|---|---|
| (a) | $t_0 \le B$ | $\vdash$ | $t_0$ | $\le T$ | | ▷(varTrans) |
| | $t_0 \le B$ | $\vdash$ | $\forall t_1.-\forall t_2 \le t_1.-t_1$ | $\le T$ | | ▷(unfold-r) (T=$\mu X.\forall t \le X.-X$) |
| | $t_0 \le B$ | $\vdash$ | $\forall t_1.-\forall t_2 \le t_1.-t_1$ | $\le \forall t_1 \le T.-T$ | | ▷($\forall$) |
| | $t_0 \le B, t_1 \le T$ | $\vdash$ | $-\forall t_2 \le t_1.-t_1$ | $\le -T$ | | ▷(-) |
| | $t_0 \le B, t_1 \le T$ | $\vdash$ | $T$ | $\le \forall t_2 \le t_1.-t_1$ | | ▷(unfold-l) |
| | $t_0 \le B, t_1 \le T$ | $\vdash$ | $\forall t_2 \le T.-T$ | $\le \forall t_2 \le t_1.-t_1$ | | ▷($\forall$dom):1; ▷($\forall$cod):2 |
| 1 | $t_0 \le B, t_1 \le T$ | $\vdash$ | $t_1$ | $\le T$ | | ▷(var) **true** |

---

[6] This underivable judgement was introduced in [Ghe]; before its discovery, it was widely believed that the algorithm described in the previous section was terminating, and so that $F_\le$ subtyping was decidable. A variant of this judgement has then been used by Pierce to encode two-counter Turing machines as $F_\le$ subtyping judgements, proving undecidability of $F_\le$ subtyping [Pie].

|     |                              |                                 |          |                              |                              |
|-----|------------------------------|---------------------------------|----------|------------------------------|------------------------------|
|  2  | $t_0 \leq B, t_1 \leq T, t_2 \leq t_1 \vdash$ | $-T$              | $\leq$   | $-t_1$                       | $\triangleright(-)$          |
|     | $t_0 \leq B, t_1 \leq T, t_2 \leq t_1 \vdash$ | $t_1$             | $\leq$   | $T$                          | $\triangleright(\text{var})$ **true** |
| (b) | $t_0 \leq B$                 | $\vdash T$                      | $\leq$   | $\forall t_1 \leq t_0.-t_0$  | $\triangleright(\text{unfold-l})$ |
|     | $t_0 \leq B$                 | $\vdash \forall t_1 \leq T.-T$  | $\leq$   | $\forall t_1 \leq t_0.-t_0$  | $\triangleright(\forall\text{dom}){:}1;\ \triangleright(\forall\text{cod}){:}2$ |
|  1  | $t_0 \leq B$                 | $\vdash t_0$                    | $\leq$   | $T$                          | $\triangleright^*$ **true** (see (a)) |
|  2  | $t_0 \leq B, t_1 \leq t_0$   | $\vdash -T$                     | $\leq$   | $-t_0$                       | $\triangleright(-)$          |
|     | $t_0 \leq B, t_1 \leq t_0$   | $\vdash t_0$                    | $\leq$   | $T$                          | $\triangleright^*$ **true** (see (a), $t_1$ plays no role) |

As already stated, the non conservativity of $\mu F_\leq$ w.r.t. $F_\leq$ implies that the undecidability of the second system does not necessarily extend to the first one. Moreover the judgement that we have shown to become derivable in $\mu F_\leq$ is exactly that judgement whose non derivability has been used by Pierce to show undecidability of $F_\leq$ subtyping. This means that it is seems rather difficult to adapt Pierce's proof to show $\mu F_\leq$ undecidability.

It could even be conceivable that adding recursive types to $F_\leq$ types makes its type-checking problem easier, in the same way as admitting rational infinite terms as solutions for the unification problem makes its solution simpler, allowing us to avoid the occur check. However, this optimistic hypothesis is not convincing, since we do not see, in practice, how a complete algorithm for $\mu F_\leq$ subtyping could be designed.

Our counterexample hints at a possible direction to explore in order to complete the algorithm. Since the structure of the recursive type T "reflects" the structure of the looping behavior of the judgement with respect to the standard type-checking algorithm (an alternation of $\forall$ and $-$ steps), we could look for an algorithm which tries to build an intermediate, possibly recursive, type whose structure is derived from the behavior of the standard algorithm.

## 4  The Compatibility Relation Defined by Diverging Judgements

We want to prove that the effects of our non-conservativity result are limited to the "pathological" set of the diverging $F_\leq$ judgements, i.e. to those judgements which make the $F_\leq$ subtype checking algorithm diverge.

To prove this partial conservativity result we first need some results about the nature of diverging judgements. To this aim, in this section we study the point of view that, when the subtype-checking algorithm is not able to prove in a finite time that T is not a subtype of U, this means that T is somehow "compatible" with U, even if it is not exactly a subtype of U, and we will study the properties of this compatibility relation.

Compatibility will be defined by negation, as the complement of the notion of being "provably not a subtype". We first model the failing runs of the subtyping algorithm using the proof system below.

(t/Top⊄)  $\Gamma \vdash \text{Top} \not\subset t$          (Top/∀⊄)  $\Gamma \vdash \text{Top} \not\subset \forall t \leq T.U$

(∀/t⊄)  $\Gamma \vdash \forall t \leq T.U \not\subset t$      (VarTrans⊄)  $\dfrac{\Gamma \vdash \Gamma(t) \not\subset T \quad T \neq t,\ T \neq \text{Top}}{\Gamma \vdash t \not\subset T}$

(∀r⊄)  $\dfrac{\Gamma, t \leq T \vdash U \not\subset U'}{\Gamma \vdash \forall t \leq T.U \not\subset \forall t \leq T.U'}$      (∀l⊄)  $\dfrac{\Gamma \vdash T \not\subset T'}{\Gamma \vdash \forall t \leq T'.U \not\subset \forall t \leq T.U'}$

*Proposition 4.1*: $\Gamma \vdash T \not\subset U$ if and only if the standard type-checking algorithm fails (in a finite time) on the judgement $\Gamma \vdash T \leq U$.

*Proof hint*: By induction, on the number of steps executed by the algorithm in one direction, and on the size of the proof of $\Gamma \vdash T \not\subseteq U$ in the other one.

In this section we will always consider all the relations written as $\Gamma \vdash T \mathrel{\mathcal{R}} U$ as ranging on the universe J of well-formed judgements:

$$J = \{ \Gamma, T, U \mid \Gamma \vdash T \text{ type}, \Gamma \vdash U \text{ type} \}$$

Hereafter, $\Gamma \vdash T \not\leq U$ denotes the intersection of the relation defined by the rules above with J; note that it would be very easy to modify the proof system of $\not\leq$ so that it only produces such judgements.

For the sake of compactness of notation, we define a function $S(T)$ ("shape of T") from $F_\leq$ types to the three element linearly ordered set $\{t < \forall < Top\}$ as follows:

for any type variable u:      $S(u) = t$
for any t,T,U:      $S(\forall t{\leq}T.U) = \forall$
     $S(Top) = Top$

Then we collect the three axioms of $\not\leq$ into one:

(Shape$\not\leq$)      $S(T) > S(U) \Rightarrow \Gamma \vdash T \not\leq U$

The compatibility relation can be now formally defined.

*Notation*: For any relation denoted as $\Gamma \vdash T \mathrel{\mathcal{R}} U$:

$\Gamma \not\vdash T \mathrel{\mathcal{R}} U \Leftrightarrow_{def} \langle\Gamma,T,U\rangle \in J$ and not $\Gamma \vdash T \mathrel{\mathcal{R}} U$.

(Note that $\not\vdash$ is just a classical negation, so that $(\text{not}(\Gamma \not\vdash T \mathrel{\mathcal{R}} U)) \Leftrightarrow \Gamma \vdash T \mathrel{\mathcal{R}} U$).

*Definition*: $\Gamma \vdash T \subseteq U$ (read: T is compatible with U w.r.t. $\Gamma$)    $\Leftrightarrow_{def} \Gamma \not\vdash T \not\leq U$

By proposition 4.1, $\Gamma \vdash T \subseteq U$ means that trying to prove $\Gamma \vdash T \leq U$ results either in success or looping. We call the $\subseteq$ relation "compatibility", since $\Gamma \vdash T \subseteq V$ means that V is so near to being a supertype of T that the standard algorithm is not able to prove the opposite.

To study the compatibility relation, we first need a "positive" connotation for it. To this aim, we first observe that compatibility shares with subtyping a set of "compatibility properties".

*Definition*: A relation $\mathcal{R}$ on J is *compatible with subtyping* (compatible, for short) if:

($\forall \mathcal{R}$)      $\Gamma \vdash \forall t{\leq}T.U \mathrel{\mathcal{R}} \forall t{\leq}T.U' \Leftrightarrow \Gamma \vdash T \mathrel{\mathcal{R}} T$ and $\Gamma, t{\leq}T \vdash U \mathrel{\mathcal{R}} U'$
(VarTrans$\mathcal{R}$) $U{\neq}Top, U{\neq}t \Rightarrow \Gamma \vdash t \mathrel{\mathcal{R}} U$    $\Leftrightarrow \Gamma \vdash \Gamma(t) \mathrel{\mathcal{R}} U$
(Top$\mathcal{R}$)      $\Gamma \vdash T \mathrel{\mathcal{R}} Top$
(Id$\mathcal{R}$)      $\Gamma \vdash t \mathrel{\mathcal{R}} t$
(Shape$\mathcal{R}$)      $\Gamma \vdash T \mathrel{\mathcal{R}} U \Rightarrow S(T) \leq S(U)$

The compatibility properties give us a different way of characterizing $\leq$ and $\subseteq$: we will prove that they are respectively the minimum and the maximum compatible relations.

First, we show that any compatible relation has no intersection with the relation $\not\leq$.

*Proposition 4.2*: If $\mathcal{R}$ is compatible and $\Gamma \vdash T \not\leq U$, then $\Gamma \not\vdash T \mathrel{\mathcal{R}} U$.

*Proof hint*: By induction on the proof of $\Gamma \vdash T \not\leq U$, and by cases on the last rule applied. Essentially, any proof of $\Gamma \vdash T \not\leq U$ corresponds to a chain of applications of ($\forall \mathcal{R}$) and (VarTrans$\mathcal{R}$) which transform $\Gamma \vdash T \not\leq U$ into a contradiction to (Shape$\mathcal{R}$).

Now the "positive characterization" of the compatibility relation can be given; the following proposition is all that is needed in the next section to prove partial conservativity.

*Proposition 4.3*: ⊆ is the maximum compatible relation on J.

*Proof hint*: First we prove that ⊆ is compatible. Since ⊆ is defined as the complement of ⊄, we first negate both sides of all the properties, to express them in terms of ⊄:

(∀⊆)              $\Gamma \vdash \forall t \leq T'.U \not\subseteq \forall t \leq T.U' \Leftrightarrow \Gamma \vdash T \not\subseteq T'$ or $\Gamma, t \leq T \vdash U \not\subseteq U'$
(VarTrans⊆) $U \neq \text{Top}, U \neq t \Rightarrow \Gamma \vdash t \not\subseteq U \Leftrightarrow \Gamma \vdash \Gamma(t) \not\subseteq U$
(Top⊆)                              $\Gamma \not\vdash T \not\subseteq \text{Top}$
(Id⊆)                                $\Gamma \not\vdash t \not\subseteq t$
(Shape⊆)        $\Gamma \vdash T \not\subseteq U$        $\Leftarrow S(T) > S(U)$

Maximality of ⊆ is a corollary of proposition 4.2: since, for any compatible $\mathcal{R}$, $\Gamma \vdash T \not\subseteq U \Rightarrow \Gamma \not\vdash T \mathcal{R} U$, then $\Gamma \vdash T \mathcal{R} U \Rightarrow \Gamma \not\vdash T \not\subseteq U$, i.e. $\Gamma \vdash T \subseteq U$.

*Proposition 4.4*: ≤ is the minimum compatible relation on J.

*Proof hint*: Compatibility and minimality of ≤ are consequences of completeness and soundness of the subtype-checking algorithm presented in Section 2.

The fact that there are at least two different compatible relations could be surprising, since the compatibility conditions are rather restrictive. They contain both positive conditions, (Top) and (Id), which impose that something is in the relation, negative conditions (Shape) imposing that something is not in the relation, and bidirectional closure conditions, (∀) and (VarTrans), which allow us to deduce both the presence and the absence of something from the presence and the absence of something else. Moreover any judgement matches at least one of the five conditions, so that it seems that you can use those conditions to decide whether any judgement belongs to $\mathcal{R}$. Obviously, this is not the case, and the problem is that trying to use consistency conditions to decide whether a pair of types is in $\mathcal{R}$ can result in an infinite loop. On the other hand, in the decidable variants of $F_\leq$, the repeated application of the compatibility conditions always produces a negative or a positive axiom, so that in those systems subtyping is really the only compatible relation. So, the existence of a full range of different compatible relations is a feature of $F_\leq$ which is strictly related to its undecidability.

Before going back to the main stream of the paper let us pause to show that the compatible relation is similar to ≤ also by enjoying the properties of transitivity, narrowing and irreflexivity (corollary 4.9).

We first prove that any compatible relation is irreflexive.

*Definition*: If t is defined in a context $\Gamma$, the definition level of t in $\Gamma$, written $\mathcal{L}(t,\Gamma)$, is the number of variable definitions in $\Gamma$ which strictly precede the definition of t:

$\mathcal{L}(t, (t \leq A, \Gamma)) = 0$
$\mathcal{L}(t, (u \leq A, \Gamma)) = \mathcal{L}(t, \Gamma)+1 \quad (t \neq u)$

The definition level of a type T well formed in $\Gamma$, written $\mathcal{L}(T,\Gamma)$, is the maximum definition level of the variables which are free in T, and is 0 if no variable is free in T.

*Fact 4.5*: In any well-formed context, $\mathcal{L}(\Gamma(t),\Gamma) < \mathcal{L}(t,\Gamma)$, since all the free variables in $\Gamma(t)$ have to be defined before t.

*Lemma 4.6*: Let $\mathcal{R}$ be a compatible relation. Then $\Gamma \not\vdash \Gamma(t) \mathcal{R} t$.

*Proof hint*: By (Shape$\mathcal{R}$), $\Gamma \vdash \Gamma(t) \mathcal{R} t$ would imply $\Gamma(t)=u$ for a certain u; by Fact 4.5, this implies $\mathcal{L}(u,\Gamma) \leq \mathcal{L}(t,\Gamma)$; so we find a contradiction by proving, by induction on $\mathcal{L}(u,\Gamma)$, that: ($\Gamma \vdash u \mathcal{R} t$ and $u \neq t$) $\Rightarrow$ ($\mathcal{L}(u,\Gamma) > \mathcal{L}(t,\Gamma)$).

**Proposition 4.7**: Any compatible $\mathcal{R}$ is irreflexive: $\Gamma \vdash T \mathcal{R} U$ and $\Gamma \vdash U \mathcal{R} T \Rightarrow T=U$.

**Proof**: By induction on the size of T and U. By (Shape$\mathcal{R}$), $\Gamma \vdash T \mathcal{R} U$ and $\Gamma \vdash U \mathcal{R} T$ implies that $\mathcal{S}(T)=\mathcal{S}(U)$. Case $\mathcal{S}(T)=t$ is managed by 4.6, the others are routine.

Now the transitivity of $\sqsubseteq$ can be proved. Since we cannot reason by induction on the $\sqsubseteq$ relation, we will reason by maximality, by proving that the transitive closure of $\sqsubseteq$ is a compatible relation which contains $\sqsubseteq$.

**Notation**: If $\mathcal{R}$ is a relation on J, its lifting $\mathcal{R}_\Gamma$ to well formed environments is defined inductively as follows:

$$0 \ \mathcal{R}_\Gamma \ 0$$
$$\Gamma, t \leq T \ \mathcal{R}_\Gamma \ \Delta, t \leq U \ \Leftrightarrow_{def} \ \Gamma \ \mathcal{R}_\Gamma \Delta \text{ and } \Gamma \vdash T \mathcal{R} U$$

Now we define the transitive closure $\sqsubseteq^+$ of the $\sqsubseteq$ relation; condition (Narrow+) is strictly needed, since narrowing is essentially another aspect of transitivity, as shown in the proof of Proposition 4.8. Conditions ($\forall$+) and (VarTrans+) are added to the definition of $\sqsubseteq^+$ to simplify the same proof.

**Definition**: $\Gamma \vdash T \sqsubseteq^+ V$ is the relation defined by the following five deduction rules:

($\sqsubseteq$)        $\Gamma \vdash T \sqsubseteq V$        $\Rightarrow \Gamma \vdash T \sqsubseteq^+ V$
(Trans+)   $\Gamma \vdash T \sqsubseteq^+ U$ and $\Gamma \vdash U \sqsubseteq^+ V$   $\Rightarrow \Gamma \vdash T \sqsubseteq^+ V$
(Narrow+)  $\Gamma \vdash T \sqsubseteq^+ U$ and $\Delta \sqsubseteq^+ \Gamma$   $\Rightarrow \Delta \vdash T \sqsubseteq^+ U$
($\forall$+)       $\Gamma \vdash T \sqsubseteq^+ T'$ and $\Gamma, t \leq T \vdash U \sqsubseteq^+ U' \Rightarrow \Gamma \vdash \forall t \leq T'.U \sqsubseteq^+ \forall t \leq T.U'$
(VarTrans+) $U \neq Top, U \neq t, \Gamma \vdash \Gamma(t) \sqsubseteq^+ U$   $\Rightarrow \Gamma \vdash t \sqsubseteq^+ U$

**Proposition 4.8**: $\sqsubseteq^+$ is equal to $\sqsubseteq$.

**Proof**: $\sqsubseteq^+ \supseteq \sqsubseteq$ is immediate by condition ($\sqsubseteq$) above. We show that $\sqsubseteq^+$ is compatible; the thesis follows by maximality of $\sqsubseteq$ among compatible relations. We show the typical case, ($\forall \sqsubseteq^+$): $\Gamma \vdash \forall t \leq T'.U \sqsubseteq^+ \forall t \leq T.U' \Leftrightarrow \Gamma \vdash T \sqsubseteq^+ T'$ and $\Gamma, t \leq T \vdash U \sqsubseteq^+ U'$

$\Leftarrow$: by ($\forall$+)

$\Rightarrow$: By induction on the size of the proof that $\Gamma \vdash \forall t \leq T'.U \sqsubseteq^+ \forall t \leq T.U'$ and by cases on the last rule applied. ($\sqsubseteq$): by ($\forall \sqsubseteq$). ($\forall$+): immediate. (VarTrans+): impossible.

(Narrow+): exists $\Delta$: $\Gamma \sqsubseteq^+ \Delta$ and $\Delta \vdash \forall t \leq T'.U \sqsubseteq^+ \forall t \leq T.U'$
$\Rightarrow$(Ind.)        $\Delta \vdash T \sqsubseteq^+ T'$ and $\Delta, t \leq T \vdash U \sqsubseteq^+ U'$
$\Rightarrow$(Narrow+) $\Gamma \vdash T \sqsubseteq^+ T'$ and $\Gamma, t \leq T \vdash U \sqsubseteq^+ U'$.

(Trans+): by (Shape$\sqsubseteq^+$) the intermediate type has the shape $\forall t \leq V.W$:
$\Gamma \vdash \forall t \leq T'.U \sqsubseteq^+ \forall t \leq V.W, \ \Gamma \vdash \forall t \leq V.W \sqsubseteq^+ \forall t \leq T.U'$
$\Rightarrow$(Ind.)        $\Gamma \vdash V \sqsubseteq^+ T'$ and $\Gamma, t \leq V \vdash U \sqsubseteq^+ W$,
            $\Gamma \vdash T \sqsubseteq^+ V$ and $\Gamma, t \leq T \vdash W \sqsubseteq^+ U'$
$\Rightarrow$(Narrow+) $\Gamma \vdash V \sqsubseteq^+ T'$ and $\Gamma, t \leq T \vdash U \sqsubseteq^+ W$,
            $\Gamma \vdash T \sqsubseteq^+ V$ and $\Gamma, t \leq T \vdash W \sqsubseteq^+ U'$
$\Rightarrow$(Trans+)  $\Gamma \vdash T \sqsubseteq^+ T'$ and $\Gamma, t \leq T \vdash U \sqsubseteq^+ U'$.

**Corollary 4.9**: $\sqsubseteq$ enjoys both transitivity and narrowing, i.e.:

$\Gamma \vdash T \sqsubseteq U$ and $\Gamma \vdash U \sqsubseteq V \Rightarrow \Gamma \vdash T \sqsubseteq V$
$\Gamma \vdash T \sqsubseteq U$ and $\Delta \sqsubseteq \Gamma \ \Rightarrow \Delta \vdash T \sqsubseteq U$

This study of compatible relations leaves two interesting problems open:

- Does compatibility imply transitivity?
- Is there some other compatible relation apart from $\leq$ and $\sqsubseteq$? Is there some other interesting compatible relation apart from $\leq$ and $\sqsubseteq$?

We have no answer for the first question, while the second will be answered in the next section.

## 5 Conservativity of $\mu F_\leq$ Over Terminating Judgements

In Section 3 we showed that $\mu F_\leq$ is not conservative over $F_\leq$, by exhibiting a non provable $F_\leq$ judgement which is provable in $\mu F_\leq$. That judgement is "divergent", which means that it makes the standard $F_\leq$ type-checking algorithm diverge. In this section we generalize that observation, by showing that every "finitely failing" judgement, i.e. which makes the standard $F_\leq$ type-checking algorithm fail, is not provable in $\mu F_\leq$, so that non conservativity is restricted to the "pathological" set of the diverging judgements.

We show this by proving that $\mu F_\leq$ subtyping is still a compatible relation, so that it cannot relate more types than the maximum compatible relation "$\sqsubseteq$".

*Definition*: $\mu J$ is the set of triples which form well-formed $\mu F_\leq$ judgements:

$$\mu J = \{ \Gamma, T, U \mid \Gamma \vdash_\mu T \text{ type}, \Gamma \vdash_\mu U \text{ type}\}$$

*Definition*: A relation $\mathcal{R}$ on $\mu J$ is $\mu$-compatible when:

($\forall \mathcal{R}$)  $\quad\Gamma \vdash \forall t \leq T'.U \; \mathcal{R} \; \forall t \leq T.U' \Leftrightarrow \Gamma \vdash T \; \mathcal{R} \; T' \text{ and } \Gamma, t \leq T \vdash U \; \mathcal{R} \; U'$
(VarTrans$\mathcal{R}$) $U \neq$ Top, $U \neq t \Rightarrow \Gamma \vdash t \; \mathcal{R} \; U \quad\Leftrightarrow \Gamma \vdash \Gamma(t) \; \mathcal{R} \; U$
(Top$\mathcal{R}$)  $\qquad\qquad\qquad\qquad\qquad\qquad\qquad\Gamma \vdash T \; \mathcal{R} \; \text{Top}$
(Id$\mathcal{R}$)  $\qquad\qquad\qquad\qquad\qquad\qquad\qquad\Gamma \vdash t \; \mathcal{R} \; t$
(Shape$\mathcal{R}$)  $\quad\Gamma \vdash T \; \mathcal{R} \; U \qquad\qquad\Rightarrow S(T) \leq S(U)$
($\mu \mathcal{R} l$)  $\qquad\Gamma \vdash \mu X.T \; \mathcal{R} \; U \quad\Leftrightarrow \Gamma \vdash [\mu X.T/X]T \; \mathcal{R} \; U$
($\mu \mathcal{R} r$)  $\qquad\Gamma \vdash T \; \mathcal{R} \; \mu X.U \quad\Leftrightarrow \Gamma \vdash T \; \mathcal{R} \; [\mu X.U/X]U$

Where, in condition (Shape$\mathcal{R}$), $S(\mu X.T) =_{\text{def}} S(T)$

*Proposition 5.1*: The relation $\Gamma \vdash_\mu T \leq U$ is $\mu$-compatible.

*Proof*: See Appendix B.

*Lemma 5.2*: Any $\mu$-compatible relation is compatible when restricted to J.

*Proof*: Easy (see the full paper).

The main theorem now follows immediately.

*Theorem 5.3*: If $\Gamma \vdash_\mu T \leq U$ and $\Gamma$, T and U do not contain $\mu$, then $\Gamma \vdash T \sqsubseteq U$

Proof: By proposition 5.1, the $\leq_\mu$ relation is $\mu$-compatible; by lemma 5.2, its restriction to $F_\leq$ is compatible. The thesis follows from the maximality of $\sqsubseteq$ among the compatible relations (proposition 4.3).

We have proved that the restriction of $\leq_\mu$ to J is strictly bigger than $F_\leq$ subtyping but is contained in $\sqsubseteq$. We can now prove that $\leq_\mu$ restricted to J is *strictly* contained in $\sqsubseteq$.

*Lemma 5.4*: The $\sqsubseteq$ relation is co-R.E. but is not R.E.

*Proof*: See Appendix B.

*Proposition 5.5*: $\mu F_\leq$ subtyping restricted to J is R.E.

Proof: All the rules of $\mu F_\leq$ subtyping are effective, so that they can be used to enumerate all the provable $\mu F_\leq$ subtyping judgements; J is a decidable subset of $\mu J$.

*Corollary*: $\leq_\mu$ restricted to J is *strictly* contained in ⊑.

Hence, $\leq_\mu$ restricted to J defines a third compatible relation, intermediate between $F_\leq$ subtyping and ⊑, and different from both of them.

This fact implies that there exists a looping judgement which is not provable in $\mu F_\leq$. So, if we knew such a judgement, we could try to use it to rephrase Pierce's proof of undecidability for $\mu F_\leq$. The problem is recognizing that a looping judgement is not provable in $\mu F_\leq$. In fact, the only technique we know to prove that a judgement is not provable in $\mu F_\leq$ is to apply backward the deduction rules to it to reduce it to a judgement violating the (Shape$\leq$) condition. However we have proved that this process never ends when is applied to a looping judgement (see proposition 4.3).

Note that compatibility of $\leq_\mu$ implies that, for every proof of $\Gamma \vdash_\mu T \leq U$, where the last rule used is transitivity, a proof exists for $\Gamma \vdash_\mu T \leq U$ where the last rule used is either (VarTrans) or a rule which is not transitivity; the key property, to prove this fact, is ($\forall\leq_\mu$). This sounds like a form of transitivity elimination. Actually, this fact only implies that, for any $n$, we can find a proof of $\Gamma \vdash_\mu T \leq U$ where full transitivity is not used in any of the last $n$ steps, but does not imply that we can find a proof where full transitivity is *never* used. The point is that, intuitively, it is possible that whenever we remove a transitivity instance from the bottom of the proof, we add some new transitivity instance to the higher levels, so that the process of transitivity elimination never ends.

## 6 Conclusion and future work

We have proved that recursive types are not a conservative extension for the $F_\leq$ system. Moreover we have shown that in the extended system $\mu F_\leq$ transitivity elimination does not hold, i.e. it is not possible to substitute the general transitivity rule with the restricted (VarTrans) version. This implies that a rich set of proofs of properties of $F_\leq$, which is based on transitivity elimination, is not valid for $\mu F_\leq$ (see, e.g., [CuGhe], [GheTh], [CaMaMiSce], [CuGhe2], [BrLo]). We have shown that the basic diverging judgement of $F_\leq$ is provable in $\mu F_\leq$, thus showing that Pierce's proof of undecidability of $F_\leq$ subtyping is not valid for $\mu F_\leq$ ([Pie]).

Non conservativity has serious practical consequences: it means that the standard type checking algorithm for $F_\leq$ is no more complete with respect to $\mu F_\leq$ judgements, even for judgements not containing recursive types.

As a positive counterpart to this set of negative results, we have shown that the difference between $F_\leq$ and $\mu F_\leq$ is restricted to the set of the diverging judgements. This means that the standard $F_\leq$ type-checking algorithm is at least sound, even if not complete, when used to check $\mu F_\leq$ subtyping on $F_\leq$ judgements, since it does not give any answer on those judgements whose provability is different in the two systems. Furthermore, the standard type-checking algorithm for $F_\leq$ "goes wrong", when applied to $\mu$-free $\mu F_\leq$ judgements, exactly in the same cases when it goes wrong with respect to $F_\leq$, i.e. on the diverging judgements. The only difference is that, in the case of $F_\leq$, the algorithm "goes wrong" by looping in some cases when it should fail, while, with $\mu F_\leq$, it loops on a set of judgements containing both provable and not provable ones. But the essential point is that those judgements have no serious possibility of appearing in a real program, so that a looping behavior of the algorithm in those cases may be tolerated. However, the actual impact of our result on the use of $F_\leq$ as a basis for language design should be more seriously taken into consideration.

Finally, conservativity over terminating judgements shows that non-conservativity of recursion is connected to the problem of undecidability of subtyping, and gives a technique to prove that recursion is conservative over the decidable variants of $F_\leq$.

We chose a very weak form of recursive types for this study. This implies that the non conservativity result has a wide applicability, since it holds for any notion of recursive types which is stronger than the one adopted. On the other hand, for the same reason, our partial conservativity result has a more limited applicability, since it immediately holds only for notions of recursive types which are weaker than the one adopted. So our next step should be the definition of a strong notion of recursive types, generalizing the [AmCa] approach to $F_\leq$, to verify whether the partial conservativity result can be extended to that system.

**Acknowledgements**

The counterexample was obtained by slightly modifying a judgement produced by Roberto Bergamini whose suggestion sparked off this work.

This work was carried out in the framework of the Galileo project, aimed at the design of a strongly typed object-oriented database language, and lead by Prof. A. Albano at the University of Pisa.

**References**

[AmCa] R. Amadio, L. Cardelli, *Subtyping Recursive Types*, DEC SRC Research Report 62, short version in Proc. of the 18th ACM Symposium on Principles of Programming Languages, 104-118, 1991.

[BrLo] K. Bruce, G. Longo, A Modest Model of Records, Inheritance and Bounded Quantification, *Information and Computation 87 (1/2)*, 196-240, 1990.

[Ca] L. Cardelli, *Fsub: the System*, note, 1991.

[CaLo] L. Cardelli, G. Longo, *A Semantic Basis for Quest*, DEC SRC Research Report 55, short version in Proc. Conf. on Lisp and Functional Programming, Nice, 1990.

[CaMaMiSce] L. Cardelli, S. Martini, J. Mitchell, A. Scedrov, An Extension of System F with Subtyping, *Proc. Conference on Theoretical Aspects of Computer Software*, Sendai, Japan, Springer-Verlag, Berlin, LNCS 526, 1991.

[CaWe] L. Cardelli, P. Wegner, On Understanding Types, Data Abstraction and Polymorphism, *ACM Computing Surveys 17 (4)*, 471-522, 1985.

[CuGhe] P.-L. Curien, G. Ghelli, Coherence of Subsumption in $F_\leq$, Minimum Typing and Type Checking, *Mathematical Structures in Computer Science 2(1)*, 55-91, 1992.

[CuGhe2] P.-L. Curien, G. Ghelli, Subtyping + Extensionality: Confluence of $\beta\eta top_\leq$ in $F_\leq$, extended abstract, *Proc. Conference on Theoretical Aspects of Computer Software*, Sendai, Japan, Springer-Verlag, Berlin, LNCS 526, 731-749, 1991.

[Ghe] G. Ghelli, *Divergence of $F_\leq$ Type-checking*, note, 1991.

[GhePie] G. Ghelli, B. Pierce, *Bounded Existentials and Minimal Typing*, note, 1992.

[GheTh] G. Ghelli, *Proof-theoretic Studies about a Minimal Type System Integrating Inclusion and Parametric Polymorphism*, PhD Thesis, TD-6/90, Univ. of Pisa, 1990.

[Ka] D. Katiyar, S. Sankar, Completely bounded quantification is decidable, Proc. of the *ACM SIGPLAN Workshop on ML and its Applications*, 1992.

[Pie] B. Pierce, Bounded Quantification is Undecidable, *Proc. of the 19th ACM Symposium on Principles of Programming Languages*, 305-315, 1992.

## Appendix A: $\mu F_{\leq}$.

*Environments* (sequences whose individual components have the form x:T or t≤T)

(∅env)    ()   env

(≤env)    $\dfrac{\Gamma \text{ env} \quad \Gamma \vdash T \text{ type} \quad t \notin \Gamma^7}{\Gamma, t \leq T \quad \text{env}}$    (:env)    $\dfrac{\Gamma \text{ env} \quad \Gamma \vdash T \text{ type} \quad x \notin \Gamma}{\Gamma, x:T \quad \text{env}}$

*Types*

(VarForm)   $\dfrac{\Gamma, t \leq T, \Gamma' \text{ env}}{\Gamma, t \leq T, \Gamma' \vdash t \text{ type}}$    (TopForm)   $\dfrac{\Gamma \text{ env}}{\Gamma \vdash \text{Top type}}$

(→ Form)   $\dfrac{\Gamma \vdash T \text{ type} \quad \Gamma \vdash U \text{ type}}{\Gamma \vdash T \rightarrow U \text{ type}}$    (∀ Form)   $\dfrac{\Gamma, t \leq T \vdash U \text{ type}}{\Gamma \vdash \forall t \leq T. U \text{ type}}$

(μForm)   $\dfrac{\Gamma \vdash [\text{Top}/X]U \text{ type}}{\Gamma \vdash \mu X.U \text{ type}}$

*Subtypes*

(Var≤)   $\dfrac{\Gamma, t \leq T, \Gamma' \text{ env}}{\Gamma, t \leq T, \Gamma' \vdash t \leq T}$    (Top≤)   $\dfrac{\Gamma \vdash T \text{ type}}{\Gamma \vdash T \leq \text{Top}}$

(→≤)   $\dfrac{\Gamma \vdash T \leq T' \quad \Gamma \vdash U \leq U'}{\Gamma \vdash T' \rightarrow U \leq T \rightarrow U'}$    (∀≤)   $\dfrac{\Gamma \vdash T \leq T' \quad \Gamma, t \leq T \vdash U \leq U'}{\Gamma \vdash \forall t \leq T'.U \leq \forall t \leq T.U'}$

(Id≤)   $\dfrac{\Gamma \vdash T \text{ type}}{\Gamma \vdash T \leq T}$    (Trans≤)   $\dfrac{\Gamma \vdash T \leq U \quad \Gamma \vdash U \leq V}{\Gamma \vdash T \leq V}$

*Recursive types*

(unfold-l)   $\dfrac{\Gamma \vdash [\mu X.T/X]T \leq U}{\Gamma \vdash \mu X.T \leq U}$    (unfold-r)   $\dfrac{\Gamma \vdash U \leq [\mu X.T/X]T}{\Gamma \vdash U \leq \mu X.T}$

*Algorithmic subtype rules: drop (Trans≤) and add:*

(VarTrans≤)   $\dfrac{\Gamma, t \leq T, \Gamma' \vdash T \leq U \quad U \neq \text{Top}, U \neq t}{\Gamma, t \leq T, \Gamma' \vdash t \leq U}$

**Appendix B**: Proofs of propositions 5.1 and 5.4.

*Proposition 5.1*: The relation $\Gamma \vdash_\mu T \leq U$ is μ-compatible.

*Proof*: We have to prove that:

| | | |
|---|---|---|
| (∀≤) | $\Gamma \vdash \forall t \leq T.U \leq \forall t \leq T.U'$ | $\Leftrightarrow \Gamma \vdash T \leq T'$ and $\Gamma, t \leq T \vdash U \leq U'$ |
| (VarTrans≤) | $U \neq \text{Top}, U \neq t \Rightarrow \Gamma \vdash t \leq U$ | $\Leftrightarrow \Gamma \vdash \Gamma(t) \leq U$ |
| (Top≤) | | $\Gamma \vdash T \leq \text{Top}$ |
| (Id≤) | | $\Gamma \vdash t \leq t$ |
| (Shape≤) | $\Gamma \vdash T \leq U$ | $\Rightarrow S(T) \leq S(U)$ |
| (μ≤l) | $\Gamma \vdash \mu X.T \leq U$ | $\Leftrightarrow \Gamma \vdash [\mu X.T/X]T \leq U$ |
| (μ≤r) | $\Gamma \vdash T \leq \mu X.U$ | $\Leftrightarrow \Gamma \vdash T \leq [\mu X.U/X]U$ |

---

[7] We write t∈Γ if t ≤ T is a component of Γ, for a certain T; similarly for x∈Γ.

(Top≤) and (Id≤) are immediate. The condition (Shape≤) is enforced by all the rules and is transmitted inductively by the (μ) rules. (VarTrans≤), in the ⇐ direction, is a consequence of the (Var≤) and (Trans≤) subtyping rules; the other ⇐ implications, including (Top≤) and (Id≤), are equivalent to the corresponding subtyping rules. Now we prove the ⇒ implications for (∀≤), (μ≤l/r) and (VarTrans≤).

(∀≤)  This is the crucial case. To prove it we strengthen the condition to the conjunction of the following four conditions (where L=μX.∀t≤T'.U and R=μY.∀t≤T.U'):

(∀1) $\Gamma \vdash \forall t \leq T'.U \leq \forall t \leq T.U'\ \Rightarrow\ \Gamma \vdash T \leq T'$ and $\Gamma, t \leq T \vdash U \leq U'$

(∀2) $\Gamma \vdash \mu X.\forall t \leq T'.U \leq \forall t \leq T.U' \Rightarrow\ \Gamma \vdash T \leq [L/X]T'$ and $\Gamma, t \leq T \vdash [L/X]U \leq U'$

(∀3) $\Gamma \vdash \forall t \leq T'.U \leq \mu Y.\forall t \leq T.U' \Rightarrow\ \Gamma \vdash [R/Y]T \leq T'$ and $\Gamma, t \leq [R/Y]T \vdash U \leq [R/Y]U'$

(∀4) $\Gamma \vdash \mu X.\forall t \leq T'.U \leq \mu Y.\forall t \leq T.U'$
    $\Rightarrow\ \Gamma \vdash [R/Y]T \leq [L/X]T'$ and $\Gamma, t \leq [R/Y]T \vdash [L/X]U \leq [R/Y]U'$

We denote the four conditions at once as follows, where (μX.)... means "optional recursion" and []U means "substitute the recursive variable in U, if needed" (this is just an informal notation to avoid repeating four times some proofs):

(∀*) $\Gamma \vdash (\mu X.)\forall t \leq T'.U \leq (\mu Y.)\forall t \leq T.U' \Rightarrow\ \Gamma \vdash []T \leq []T'$ and $\Gamma, t \leq []T \vdash []U \leq []U'$

We prove the four conditions by induction on the minimum depth of a proof proving

(*) $\Gamma \vdash_\mu (\mu X.)\forall t \leq T'.U \leq (\mu Y.)\forall t \leq T.U'$.

(*) can be proved by transitivity, by (∀) or by (unfold-l/r) (the (Shape≤) condition leaves no other possibility).

(Trans): (the shape of the intermediate type is determined by (Shape≤)):

$\Gamma \vdash_\mu (\mu X.)\forall t \leq T'.U \leq (\mu Z.)\forall t \leq R.S \qquad \Gamma \vdash_\mu (\mu Z.)\forall t \leq R.S \leq (\mu Y.)\forall t \leq T.U'$

⇒(by induction)   $\Gamma \vdash []R \leq []T'$       $\Gamma \vdash []T \leq []R$
                  $\Gamma, t \leq []R \vdash_\mu []U \leq []S$   $\Gamma, t \leq []T \vdash_\mu []S \leq []U'$
⇒(by narrowing[8]) $\Gamma, t \leq []T \vdash_\mu []U \leq []S$
⇒(by transitivity) $\Gamma \vdash []T \leq []T',\ \Gamma, t \leq []T \vdash_\mu []U \leq []U'$

(∀): The thesis is immediate

(unfold-r):  $\dfrac{\Gamma \vdash_\mu \forall t \leq []T'.[]U \leq (\mu Y.)\forall t \leq T.U'}{\Gamma \vdash_\mu \mu X.\forall t \leq T'.U \leq (\mu Y.)\forall t \leq T.U'}$

By induction: $\Gamma \vdash []T \leq []T',\ (\Gamma, t \leq []T) \vdash_\mu []U \leq []U'$; (unfold-r) is identical.

(μ≤l/r)  $\Gamma \vdash \mu X.V \leq U \Rightarrow\ \Gamma \vdash [\mu X.V/X]V \leq U$

Let $\Gamma \vdash \mu X.V \leq U$. Either it has been proved by (unfold-l), and the thesis follows immediately, or it has been proved by transitivity from:

$\Gamma \vdash \mu X.V \leq T \qquad \Gamma \vdash T \leq U$
⇒(induction)  $\Gamma \vdash [\mu X.V/X]V \leq T \qquad \Gamma \vdash T \leq U$
⇒(Trans)      $\Gamma \vdash [\mu X.V/X]V \leq U$

---

[8] Narrowing: $\Delta \vdash T \leq U$ and $\Gamma \leq_\Gamma \Delta\ \Rightarrow\ \Gamma \vdash T \leq U$; proof hint: substitute any use of an axiom $\Delta, \Delta' \vdash t \leq \Delta(t)$ in the proof of $\Delta \vdash T \leq U$ with the transitivity composition of an axiom $\Gamma, \Delta' \vdash t \leq \Gamma(t)$ and a proof of $\Gamma, \Delta' \vdash \Gamma(t) \leq \Delta(t)$.

or it has been proved by (unfold-r) (U=μY.T), from:

$\Gamma \vdash \mu X.V \leq [\mu Y.T/Y]T$
⇒(induction)   $\Gamma \vdash [\mu X.V/X]V \leq [\mu Y.T/Y]T$
⇒(unfold-r)    $\Gamma \vdash [\mu X.V/X]V \leq \mu Y.T(=U)$

(VarTrans≤)   U≠Top, U≠t, $\Gamma \vdash t \leq U$  ⇒  $\Gamma \vdash \Gamma(t) \leq U$

Let $\Gamma \vdash t \leq U$. Either $\Gamma \vdash t \leq U$ is proved by VarTrans (trivial) or by transitivity:
$\Gamma \vdash t \leq V$   $\Gamma \vdash V \leq U$;

If V=t, the thesis is immediate; otherwise, by (Shape), V≠Top:
⇒(induction)   $\Gamma \vdash \Gamma(t) \leq V$   $\Gamma \vdash V \leq U$
⇒(Trans)       $\Gamma \vdash \Gamma(t) \leq U$

or $\Gamma \vdash t \leq U$ is proved by (unfold-r) (U=μY.T), from:

$\Gamma \vdash t \leq [\mu Y.T/Y]T$
⇒(induction)   $\Gamma \vdash \Gamma(t) \leq [\mu Y.T/Y]T$
⇒(unfold-r)    $\Gamma \vdash \Gamma(t) \leq \mu Y.T(=U)$

□

*Lemma 5.4*: The ⊑ relation is co-R.E. but is not R.E.

*Proof hint*: The ⊑ relation is co-R.E. since its complement ⋢ is semidecidable, because it is semidecided by the finite failure of the standard type-checking algorithm. Since ⊑ is co-R.E, we prove that it is not R.E. by proving that it is not decidable. This can be proved using, essentially, the same proof used in [Pie] to show the undecidability of ≤. We cannot rephrase here the whole Pierce's proof, but we will just give an outline.

Pierce defines a state machine, called the "rowing machine", whose halting problem is undecidable, since it is equivalent to the halting problem of two-counter Turing machines. Then he considers the following subset $J^F$ of J (F stands for "Flattened")[9]:

$T^+ = \text{Top} \mid \forall x_1 \leq T^-_1 ... \forall x_n \leq T^-_n.T^-$
$T^- = x \mid \forall x_1 ... \forall x_n.T^+$
$J^F = \vdash T^- \leq T^+$

he shows that the following two-rule reduction system is correct and complete for $F_\leq$ subtyping restricted to $J^F$ triples:

(FTop)   $\vdash T^- \leq \text{Top}$                                          ▷ **true**
(F∀-)    $\vdash \forall x_1 ... \forall x_n.T^+ \leq \forall x_1 \leq T^-_1 ... \forall x_n \leq T^-_n.T^-$ ▷
         $\vdash [T^-_1/x_1 ... T^-_n/x_n]T^- \leq [T^-_1/x_1 ... T^-_n/x_n]T^+$

finally he shows that every rowing machine can be encoded as a $J^F$ triple which terminates if and only if the rowing machine does, which implies that termination of the two-rule system on $J^F$ triples is undecidable, hence that the provability of $J^F$ judgements is undecidable, and that subtyping on the whole $F_\leq$ is undecidable.

Note that every $J^F$ triple is either a provable or a diverging judgement; to prove undecidability of finite failure we just substitute $J^F$ with a different subset $J^G$ of J which contains only finitely failing or diverging triples, by susbstituting Top with a fresh variable y:

---

[9] The actual definition of $J^F$ is slightly more complex: first, in every type $\forall x1 \leq T1...xn \leq Tn.T$, no xi can be free in any Ti+j; furthermore, all the quantifier prefixes $\forall x1 \leq T1...xn \leq Tn$ appearing in a unique judgement must have the same length.

$T^+ = y \mid \forall x_1 \leq T^-_1 ... \forall x_n \leq T^-_n.\text{-}T^-$
$T^- = x$ (with $x \neq y$) $\mid \forall x_1 ... \forall x_n.\text{-}T^+$
$J^G = y \leq \text{Top} \vdash T^- \leq T^+$

Now we can rephrase Pierce's proof by using the following reduction system, correct and complete on $J^G$ with respect to finite failure:

(Gy)  $y \leq \text{Top} \vdash T^- \not\leq y$  ▷  **true**

(G∀-)  $y \leq \text{Top} \vdash \forall x_1 ... \forall x_n.\text{-}T^+ \not\leq \forall x_1 \leq T^-_1 ... \forall x_n \leq T^-_n.\text{-}T^-$  ▷
$\qquad\qquad y \leq \text{Top} \vdash [T^-_1/x_1 ... T^-_n/x_n]T^- \not\leq [T^-_1/x_1 ... T^-_n/x_n]T^+$

Now, by encoding rowing machines over $J^G$ triples, we prove that finite failure of subtyping is undecidable on $J^G$, and hence is undecidable on the whole J.

# The Conservation Theorem revisited

Philippe de Groote

INRIA-Lorraine – CRIN – CNRS
Campus Scientifique - B.P. 239
54506 Vandœuvre-les-Nancy Cedex – FRANCE

**Abstract.** This paper describes a method of proving strong normalization based on an extension of the conservation theorem. We introduce a structural notion of reduction that we call $\beta_S$, and we prove that any $\lambda$-term that has a $\beta_I\beta_S$-normal form is strongly $\beta$-normalizable. We show how to use this result to prove the strong normalization of different typed $\lambda$-calculi.

## 1 Introduction

We present a method of proving strong normalization for several typed $\lambda$-calculi. This method is based on a extension of the conservation theorem.

The conservation theorem for $\lambda I$ [1, CHAP. 11, §3.]), says that all the $\beta$-reducts of a $\lambda I$-term that is not strongly $\beta$-normalizable are not strongly $\beta$-normalizable. As a corollary, any $\lambda I$-term that has a $\beta$-normal form is strongly $\beta$-normalizable. This is one of the few results relating strong normalization to (weak) normalization.

The above property, of course, fails for $\lambda K$-terms. The conservation theorem, when formulated for $\lambda K$, concerns only the $\beta_I$-reducts of the $\lambda K$-terms and not all its $\beta$-reducts. Hence, we lose the corollary and we cannot use the theorem to turn a proof of normalization into a proof of strong normalization.

In this paper, we state and prove a version of the corollary that holds for $\lambda K$-terms. To this end, we introduce a new notion of reduction ($\beta_S$) that allows the contraction of the $\beta_K$-redexes to be delayed. We prove that any $\lambda$-term that has a $\beta_I\beta_S$-normal form is strongly $\beta_I\beta_S$-normalizable. Then, it turns out by postponement that any $\lambda$-term that has a $\beta_I\beta_S$-normal form is strongly $\beta$-normalizable. This is the central result of the paper and we show how to use it to prove the strong normalization of different typed $\lambda$-calculi.

The typed calculi we consider are calculi à la Church [3]. Yet we do not consider that the type discipline is part of the term formation rules. We rather consider that the typing rules allow well-typed terms to be singled out from the set of raw terms. Therefore our technical framework is the untyped $\lambda$-calculus and our strong normalization proofs rely on the so-called erasing trick. Nevertheless, we show in Section 6 how to extend our results to Barendregt's set $\mathcal{T}$ of pseudo-terms.

The paper is organized as follows. Section 2 is reminder of well-known definitions concerning the untyped $\lambda$-calculus. In Section 3, we introduce the notion of reduction $\beta_S$ and we establish the postponement property of $\beta_K$-contractions with respect to $\beta_S$- and $\beta_I$-contractions. In Section 4, we prove the conservation theorem for $\beta_I\beta_S$-reductions. To this end, we use labeled $\lambda$-terms à la Lévy. In Section 5, we obtain, as a corollary of the postponement and conservation properties, that $\beta_I\beta_S$-normalization implies strong $\beta$-normalization. This result is used in section Section 6 to prove the strong normalization of Church's simply typed $\lambda$-calculus. Then we show how to extend the proof to Barendregt's $\lambda$-cube. We present our conclusions in Section 7.

## 2 Basic Definitions

In this section we remind the reader of some basic notions about the type-free $\lambda$-calculus. The definitions we give, which are taken from [1], concern mainly the concepts of reduction and normalization. The reader familiar with this material may proceed directly to Section 3.

Type-free $\lambda$-terms are built up on an infinite numerable set of variable $\mathcal{V}$ according to the following definition.

**Definition 1.** The set $\Lambda$ of $\lambda$-terms is inductively defined as follows:

i. $x \in \mathcal{V} \Rightarrow x \in \Lambda$,
ii. $x \in \mathcal{V}, M \in \Lambda \Rightarrow (\lambda x.M) \in \Lambda$,
iii. $M, N \in \Lambda \Rightarrow (M N) \in \Lambda$.

The symbol $\lambda$ is a binding operator and the notions of free and bound occurrences of a variable are as usual in logic. In particular, the free occurrences of $x$ in $M$ are bound in $(\lambda x.M)$. The set of variables occurring free in a $\lambda$-term $M$ is denoted FV($M$). The $\lambda$-terms that can be transformed into each other by renaming their bound variables are identified. We also consider that some variable convention prevents us from clashes of variables (see [1, page 26], also [5] for a formal treatment).

**Definition 2.** Any binary relation $R \subset \Lambda \times \Lambda$ is called a notion of reduction. If $R$ is a notion of reduction and $(M, N) \in R$, $M$ is called a $R$-redex and $N$ is called the contractum of $M$.

Given some notion of reduction $R$, one defines the following binary relations between $\lambda$-terms: the relation of $R$-contraction $(\rightarrow_R)$, the relation of $R$-reduction $(\twoheadrightarrow_R)$, the relation of strict $R$-reduction $(\xrightarrow{+}_R)$, and the relation of $R$-conversion $(\leftrightarrow\!\!\!\rightarrow_R)$.

**Definition 3.** Let $R$ be a notion of reduction. The corresponding contraction relation is the least relation containing $R$, and compatible with the $\lambda$-term formation rules. This relation is inductively defined by the following rules:

i. $M \rightarrow_R N$ if $(M, N) \in R$,   ii. $\dfrac{M \rightarrow_R N}{(\lambda x.M) \rightarrow_R (\lambda x.N)}$,

iii. $\dfrac{M \rightarrow_R N}{(M O) \rightarrow_R (N O)}$,   iv. $\dfrac{M \rightarrow_R N}{(O M) \rightarrow_R (O N)}$.

The relation of $R$-reduction $(\twoheadrightarrow_R)$, is the transitive, reflexive closure of the relation of $R$-contraction; the relation of strict $R$-reduction $(\xrightarrow{+}_R)$ is the transitive closure of the relation of $R$-contraction; the relation of $R$-conversion $(\leftrightarrow\!\!\!\rightarrow_R)$ is the transitive, reflexive, symmetric closure of the relation of $R$-contraction.

Any notion of reduction induces also the corresponding concepts of normal form, normalization, and strong normalization.

**Definition 4.** Let $R$ be a notion of reduction. A $\lambda$-term $M$ is called a $R$-normal form if and only if there does not exist $N \in \Lambda$ such that $M \rightarrow_R N$.

**Definition 5.** Let $R$ be a notion of reduction. A $\lambda$-term $M$ is called $R$-normalizable if and only if there exists a $R$-normal form $N$ such that $M \twoheadrightarrow_R N$.

**Definition 6.** Let $R$ be a notion of reduction. A $\lambda$-term $M$ is called strongly $R$-normalizable if and only if there exists an upper bound to the length $n$ of any sequence of $R$-contractions starting in $M$:

$$M \equiv M_0 \to_R M_1 \to_R \ldots \to_R M_n.$$

$M[x:=N]$ denotes the result of substituting a $\lambda$-term $N$ for the free occurrences of a variable $x$ in a $\lambda$-term $M$. This operation is defined as follows.

**Definition 7.**

i. $x[x:=N] \equiv N$,
ii. $y[x:=N] \equiv y$   if $x \not\equiv y$,
iii. $(\lambda y. M)[x:=N] \equiv (\lambda y. M[x:=N])$,
iv. $(M\,O)[x:=N] \equiv (M[x:=N]\,O[x:=N])$.

The principal notion of reduction of the $\lambda$-calculus is the notion of reduction $\beta$.

**Definition 8.** The notion of reduction $\beta$ is defined by the following contraction rule:

$\beta:\quad ((\lambda x.M)\,N) \to M[x:=N]$.

## 3 The Notions of Reduction $\beta_I$, $\beta_K$, and $\beta_S$

A $\beta$-redex $((\lambda x.M)\,N)$ is called an $I$-redex if $x \in \mathrm{FV}(M)$ and a $K$-redex otherwise. This distinction allows the notion of $\beta$-reduction to be split into the two notions of $\beta_I$- and $\beta_K$-reduction.

**Definition 9.** The notions of reduction $\beta_I$ and $\beta_K$ are respectively defined by the following contraction rules:

$\beta_I:\quad ((\lambda x.M)\,N) \to M[x:=N]$   if $x \in \mathrm{FV}(M)$,
$\beta_K:\quad ((\lambda x.M)\,N) \to M$   if $x \notin \mathrm{FV}(M)$.

A possible strategy to prove strong normalization when dealing with two different notions of reduction is to take advantage of some postponement property. For instance, the postponement of $\eta$-contractions with respect to $\beta$-contractions is a well known property (see [1, page 386]) from which it follows that any strongly $\beta$-normalizable $\lambda$-term is strongly $\beta\eta$-normalizable.

The postponement strategy may be used when two notions of reduction, say $R_1$ and $R_2$, are such that

1. any term of interest is strongly $R_1$-normalizable,
2. any term of interest is strongly $R_2$-normalizable,
3. the contraction of a $R_2$-redex cannot create a $R_1$-redex.

In the case of $\beta_I$ and $\beta_K$, Conditions 1 and 2 are satisfied. Indeed, we have that

1. any $M \in \Lambda$ that is $\beta_I$-normalizable is strongly $\beta_I$-normalizable—this is an easy consequence of the conservation theorem for $\lambda I$ (see [1, Chap. 11, §3.]),
2. any $M \in \Lambda$ is strongly $\beta_K$-normalizable—this is obvious because the length of any $\beta_K$-contractum is strictly less than the length of the corresponding $\beta_K$-redex.

Unfortunately, neither $\beta_I$-contractions nor $\beta_K$-contractions may be postponed. This is shown by the following counterexamples:

$((\lambda x.(x\,M))(\lambda y.z)) \to_{\beta_I} ((\lambda y.z)\,M)$    a $\beta_K$-redex is created,
$(((\lambda y.(\lambda x.x))\,M)\,N) \to_{\beta_K} ((\lambda x.x)\,N)$    a $\beta_I$-redex is created.

In order to fix this problem, we are going to introduce a third notion of reduction, namely $\beta_S$. This notion of reduction is such that

1. if $M \twoheadrightarrow_{\beta_S} N$ then $M \leftrightarrow\!\!\!\twoheadrightarrow_\beta N$,
2. $\beta_K$-contractions may be postponed with respect of $\beta_I\beta_S$-contractions,
3. the conservation theorem holds for $\beta_I$ and $\beta_S$.

The first of these three properties is immediate. The second one and the third one will be established respectively as Lemma 11 and Theorem 21.

**Definition 10.** The notion of reduction $\beta_S$ is defined by the following contraction rule:

$$\beta_S: \quad (((\lambda x.M)\,N)\,O) \to ((\lambda x.(M\,O))\,N) \quad \text{if } x \notin \mathrm{FV}(M).$$

Notice that we have by the variable convention that $x \notin \mathrm{FV}(O)$.

**Lemma 11.** (postponement of $\beta_K$-contractions) *Let* $R \in \{\beta_I, \beta_S\}$. *Let* $M, N, O \in \Lambda$ *be such that*

$$M \to_{\beta_K} N \quad \text{and} \quad N \to_R O$$

*Then there exists* $P \in \Lambda$ *such that*

$$M \xrightarrow{+}_{\beta_I\beta_S} P \quad \text{and} \quad P \twoheadrightarrow_{\beta_K} O$$

**Proof.** The proof is by induction on the derivation of $M \to_{\beta_K} N$, distinguishing subcases according to the way $N \to_R O$. The details are given in Appendix A. □

## 4 The Conservation Theorem for $\beta_I$ and $\beta_S$

In this section we establish the main technical result of this paper, namely that any $\beta_I\beta_S$-normalizable $\lambda$-term is strongly $\beta_I\beta_S$-normalizable. To this end we use labeled $\lambda$-terms à la Lévy (see [1, CHAP. 14]).

**Definition 12.** The set $\Lambda^{\mathbb{N}}$ of labeled $\lambda$-terms is inductively defined as follows:

i. $n \in \mathbb{N}, x \in \mathcal{V} \Rightarrow (x)^n \in \Lambda^{\mathbb{N}}$,
ii. $n \in \mathbb{N}, x \in \mathcal{V}, M \in \Lambda^{\mathbb{N}} \Rightarrow (\lambda x.M)^n \in \Lambda^{\mathbb{N}}$,
iii. $n \in \mathbb{N}, M \in \Lambda^{\mathbb{N}}, N \in \Lambda^{\mathbb{N}} \Rightarrow (M\,N)^n \in \Lambda^{\mathbb{N}}$.

We use $M, N, O, \ldots$ as metavariable ranging on labeled $\lambda$-terms. We use the notation $M^n$ or $(M)^n$ to stress that the outermost label of a labeled $\lambda$-term $M$ is $n$. Thus, according to this last convention, the meta-expressions $M$, $M^n$, and $(M)^n$ used in a same context stand exactly for the same labeled $\lambda$-term. We also write $M^{+m}$ or $(M)^{+m}$ for the labeled $\lambda$-term obtained by adding $m$ to the outermost label of a labeled $\lambda$-term $M$. Thus, if the outermost label of $M$ is $n$, then $(M)^{+m}$ denotes the same term than $(M)^{n+m}$.

Let $M \in \Lambda^{\mathbb{N}}$. We write $|M|$ for the (unlabeled) $\lambda$-term obtained by erasing all the labels in $M$. Therefore, for $M \in \Lambda^{\mathbb{N}}$, we have $|M| \in \Lambda$.

Now, let $M \in \Lambda$. We identify $M$ with the labeled $\lambda$-term $M'$ such that (i) $|M'| \equiv M$, (ii) all the labels in $M'$ are 0. Therefore, we have that $\Lambda \subset \Lambda^{\mathbb{N}}$.

Labels will be used as counters to record the number of contracted redexes when reducing a term. This idea motivates the next two definitions. The first one generalizes the operation of substitution on labeled $\lambda$-terms. The second one introduces labeled versions of the notions of reduction $\beta_I$ and $\beta_S$.

**Definition 13.** The substitution $M[x := N]$ of a labeled $\lambda$-term $M$ for the free occurrences of a variable $x$ in a labeled $\lambda$-term $N$ is defined as follows.

i. $(x)^m[x := N^n] \equiv (N)^{m+n}$,
ii. $(y)^m[x := N^n] \equiv (y)^m$    if $x \not\equiv y$,
iii. $(\lambda y. M)^m[x := N^n] \equiv (\lambda y. M[x := N^n])^m$,
iv. $(M\,O)^m[x := N^n] \equiv (M[x := N^n]\,O[x := N^n])^m$.

The labeled versions of the notions of reduction $\beta_I$ and $\beta_S$ are called respectively $\beta_I^+$ and $\beta_S^+$.

**Definition 14.** The notions of reduction $\beta_I^+$ and $\beta_S^+$ are respectively defined by the following contraction rules:

$\beta_I^+ : \quad ((\lambda x. M)^m\, N)^n \rightarrow (M[x := N])^{+(m+n+1)}$    if $x \in \text{FV}(M)$,
$\beta_S^+ : \quad (((\lambda x. M)^m\, N)^n\, O)^o \rightarrow ((\lambda x.(M\,O))\, N)^{m+n+o+1}$    if $x \notin \text{FV}(M)$.

To adapt the definition of contraction is straightforward.

**Definition 15.** Let $R \subset \Lambda^{\mathbb{N}} \times \Lambda^{\mathbb{N}}$. The relation of $R$-contraction is inductively defined by the following rules:

i.   $M \rightarrow_R N$   if $(M, N) \in R$,      ii.   $\dfrac{M \rightarrow_R N}{(\lambda x. M)^n \rightarrow_R (\lambda x. N)^n}$,

iii.   $\dfrac{M \rightarrow_R N}{(M\,O)^n \rightarrow_R (N\,O)^n}$,      iv.   $\dfrac{M \rightarrow_R N}{(O\,M)^n \rightarrow_R (O\,N)^n}$.

The definitions of $R$-reduction, strict $R$-reduction, and $R$-conversion are unchanged.

The notion of reduction $\beta_I^+ \cup \beta_S^+$ satisfies the Church-Rosser property.

**Theorem 16.** (Church-Rosser) *Let $M, N, O \in \Lambda^{\mathbb{N}}$ be such that*

$$M \twoheadrightarrow_{\beta_I^+ \beta_S^+} N \text{ and } M \twoheadrightarrow_{\beta_I^+ \beta_S^+} O.$$

*Then there exists $P \in \Lambda^{\mathbb{N}}$ such that*

$$N \twoheadrightarrow_{\beta_I^+ \beta_S^+} P \text{ and } O \twoheadrightarrow_{\beta_I^+ \beta_S^+} P.$$

*Proof.* One uses the lemma of Hindley-Rossen: first one establishes that $\beta_I^+$ and $\beta_S^+$ are individually Church-Rosser; then one shows that $\beta_I^+$ and $\beta_S^+$ commute. This can be done using the method of Tait and Martin-Löf. The details are given in Appendix B. □

The next step is to define the weight of a term as the sum of all its labels.

**Definition 17.** The weight $\Theta[M]$ of a labeled $\lambda$-term $M$ is defined as follows.

i. $\Theta[(x)^n] = n$,
ii. $\Theta[(\lambda y.\,M)^n] = n + \Theta[M]$,
iii. $\Theta[(M\,N)^n] = n + \Theta[M] + \Theta[N]$.

We are now in the position of proving the conservation theorem for $\beta_I$ and $\beta_S$. The proof consists of three easy lemmas.

**Lemma 18.** *Let $R \in \{\beta_I^+, \beta_S^+\}$, and let $M, N \in \Lambda^{\mathbb{N}}$ be such that $M \xrightarrow{+}_R N$. Then $\Theta[M] < \Theta[N]$.*

*Proof.* The statement is proven for one-step reduction by a straightforward induction on the definition of contraction. Notice, in the case of $\beta_I^+$, the part played by the proviso $x \in \mathrm{FV}(M)$. □

**Lemma 19.** *(Conservation for $\beta_I^+$ and $\beta_S^+$) Let $M \in \Lambda^{\mathbb{N}}$ be $\beta_I^+\beta_S^+$-normalizable. Then $M$ is strongly $\beta_I^+\beta_S^+$-normalizable.*

*Proof.* Since $M$ is $\beta_I^+\beta_S^+$-normalizable, it has at least one $\beta_I^+\beta_S^+$-normal form. On the other hand, by the Church-Rosser property, $M$ has at most one $\beta_I^+\beta_S^+$-normal form. So, let $M^*$ be the unique $\beta_I^+\beta_S^+$-normal form of $M$. Then, according to Lemma 18, we have that the length of any sequence of $\beta_I^+\beta_S^+$-reduction starting in $M$ is bounded by $\Theta[M^*] - \Theta[M]$. □

**Lemma 20.** *Let $M, N \in \Lambda$ be such that $M \to_{\beta_I\beta_S} N$. Then there exist $M^*, N^* \in \Lambda^{\mathbb{N}}$ such that $|M^*| \equiv M$, $|N^*| \equiv N$, and $M^* \to_{\beta_I^+\beta_S^+} N^*$.*

*Proof.* Follows from the fact that $\Lambda \subset \Lambda^{\mathbb{N}}$ and that, if $P \to_{\beta_I^+\beta_S^+} Q$, then $|P| \to_{\beta_I\beta_S} |Q|$. □

**Theorem 21.** *(Conservation for $\beta_I$ and $\beta_S$) Let $M \in \Lambda$ be $\beta_I\beta_S$-normalizable. Then $M$ is strongly $\beta_I\beta_S$-normalizable.*

*Proof.* Follows from Lemma 19 and Lemma 20. □

## 5 Main Result

Our goal is to take advantage of the result established in the previous section when dealing with the notion of reduction $\beta$.

Theorem 21 may be seen as a generalized version of the conservation theorem. Indeed the usual conservation theorem [1, CHAP. 11, §3.] appears as a particular case of Theorem 21. Nevertheless, we may go further in the generalization and state the following theorem, which is the central result of this paper.

**Theorem 22.** *(Generalized conservation) Let $M \in \Lambda$ be $\beta_I\beta_S$-normalizable. Then $M$ is strongly $\beta$-normalizable.*

*Proof.* Suppose that there exists a λ-term $M \in \Lambda$ that is $\beta_I\beta_S$-normalizable but that is not strongly $\beta$-normalizable. Then there exists an infinite sequence of terms $M_0, M_1, M_2, \ldots \in \Lambda$ such that

$$M \equiv M_0 \text{ and } \forall i \in \mathbb{N}, M_i \to_\beta M_{i+1}.$$

This sequence must be such that

$$\forall k \in \mathbb{N}, \exists l \in \mathbb{N}, l \geq k \text{ and } M_l \to_{\beta_I} M_{l+1}$$

otherwise there would exist an infinite sequence of $\beta_K$-contractions, which is absurd. But then, by Lemma 11, it is possible to construct an infinite sequence of $\beta_I\beta_S$-contractions starting in $M$, and this contradicts the fact that, by Theorem 21, $M$ is strongly $\beta_I\beta_S$-normalizable. □

## 6 Application to Typed λ-Calculi

In this section we show how to use Theorem 22 to prove the strong normalization of several typed λ-calculi. We first establish the strong normalization of Church's simply typed λ-calculus. Then we discuss how to extend the proof to Barendregt's λ-cube by following the edges of the cube in the three possible directions.

### 6.1 Church's Simply Typed λ-Calculus

We first define the raw syntax of simple types and simply typed λ-terms.

Let $\mathcal{A}$ be a set of symbols called atomic types. The set $\mathcal{S}$ of simple types is inductively defined as follows.

**Definition 23.**

i. $a \in \mathcal{A} \Rightarrow a \in \mathcal{S}$,
ii. $\alpha, \beta \in \mathcal{S} \Rightarrow (\alpha \to \beta) \in \mathcal{S}$

Let $\mathcal{V}$ be the set of variables of type $\alpha$. We define the set $\Lambda^\to$ of raw simply typed λ-terms.

**Definition 24.** The set $\Lambda^\to$ of raw simply typed λ-terms is inductively defined as follows:

i. $x \in \mathcal{V} \Rightarrow x \in \Lambda^\to$,
ii. $x \in \mathcal{V}, \alpha \in \mathcal{S}, M \in \Lambda^\to \Rightarrow (\lambda x : \alpha . M) \in \Lambda^\to$,
iii. $M, N \in \Lambda^\to \Rightarrow (M N) \in \Lambda^\to$.

If $M \in \Lambda^\to$ and $\alpha \in \mathcal{S}$, an expression of the form $M : \alpha$ is called a statement. $M$ is called the subject of the statement and $\alpha$ is called the predicate. A statement whose subject is a variable is called a declaration. A sequence of declarations whose subjects are all distinct is called a typing context. We will use $\Gamma, \Delta, \ldots$ as metavariables ranging over typing contexts.

The notion of well-typed term is defined by providing a proof system to derive typing judgements of the shape

$$\Gamma \vdash M : \alpha$$

where $\Gamma$ is a typing context, $M \in \Lambda^\rightarrow$, and $\alpha \in \mathcal{S}$.

**Definition 25.**

$$\Gamma \vdash x : \alpha \quad \text{if } x : \alpha \in \Gamma \qquad \text{(variable)}$$

$$\frac{\Gamma, x : \alpha \vdash M : \beta}{\Gamma \vdash (\lambda x : \alpha. M) : (\alpha \rightarrow \beta)} \qquad \text{(abstraction)}$$

$$\frac{\Gamma \vdash M : (\alpha \rightarrow \beta) \quad \Gamma \vdash N : \alpha}{\Gamma \vdash (M N) : \beta} \qquad \text{(application)}$$

We now establish the normalization of Church's simply typed $\lambda$-calculus by giving a proof due to Turing [9].

**Theorem 26.** (Normalization) *Let $\Gamma$ be a context, and let $M \in \Lambda^\rightarrow$ and $\alpha \in \mathcal{S}$ be such that*

$$\Gamma \vdash M : \alpha$$

*then $M$ has a $\beta$-normal form.*

*Proof.* One defines the order of a $\beta$-redex $((\lambda x : \alpha. M) N)$ as the length of the type assigned to $(\lambda x : \alpha. M)$. Now, consider some $\beta$-contraction $P \rightarrow_\beta Q$ where the contracted redex is $((\lambda x : \alpha. M) N)$. The redexes in $Q$ are of six kinds:

1. redexes occurring in $P$ disjointly with $((\lambda x : \alpha. M) N)$; these redexes are unchanged;
2. redexes occurring in $M$, and possibly modified by the substitution $M[x := N]$; their orders are unchanged;
3. new redexes $((\lambda y : \alpha_1. N_1) O_i)$, if $N \equiv (\lambda y : \alpha_1. N_1)$ and $(x O_i)$ occurs in $M$; the order of these redexes is the length of the type assigned to $N$, which is less than the order of $((\lambda x : \alpha. M) N)$;
4. a new redex $((\lambda y : \alpha_1. M_1[x := N]) O)$, if $M \equiv (\lambda y : \alpha_1. M_1)$ and $((\lambda x : \alpha. M) N)$ occurs in $P$ as the left subterm of an application $(((\lambda x : \alpha. M) N) O)$; the order of this redex is the length of the type assigned to $M$, which is less than the order of $((\lambda x : \alpha. M) N)$;
5. a new redex $((\lambda y : \alpha_1. N_1) O)$, if $N \equiv (\lambda y : \alpha_1. N_1)$, $M \equiv x$, and $((\lambda x : \alpha. x) N)$ occurs in $P$ as the left subterm of an application $(((\lambda x : \alpha. x) N) O)$; the order of this redex is the length of the type assigned to $N$, which is less than the order of $((\lambda x : \alpha. M) N)$;
6. redexes occurring in $N$ and possibly multiplied by the substitution $M[x := N]$; their orders are kept unchanged.

The normalization procedure runs as follows: contract one of the $\beta$-redexes of highest order that is innermost, and repeat this process until no more redex remains.

Since the contracted redex, say $((\lambda x : \alpha. M) N)$, is chosen innermost, the order of any redex occurring in $N$ is strictly less than the order of the contracted redex. Therefore, at each step, the orders of the created or multiplied redexes (of kind 3, 4, 5, and 6) are strictly less than the order of the redex that disappears. The theorem follows by an induction up to $\omega^2$. $\square$

Normalization concerns only the existence of normal forms. Hence, to prove a normalization theorem, it sufficient to exhibit one normalization procedure. Strong normalization is, in general, more complex to establish. One has to show that any reduction strategy is normalizing. This is why Theorem 22 is interesting: it allows one to reduce a proof of strong normalization to a proof of normalization. This is illustrated by the proof of the next theorem.

**Theorem 27.** (Strong normalization) *Let $\Gamma$ be a context, and let $M \in \Lambda^\rightarrow$ and $\alpha \in \mathcal{S}$ be such that*
$$\Gamma \vdash M : \alpha$$
*then $M$ is strongly $\beta$-normalizable.*

*Proof.* According to Theorem 22, it is sufficient to show that $M$ has a $\beta_I \beta_S$-normal form.

First one defines the order of a $\beta_S$-redex $(((\lambda x{:}\alpha.\,M)\,N)\,O)$ as the length of the type assigned to $(\lambda x{:}\alpha.\,M)$. Then, it is straightforward to replay the proof of Theorem 26. □

A technical advantage of the above proof, as opposed to a proof à la Tait (see [15, APP. 2]) is that our proof is arithmetizable. We will come back to this point in the conclusions.

## 6.2 Barendregt's $\lambda$-Cube

Barendregt's $\lambda$-cube is a system of eight typed $\lambda$-calculi ordered by inclusion. The simplest of these calculi is Church's simply typed $\lambda$-calculus and the more complex is Coquand's Constructions [4]. The others system include Girard's systems F [12, 13], its higher order version $F_\omega$, and the system LF [14], which is a variant of AUT-QE [6].

A complete description of the $\lambda$-cube, together with examples, may be found in [3, 2]. We just give briefly the main definitions.

The set $\mathcal{T}$ of raw types and of raw typed $\lambda$-terms is defined at once. This set is called the set of pseudo-terms.

**Definition 28.** The set $\mathcal{T}$ of pseudo-terms is inductively defined as follows:

i. $\square \in \mathcal{T}$,
ii. $* \in \mathcal{T}$,
iii. $x \in \mathcal{V} \Rightarrow x \in \mathcal{T}$,
iv. $x \in \mathcal{V}, M, N \in \mathcal{T} \Rightarrow (\lambda x{:}M.\,N) \in \mathcal{T}$,
v. $x \in \mathcal{V}, M, N \in \mathcal{T} \Rightarrow (\Pi x{:}M.\,N) \in \mathcal{T}$,
vi. $M, N \in \mathcal{T} \Rightarrow (M\,N) \in \mathcal{T}$.

The two constants $\square$ and $*$ are called sorts. We let $s$, $s_1$ and $s_2$ range over sorts. The statements are of the form $M : N$, with $M, N \in \mathcal{T}$. The notions of declaration and typing context are defined as previously. Type assignment is defined as follows.

**Definition 29.**

i. $\vdash * : \square$,

ii. $\dfrac{\Gamma \vdash M : s}{\Gamma, x{:}M \vdash x : M}$,

iii. $\dfrac{\Gamma \vdash M : N \quad \Gamma \vdash O : s}{\Gamma, x{:}O \vdash M : N}$,

iv. $\dfrac{\Gamma \vdash M : (\Pi x{:}O.\,P) \quad \Gamma \vdash N : O}{\Gamma \vdash (M\,N) : P[x := N]}$,

v. $\dfrac{\Gamma \vdash M : N \quad \Gamma \vdash O : s}{\Gamma \vdash M : O}$ if $N \leftrightarrow\!\!\!\twoheadrightarrow_\beta O$,

vi. $\dfrac{\Gamma \vdash M : s_1 \quad \Gamma, x : M \vdash N : s_2}{\Gamma \vdash (\Pi x : M . N) : s_2}$,

vii. $\dfrac{\Gamma \vdash M : s_1 \quad \Gamma, x : M \vdash N : O \quad \Gamma, x : M \vdash O : s_2}{\Gamma \vdash (\lambda x : M . N) : (\Pi x : M . O)}$.

Rules vi and vii are parametrized by the pair of sorts $(s_1, s_2)$. By taking this pair to be $(*, *)$, one gets a new formulation of Church's simply typed $\lambda$-calculus ($\lambda^\rightarrow$). To see this, one defines $(\alpha \rightarrow \beta)$ as $(\Pi x : \alpha . \beta)$ where the variable $x$ does not occur in $\beta$. For a precise correspondence between this new formulation and Definition 25, see [3].

By taking $(s_1, s_2)$ to be $(\Box, *)$, $(\Box, \Box)$, or $(*, \Box)$, one gets respectively *terms depending on types*, *types depending on types*, and *types depending on terms*. This corresponds to the three possible directions of the edges of the $\lambda$-cube.

**Terms Depending on Types.** By allowing for terms depending on types, one obtains systems containing Girard's system F. Therefore it is vain to seek for some arithmetizable proof [13]. The method of reducibility candidates [8] is somehow required.

This method, due to Girard, extends Tait's technique and yields strong normalization proofs. Therefore, it may be unclear how one can take advantage of Theorem 22 in this context. Nevertheless, when one is simply interested in normalization (as opposed to strong normalization), there exists a simplification of the method, which is due to Scedrov [18].

In the course of (one version of) the strong normalization proof of system F, one defines the notion of saturated set [8]. A saturated set $S$ is a set of strongly normalizable $\lambda$-terms such that[1]:

1. if $x$ is a variable and $M_1, \ldots, M_n$ are strongly normalizable $\lambda$-terms, then $((x M_1) \ldots M_n) \in S$,
2. if $N_1$ is a strongly normalizable $\lambda$-term and if $((M[x := N_1] N_2) \ldots N_n) \in S$, then $((((\lambda x . M) N_1) N_2) \ldots N_n) \in S$.

Scedrov notices that normalizable $\lambda$-terms, as opposed to strongly normalizable $\lambda$-terms, are closed under $\beta$-expansion. Therefore, when adapting the above definition to the case of normalization, one may drop Condition 2. The same idea may be used to establish $\beta_I \beta_S$-normalization.

**Types Depending on Types.** Our proof technique relies on the so-called erasing trick. To prove that a typed $\lambda$-term $M$ is strongly normalizable, we prove that the untyped $\lambda$-term that is obtained by erasing the type information from $M$ is strongly-normalizable.

Types depending on types introduces redexes at the type level. For instance, one may derive the following:

$$\vdash (\Pi x : * . *) : \Box$$
$$\vdash (\lambda a : * . a) : (\Pi x : * . *)$$
$$b : * \vdash (\lambda x : ((\lambda a : * . a) b) . x) : (\Pi x : b . b)$$

---

[1] The conditions that we give are due to Mitchell. For a comparison between the different versions of Girard's method see the comprehensive article of Gallier [8].

Nevertheless, the erasing trick may still be used because the strong normalization at the type level may be established independently of the strong normalization at the term level. This is even true for $F_\omega$ [8].

**Types Depending on Terms.** With types depending on terms, the erasing trick fails. Thus we cannot reason in the framework of untyped $\lambda$-calculus any more. We have to work with the complete set of pseudo-terms $\mathcal{T}$.

To adapt Theorem 22 to the set $\mathcal{T}$ is not straightforward. The problem is that even the usual version of the conservation theorem fails for $\mathcal{T}$. There exist pseudo-terms that are $\beta_I$-normalizable but not strongly $\beta_I$-normalizable as shown by the following counterexample:

$$((\lambda x : \Omega . x) y) \to_{\beta_I} y$$

where $\Omega \equiv ((\lambda x : y . x\, x)(\lambda x : y . x\, x))$.

Nevertheless, a version of Theorem 22 may be stated for pseudo-terms by following an idea due to Nederpelt [17]. This idea consists in adapting the notion of reduction $\beta_I$ as follows.

$\beta_I^{\mathcal{T}} : ((\lambda x : M . N) O) \to ((\lambda x : M . N[x := O]) O) \qquad$ if $x \in \mathrm{FV}(M)$.

On the other hand, our notion of reduction $\beta_S$ may be kept unchanged (except for the syntax).

$\beta_S^{\mathcal{T}} : (((\lambda x : M . N) O) P) \to ((\lambda x : M . (N\, P)) O) \qquad$ if $x \notin \mathrm{FV}(N)$.

With this new definitions, Theorem 22 holds for pseudo-terms.

# 7 Conclusions

The technique we have developed in this paper yields rather transparent proofs of strong normalization (transparency, of course, is chiefly a matter of style—or even a matter of taste).

As we already mentioned, the strong normalization proof that we have given for Church's simply typed $\lambda$-calculus is arithmetizable. Other arithmetizable proofs exist, among which the one by Gandy [10] is may be the best known. Gandy's proof is based on a semantic interpretation of the simply typed $\lambda$-terms and is, therefore, quite different form our proof. Nevertheless, it is interesting to note that he transforms $\beta_K$-redexes into $\beta_I$-redexes.

The untyped $\lambda$-terms typable in Nederpelt's calculus correspond exactly to the simply typable $\lambda$-terms [7]. Therefore Nederpelt's proof [17] may also be seen as an arithmetizable proof for Church's simply typed $\lambda$-calculus. As we explained in Section 6, our proof technique is close to the one of Nederpelt. The main difference is that Nederpelt does not use the notion of reduction $\beta_S$, but generalizes further the notion of reduction $\beta_I^{\mathcal{T}}$. This yields some technical complications when proving the Church-Rosser property. Another difference is that Nederpelt's proof is tailormade for his own calculus. Nevertheless, Nederpelt says in his thesis [17] that his proof may be turn in a general method of proving strong normalization from normalization. This statement is worked out by Klop in [16]. Indeed Klop provides a generalization of Nederpelt's method for a large class of reduction systems. The present work may also be seen as an exploration of Nederpelt's statement.

Van Daalen, in his thesis [19], gives also an arithmetizable proof of strong normalization for simply typed $\lambda$-calculus. His proof, which is totally syntactic, does not use any other notion of reduction than $\beta$, and is based on a strong substitution lemma.

We have briefly explained how to extend the strong normalization proof of simply typed $\lambda$-calculus to Barendregt's $\lambda$-cube. The ideas that we have developed are immediately applicable to the direct successors of $\lambda^{\rightarrow}$, namely $\lambda 2$, $\lambda\underline{\omega}$, and $\lambda P$. To put these ideas together into a modular proof of strong normalization for the calculus of Constructions (in the spirit of [11]) will be the subject of future work.

# References

[1] H.P. Barendregt. *The lambda calculus, its syntax and semantics*. North-Holland, revised edition, 1984.

[2] H.P. Barendregt. Introduction to Generalised Type Systems. *Journal of Functional Programming*, 1(2):125–154, 1991.

[3] H.P. Barendregt. Lambda calculi with types. In S. Abramsky, D. Gabbai, and T. Maibaum, editors, *Handbook of Logic in Computer Science*. Oxford University Press, 1992.

[4] Th. Coquand. Metamathematical investigations of a calculus of constructions. In P. Odifreddi, editor, *Logic and Computer Science*, pages 91–122. Academic Press, 1990.

[5] N.G. de Bruijn. Lambda calculus notations with nameless dummies, a tool for automatic formula manipulation, with an application to the Church-Rosser theorem. *Indigationes Mathematicae*, 34:381–392, 1972.

[6] N.G. de Bruijn. A survey of the project Automath. In J.P. Seldin and J.R. Hindley, editors, *to H. B. Curry: Essays on Combinatory Logic, Lambda Calculus and Formalism*, pages 579–606. Academic Press, 1980.

[7] Ph. de Groote. *Définition et Propriétés d'un métacalcul de représentation de théories*. PhD thesis, Université Catholique de Louvain, Unité d'Informatique, 1991.

[8] J.H. Gallier. On Girard's "Candidats de Réductibilité". In P. Odifreddi, editor, *Logic and Computer Science*, pages 123–203. Academic Press, 1990.

[9] R.O. Gandy. An early proof of normalization by A.M. Turing. In J. P. Seldin and J. R. Hindley, editors, *to H. B. Curry: Essays on Combinatory Logic, Lambda Calculus and Formalism*, pages 453–455. Academic Press, 1980.

[10] R.O. Gandy. Proofs of strong normalization. In J. P. Seldin and J. R. Hindley, editors, *to H. B. Curry: Essays on Combinatory Logic, Lambda Calculus and Formalism*, pages 457–478. Academic Press, 1980.

[11] H. Geuvers and M.-J. Nederhof. Modular proof of strong normalization for the calculus of construction. *Journal of Functional Programming*, 1(2):155–189, 1991.

[12] J.-Y. Girard. The system F of variable types, fifteen years later. *Theoretical Computer Science*, 45:159–192, 1986.

[13] J.-Y. Girard, Y. Lafont, and P. Taylor. *Proofs and Types*, volume 7 of *Cambridge Tracts in Theoretical Computer Science*. Cambridge University Press, 1989.

[14] R. Harper, F. Honsel, and G. Plotkin. A framework for defining logics. In *Proceedings of the second annual IEEE symposium on logic in computer science*, pages 194–204, 1987.

[15] J.R. Hindley and J.P. Seldin. *Introduction to combinators and $\lambda$-calculus*. London Mathematical Society Student Texts. Cambridge University Press, 1986.

[16] J.W. Klop. *Combinatory Reduction Systems*. PhD thesis, CWI, Amsterdam, Mathematical Centre Tracts Nr. 127, 1980.
[17] R.P. Nederpelt. *Strong normalization in a typed lambda calculus with lambda structured types*. PhD thesis, Technische hogeschool Eindhoven, 1973.
[18] A. Scedrov. Normalization revisited. In J.W. Gray and A. Scedrov, editors, *Proceedings of the AMS research conference*, pages 357–369. American Mathematical Society, 1987.
[19] D.T. van Daalen. *The language theory of Automath*. PhD thesis, Technische hogeschool Eindhoven, 1980.

## A  Postponement of $\beta_K$-Contractions

**Lemma 30.** *Let $M, N, O \in \Lambda$ be such that $M \to_{\beta_K} N$. Then $M[x:=O] \to_{\beta_K} N[x:=O]$.*

*Proof.* By induction on the derivation of $M \to_{\beta_K} N$. □

**Lemma 31.** *Let $M, N, O \in \Lambda$ be such that $M \to_{\beta_K} N$. Then $O[x:=M] \twoheadrightarrow_{\beta_K} O[x:=N]$.*

*Proof.* By induction on the structure of $O$. □

**Proof of Lemma 11.**

We treat the case $R = \beta_I$, the other case being similar. The proof is by induction on the derivation of $M \to_{\beta_K} N$.

1. $M \equiv ((\lambda x. M_1) M_2)$, $N \equiv M_1$, and $x \notin \mathrm{FV}(M_1)$
   Take $P \equiv ((\lambda x. O) M_2)$.
2. $M \equiv (\lambda x. M_1)$, $N \equiv (\lambda x. N_1)$ and $M_1 \to_{\beta_K} N_1$.
   $O$ must be of the form $(\lambda x. O_1)$, with $N_1 \to_{\beta_I} O_1$. Apply the induction hypothesis.
3. $M \equiv (M_1 M_2)$, $N \equiv (N_1 M_2)$ and $M_1 \to_{\beta_K} N_1$.
   There are three subcases according to the way $N \to_{\beta_I} O$.
   (a) $O \equiv (O_1 M_2)$ and $N_1 \to_{\beta_I} O_1$.
       Apply the induction hypothesis.
   (b) $O \equiv (N_1 O_2)$ and $M_2 \to_{\beta_I} O_2$.
       Take $P \equiv (M_1 O_2)$.
   (c) $N_1 \equiv (\lambda x. N_{11})$ and $O \equiv N_{11}[x:=M_2]$.
       There are two subcases according to the form of $M_1$.
       i. $M_1 \equiv (\lambda x. M_{11})$ and $M_{11} \to_{\beta_K} N_{11}$.
          Take $P \equiv M_{11}[x := M_2]$. Indeed, $M_{11}[x := M_2] \to_{\beta_K} N_{11}[x := M_2]$, by Lemma 30.
       ii. $M_1 \equiv ((\lambda y. (\lambda x. N_{11})) M_{11})$ and $y \notin \mathrm{FV}((\lambda x. N_{11}))$.
          Take $P \equiv ((\lambda y. N_{11}[x:=M_2]) M_{11})$. Indeed we have:
          $$\begin{aligned}(((\lambda y.(\lambda x. N_{11})) M_{11}) M_2) &\to_{\beta_S} ((\lambda y. ((\lambda x. N_{11}) M_2)) M_{11}) \\ &\to_{\beta_I} ((\lambda y. N_{11}[x:=M_2]) M_{11}) \\ &\to_{\beta_K} N_{11}[x:=M_2].\end{aligned}$$
4. $M \equiv (M_1 M_2)$, $N \equiv (M_1 N_2)$ and $M_2 \to_{\beta_K} N_2$.
   This case is similar to case (3), using Lemma 31 instead of Lemma 30.  □

## B   The Church-Rosser Property for $\beta_I^+$ and $\beta_S^+$

### The Church-Rosser Property for $\beta_S^+$

The contraction of a $\beta_S^+$-redex does not multiply the other $\beta_S^+$-redexes. For this reason, the relation of $\beta_S^+$-contraction satisfies the Church-Rosser property.

**Lemma 32.** *Let* $M, N, O \in \Lambda^{\mathbb{N}}$ *be such that* $M \to_{\beta_S^+} N$ *and* $M \to_{\beta_S^+} O$. *Then there exists* $P \in \Lambda^{\mathbb{N}}$ *such that* $N \to_{\beta_S^+} P$ *and* $O \to_{\beta_S^+} P$.

*Proof.* By induction on the derivation of $M \to_{\beta_S^+} N$, distinguishing subcases according to the way $M \to_{\beta_S^+} O$. □

**Lemma 33.** *Let* $M, N, O \in \Lambda^{\mathbb{N}}$ *be such that* $M \twoheadrightarrow_{\beta_S^+} N$ *and* $M \twoheadrightarrow_{\beta_S^+} O$. *Then there exists* $P \in \Lambda^{\mathbb{N}}$ *such that* $N \twoheadrightarrow_{\beta_S^+} P$ *and* $O \twoheadrightarrow_{\beta_S^+} P$.

*Proof.* By a diagram chase, using Lemma 32. □

### The Church-Rosser Property for $\beta_I^+$

To prove that $\beta_I^+$ is Church-Rosser, we use the Tait–Martin-Löf method. We first define the binary relation $\twoheadrightarrow_1$ on $\Lambda^{\mathbb{N}}$.

**Definition 34.** The binary relation $\twoheadrightarrow_1$ is defined on $\Lambda^{\mathbb{N}}$ by the following system.

i. $M \twoheadrightarrow_1 M$,

ii. $\dfrac{M \twoheadrightarrow_1 N}{(\lambda x.\, M)^n \twoheadrightarrow_1 (\lambda x.\, N)^n}$,

iii. $\dfrac{M \twoheadrightarrow_1 O \quad N \twoheadrightarrow_1 P}{(M\, N)^n \twoheadrightarrow_1 (O\, P)^n}$,

iv. $\dfrac{M \twoheadrightarrow_1 O \quad N \twoheadrightarrow_1 P}{((\lambda x.\, M)^m\, N)^n \twoheadrightarrow_1 (O[x := P])^{+(m+n+1)}}$   if $x \in \mathrm{FV}(M)$.

**Lemma 35.** *Let* $M, N \in \Lambda^{\mathbb{N}}$ *be such that* $M \twoheadrightarrow_1 N$. *Then* $M^{+n} \twoheadrightarrow_1 N^{+n}$.

*Proof.* By induction on the derivation of $M \twoheadrightarrow_1 N$. □

**Lemma 36.** *Let* $M, N, O \in \Lambda^{\mathbb{N}}$ *be such that* $M \twoheadrightarrow_1 N$. *Then* $O[x := M] \twoheadrightarrow_1 O[x := N]$.

*Proof.* By induction on the structure of $O$, using Lemma 35 when $O \equiv (x)^n$. □

**Lemma 37.** *Let* $M, N, O, P \in \Lambda^{\mathbb{N}}$ *be such that* $M \twoheadrightarrow_1 N$ *and* $O \twoheadrightarrow_1 P$. *Then* $M[x := O] \twoheadrightarrow_1 N[x := P]$.

*Proof.* By induction on the derivation of $M \twoheadrightarrow_1 N$, using Lemma 36 for the case $M \twoheadrightarrow_1 M$. □

**Lemma 38.** Let $M, N, O \in \Lambda^{\mathbb{N}}$ be such that $M \twoheadrightarrow_1 N$ and $M \twoheadrightarrow_1 O$. Then there exists $P \in \Lambda^{\mathbb{N}}$ such that $N \twoheadrightarrow_1 P$ and $O \twoheadrightarrow_1 P$.

*Proof.* By induction on the derivation of $M \twoheadrightarrow_1 N$.

1. $M \equiv N$.
   Take $P \equiv O$.
2. $M \equiv (\lambda x. M_1)^n$, $N \equiv (\lambda x. N_1)^n$, and $M_1 \twoheadrightarrow_1 N_1$.
   We must have $O \equiv (\lambda x. O_1)^n$, with $M_1 \twoheadrightarrow_1 O_1$. Apply the induction hypothesis.
3. $M \equiv (M_1 M_2)^n$, $N \equiv (N_1 N_2)^n$, $M_1 \twoheadrightarrow_1 N_1$, and $M_2 \twoheadrightarrow_1 N_2$.
   There are two subcases according to the way $M \twoheadrightarrow_1 O$.
   (a) $O \equiv (O_1 O_2)^n$, $M_1 \twoheadrightarrow_1 O_1$, and $M_2 \twoheadrightarrow_1 O_2$.
       Apply the induction hypothesis.
   (b) $M_1 \equiv (\lambda x. M_{11})^m$, $O \equiv (O_{11}[x := O_2])^{+(m+n+1)}$, $M_{11} \twoheadrightarrow_1 O_{11}$, and $M_2 \twoheadrightarrow_1 O_2$.
       Then we must have $N_1 \equiv (\lambda x. N_{11})^m$, with $M_{11} \twoheadrightarrow_1 N_{11}$. Therefore, by induction hypothesis, there exists $P_{11}, P_2 \in \Lambda^{\mathbb{N}}$ such that $N_{11} \twoheadrightarrow_1 P_{11}$, $O_{11} \twoheadrightarrow_1 P_{11}$, $N_2 \twoheadrightarrow_1 P_2$ and $O_2 \twoheadrightarrow_1 P_2$. Hence, by Lemmas 35 and 37, we may take $P \equiv (P_{11}[x := P_2])^{+(m+n+1)}$.
4. $M \equiv ((\lambda x. M_1)^m M_2)^n$, $N \equiv (N_1[x := N_2])^{+(m+n+1)}$, $M_1 \twoheadrightarrow_1 N_1$, and $M_2 \twoheadrightarrow_1 N_2$.
   There are two subcases according to the way $M \twoheadrightarrow_1 O$.
   (a) $O \equiv ((\lambda x. O_1)^m O_2)^n$, $M_1 \twoheadrightarrow_1 O_1$, and $M_2 \twoheadrightarrow_1 O_2$.
       This case is symmetric to Case (3.b).
   (b) $N \equiv (O_1[x := O_2])^{+(m+n+1)}$, $M_1 \twoheadrightarrow_1 O_1$, and $M_2 \twoheadrightarrow_1 O_2$.
       Apply the induction hypothesis and use Lemmas 35 and 37. □

**Lemma 39.** Let $M, N, O \in \Lambda^{\mathbb{N}}$ be such that $M \twoheadrightarrow_{\beta_I^+} N$ and $M \twoheadrightarrow_{\beta_I^+} O$. Then there exists $P \in \Lambda^{\mathbb{N}}$ such that $N \twoheadrightarrow_{\beta_I^+} P$ and $O \twoheadrightarrow_{\beta_I^+} P$.

*Proof.* Because $\twoheadrightarrow_{\beta_I^+}$ is the transitive closure of $\twoheadrightarrow_1$. □

## Commutation of $\beta_I^+$ and $\beta_S^+$

**Lemma 40.** Let $M, N \in \Lambda^{\mathbb{N}}$ be such that $M \to_{\beta_S^+} N$. Then $M^{+n} \to_{\beta_S^+} N^{+n}$.

*Proof.* By induction on the derivation of $M \to_{\beta_S^+} N$. □

**Lemma 41.** Let $M, N, O \in \Lambda^{\mathbb{N}}$ be such that $M \to_{\beta_S^+} N$. Then $M[x := O] \to_{\beta_S^+} N[x := O]$.

*Proof.* By induction on the derivation of $M \to_{\beta_S^+} N$. □

**Lemma 42.** Let $M, N, O \in \Lambda^{\mathbb{N}}$ be such that $M \to_{\beta_S^+} N$. Then $O[x := M] \to_{\beta_S^+} O[x := N]$.

*Proof.* By induction on the structure of $O$, using Lemma 40 for the case $O \equiv (x)^n$. □

To establish the commutation of $\beta_I^+$- and $\beta_S^+$-reductions, we prove what some authors call the trapezium property, by analogy with the diamond property.

**Lemma 43.** *Let $M, N, O \in \Lambda^{\mathbb{IN}}$ be such that $M \to_{\beta_I^+} N$ and $M \to_{\beta_S^+} O$. Then there exists $P \in \Lambda^{\mathbb{IN}}$ such that $N \twoheadrightarrow_{\beta_S^+} P$ and $O \to_{\beta_I^+} P$.*

*Proof.* By induction on the derivation of $M \to_{\beta_I^+} N$.

1. $M \equiv ((\lambda x. M_1)^m M_2)^n$ and $N \equiv (M_1[x := M_2])^{+(m+n+1)}$.
   There are two subcases according to the way $M \to_{\beta_S^+} O$.
   (a) $O \equiv ((\lambda x. O_1)^m M_2)^n$ and $M_1 \to_{\beta_S^+} O_1$.
       By lemma 41, we may take $P \equiv (O_1[x := M_2])^{+(m+n+1)}$
   (b) $O \equiv ((\lambda x. M_1)^m O_2)^n$ and $M_2 \to_{\beta_S^+} O_2$.
       By lemma 42, we may take $P \equiv (M_1[x := O_2])^{+(m+n+1)}$
2. $M \equiv (\lambda x. M_1)^n$, $N \equiv (\lambda x. N_1)^n$ and $M_1 \to_{\beta_I^+} N_1$.
   $O$ must be of the form $(\lambda x. O_1)^n$, with $M_1 \to_{\beta_S^+} O_1$. Apply the induction hypothesis.
3. $M \equiv (M_1 M_2)^o$, $N \equiv (N_1 M_2)^o$ and $M_1 \to_{\beta_I^+} N_1$.
   There are three subcases according to the way $M \to_{\beta_S^+} O$.
   (a) $O \equiv (O_1 M_2)^o$ and $M_1 \to_{\beta_S^+} O_1$.
       Apply the induction hypothesis.
   (b) $O \equiv (M_1 O_2)^o$ and $M_2 \to_{\beta_S^+} O_2$.
       Take $P \equiv (N_1 O_2)^o$.
   (c) $M_1 \equiv ((\lambda x. M_{11})^m M_{12})^n$, $x \notin \mathrm{FV}(M_{11})$, and $O \equiv ((\lambda x. (M_{11} M_2)) M_{12})^{m+n+o+1}$.
       We must have $N_1 \equiv ((\lambda x. N_{11})^m N_{12})^n$ with $M_{11} \to_{\beta_I^+} N_{11}$ and $M_{12} \equiv N_{12}$, or with $M_{11} \equiv N_{11}$ and $M_{12} \to_{\beta_I^+} N_{12}$. Take $P \equiv ((\lambda x. (N_{11} M_2)) N_{12})^{m+n+o+1}$.
4. $M \equiv (M_1 M_2)^o$, $N \equiv (M_1 N_2)^o$ and $M_2 \to_{\beta_I^+} N_2$.
   This case is similar to Case (3).

□

**Lemma 44.** *Let $M, N, O \in \Lambda^{\mathbb{IN}}$ be such that $M \twoheadrightarrow_{\beta_I^+} N$ and $M \twoheadrightarrow_{\beta_S^+} O$. Then there exists $P \in \Lambda^{\mathbb{IN}}$ such that $N \twoheadrightarrow_{\beta_S^+} P$ and $O \twoheadrightarrow_{\beta_I^+} P$.*

*Proof.* By a diagram chase, using Lemma 43. □

**Proof of Theorem 16.**

By Lemma 33, Lemma 39, and Lemma 44. □

# Modified Realizability Toposes

# and

# Strong Normalization Proofs

(Extended Abstract)

J. M. E. Hyland[1]     C.-H. L. Ong[2]

University of Cambridge, England

**Abstract**

This paper is motivated by the discovery that an appropriate quotient $SN_*$ of the strongly normalising untyped $\lambda*$-terms (where $*$ is just a formal constant) forms a partial applicative structure with the inherent application operation. The quotient structure satisfies all but one of the axioms of a partial combinatory algebra (PCA). We call such partial applicative structures *conditionally partial combinatory algebras* (C-PCA). Remarkably, an arbitrary *right-absorptive* C-PCA gives rise to a *tripos* provided the underlying intuitionistic predicate logic is given an interpretation in the style of Kreisel's *modified realizability*, as opposed to the standard Kleene-style realizability. Starting from an arbitrary right-absorptive C-PCA $U$, the tripos-to-topos construction due to Hyland *et al.* can then be carried out to build a *modified realizability topos* $\mathbf{TOP}_m(U)$ of non-standard sets equipped with an equality predicate. Church's Thesis is internally valid in $\mathbf{TOP}_m(K_1)$ (where the PCA $K_1$ is "Kleene's first model" of natural numbers) but not Markov's Principle. There is a topos inclusion of **SET** — the "classical" topos of sets — into $\mathbf{TOP}_m(U)$; the image of the inclusion is just sheaves for the $\neg\neg$-topology. Separated objects of the $\neg\neg$-topology are characterized. We identify the appropriate notion of PERs (partial equivalence relations) in the modified realizability setting and state its completeness properties. The topos $\mathbf{TOP}_m(U)$ has enough completeness property to provide a category-theoretic semantics for a family of higher type theories which include Girard's System F and the Calculus of Constructions due to Coquand and Huet. As an important application, by interpreting type theories in the topos $\mathbf{TOP}_m(SN_*)$, a clean semantic explanation of the Tait-Girard style strong normalization argument is obtained. We illustrate how a strong normalization proof for an impredicative and dependent type theory may be assembled from two general "stripping arguments" in the framework of the topos $\mathbf{TOP}_m(SN_*)$. This opens up the possibility of a "generic" strong normalization argument for an interesting class of type theories.

## 1 Introduction

A celebrated result at the junction of proof theory and theoretical computer science is the strong normalization of System F [Gir72, Gir86]. This result is notable not only because of the impact it has in proof theory (e.g. proof of the syntactic version of Takeuti's conjecture) but also because the proof is notoriously hard. To get a feel of how intrinsically difficult it is, it is helpful to recall Girard's observation that the strong normalization (s.n.) of System F implies the consistency of $\mathbf{PA_2}$ (second order Peano Arithmetic). The system $\mathbf{PA_2}$ is highly expressive. In Girard's words, "it suffices for everyday mathematics". By Gödel's Second Incompleteness Theorem, the consistency of $\mathbf{PA_2}$ is not provable in $\mathbf{PA_2}$. In view of Girard's observation, the s.n. of System F is therefore also *not* provable in

---
[1] Dept. of Pure Maths. and Mathematical Statistics, 16 Mill Lane, Cambridge CB2 1SB, UK; jmeh1@phx.cam.ac.uk.
[2] Computer Laboratory, Pembroke Street, Cambridge CB2 3QG, UK; chlo1@cl.cam.ac.uk. On leave from the National University of Singapore and supported by a fellowship from Trinity College, Cambridge.

$PA_2$. This says something about the complexity of the s.n. proof. Remarkably, Girard succeeded in proving it by using a powerful induction technique known as *reducibility candidates* which is based on an earlier method of Tait [Tai67, Tai75]; see [Gal90] for a careful exposition and [Sce89] for a semantic approach. System F, also known as Second Order Polymorphic Lambda Calculus was reinvented by Reynolds [Rey74].

Other higher type theories have emerged over the past two decades. Martin-Löf introduced a series of intuitionistic dependent type theories *e.g.* [ML73, ML84]. More recently, Coquand and Huet introduced the Calculus of Constructions [CH88] which is both a dependent as well as impredicative type theory. Strong normalization holds for all of these theories. The proof in each case, being essentially an appropriate extension of the Tait-Girard method, is invariably hard. It would therefore be highly desirable if a "generic" s.n. proof could be invented which not only sheds light on the Tait-Girard argument but also reduces the s.n. property of a class of type theories to a couple of sufficient conditions. It is this dream of a "generic" s.n. argument which provided the initial motivation to our work.

**A Key Idea** It is well known (see *e.g.* [LS86]) that there is an equivalence between cartesian closed categories (CCC) and the simply-typed $\lambda$-calculi with surjective pairing. Consider the interpretation of $\lambda^{\rightarrow}$, the pure simply-typed $\lambda$-calculus in a particular (class of) CCC $\mathbf{PER}(U)$, the category of partial equivalence relations (PERs) over a partial combinatory algebra (PCA) $U$. Recall that objects of this category are just PERs *i.e.* symmetric and transitive relations over $U$. For any PERs $R$ and $S$ over $U$, a morphism $F : R \rightarrow S$ is a function $F$ from $[R]$ (the $R$-equivalence classes) to $[S]$ which is *realised*[3] (or *tracked*) by some element of $U$; that is to say, for some realiser $f \in U$, whenever $r\,R\,r'$, then $(fr)\,S\,(fr')$ and $F([r]_R) = [fr]_S$. (We assume throughout this paper that relations defined over a partial applicative structure are *strict* i.e. whenever we write $(fr)\,S\,(fr')$, we are implicitly asserting that $fr$ and $fr'$ are defined.) Categories constructed in this way have morphisms which are *by definition* realised by elements of the underlying PCA. We shall refer to such categories informally as "realizability categories".

Consider the standard interpretation of $\lambda^{\rightarrow}$ in $\mathbf{PER}(U)$ (see *e.g.* [LS86]). For any derivable type-assignment sequent $x_1 : \sigma_1, \cdots, x_n : \sigma_n \vdash s : \tau$ (assuming $\{\vec{x}\} \supseteq \mathrm{FV}(s)$, the free variables of $s$), the types $\sigma_1, \cdots, \sigma_n, \tau$ are interpreted as objects of $\mathbf{PER}(U)$, and $s$ the term in question is interpreted as the morphism $[\![s]\!] : [\![1]\!] \times [\![\sigma_1]\!] \times \cdots \times [\![\sigma_n]\!] \rightarrow [\![\tau]\!]$ which is realised by some element of the underlying PCA $U$. This interpretation makes sense in the class of CCCs $\mathbf{PER}(U)$ where $U$ is *any* PCA. Now, *suppose* the following statement is valid.

**Assumption 0** *An appropriate quotient of the s.n. untyped $\lambda$-terms yields a* PCA, *call it* SN. □

As before, we may interpret the calculus $\lambda^{\rightarrow}$ in the CCC $\mathbf{PER}(\mathsf{SN})$. We can now demonstrate how a s.n. "proof" may be assembled from the following **"stripping" arguments**.

1 **Realiser** *For any derivable sequent $x_1 : \sigma_1, \cdots, x_n : \sigma_n \vdash s : \tau$, the morphism $[\![\vec{x}:\vec{\sigma} \vdash s:\tau]\!]$ is realised by $\lambda\xi.\lambda x_1\cdots x_n.\lceil s\rceil$ where $\lceil s\rceil$ is obtained from $s$ by "stripping off" (or erasing) all embedded type expressions. (Since $\lceil s\rceil \in \mathsf{SN}$, $\lceil s\rceil$ is s.n.)* □

2 **Reflection** *For any well-typed term $s$, if $\lceil s\rceil$ is s.n. in $\Lambda$ then $s$ is s.n. in the typed regime.* □

The two stripping arguments above are easily seen to be valid in the case of the calculus $\lambda^{\rightarrow}$. *Provided* Assumption 0 is valid, the proof of the s.n. of $\lambda^{\rightarrow}$ is now complete.

---
[3]*Realizability* is a notion primarily associated with proofs of propositions. We use it in a broader sense of "witnessing" a property, not necessarily in the context of formal logic.

**A "Generic" Strong Normalization Argument** Does the argument "scale up"? Does an appropriate extrapolation of this argument establish s.n. of say, System F or even better, the Calculus of Constructions? This approach hinges upon two things:

- there is a realizability category **C** with the untyped s.n. $\lambda$-terms as realisers,
- this category **C** is a model of the type theory in question.

We know from the works of Seely [See87], Pitts [Pit87] and Hyland [HP89] that the category-theoretic interpretation of such sophisticated type theories as System F or the Calculus of Constructions places heavy demands on the structure of categories. For example, in the case of System F, we essentially need a cloven fibration $\begin{smallmatrix} \mathbf{E} \\ \downarrow p \\ \mathbf{B} \end{smallmatrix}$ (see e.g. [B85]) such that:

(i) the base category **B** has finite products,

(ii) $\begin{smallmatrix} \mathbf{E} \\ \downarrow p \\ \mathbf{B} \end{smallmatrix}$ is a fibred CCC *i.e.* each fibre is a CCC and reindexing is a *strict* CCC-morphism,

(iii) there is a generic object $G \in \mathbf{E}_P$ for some distinguished object $P \in \mathbf{B}$,

(iv) $\begin{smallmatrix} \mathbf{E} \\ \downarrow p \\ \mathbf{B} \end{smallmatrix}$ has $P$-indexed product *i.e.* for each $I \in \mathbf{B}$, the reindexing functor $\pi^*_{P,I}$ has a right-adjoint $\Pi_{P,I}$ satisfying the Beck-Chevalley condition where $\pi_{P,I} : P \times I \to I$ is the projection morphism.

Happily, realizability categories satisfying the above requirements are available as appropriate substructures of *realizability toposes* [Hyl82, Hyl88] (see also [HRR90]). Hyland shows that any PCA $U$ gives rise in a systematic way to a Kleene-style realizability topos $\mathbf{TOP}_s(U)$ which has more than sufficient completeness properties for interpreting at least the class of impredicative and dependent type theories [Pit87, HP89].

To carry our programme through, the first step is to verify Assumption 0.

## 2  Strongly Normalising Untyped $\lambda*$-Terms

Our immediate task is the following:

> To construct a PCA of strongly normalising untyped $\lambda$-terms (or an appropriate quotient thereof) using the inherent application operator of the calculus.

We begin by searching for an equivalence relation on SN (the collection of s.n. pure untyped $\lambda$-terms), say $\sim$ which is *compatible* with application (so that the associated quotient structure has a well-defined partial application). That is to say, for $M, M', N, N' \in$ SN, whenever $M \sim M'$ and $N \sim N'$, then

- *either* both $MN$ and $M'N'$ are not s.n.,
- *or* both are s.n. and $MN \sim M'N'$.

**First Attempt** The most natural candidate for $\sim$ is $\beta$-equivalence (which we denote as $=_\beta$). Unfortunately, it is not compatible with application. It is instructive to see why this is so. Consider the s.n. $\lambda$-terms: $M \equiv \lambda z.(\lambda x.\mathbf{i})(z(\lambda y.yy)) =_\beta \lambda z.\mathbf{i} \equiv N$ where $\mathbf{i}$ is the identity. Take $P \equiv (\lambda y.yy)$. Clearly $NP$ is s.n. However $MP$ is not s.n. because the free occurrence of the variable $z$ in the redex-subterm $\Delta \equiv (\lambda x.\mathbf{i})(z(\lambda y.yy))$ of $M$ allows "offending" terms (like $P$ in this case) to be introduced as a result of the application $MP$. This situation will not arise in the case of $NP$ simply because $z$ does not occur free in $N$.

The above example leads us to consider the equivalence relation generated by a reduction scheme that contracts *only* those $\beta$-redexes which do not contain any free variables. More formally, we define *closed $\beta$-reduction* on $\Lambda_*$ (the collection of $\lambda$-terms generated from a distinguished formal constant $*$) denoted $\to_{cl}$, as the relation inductively defined by the following rules:

$$\overline{(\lambda x.P)Q \to_{cl} P[Q/x]} \quad \mathsf{FV}((\lambda x.P)Q) = \emptyset$$

$$\frac{M \to_{cl} N}{PM \to_{cl} PN \quad MP \to_{cl} NP \quad \lambda x.M \to_{cl} \lambda x.N}$$

We further define $\twoheadrightarrow_{cl}$ as the reflexive, transitive closure of $\to_{cl}$, and $\sim_{cl}$ as the symmetric closure of $\twoheadrightarrow_{cl}$. For example, $\lambda z.z((\lambda x.z)\mathrm{i}) \not\to_{cl} \lambda z.zz$ but $\lambda z.z((\lambda x.x)\mathrm{i}) \to_{cl} \lambda z.z\mathrm{i}$.

The main result of this section is the following theorem. We define $\mathrm{SN}_*$ to be the subset of $\Lambda_*$ consisting of all s.n. terms. Note that there are no $\delta$-rules associated with the constant $*$.

**Theorem 2.1 (Compatibility)** *Let $M, M', N, N' \in \mathrm{SN}_*$ with $M \sim_{cl} M'$ and $N \sim_{cl} N'$. Then,*

*(i)* $MN \in \mathrm{SN}_* \iff M'N' \in \mathrm{SN}_*$,

*(ii)* $MN \in \mathrm{SN}_* \implies MN \sim_{cl} M'N'$.

*Hence, the quotient structure $\langle \mathrm{SN}_*/\sim_{cl}, \cdot/\sim_{cl} \rangle$ is a well-defined partial applicative structure.* □

We sketch an outline of the proof. In general, we follow the notational conventions of [Bar84] with the notable exception of substitution; we write $M[N/x]$ to mean "in $M$, substitute $N$ for every occurrence of $x$" taking care to follow the substitution convention in [Bar84, p. 26]. We let $\Delta, \Delta_i$ range over $\beta$-redexes. Let $\Delta \equiv (\lambda x.A)B$. By convention, we write $\Delta \equiv A[B/x]$ (similarly for $\Delta_i$). The binary relation $\beta$ on $\Lambda_*$ is defined as the collection of pairs of the shape $\langle (\lambda x.P)Q, P[Q/x] \rangle$ with $P, Q$ ranging over $\Lambda_*$. We write $\to$ as the compatible closure of $\beta$ and $\twoheadrightarrow$ the reflexive, transitive closure of $\to$.

**Multi-holed Linear Contexts** In the following, we shall work with a sub-collection of what Barendregt calls "multiple numbered contexts" (see [Bar84, p. 375]) known as *multi-holed linear contexts*. We write them typically as $C[[]_1, \cdots, []_n]$ where the $[]_i$'s serve as place-markers for the "holes". Note that all $[]_i$'s are distinct and each hole $[]_i$ occurs exactly once (hence the adjective "linear").

**Definition 2.2** For $U, V \in \Lambda_*$, we say that $U \twoheadrightarrow V$ is an *innocuous reduction* if for some $n \geq 1$ there are redexes $\Delta_i$ for $1 \leq i \leq n$ and an $n$-holed linear context $C[[]_1, \cdots, []_n]$ satisfying the *non-interference* condition

"no free variables in any of the $\Delta_i$ becomes bound in $C[\Delta_1, \cdots, \Delta_n]$"

and that $U \equiv C[\Delta_1, \cdots, \Delta_n]$ and $V \equiv C[\bar\Delta_1, \cdots, \bar\Delta_n]$. The point of the non-interference condition is to ensure that for fresh variables $\xi_1, \cdots, \xi_n$, the term $C[\Delta_1, \cdots, \Delta_n]$ is the same as $C[\xi_1, \cdots, \xi_n][\vec{\Delta}/\vec{\xi}]$.

For example, with $C[] \equiv \lambda z.[]$ and $\Delta \equiv (\lambda x.\mathrm{i})(z(\lambda y.yy))$, the reduction $C[\Delta] \to C[\bar\Delta]$ is not innocuous because the non-interference condition is violated. Closed reduction is immediately seen to be innocuous *i.e.* for $U, V \in \Lambda_*$, whenever $U \to_{cl} V$, there is a 1-holed linear context $C[]$ and a *closed* redex $\Delta$ such that $U \equiv C[\Delta] \to_{cl} C[\bar\Delta] \equiv V$.

It is easy to see that $\beta$-reduction preserves s.n. property *i.e.* assuming $U \to V$, if $U \in \mathrm{SN}_*$ then $V \in \mathrm{SN}_*$. However, the converse *i.e.*

(†) $\qquad\qquad\qquad U \to V \ \& \ V \in \mathrm{SN}_* \quad \Longrightarrow \quad U \in \mathrm{SN}_*$

is not true in general; just consider $U \equiv (\lambda x.y)\Omega \to y \equiv V$ where $\Omega$ is any unsolvable term. We establish sufficient conditions for the above implication (†) in the following proposition. This result is also a crucial step (but stronger than is necessary) in the proof of the Compatibility Theorem.

**Proposition 2.3 (Crucial)** *For $U, V \in \Lambda_*$ such that $U \equiv C[\Delta_1,\cdots,\Delta_n] \twoheadrightarrow C[\Delta_1,\cdots,\Delta_n] \equiv V$ is an innocuous reduction with $\Delta_1,\cdots,\Delta_n \in \mathrm{SN}_*$. If $V \in \mathrm{SN}_*$, then $U \in \mathrm{SN}_*$.* $\qquad\square$

The proposition is a corollary of a Technical Lemma which we omit. We are now in a position to prove the Compatibility Theorem which is actually valid for any $\lambda$-calculus (including the *pure* calculus) generated from a set (possibly empty) of formal constants with no $\delta$-rules.

**Proof of the Theorem** (i) Since s.n. is preserved by closed $\beta$-reduction, it suffices to prove: for $M, N, P \in \mathrm{SN}_*$,

(1) $M \to_{\mathrm{cl}} N \ \& \ NP \in \mathrm{SN}_* \quad \Longrightarrow \quad MP \in \mathrm{SN}_*$;

(2) $M \to_{\mathrm{cl}} N \ \& \ PN \in \mathrm{SN}_* \quad \Longrightarrow \quad PM \in \mathrm{SN}_*$.

Now, to prove (1), note that $MP \to_{\mathrm{cl}} NP$. For some 1-holed linear context $C[\ ]$ and for some closed redex $\Delta$, $MP \equiv C[\Delta] \to C[\Delta] \equiv NP$. $\Delta$ is s.n. since it is a subterm of $M$ — a s.n. term. The result then follows by an appeal to the Crucial Proposition. The argument for (2) is entirely similar. Part (ii) of the theorem is an easy consequence of the fact that the equivalence relation $\sim_{\mathrm{cl}}$ is by definition a compatible closure. $\qquad\square$

In the following, we shall often need to reason with the s.n. property of $\lambda *$-terms. We gather some useful arguments in the following proposition to this end.

**Proposition 2.4 (Strong Normalization Arguments)** *The following arguments are valid:*

(1) **subterm**     If $M \in \mathrm{SN}_*$ and $N$ is a subterm of $M$, then $N \in \mathrm{SN}_*$;

(2) **reduction**     If $M \in \mathrm{SN}_*$ and $M \twoheadrightarrow N$, then $N \in \mathrm{SN}_*$;

(3) **abstraction**     $M \in \mathrm{SN}_*$ iff $\lambda x.M \in \mathrm{SN}_*$;

(4) **substitution**     If $M[N/x] \in \mathrm{SN}_*$ for some $N$, then $M \in \mathrm{SN}_*$;

(5) **redex**     $Q, P[Q/x] \in \mathrm{SN}_*$ iff $(\lambda x.P)Q \in \mathrm{SN}_*$;

(6) **head variable**     Let $\nu \equiv *$ or a variable. Then, $\vec{L} \subseteq \mathrm{SN}_*$ iff $\nu\vec{L} \in \mathrm{SN}_*$;

(7) **application**     Let $\nu \equiv *$ or a variable. Then, $M, \vec{L} \subseteq \mathrm{SN}_*$ iff $M(\nu\vec{L}) \in \mathrm{SN}_*$;

(8) **anti-reduction**     If $M, N \in \mathrm{SN}_*$ and $MN \equiv C[\Delta] \to C[\Delta] \in \mathrm{SN}_*$ with $\Delta$ and $C[\ ]$ satisfying the non-interference condition, then $MN \in \mathrm{SN}_*$.

*The converse of each of the above arguments, where applicable, is not valid.* $\qquad\square$

**Is SN$_*$ a partial combinatory algebra?** Let SN$_*$ denote the quotient structure $\langle \text{SN}^o_*/\sim_{\text{cl}}, \cdot/\sim_{\text{cl}} \rangle$ where SN$^o_*$ is the collection of closed s.n. untyped $\lambda*$-terms. We shall very often confuse SN$_*$ with the underlying set. Let **s** and **k** denote the (respective $\sim_{\text{cl}}$-equivalence classes of the) standard combinators $\lambda xyz.xz(yz)$ and $\lambda xy.x$. For any partial applicative structure $U$, denote the formal applicative algebra freely generated from $U$ and Var as $\mathcal{T}(U)$. We shall call elements of $\mathcal{T}(U)$ *polynomials* over $U$.

**Definition 2.5** A *conditionally partial combinatory algebra* (C-PCA) is a partial applicative structure $\langle U, \cdot \rangle$ where $U$ is a set with at least two elements; there are distinguished elements **k**, **s** $\in U$ satisfying the following axioms: for any $f, g, a \in U$

$$\text{(S)} \qquad sfga \simeq fa(ga),$$
$$\text{(K)} \qquad kab = a,$$

where $t\!\downarrow$ means "$t$ is defined" and $\simeq$ is *Kleene equality* i.e. either both sides are defined and are equal, or both are undefined. We shall assume that application is strict i.e. to assert $st\!\downarrow$ is to assert implicitly both $s\!\downarrow$ and $t\!\downarrow$ for polynomials $s$ and $t$. Also, the assertion $u = v$ has the force of $u\!\downarrow$ and $v\!\downarrow$. Note that the following axioms are valid in a C-PCA: for any $f, g, a \in U$, we have (K$_1$): $ka\!\downarrow$ and

$$(\text{S}_2^-) \qquad \exists a \in U. fa(ga)\!\downarrow \implies sfg\!\downarrow.$$

It is easy to check that **skk** which we shall call **i** is always defined in a C-PCA. The stipulation **s** $\neq$ **k** is equivalent to $\exists a, b \in U. a \neq b$. If, in addition, the axiom (S$_2$): $\forall f, g \in U. sfg\!\downarrow$ holds in a C-PCA $U$, then $U$ is by definition, a *partial combinatory algebra*. Of course, if the axiom (S$_2$) holds, then so does the axiom (S$_1$): $\forall f \in U. sf\!\downarrow$; but the converse is not true.

As an important corollary of Proposition 2.4, we have the following theorem.

**Theorem 2.6 (S.N. Realisers)** *The quotient structure* SN$_*$ *is a C-PCA in which* (S$_1$) *is valid but not* (S$_2$). *Hence,* SN$_*$ *is not a* PCA.

**Proof** (Sketch) To see why the axiom (S$_2$) fails[4]: just consider $f = g = \lambda z.(\lambda x.xx)$. We will just show the validity of (S) for illustration. First for "$\Rightarrow$", for any $f, g, a \in$ SN$_*$, observe that:

$$sfga \to_{\text{cl}} [\lambda yz.fz(yz))g]_2 a \to_{\text{cl}} [(\lambda z.fz(gz))a]_1 \to_{\text{cl}} fa(ga).$$

If $sfga \in$ SN$_*$, then so does $fa(ga)$, by the reduction argument. To prove the other direction "$\Leftarrow$", suppose $fa(ga) \in$ SN$_*$. Note that $(fz(gz))[a/z] \equiv fa(ga)$. By the redex argument, $[(\lambda z.fz(gz))a]_1 \in$ SN$_*$. Now, by the subterm argument, $(\lambda z.fz(gz)) \in$ SN$_*$; and so, by the redex argument, the closed redex $[(\lambda yz.fz(yz))g]_2 \in$ SN$_*$. Take $C[] \equiv []a$, by the Crucial Proposition, $[\lambda yz.fz(yz))g]_2 a \equiv C[\lambda yz.fz(yz))g] \in$ SN$_*$. By an entirely similar argument, we have $sfga \in$ SN$_*$. □

The relevance of PCAs to constructive logic is well-known, see e.g. [Bee85, Ch. VI]; not so in the case of C-PCAs. As far as we know, the notion of C-PCA is new. Schönfinkel showed that combinatory algebras may be characterised as precisely those (total) applicative structures $U$ which are *combinatory complete* i.e. every polynomial $t$ over $U$ is *internally representable* which means that for any fixed $\{\vec{x}\}$ containing Var($t$), we have

$$\exists u \in U. \forall \vec{a} \subseteq U. t[\vec{a}/\vec{x}] = u\vec{a}.$$

Of course, for the simultaneous substitution $(-)[\vec{a}/\vec{x}]$ to make sense, we require the two sequences $\vec{a}$ and $\vec{x}$ to be compatible i.e. they have the same length. In a similar vein, Bethke [Bet87] recently

---
[4] We are grateful to E. Robinson for pointing this example out to us.

established a characterization result for PCAs. She showed that a partial applicative structure $U$ is a PCA iff for any $t \in \mathcal{T}(U)$ and for $\text{Var}(t) \subseteq \{x_0, \cdots, x_{n+1}\}$, $U$ satisfies the following:

$$\exists y \in U. \forall a_0, \cdots, a_{n+1} \subseteq U.[ta_0 \cdots a_n \downarrow \ \& \ ya_0 \cdots a_{n+1} \simeq t[a_0, \cdots, a_{n+1}/x_0, \cdots, x_{n+1}]].$$

Can C-PCAs be characterised along similar lines? For any partial applicative structure $U$, we say that $U$ is *conditionally combinatory complete* iff for any $t \in \mathcal{T}(U)$ with $\text{Var}(t) \subseteq \{\vec{x}\}$, we have:

(c-cc) $\qquad\qquad \exists \vec{b} \subseteq U. t[\vec{b}/\vec{x}] \downarrow \quad \Longrightarrow \quad \exists y \in U. \forall \vec{a} \subseteq U. y\vec{a} \simeq t[\vec{a}/\vec{x}].$

For $t \in \mathcal{T}(U)$ and $x$ a variable, define $\lambda^* x.t$ by structural induction as follows:

$$\lambda^* x.x \stackrel{\text{def}}{=} \text{skk};$$
$$\lambda^* x.t \stackrel{\text{def}}{=} \text{k}t \qquad \text{if } x \notin \text{FV}(t);$$
$$\lambda^* x.st \stackrel{\text{def}}{=} \text{s}(\lambda^* x.s)(\lambda^* x.t). \quad \text{if } x \in \text{FV}(st).$$

We write $\lambda^* \vec{x}.t$ for $\lambda^* x_1.(\cdots(\lambda^* x_n.t)\cdots)$. This algorithm is known as the *Curry abstraction algorithm*. Note that for any $t \in \mathcal{T}(U)$ with $\text{Var}(t) \subseteq \{x\}$, the polynomial $\lambda^* x.t$ is not necessarily defined in $U$. For example, in any C-PCA $U$ where there are elements $a, b \in U$ such that $a \cdot b$ is undefined, the polynomial $\lambda^* x.ab \stackrel{\text{def}}{=} \text{k}(ab)$ is undefined. We can show:

**Theorem 2.7 (Constructive Characterization)** *A partial applicative structure $U$ is a C-PCA iff conditional combinatory completeness is valid in $U$.* □

As an important corollary, we obtain an axiom useful for many "realiser calculations" in the sequel:

(abs) $\qquad\qquad \exists a_1, \cdots, a_n \in U. t[\vec{a}/\vec{x}] \downarrow \quad \Longrightarrow \quad \lambda^* \vec{x}.t \downarrow.$

## 3 Right-Absorptive C-PCA and Modified Realizability

Recall that our programme to produce a "generic" s.n. proof relies crucially on Hyland's construction which builds a Kleene-style realizability topos $\text{TOP}_s(U)$ out of an arbitrary PCA $U$. It is helpful to understand the construction from a logical perspective. The topos $\text{TOP}_s(U)$ may be thought of as a constructive set theory according to the Kleene-style realizability interpretation of truth, and the underlying PCA $U$ is just the collection of "realisers". For instance, in the case of the *effective topos* $\mathcal{E}\!f\!f$, the "canonical" realizability topos over the PCA $K_1$ — "Kleene's first model" of natural numbers, a sentence of Heyting Arithmetic is recursively realised if and only if it is "internally" true of the natural number object of the topos $\mathcal{E}\!f\!f$ (see [Hyl82]).

Hyland's PCA-to-topos construction is actually an instance of a more general construction that yields a class of toposes as studied in Tripos Theory [HJP80]. Formally, a *tripos* (which is an acronym for "Topos Representing Indexed Pre-Ordered Sets") is a structure couched in the language of category theory which ressembles Lawvere's hyperdoctrines. It provides a semantics for typed intuitionistic predicate logic *without* equality. The Fundamental Theorem of Tripos Theory states that:

> Every tripos P gives rise to a topos P-SET of non-standard sets.

Logically, the passage from a tripos P to the associated topos P-SET corresponds to the addition of equality and the axiom of extensionality. In essence and this can be made precise, the logic of a tripos P is an "external" (and hence arguably more convenient) but equivalent presentation of the *internal* logic of the associated topos P-SET. In light of Tripos Theory, the PCA-to-topos construction may be re-organised in terms of two stages:

I. Every PCA $U$ yields a "standard" Kleene-style realizability tripos $P_s(U)$ (subscript "s" for "standard").

II. Every standard realizability tripos $P_s(U)$ gives rise to an associated standard realizability topos $TOP_s(U)$, by the Fundamental Theorem of Tripos Theory.

We have established that $SN_*$ — the quotient structure of central interest to our programme — is a C-PCA but not a PCA. If the prescription of stage I above is applied to $SN_*$, does the process of construction yield a tripos (and in turn, a topos)? Unfortunately, the answer is no and this is because we need the full "structure" of a PCA in order to use its elements as "realisers" for a *Kleene*-style realizability interpretation of intuitionistic predicate logic (which is what stage I is all about). We will now explain why this is so.

**Scott Implication** Let $U$ be a C-PCA. We define *Scott implication*, a binary operation $\to$ on the powerset of $U$ as follows: for $P, Q \subseteq U$,

$$(P \to Q) \stackrel{\text{def}}{=} \{ u \in U : \forall a \in P . ua \downarrow \ \& \ ua \in Q \}.$$

Reading subsets of $U$ as "propositions", Scott implication is the precise semantic counterpart of Kleene's realizability interpretation of implication. Now assume that the axiom $(S_2)$ is not valid in $U$. We claim that $U$ is inadequate as realisers for a Kleene-style realizability interpretation of intuitionistic predicate logic. To show this, it suffices to show that Scott implication over $U$ does not model minimal logic. Suppose, for a contradiction, it does. Then for any $P, Q, R \subseteq U$, the following subset

$$V \stackrel{\text{def}}{=} (P \to Q \to R) \to (P \to Q) \to (P \to R)$$

is non-empty and contains the element $\mathbf{s}$. Take $P \equiv \emptyset$ and $V$ degenerates to $U \to U \to U$. Clearly, $\mathbf{s} \in U \to U \to U$ if and only if $(S_2)$ is valid. The following lemma gives further properties of Scott implication over C-PCAs: note the extent to which they are weaker (*vide* non-emptyness assumption in (iii)) than those of the same notion defined over PCAs.

**Lemma 3.1** *For any* C-PCA $U$ *and for any* $P, Q, R \subseteq U$,

(i) $P = \emptyset \implies (P \to Q) = U$. *Suppose there are some* $a, b \in U$ *for which* $ab$ *is undefined and that the axiom* $(S_2)$ *holds, then the converse is also valid.*

(ii) $\mathbf{k} \in P \to (Q \to P)$.

(iii) $P, Q \neq \emptyset \implies \mathbf{s} \in (P \to (Q \to R)) \to ((P \to Q) \to (P \to R))$.

*The non-emptyness assumption in (iii) is indispensable.* □

This above observation leads us to consider a more intensional variant of the "standard" realizability interpretation; that of *modified realizability* in the style of Kreisel [Kre59, Tro73]. In the following, we shall present modified realizability in the general framework of a *right-absorptive* C-PCA.

**Definition 3.2** Let $\langle U, \cdot \rangle$ be a partial applicative structure. For any non-empty subset $\Theta$ of $U$, we say that $\Theta$ is *right-absorptive* if $\forall \theta \in \Theta . \forall u \in U . \theta u \downarrow \ \& \ \theta u \in \Theta$. Informally, the subset $\Theta$ is a right-ideal of $U$ i.e. $\Theta \cdot U \subseteq \Theta$. We call a C-PCA $U$ *right-absorptive* if $U$ has a right-absorptive subset.

The point of right-absorptiveness is this: supersets of any right-absorptive subset $\Theta$ are closed under the operation of Scott implication. More precisely, let $\Theta$ be a non-empty subset of a C-PCA $U$. Then

$$\Theta \text{ is right-absorptive} \iff \forall Q, P \subseteq U.[\Theta \subseteq Q \implies \Theta \subseteq (P \to Q)].$$

Clearly, if a C-PCA $U$ has a *right-absorptive element* $\theta$ i.e. $\forall u \in U.\theta u\downarrow$ & $\theta u = \theta$; then the singleton set $\{\theta\}$ is a right-absorptive subset of $U$.

**Example** A C-PCA may have more than one right-absorptive subset or it may have none. Consider the C-PCA of closed $\beta$-equivalent untyped $\lambda$-terms (which is actually a $\lambda$-model and so, is *a fortiori* a total combinatory algebra): it has a right-absorptive element **YK** where **Y** is the paradoxical combinator $\lambda f.(\lambda x.f(xx))(\lambda x.f(xx))$. It may be helpful to think of **YK** as the "solution" to the recursive equation $X = \lambda x.X$ in meta-variable X. The collection of all elements with *unbounded order*[5] forms a further right-absorptive subset.

Not every C-PCA is right-absorptive.

**Proposition 3.3** *(i) The pure untyped s.n. $\lambda$-terms quotiented with $\sim_{cl}$ is not right-absorptive.*

*(ii) $\mathsf{SN}_*$ has a right-absorptive subset $\Theta \stackrel{\text{def}}{=} \{*\vec{u} : \vec{u} \subseteq \mathsf{SN}_*\}$.* □

**Definition 3.4** Given a C-PCA $U$ with a right-absorptive subset $\Theta$, a *proof-extension pair* is a pair $P \equiv \langle P_0, P_1 \rangle$ with $P_0, P_1 \subseteq U$ satisfying $\Theta \subseteq P_1$ and $P_0 \subseteq P_1$. Note that by definition $\Theta \neq \emptyset$, and so $P_1 \neq \emptyset$.

We may read a proof-extension pair $P$ as a non-standard proposition with the first component $P_0$ containing the set of (codes for the) "actual proofs" of the proposition $P$ and the second component $P_1$ the set of (codes for the) "potential proofs" or "proof extensions" of the proposition. The notion of right-absorptiveness was introduced because the failure of axiom ($S_2$) places a serious restriction on the availability and use of the crucial realiser s. This constraint is connected with the non-emptyness assumption in propositions of certain shapes: see Lemma 3.1(iii). We circumvent this problem by modifying the way we model propositions: instead of a set of realisers we think of a proposition as a *pair* of sets of realisers, and *crucially* the second component of each pair is designed to be a superset of $\Theta$ and so, it is necessarily non-empty. For this reason, we are forced to consider *right-absorptive* C-PCAs. Now the pure s.n. $\lambda$-terms are not right-absorptive whereas the s.n. $\lambda*$-terms are (Proposition 3.3) — this is precisely why our programme is built on a quotient of s.n. $\lambda*$-terms (as opposed to the pure terms).

Let $\mathsf{Prf}(U)$ denote the collection of proof-extension pairs over a right-absorptive C-PCA $U$ (w.r.t. a fixed $\Theta$) and we use meta-variables $P, Q, R$ etc. to range over $\mathsf{Prf}(U)$. An important property of proof-extension pairs is that they are closed under (component-wise) intersection. For any collection of proof-extensions $\{A(i) \in \mathsf{Prf}(U) : i \in \mathcal{I}\}$, we define the *intersection* as:

$$\bigcap_{i \in \mathcal{I}} A(i) \stackrel{\text{def}}{=} \langle \bigcap_{i \in \mathcal{I}} A(i)_0, \bigcap_{i \in \mathcal{I}} A(i)_1 \rangle.$$

It is easy to check that $\bigcap_{i \in \mathcal{I}} A(i)$ thus defined is a proof-extension pair.

**Definition 3.5** Let $U$ be a right-absorptive C-PCA. We define *Kreisel implication* $\stackrel{\circ}{\to}$ which is a binary operation on $\mathsf{Prf}(U)$ as follows: for any two proof-extension pairs $P$ and $Q$,

$$(P \stackrel{\circ}{\to} Q) \stackrel{\text{def}}{=} \langle (P_0 \to Q_0) \cap (P_1 \to Q_1), \ (P_1 \to Q_1) \rangle.$$

It is a property of the right-absorptive subset that $(P \stackrel{\circ}{\to} Q)$ thus defined is a proof-extension pair.

---

[5] A $\lambda$-term $M$ has *unbounded order* if for any natural number $n$, however large, there exists a term $N$ such that $M =_\beta \lambda x_1 \cdots x_n.N$. Such a term is always unsolvable, see e.g. [Bar84].

Kreisel implication is the precise semantic counterpart of the Kreisel-style modified realizability interpretation of implication, of which more anon.

**Modified Realizability Triposes** Our aim for the rest of this section is to show that a new class of *modified* realizability triposes $P_m(U)$ may be systematically constructed from an arbitrary right-absorptive C-PCA $U$. We shall assume familiarity with the basic notions of Tripos Theory (including the respective definitions of a tripos P and its associated topos P-SET of non-standard sets) of which the most accessible reference is [HJP80]. For any set $I$, define $P_m(I)$ to be $\langle \text{Prf}(U)^I, \vdash_I \rangle$, the set of all functions from $I$ to $\text{Prf}(U)$ which is to be thought of as a collection of proof-extension pairs indexed over the set $I$. The pre-order $\vdash_I$ is defined as

$$A \vdash_I B \stackrel{\text{def}}{=} \bigcap_{i \in I}(A(i) \stackrel{\circ}{\to} B(i))_0 \neq \emptyset.$$

In the case of $I \equiv \emptyset$, we decree that the intersection of a collection of proof-extension pairs indexed over the empty set is $\langle U, U \rangle$, the largest (w.r.t. inclusion) such pair, as is consistent with convention. If $a$ is a member of the above intersection, we say that $a$ *realises* or *witnesses* $A \vdash_I B$.

We define a binary operation $\stackrel{\circ}{\to}$ on $\text{Prf}(U)^I$ by point-wise extension of $\stackrel{\circ}{\to}$: given $A, B \in \text{Prf}(U)^I$, the map $A \stackrel{\circ}{\to} B$ is defined as $i \mapsto (A(i) \stackrel{\circ}{\to} B(i))$. The structure $\langle \text{Prf}(U)^I, \vdash_I \rangle$ is a pre-ordered set: we prove this fact as a corollary of the following lemma.

**Lemma 3.6** *Kreisel implication on proof-extension pairs of a right-absorptive* C-PCA *models minimal logic i.e. intuitionistic implication.* □

A bottom element of the preorder $\langle \text{Prf}(U)^I, \vdash_I \rangle$ is a map $i \mapsto \langle \emptyset, X_i \rangle$ for any $X_i \supseteq \Theta$ and for any $i \in I$; a top element is the constant map $i \mapsto \langle U, U \rangle$; another is the map $i \mapsto \langle \{i\}, \{i\} \cup \Theta \rangle$.

We will next show how the universal quantifier is to be interpreted. The interpretation of the other connectives of the logic may then be defined in terms of the interpretation for the implication and the universal quantifier, using a definability result in [HJP80, Theorem 1.4]. Though cast in category-theoretic language, the result in *op. cit.* is essentially the inter-definability result of second order logical connectives which is attributed to Russell, see *e.g.* [Pra65].

**Definition 3.7** Let $f : I \to J$ be a map between sets. Recall that $P_m(I) \stackrel{\text{def}}{=} \langle \text{Prf}(U)^I, \vdash_I \rangle$. We define functors $P_m(f) : P_m(J) \to P_m(I)$ and $\forall f : P_m(I) \to P_m(J)$ as follows: for any $A \in \text{Prf}(U)^J$ and $B \in \text{Prf}(U)^I$, $P_m(f)$ is just composition with $f$ i.e. $(P_m(f))A(i) \stackrel{\text{def}}{=} A(f(i))$ for any $i \in I$. For any $j \in J$, $(\forall f)B(j)$ is defined as:

$$(\forall f)B(j) \stackrel{\text{def}}{=} \bigcap_i [|f(i) = j| \stackrel{\circ}{\to} B(i)];$$

where

$$|f(i) = j| \stackrel{\text{def}}{=} \begin{cases} \langle \{i\}, \Theta \cup \{i\} \rangle & \text{if } f(i) = j, \\ \langle \emptyset, \Theta \rangle & \text{else.} \end{cases}$$

For the case of $I \equiv \emptyset$, $P_m(I)$ is just the singleton set, note that $(\forall f)B(j) \stackrel{\text{def}}{=} \langle U, U \rangle$ by convention. It is easy to check that $(\forall f)B(j)$ defines a proof-extension pair for each $j$. We can show that the conditions of Theorem 1.4 in [HJP80] are satisfied. Hence, we have the following theorem.

**Theorem 3.8 (Modified Realizability Topos)** *Given any right-absorptive* C-PCA $U$, $P_m(U)$ *thus defined is a (canonically presented) tripos. Applying the Fundamental Theorem of Tripos Theory, we can then construct the associated modified realizability topos* $\text{TOP}_m(U)$. □

We give a summary of the various constructions that make up a modified realizability tripos $\mathbf{P}_m(U)$ over a right-absorptive C-PCA $U$. Let $I$ be a set, for any $A, B \in \operatorname{Prf}(U)^I$ and any set-theoretic function $f: I \to J$:

$$(A \times B)(i) \stackrel{\text{def}}{=} \bigcap_{X \in \operatorname{Prf}(U)}((A(i) \stackrel{\circ}{\to} B(i) \stackrel{\circ}{\to} X) \stackrel{\circ}{\to} X)$$

$$(A + B)(i) \stackrel{\text{def}}{=} \bigcap_{X \in \operatorname{Prf}(U)}(A(i) \stackrel{\circ}{\to} X) \stackrel{\circ}{\to} (B(i) \stackrel{\circ}{\to} X) \stackrel{\circ}{\to} X),$$

$$(A \stackrel{\circ}{\to} B)(i) \stackrel{\text{def}}{=} A(i) \stackrel{\circ}{\to} B(i),$$

$$(\forall f)A(j) \stackrel{\text{def}}{=} \bigcap_i [|f(i) = j| \stackrel{\circ}{\to} A(i)],$$

$$(\exists f)A(j) \stackrel{\text{def}}{=} \bigcup_i [|f(i) = j| \times A(i)].$$

Since the meaning of $(\exists f)A(j)$ involves indexed union, we decree that union of proof-extension pairs indexed over the emptyset is the least (w.r.t. inclusion) proof-extension pair $\langle \varnothing, \Theta \rangle$.

## 4 Modified Realizability Toposes

We already know some general properties of P-SET, the topos of non-standard sets associated with an arbitrary tripos P from [HJP80]. In addition, an in-depth study of a particular realizability topos $\mathcal{E}\!f\!f$, the *effective topos* has been carried out in [Hyl82]. This section presents a summary of the properties of $\textbf{TOP}_m(U)$ where $U$ is any right-absorptive C-PCA.

**Embeddings of SET into $\textbf{TOP}_m(U)$** The topos $\textbf{TOP}_m(U)$ is a category of sets equipped with a $\operatorname{Prf}(U)$-valued equality predicate. Unlike the standard realizability toposes, there are two canonical embeddings of SET into $\textbf{TOP}_m(U)$ which we shall refer to as type I and type II embeddings respectively. In particular, the type II embedding is a topos inclusion.

**Type I Embedding:** $\Delta : \textbf{SET} \to \textbf{TOP}_m(U)$  Write $\mathsf{T} \stackrel{\text{def}}{=} \langle \{i\}, \{i\} \cup \Theta \rangle$ and $\mathsf{F} \stackrel{\text{def}}{=} \langle \varnothing, \Theta \rangle$. For any set $X$, the image $\Delta X$ has $X$ as the underlying set; the equality predicate = is defined as follows: for any $x, y \in X$,

$$[x = y] \stackrel{\text{def}}{=} \begin{cases} \mathsf{T} & \text{if } x = y, \\ \mathsf{F} & \text{else.} \end{cases}$$

For any set-theoretic function $f: X \to Y$, we define a functional relation $\Delta f: X \times Y \to \operatorname{Prf}(U)$ by: for any $x \in X$ and $y \in Y$,

$$\Delta f(x, y) \stackrel{\text{def}}{=} \begin{cases} \mathsf{T} & \text{if } f(x) = y, \\ \mathsf{F} & \text{else.} \end{cases}$$

Note that the designated truth values $\mathsf{T}$ and $\mathsf{F}$ are identical to those in Definition 3.7. The data associated with $\Delta$ may be cast in tripos-theoretic language: for any map $f: X \to Y$ in SET, $\Delta X$ is just the object $\langle X, \exists \Delta_X(\mathsf{T}_X) \rangle$ where $\Delta_X : X \to X \times X$ is the diagonal map, $\mathsf{T}_X$ is the constant map sending $x \in X$ to $\langle U, U \rangle$; and $\Delta f : \Delta X \to \Delta Y$ is equivalent to $\exists (1_X, f)(\mathsf{T}_X) \in \operatorname{Prf}(U)^{X \times Y}$.

**Proposition 4.1** *The functor* $\Delta: \textbf{SET} \to \textbf{TOP}_m(U)$ *is full and faithful, and left-exact. Further, every object* $\langle X, = \rangle$ *of the topos* $\textbf{TOP}_m(U)$ *is a quotient of a subobject of the object* $\Delta X$. □

**Type II Embedding: topos inclusion** $\nabla : \textbf{SET} \to \textbf{TOP}_m(U)$  We spell out the inverse and direct image functors of the geometric morphism: $\Gamma \dashv \nabla$.

**Direct Image Functor:** $\nabla : \mathbf{SET} \hookrightarrow \mathbf{TOP}_m(U)$ is defined in exactly the same way as the functor $\Delta$ except in place of $\mathsf{F}$, we use $\mathsf{F}^+ \stackrel{\text{def}}{=} \langle \emptyset, \Theta \cup \{i\} \rangle$. Since adjunction is defined up to isomorphism, replacing the pair $(\mathsf{T}, \mathsf{F}^+)$ in the definition by another equivalent (in $\langle \text{Prf}(U)^2, \vdash_2 \rangle$) pair of proof-extension pairs say $(\langle U, U \rangle, \langle \emptyset, U \rangle)$, defines the same direct image functor.

**Inverse Image Functor:** $\Gamma : \mathbf{TOP}_m(U) \to \mathbf{SET}$ is just the global sections functor:

- **objects:** for any $\langle X, = \rangle \in \mathbf{TOP}_m(U)$, the image under $\Gamma$ is the set $\Gamma\langle X, = \rangle \stackrel{\text{def}}{=} \{ x : \llbracket x \rrbracket \neq \emptyset \}/\simeq$ where $x \simeq x' \stackrel{\text{def}}{=} \llbracket x = x' \rrbracket \neq \emptyset$. We write $[x]_\simeq$ for the $\simeq$-equivalence class of $x$ as an element of $\Gamma\langle X, = \rangle$.

- **morphisms:** for any morphism $F : \langle X, = \rangle \to \langle Y, = \rangle \in \mathbf{TOP}_m(U)$, its image under $\Gamma$ is the set-theoretic function $\Gamma F : \Gamma\langle X, = \rangle \to \Gamma\langle Y, = \rangle$ with $\Gamma F([x]_\simeq) \stackrel{\text{def}}{=} [y]_\simeq$ for any $y$ such that $F(x,y) \neq \emptyset$. Such a $y$ is guaranteed to exist by the requirement of totality on the functional relation $F$.

**Theorem 4.2** *The functor* $\nabla : \mathbf{SET} \hookrightarrow \mathbf{TOP}_m(U)$ *is a topos inclusion.* □

The above topos inclusion is the canonical inclusion of $\mathbf{SET}$ into the modified realizability topos $\mathbf{TOP}_m(U)$. We know from the work of Hyland et al. ([HJP80, Corollary 4.6]) that in general, any arbitrary tripos which is $\exists$-*standard* (which $\mathsf{P}_m(U)$ is) gives rise to just such a topos inclusion which is defined entirely by the logical properties of the tripos.

**Double-Negation Topology** Given the topos inclusion of $\mathbf{SET}$ into $\mathbf{TOP}_m(U)$, by a theorem of Lawvere and Tierney (see e.g. Theorem 4.14 and Proposition 4.15 in [Joh77, pp. 104 – 105]), there is a unique Lawvere-Tierney topology $j$ in $\mathbf{TOP}_m(U)$ such that $\mathbf{SET} \simeq \mathrm{Shv}_j(\mathbf{TOP}_m(U))$, where $\mathrm{Shv}_j(\mathbf{C})$ denotes the category of $j$-sheaves of the category $\mathbf{C}$. The topology $j$ in question is the double negation topology.

**Proposition 4.3** *For any right-absorptive* C-PCA, $\mathbf{SET} \simeq \mathrm{Shv}_{\neg\neg}(\mathbf{TOP}_m(U))$. □

**¬¬-Seperated Objects** Let $j$ be a Lawvere-Tierney topology over a topos $\mathbf{E}$. Recall the result that the inclusion functor $\nabla : \mathrm{Shv}_j(\mathbf{E}) \hookrightarrow \mathbf{E}$ has a left-adjoint $\mathbf{a} : \mathbf{E} \to \mathrm{Shv}_j(\mathbf{E})$ called the *sheafification functor* which is left-exact.

Recall also the following equivalent characterizations of a $j$-separated object $A$.

(1) the diagonal $\Delta_A : A \rightarrowtail A \times A$ is a $j$-closed subobject of $A \times A$,

(2) the unit of the adjunction $\eta_A : A \to \nabla(\mathbf{a}A)$ is monic,

(3) for any $j$-dense subobject $E \rightarrowtail X$, whenever the partial map $X \leftarrowtail E \to A$ extends to a total map $X \to A$, the extension is unique.

Let $X$ be a map. A map $R \in \text{Prf}(U)^X$ is said to be *1-stable*, or simply *stable* if for any $x, y \in X$, $R(x)_1 = R(y)_1$. Stable maps characterize strict relations over sheaves in the sense which we will now clarify. For an arbitrary strict relation $R \in \text{Prf}(U)^X$ over the sheaf $\nabla X$, define a new map $\overline{R} \in \text{Prf}(U)^X$ from $R$ by fixing the second component: for each $x \in X$,

$$\overline{R}(x) \stackrel{\text{def}}{=} \langle R(x)_0, \bigcup_{y \in X} R(y)_1 \rangle.$$

We claim: $P_m(U) \vDash \forall x \in X.R(x) \leftrightarrow \overline{R}(x)$. Since there is a 1-1 correspondence between monics into an object and strict relations over it, we see that there is a 1-1 correspondence between subobjects of the sheaf $\nabla X$ and stable maps $R \in \text{Prf}(U)^X$ for any set $X$.

**Definition 4.4** An object $\langle X, = \rangle$ of the topos $\text{TOP}_m(U)$ is said to be *canonically separated* if there is a stable map $R \in \text{Prf}(U)^X$ such that for any $x, y \in X$

$$[x = y] \stackrel{\text{def}}{=} \begin{cases} R(x) & \text{if } x = y, \\ \langle \emptyset, R(x)_1 \rangle & \text{else.} \end{cases}$$

Note that by definition of stability, $R(x)_1$ is constant as $x$ varies over $X$. It is not difficult to check that the above data specifies a well-defined object of $\text{TOP}_m(U)$.

**Proposition 4.5** *An object of $\text{TOP}_m(U)$ is $\neg\neg$-separated iff it is isomorphic to a canonically separated object.* □

**Validity of Constructive Principles** Modified realizability was introduced by Kreisel [Kre59, Tro73] as an intensional variant of the Kleene-style realizability interpretation. The most distinctive feature of modified realizability is that it provides a setting in which the *Markov Principle* ($\text{MP}_{pr}$)

$(\text{MP}_{pr})$ $\qquad\qquad \neg\neg\exists x \in \mathbf{N}.A \quad \to \quad \exists x \in \mathbf{N}.A \qquad A$ is primitive recursive.

is invalidated.

**Theorem 4.6** *(i) The* Independence of Premise (IP) *axiom is internally valid in $\text{TOP}_m(U)$ for any right-absorptive* C-PCA $U$: *for any $B$, and $A$ in which $y$ is not free,*

$(\text{IP}) \qquad\qquad \neg A \to \exists y \in \mathbf{N}.B \quad \to \quad \exists y \in \mathbf{N}.(\neg A \to B)$

*(ii)* Church's Thesis ($\text{CT}_0$) *is internally valid in $\text{TOP}_m(K_1)$, where the PCA $K_1$ is "Kleene's first model" of natural numbers and where T and U are Kleene's T-predicate and output function respectively:*

$(\text{CT}_0) \qquad\qquad \forall x.\exists y.B(x,y) \quad \to \quad \exists e.\forall x.\exists z.[\text{T}(e,x,z) \wedge B(x,\text{U}(z))]$

□

Since $\text{HA} + \text{IP} + \text{CT}_0 + \text{MP}_{pr}$ is inconsistent (see [Tro73]), we infer that $\text{MP}_{pr}$ is invalid in $\text{TOP}_m(\mathbf{N})$. Here, it is appropriate to mention an unpublished manuscript of Grayson [Gra81] which provides sketchy details of a modified realizability topos (based on the tripos construction) and the more recent work [vO91, Str92].

**PERS-Extension Pairs** The collection $\text{PER}(U)$ (as defined in the Introduction) of PERs over a *proper* C-PCA $U$ (i.e. one in which $(S_2)$ fails) does not form a category since the axiom $(S_2)$ is needed to establish closure of composition. What then is the right notion of "PERs" (and "modest sets") in the modified realizability setting?

A PER-*extension pair* over a right-absorptive C-PCA $U$ is a pair $\langle R, \overline{R} \rangle$ where $R$ is a PER over $U$ and $\text{dom}(R) \subseteq \overline{R} \supseteq \Theta$. We define $\mathbf{P}_{ext}(U)$, the category of PER-extension pairs over $U$ as follows:

- *objects*: PER-extension pairs $\langle R, \overline{R} \rangle$,

- *morphisms*: $F : \langle R, \overline{R} \rangle \to \langle S, \overline{S} \rangle$ where $F$ is a function from $[R]$ ($R$-equivalence classes) to $[S]$ such that $F$ is *realised* by some $f \in U$ i.e. for any $r, r' \in U$, whenever $r \, R \, r'$ then $(fr) \, S \, (fr')$ and $F[r]_R = [fr]_S$, and $f \in (\overline{R} \to \overline{S})$. Two such realisers $f$ and $g$ are equivalent, written $f \sim g$ iff for any $r \in \mathrm{dom}(R)$, $(fr) \, S \, (gr)$, so $F$ is characterised by $[f]_\sim$.

**Lemma 4.7** *The above data defines a category* $\mathbf{P}_{\mathrm{ext}}(U)$ *which is Cartesian closed.* □

For instance, for any PER-extension pairs $\langle R, \overline{R} \rangle$ and $\langle S, \overline{S} \rangle$, the exponential $\langle S, \overline{S} \rangle^{\langle R, \overline{R} \rangle}$ has as the first component a PER $T$ whose domain is $\{ e \in U \, : \, e \text{ realises a morphism } R \to S \}$ and $e \, T \, e'$ iff $e \sim e'$; the second component is just $\overline{R} \to \overline{S}$.

**Fibration of $\mathbf{P}_{\mathrm{ext}}(U)$ over SET** For each set $I$, define the following category $(\mathbf{P}_{\mathrm{ext}}(U))_I$:

- *objects*: $I$-indexed families of PER-extension pairs $\{ \langle R_i, \overline{R_i} \rangle \}_{i \in I}$,

- *morphisms*: $[e]_\sim : \{ \langle R_i, \overline{R_i} \rangle \}_{i \in I} \to \{ \langle S_i, \overline{S_i} \rangle \}_{i \in I}$ where for each $i \in I$, $e$ realises a morphism $\langle R_i, \overline{R_i} \rangle \to \langle S_i, \overline{S_i} \rangle$; and $e \sim e'$ iff for each $i \in I$, $e \sim e'$ as realisers of morphisms from $\langle R_i, \overline{R_i} \rangle$ to $\langle S_i, \overline{S_i} \rangle$.

**Theorem 4.8** *The above data defines a cloven fibration* $\begin{array}{c} \mathbf{P}_{\mathrm{ext}}(U) \\ \downarrow \\ \mathrm{SET} \end{array}$ *which is complete i.e.*

(i) *each fibre* $(\mathbf{P}_{\mathrm{ext}}(U))_I$ *has finite limits and reindexing functors preserve limits,*

(ii) *for each morphism* $\phi : I \to J$ *in* SET, $\phi^*$ *has a right adjoint* $\Pi_\phi$ *satisfying the Beck-Chevalley condition,*

*and it has a generic object.* □

## 5 "Generic" S.N. Argument: an application

As an application of the machinery which we have set up, consider the s.n. argument of System F. First, note that the fibration $\begin{array}{c} \mathbf{P}_{\mathrm{ext}}(U) \\ \downarrow \\ \mathrm{SET} \end{array}$ satisfies all the structural requirements of a category-theoretic model for System F (as spelt out in the Introduction). For any derivable type-assignment sequent of the form $x_1 : \sigma_1, \cdots, x_n : \sigma_n \vdash_{\vec{X}} s : \tau$ where the free type variables of $\sigma_1, \cdots, \sigma_n, \tau$ are a subset of $\{ \vec{X} \}$, we can establish the *realiser argument* (the first of the "stripping arguments") by a straightforward induction:

**Lemma 5.1 (Realiser)** *For any derivable sequent* $x_1 : \sigma_1, \cdots, x_n : \sigma_n \vdash_{\vec{X}} s : \tau$, *the untyped term* $\lambda \xi. \lambda x_1 \cdots x_n. \lceil s \rceil$ *(where* $\lceil s \rceil$ *is obtained from $s$ by stripping off all embedded type expression) realises the following morphism*

$$[\vec{x} : \vec{\sigma} \vdash_{\vec{X}} s : \tau] \quad : \quad [\vec{X}; 1] \times [\vec{X}; \sigma_1] \times \cdots \times [\vec{X}; \sigma_n] \quad \to \quad [\vec{X}; \tau]$$

*in the fibre over* $(\mathbf{P}_{\mathrm{ext}}(U))^m$ *where $m$ is the length of $\vec{X}$.* □

To see the validity of the *reflection argument*, note that the term-$\beta$ reduction in System F corresponds precisely to $\beta$-reduction of the stripped terms. Moreover, the type-$\beta$ reduction leaves the corresponding stripped terms unchanged. The argument for the s.n. of System F is now complete.

**Further Directions** Does the above argument establish s.n. of the Calculus of Constructions? We can show that the category of PER-extension pairs is strongly complete (see *e.g.* [HRR90]) as a fibration over an appropriate category of *assemblies* (*not* equivalent to the category of ¬¬-separated objects) thus giving rise to a model of the calculus and much more. The stripping arguments therein are valid and the details will be presented elsewhere. The challenge that remains is to show that this approach is systematically applicable to the family of Generalized Type Systems *i.e.* Barendregt's "cube" [Bar91].

Until and unless we can establish the general applicability of the two "stripping arguments", or demonstrate that they are easily verifiable for a significant class of type theories, we cannot properly claim to have a generic s.n. proof.

# References

[Bé85] J. Bénabou. Fibred categories and the foundations of naïve category theory. *J. Symb. Logic*, pages 10–37, 1985.

[Bar84] H. Barendregt. *The Lambda Calculus*. North-Holland, revised edition, 1984.

[Bar91] H. Barendregt. Introduction to generalized type systems. *J. Functional Prog.*, 1:125–154, 1991.

[Bee85] M. J. Beeson. *Foundations of Constructive Mathematics*. Springer-Verlag, 1985.

[Bet87] I. Bethke. On the existence of extensional partial combinatory algebras. *J. Symb. Logic*, pages 819–833, 1987.

[CH88] T. Coquand and G. Huet. The Calculus of Constructions. *Info. and Comp.*, 76:95–120, 1988.

[Gal90] J. Gallier. On Girard's "Candidats de Réductibilité". In P. Odifreddi, editor, *Logic and Computer Science*. Academic Press, 1990.

[Gir72] J.-Y. Girard. Interprétation fonctionelle et elimination des coupures dans l'arithmétique d'order supérieur. Thèse de Doctorat d'Etat, Paris, 1972.

[Gir86] J.-Y. Girard. The system $F$ of variable types, fifteen years later. *Theoretical Computer Science*, 45:159–192, 1986.

[Gra81] R. Grayson. Modified realizability toposes. unpublished manuscript, 1981.

[HJP80] J. M. E. Hyland, P. T. Johnstone, and A. M. Pitts. Tripos theory. *Math. Proc. Camb. Phil. Soc.*, 88:205–232, 1980.

[HP89] J. M. E. Hyland and A. M. Pitts. The theory of constructions: Categorical semantics and topos-theoretic models. *Contemporary Mathematics*, 92:137–199, 1989.

[HRR90] J. M. E. Hyland, E. P. Robinson, and G. Rosolini. The discrete objects in the effective topos. *Proc. London Math. Soc. (3)*, 60:1–36, 1990.

[Hyl82] J. M. E. Hyland. The effective topos. In *The L. E. J. Brouwer Centenary Symposium*, pages 165–216. North-Holland, 1982.

[Hyl88]  J. M. E. Hyland. A small complete category. *Annals of Pure and Applied Logic*, 40:135–165, 1988.

[Joh77]  P. T. Johnstone. *Topos Theory*. Academic Press, 1977. L.M.S. Monograph No. 10.

[Kre59]  G. Kreisel. Interpretation of analysis by means of constructive functionals of finite type. In A Heyting, editor, *Constructivity in Mathematics*. North-Holland, 1959.

[LS86]  J. Lambek and P. J. Scott. *Introduction to Higher Order Categorical Logic*. Cambridge Studies in Advanced Mathematics No. 7. Cambridge University Press, 1986.

[ML73]  P. Martin-Löf. An intuitionistic theory of types: Predicative part. In Rose and Shepherdson, editors, *Logic Colloquium '73*. North-Holland, 1973.

[ML84]  P. Martin-Löf. *Intuitionistic Type Theory*. Bibliopolis, 1984. Studies in Proof Theory Series.

[Pit87]  A. M. Pitts. Polymorphism is set-theoretical, constructively. In D. H. Pitt et al., editor, *Proc. Conf. Category Theory and Computer Science, Edinburgh*, Berlin, 1987. Springer-Verlag. LNCS. Vol. 287.

[Pra65]  D. Prawitz. *Natural Deduction*. Almqvist and Wiksell, 1965. Stockholm Studies in Philosophy 3.

[Rey74]  J. C. Reynolds. Towards a theory of type structure. In B. Robinet, editor, *Colloque sur la Programmation*, pages 405–425. Springer-Verlag, 1974. Lecture Notes in Computer Science Vol. 19.

[Sce89]  A. Scedrov. Normalization revisited. In J. W. Gray and A. Scedrov, editors, *Categories in Computer Science and Logic*, pages 357 – 369. AMS, 1989.

[See87]  R. A. G. Seely. Categorical semantics for higher order polymorphic lambda calculus. *J. Symb. Logic*, 52:969–989, 1987.

[Str92]  T. Streicher. Truly intensional models of type theory arising from modified realizability. dated 25 May '92 mailing list at CATEGORIES@mta.ca, 1992.

[Tai67]  W. W. Tait. Intensional interpretation of functionals of finite type i. *J. Symb. Logic*, 32:198–212, 1967.

[Tai75]  W. W. Tait. A realizability interpretation of the theory of species. In *Logic Colloquium*. Springer-Verlag, 1975. Lecture Notes in Mathematics Vol. 453.

[Tro73]  A. Troelstra. *Metamathematical Investigation of Intuitionistic Arithmetic and Analysis*. Springer, 1973. Springer Lecture Notes in Mathematics 344.

[vO91]  J. van Oosten. *Exercises in Realizability*. PhD thesis, University of Amsterdam, 1991.

# Semantics of lambda-I
## and of other substructure lambda calculi[1]

### Bart Jacobs[2]

*The ordinary untyped $\lambda$-calculus (the main object of study in [3]) will be denoted here by $\lambda K$. Church originally introduced the $\lambda I$-calculus, which can be understood as the $\lambda K$-calculus without weakening: one cannot throw away variables. Similarly there is a affine calculus $\lambda A$ without contraction: there, one cannot duplicate variables. There is also a linear calculus $\lambda L$ in which one has neither weakening nor contraction. In $\lambda L$ variables occur precisely once.*

*We give a systematic description of the semantics of these four calculi. It starts with two sorts of domain theoretic models: graph models and filter models (of intersection types) are constructed for each of these calculi. Later on, we describe an appropriate categorical way to capture such structures in terms of monoidal categories (with diagonals or projections).*

## 1 Introduction

Logical systems in which the use of structural rules is restricted are called *substructure logics*. In the $\lambda I$-calculus one cannot use weakening. In a similar way, there is a calculus $\lambda A$ without contraction and a *linear* calculus $\lambda L$ in which one has neither weakening nor contraction. Therefore, $\lambda I$, $\lambda L$ and $\lambda A$ will be called substructure $\lambda$-calculi.

A bit more concretely, let $M$ be a term from the ordinary $\lambda$-calculus (denoted by $\lambda K$). Let $V(M)$ denote the union of the sets of free and bound variables in $M$; by convention, this union is a disjoint one. Intuitively,

$M$ is in $\lambda I$    $\Leftrightarrow$    every $x \in V(M)$ occurs *at least* once
$M$ is in $\lambda A$    $\Leftrightarrow$    every $x \in V(M)$ occurs *at most* once.
$M$ is in $\lambda L$    $\Leftrightarrow$    every $x \in V(M)$ occurs *precisely* once

A precise syntax using sequents will be presented in the next section.

The subsequent third section deals with an order theoretic description. It will be shown that one can interprete

| | | | |
|---|---|---|---|
| $\lambda K$-terms | as | *continuous* functions | (preserving directed sups) |
| $\lambda I$-terms | as | *strict* functions | (preserving directed sups and bottom) |
| $\lambda A$-terms | as | *affine* functions | (preserving all non-empty sups) |
| $\lambda L$-terms | as | *linear* functions | (preserving all sups) |

For all four cases concrete models $D$ will be described where the set of appropriate endofunctions $D \to D$ is a retract of $D$. The case of the ordinary $\lambda K$-calculus is of course well-known.

Thus far, no categorical tools are used. Given the fact that in the $\lambda I$-calculus one has "duplication" and in the $\lambda A$-calculus one has "projections" we are lead to describe these

---

[1]The research reported here was done during the academic year '91–'92 at the Pure Maths Dep., Univ. of Cambridge, UK. An early version is [5].

[2]Mathematical Institute RUU, P.O.Box 80.010, 3508 TA Utrecht, NL. bjacobs@math.ruu.nl

notions of duplication and projection in monoidal categories. It turns out that the well-known smash product in the category of complete lattices with strict functions is a "tensor with diagonals". Similarly there is a "tensor with projections" in the category of complete lattices with affine functions.

In the subsequent section the notion of a (categorical) $\lambda K$-algebra is recalled. A $\lambda L$-algebra is described analogously by replacing the finite products in the definition of a $\lambda K$-algebra by a monoidal structure. A $\lambda I$-algebra (resp. $\lambda A$-algebra) is then defined as a $\lambda L$-algebra plus diagonals (resp. projections). Finally, a uniform construction is given to obtain such categories from set theoretic examples.

## 2 Syntax

The formation of terms in the substructure $\lambda$-calculi is most adequately described by sequent rules. Let *Var* be an infinite collection of (term) variables. Contexts $\Gamma$ are finite sequences of such variables, in which multiple occurrences are allowed. A sequent $\Gamma \vdash M$ expresses that $M$ is a well-formed term in context $\Gamma$.

In all systems $\lambda K$, $\lambda I$, $\lambda A$ and $\lambda L$ one has the following rules.

$$\frac{}{x \vdash x} \ (x \in \mathit{Var}) \qquad \frac{\Gamma \vdash M}{\Gamma' \vdash M} \ (\Gamma' \text{ is a permutation of } \Gamma) \qquad \frac{\Gamma, x \vdash M}{\Gamma \vdash \lambda x.M} \ (x \text{ not in } \Gamma)$$

Only in $\lambda K$ and $\lambda A$ one has weakening: $\quad \dfrac{\Gamma \vdash M}{\Gamma, x \vdash M}$

and only in $\lambda K$ and $\lambda I$ one has contraction: $\quad \dfrac{\Gamma, x, x \vdash M}{\Gamma, x \vdash M.}$

Application in $\lambda L$ and $\lambda A$ is described by $\quad \dfrac{\Gamma \vdash M \quad \Delta \vdash N}{\Gamma, \Delta \vdash MN} \ (\Gamma, \Delta \text{ disjoint})$

and in $\lambda K$ and in $\lambda I$ by $\quad \dfrac{\Gamma, \Theta \vdash M \quad \Theta, \Delta \vdash N}{\Gamma, \Theta, \Delta \vdash MN} \ (\Gamma, \Delta \text{ disjoint})$

Notice that there are the following inclusions (of terms).

$$\begin{array}{ccc} \lambda L & \subset & \lambda I \\ \cap & & \cap \\ \lambda A & \subset & \lambda K \end{array}$$

In fact, $\lambda L$ can be understood as the intersection $\lambda I$ and $\lambda A$, and $\lambda K$ as their union.

**2.1. Example.** The combinators $I \equiv \lambda x.x$ (identity), $B \equiv \lambda xyz.x(yz)$ (composition) and $C \equiv \lambda xyz.xzy$ (exchange) can be formed in $\lambda L$. One has $S \equiv \lambda xyz.xz(yz)$ (duplication) in $\lambda I$ and $K \equiv \lambda xy.x$ (projection) in $\lambda A$.

Put for $n \in \mathbb{N}$, $\mathbf{1}_n \equiv \lambda zx_1\ldots x_n.\, zx_1\ldots x_n$. Then $\mathbf{1}_0 = I$ and $\mathbf{1}_{n+1} = B\mathbf{1}_n$; these $\mathbf{1}_n$ exist in $\lambda L$ (and hence in all systems).

## 3 Domain models

This section is organized as follows. We start with some general domain theoretic notions, which will be used subsequently for two specific constructions. The first one generalizes Engeler's graph models (see e.g. [3]) to all four of the present $\lambda$-calculi, and the second one generalizes the filter models from [4].

**3.1. Definition.** Let $D, D'$ be complete lattices (i.e. posets with suprema of all subsets). A function $f: D \to D'$ is called

*continuous* if $f(\bigvee a) = \bigvee f(a)$ for all *directed* $a \subseteq D$,
(i.e. for non-empty $a$ satisfying $\forall x, y \in a.\ \exists z \in a.\ x \leq z\ \&\ y \leq z$)
*strict* if $f(\bigvee a) = \bigvee f(a)$ for all directed or empty $a \subseteq D$
*affine* if $f(\bigvee a) = \bigvee f(a)$ for all non-empty $a \subseteq D$
*linear* if $f(\bigvee a) = \bigvee f(a)$ for all $a \subseteq D$

Thus a continuous function $f$ is strict if and only if $f(\bot) = \bot$, where $\bot = \bigvee \emptyset$ is the bottom element. Similarly an affine function $f$ is linear if and only if $f(\bot) = \bot$. Let's write $\mathbf{CL}_{cont}$, $\mathbf{CL}_{str}$, $\mathbf{CL}_{aff}$ and $\mathbf{CL}_{lin}$ for the categories of complete lattices with continuous, strict, affine and linear functions. There are inclusion functors

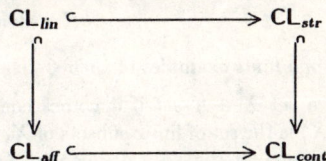

The associated function spaces will be denoted as follows; here we write

$[D \to_K D']$ for the set of *continuous* functions $D \to D'$
$[D \to_I D']$ for the set of *strict* functions $D \to D'$
$[D \to_A D']$ for the set of *affine* functions $D \to D'$
$[D \to_L D']$ for the set of *linear* functions $D \to D'$

These function spaces are complete lattices again under pointwise ordering. Finally, for $\Diamond \in \{K, I, L, A\}$ we call a complete lattice $D$ $\Diamond$-*reflexive* if it comes equipped with two functions

$F \in [D \to_\Diamond [D \to_\Diamond D]]$ for application
$G \in [[D \to_\Diamond D] \to_\Diamond D]$ for abstraction

satisfying $F \circ G = id$. In that case one has that $[D \to_\Diamond D]$ is a retract of $D$, which is often written as $[D \to_\Diamond D] \triangleleft D$.

We claim that for $\Diamond \in \{K, I, A, L\}$ one has that a $\Diamond$-reflexive complete lattice (with $F, G$ as described above) forms a model for the $\lambda\Diamond$-calculus. This goes as follows. For a valuation (or environment) $\rho: \mathrm{Var} \to D$ one defines the interpretation $[\![M]\!]_\rho$ of a $\lambda\Diamond$-term $M$ in the standard way, i.e.

$[\![x]\!]_\rho = \rho(x)$
$[\![MN]\!]_\rho = F([\![M]\!]_\rho)([\![N]\!]_\rho)$
$[\![\lambda x.M]\!]_\rho = G(\lambda d \in D.\, [\![M]\!]_{\rho(x:=d)})$

**3.2. Lemma.** *The above interpretation is well-defined: for a $\lambda\Diamond$-term of the form $\lambda x.M$ one has that $f = \lambda d \in D.\, [\![M]\!]_{\rho(x:=d)}$ is in $[D \to_\Diamond D]$.*

*Proof.* By induction on $M$. We shall do the application case $M \equiv M_1 M_2$ for $\Diamond = L, A, I$.

$\Diamond = L$ Assume without loss of generality that $x \in FV(M_1)$ and $x \notin FV(M_2)$. Then

$\begin{aligned}
f(\bigvee_i d_i) &= F([\![M_1]\!]_{\rho(x:=\bigvee_i d_i)})([\![M_2]\!]_\rho) \\
&= F(\bigvee_i [\![M_1]\!]_{\rho(x:=d_i)})([\![M_2]\!]_\rho) \quad \text{by induction hypothesis} \\
&= \bigvee_i F([\![M_1]\!]_{\rho(x:=d_i)})([\![M_2]\!]_\rho) \quad F \text{ is linear (in both places)} \\
&= \bigvee_i f(d_i).
\end{aligned}$

$\diamond = A$  The interesting remaining case is when $x \notin FV(M_1)$ and $x \notin FV(M_2)$. Then for a non-empty collection $\{d_i\}$,

$$\begin{aligned}
f(\vee_i d_i) &= F([\![M_1]\!]_\rho)([\![M_2]\!]_\rho) \\
&= f(d_{i_0}) \qquad \text{for some } i_0 \\
&= \vee_i f(d_i).
\end{aligned}$$

$\diamond = I$  Assume now (without loss of generality) that $x \in FV(M_1)$. The $\bot$-case is the most interesting:

$$\begin{aligned}
f(\bot) &= F([\![M_1]\!]_{\rho(x:=\bot)})([\![M_2]\!]_{\rho(x:=\bot)}) \\
&= F(\bot)([\![M_2]\!]_{\rho(x:=\bot)}) \qquad \text{by induction hypothesis} \\
&= \bot \qquad\qquad\qquad\qquad\quad F \text{ is strict (in both places).} \qquad \square
\end{aligned}$$

In the rest of this section one finds examples of such $\diamond$-reflexive complete lattices.

**Graph models.** Let's call a set $X$ a *K-set* if it comes equipped with an injection $\langle -, - \rangle : \mathcal{P}_f(X) \times X \rightarrowtail X$, where $\mathcal{P}_f(X)$ is the set of finite subsets of $X$. As is well-known, the powerset $\mathcal{P}(X)$ of a $K$-set $X$ yields a model for the $\lambda K$-calculus with application and abstraction

$$\begin{aligned}
F(a)(b) &= \{x \in X \mid \exists \beta \in \mathcal{P}_f(X).\ \beta \subseteq b \text{ and } \langle \beta, x \rangle \in a\} \\
G(f) &= \{\langle \beta, x \rangle \mid \beta \in \mathcal{P}_f(X) \text{ and } x \in f(\beta)\}
\end{aligned}$$

using that a continuous function on a powerset lattice is determined by its behaviour on finite sets. One obtains that $\mathcal{P}(X)$ is a $K$-reflexive complete lattice. The first example of such a $K$-set is the set $\omega$ of natural numbers (as noticed by Scott and Plotkin, see [3], 18.1.3). Later, Engeler constructed code-free variants of this model.

In order to code up a strict function $f : \mathcal{P}(X) \to \mathcal{P}(X)$ it suffices to code up its behaviour on *non-empty* finite sets — since $f(\emptyset) = \emptyset$. Hence we call $X$ an *I-set* if it comes with an injection $\mathcal{P}_f^+(X) \times X \rightarrowtail X$, where $\mathcal{P}_f^+(X)$ is the set of non-empty finite subsets of $X$. Then one can define application and abstraction functions as above — with $\mathcal{P}_f$ replaced by $\mathcal{P}_f^+$ — and verify that the powerset $\mathcal{P}(X)$ becomes an $I$-reflexive complete lattice (and thus a model of the $\lambda I$-calculus).

Next we notice that a linear function $f : \mathcal{P}(X) \to \mathcal{P}(X)$ is determined by its behaviour on singletons. Hence we define an *L-set* to be a set $X$ together with an injection $X \times X \rightarrowtail X$. The powerset of an $L$-set is an $L$-reflexive complete lattice. Specifically, it was noted by Abramsky and by Phoa that the pairing on $\omega$ makes $\mathcal{P}\omega$ a model of the linear $\lambda$-calculus.

Finally, an affine function $f : \mathcal{P}(X) \to \mathcal{P}(X)$ is determined by its behaviour on the set $\bot(X)$ of what are usually called *subsingletons* of $X$. It is given by

$$\begin{aligned}
\bot(X) &= \{\alpha \in \mathcal{P}(X) \mid \forall x, y \in \alpha.\ x = y\} \\
&= \{\{x\} \mid x \in X\} \cup \{\emptyset\}.
\end{aligned}$$

Hence an *A-set* consists of a set $X$ together with an injection $\bot(X) \times X \rightarrowtail X$. Its powerset $\mathcal{P}(X)$ is a $A$-reflexive complete lattice. We summarize

|  | K-set | I-set | A-set | L-set |
|---|---|---|---|---|
| definition | $\mathcal{P}_f(X) \times X \rightarrowtail X$ | $\mathcal{P}_f^+(X) \times X \rightarrowtail X$ | $\bot(X) \times X \rightarrowtail X$ | $X \times X \rightarrowtail X$ |
| $\mathcal{P}(X)$ is .. | $K$-reflexive | $I$-reflexive | $A$-reflexive | $L$-reflexive |
| .. and model of | $\lambda K$-calculus | $\lambda I$-calculus | $\lambda A$-calculus | $\lambda L$-calculus |

Finally, we mention how to obtain such sets. Let $A$ be a (non-empty) set of atoms. The least $L$-set containing $A$ is $\bigcup_{n \in \mathbb{N}} \cdot A_n$, where

$$A_0 = A, \qquad A_{n+1} = A_n \cup (A_n \times A_n).$$

A similar construction applies to the other cases. For example, to obtain an $I$-set one takes

$$A_{n+1} = A_n \cup (\mathcal{P}_f^+(A_n) \times A_n).$$

In such a way, one also obtains Engeler's $D_A$ as the least $K$-set containing $A$, see e.g. [3].

**Filter models.** The intersection type discipline introduced by the Turin school captures finite forms of polymorphism via finite intersections of types $\sigma \wedge \tau$ and $\omega$, where $\omega$ stands for the empty intersection, i.e. the top element. The intention is that $M : \sigma \wedge \tau$ if and only if both $M : \sigma$ and $M : \tau$ and that $M : \omega$ always holds. Thus one describes overloading of addition by

$$+ : (\text{int} \to \text{int} \to \text{int}) \wedge (\text{real} \to \text{real} \to \text{real})$$

The relevance for our present investigation lies in the fact that the finite intersections allow us to introduce types in the contraction and weakening rules. More specifically, $\wedge$ is used for contraction and $\omega$ for weakening:

$$\frac{\Gamma, x : \sigma, x : \tau \vdash M : \rho}{\Gamma, x : \sigma \wedge \tau \vdash M : \rho} \ (C) \qquad \frac{\Gamma \vdash M : \rho}{\Gamma, x : \omega \vdash M : \rho} \ (W)$$

We are thus led to adapt the filter models of the $\lambda K$-calculus described in [4] to the other $\lambda$-calculi by using

> at least one intersection   for   contraction
> at most one intersection   for   weakening

This will be worked out below. The 'K' case covers what was done in [4].

Let $TypeVar$ be a non-empty set of type variables. The sets of types $T_K$, $T_I$, $T_A$ and $T_L$ are the least containing $TypeVar$ and closed under $\to$ such that the following extra requirements are satisfied.

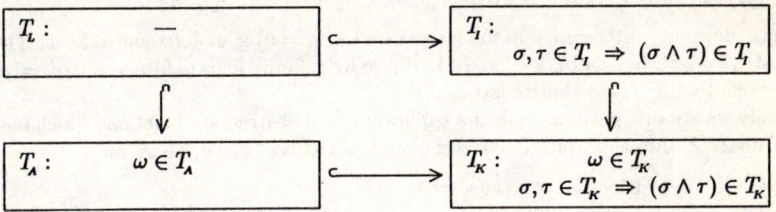

where $\omega$ is a constant type (not in $TypeVar$). Further, consider the following conditions for an order relation $\leq$ on types.

$$(L) \begin{cases} \sigma \leq \sigma \\ \sigma \leq \tau, \tau \leq \rho \Rightarrow \sigma \leq \rho \\ \sigma' \leq \sigma, \tau \leq \tau' \Rightarrow \sigma \to \tau \leq \sigma' \to \sigma' \end{cases} \quad (I) \begin{cases} \sigma \leq \sigma \wedge \sigma \\ \sigma \wedge \tau \leq \sigma \\ \sigma \wedge \tau \leq \tau \\ (\sigma \to \tau) \wedge (\sigma \to \rho) \leq \sigma \to (\tau \wedge \rho) \\ \sigma \leq \sigma', \tau \leq \tau' \Rightarrow \sigma \wedge \tau \leq \sigma' \wedge \tau' \end{cases}$$

$$(A) \begin{cases} \sigma \leq \omega \\ \omega \leq \omega \to \omega \end{cases}$$

We then define:

$\leq_L$ is the least relation on $T_L$ satisfying $(L)$
$\leq_A$ is the least relation on $T_A$ satisfying $(L) + (A)$
$\leq_I$ is the least relation on $T_I$ satisfying $(L) + (I)$
$\leq_K$ is the least relation on $T_K$ satisfying $(L) + (A) + (I)$

and write $\sigma \sim_\diamond \tau$ for $\sigma \leq_\diamond \tau$ and $\tau \leq_\diamond \sigma$ with $\diamond \in \{L, A, I, K\}$.

The following is essentially Lemma 2.4 in [4]; the proof is similar and left to the interested reader. It gives some technical properties of the preorders $(T_\diamond, \leq_\diamond)$ which are needed later.

**3.3. Lemma.** (i) *Let $\diamond$ be $A$ or $K$. Then $\sigma \to \tau \sim_\diamond \omega \Leftrightarrow \tau \sim_\diamond \omega$.*

(ii) *In $(T_L, \leq_L)$ one has $\mu \to \nu \leq_L \sigma \to \tau \Rightarrow \sigma \leq_L \mu \ \& \ \nu \leq_L \tau$.*
*In $(T_A, \leq_A)$, one has $\mu \to \nu \leq_A \sigma \to \tau \ \& \ \tau \not\sim_A \omega \Rightarrow \sigma \leq_A \mu \ \& \ \nu \leq_A \tau$.*
*In $(T_I, \leq_I)$, one has that $(\mu_1 \to \nu_1) \wedge \cdots \wedge (\mu_n \to \nu_n) \leq_I \sigma \to \tau$ implies*
$\exists i_1, \ldots, i_m \in \{1, \ldots, n\}. \ \sigma \leq_I \mu_{i_1} \wedge \cdots \wedge \mu_{i_m} \ \& \ \nu_{i_1} \wedge \cdots \wedge \nu_{i_m} \leq_I \tau.$
*In $(T_K, \leq_K)$, one has that $(\mu_1 \to \nu_1) \wedge \cdots \wedge (\mu_n \to \nu_n) \leq_K \sigma \to \tau \ \& \ \tau \not\sim_K \omega$ implies*
$\exists i_1, \ldots, i_m \in \{1, \ldots, n\}. \ \sigma \leq_K \mu_{i_1} \wedge \cdots \wedge \mu_{i_m} \ \& \ \nu_{i_1} \wedge \cdots \wedge \nu_{i_m} \leq_K \tau.$ □

In a preorder an *upset* is a subset $a$ which is upwardly closed, i.e. satisfies $x \geq y \in a \Rightarrow x \in a$. We then define an *L-filter* to be an upset in the preorder $(T_L, \leq_L)$.

Next, an *A-filter* is an upset $a$ in $(T_A, \leq_A)$ satisfying $\omega \in a$, i.e. A-filters are non-empty L-filters. Further, an *I-filter* is an upset $a$ in $(T_I, \leq_I)$ which satisfies $\sigma, \tau \in a \Rightarrow \sigma \wedge \tau \in a$ and a *K-filter* is an upset $a$ in $(T_K, \leq_K)$ satisfying both $\omega \in a$ and $\sigma, \tau \in a \Rightarrow \sigma \wedge \tau \in a$.

It is then easy to verify that for $\diamond \in \{L, A, I, K\}$ one has that the set $\mathcal{F}_\diamond$ of $\diamond$-filters is closed under arbitrary intersections. For an arbitrary $b \subseteq T_\diamond$ put

$$\Uparrow_\diamond b = \bigcap \{a \in \mathcal{F}_\diamond \mid a \supseteq b\}$$

for the least $\diamond$-filter containing $b$. As usual, one writes $\Uparrow_\diamond \sigma$ for $\Uparrow_\diamond \{\sigma\}$. One then easily verifies that

$$\Uparrow_L b = \{\sigma \in T_L \mid \exists \tau \in b. \ \tau \leq_L \sigma\}$$
$$\Uparrow_A b = \{\sigma \in T_A \mid \exists \tau \in b. \ \tau \leq_A \sigma\} \cup \{\omega\}$$
$$\Uparrow_I b = \{\sigma \in T_I \mid \exists n \geq 1. \ \exists \tau_1, \ldots, \tau_n \in b. \ \tau_1 \wedge \cdots \wedge \tau_n \leq_I \sigma\}$$
$$\Uparrow_K b = \{\sigma \in T_K \mid \exists n \geq 0. \ \exists \tau_1, \ldots, \tau_n \in b. \ \tau_1 \wedge \cdots \wedge \tau_n \leq_K \sigma\}$$

whereby the empty intersection in the latter case (for $n = 0$) is understood to be $\omega$. Thus one has that $\Uparrow_A b = \Uparrow_L b \cup \{\omega\}$ and $\Uparrow_K b = \Uparrow_I b \cup \{\omega\}$. Hence by defining joins of filters as $\bigvee_i a_i = \Uparrow_\diamond (\bigcup_i a_i)$ one obtains that $\mathcal{F}_\diamond$ is a complete lattice.

Finally we are in a position to define application and abstraction functions which make the sets of filters $\mathcal{F}_\diamond$ into appropriate reflexive complete lattices. For $a, b \in \mathcal{F}_\diamond$ put

$$F(a)(b) = \{\tau \in T_\diamond \mid \exists \sigma \in b. \ \sigma \to \tau \in a\}$$

and for a function $f : \mathcal{F}_\diamond \to \mathcal{F}_\diamond$ put

$$G(f) = \Uparrow_\diamond \{\sigma \to \tau \mid \sigma \in T_\diamond \text{ and } \tau \in f(\Uparrow_\diamond \sigma)\}.$$

It is then not difficult to verify that indeed $F$ is in $[\mathcal{F}_\diamond \to_\diamond [\mathcal{F}_\diamond \to_\diamond \mathcal{F}_\diamond]]$ and that $G$ is in $[[\mathcal{F}_\diamond \to_\diamond \mathcal{F}_\diamond] \to_\diamond \mathcal{F}_\diamond]$ and with a little more effort that $F \circ G = id$, using the above Lemma and characterization of $\Uparrow_\diamond b$. Thus $\mathcal{F}_\diamond$ is a model of the $\lambda \diamond$-calculus.

In [2] one finds various (other) subsystems of the intersection type discipline. It was already noticed there that a filter model of intersection types without $\omega$ yields a model of the $\lambda I$-calculus. In [8] the intersection $\wedge$ is generalized to second order quantification $\forall$. The resulting filters model only the affine $\lambda A$-calculus, because one cannot type the contraction rule with $\forall$.

## 4 Towards a categorical understanding

A *monoidal category* is a category $\mathsf{B}$ provided with a *tensor product* $\otimes$ with a neutral element $I$. A bit more formally, $\otimes$ is a functor $\mathsf{B} \times \mathsf{B} \to \mathsf{B}$ which comes equipped with natural isomorpisms

$$X \otimes (Y \otimes Z) \xrightarrow[\sim]{\alpha} (X \otimes Y) \otimes Z \qquad I \otimes X \xrightarrow[\sim]{\lambda} X \xleftarrow[\sim]{\rho} X \otimes I$$

satisfying some equations, see e.g. [9]. The monoidal category is called *symmetric* if there are additional isomorphism

$$X \otimes Y \xrightarrow[\sim]{\gamma} Y \otimes X$$

satisfying some additional equations. A monoidal category is called *closed* if it comes equipped with "function spaces" or "internal homs". That is, if for each pair of objects $X, Y$, there is an object $X \multimap Y$ together with an evaluation map $ev : (X \multimap Y) \otimes X \to Y$ such that for every $f : Z \otimes X \to Y$ there is a unique $\Lambda(f) : Z \to (X \multimap Y)$ satisfying $ev \circ \Lambda(f) \otimes id = f$.

We shall abbreviate 'symmetric monoidal category' as SMC and 'symmetric monoidal closed category' as SMCC. The category of abelian groups is the standard example of an SMCC.

Suppose now $\Omega$ is an object in an SMC $\mathsf{B}$. The idea is to interpret a sequent $x_1, \ldots, x_n \vdash M$ as a morphism in $\mathsf{B}$, $\Omega^n \to \Omega$, where $\Omega^0 = I$ and $\Omega^{i+1} = \Omega^i \otimes \Omega$. Because the tensor is associative, the precise bracketing in $\Omega^n$ can be ignored. Since we have symmetry as well, the components in $\Omega^n$ can be exhanged. This gives the validity of the context permutation rule. An elementary but crucial observation is that one needs *diagonals* $X \to (X \otimes X)$ for the validity of the contraction rule and *projections* $X \leftarrow (X \otimes Y) \to Y$ for the weakening rule. This can be seen as follows. Suppose we have a term $\Gamma, x, x \vdash M$ interpreted as a morphism $f : \Omega^n \otimes \Omega \otimes \Omega \to \Omega$ then by using $\delta : \Omega \to \Omega \otimes \Omega$ one can form the arrow which corresponds to the outcome $\Gamma, x \vdash M$ of the contraction rule by $f \circ id_{\Omega^n} \otimes \delta : \Omega^n \otimes \Omega \to \Omega^n \otimes \Omega \otimes \Omega \to \Omega$. For weakening, suppose we have $\Gamma \vdash M$ interpreted as $g : \Omega^n \to \Omega$, then one can use the (first) projection $\pi : (\Omega^n) \otimes \Omega \to \Omega^n$ to interprete $\Gamma, x \vdash M$ as $g \circ \pi : \Omega^n \otimes \Omega \to \Omega^n \to \Omega$.

These observations lead to the following Definition.

**4.1. Definition.** Let $(\mathsf{B}, \otimes, I, \alpha, \lambda, \rho, \gamma)$ be an SMC as described above. We say that $\mathsf{B}$ has

(i) *diagonals* if there is a natural transformation $\delta : Id \overset{\cdot}{\to} \Delta$, where the latter functor is described by $\Delta(X) = X \otimes X$. In addition, commutation of the following diagrams is required.

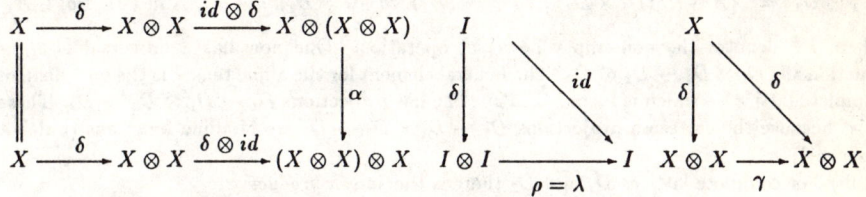

(ii) *projections* if the neutral element $I$ is terminal.

The word 'projections' used in the second point of the above Definition is appropriate because in case the neutral element is terminal, one obtains two natural transformations $\pi : \otimes \overset{\cdot}{\to} Fst$ and $\pi' : \otimes \overset{\cdot}{\to} Snd$ with components

$$\begin{array}{rcl}
\pi & = & \rho^{-1} \circ id \otimes\, ! \; : \; X \otimes Y \to X \otimes I \overset{\sim}{\to} X \\
\pi' & = & \lambda^{-1} \circ\, ! \otimes id \; : \; X \otimes Y \to I \otimes Y \overset{\sim}{\to} Y
\end{array}$$

These projections satisfy
$$\pi \circ \pi \circ \alpha = \pi \qquad \pi \circ \gamma = \pi'$$
$$\pi' \circ \alpha = \pi' \circ \pi' \qquad \pi' \circ \gamma = \pi$$
$$\pi' \circ \pi \circ \alpha = \pi \circ \pi'$$

Notice that if a (symmetric) monoidal category has both diagonals and projections as described above and additionally the following equations hold,
$$\pi \circ \delta = id \qquad \pi \circ \delta = id \qquad \delta \circ (\pi \otimes \pi') = id$$
then $\otimes$ becomes a cartesian product with pairing $\langle f, g \rangle = (f \otimes g) \circ \delta$. Moreover, in that case one has $\alpha = \langle \langle \pi, \pi \circ \pi' \rangle, \pi' \circ \pi' \rangle$ and $\gamma = \langle \pi', \pi \rangle$.

In the rest of this section it will be shown that the categories of complete lattices introduced in Definition 3.1 give examples of the above monoidal structures. Much of the next result is standard; maybe points (ii) and (iii) are not so well-known.

**4.2. Proposition.** *The category of complete lattices with*
  (i) *linear functions (i.e.* $\mathbf{CL}_{lin}$) *is an SMCC;*
  (ii) *affine functions (i.e.* $\mathbf{CL}_{aff}$) *is an SMCC with projections;*
  (iii) *strict functions (i.e.* $\mathbf{CL}_{str}$) *is an SMCC with diagonals;*
  (iv) *continuous functions (i.e.* $\mathbf{CL}_{cont}$) *is a CCC (cartesian closed category; it thus has both projections and diagonals).*

*Proof.* (i) For complete lattices $D_1$ and $D_2$ one defines a linear tensor
$$D_1 \otimes D_2 = \{A \in \mathcal{P}(D_1 \times D_2) \mid \forall a \in \mathcal{P}(D_1). \forall b \in \mathcal{P}(D_2). \, a \times b \subseteq A \text{ iff } (\bigvee a, \bigvee b) \in A\}$$
It comes together with a map $\mathrm{lin} : D_1 \times D_2 \to D_1 \otimes D_2$ given by
$$(x, y) \mapsto \downarrow(x, y) = \{(x', y') \mid x' \leq x \text{ and } y' \leq y\}$$
This map is *bi-linear*, i.e. linear in each coordinate separately. Even more, it is a universal bi-linear function with domain $D_1 \times D_2$: for every bi-linear function $f : D_1 \times D_2 \to D_3$ there is a unique linear function $\overline{f} : D_1 \otimes D_2 \to D_3$ with $\overline{f} \circ \mathrm{lin} = f$. This shows that the category $\mathbf{CL}_{lin}$ is closed (i.e. has internal homs). The neutral element for the linear tensor is the two-element complete lattice.

(ii) The affine tensor of $D_1$ and $D_2$ is
$$D_1 \otimes D_2 = \{A \in \mathcal{P}^+(D_1 \times D_2) \mid \forall a \in \mathcal{P}^+(D_1). \forall b \in \mathcal{P}^+(D_2). \, a \times b \subseteq A \text{ iff } (\bigvee a, \bigvee b) \in A\}$$
where $\mathcal{P}^+$ denotes the non-empty powerset operation. One now has a universal *bi-affine* function $\mathrm{aff} : D_1 \times D_2 \to D_1 \otimes D_2$. The neutral element for the affine tensor is the one-element complete lattice — which is terminal. Thus one has projections $D_1 \leftarrow D_1 \otimes D_2 \to D_2$. These exist because the cartesian projections $D_1 \leftarrow D_1 \times D_2 \to D_2$ are bi-affine functions (but bot bi-linear!).

(iii) For complete lattices $D_1$ and $D_2$ there is the *smash product*
$$D_1 \otimes D_2 = \{(x, y) \in D_1 \times D_2 \mid x = \bot \text{ iff } y = \bot\}$$
which comes equipped with a universal *bi-strict* function $\mathrm{str} : D_1 \times D_2 \to D_1 \otimes D_2$, given by
$$(x, y) \mapsto \text{if } x \neq \bot \wedge y \neq \bot \text{ then } (x, y) \text{ else } (\bot, \bot).$$
The 2-element complete lattice is neutral element for $\otimes$. There are obvious diagonals $\delta : D \to D \otimes D$ by $x \mapsto (x, x)$.

(iv) Well-known. □

# 5 Appropriate categorical notions

Following earlier work by Obtulowicz, it is argued in [6] that a categorical structure for the $\lambda K$-calculus looks as follows.

**5.1. Definition.** (i) A *(categorical) $\lambda K$-algebra* is given by a category **B** with finite products, containing a distinguished non-empty object $\Omega$ (i.e. the homset $\mathbf{B}(1, \Omega) \neq \emptyset$ where 1 is the terminal object) provided with

- an application map $app : \Omega \times \Omega \to \Omega$;
- abstraction operations $\lambda(-) : \mathbf{B}(X \times \Omega, \Omega) \to \mathbf{B}(X, \Omega)$,

such that
$$app \circ \lambda(f) \times id = f \qquad \lambda(f \circ g \times id) = \lambda(f) \circ g.$$
(ii) A *$\lambda \eta K$-algebra* is a $\lambda K$-algebra in which $\lambda(app) = id$.

The second equation $\lambda(f \circ g \times id) = \lambda(f) \circ g$ above is a naturality condition which provides the proper distribution of substitution over abstraction. The first equation is of course $\beta$; and the equation $\lambda(app) = id$ in (ii) is $\eta$.

**5.2. Examples.** (i) Suppose $\langle D, \cdot \rangle$ is a set theoretic $\lambda$-algebra as in [3]. For $n \in \mathbb{N}$, put
$$(D^n \to D) = \{a \in D \mid 1_n \cdot a = a\}.$$
Define a category **D** with $n \in \mathbb{N}$ as objects. Morphisms $n \to m$ in **D** are $m$-tuples of elements from $(D^n \to D)$. The object 0 is then terminal and $n + m$ is the cartesian product of $n$ and $m$. As distinguished object ('$\Omega$') one takes $1 \in \mathbf{D}$. One has $app = \lambda xy.xy : 1 + 1 \to 1$ and for $a : n + 1 \to 1$ one has $\lambda(a) = a : n \to 1$.

(ii) D. Scott first constructed a complete lattice $D_\infty$ for which the complete lattice $[D_\infty \to_K D_\infty]$ of continuous endofunctions is isomorphic to $D_\infty$ itself (say via $F : D_\infty \to [D_\infty \to_K D_\infty]$ and $G : [D_\infty \to_K D_\infty] \to D_\infty$). A category **B** is formed with $n \in \mathbb{N}$ as objects. A morphism $n \to m$ is an $m$-tuple $(f_1, \ldots, f_m)$ of continuous functions $f_i : (D_\infty)^n \to D_\infty$. Again 0 is terminal, $n + m$ is cartesian product and 1 serves as distinguished object. Then
$$app = \lambda(x,y). F(x)(y) : 1 + 1 \to 1$$
and for $f : n + 1 \to 1$ one has $\lambda(f) : n \to 1$ described by
$$\lambda(f) = \boldsymbol{\lambda}(x_1, \ldots, x_n). G(\lambda y. f(x_1, \ldots, x_n, y))$$
where the boldface $\boldsymbol{\lambda}$ is used for meta abstraction. One obtains a $\lambda \eta K$-algebra.

D. Scott used in [10] a *reflexive object* in a cartesian closed category (CCC) to describe the semantics of the untyped $\lambda$-calculus (like in Definition 3.1). This notion can be described in an arbitrary symmetric monoidal closed category (SMCC). Actually, symmetry is not needed.

**5.3. Definition.** Let **B** be an SMCC. An object $\Omega \in \mathbf{B}$ will be called *reflexive* if it comes equipped with two morphisms
$$F : \Omega \to (\Omega \multimap \Omega) \qquad\qquad G : (\Omega \multimap \Omega) \to \Omega$$
satisfying $F \circ G = id$. It will be called *extensional* if additionally $G \circ F = id$.

The complete lattice $D_\infty$ in the above Example (ii) is an example of an extensional reflexive object in the cartesian closed category $\mathbf{CL}_{cont}$ of complete lattices and continuous functions. The following is now an easy result.

**5.4. Lemma.** *Suppose* B *is a CCC with a reflexive object* $\Omega$. *One obtains a* $\lambda K$*-algebra by putting*

$$app \;=\; ev \circ F \times id \qquad and \qquad \lambda(f) \;=\; G \circ \Lambda(f).$$

*One obtains a* $\lambda \eta K$*-algebra if and only if the reflexive object* $\Omega$ *is extensional.* □

Notice that a categorical description of untyped $\lambda$-calculi in terms of reflexive objects requires an ambient category with *all* exponents, whereas for $\lambda K$-algebras (in Definition 5.1) we use a weaker ambient category with essentially only one exponent. This more economical formulation gives the straightforward descriptions in Examples 5.2 above. Below one finds how every $\lambda K$-algebra can be fully embedded into a CCC with a reflexive object.

The following is a straightforward generalization of Definition 5.1.

**5.5. Definition.** (i) A $\lambda L$*-algebra* is given by a symmetric monoidal category (SMC) B containing a distinguished non-empty object $\Omega$ provided with a map $app: \Omega \otimes \Omega \to \Omega$ and operations $\lambda(-): \mathsf{B}(X \otimes \Omega, \Omega) \to \mathsf{B}(X, \Omega)$ such that $app \circ \lambda(f) \otimes id = f$ and $\lambda(f \circ g \otimes id) = \lambda(f) \circ g$.
(ii) A $\lambda \eta L$*-algebra* is a $\lambda L$-algebra in which $\lambda(app) = id$.

The next step is of course to add diagonals and projections.

**5.6. Definition.** (i) A $\lambda(\eta)I$*-algebra* is a $\lambda(\eta)L$-algebra with diagonals.
(ii) A $\lambda(\eta)A$*-algebra* is a $\lambda(\eta)L$-algebra with projections.

**5.7. Interpretation.** Let $(\mathsf{B}, \Omega)$ be a $\lambda L$-algebra and let $\Gamma$ be a context of length $n$; put $[\![\Gamma]\!] = \Omega^n$, where $\Omega^0 = I$ and $\Omega^{n+1} = \Omega^n \otimes \Omega$. A term $\Gamma \vdash M$ can be interpreted as a morphism $[\![M]\!]_\Gamma : [\![\Gamma]\!] \to \Omega$. We only describe abstraction and application. How to interpret the contraction and weakening rules may be found before Definition 4.1.

Assume $\Gamma, x \vdash M$ interpreted as $[\![M]\!]_{\Gamma,x} : [\![\Gamma]\!] \otimes \Omega \to \Omega$. One takes

$$[\![\lambda x.M]\!]_\Gamma \;=\; \lambda([\![M]\!]_{\Gamma,x}) \;:\; [\![\Gamma]\!] \;\to\; \Omega$$

Let's turn to application. Given terms $\Gamma \vdash M$ and $\Delta \vdash N$ in $\lambda L$ or $\lambda A$ with $\Gamma, \Delta$ disjoint, one puts

$$[\![MN]\!]_{\Gamma,\Delta} \;=\; app \circ [\![M]\!]_\Gamma \otimes [\![N]\!]_\Delta \;:\; [\![\Gamma]\!] \otimes [\![\Delta]\!] \;\to\; \Omega \otimes \Omega \;\to\; \Omega.$$

For the interpretation of application in $\lambda K$ or $\lambda I$ one needs diagonals: given terms $\Gamma, \Theta \vdash M$ and $\Theta, \Delta \vdash N$ with $\Gamma, \Delta$ disjoint, one takes

$$[\![MN]\!]_{\Gamma,\Theta,\Delta} \;=\; app \circ [\![M]\!]_{\Gamma,\Theta} \otimes [\![N]\!]_{\Theta,\Delta} \circ id_{[\![\Gamma]\!]} \otimes \delta_{[\![\Theta]\!]} \otimes id_{[\![\Delta]\!]} \;:$$
$$[\![\Gamma]\!] \otimes [\![\Theta]\!] \otimes [\![\Delta]\!] \;\to\; [\![\Gamma]\!] \otimes [\![\Theta]\!] \otimes [\![\Theta]\!] \otimes [\![\Delta]\!] \;\to\; \Omega \otimes \Omega \;\to\; \Omega.$$

**5.8. Examples.** In analogy with Lemma 5.4 one obtains that a reflexive object in an SMCC (with diagonals/projections) yields an example of a $\lambda L$-algebra ($\lambda I$-/$\lambda A$-algebra). Thus Proposition 4.2 provides the setting for the following examples.

| Examples | $\mathcal{P}(X)$, $X$ L-set $\mathcal{F}_L$ | $\mathcal{P}(X)$, $X$ A-set $\mathcal{F}_A$ $\mathcal{F}_v$ | $\mathcal{P}(X)$, $X$ I-set $\mathcal{F}_I$ | $\mathcal{P}(X)$, $X$ K-set $\mathcal{F}_K$ $D_\infty$ |
|---|---|---|---|---|
| reflexive object in | **CL**$_{lin}$ | **CL**$_{aff}$ | **CL**$_{str}$ | **CL**$_{cont}$ |
| model of | $\lambda L$-calculus | $\lambda A$-calculus | $\lambda I$-calculus | $\lambda K$-calculus |

where $\mathcal{F}$ is the filter model of ∀-types from [8].

The following result occurs in [6], 4.9 (iii). It says that the notion of a $\lambda K$-algebra can be described via a reflexive object in a richer environment, namely in a topos of presheaves. Later, we give similar results for the other calculi. We assume the CCC structure of a category of presheaves to be known, but will provide the SMCC definitions.

**5.9. Proposition.** *Let* $\mathsf{B}$ *be a category with finite products and* $\Omega$ *a non-empty object in* $\mathsf{B}$. *Then*

$$(\mathsf{B},\Omega) \text{ is a } \lambda K\text{-algebra} \Leftrightarrow \mathsf{B}(\_,\Omega) \text{ is a reflexive object in } \hat{\mathsf{B}} = \mathit{Sets}^{\mathsf{B}^{op}}$$

*Proof.* Let's write $H = \mathsf{B}(\_,\Omega) : \mathsf{B}^{op} \to \mathit{Sets}$. The exponent $H \Rightarrow H$ in $\hat{\mathsf{B}}$ is described by

$$\begin{aligned}(H \Rightarrow H)(X) &= \mathit{Nat}\bigl(\mathsf{B}(\_,X) \times H, H\bigr) \\ &\cong \mathit{Nat}\bigl(\mathsf{B}(\_,X \times \Omega), \mathsf{B}(\_,\Omega)\bigr) \\ &\cong \mathsf{B}(X \times \Omega, \Omega)\end{aligned}$$

The latter by Yoneda. Thus $H = \mathsf{B}(\_,\Omega)$ is a reflexive object in $\hat{\mathsf{B}}$ if and only if there are natural transformations $F, G$ with components

$$\mathsf{B}(X \times \Omega, \Omega) \underset{F_X}{\overset{G_X}{\rightleftarrows}} \mathsf{B}(X, \Omega)$$

such that $F \circ G = id$.

Now assume that $(\mathsf{B},\Omega)$ is a $\lambda K$-algebra with $app$ and $\lambda(\_)$. Put

$$\begin{aligned}F_X(f : X \to \Omega) &= app \circ f \times id \\ G_X(g : X \times \Omega \to \Omega) &= \lambda(f)\end{aligned}$$

Conversely, given natural transformations $F, G$ as above, take

$$\begin{aligned}app &= F_\Omega(id_\Omega) \\ \lambda(f : X \times \Omega \to \Omega) &= G_X(f)\end{aligned} \qquad \square$$

One easily extends this result to a correspondence between $\lambda\eta K$-algebras and extensional reflexive objects. For the other calculi, we need the following basic result, due to Day. A clear exposition may be found in [1], 4.1.3. The second point is added here.

**5.10. Lemma.** *Let* $\mathsf{B}$ *be an SMC.*
 (i) *The category of presheaves* $\hat{\mathsf{B}} = \mathit{Sets}^{\mathsf{B}^{op}}$ *is then an SMCC and the Yoneda embedding* $\mathcal{Y} : \mathsf{B} \to \hat{\mathsf{B}}$ *preserves the SMC structure.*
 (ii) *If* $\mathsf{B}$ *has diagonals/projections, then so has* $\hat{\mathsf{B}}$.

*Proof.* (i) The tensor of presheaves $A, B : \mathsf{B}^{op} \to \mathit{Sets}$ is given (at $X$) as coend in $\mathit{Sets}$

$$(A \otimes B)(X) = \int^{(Y,Z)} A(Y) \times B(Z) \times \mathsf{B}(X, Y \otimes Z)$$

and the exponent by the end

$$\begin{aligned}(A \multimap B)(X) &= \int_Z B(Z \otimes X)^{A(X)} \\ &\cong \mathit{Nat}\bigl(A, B(X \otimes \_)\bigr)\end{aligned}$$

Finally the presheaf $\mathcal{Y}(I) = \mathsf{B}(\_,I)$ is neutral for this tensor on $\hat{\mathsf{B}}$.

(ii) If $I$ is terminal in $\mathsf{B}$, then so is $\mathcal{Y}(I)$ in $\hat{\mathsf{B}}$. This settles the case of projections. For diagonals, notice that the above coend comes together with a universal wedge

$$A(Y) \times B(Z) \times \mathsf{B}(X,\, Y \otimes Z) \xrightarrow{\omega(X)_{(Z,Y)}} (A \otimes B)(X)$$

which enables us to form a diagonal in $\hat{\mathsf{B}}$ with components

$$\lambda a.\, \omega(X)_{(X,X)}(a, a, \delta_X) \;:\; A(X) \;\to\; (A \otimes A)(X). \qquad \square$$

**5.11. Proposition.** *Let $\mathsf{B}$ be an SMC and $\Omega \in \mathsf{B}$ a non-empty object. Reflexive objects below are with respect to the above implication $\multimap$ on $\hat{\mathsf{B}} = \mathsf{Sets}^{\mathsf{B}^{op}}$.*
 (i) *One has that*
$$(\mathsf{B}, \Omega) \text{ is a } \lambda L\text{-algebra} \;\Leftrightarrow\; \mathsf{B}(\_, \Omega) \text{ is a reflexive object in } \hat{\mathsf{B}}$$
 (ii) *The same holds for $\lambda A$- and $\lambda I$-algebras.*

*Proof.* (i) Note that $\mathsf{B}(\_, \Omega)$ is reflexive in $\hat{\mathsf{B}}$ if and only if there are natural transformations $F, G$ with components

$$\mathrm{Nat}\bigl(\mathsf{B}(\_, \Omega),\, \mathsf{B}(X \otimes \_, \Omega)\bigr) \;\underset{F_X}{\overset{G_X}{\rightleftarrows}}\; \mathsf{B}(X, \Omega)$$

satisfying $F \circ G = id$. These can be obtained from $app$ and $\lambda(\_)$ by

$$\begin{aligned}
\bigl(F_X(f)\bigr)_Y(g) &= app \circ f \otimes g \\
G_X(\sigma) &= \lambda\bigl(\sigma_\Omega(id_\Omega)\bigr)
\end{aligned}$$

Conversely, given $F$ and $G$ one obtains $app$ and $\lambda(\_)$ by

$$\begin{aligned}
app &= \bigl(F_\Omega(id_\Omega)\bigr)_\Omega(id_\Omega) \\
\lambda(f : X \otimes \Omega \to \Omega) &= G_X(\lambda g.\, f \circ id \otimes g)
\end{aligned}$$

(ii) By (ii) in the previous Lemma and (i) above using that the implication $\multimap$ remains the same. $\qquad\square$

In case the SMC structure on $\mathsf{B}$ is cartesian, the induced SMCC structure on $\hat{\mathsf{B}}$ described in Lemma 5.10 becomes the familiar CCC structure used in Proposition 5.9. Thus the results 5.9 and 5.11 are completely uniform.

# 6 Set theoretic versus categorical

The more set theoretic oriented reader may want to know what *set theoretic* $\lambda I$-, $\lambda A$- or $\lambda L$-algebras are. One expects definitions similar to Definition 5.2.2 in [3] of (set theoretic) $\lambda K$-algebras. Since the latter is aesthetically so unappealing, we don't bother about this and contend ourselves with the clear and concise categorical definitions of $\lambda I$-, $\lambda A$- and $\lambda L$-algebras.

In order to massage away remaining feelings of (set theoretic) discomfort we intend to do the following. Let's assume that we do understand what a set theoretic $\lambda I$-, $\lambda L$- or $\lambda A$-algebra is. We suppose it is an applicative structure $\langle D, \cdot \rangle$ with appropriate combinators ($I, B, C$ for $\lambda L$; additionally $S$ for $\lambda I$ and $K$ for $\lambda A$) in which one can interpret the terms of the corresponding calculi. As usual, we reason directly in $D$. The following construction produces categorical $\lambda I$-, $\lambda A$- and $\lambda L$-algebras from such data; it generalizes Example 5.2 (i).

**6.1. Definition.** Let $\Diamond \in \{L, A, I, K\}$ and $\langle D, \cdot \rangle$ be a "set theoretic $\lambda\Diamond$-algebra". We form a category $\mathbf{B}_\Diamond$ with $n \in \mathbf{N}$ as objects. A morphism $n \to m$ in $\mathbf{B}_\Diamond$ consists of an element $a$ of $D$ of the form
$$a = \lambda z x_1 \ldots x_n. \, z\big(a_1(\vec{x})_1\big) \ldots \big(a_m(\vec{x})_m\big) \tag{*}$$
where $a_1, \ldots a_m$ are elements of $D$ called *factors* and $(\vec{x})_1 \ldots (\vec{x})_m$ are subsequences of $x_1, \ldots, x_n$ in such a way that the term (*) exists in the $\lambda\Diamond$-calculus. Composition of $a: n \to m$ and $b: m \to k$ is given by $b \circ a = Bab = \lambda z. \, a(bz)$. It is thus associative. The identity on $n$ is $\mathbf{1}_n = \lambda z x_1 \ldots x_n. \, z x_1 \ldots x_n$ (with factors $I$).

**6.2. Lemma.** (i) *Each category $\mathbf{B}_\Diamond$ has a (strict) monoidal structure with $n \otimes m = n + m$.*

(ii) *The monoidal structure has projections in $\mathbf{B}_A$, has diagonals in $\mathbf{B}_I$ and is cartesian in $\mathbf{B}_K$.*

*Proof.* (i) For
$$a = \lambda z x_1 \ldots x_n. \, z\big(a_1(\vec{x})_1\big) \ldots \big(a_l(\vec{x})_l\big) \; : \; n \to l$$
$$b = \lambda z y_1 \ldots y_m. \, z\big(b_1(\vec{y})_1\big) \ldots \big(b_k(\vec{y})_k\big) \; : \; m \to k$$
define $a \otimes b: n \otimes m \to l \otimes k$ by
$$a \otimes b = \lambda z x_1 \ldots x_n y_1 \ldots y_m. \, z\big(a_1(\vec{x})_1\big) \ldots \big(a_l(\vec{x})_l\big)\big(b_1(\vec{y})_1\big) \ldots \big(b_k(\vec{y})_k\big)$$
i.e. $a \otimes b = \lambda z \vec{x} \vec{y}. \, b(a z \vec{x}) \vec{y}$. The object 0 is neutral element for this tensor. Further, symmetry is given by
$$\lambda z x_1 \ldots x_n y_1 \ldots y_m. \, z y_1 \ldots y_m x_1 \ldots x_n \; : \; n \otimes m \to m \otimes n.$$

(ii) We need to check that 0 is terminal in $\mathbf{B}_A$. The unique arrow $n \to 0$ in $\lambda z x_1 \ldots x_n. z$. Notice that weakening is used in an essential way to form this morphism. The resulting projections $\pi: n \otimes m \to n$ and $\pi': n \otimes m \to m$ are
$$\pi = \lambda z x_1 \ldots x_n y_1 \ldots y_m. \, z x_1 \ldots x_n \qquad \pi' = \lambda z x_1 \ldots x_n y_1 \ldots y_m. \, z y_1 \ldots y_m.$$
Further, the category $\mathbf{B}_I$ has diagonals because one can use contraction:
$$\lambda z x_1 \ldots x_n. \, z x_1 \ldots x_n x_1 \ldots x_n \; : \; n \to n \otimes n.$$
In the category $\mathbf{B}_K$ these projections and diagonals both exist and yield cartesian products. □

**6.3. Proposition.** *For $\Diamond \in \{L, A, I, K\}$ one has that the pair $(\mathbf{B}_\Diamond, 1)$ is a $\lambda\Diamond$-algebra.*

*Proof.* In all four cases one has
$$\mathrm{app} = B = \lambda z x y. \, z(x y) \; : \; 1 \otimes 1 \to 1$$
with factor $\mathbf{1}_1 = \lambda x y. \, x y$. For $a: n \otimes 1 \to 1$ of the form $a = \lambda z x_1 \ldots x_n x_{n+1}. \, z\big(a_1(\vec{x})_1\big)$ one obtains an abstraction morphism
$$\lambda(a) = \lambda z x_1 \ldots x_n. z\big(\lambda x_{n+1}. \, a_1(\vec{x})_1\big) \; : \; n \to 1. \qquad \Box$$

One easily establishes that the above category $\mathbf{B}_K$ is isomorphic to the category $\mathbf{D}$ described in Example 5.2 (i).

## 7 Final Remarks

Soon after investigating the matters reported here, I realized that the phenomena involved have a wider significance and give rise to a refinement of some aspect of Girard's linear logic, see [7]. The crucial "shriek" operation ! of linear logic reinstates weakening and contraction in annotated form. One can read $!A$ as "$A$, as many times as you like". This ! can be split into two operations: $!_w$ for weakening and $!_c$ for contraction. These can be read as

$!_w A$ means $A$, at *most* once $\qquad\qquad$ $!_c A$ means $A$, at *least* once

This distinction at least once/at most once underlies all models given above. In a linear setting where one has a linear implication $\multimap$, one can describe a model $\Omega$ for one of our untyped substructure $\lambda$-calculi as a retract of the following form.

$$\begin{array}{llll} \lambda K & (!\Omega \multimap \Omega) \triangleleft \Omega & \lambda A & (!_w \Omega \multimap \Omega) \triangleleft \Omega \\ \lambda I & (!_c \Omega \multimap \Omega) \triangleleft \Omega & \lambda L & (\Omega \multimap \Omega) \triangleleft \Omega \end{array}$$

For more details on $!_w$ and $!_c$, see [7].

Interesting operational aspects crop up at this point. Under the above reading, a function $f : !_w A \to B$ uses its argument at *most* once. Hence in computing an application $f a$ one better first fully evaluates $f$—instead of $a$—because $a$ may be thrown away. Similarly, a function $g : !_c A \to B$ uses its argument at *least* once. So in computing $g a$ one better computes $a$ first. Such a type system involving $!_w$ and $!_c$ contains information about the reduction strategy to be followed.

### Acknowledgements

Thanks are due to Martin Hyland, Valeria de Paiva and Alex Simpson for helpful discussions.

## References

[1] , S.J. Ambler, 'First Order Linear Logic in Symmetric Monoidal Closed Categories', PhD. Thesis, Univ. of Edinburgh 1992. Techn. Rep. CST-87-92.

[2] S. van Bakel, 'Complete restrictions of the intersection type discipline', *Theor. Computer Science*, vol. 102-2 (1992), pp. 135–163.

[3] H.P. Barendregt, *The Lambda Calculus. Its Syntax and Semantics*, 2$^{nd}$ rev. ed. (North-Holland, Amsterdam, 1984).

[4] H.P. Barendregt, M. Coppo and M. Dezani-Ciancaglini, 'A filter lambda model and the completeness of type assignment', *Journ. Symbolic Logic* 48-4 (1983), pp. 931–940.

[5] B. Jacobs, Notes of a talk held at the 48$^{th}$ PSLL, Edinburgh, 9/10 nov. 1991.

[6] B. Jacobs, 'Simply Typed and Untyped Lambda Calculus Revisited', in *Applications of Categories in Computer Science* edited by M.P. Fourman and P.T. Johnstone and A.M. Pitts, LMS vol. 177 (Cambridge Univ. Press, 1992), pp. 119–142.

[7] B. Jacobs, 'Semantics of Weakening and Contraction', manuscript, May 1992.

[8] B. Jacobs, I. Margaria and M. Zacchi, 'Filter models with polymorphic types', *Theor. Computer Science* 95-2 (1992), pp. 143–158.

[9] S. Mac Lane *Categories for the Working Mathematician*, (Springer, Berlin, 1971).

[10] D.S. Scott, 'Relating theories of the $\lambda$-calculus', in *To H.B. Curry: Essays on Combinatory Logic, Lambda Calculus and Formalism* edited by J.R. Hindley and J.P. Seldin (Academic Press, New York and London, 1980), pp. 403–450.

# Translating Dependent Type Theory into Higher Order Logic[1]

Bart Jacobs[2] and Tom Melham[3]

**Abstract.** *This paper describes a translation of the complex calculus of dependent type theory into the relatively simpler higher order logic originally introduced by Church. In particular, it shows how type dependency as found in Martin-Löf's Intuitionistic Type Theory can be simulated in the formulation of higher order logic mechanized by the HOL theorem-proving system. The outcome is a theorem prover for dependent type theory, built on top of HOL, that allows natural and flexible use of set-theoretic notions. A bit more technically, the language of the resulting theorem-prover is the internal language of a (boolean) topos (as formulated by Phoa).*

## 1  Introduction

In dependent type theory (DTT, for short) terms may occur in types. A typical example is the dependent type $Nat[n]$ of natural numbers less than $n$. This gives DTT greater expressive power than simple type theory, where types cannot depend on the values of terms but only on other types. This advantage of DTT will be illustrated by examples in section 3.2 below.

Church's higher order logic consists of simply typed $\lambda$-calculus with the type constructors $\to$ for function space and $\times$ for cartesian product and with a special type *bool* of 'truth values' or 'booleans'. It comes equipped with constants $T : bool$ and $F : bool$ for true and false and the usual connectives $\supset$, $\wedge$, $\vee$, and $\neg$. Formulas are just boolean terms $\varphi : bool$, and the logic involved is classical. The logic can conveniently be formulated with sequents of the form

$$\Gamma \mid \varphi_1, \ldots, \varphi_m \vdash \psi \tag{1}$$

which should be read as follows: 'in a context $\Gamma$ containing type declarations of all the term variables free in $\varphi_1, \ldots, \varphi_m, \psi$, the formula $\psi$ is true under the assumption that the formulas $\varphi_1, \ldots, \varphi_m$ are true'. The context $\Gamma$ in such a sequent has the form $\{x_1 : \sigma_1, \ldots, x_n : \sigma_n\}$. The sign '|' is used as a separator.

Gordon's HOL theorem prover [4, 5] implements a specific formulation of Church's higher order logic, which will be described in some detail in section 2 below. In HOL, as in Church's original logic, the context $\Gamma$ in a sequent is omitted; types are instead attached to the variables themselves. But in the background, the context can be regarded as still being there, because the information it contains can be reconstructed from the formulas of a sequent. We shall therefore feel free to refer to a formulation of HOL with explicit contexts when this has technical or notational advantages.

For example, higher order logic includes universal and existential quantification over the values of each type. Using a formulation based on sequents with an explicit context, the introduction rules for these take the form

$$\frac{\Gamma, x:\sigma \mid \varphi_1, \ldots, \varphi_m \vdash \psi}{\Gamma \mid \varphi_1, \ldots, \varphi_m \vdash \forall x{:}\sigma.\, \psi} \quad (x \text{ not in } \varphi_1, \ldots, \varphi_m)$$

---

[1] Research carried out during 1991–92 at the University of Cambridge under SERC grant GR/F 36675.
[2] Mathematical Institute RUU, P.O.Box 80.010, 3508 TA Utrecht, NL. bjacobs@math.ruu.nl
[3] Computer Laboratory, University of Cambridge, New Museums Site, Pembroke Street, Cambridge CB2 3QG, UK. tfm@cl.cam.ac.uk

$$\frac{\Gamma \vdash a : \sigma \quad \Gamma \mid \varphi_1, \ldots, \varphi_m \vdash \psi[a/x]}{\Gamma \mid \varphi_1, \ldots, \varphi_m \vdash \exists x{:}\sigma.\,\psi}$$

Note that the side condition '$x$ not in $\varphi_1, \ldots, \varphi_m$' appears natural in this formulation, because in forming $\forall x{:}\sigma.\,\psi$ one has to lift $x$ over the $\varphi_1, \ldots, \varphi_m$.

Another advantage of using explicit contexts is that this highlights the fact that higher order logic is a system of (classical) logic in an environment with *simple types*; a context is just an assignment of simple types to the free variables of a proposition and its assumptions. In this paper, we describe a translation into HOL of a system of classical logic in an environment with *dependent types*. This also has sequents of the form illustrated by (1) above, but the context $\Gamma$ may now contain dependent types. This gives the logic increased expressive power. Our favourite example is the result which says that every injective endofunction on a finite set is surjective. Using such a 'dependent logic', it can be expressed simply as

$$n :: \mathit{Nat}, f :: \mathit{Nat}[n] \rightarrow \mathit{Nat}[n] \mid \mathsf{injective}(f) \vdash \mathsf{surjective}(f)$$

where we use :: for inhabitation in DTT.[4] Translating such assertions into the mechanized HOL logic gives us a theorem prover with great expressive power, having a logic with dependent types but based on the underlying HOL theory of simple types.

Our main aim in this work is to obtain this expressive advantage, rather than to use 'propositions-as-types'. Moreover, we are not so much interested in *programming* in DTT, as discussed in [13], as in using the logic of type theory for specification and reasoning. We wish the logic we use to be as expressive and natural as possible. Our motivation for using DTT comes, for example, from the work of Hanna, Daeche and Longley [7], who have made a strong case for the utility of dependent types for hardware specification and verification. Leeser [9] has also shown how the dependent types of the Nuprl theorem-prover can be used to structure the information content of the theorems involved in mechanized reasoning about hardware. The main advantage over working in simple type theory is that typing judgements DTT have increased information content; an inhabitation judgement, for example, can bear the information that a term meets a partial specification.

The translation of DTT into HOL uses the more or less familiar idea of sending a dependent type to a predicate. It is also used in [2] and on a more abstract level in section 4.3.5 of [8]; the translation is extracted from the interpretation of DTT in a topos. A detailed description of this translation forms the basis for an actual implementation, which is briefly described in section 6. Our practical experience with the translated version of DTT is still at the level of playing with examples. In particular, detailed investigation of how best to reason in a mixture of both HOL and DTT has been left for future work.

## 2  Higher order logic and HOL

The higher order logic mechanized by the HOL theorem prover is based on Church's formulation of simple type theory [1]. Gordon's machine-oriented formulation, which we shall call the HOL logic, or just HOL, extends Church's theory in two significant ways: the syntax of types includes the polymorphic type discipline developed by Milner for the LCF logic PP$\lambda$ [6], and the primitive basis of the logic includes formal rules of definition for extending the logic with new constants and new types. The following section gives a quick overview, mainly of the notation we'll be using; see [5] for a complete description, including a set-theoretic semantics.

---

[4]Strictly speaking, the propositions injective($f$) and surjective($f$) will also make reference to the domain and codomian of $f$. We have left this implicit here.

**Types.** The syntax of types in HOL is given by

$$\sigma ::= c \mid v \mid (\sigma_1, \ldots, \sigma_n)op$$

where $\sigma, \sigma_1, \ldots, \sigma_n$ range over types, $c$ ranges over type *constants*, $v$ ranges over type *variables*, and *op* ranges over $n$-ary type *operators* (for $n \geq 1$).

In fact, the basic type system is very small; it contains only the primitive types *bool* (the two-element set of truth-values) and *ind* (an infinite set of 'individuals') and one type operator, namely $\rightarrow$ for function space. All other types are formally defined in terms of these primitive ones using one of the rules of definition mentioned above.

Among the types definable in HOL is a singleton type *one* whose sole element will be written $*$. A cartesian product type $\sigma \times \tau$ is also definable, with (surjective) pairing written as $\langle -, - \rangle$ and with projections $\pi, \pi'$. In this paper, we also use a type *num* constant of natural numbers and a polymorphic type constructor $(\alpha)list$, both of which are also formally definable in the HOL logic. Details of the definitions of all these types can be found in [3] or [12].

**Terms.** The syntax of (untyped) terms in the HOL logic is given by

$$M ::= c \mid v \mid (M\,N) \mid \lambda v.M$$

where $c$ ranges over constants, $v$ ranges over variables, and M and N range over terms. Sans serif identifiers (e.g. a, b, c, Const) and non-alphabetical symbols (e.g. $\supset, =, \forall$) are generally used for constants, and italic identifiers (e.g. $v, x, x_1, f$) are used for variables.

Every well-formed term in higher order logic must be *well-typed*. Writing $M : \sigma$ indicates explicitly that the term M is well-typed with type $\sigma$. Typing of terms takes place within the context of an assignment of types to constants. Each constant $c$ has fixed *generic* type $Gen(c)$ associated with it. A constant $c$ with a polymorphic generic type $\sigma$ is well typed with any substitution instance of $\sigma$ obtainable by substituting types for type variables. Given an assignment of generic types to constants, the well-typed terms of HOL are defined inductively by the following typing rules.

$$\text{var} \frac{}{v : \sigma} \qquad \text{con} \frac{}{c : \sigma} \quad (\sigma \text{ a substitution instance of } Gen(c))$$

$$\text{abs} \frac{v : \sigma \quad M : \tau}{(\lambda v.M) : \sigma \rightarrow \tau} \qquad \text{app} \frac{M : \sigma \rightarrow \tau \quad N : \sigma}{(M\,N) : \tau}$$

It follows from these rules that the type of a term is uniquely determined by the types of constants and variables it contains.

**Definitions and Axioms.** The HOL logic is based on the following fundamental definitions for quantifiers and connectives, which we simply list here without further comment.

**D1** $\vdash T = ((\lambda x{:}bool.\,x) = (\lambda x{:}bool.\,x))$
**D2** $\vdash \forall = \lambda P{:}\alpha{\rightarrow}bool.\,P = (\lambda x.\,T)$
**D3** $\vdash \exists = \lambda P{:}\alpha{\rightarrow}bool.\,P(\varepsilon\,P)$
**D4** $\vdash F = \forall b{:}bool.\,b$
**D5** $\vdash \neg = \lambda b.\,b \supset F$
**D6** $\vdash \wedge = \lambda b_1\,b_2.\forall b.\,(b_1 \supset (b_2 \supset b)) \supset b$
**D7** $\vdash \vee = \lambda b_1\,b_2.\forall b.\,(b_1 \supset b) \supset ((b_2 \supset b) \supset b)$

Note that Hilbert's $\varepsilon$-operator, a primitive constant of HOL, is used in **D3** to define existential quantification. Informally, the semantics of this operator is as follows. If $P{:}\sigma{\rightarrow}bool$ is a predicate on values of type $\sigma$, then the application '$\varepsilon$ P' denotes a value of type $\sigma$ for which P is true. If there is no such value, then the term '$\varepsilon$ P' denotes a fixed but unknown value of type $\sigma$. A consequence is that all types in HOL must be non-empty, since for any predicate $P{:}\sigma{\to}bool$, the term $\varepsilon$ P always denotes a value of type $\sigma$. For further discussion of $\varepsilon$, see [10].

Some typical examples of formulas written using this defined logical notation are the five axioms of the HOL logic, which are shown below.

A1 $\vdash \forall b.(b = \mathsf{T}) \lor (b = \mathsf{F})$
A2 $\vdash \forall b_1 b_2.(b_1 \supset b_2) \supset (b_2 \supset b_1) \supset (b_1 = b_2)$
A3 $\vdash \forall f{:}\alpha{\to}\beta.(\lambda x. f\, x) = f$
A4 $\vdash \forall P{:}\alpha{\to}bool.\forall x.\, P\, x \supset P(\varepsilon\, P)$
A5 $\vdash \exists f{:}ind{\to}ind.\,(\forall x\, y.\,(fx = fy) \supset (x = y)) \land \neg\forall x.\exists y.\, x = f\, y$

Together with the primitive inference rules of HOL, the axioms **A1**, **A2** and **A3** define (classical) higher order propositional and functional calculus. The additional axioms **A4** and **A5** are the axiom of choice and axiom of infinity.

**Inference Rules.** The style of proof used in Gordon's formulation of higher order logic is a form of natural deduction in which sequents are used to keep track of assumptions. A sequent is written $\varphi_1,\ldots,\varphi_n \vdash \psi$, where $\varphi_1,\ldots,\varphi_n$ is a sequence of boolean terms called the assumptions and $\psi$ is a boolean term called the conclusion.[5] This sequent notation can be read as the metalinguistic assertion that there exists a natural deduction proof of the conclusion $\psi$ from the assumptions in $\varphi_1,\ldots,\varphi_n$. When there are no assumptions, the notation $\vdash \psi$ is used. In this case, $\psi$ is a formal theorem of the logic.

The inference rules of HOL are without surprises. For example, one has the following rules introduction and elimination rules for implication.

$$\dfrac{\varphi_1,\ldots,\varphi_n,\psi \vdash \chi}{\varphi_1,\ldots,\varphi_n \vdash \psi \supset \chi} \qquad \dfrac{\varphi_1,\ldots,\varphi_n \vdash \psi \supset \chi \quad \varphi_1,\ldots,\varphi_n \vdash \psi}{\varphi_1,\ldots,\varphi_n \vdash \chi}$$

Note that these rules do not explicitly mention contexts, as discussed above on in section 1. For a complete account of the HOL inference rules, both primitive and derived, see [5].

## 3 Dependent type theory

In this section we do not intend to give a complete and systematic description of dependent type theory (DTT)—for this, refer to [11] or [16]. Rather it is our aim to get across just the main ideas and features of such a type theory, especially the dependent product $\Pi$ and the dependent sum $\Sigma$. Our secondary aim is to point out some subtleties in the presentation of the calculus.

The language we use is rather informal. We shall use :: for inhabitation in DTT in order to distinguish it from the typing symbol : used in HOL.

### 3.1 Informal description

In HOL a term variable $x$ may occur in a term (e.g. in $x + 3 : num$) but not in a type. This is different in dependent type theory, where for example one can have a term $n$ that denotes a natural number and occurs in types. Some typical examples are

$Nat[n]$ the type of natural numbers less than $n$, i.e. $\{0, 1,\ldots, n{-}1\}$
$List[n]$ the type of lists of length $n$
$Matrix[n, m]$ the type of $n \times m$ matrices

---

[5]In fact, not sequences but *sets* of assumptions are used in the formulations of HOL presented in [3, 5].

One can then have term judgements that involve these types, for example:

$$n :: \textit{Nat}, m :: \textit{Nat}[n] \;\vdash\; \text{suc}\, m :: \textit{Nat}[\text{suc}\, n]$$
$$n :: \textit{Nat}, m :: \textit{Nat}, z :: \textit{Matrix}[n,m] \;\vdash\; \text{firstrow}\, z :: \textit{List}[n]$$

where suc:$\textit{num} \to \textit{num}$ is the successor function on natural numbers in HOL, and firstrow is a HOL function with the obvious meaning.

A first thing to note about DTT is that a dependent type makes sense only in a context which contains declarations of its free variables. For example one can have

$$n :: \textit{Nat} \vdash \textit{Nat}[n] :: \textit{DType} \qquad\qquad n :: \textit{Nat}[5] \vdash \textit{Nat}[n] :: \textit{DType}$$

where it is clear that the sets denoted by the type $\textit{Nat}[n]$ in the judgement on the left can be quite different from the ones denoted by $\textit{Nat}[n]$ in the judgement on the right.[6] Thus the most basic judgements of DTT have the form

$$\Gamma \vdash A :: \textit{DType}$$

where $\Gamma$ is a context of the form $x_1 :: A_1, \ldots, x_n :: A_n$. This says that $A$ is a (dependent) type in context $\Gamma$. Alternatively one can write

$$x_1 :: A_1,\; x_2 :: A_2[x_1], \ldots,\; x_n :: A_n[x_1, \ldots, x_{n-1}] \vdash A[x_1, \ldots, x_n] :: \textit{DType}$$

where the possible occurrences of variables are made explicit. The second form of judgement in dependent type theory is

$$\Gamma \vdash A = B :: \textit{DType}$$

This says that $A$ and $B$ are equal dependent types, where it is therefore understood that $\Gamma \vdash A :: \textit{DType}$ and $\Gamma \vdash B :: \textit{DType}$. Notice that such a type equality judgement can have a computational content, since terms may occur in types. As a trivial example, consider

$$\vdash \textit{Nat}[25] = \textit{Nat}[\text{square}\, 5] :: \textit{DType}$$

Thirdly, for a dependent type $\Gamma \vdash A :: \textit{DType}$ we can say that $a$ is a term of type $A$ by writing

$$\Gamma \vdash a :: A$$

where it is understood that all free variables of $a$ occur in $\Gamma$. Finally, one wants judgements about equality of terms. These take the form

$$\Gamma \vdash a = b :: A$$

for terms $\Gamma \vdash a :: A$ and $\Gamma \vdash b :: A$. As a typical example one can state the following in DTT.

$$n :: \textit{Nat}[2] \vdash n^4 = n^5 :: \textit{Nat}[2]$$

Notice how much information is contained in this judgement. Indeed, the facility for compact expression of complex facts is what makes DTT so attractive.

The above judgements are the four kinds of judgement distinguished by Martin-Löf [11].

---

[6]Notice, by the way, that on the right-hand side $n :: \textit{Nat}[5]$ is implicitly coerced to $n :: \textit{Nat}$.

**Products and sums.** The most often used type constructors in HOL are function space $\to$ and cartesian product $\times$. In DTT, these generalize to the dependent product $\Pi$ and dependent sum $\Sigma$. The formation rules for these types are

$$\frac{\Gamma, x :: A \vdash B :: DType}{\Gamma \vdash \Pi_{x::A}.B :: DType} \qquad \frac{\Gamma, x :: A \vdash B :: DType}{\Gamma \vdash \Sigma_{x::A}.B :: DType}$$

That is, given a dependent type $B$ (possibly) containing a variable of type $A$, then one can form the dependent product $\Pi_{x::A}.B$ and the dependent sum $\Sigma_{x::A}.B$. Denotationally one thinks about these in the following way.

$$[\![\Pi_{x::A}.B]\!] = \begin{cases} \text{the set of functions } f \text{ with domain } [\![A]\!], \text{ such that for each} \\ a \in [\![A]\!] \text{ one has } f(a) \in [\![B[a/x]]\!]; \end{cases}$$

$[\![\Sigma_{x::A}.B]\!] =$ the set of 'dependent' pairs $(a,b)$ where $a \in [\![A]\!]$ and $b \in [\![B[a/x]]\!]$.

Thus in case a variable $x$ does not occur in $B$ one has that $\Pi_{x::A}.B$ is $A \to B$ and that $\Sigma_{x::A}.B$ is $A \times B$. In this way $\Pi$ and $\Sigma$ generalize $\to$ and $\times$.

In line with the interpretations given above, one has the following introduction, elimination and conversion rules for $\Pi$ and $\Sigma$.

$$\frac{\Gamma, x :: A \vdash b :: B}{\Gamma \vdash \lambda x :: A.b :: \Pi_{x::A}.B} \text{ (}\Pi\text{-I)} \qquad \frac{\Gamma \vdash c :: \Pi_{x::A}.B \quad \Gamma \vdash a :: A}{\Gamma \vdash c\,a :: B[a/x]} \text{ (}\Pi\text{-E)}$$

$$\frac{\Gamma, x :: A \vdash b :: B \quad \Gamma \vdash a :: A}{\Gamma \vdash (\lambda x :: A.b)\,a = b[a/x] :: B[a/x]} \qquad \frac{\Gamma \vdash c :: \Pi_{x::A}.B}{\Gamma \vdash \lambda x :: A.c\,x = c :: \Pi_{x::A}.B}$$

$$\frac{\Gamma \vdash a :: A \quad \Gamma \vdash b :: B[a/x]}{\Gamma \vdash \langle a,b\rangle :: \Sigma_{x::A}.B} \text{ (}\Sigma\text{-I)}$$

$$\frac{\Gamma \vdash c :: \Sigma_{x::A}.B}{\Gamma \vdash \pi c :: A} \text{ (}\Sigma\text{-E}_1\text{)} \qquad \frac{\Gamma \vdash c :: \Sigma_{x::A}.B}{\Gamma \vdash \pi' c :: B[\pi c/x]} \text{ (}\Sigma\text{-E}_2\text{)}$$

$$\frac{\Gamma \vdash a :: A \quad \Gamma \vdash b :: B[a/x]}{\Gamma \vdash \pi\langle a,b\rangle = a :: A} \qquad \frac{\Gamma \vdash a :: A \quad \Gamma \vdash b :: B[a/x]}{\Gamma \vdash \pi'\langle a,b\rangle = b :: B[a/x]}$$

$$\frac{\Gamma \vdash c :: \Sigma_{x::A}.B}{\Gamma \vdash \langle \pi c, \pi' c\rangle = c :: \Sigma_{x::A}.B}$$

The $\Sigma$ types we are describing are the so-called 'strong sums', which come equipped with both projections. See [8] for more details about strong and weak sums.

## 3.2 Examples

Let's turn to some examples. The first one is taken from a paper by Hanna, Daeche, and Longley [7]. On first thought one may view a type *date* as $Nat \times Nat \times Nat$, where the first component represents the year, the second the month and the third the day. This can obviously be done in a far more precise way, since the number of months in a year does not exceed 12 and the number of days in a month does not come above 31. So a second try is $Nat \times Nat[12] \times Nat[31]$, which is already much better. But not every month has 31 days; even worse, the length of the month of February depends on the year. So the best representation is

$$date = \Sigma_{y::Nat}. \Sigma_{m::Nat[12]}. Nat[\text{length of month } m \text{ in year } y]$$

where the term 'length of month $m$ in year $y$' is defined by cases in an obvious way. A typical term of type $date$ is $\langle 1992, \langle 5, 16 \rangle \rangle$.

This example nicely illustrates an important point made in [7]: by using dependent types one has a precise and concise typing discipline. This is especially convenient in hardware verification, where there are many kinds of values that are parameterized by size—for example bit-vectors, or integers mod $n$.

The second example comes from [11]. It shows that the axiom of choice (AC) holds in dependent type theory. Under the 'propositions-as-types' reading one views a type $A$ as a proposition and a term $a :: A$ as a proof of $A$. Then one further reads

$\Pi_{x::A}.B$ as universal quantification: for all $x$ in $A$ one has $B$

$\Sigma_{x::A}.B$ as existential quantification: there is an $x$ in $A$ for which $B$

Indeed a proof $c :: \Pi_{x::A}.B$ is then seen as a function which gives for each element $a$ of $A$ a proof $c\,a$ of $B[a/x]$. And a proof $c :: \Sigma_{x::A}.B$ consists of an element $\pi c$ of $A$ and a proof $\pi' c$ of $B[\pi c/x]$. This is the so-called 'Brouwer-Heyting-Kolmogorov' interpretation, see [17].

Thus the fact that (AC) holds in DTT amounts to the existence of a term ac such that

$$\vdash \text{ac} :: (\Pi_{x::A}.\Sigma_{y::B}.C[x,y]) \to (\Sigma_{f::(\Pi_{x::A}.B)}.\Pi_{x::A}.C[x,f\,x])$$

We reason informally; suppose we are given a term $z :: \Pi_{x::A}.\Sigma_{y::B}.C[x,y]$. That is, suppose $z$ is a proof of 'for all $x$ in $A$ there is a $y$ in $B$ such that $C[x,y]$'. The aim is then to construct a function $f$ which gives such a $y$ in $B$ for each $x$ in $A$. Notice that for every $x :: A$ one has $z\,x :: \Sigma_{y::B}.C[x,y]$ and thus $\pi(z\,x) :: B$ and $\pi'(z\,x) :: C[x, \pi(z\,x)]$. So if we take $f$ to be the term $\lambda x :: A.\ \pi(z\,x)$ of type $\Pi_{x::A}.B$, then $\pi'(z\,x) :: C[x, f\,x]$, which yields $\lambda x :: A.\ \pi'(z\,x) :: \Pi_{x::A}.C[x, f\,x]$. Hence we have

$$\text{ac} = \lambda z :: (\Pi_{x::A}.\Sigma_{y::B}.C[x,y]).\ \langle \lambda x :: A.\ \pi(z\,x),\ \lambda x :: A.\ \pi'(z\,x) \rangle$$

Our third and final example is taken from [15]. The perspective is again 'propositions-as-types'. The aim of predicate logic is to formalize inferences like the following: 'All men are mortal; Socrates is a man; hence Socrates is mortal.' But not everything can be formulated in predicate logic. A famous elusive phrase is the so-called *donkey sentence*:

every man who owns a donkey beats it.

Try to formalize it! You'll find that the problem lies in 'it' referring back to the donkey. One might want to try

$$\forall d \in Donkey.\ \forall m \in Man.\ \text{Owns}(m,d) \supset \text{Beats}(m,d)$$

But the quantification here is over all donkeys instead of over men owning a donkey. This problem is solved in dependent type theory in the following elegant way.

$$\Pi_{m::Man}.\ \Pi_{x::(\Sigma_{d::Donkey}.\text{Owns}(m,d))}.\ \text{Beats}(m, \pi x)$$

### 3.3 Some formalities

The possibility in dependent type theory of term variables occurring in types has advantages when it comes to expressive power. A definite disadvantage is that it becomes rather cumbersome to formulate the calculus in a precise way. This is because one cannot first give the rules for types and then for terms; they depend on each other, and so one has to present them in a simultaneous induction.

A further point is that one needs certain predefined dependent types (like $Nat[n]$) to start with. Otherwise one doesn't get off the ground. Below we present the basic rules, mainly having to do with contexts. In these rules, $\mathcal{J}$ stands for one of the four judgements $A :: DType$, $a :: A$, $A = B :: DType$ and $a = b :: A$.

$$\text{start } \frac{}{\Gamma \vdash A :: DType} \text{ (for predefined } A \text{ and } \Gamma)$$

$$\text{projection } \frac{\Gamma \vdash A :: DType}{\Gamma, x :: A \vdash x :: A} \text{ (with } x \text{ a fresh variable)}$$

$$\text{exchange } \frac{\Gamma, x :: A, y :: B, \Delta \vdash \mathcal{J}}{\Gamma, y :: B, x :: A, \Delta \vdash \mathcal{J}} \text{ (if } x \text{ not free in } B)$$

$$\text{weakening } \frac{\Gamma, \Delta \vdash \mathcal{J} \quad \Gamma \vdash A :: DType}{\Gamma, x :: A, \Delta \vdash \mathcal{J}} \text{ (with } x \text{ a fresh variable)}$$

$$\text{substitution } \frac{\Gamma, x :: A, \Delta \vdash \mathcal{J} \quad \Gamma \vdash a :: A}{\Gamma, \Delta[a/x] \vdash \mathcal{J}[a/x]}$$

We shall assume that we always have some start rules together with these basic context rules, in addition to whatever other rules are present in the theory.

## 4 Translating type dependency

The typical reaction of an experienced HOL user to the examples in section 3.2 is: 'But that can all be done in HOL; just use some suitable predicates to mimic dependent types'. The translation we are about to describe can be seen as a systematic elaboration of such a reaction.

### 4.1 Prejudgements

The table below gives a first description of the intended translation of dependent type theory into higher order logic.

| DTT | becomes in HOL |
|---|---|
| type, in isolation | predicate |
| declaration of variable $v$ | assumption of a predicate being true of $v$ |
| context | 'well-formed' list of assumptions |
| equality of types, in context | equivalence of predicates, with a list of assumptions |
| inhabitation judgement, in context | theorem, with a list of assumptions |
| equality of terms, in context | equality of terms, with a list of assumptions |

This table will be explained in more detail. The HOL predicates that will be used for the translation have the following form.

$$P : \sigma \to bool$$

Such a predicate $P$ need not be closed; it may contain free term variables, say $x_1, \ldots, x_n$. In case we want to have these explicit, we write $P[x_1, \ldots, x_n]$. Of course we could work with closed terms $P : (\sigma_1 \times \cdots \times \sigma_n) \to \sigma \to bool$ or simply with propositions $P[x_1, \ldots, x_n, y] : bool$, but the above form makes the most efficient use of the underlying HOL mechanism for handling variables.

Informally we'll say that $a :: P$ if $a$ is a HOL term of type $\sigma$ for which $P\,a$ is true. For example, one can take $Nat[n]$ to be the predicate $\lambda m.\ m < n$ of HOL type $num \to bool$. Then $m :: Nat[n]$ if and only if $m : num$ in HOL and $m < n$.

**Definition 4.1** *A pseudo context is a sequence $x_1 :: P_1, \ldots, x_n :: P_n$ where*

- $P_1, \ldots, P_n$ *are HOL predicates of the form* $P_i : \sigma_i \to bool$;
- $x_1, \ldots, x_n$ *are HOL variables with* $x_i : \sigma_i$;
- *the free variables of $P_i$ are among $x_1, \ldots, x_{i-1}$. $P_1$, in particular, is closed.*

Next we describe pseudo versions of the four forms of judgement in dependent type theory.

**Definition 4.2** *Let $\Gamma = x_1 :: P_1, \ldots, x_n :: P_n$ be a pseudo context.*

(i) *$Q$ is a pseudo type in context $\Gamma$ if $Q$ is of the form $\tau \to bool$ and all its free variables are among $x_1, \ldots, x_n$.*

(ii) *$a$ is a pseudo term of type $Q$ in context $\Gamma$ if $Q$ is a pseudo type in $\Gamma$ as in (i) and $a$ is a HOL term of type $\tau$ with free variables among $x_1, \ldots, x_n$ for which there is a HOL theorem (with assumptions)*

$$P_1\,x_1,\ \ldots,\ P_n\,x_n \vdash Q\,a$$

(iii) *$Q$ and $R$ are equal pseudo types in context $\Gamma$ if both $Q$ and $R$ are pseudo types in context $\Gamma$ for which the following is a HOL theorem*

$$P_1\,x_1,\ \ldots,\ P_n\,x_n \vdash \forall y.\, Q\,y = R\,y$$

*where $=$ on booleans means logical equivalence in HOL.*

(iv) *$a$ and $b$ are equal pseudo terms of type $Q$ in context $\Gamma$ if both $a$ and $b$ are pseudo terms of type $Q$ in $\Gamma$ for which one has in HOL*

$$P_1\,x_1,\ \ldots,\ P_n\,x_n \vdash a = b$$

For example, with the formulation of $Nat[n]$ given above one has a 'pseudo term in context' $m :: Nat[10] \vdash m^2 :: Nat[82]$ and an 'equality of pseudo terms' $m :: Nat[2] \vdash m^4 = m^5 :: Nat[2]$. It is easy to work out the meanings of these 'pseudo judgements' from the above definition.

## 4.2 Validity of DTT rules

In order to prevent confusion about the calculus in which we are working, we shall from now on use $\vdash_H$ for deducibility in HOL and $\vdash_D$ for deducibility in DTT. It will be shown that all the DTT rules are valid under the interpretation described above. For a pseudo judgement $\mathcal{J}$ in context $\Gamma$, we shall write $\Gamma \vdash_D \mathcal{J}$ if this is a judgement of DTT that can be derived from the underlying translation.

**Lemma 4.3** *Start rules are available. First of all, for any HOL type $\sigma$, define*

$$D(\sigma) = \lambda x.\,\mathsf{T}\ :\ \sigma \to bool$$

*where $\mathsf{T} : bool$ is the truth value 'true'. Then one has in the empty context $\vdash_D D(\sigma) :: DType$. We'll write $Nat = D(num)$ and $Bool = D(bool)$ for the resulting 'dependent' naturals and booleans.*

*Here are some more primitive dependent types.*

$$\begin{array}{rcl}
\text{Nat}[n] & = & \lambda m.\, m < n \;:\; num \to bool \\
\text{List}[n] & = & \lambda l.\, \text{length}(l) = n \;:\; (num)list \to bool \\
\text{Matrix}[n,m] & = & \lambda z.\, \text{length}(z) = n \wedge \forall i \leq n.\, \text{depth}(i,z) = m \;:\; ((num)list)list \to bool
\end{array}$$

where $\text{depth}(i, z)$ is a HOL term which gives the length of the i-th (list) element of a list of lists z. We also have the following example judgements that involve these primitive types

$$\begin{array}{rcl}
n :: \text{Nat} & \vDash_{\text{D}} & \text{Nat}[n] :: Dtype \\
n :: \text{Nat} & \vDash_{\text{D}} & \text{List}[n] :: Dtype \\
n :: \text{Nat}, m :: \text{Nat} & \vDash_{\text{D}} & \text{Matrix}[n,m] :: Dtype
\end{array}$$

**Proof.** Obvious, by definition 4.2 (i). □

**Lemma 4.4** *The context rules 'projection', 'exchange', 'weakening' and 'substitution' are all valid.*

**Proof.** Suppose that $\Gamma = x_1 :: P_1, \ldots, x_n :: P_n$ is a context with $\Gamma \vDash_{\text{D}} Q :: DType$. Then $\Gamma, y :: Q \vDash_{\text{D}} y :: Q$ translates to the sequent $P_1 x_1, \ldots, P_n x_n, Q y \vDash_{\text{H}} Q y$ being a theorem. The $P_i x_i$ are simply superfluous assumptions. In an equally simple way one has that the exchange rule involves changing the order of HOL assumptions and that weakening involves adding another HOL assumption. This can all be done without problems. Finally for the substitution rule, suppose we have $\Gamma \vDash_{\text{D}} a :: Q$ and $\Gamma, y :: Q, \Delta \vDash_{\text{D}} \mathcal{J}$. Then we have a HOL theorem $P_1 x_1, \ldots P_n x_n \vDash_{\text{H}} Q a$. Thus in the HOL interpretation of the judgement $\Gamma, y :: Q, \Delta \vDash_{\text{D}} \mathcal{J}$ the assumption $Q y$ can be removed by first substituting $a$ for $y$ and then performing a cut on $Q a$. So one obtains $\Gamma, \Delta[a/x] \vDash_{\text{D}} \mathcal{J}[a/x]$ as required. □

**Lemma 4.5** *The rules for the dependent product $\Pi$ are all valid.*

**Proof.** Let $\Gamma = x_1 :: P_1, \ldots, x_n :: P_n$ be a context provided with a type $\Gamma, y :: Q \vDash_{\text{D}} R :: DType$, say with $Q$ of HOL type $\sigma \to bool$ and $R$ of HOL type $\tau \to bool$. We define a new predicate $\Pi_{y::Q}. R$ of HOL type $(\sigma \to \tau) \to bool$ by

$$\Pi_{y::Q}. R \;=\; \lambda f.\, \forall y{:}\sigma.\, Q y \supset R(f y)$$

Then $\Gamma \vDash_{\text{D}} \Pi_{y::Q}. R :: DType$ because all the free variables of $\Pi_{y::Q}. R$ are among $x_1, \ldots, x_n$. Note that $y$ may, of course, occur in $R$.

Next we have to establish the validity of the introduction and elimination rules. Therefore assume we have $\Gamma, y :: Q \vDash_{\text{D}} b :: R$. That is, assume that $b$ is a HOL term of type $\tau$ with free variables among $x_1, \ldots, x_n, y$ for which there is a HOL theorem $P_1 x_1, \ldots, P_n x_n, Q y \vDash_{\text{H}} R b$. Then we deduce in HOL,

$$\frac{\displaystyle \frac{\displaystyle \frac{\displaystyle \frac{\displaystyle \frac{P_1 x_1, \ldots, P_n x_n, Q y \vDash_{\text{H}} R b}{P_1 x_1, \ldots, P_n x_n, Q y \vDash_{\text{H}} R((\lambda y.\, b)\, y)}}{P_1 x_1, \ldots, P_n x_n \vDash_{\text{H}} Q y \supset R((\lambda y.\, b)\, y)}}{P_1 x_1, \ldots, P_n x_n \vDash_{\text{H}} \forall y{:}\sigma.\; Q y \supset R((\lambda y.\, b)\, y)}}{P_1 x_1, \ldots, P_n x_n \vDash_{\text{H}} (\lambda f.\, \forall y{:}\sigma.\; Q y \supset R(f y))\,(\lambda y.\, b)}}{P_1 x_1, \ldots, P_n x_n \vDash_{\text{H}} (\Pi_{y::Q}. R)\,(\lambda y.\, b)}$$

and thus by defining $\lambda y :: Q.\, b = \lambda y.\, b$ we obtain

$$\Gamma \vdash_{\overline{D}} \lambda y :: Q.\, b :: \Pi_{y::Q}.\, R$$

Hence abstraction in our translation of DTT is simply abstraction in HOL. Similarly for application: given $\Gamma \vdash_{\overline{D}} c :: \Pi_{y::Q}.\, R$ and $\Gamma \vdash_{\overline{D}} a :: Q$ we use the following deduction in HOL

$$\cfrac{\cfrac{\cfrac{P_1 x_1, \ldots, P_n x_n \vdash_{\overline{H}} (\Pi_{y::Q}.\, R)\, c}{P_1 x_1, \ldots, P_n x_n \vdash_{\overline{H}} \forall y{:}\sigma.\ Q\, y \supset R(c\, y)}}{P_1 x_1, \ldots, P_n x_n \vdash_{\overline{H}} Q\, a \supset R[a/y](c\, a)} \qquad P_1 x_1, \ldots, P_n x_n \vdash_{\overline{H}} Q\, a}{P_1 x_1, \ldots, P_n x_n \vdash_{\overline{H}} R[a/y](c\, a)}$$

which yields that $\Gamma \vdash_{\overline{D}} c\, a :: R[a/y]$. The associated $\beta$ and $\eta$ rules are obviously valid. □

**Lemma 4.6** *The rules for the dependent sum $\Sigma$ are valid.*

*Proof.* Assume again we have a context $\Gamma = x_1 :: P_1, \ldots, x_n :: P_n$ and a type $\Gamma, y :: Q \vdash_{\overline{D}} R :: DType$ with $Q : \sigma \to bool$ and $R : \tau \to bool$. We define $\Gamma \vdash_{\overline{D}} \Sigma_{y::Q}.\, R :: DType$ by

$$\Sigma_{y::Q}.\, R \;=\; \lambda z.\, Q\,(\pi z) \wedge R[\pi z/y](\pi' z) \;:\; (\sigma \times \tau) \to bool$$

If we have terms $\Gamma \vdash_{\overline{D}} a :: Q$ and $\Gamma \vdash_{\overline{D}} b :: R[a/y]$, then we deduce in HOL

$$\cfrac{\cfrac{P_1 x_1, \ldots, P_n x_n \vdash_{\overline{H}} Q\, a \qquad P_1 x_1, \ldots, P_n x_n \vdash_{\overline{H}} R[a/y]\, b}{P_1 x_1, \ldots, P_n x_n \vdash_{\overline{H}} Q\, a \wedge R[a/y]\, b}}{P_1 x_1, \ldots, P_n x_n \vdash_{\overline{H}} (\Sigma_{y::Q}.\, R)(\langle a, b \rangle)}$$

which yields $\Gamma \vdash_{\overline{D}} \langle a, b \rangle :: \Sigma_{y::Q}.\, R$, where $\langle -, - \rangle$ denotes the pairing from HOL. Similarly from $\Gamma \vdash_{\overline{D}} c :: \Sigma_{y::Q}.\, R$ one obtains $\Gamma \vdash_{\overline{D}} \pi c :: Q$ and $\Gamma \vdash_{\overline{D}} \pi' c :: R[\pi c/y]$. □

The identity types from [11] are also supported by our interpretation.

**Lemma 4.7** *The following rules for identity types are valid.*

$$\cfrac{\Gamma \vdash_{\overline{D}} a :: Q \qquad \Gamma \vdash_{\overline{D}} b :: Q}{\Gamma \vdash_{\overline{D}} \mathsf{Eq}_Q(a,b) :: DType} \qquad\qquad \cfrac{\Gamma \vdash_{\overline{D}} a = b :: Q}{\Gamma \vdash_{\overline{D}} * :: \mathsf{Eq}_Q(a,b)}$$

$$\cfrac{\Gamma \vdash_{\overline{D}} c :: \mathsf{Eq}_Q(a,b)}{\Gamma \vdash_{\overline{D}} a = b :: Q} \qquad\qquad \cfrac{\Gamma \vdash_{\overline{D}} c :: \mathsf{Eq}_Q(a,b)}{\Gamma \vdash_{\overline{D}} c = * :: \mathsf{Eq}_Q(a,b)}$$

That is, $\mathsf{Eq}_Q(a,b)$ is inhabited if and only if $a = b$ and its only possible inhabitant is $*$.

*Proof.* Define $\mathsf{Eq}_Q(a,b) = \lambda z.\, a = b \,:\, one \to bool$ with $z$ not in $a$ or $b$. □

**Theorem 4.8** *Dependent type theory can be translated into HOL.*

*Proof.* By the above series of lemmas. □

**Remark 4.9** The previous theorem deals with only the essential aspects of DTT. Here are some more details about the translation. One easily verifies that the following rules hold.

$$\frac{\Gamma \vdash_D Q = R :: DType \quad \Gamma \vdash_D a :: Q}{\Gamma \vdash_D a :: R} \qquad \frac{\Gamma \vdash_D a :: Q \quad \Gamma \vdash_D a = b :: Q}{\Gamma \vdash_D b :: Q}$$

We further mention that the translation yields a so-called 'extensional' version of dependent type theory, see section 11–5 of [17]. Rules like the following are valid.

$$\frac{\Gamma \vdash_D Q = Q' :: DType \quad \Gamma, y :: Q \vdash_D R = R' :: DType}{\Gamma \vdash_D \Pi_{y::Q}.R \; = \; \Pi_{y::Q'}.R' :: DType} \qquad \frac{\Gamma, x :: Q \vdash_D b = b' :: R}{\Gamma \vdash_D \lambda x.b = \lambda x.b' :: \Pi_{y::Q}.R}$$

Finally we'll have a closer look at the embedding $D$ : HOL $\to$ DTT from lemma 4.3. We show that it trivially extends to terms and preserves $\times$ and $\to$.

**Theorem 4.10** (i) *HOL terms* $M : \tau$ *with free variables* $x_1 : \sigma_1, \ldots, x_n : \sigma_n$ *are the same as terms* $x_1 :: D(\sigma_1), \ldots, x_n :: D(\sigma_n) \vdash_D M :: D(\tau)$ *in our translated* DTT.

(ii) *For* HOL *types* $\sigma, \tau$ *one has equalities of types*

$$\vdash_D D(\sigma \times \tau) = D(\sigma) \times D(\tau) :: DType \qquad \text{and} \qquad \vdash_D D(\sigma \to \tau) = D(\sigma) \to D(\tau) :: DType$$

**Proof.** (i) Obvious; the predicates involved in $D(-)$ are always true and hence irrelevant.

(ii) By definition one has logical equivalences

$$\begin{aligned}
\vdash_H (D(\sigma) \times D(\tau)) z &= (\Sigma_{y::D(\sigma)}.D(\tau)) z & \text{with } y \text{ not in } D(\tau) \\
&= D(\sigma)(\pi z) \land D(\tau)[\pi z/y](\pi' z) \\
&= T \land T = T = D(\sigma \times \tau) z.
\end{aligned}$$

Thus $\vdash_D D(\sigma \times \tau) = D(\sigma) \times D(\tau) :: DType$ by definition 4.2 (iii). Similarly for $\to$ we have

$$\begin{aligned}
\vdash_H (D(\sigma) \to D(\tau)) f &= (\Pi_{y::D(\sigma)}.D(\tau)) f & \text{with } y \text{ not in } D(\tau) \\
&= \forall y{:}\sigma.\; D(\sigma) y \supset D(\tau)(f y) \\
&= \forall y{:}\sigma.\; T \supset T = T = D(\sigma \to \tau) f.
\end{aligned}$$
□

## 4.3 Translation of logic

Having seen the above translation of dependent type theory into HOL one asks how much of the logic of HOL can be used in the translated DTT. After all, by lemma 4.3 there is a dependent type $Bool = D(bool)$. But note that if we go down this road in seeking an answer, we depart from a 'propositions-as-types' perspective, in that we will be using terms $\varphi :: Bool$ as logical formulas instead of types $A :: DType$ (as in the last two examples in section 3.2).

Recall that in a formulation with explicit contexts of term variables, the logic of HOL can be given in terms of sequents $\Gamma \mid \varphi_1, \ldots, \varphi_m \vdash_H \psi$, where $\varphi_1, \ldots, \varphi_n, \psi$ are terms of type $bool$ with all to their free term variables declared in the context $\Gamma$. We hope that this notation clearly conveys the essential point that with HOL one has (classical) logic in a simply typed ambience.

The same notation, only this time with the annotated turnstile $\vdash_D$, will be used to describe the logic we find in our translated dependent type theory. This expresses the fact that we now have (classical) logic in a dependently typed environment. Thus we write

$$\Gamma \mid \varphi_1, \ldots, \varphi_m \vdash_D \psi$$

where $\Gamma = x_1 :: P_1, \ldots, x_n :: P_n$ is a (translated) dependent context as before and $\varphi_1, \ldots, \varphi_m, \psi$ are terms of dependent type *Bool* in context $\Gamma$, if there is a HOL theorem

$$P_1 x_1, \ldots, P_n x_n, \varphi_1, \ldots, \varphi_m \vDash_H \psi$$

We shall call the resulting system *dependent logic*.

**Proposition 4.11** *In dependent logic one has universal and existential quantification over the values of dependent types. These are given by the following rules.*

$$\frac{\Gamma, y :: Q \mid \varphi_1, \ldots, \varphi_m \vDash_D \psi}{\Gamma \mid \varphi_1, \ldots, \varphi_m \vDash_D \forall y :: Q. \psi} \quad (y \text{ not in } \varphi_1, \ldots, \varphi_m)$$

$$\frac{\Gamma \vDash_D a :: Q \quad \Gamma \mid \varphi_1, \ldots, \varphi_m \vDash_D \forall y :: Q. \psi}{\Gamma \mid \varphi_1, \ldots, \varphi_m \vDash_D \psi[a/y]} \qquad \frac{\Gamma \vDash_D a :: Q \quad \Gamma \mid \varphi_1, \ldots, \varphi_m \vDash_D \psi[a/y]}{\Gamma \mid \varphi_1, \ldots, \varphi_m \vDash_D \exists y :: Q. \psi}$$

$$\frac{\Gamma \mid \varphi_1, \ldots, \varphi_m \vDash_D \exists y :: Q. \psi \quad \Gamma, y :: Q \mid \varphi_1, \ldots, \varphi_m, \psi \vDash_D \chi}{\Gamma \mid \varphi_1, \ldots, \varphi_m \vDash_D \chi} \quad (y \text{ not in } \varphi_1, \ldots, \varphi_m, \chi)$$

*Proof.* Suppose $Q$ is of HOL type $\tau \to bool$. Then take

$$\forall y :: Q. \psi \ = \ \forall y{:}\tau.\ Q\,y \supset \psi \qquad \text{and} \qquad \exists y :: Q. \psi \ = \ \exists y{:}\tau.\ Q\,y \land \psi$$

The above rules then follow from some easy deductions in the HOL logic. □

**Proposition 4.12** *Dependent logic has separation of the following form.*

$$\frac{\Gamma, y :: Q \vDash_D \psi :: Bool}{\Gamma \vDash_D \{y :: Q \mid \psi\} :: DType} \text{(sep)} \qquad \frac{\Gamma \vDash_D a :: Q \quad \Gamma \vDash_D \psi[a/x]}{\Gamma \vDash_D a :: \{y :: Q \mid \psi\}}$$

$$\frac{\Gamma, y :: Q \vDash_D \psi :: Bool}{\Gamma, z :: \{y :: Q \mid \psi\} \vDash_D z :: Q} \qquad \frac{\Gamma, y :: Q \mid \varphi_1, \ldots, \varphi_n, \psi \vDash_D \chi}{\Gamma, z :: \{y :: Q \mid \psi\} \mid \varphi_1[z/y], \ldots, \varphi_n[z/y] \vDash_D \chi[z/y]}$$

*Proof.* Take $\{y :: Q \mid \psi\} = \lambda y.\ Q\,y \land \psi$. □

**Example 4.13** (i) The above rule (sep) is quite useful for introducing new types. For example one can make *Nat[n]* an abbreviation for the type introduced by

$$\frac{n :: Nat, m :: Nat \vDash_D m < n :: Bool}{n :: Nat \vDash_D \{m :: Nat \mid m < n\} :: Dtype}$$

and the interval $Int[m,n] = \{m, m+1, \ldots, n\}$ is defined by

$$\frac{n :: Nat, m :: Nat[n+1], k :: Nat[n+1] \vDash_D m \leq k :: Bool}{n :: Nat, m :: Nat[n+1] \vDash_D \{k :: Nat[n+1] \mid m \leq k\} :: DType}$$

(ii) There are dependent equality predicates $=_P$ for dependent types $P$ for which one has rules in both directions

$$\frac{\Gamma \vDash_D a = b :: P}{\Gamma \mid \emptyset \vDash_D a =_P b} \qquad \frac{\Gamma \mid \emptyset \vDash_D a =_P b}{\Gamma \vDash_D a = b :: P}$$

where, for clarity, the empty logical context is written explicitly. One simply defines $a =_P b$ to be '$P\,a \land P\,b \land a = b$'.

**Remark 4.14** The above dependent logic with sequents $\Gamma \mid \varphi_1, \ldots, \varphi_m \vDash_D \psi$ corresponds to the internal language of a topos as described in detail in [14]. Because our language is based on the classical logic of HOL, we get the language of a *boolean* topos—that is, of a topos with classical logic.

## 4.4 Translation of some additional type constructors

In this section it will be shown how some extra features of HOL extend to the translated DTT. First, we mention that a consequence of theorem 4.10 is the following (expected) result.

**Proposition 4.15** *The type Nat $= D(\text{num})$ is a natural numbers object in DTT, i.e. it comes with the following rules*

$$\dfrac{}{\vdash_{\text{D}} 0 :: \text{Nat}} \qquad \dfrac{\Gamma \vdash_{\text{D}} n :: \text{Nat}}{\Gamma \vdash_{\text{D}} \text{suc}\, n :: \text{Nat}}$$

$$\dfrac{\Gamma \vdash_{\text{D}} a :: Q[0/z] \quad \Gamma, x :: \text{Nat}, y :: Q[x/z] \vdash_{\text{D}} b :: Q[\text{suc}\, x/z]}{\Gamma, n :: \text{Nat} \vdash_{\text{D}} \text{R}_{x,y}(n,a,b) :: Q[n/z]}$$

*where $\text{R}_{x,y}$ is a constant which binds the variables $x,y$ in $b$. It satisfies*

$$\begin{aligned} \text{R}_{x,y}(0,a,b) &= a \\ \text{R}_{x,y}(\text{suc}\, n,a,b) &= b[n/x, \text{R}_{x,y}(n,a,b)/y] \end{aligned}$$

*These are the rules for dependent naturals as described in [11].*

**Proof.** Assume $\Gamma = x_1 :: P_1, \ldots, x_n :: P_n$ and $Q : \sigma \to \text{bool}$ where the variable $z :: \text{Nat}$ in $Q$ is one of the $x_i$. The validity of the two introduction rules follows from (i) in theorem 4.10. For the elimination rule we make use the following HOL theorem, see [3] or [12].

$$\vdash_{\text{H}} \forall x f.\, \exists! g.\, (g\, 0 = x) \wedge \forall n.\, g(\text{suc}\, n) = f\,(g\, n)\, n$$

Applying this to $a$ and $\lambda xy.b$, we obtain the existence of a term $g : \text{num} \to \sigma$ with

$$\vdash_{\text{H}} g\, 0 = a \qquad \text{and} \qquad \vdash_{\text{H}} g(\text{suc}\, n) = b[n/x,\, g\, n/y]$$

From the premises

$$P_1\, x_1, \ldots, P_n\, x_n \vdash_{\text{H}} Q[0/x]\, a \qquad \text{and} \qquad P_1\, x_1, \ldots, P_n\, x_n, Q[x/z]\, y \vdash_{\text{H}} Q[\text{suc}\, x/z]\, b$$

one obtains by an induction proof in HOL that $P_1\, x_1, \ldots, P_n\, x_n \vdash_{\text{H}} Q[n/z]\,(g\, n)$. So if we put $\text{R}_{x,y}(n,a,b) = g\, n$, then $\Gamma, n :: \text{Nat} \vdash_{\text{D}} \text{R}_{x,y}(n,a,b) :: Q[n/z]$. □

As discussed in [12], a coproduct (or sum) type $\sigma + \tau$ with the following introduction and elimination rules is definable in HOL.

$$\dfrac{M : \sigma}{\text{inl}\, M : \sigma + \tau} \qquad \dfrac{N : \tau}{\text{inr}\, N : \sigma + \tau} \qquad \dfrac{L : \sigma + \tau \quad P[x] : \rho \quad Q[y] : \rho}{\text{case}_{x,y}(L, P, Q) : \rho}$$

where the variables $x : \sigma$ in $P$ and $y : \tau$ in $Q$ become bound in $\text{case}_{x,y}(L,P,Q)$. The associated conversions are

$$\begin{aligned} \text{case}_{x,y}(\text{inl}\, M, P, Q) &= P[M/x] \\ \text{case}_{x,y}(\text{inr}\, N, P, Q) &= Q[N/y] \\ \text{case}_{x,y}(L, R[(\text{inl}\, x)/z], R[(\text{inr}\, y)/z]) &= R[L/z] \end{aligned}$$

**Proposition 4.16** *Coproducts can be defined in (our version of)* DTT *with rules*

$$\frac{\Gamma \vdash_{\mathrm{D}} P :: DType \quad \Gamma \vdash_{\mathrm{D}} Q :: DType}{\Gamma \vdash_{\mathrm{D}} P + Q :: DType}$$

$$\frac{\Gamma \vdash_{\mathrm{D}} a :: P}{\Gamma \vdash_{\mathrm{D}} \mathsf{inl}\, a :: P + Q} \qquad \frac{\Gamma \vdash_{\mathrm{D}} b :: Q}{\Gamma \vdash_{\mathrm{D}} \mathsf{inr}\, b :: P + Q}$$

$$\frac{\Gamma, x :: P \vdash_{\mathrm{D}} c :: R \quad \Gamma, y :: Q \vdash_{\mathrm{D}} d :: R}{\Gamma, z :: P + Q \vdash_{\mathrm{D}} \mathsf{case}_{x,y}(z, c, d) :: R} \text{ (with } x, y \text{ not in } R\text{)}$$

*Moreover, the embedding* $D : \mathrm{HOL} \to \mathrm{DTT}$ *preserves coproducts.*

**Proof.** Assume $\Gamma = x_1 :: P_1, \ldots, x_n :: P_n$ with $P : \sigma \to bool$ and $Q : \tau \to bool$. Define

$$P + Q \;=\; \lambda z.\, (\exists x{:}\sigma.\; z = \mathsf{inl}\, x \;\wedge\; P\, x) \;\vee\; (\exists y{:}\tau.\; z = \mathsf{inr}\, y \;\wedge\; Q\, y) \;:\; (\sigma + \tau) \to bool$$

The above introduction rules are then clearly valid. For the elimination rule we deduce in HOL from the first premisse

$$\frac{\dfrac{\dfrac{P_1\, x_1, \ldots, P_n\, x_n, P\, x \vdash_{\mathrm{H}} R\, c}{P_1\, x_1, \ldots, P_n\, x_n, (z = \mathsf{inl}\, x \;\wedge\; P\, x) \vdash_{\mathrm{H}} R\, c}}{P_1\, x_1, \ldots, P_n\, x_n, (z = \mathsf{inl}\, x \;\wedge\; P\, x) \vdash_{\mathrm{H}} R(\mathsf{case}_{x,y}(z, c, d))}}{P_1\, x_1, \ldots, P_n\, x_n, \exists x{:}\sigma.\; (z = \mathsf{inl}\, x \;\wedge\; P\, x) \vdash_{\mathrm{H}} R(\mathsf{case}_{x,y}(z, c, d))}$$

In a similar way one obtains from the second premisse a theorem

$$P_1\, x_1, \ldots, P_n\, x_n, \exists y{:}\tau.\; (z = \mathsf{inl}\, y \;\wedge\; Q\, y) \vdash_{\mathrm{H}} R(\mathsf{case}_{x,y}(z, c, d))$$

By unfolding the definition of $P + Q$ and using $\vee$-elimination in HOL one gets the theorem

$$P_1\, x_1, \ldots, P_n\, x_n, (P + Q)\, z \vdash_{\mathrm{H}} R(\mathsf{case}_{x,y}(z, c, d))$$

That is, one has in DTT that $\Gamma, z :: P + Q \vdash_{\mathrm{D}} \mathsf{case}_{x,y}(z, c, d) :: R$.

The embedding $D : \mathrm{HOL} \to \mathrm{DTT}$ preserves coproducts because for HOL types $\sigma, \tau$ one has

$$\begin{aligned}
\vdash_{\mathrm{H}} (D(\sigma) + D(\tau))\, z &= (\exists x{:}\sigma.\; z = \mathsf{inl}\, x \;\wedge\; D(\sigma)\, x) \;\vee\; (\exists y{:}\tau.\; z = \mathsf{inr}\, y \;\wedge\; D(\tau)\, y) \\
&= (\exists x{:}\sigma.\; z = \mathsf{inl}\, x) \;\vee\; (\exists y{:}\tau.\; z = \mathsf{inr}\, y) \\
&\stackrel{*}{=} \mathsf{T} = D(\sigma + \tau)\, z.
\end{aligned}$$

in which the equivalence $\stackrel{*}{=}$ holds by the following argument. For $z : \sigma + \tau$ abbreviate

$$R[z] \;=\; (\exists x{:}\sigma.\; z = \mathsf{inl}\, x) \;\vee\; (\exists y{:}\tau.\; z = \mathsf{inr}\, y)$$

Then, using the fact that the HOL types $\sigma$ and $\tau$ are non-empty we have $\vdash_{\mathrm{H}} R[(\mathsf{inl}\, x)/z] = \mathsf{T}$ and $\vdash_{\mathrm{H}} R[(\mathsf{inr}\, y)/z] = \mathsf{T}$. And thus

$$\begin{aligned}
R[z] &= \mathsf{case}_{x,y}(z,\, R[(\mathsf{inl}\, x)/z],\, R[(\mathsf{inr}\, y)/z]) \\
&= \mathsf{case}_{x,y}(z,\, \mathsf{T},\, \mathsf{T}) \\
&= \mathsf{case}_{x,y}(z,\, \mathsf{T}[(\mathsf{inl}\, x)/z],\, \mathsf{T}[(\mathsf{inr}\, y)/z]) \\
&= \mathsf{T}[z/z] = \mathsf{T}. \qquad \Box
\end{aligned}$$

Preliminary work indicates that all the type constructors definable using the HOL system's recursive types package [12] (e.g. the polymorphic list type $(\alpha)\mathit{list}$) can also be extended to DTT in a uniform and straightforward way. The details have, however, been left for future work.

# 5 Examples

In this section we describe three illustrative examples. The first one concerns the typical difference between programming in simple type theory and in dependent type theory. The second one involves the dependent logic described in section 4.3 and focusses on the use of the dependent and logical contexts. The third example gives some details of the proof of the pigeon hole principle in dependent logic.

Let's return to the first example in section 3.2. Consider the types of dates

$$date_1 = Nat \times Nat \times Nat$$
$$date_2 = \Sigma_{y::Nat}. \Sigma_{m::Nat[12]}. Nat[\text{length of month } m \text{ in year } y]$$

together with functions

$$\text{dayadder}_i : Nat \times date_i \to date_i$$

which add a number of days to a given date (for $i = 1, 2$). The rather complicated precise form of such functions is not of much interest at this point. What we do want to emphasize is that the well-typedness of dayadder$_1$ is a trivial matter, whereas proving the well-typedness of dayadder$_2$ involves showing that it yields a triple whose components lie in the appropriate range (viz. in $Nat[12]$ and $Nat[\text{length of month } m \text{ in year } y]$). One of the main advantages of typing in general is that it provides partial specification. We conclude that functions that are well-typed in dependent type theory are more likely to be correct than functions that are well-typed in simple type theory.

Of course there is a price to pay; in DTT a type of a term must be provided by the programmer, together with a proof of well-typedness. This becomes particularly clear in our implementation, where such a proof is done in HOL. In contrast, type information in simple type theory can be inferred automatically.

Once it has been shown in our version of DTT that a term is well-typed, the typing is preserved even though the term itself is transformed by deductions in the underlying HOL logic (for example, by $\beta$-reduction). Thus there is a similarity—but on a different level—with functional languages whose programs are implemented as untyped combinators. Running a well-typed program means running it as a combinator term which is stripped of all type information. Similarly a well-typed DTT term can be run as a HOL term (i.e. term of the simply typed $\lambda$-calculus) in the underlying HOL logic.

The second example involves the logic of our dependent type theory. We introduce

$$\frac{\Gamma \vdash_{\overline{D}} P :: DType}{\Gamma \vdash_{\overline{D}} \mathcal{A}s(P) :: DType}$$

where $\mathcal{A}s(P)$ is an abbreviation defined by

$$\mathcal{A}s(P) = \Sigma_{m::P \to P \to P}. \Sigma_{u::P}. \Sigma_{v::P}. \Pi_{x::P}. \mathsf{Eq}_P(m\,u\,x, x) \wedge \mathsf{Eq}_P(m\,x\,v, x)$$

This type expresses the property that $P$ carries an applicative structure with left and right units. For an element $z :: \mathcal{A}s(P)$ we abbreviate

$$\mathsf{m}\,z = \pi\,z, \qquad \mathsf{lu}\,z = \pi\,(\pi'\,z), \qquad \text{and} \qquad \mathsf{ru}\,z = \pi\,(\pi'\,(\pi'\,z))$$

for the multiplication, left unit, and right unit involved. The following formula states that such an applicative structure $z :: \mathcal{A}s(P)$ is commutative.

$$\mathsf{comm}\,z = \forall x :: P.\,\forall y :: P.\,\mathsf{m}\,z\,x\,y =_P \mathsf{m}\,z\,y\,x$$

An easy result in our logic is the theorem

$$\Gamma, z :: \mathcal{A}s(P) \mid \text{comm}\, z \vdash_{\overline{D}} \text{lu}\, z =_p \text{ru}\, z$$

which states a basic fact about such applicative structures.

This example illustrates the interaction between declarations in the *dependent* context of a sequent (the part before the separator '|') and the formulas in the *logical* context (the part after '|'). Similarly, one can define a type $\mathcal{G}(P)$ expressing that $P$ carries a group structure. If all the theorems one proves in dependent logic about a group $z :: \mathcal{G}(P)$ involve the logical assumption comm $z$ stating that $z$ is commutative, then it is better to put this requirement into the dependent context by assuming $z :: \mathcal{AG}(P)$, where $\mathcal{AG}(P)$ means that $P$ carries an *abelian* group structure.

The third example is about the so-called pigeon hole principle. It says that if you distribute $n+2$ items over $n+1$ pigeon holes (for $n \geq 0$), then there is at least one pigeon hole containing two items. In a more mathematical formulation, it says that there is no injective function $Nat[n+2] \to Nat[n+1]$:

$$n :: Nat, f :: Nat[n+2] \to Nat[n+1] \mid \text{injective}\, f \vdash_{\overline{D}} \mathsf{F}$$

where F denotes false. The principle can be proved by induction on $n$. We reason semi-informally in dependent logic and use the following lemma.

**Lemma 5.1** *For $n :: Nat$ define the function $g_n :: Nat[n+2] \to Nat[n+2] \to Nat[n+1]$ by*

$$g_n = \lambda m :: Nat[n+2].\, \lambda k :: Nat[n+2].\, \text{if}\ k{<}m\ \text{then}\ k\ \text{elseif}\ k{>}m\ \text{then}\ k{-}1\ \text{else}\ 0\ \text{fi}$$

*Then one has*

$$n :: Nat, m, k_1, k_2 :: Nat[n+2] \mid m \neq k_1, m \neq k_2 \vdash_{\overline{D}} g_n\, m\, k_1 = g_n\, m\, k_2 \supset k_1 = k_2$$

*Proof.* Distinguish the cases

- $k_1 < m, k_2 < m$. Then $k_1 = g_n\, m\, k_1 = g_n\, m\, k_2 = k_2$.
- $k_1 > m, k_2 > m$. Then $k_1 = (g_n\, m\, k_1) + 1 = (g_n\, m\, k_2) + 1 = k_2$.
- $k_1 < m, k_2 > m$. This is impossible, since it yields $m > g_n\, m\, k_1 = g_n\, m\, k_2 \geq m$.
- $k_1 > m, k_2 < m$. As before. □

We now give the induction proof.
Case $n = 0$. Assume $f :: Nat[2] \to Nat[1]$. Then $f\, 0 = f\, 1 :: Nat[1]$, which, together with the assumption injective $f$, yields the desired contradiction.
Case $n > 0$. Assume $f :: Nat[n+3] \to Nat[n+2]$ and put $m = f\,(n+2) :: Nat[n+2]$. Then by injective $f$ one has $k :: Nat[n+2] \vdash_{\overline{D}} f\, k \neq m$. Define $f' = \lambda k :: Nat[n+2].\, g_n\, m\,(f\, k)$ where $g_n$ is the function defined in the above lemma. We claim that $f' :: Nat[n+2] \to Nat[n+1]$ is injective; for $k_1, k_2 :: Nat[n+2]$ one derives using the above lemma

$$\frac{\dfrac{f'\, k_1 = f'\, k_2}{g_n\, m\,(f\, k_1) = g_n\, m\,(f\, k_2)} \qquad f\, k_1 \neq m \qquad f\, k_2 \neq m}{\dfrac{f\, k_1 = f\, k_2}{k_1 = k_2}}$$

Hence the induction hypothesis applied to $f'$ yields the required contradiction.

# 6 Implementation

The HOL system is based on the LCF approach to interactive theorem proving, originally due to Milner [6]. As in LCF, the system is based on the strongly-typed functional programming language ML.[7] Propositions and theorems of the logic are represented by ML abstract data types, and theorem-proving takes place by executing ML programs that operate on values of these data types. Because ML is a programming language, the user can write arbitrarily complex programs to implement proof strategies. Furthermore, because of the way the logic is represented in ML, such user-defined proof strategies are guaranteed to perform only valid logical inferences.

This approach is explained in more detail as follows. ML is a strongly-typed language; all expressions have types, and only consistently-typed expressions are syntactically well-formed. This type discipline is the basis for the soundness of proofs in HOL. The HOL system is built on top of ML by adding an abstract data type thm, values of which are theorems of higher order logic. The only predefined values of type thm are those which correspond to the five axioms of higher order logic listed in section 2. Furthermore, the only way of creating new values of type thm is by using certain built-in ML functions that take theorems as arguments and return theorems as results. Each of these corresponds to one of the primitive inference rules of the logic and returns only theorems that are deducible using this rule. Any value of type thm obtained in HOL must therefore be either an axiom or have been generated using the functions that represent the primitive inference rules of the logic—i.e. the only way to generate a theorem is to prove it.

In addition to the primitive inference rules, there are many derived inference rules available in the HOL system. These are ML procedures that perform commonly-used sequences of primitive inferences by calling the appropriate sequence of ML functions. Derived inference rules allow the user of HOL to do proofs in bigger steps, omitting explicit mention of all the necessary primitive inferences. The ML code for a derived rule can be arbitrarily complex, but it will never produce a theorem that does not follow by valid logical inference.

The LCF methodology just described is also the basis for the implementation of our translation of DTT into higher order logic. The judgements of DTT are represented by the values of certain abstract data types in ML, and these are defined so that the only admissible operations over them are ones that correspond to valid inferences of dependent type theory. The table below gives a sketch of the ML data types involved. The ML type term in the third column is a predefined HOL type whose elements are the well-typed terms of higher order logic.

| Judgement | ML type | Representation |
|---|---|---|
| $\Gamma \vdash_D A :: DType$ | dtype | (term)list × term |
| $\Gamma \vdash_D A = B :: DType$ | dthm | thm × dtype × dtype |
| $\Gamma \vdash_D a :: A$ | dthm | thm × dtype |
| $\Gamma \vdash_D a = b :: A$ | dthm | thm × dthm × dthm |

The left column of this table lists the four forms of judgement in DTT, and the middle column shows the names of the ML abstract data types whose values represent these judgements. In ML, one defines an abstract data type by giving its values appropriate representations in an already existing data type. The third column of the table shows the representing types used in defining the abstract data types dtype and dthm. As was pointed out in section 3.3, the definitions of these two data types must be recursive, since the DTT rules for types and terms depend on each other.

---

[7]Not Standard ML, but an earlier version of the language; see [5].

As can be seen from this table, the HOL implementation of dependent type theory closely follows the interpretations of the four forms of DTT judgement given by definition 4.2. For example, a type-in-context $\Gamma \vdash_D A :: Dtype$ (i.e. an element of dtype) is represented by a pair consisting of a list of terms representing the context $\Gamma$ together with a term representing the type $A$. This corresponds directly to the pseudo-context and predicate in the first clause of definition 4.2. Likewise, an inhabitation judgement $\Gamma \vdash_D a :: A$ is represented by a type in context together with the appropriate HOL theorem (i.e. an element of thm).

Once the representation for an ML abstract data type has been specified, operations on the values of the abstract data type can then be defined in terms of corresponding operations on this representation. In ML, this is done in such a way that both the representation and the operations over it are hidden once the definition of an abstract type is completed, so that only the abstract values and operations are then available to the user. In the LCF approach to theorem proving, where abstract data types represent the judgements of a logic, these abstract operations correspond to primitive inference rules and are the only means of constructing judgements.

Our implementation of DTT also follows this general approach, with one important modification. In HOL there is a small fixed set of primitive inference rules, and so these can be taken as the only operations of the abstract type thm of HOL theorems. The set of inference rules for DTT, however, is rather open-ended; one often extends the system by postulating new rules for additional type constructors. We have therefore defined the operations on our abstract type dthm not to be inference rules, but rather mappings into the interpretation of DTT judgements in higher order logic. This allows new rules to be added just by programming new *derived* rules, rather than modifying the definition of the abstract type itself.[8]

For inhabitation judgements $\Gamma \vdash_D a :: A$, for example, the definition of dthm provides a single operation, namely the ML function

    INHAB : dtype → thm → dthm

This function is expected to be applied to an element of dtype

$$x_1 :: P_1, \ldots, x_n :: P_n \vdash_D A :: Dtype$$

representing a dependent type in context and a corresponding HOL theorem

$$P_1\, x_1, \ldots, P_n\, x_n \vdash_H A\, a$$

where all the free variables in $a$ occur in $\{x_1, \ldots, x_n\}$. When applied to these values, the ML function INHAB produces an element of dthm that represents the inhabitation judgement

$$x_1 :: P_1, \ldots, x_n :: P_n \vdash_D a :: A$$

This is just a direct implementation of clause (ii) of definition 4.2, which gives the interpretation of inhabitation judgements from DTT in the HOL logic. Similar mappings into dthm are defined for type equality judgements $\Gamma \vdash_D A = B :: Dtype$ and term equality judgements $\Gamma \vdash_D a = b :: A$. Note that a run-time error occurs if any of these functions is applied to inappropriate arguments; this ensures that dthm contains only judgements that are valid under our interpretation.

The implementation also provides functions that map DTT judgements (i.e. elements of dthm) to their interpretations. For example, for inhabitation judgements we have ML functions

    HOL : dthm → thm     and     DTYPE : dthm → dtype

---
[8]This is in fact not quite true; see below.

which extract from an inhabitation judgement its HOL theorem and dependent type components, respectively. Similar functions are provided for type equality and term equality judgements. These functions, together with the mappings into the interpretation explained above, allow us to program derived inference rules for DTT in the system. For example, the derivation given in section 4.2 for the dependent product rule

$$\frac{\Gamma, x :: A \vdash_{\overline{D}} b :: B}{\Gamma \vdash_{\overline{D}} \lambda x :: A.\, b :: \Pi_{x::A}.\, B} \quad (\Pi\text{-}I)$$

is implemented by an ML function

```
PRODUCT_INTRO : dthm → dthm
```

that extracts the HOL component of the judgement above the line, carries out the little HOL proof shown in the proof of lemma 4.5, and then injects the result back into the type dthm to get the judgement below the line.

The scheme described above applies only to the definition of the abstract type dthm, whose elements are the type equality, inhabitation, and term equality judgements of DTT. Judgements of the remaining form, namely dependent types in context, are represented by elements of the abstract type dtype. In contrast to the approach taken to defining dthm, the abstract type dtype is defined so that its operations form a set of 'primitive inference rules' for inferring judgements of the form $\Gamma \vdash_{\overline{D}} A :: Dtype$. In this case, we have chosen *not* to make available direct mappings into the interpretation, since otherwise any HOL predicate (in a suitable context) could become a dependent type. Instead, we wish dtype to contain only judgements that follow from explicitly-stated type formation rules together with the context rules in section 3.3.

Note that the construction of an element of dthm always requires an element of dtype, so by adopting the method just described we are also suitably restricting the dthm judgements we can generate to ones that involve only the dependent types we choose to make available. On the other hand, this scheme also means that we must extend the ML definition of dtype with new type formation rules whenever we wish to add a new type constructor; the rules cannot just be derived from existing ones. Furthermore, derived rules for dthm must carry out actual proofs of the required dtype judgements, in addition to doing the HOL proofs involved.

The ML types and functions described above form only the most primitive basis required for a theorem prover for DTT. To make the system practical, a very considerable infrastructure will have to be built on top of this basis—this still remains to be done. Future work will also involve an investigation of how best to mix reasoning in both DTT and HOL.

# References

[1] A. Church, 'A Formulation of the Simple Theory of Types', *The Journal of Symbolic Logic*, vol. 5 (1940), pp. 56–68.
[2] A. Felty, 'Encoding Dependent Types in an Intuitionistic Logic', in *Logical Frameworks*, edited by G. Huet and G. Plotkin (Cambridge University Press, 1991), pp. 215–251.
[3] M. Gordon, 'HOL: A Machine Oriented Formulation of Higher Order Logic', Technical Report 68, Computer Laboratory, University of Cambridge, revised version (July 1985).
[4] M. J. C. Gordon, 'HOL: A Proof Generating System for Higher-Order Logic', in *VLSI Specification, Verification and Synthesis*, edited by G. Birtwistle and P. A. Subrahmanyam, Kluwer International Series in Engineering and Computer Science (Kluwer, 1988), pp. 73–128.

[5] M. J. C. Gordon and T. F. Melham (eds.), *Introduction to HOL: A theorem proving environment for higher order logic*, forthcoming book (Cambridge University Press, 1993).

[6] M. J. Gordon, A. J. Milner, and C. P. Wadsworth, *Edinburgh LCF: A Mechanised Logic of Computation*, Lecture Notes in Computer Science, vol. 78 (Springer-Verlag, 1979).

[7] F. K. Hanna, N. Daeche, and M. Longley, 'Specification and Verification Using Dependent Types', *IEEE Transactions on Software Engineering*, vol. 16, no. 9 (September 1990), pp. 949–964.

[8] B. P. F. Jacobs, *Categorical Type Theory* (Ph.D. dissertation, University of Nijmgen, 1991).

[9] M. Leeser, 'Using Nuprl for the verification and synthesis of hardware', in *Mechanized Reasoning and Hardware Design: a Discussion Meeting held at the Royal Society, October 1991*, edited by C. A. R. Hoare and M. J. C. Gordon, Prentice Hall International Series in Computer Science (Prentice Hall, 1992), pp. 49–68.

[10] A. C. Leisenring, *Mathematical Logic and Hilbert's $\varepsilon$-Symbol*, University Mathematical Series (Macdonald & Co., 1969).

[11] P. Martin-Löf, *Intuitionistic Type Theory* (Bibliopolis, Naples, 1984).

[12] T. F. Melham, 'Automating Recursive Type Definitions in Higher Order Logic', in *Current Trends in Hardware Verification and Automated Theorem Proving*, edited by G. Birtwistle and P. A. Subrahmanyam (Springer-Verlag, 1989), pp. 341–386.

[13] B. Nordström, K. Petersson, and J. M. Smith, *Programming in Martin-Löf's Type Theory: An Introduction*, International Series of Monographs on Computer Science 7 (Oxford University Press, 1990).

[14] W. Phoa, 'An introduction to fibrations, topos theory, the effective topos and modest sets', Technical Report ECS-LFCS-92-208, LFCS, Department of Computer Science, University of Edinburgh (April 1992).

[15] G. Sundholm, 'Proof Theory and Meaning', in *Alternatives in Classical Logic*, vol. 3 of *Handbook of Philosophical Logic*, edited by D. Gabbay and F. Guenthner, 4 vols. (D. Reidel, 1983–9), pp. 471–506.

[16] A. S. Troelstra, 'On the Syntax of Martin-Löf's Theories', *Theoretical Computer Science*, vol. 51, nos. 1–2 (1987), pp. 1–26.

[17] A. S. Troelstra and D. van Dalen, *Constructivism in Mathematics: An Introduction*, Studies in Logic and the Foundations of Mathematics, 2 vols. (North Holland, 1988).

# Studying the Fully Abstract Model of PCF within its Continuous Function Model

Achim Jung[1] and Allen Stoughton[2]

**Abstract.** We give a concrete presentation of the inequationally fully abstract model of PCF as a continuous projection of the inductively reachable subalgebra of PCF's continuous function model.

## 1 Introduction

As is well known, the continuous function model $\mathcal{E}$ of the applied typed lambda calculus PCF fails to be inequationally fully abstract [Plo77], but PCF has a unique inequationally fully abstract, order-extensional model $\mathcal{F}$ [Mil77, Sto90], where models are required to interpret the ground type $\iota$ as the flat cpo of natural numbers. Two attempts at finding connections between $\mathcal{E}$ and $\mathcal{F}$ have been made in the literature.

Mulmuley's idea was to connect complete lattice versions of $\mathcal{E}$ and $\mathcal{F}$ [Mul87]. Using a syntactically defined inductive (inclusive) predicate, he defines an inequationally fully abstract, order-extensional model $\mathcal{F}'$ as the image of a continuous closure (retraction that is greater than the identity function) of the complete lattices version $\mathcal{E}'$ of the continuous function model. The use of complete lattices is essential in this construction, and, e.g., parallel or is mapped to ⊤. Very pleasingly, $\mathcal{F}'$ inherits both its ordering relation and function application operation from $\mathcal{E}'$. Thus some of PCF's operations must be sent by the closure to strictly greater functions. Although the closure isn't a homomorphism between $\mathcal{E}'$ and $\mathcal{F}'$ (since it doesn't preserve application in general), it does preserve the meaning of terms. $\mathcal{F}'$ isn't a combinatory algebra, since all functions of $\mathcal{F}'$ preserve ⊤, and thus the usual axiom for the K combinator cannot hold. Finally, Mulmuley is able to recover $\mathcal{F}$ from $\mathcal{F}'$ simply by removing ⊤ at all types.

A more algebraic connection between $\mathcal{E}$ and $\mathcal{F}$ was subsequently developed by the second author [Sto88]. Here one begins by forming the inductively reachable subalgebra $R(\mathcal{E})$ of $\mathcal{E}$, which in this case simply consists of those elements of $\mathcal{E}$ that are lub's of directed sets of denotable elements. $R(\mathcal{E})$ is then continuously quotiented by a syntactically defined inductive pre-ordering, producing $\mathcal{F}$. Furthermore, in contrast to the situation with $\mathcal{E}'$ and $\mathcal{F}'$ above, there is a continuous homomorphism from $R(\mathcal{E})$ to $\mathcal{F}$.

---

[1] Fachbereich Mathematik, Technische Hochschule Darmstadt, Schloßgartenstraße 7, D-6100 Darmstadt, Germany, e-mail: jung@mathematik.th-darmstadt.de.
[2] School of Cognitive and Computing Sciences, University of Sussex, Falmer, Brighton BN1 9QH, UK, e-mail: allen@cogs.sussex.ac.uk.

The purpose of this paper is to give a concrete presentation of this construction of $\mathcal{F}$ from $R(\mathcal{E})$. We define a model $N(\mathcal{E})$ as the image of a syntactically defined continuous projection over $R(\mathcal{E})$, and show that $N(\mathcal{E})$ is inequationally fully abstract and order-extensional, and is thus order-isomorphic to $\mathcal{F}$. As in Mulmuley's construction, the ordering relation of $N(\mathcal{E})$ is inherited from $\mathcal{E}$ (and $R(\mathcal{E})$). On the other hand we prove that the application operation of $N(\mathcal{E})$ cannot be inherited from $\mathcal{E}$. There is, of course, a continuous homomorphism from $R(\mathcal{E})$ to $N(\mathcal{E})$.

In the final section of the paper, we consider the relationship between the full abstraction problem, lambda definability and our presentation of $\mathcal{F}$, and propose a minimal condition that any "solution" to the full abstraction problem should satisfy.

## 2   Background

The reader is assumed to be familiar with such standard domain-theoretic concepts as (directed) complete partial orders (cpo's), (directed) continuous functions, and $\omega$-algebraic, strongly algebraic (SFP) and consistently complete cpo's.

If $X$ is a subset of a poset $P$, then we write $\bigsqcup X$ and $\bigsqcap X$ for the lub and glb, respectively, of $X$ in $P$, when they exist. We abbreviate $\bigsqcup\{x,y\}$ and $\bigsqcap\{x,y\}$ to $x \sqcup y$ and $x \sqcap y$, respectively. We write $\omega_\perp$ for the flat cpo of natural numbers. Given cpo's $P$ and $Q$, we write $P \xrightarrow{c} Q$ for the cpo of all continuous functions from $P$ to $Q$, ordered pointwise. A cpo $P$ is a *subcpo* of a cpo $Q$ iff $P \subseteq Q$, $\sqsubseteq_P$ is the restriction of $\sqsubseteq_Q$ to $P$, $\perp_P = \perp_Q$ and $\bigsqcup_P D = \bigsqcup_Q D$ for all directed $D \subseteq P$. A pre-ordering $\leq$ over a cpo $P$ is *inductive* iff $\sqsubseteq_P \subseteq \leq$ and, whenever $D$ is a directed set in $\langle P, \sqsubseteq_P \rangle$ and $p$ is an ub of $D$ in $\langle P, \leq \rangle$, the lub of $D$ in $\langle P, \sqsubseteq_P \rangle$ is $\leq p$.

In the remainder of this section, we briefly recall the definitions and results from [Sto88] that will be required in the sequel.

The reader is assumed to be familiar with many-sorted signatures $\Sigma$ over sets of sorts $S$, as well as algebras over such signatures, i.e., $\Sigma$-algebras. Signatures are assumed to contain distinguished constants $\Omega_s$ at each sort $s$, which intuitively stand for divergence. Many operations and concepts extend naturally from sets to $S$-indexed families of sets, in a pointwise manner. For example, if $A$ and $B$ are $S$-indexed families of sets, then a function $f: A \to B$ is an $S$-indexed family of functions $f_s: A_s \to B_s$. We will make use of this and other such extensions without explicit comment. We use uppercase script letters ($\mathcal{A}$, $\mathcal{B}$, etc.) to denote algebras and the corresponding italic letters ($A$, $B$, etc.) to stand for their carriers.

We write $\mathcal{T}_\Sigma$ (or just $\mathcal{T}$) for the initial (term) algebra, so that $T_s$ is the set of terms of sort $s$. Given an algebra $\mathcal{A}$ and a term $t$ of sort $s$, $[\![t]\!]_\mathcal{A}$ (or just $[\![t]\!]$) is the meaning of $t$ in $A_s$, i.e., the image of $t$ under the unique homomorphism from $\mathcal{T}$ to $\mathcal{A}$. Sometimes we write $t_\mathcal{A}$ (or even just $t$) for $[\![t]\!]_\mathcal{A}$.

An algebra is *reachable* iff all of its elements are denotable (definable) by terms. A pre-ordering over an algebra is *substitutive* iff it is respected by all of the operations of that algebra. Substitutive equivalence relations are called *congruences*, as usual. The congruence over $\mathcal{T}$ that is induced by an algebra $\mathcal{A}$ is called $\approx_\mathcal{A}$: two terms are congruent when they are mapped to the same element of $A$. When we say that $c[v_1, \ldots, v_n]$ is a *derived operator* of type $s_1 \times \cdots \times s_n \to s'$, this means that $c$ is a context of sort $s'$ over context variables $v_i$ of sort $s_i$. We write $c_\mathcal{A}$ for the corresponding *derived operation* over an algebra $\mathcal{A}$.

The reader is also assumed to be familiar with *ordered algebras*, i.e., algebras $\mathcal{A}$ whose carriers are $S$-indexed families of posets $A_s = \langle A_s, \sqsubseteq_s \rangle$ with least elements $\perp_s$ denoted by the $\Omega_s$ constants, and whose operations are monotone functions. Such an algebra is called *complete* when its carrier is a cpo and operations are continuous. A homomorphism over complete ordered algebras is called *continuous* when it is continuous on the underlying cpo's. We write $\mathcal{OT}_\Sigma$ (or just $\mathcal{OT}$) for the initial ordered algebra, which consists of $\mathcal{T}$ with the "$\Omega$-match" ordering: one term is less than another when the second can be formed by replacing occurrences of $\Omega$ in the first by terms. The substitutive pre-ordering over $\mathcal{T}$ that is induced by an ordered algebra $\mathcal{A}$ is called $\preceq_\mathcal{A}$: one term is less than another when the meaning of the first is less than that of the second in $\mathcal{A}$.

Given complete ordered algebras $\mathcal{A}$ and $\mathcal{B}$, we say that $\mathcal{A}$ is an *inductive subalgebra* of $\mathcal{B}$ (written $\mathcal{A} \preceq \mathcal{B}$) iff $\mathcal{A}$ is a subalgebra of $\mathcal{B}$ and $A$ is a subcpo of $B$. Given a complete ordered algebra $\mathcal{A}$, we write $R(\mathcal{A})$ for the $\preceq$-least inductive subalgebra of $\mathcal{A}$. Its carrier $R(A)$ is the least set containing all denotable elements and closed under lub's of directed sets; thus we are able to carry out proofs by induction on $R(A)$. A complete ordered algebra $\mathcal{A}$ is *inductively reachable* iff $\mathcal{A} = R(\mathcal{A})$. Hence $R(\mathcal{A})$ is inductively reachable for all complete ordered algebras $\mathcal{A}$. Complete ordered algebras whose carriers are $\omega$-algebraic are inductively reachable iff all of their isolated elements are denotable.

If $\mathcal{A}$ is an algebra and $R$ is a pre-ordering over $A$, then $R$ is *unary-substitutive* iff all unary-derived operations respect $R$: for all derived operators $c[v]$ of type $s \to s'$ and $a, a' \in A_s$, if $a R_s a'$, then $c(a) R_{s'} c(a')$. Unary-substitutive pre-orderings can fail to be substitutive; see Lemma 2.2.27 of [Sto88] and Counterexample 4.7.

If $P \subseteq S$, $\mathcal{A}$ is an algebra and $R$ is a pre-ordering over $A|P$, then $R^c$, the *contextualization* of $R$, is the relation over $A$ defined by: $a R^c_s a'$ iff $c(a) R_p c(a')$, for all derived operators $c[v]$ of type $s \to p$, $p \in P$.

**Lemma 2.1** *If $P \subseteq S$, $\mathcal{A}$ is an algebra and $R$ is a pre-ordering (respectively, equivalence relation) over $A|P$, then $R^c$ is the greatest unary-substitutive pre-ordering (respectively, equivalence relation) over $\mathcal{A}$ whose restriction to $P$ is included in $R$.*

**Proof.** See Lemma 2.2.25 of [Sto88]. □

**Lemma 2.2** *If $P \subseteq S$, $\mathcal{A}$ is a complete ordered algebra and $\leq$ is an inductive pre-ordering over $A|P$, then $\leq^c$ is a unary-substitutive, inductive pre-ordering over $\mathcal{A}$.*

**Proof.** See Lemma 2.3.14 of [Sto88]. □

**Lemma 2.3** (i) *Unary substitutive pre-orderings over reachable algebras are substitutive.*

(ii) *Unary substitutive, inductive pre-orderings over inductively reachable, complete ordered algebras are substitutive.*

**Proof.** See Lemmas 2.2.29 and 2.3.35 of [Sto88]. □

## 3 Syntax and Semantics of PCF

For technical simplicity, we have chosen to work with a combinatory logic version of PCF with a single ground type $\iota$, whose intended interpretation is the natural numbers. From the viewpoint of the conditional operations, non-zero and zero are interpreted as true and false, respectively.

The syntax of PCF is specified by a signature, the sorts of which consist of PCF's types. The set of sorts $S$ is least such that
  (i) $\iota \in S$, and
  (ii) $s_1 \to s_2 \in S$ if $s_1 \in S$ and $s_2 \in S$.
As usual, we let $\to$ associate to the right. Define $s^n$, for $n \in \omega$, by: $s^0 = s$ and $s^{n+1} = s \to s^n$. The signature $\Sigma$ over $S$ has binary (application) operators $\cdot_{s_1,s_2}$ of type $(s_1 \to s_2) \times s_1 \to s_2$ for all $s_1, s_2 \in S$, as well as the following constants (nullary operators) for all $s_1, s_2, s_3 \in S$:
  (i) $\Omega_s$ of sort $s$,
  (ii) $K_{s_1,s_2}$ of sort $s_1 \to s_2 \to s_1$,
  (iii) $S_{s_1,s_2,s_3}$ of sort $(s_1 \to s_2 \to s_3) \to (s_1 \to s_2) \to s_1 \to s_3$,
  (iv) $Y_s$ of sort $s^1 \to s$,
  (v) $n$ of sort $\iota$, for $n \in \omega$,
  (vi) Succ and Pred of sort $\iota^1$, and
  (vii) If$_s$ of sort $\iota \to s^2$.
We usually abbreviate $x \cdot y$ to $x\,y$, and let application associate to the left.

A *model* $\mathcal{A}$ is a complete ordered algebra such that the following conditions hold:
  (i) $A_\iota = \{\bot_\iota, 0_\mathcal{A}, 1_\mathcal{A}, \ldots\}$, where $\bot_\iota \sqsubseteq n_\mathcal{A}$ for all $n \in \omega$ and $n_\mathcal{A}$ and $m_\mathcal{A}$ are incomparable whenever $n \neq m$ (we often confuse $A_\iota$ with $\omega_\bot$ below);
  (ii) For all $x \in A_{s_1}$ and $y \in A_{s_2}$, $K_{s_1,s_2}\,x\,y = x$;
  (iii) For all $x \in A_{s_1 \to s_2 \to s_3}$, $y \in A_{s_1 \to s_2}$ and $z \in A_{s_1}$, $S_{s_1,s_2,s_3}\,x\,y\,z = x\,z\,(y\,z)$;
  (iv) For all $x \in A_{s^1}$, $Y_s\,x$ is the least fixed point of the continuous function over $A_s$ that $x$ represents;
  (v) For all $x \in A_\iota$, Succ $x$ is equal to $\bot$, if $x = \bot$, and is equal to $x + 1$, if $x \in \omega$;
  (vi) For all $x \in A_\iota$, Pred $x$ is equal to $\bot$, if $x = \bot$, is equal to 0, if $x = 0$, and is equal to $x - 1$, if $x \in \omega - \{0\}$; and
  (vii) For all $x \in A_\iota$ and $y, z \in A_s$, If$_s\,x\,y\,z$ is equal to $\bot$, if $x = \bot$, is equal to $y$, if $x \in \omega - \{0\}$, and is equal to $z$, if $x = 0$.

A model $\mathcal{A}$ is *extensional* iff, for all $x_1, x_2 \in A_{s_1 \to s_2}$, if $x_1\,y = x_2\,y$ for all $y \in A_{s_1}$, then $x_1 = x_2$, and *order-extensional* iff, for all $x_1, x_2 \in A_{s_1 \to s_2}$, if $x_1\,y \sqsubseteq x_2\,y$ for all $y \in A_{s_1}$, then $x_1 \sqsubseteq x_2$. Finally, *morphisms* between models are simply continuous homomorphisms between the complete ordered algebras.

Application is left-strict in all models $\mathcal{A}$ since $\bot_{s_1 \to s_2} \sqsubseteq_{s_1 \to s_2} K_{s_2,s_1}\,\bot_{s_2}$, and thus $\bot_{s_1 \to s_2}\,x \sqsubseteq_{s_2} K_{s_2,s_1}\,\bot_{s_2}\,x = \bot_{s_2}$, for all $x \in A_{s_1}$.

The *continuous function model* $\mathcal{E}$ is the unique model $\mathcal{E}$ such that $E_\iota = \omega_\bot$, $E_{s_1 \to s_2} = E_{s_1} \overrightarrow{c} E_{s_2}$ for all $s_1, s_2 \in S$, application is function application and $n_\mathcal{A} = n$ for all $n \in \omega$. $\mathcal{E}$ is clearly order-extensional. The parallel or operation por $\in E_{\iota^2}$ is defined by: por $x\,y = 1$, if $x \in \omega - \{0\}$ or $y \in \omega - \{0\}$, por $x\,y = 0$, if $x = 0$ and $y = 0$, and por $x\,y = \bot$, otherwise.

**Lemma 3.1** *If $\mathcal{A}$ is a model, then so is $R(\mathcal{A})$.*

**Proof.** Follows easily from the fact that $R(\mathcal{A})$ is an inductive subalgebra of $\mathcal{A}$. □

For $s \in S$, we write $I_s$ for the term $S_{s,s^1,s} K_{s,s^1} K_{s,s}$ of sort $s^1$. I is the identity operation in all models. We code lambda abstractions in terms of the S, K and I combinators, in the standard way.

For $s \in S$, define approximations $Y_s^n$ to $Y_s$ of sort $s^1 \to s$ by $Y_s^0 = \Omega_{s^1 \to s}$ and $Y_s^{n+1} = S_{s^1,s,s} I_{s^1} Y_s^n$, so that $Y_s^n$ is an $\omega$-chain in the initial ordered algebra, and thus in all models.

Following [Mil77, BCL85], we can define syntactic projections $\Psi_s^n$ of sort $s^1$, for all $n \in \omega$ and $s \in S$, by $\Psi_\iota^n = Y_\iota^n F$ and $\Psi_{s_1 \to s_2}^n = \lambda x y. \Psi_{s_2}^n(x(\Psi_{s_1}^n y))$, where $F$ of sort $\iota^1 \to \iota^1$ is $\lambda x y.$ If $y$ (Succ$(x($Pred $y))) 0$. Expanding the abstractions, one can see that the $\Psi_s^n$ form an $\omega$-chain in the initial ordered algebra, and thus in all models. Given a model $\mathcal{A}$, we write $A_s^n$ for the subposet of $A_s$ whose elements are $\{\Psi_s^n x \mid x \in A_s\}$. Clearly $A_\iota^n = \{\bot, 0, 1, \ldots, n-1\}$ for all $n \in \omega$.

**Lemma 3.2 (Milner/Berry)** *Suppose $\mathcal{A}$ is an extensional model and $s \in S$. The $\Psi_s^n$ represent an $\omega$-chain of continuous projections with finite image over $A_s$ whose lub is the identify function. Hence $x \in A_s^n$ iff $x = \Psi_s^n x$, $A_s^n \subseteq A_s^m$ whenever $n \le m$, and $A_s$ is a strongly algebraic cpo whose set of isolated elements is $\bigcup_{n \in \omega} A_s^n$.*

**Proof.** The $\Psi_s^n$ obviously represent an $\omega$-chain of continuous functions. Inductions on $S$ suffice to show that they are retractions, have finite image and that their lub is the identity function. But then each $\Psi_s^n$ is less than the identity function. The rest follows easily. □

**Lemma 3.3** *Suppose $\mathcal{A}$ is an extensional model and $s \in S$. The $\Psi_s^n$ also represent an $\omega$-chain of continuous projections with finite image over $R(A)_s$ whose lub is the identity function. Hence $R(A)_s$ is a strongly algebraic cpo and, for all $x \in A_s$,*

(i) *if $x \in R(A)_s$, then $x$ is isolated in $R(A)_s$ iff $x$ is isolated in $A_s$;*

(ii) *if $x \in R(A)_s$ is isolated, then $x$ is denotable; and*

(iii) *$x \in R(A)_s$ iff $\Psi^n x$ is denotable for all $n \in \omega$.*

**Proof.** Everything except (ii) and the "only if" direction of (iii) follows by Lemma 3.2 and the fact that $R(\mathcal{A})$ is an inductive subalgebra of $\mathcal{A}$. (ii) follows by induction on $R(A)_s$, and the "only if" direction of (iii) follows from (ii). □

We write por$^2$ for $\Psi^2$ por. It is easy to see that por$^2$ and por are interdefinable elements of $E_{\iota^2}$.

Let the equality test Eq of sort $\iota^2$ be

$$Y(\lambda z x y. \text{If } x \,(\text{If } y \,(z(\text{Pred } x)(\text{Pred } y)) \, 0) \,(\text{Not } y)),$$

where Not of sort $\iota^1$ is $\lambda x.$ If $x \, 0 \, 1$.

For $n \in \omega$, define operators And$_n$ of sort $\iota^n$ by: And$_0 = 1$ and

$$\text{And}_{n+1} = \lambda x y_1 \cdots y_n. \text{If } x \,(\text{And}_n \, y_1 \cdots y_n) \, 0.$$

Also following [Mil77, BCL85], define glb operators Inf$_s^n$ of sort $s^n$, for $n \ge 1$, by:

$$\text{Inf}_\iota^n = \lambda x_1 \cdots x_n. \text{If } (\text{And}_{n-1} \,(\text{Eq } x_1 \, x_2) \cdots (\text{Eq } x_1 \, x_n)) \, x_1 \, \Omega$$
$$\text{Inf}_{s_1 \to s_2}^n = \lambda x_1 \cdots x_n y. \text{Inf}_{s_2}^n \,(x_1 \, y) \cdots (x_n \, y).$$

**Lemma 3.4 (Milner)** *If $\mathcal{A}$ is an order-extensional model, $x_1, \ldots, x_n \in A_s$, $n \geq 1$ and $s \in S$, then $\text{Inf}^n\, x_1 \cdots x_n$ is the glb of $\{x_1, \ldots, x_n\}$ in $A_s$—and also in $R(A)_s$, if the $x_i$ are in $R(A)_s$.*

**Proof.** By induction on $S$. □

**Lemma 3.5** *Suppose $\mathcal{A}$ is an order-extensional model, $X \subseteq A_s$ is nonempty and $s \in S$. Then $\bigsqcup_{n \in \omega}(\bigsqcap(\Psi^n X))$ is the glb of $X$ in $A_s$—and also in $R(A)_s$, if $X \subseteq R(A)_s$. Hence $A_s$ and $R(A)_s$ are consistently complete, $\omega$-algebraic cpo's.*

**Proof.** Suppose $X \subseteq A_s$ is nonempty. Then the $\bigsqcap(\Psi^n X)$ are well-defined and form an $\omega$-chain by Lemmas 3.2 and 3.4. Let $x \in X$. Then $\bigsqcap(\Psi^n X) \sqsubseteq x$ for all $n \in \omega$, and thus $\bigsqcup_{n \in \omega}(\bigsqcap(\Psi^n X)) \sqsubseteq x$. Now, let $y$ be a lb of $X$. Then $\Psi^n y \sqsubseteq \bigsqcap(\Psi^n X)$ for all $n \in \omega$, so that $y = \bigsqcup_{n \in \omega}(\Psi^n y) \sqsubseteq \bigsqcup_{n \in \omega}(\bigsqcap(\Psi^n X))$, completing the proof that $\bigsqcup_{n \in \omega}(\bigsqcap(\Psi^n X))$ is the glb of $X$ in $A_s$. But, if $X \subseteq R(A)_s$, then each $\bigsqcap(\Psi^n X) \in R(A)_s$ by Lemma 3.4, so that $\bigsqcup_{n \in \omega}(\bigsqcap(\Psi^n X)) \in R(A)_s$, as required. The rest follows by Lemmas 3.2 and 3.3. □

The preceding lemma allows us to conclude that both $E$ and $R(E)$ are consistently complete, $\omega$-algebraic cpo's.

**Lemma 3.6** *If $\mathcal{A}$ is an order-extensional model, then $\Psi^n(\bigsqcap X) = \bigsqcap(\Psi^n X)$, for all $n \in \omega$ and finite, nonempty $X \subseteq A_s$.*

**Proof.** By induction on $S$, using the fact (Lemma 3.4) that finite, nonempty glb's are determined pointwise. □

Since glb's of infinite subsets of $E$ are not always determined pointwise, it is somewhat surprising that we have an infinitary version of the preceding lemma.

**Lemma 3.7** *If $\mathcal{A}$ is an order-extensional model, then $\Psi^n(\bigsqcap X) = \bigsqcap(\Psi^n X)$, for all $n \in \omega$ and nonempty $X \subseteq A_s$.*

**Proof.** For all $x \in X$, we have that $\Psi^n(\bigsqcap X) \sqsubseteq \Psi^n x$. Thus $\Psi^n(\bigsqcap X) \sqsubseteq \bigsqcap(\Psi^n X)$. For the other direction, $\bigsqcap(\Psi^n X) = \bigsqcap(\Psi^n(\Psi^n X)) = \Psi^n(\bigsqcap(\Psi^n X)) \sqsubseteq \Psi^n(\bigsqcap X)$ by Lemma 3.6 and the fact that $\bigsqcap(\Psi^n X) \sqsubseteq \bigsqcap X$. □

Following [Plo80], we say that an *n-ary logical relation* $L$ over a model $\mathcal{A}$, for $n \in \omega$, is an $n$-ary relation over $A$ such that $\langle x_1, \ldots, x_n \rangle \in L_{s_1 \to s_2}$ iff $\langle x_1 y_1, \ldots, x_n y_n \rangle \in L_{s_2}$ for all $\langle y_1, \ldots, y_n \rangle \in L_{s_1}$. Given such an $L$ and $\mathcal{A}$, we say that an element $x \in A_s$ *satisfies* $L$ iff $\langle x, \ldots, x \rangle \in L_s$.

**Lemma 3.8** *Suppose $L$ is an n-ary logical relation over a model $\mathcal{A}$, $s \in S$ and $D_1, \ldots, D_n \subseteq A_s$ are directed sets such that, for all $x_i \in D_i$, $1 \leq i \leq n$, there are $y_i \in D_i$, $1 \leq i \leq n$, such that $x_i \sqsubseteq y_i$ for all $i$ and $\langle y_1, \ldots, y_n \rangle \in L_s$. Then $\langle \bigsqcup D_1, \ldots, \bigsqcup D_n \rangle \in L_s$.*

**Proof.** By induction on $S$. □

**Lemma 3.9** *Suppose $L$ is an n-ary logical relation over a model $\mathcal{A}$. If $L$ is satisfied by $\Omega_\iota$, $n$, for all $n \in \omega$, Succ, Pred and $\text{If}_\iota$, then all elements of $R(\mathcal{A})$ satisfy $L$.*

**Proof.** First we must show that the remaining constants satisfy $L$. The satisfaction of $L$ by K and S at all sorts follows as usual. One shows that $\Omega$ satisfies $L$ at all sorts by induction on $S$, using the fact that application is strict in its left argument. The proof that If satisfies $L$ at all sorts also proceeds by induction on $S$, using the fact that $\text{If}_{s_1 \to s_2} x\,y\,z\,w = \text{If}_{s_2} x\,(y\,w)\,(z\,w)$ for all $x \in A_\iota$, $y, z \in A_{s_1 \to s_2}$ and $w \in A_{s_1}$. Finally, the satisfaction of $L$ by Y at all sorts follows using Lemma 3.8.

A simple induction on $T$ then shows that all denotable elements of $A$ satisfy $L$, following which we use Lemma 3.8 again to show, by induction on $R(A)$, that $L$ is satisfied by all elements of $R(A)$. □

**Lemma 3.10 (Plotkin)** *There is no $f \in R(\mathcal{E})_{\iota^2}$ such that $f \sqsupseteq \text{por}^2$.*

**Proof.** Following [Sie92], let $L$ be the ternary logical relation over $\mathcal{E}$ such that $\langle x_1, x_2, x_3 \rangle \in L_\iota$ iff either $x_i = \bot$ for some $i$ or $x_1 = x_2 = x_3$. It is easy to see that $L$ satisfies the hypotheses of Lemma 3.9, and thus all elements of $R(E)$ satisfy $L$. Clearly, $\langle 1, \bot, 0 \rangle \in L_\iota$ and $\langle \bot, 1, 0 \rangle \in L_\iota$. Thus, if there were such an $f$, then we would have that $\langle x_1, x_2, x_3 \rangle \in L_\iota$, where $x_1 = f\,1\,\bot$, $x_2 = f\,\bot\,1$ and $x_3 = f\,0\,0$. But $x_1 = 1$, $x_2 = 1$ and $x_3 = 0$—contradicting the definition of $L$. □

The following theorem allows us to define the meaning $[\![M]\!] \in \omega_\bot$ of a term $M$ of sort $\iota$ to be $[\![M]\!]_\mathcal{A}$, for an arbitrary model $\mathcal{A}$.

**Theorem 3.11 (Plotkin)** *For all models $\mathcal{A}$ and $\mathcal{B}$ and terms $M$ of sort $\iota$, $[\![M]\!]_\mathcal{A} = [\![M]\!]_\mathcal{B}$.*

**Proof.** See Theorem 3.1 of [Plo77]. □

We now define notions of program ordering and equivalence for PCF. Define a pre-ordering $\sqsubseteq$ over $T|\{\iota\}$ by $M \sqsubseteq_\iota N$ iff $[\![M]\!] \sqsubseteq [\![N]\!]$, and let $\approx$ be the equivalence relation over $T|\{\iota\}$ induced by $\sqsubseteq$. By Lemmas 2.1 and 2.3(i), $\sqsubseteq^c$ is a substitutive pre-ordering over $T$ and $\approx^c$ is a congruence over $T$. It is easy to see that $\sqsubseteq^c$ induces $\approx^c$. We say that a model $\mathcal{A}$ is *inequationally fully abstract* iff $\preceq_\mathcal{A} = \sqsubseteq^c$. From [Plo77], we know that $\mathcal{E}$ is not inequationally fully abstract. On the other hand, by [Mil77], there exists a unique (up to order-isomorphism) inequationally fully abstract, order-extensional model.

Finally, we recall Milner's important result concerning the order-extensional nature of $\sqsubseteq^c$ and the extensional nature of $\approx^c$ [Mil77].

**Lemma 3.12 (Milner)** (i) $\sqsubseteq^c_\iota = \sqsubseteq_\iota$ and $\approx^c_\iota = \approx_\iota$.
 (ii) *If $M_1\,N \sqsubseteq^c_{s_2} M_2\,N$ for all $N \in T_{s_1}$, then $M_1 \sqsubseteq^c_{s_1 \to s_2} M_2$.*
 (iii) *If $M_1\,N \approx^c_{s_2} M_2\,N$ for all $N \in T_{s_1}$, then $M_1 \approx^c_{s_1 \to s_2} M_2$.*

**Proof.** See Lemma 4.1.11 of [Cur86]. □

From Lemma 3.12(i), we know that, for all terms $M$ of sort $\iota$, either $M \approx^c_\iota \Omega$ or $M \approx^c_\iota n$ for some $n \in \omega$.

## 4  Normalization of $R(\mathcal{E})$

In this section, we focus on $\mathcal{E}$. Apart from the counterexamples, however, we could just as well work with any other order-extensional model, such as the bidomains model [BCL85]. We begin by defining semantic analogues of $\sqsubseteq^c$ and $\approx^c$.

**Definition 4.1**  Define an inductive pre-ordering $\leq$ over $E|\{\iota\}$ by $x \leq_\iota y$ iff $x \sqsubseteq y$, and let $\equiv$ be the equivalence relation over $E|\{\iota\}$ induced by $\leq$.

Clearly, $\equiv$ is just the identity relation over $E|\{\iota\}$.

**Lemma 4.2**  (i) $\leq^c$ is a unary-substitutive, inductive pre-ordering over $\mathcal{E}$.
(ii) $\equiv^c$ is the unary-substitutive equivalence relation over $\mathcal{E}$ induced by $\leq^c$.
(iii) For all $M, N \in T_s$, $M \sqsubseteq^c_s N$ iff $[\![M]\!] \leq^c_s [\![N]\!]$.
(iv) For all $M, N \in T_s$, $M \approx^c_s N$ iff $[\![M]\!] \equiv^c_s [\![N]\!]$.

**Proof.**  (i) follows from Lemma 2.2, (ii) is by Lemma 2.1 and an easy calculation, (iii) can be shown by another simple calculation, and (iv) follows from (ii) and (iii). □

**Lemma 4.3**  (i) The restriction of $\leq^c$ to $R(E)$ is a substitutive, inductive pre-ordering over $R(\mathcal{E})$.
(ii) The restriction of $\equiv^c$ to $R(E)$ is a congruence over $R(\mathcal{E})$.

**Proof.**  (i) follows by Lemma 2.3(ii) and the fact that $R(\mathcal{E})$ is an inductive subalgebra of $\mathcal{E}$, and (ii) follows from (i). □

**Lemma 4.4**  (i) $\leq^c_\iota = \leq_\iota$ and $\equiv^c_\iota = \equiv_\iota$.
(ii) For all $x_1, x_2 \in R(E)_{s_1 \to s_2}$, if $x_1 y \leq^c x_2 y$ for all $y \in R(E)_{s_1}$, then $x_1 \leq^c x_2$.
(iii) For all $x_1, x_2 \in R(E)_{s_1 \to s_2}$, if $x_1 y \equiv^c x_2 y$ for all $y \in R(E)_{s_1}$, then $x_1 \equiv^c x_2$.

**Proof.**  (i) follows from Lemma 3.12(i). For (ii), it suffices to show that $\Psi^n x_1 \leq^c \Psi^n x_2$ for all $n \in \omega$, since $\leq^c$ is inductive. But isolated elements of $R(E)$ are denotable, and thus, by Lemma 3.12(ii), it is sufficient to show that $\Psi^n x_1 y \leq^c \Psi^n x_2 y$ for all $y \in R(E)_{s_1}$. But $\Psi^n(x_1(\Psi^n y)) \leq^c \Psi^n(x_2(\Psi^n y))$ follows from the hypothesis and Lemma 4.3. Finally, (iii) follows from (ii). □

The following term features prominently below and is a generalization of the parallel or tester introduced in [Plo77].

**Definition 4.5**  Let the term Test of sort $\iota \to \iota \to \iota^2 \to \iota$ be

$$\lambda xyf.\, \text{If}\, (\text{Eq}\,(f\,1\,y)\,1)$$
$$\qquad (\text{If}\,(\text{Eq}\,(f\,x\,1)\,1)$$
$$\qquad\qquad (\text{If}\,(f\,0\,0)\,\Omega\,0)$$
$$\qquad\qquad \Omega)$$
$$\qquad \Omega.$$

**Lemma 4.6**  For all $f \in E_{\iota^2}$, $\text{Test} \perp \perp f$ is $0$, if $f \sqsupseteq \text{por}^2$, and $\perp$, otherwise. □

The following is a counterexample to $\equiv^c$ (and thus $\leq^c$) being substitutive.

**Counterexample 4.7** Test $\bot \bot \equiv^c \bot$, *but* Test $\bot \bot$ por $\not\equiv^c \bot$ por.

**Proof.** By Lemmas 4.4, 4.6 and 3.10, we have Test $\bot \bot \equiv^c \bot$. But Test $\bot \bot$ por $= 0$, and thus Test $\bot \bot$ por $\not\equiv^c \bot$ por. □

We do, however, have:

**Lemma 4.8** (i) *For all* $x_1, x_2 \in E_{s_1 \to s_2}$ *and* $y \in R(E)_{s_1}$, *if* $x_1 \leq^c x_2$, *then* $x_1 y \leq^c x_2 y$.
(ii) *For all* $x \in R(E)_{s_1 \to s_2}$ *and* $y_1, y_2 \in E_{s_2}$, *if* $y_1 \leq^c y_2$, *then* $x y_1 \leq^c x y_2$.

**Proof.** For (i), since application is continuous and $\leq^c$ is inductive, it suffices to show $x_1 y \leq^c x_2 y$ when $y$ is isolated. But this follows since all isolated elements of $R(E)_{s_1}$ are denotable and $\leq^c$ is unary-substitutive. (ii) follows similarly. □

The following result shows that we cannot allow $x_1, x_2$ to range over $E_{s_1 \to s_2}$ in parts (ii) and (iii) of Lemma 4.4. This raises the question (which we leave unanswered) of when nondenotable elements are related by $\leq^c$ and $\equiv^c$.

**Counterexample 4.9** *Define* $G_1, G_2 \in E_{\iota^2 \to \iota^2}$ *by*

$$G_1 = \lambda f.\text{If}\,(f\,0\,0)\,\text{por}\,(\lambda xy.\,\text{Test}\,\Omega\,\Omega\,f), \qquad G_2 = \lambda f.\text{If}\,(f\,0\,0)\,\text{por}\,\Omega.$$

*Then* $G_1 f \equiv^c G_2 f$ *for all* $f \in R(E)_{\iota^2}$, *but* $G_1 \not\equiv^c G_2$.

**Proof.** It is easy to see that $G_1 f \equiv^c G_2 f$ for all $f \in R(E)_{\iota^2}$. But $c\langle G_1 \rangle = 0$ and $c\langle G_2 \rangle = \bot$, where the derived operator $c[v]$ of type $(\iota^2 \to \iota^2) \to \iota$ is $v\,(v\,(\lambda xy.\,1))\,\Omega\,\Omega$. Thus $G_1 \not\equiv^c G_2$. □

We are now ready to define our continuous projection over $R(E)$.

**Definition 4.10** The function norm: $R(E) \to R(E)$ is defined by

$$\text{norm}_s\,x = \bigsqcap \{\,x' \in R(E)_s \mid x' \equiv^c x\,\}.$$

By Lemma 3.7, $\Psi^n(\text{norm}\,x) = \bigsqcap \{\,\Psi^n\,x' \mid x' \equiv^c x \text{ and } x' \in R(E)_s\,\}$ for all $n \in \omega$ and $x \in R(E)_s$, $s \in S$. We write $x \sqsubseteq \equiv^c y$ for $x \sqsubseteq y$ and $x \equiv^c y$.

**Lemma 4.11** *If $X$ is a finite subset of $R(E)_s$ and $x' \in X$ is such that $x' \leq^c x$ for all $x \in X$, then $x' \equiv^c \bigsqcap X$.*

**Proof.** By induction on $S$. □

**Lemma 4.12** *Let $x, y \in R(E)_s$, $s \in S$, and $n \in \omega$.*
(i) $\text{norm}\,x \sqsubseteq x$.
(ii) $\text{norm}\,x \equiv^c x$.
(iii) *If* $x \sqsubseteq \equiv^c \text{norm}\,y$, *then* $x = \text{norm}\,y$.
(iv) $x \leq^c y$ *iff* $\text{norm}\,x \sqsubseteq \text{norm}\,y$.

(v) $x \equiv^c y$ iff $\operatorname{norm} x = \operatorname{norm} y$.
(vi) If $x \sqsubseteq y$, then $\operatorname{norm} x \sqsubseteq \operatorname{norm} y$.
(vii) $\operatorname{norm}(\operatorname{norm} x) = \operatorname{norm} x$.
(viii) $\operatorname{norm}(\Psi^n x) \sqsubseteq \equiv^c \Psi^n(\operatorname{norm} x)$.
(ix) $\Psi^n(\operatorname{norm}(\Psi^n x)) = \operatorname{norm}(\Psi^n x)$.
(x) $\operatorname{norm} x = \bigsqcup_{n \in \omega} \operatorname{norm}(\Psi^n x)$.
(xi) $\operatorname{norm}_s$ *is continuous.*

**Proof.** (i) Immediate from the reflexivity of $\equiv^c$.

(ii) By Lemma 4.11, we have that $\Psi^n x \equiv^c \bigsqcap \{ \Psi^n x' \mid x' \equiv^c x$ and $x' \in R(E)_s \} \sqsubseteq \operatorname{norm} x$, for all $n \in \omega$. Thus $x \leq^c \operatorname{norm} x$, since $\leq^c$ is inductive. The result then follows by (i).

(iii) If $x \sqsubseteq \equiv^c \operatorname{norm} y$, then $x \equiv^c \operatorname{norm} y \equiv^c y$ by (ii), so that $\operatorname{norm} y \sqsubseteq x$. But then $x = \operatorname{norm} y$, since $x \sqsubseteq \operatorname{norm} y$.

(iv) The "if" direction follows from (ii) and the fact that $\sqsubseteq_s \subseteq \leq^c_s$. For the "only if" direction, suppose that $x \leq^c y$. Let $y' \in R(E)_s$ be such that $y' \equiv^c y$. Then $x \leq^c y'$, so that $x \sqcap y' \equiv^c x$ by Lemma 4.11. But then $\operatorname{norm} x \sqsubseteq x \sqcap y' \sqsubseteq y'$. Thus $\operatorname{norm} x \sqsubseteq \operatorname{norm} y$.

(v) Immediate from (iv).
(vi) Follows from (iv), since $\sqsubseteq_s \subseteq \leq^c_s$.
(vii) Follows by (ii) and (v).
(viii) Follows by (i), (ii) and (v).

(ix) Since $\operatorname{norm}(\Psi^n x) \equiv^c \Psi^n x$, we have $\Psi^n(\operatorname{norm}(\Psi^n x)) \equiv^c \Psi^n(\Psi^n x) = \Psi^n x \equiv^c \operatorname{norm}(\Psi^n x)$, and thus $\Psi^n(\operatorname{norm}(\Psi^n x)) \equiv^c \operatorname{norm}(\Psi^n x)$. The result then follows by (iii), since $\Psi^n(\operatorname{norm}(\Psi^n x)) \sqsubseteq \operatorname{norm}(\Psi^n x)$.

(x) By (vi) and (viii), $\operatorname{norm}(\Psi^n x) \sqsubseteq \operatorname{norm} x$ and $\bigsqcup_{n \in \omega} \operatorname{norm}(\Psi^n x) \geq^c \Psi^n(\operatorname{norm} x)$, for all $n \in \omega$. Thus $\bigsqcup_{n \in \omega} \operatorname{norm}(\Psi^n x) \sqsubseteq \equiv^c \operatorname{norm} x$, since $\leq^c$ is inductive. The result then follows by (iii).

(xi) Follows from (x). □

**Lemma 4.13** $\operatorname{norm}(\operatorname{Test} \bot \bot) = \bot$.

**Proof.** Follows from Counterexample 4.7. □

**Lemma 4.14** $\operatorname{norm}_\iota$ *is the identity function on* $R(E)_\iota$.

**Proof.** Immediate by Lemma 4.4(i). □

The following counterexample shows that Lemma 4.12(viii) cannot be strengthened to an identity.

**Counterexample 4.15** $\operatorname{norm}(\Psi^2(\operatorname{Test} 2\, 2)) \neq \Psi^2(\operatorname{norm}(\operatorname{Test} 2\, 2))$.

**Proof.** Let the term $A$ of sort $\iota^2$ be

$$\lambda xy.\, \text{If}\, (\text{And}_2\, (\text{Eq}\, x\, 1)\, (\text{Eq}\, y\, 2))$$
$$1$$
$$(\text{If}\, (\text{And}_2\, (\text{Eq}\, x\, 2)\, (\text{Eq}\, y\, 1))$$
$$1$$
$$(\text{If}\, (\text{And}_2\, (\text{Eq}\, x\, 0)\, (\text{Eq}\, y\, 0))\, 0\, \Omega)),$$

so that $A \sqsubseteq \text{por}^2$. Since $\text{Test}\,2\,2\,A = 0$, it follows that $(\text{norm}(\text{Test}\,2\,2))A = 0$, and thus that $(\text{norm}(\text{Test}\,2\,2))\text{por}^2 = 0$. But then

$$\Psi^2\,(\text{norm}(\text{Test}\,2\,2))\,\text{por} = \Psi^2((\text{norm}(\text{Test}\,2\,2))\text{por}^2) = \Psi^2\,0 = 0,$$

showing that $\Psi^2(\text{norm}(\text{Test}\,2\,2)) \neq \bot$. On the other hand, it is easy to show that $\Psi^2(\text{Test}\,2\,2) = \text{Test} \bot \bot$, and thus $\text{norm}(\Psi^2(\text{Test}\,2\,2)) = \bot$ by Lemma 4.13. □

In preparation for three key counterexamples, we now define the following or operations of sort $\iota^2$, where the "L", "R" and "D" stand for "Left", "Right" and "Divergent", respectively:

$$\text{LOr} = \lambda xy.\,\text{If}\,x\,1\,(\text{If}\,y\,1\,0)$$
$$\text{ROr} = \lambda xy.\,\text{If}\,y\,1\,(\text{If}\,x\,1\,0)$$
$$\text{DOr} = \lambda xy.\,\text{If}\,x\,(\text{If}\,y\,\Omega\,1)\,(\text{If}\,y\,1\,0).$$

**Lemma 4.16** *There is no $h \in R(E)_{\iota \to \iota \to \iota^2 \to \iota}$ such that*

$$h \bot \bot \text{por} = \bot, \qquad h\,0 \bot \text{LOr} = 0, \qquad h \bot 0\,\text{ROr} = 0, \qquad h\,0\,0\,\text{DOr} = 0.$$

**Proof.** Suppose, toward a contradiction, that such an $h$ does exist.

Let $L$ be the 4-ary logical relation over $\mathcal{E}$ such that $\langle x_1, x_2, x_3, x_4 \rangle \in L_\iota$ iff $\{x_1, x_2, x_3, x_4\} \subseteq \{\bot, n\}$ for some $n \in \omega$ and, if $x_1 = \bot$, then one of $x_2, x_3, x_4$ is also $\bot$. Clearly, $\Omega_\iota$ and all $n \in \omega$ satisfy $L$. Furthermore, Succ and Pred satisfy $L$ since it is satisfied by all elements of $E_{\iota^1}$. Finally, it is easy to show that $L$ is satisfied by $\text{If}_\iota$. Hence $h$ satisfies $L$, by Lemma 3.9.

Next, we show that $\langle \text{por}, \text{LOr}, \text{ROr}, \text{DOr} \rangle \in L_{\iota^2}$. Suppose that $\langle x_1, x_2, x_3, x_4 \rangle \in L_\iota$ and $\langle y_1, y_2, y_3, y_4 \rangle \in L_\iota$. We must show that $\langle z_1, z_2, z_3, z_4 \rangle \in L_\iota$, where $z_1 = \text{por}\,x_1\,y_1$, $z_2 = \text{LOr}\,x_2\,y_2$, $z_3 = \text{ROr}\,x_3\,y_3$ and $z_4 = \text{DOr}\,x_4\,y_4$. Clearly, each $z_i \in \{\bot, 0, 1\}$. Furthermore, if $z_i = 0$ for some $i$, then both $x_i$ and $y_i$ must be 0, so that no $x_j$ or $y_j$ is a nonzero element of $\omega$, and thus no $z_j = 1$. Now, suppose that $z_1 = \bot$. We must show that one of $z_2, z_3, z_4$ is $\bot$. Either $x_1$ or $y_1$ must be $\bot$, and we consider the case when $x_1 = \bot$, the other case being dual. Since LOr and DOr are strict in their first arguments, if $x_i = \bot$ for some $i \in \{2, 4\}$, then $z_i = \bot$. Otherwise, we must have that $x_3 = \bot$ and $x_2 = x_4 \neq \bot$. Now, if $y_3 \in \{\bot, 0\}$, then $z_3 = \bot$. Otherwise, $y_3 \in \omega - \{0\}$ and $y_1, y_2, y_4 \in \{\bot, y_3\}$. But then $y_1 = \bot$ (otherwise $z_1 = 1$), and thus either $y_2 = \bot$ or $y_4 = \bot$. Since DOr is also strict in its second argument, if $y_4 = \bot$, then $z_4 = \bot$. Otherwise, $y_2 = \bot$ and $y_4 = y_3$. Now, if $x_2 = 0$, then $z_2 = \bot$. Otherwise, we have that $x_2 = x_4 \in \omega - \{0\}$. But then $z_4 = \bot$, since both $x_4, y_4 \in \omega - \{0\}$.

Summarizing, we have that $h$ satisfies $L$ and $\langle \text{por}, \text{LOr}, \text{ROr}, \text{DOr} \rangle \in L_{\iota^2}$. Furthermore, $\langle \bot, 0, \bot, 0 \rangle \in L_\iota$ and $\langle \bot, \bot, 0, 0 \rangle \in L_\iota$, so that $\langle z_1, z_2, z_3, z_4 \rangle \in L_\iota$, where $z_1 = h \bot \bot \text{por}$, $z_2 = h\,0\,\bot\,\text{LOr}$, $z_3 = h \bot 0\,\text{ROr}$ and $z_4 = h\,0\,0\,\text{DOr}$. But $z_1 = \bot$ and $z_2 = z_3 = z_4 = 0$, contradicting the definition of $L$. □

The following counterexample shows that application is not preserved by norm.

**Counterexample 4.17** $\text{norm}(\text{Test} \bot) \neq (\text{norm}\,\text{Test})(\text{norm}\,\bot)$.

**Proof.** By Lemma 4.4(iii) and Counterexample 4.7, we have that

$$\text{Test} \perp \equiv^c \lambda y. \text{If } y \, (\text{Test} \perp y) \, (\text{Test} \perp y),$$

so that $(\text{norm}(\text{Test} \perp)) \perp \text{por} = \perp$. Since $\text{norm} \perp = \perp$, it is thus sufficient to show that $h \perp \perp \text{por} \neq \perp$, where $h = \text{norm Test}$. But

$$h\, 0 \perp \text{LOr} = 0, \qquad h \perp 0\, \text{ROr} = 0, \qquad h\, 0\, 0\, \text{DOr} = 0,$$

since $h \equiv^c \text{Test}$, and thus $h \perp \perp \text{por} \neq \perp$ by Lemma 4.16. □

Since $\text{norm}((\text{norm Test})(\text{norm} \perp)) = \text{norm}(\text{Test} \perp)$, it follows from the preceding counterexample that the image of norm is not closed under application.

**Counterexample 4.18** *There is no* $\text{norm}' \in R(E)_{(\iota \to \iota^2 \to \iota)^1}$ *such that* $\text{norm}' x = \text{norm } x$ *for all* $x \in R(E)_{\iota \to \iota^2 \to \iota}$.

**Proof.** Suppose, toward a contradiction, that such a $\text{norm}'$ does exist. Then, for all $y \in R(E)_\iota$,

$$\text{Test } y \equiv^c \text{norm}(\text{Test } y) = \text{norm}'(\text{Test } y) = (\lambda y.\, \text{norm}'(\text{Test } y))\, y,$$

so that $\text{Test} \equiv^c \lambda y.\, \text{norm}'(\text{Test } y)$. Then,

$$\begin{aligned}
(\text{norm Test})(\text{norm} \perp) &= (\text{norm}(\lambda y.\, \text{norm}'(\text{Test } y))) \perp \\
&\sqsubseteq^c (\lambda y.\, \text{norm}'(\text{Test } y)) \perp \\
&= \text{norm}'(\text{Test} \perp) \\
&= \text{norm}(\text{Test} \perp).
\end{aligned}$$

But then $(\text{norm Test})(\text{norm} \perp) = \text{norm}(\text{Test} \perp)$, contradicting Counterexample 4.17. □

The following counterexample shows that denotable elements can be contextually equivalent to nondenotable ones.

**Counterexample 4.19** $h \equiv^c \text{Test}$ *does not imply that* $h \in R(E)$.

**Proof.** Let $h \in E_{\iota \to \iota \to \iota^2 \to \iota}$ be $\lambda x y.$ If $(\text{pcon } x\, y)(\text{Test } x\, y)\,\Omega$, where the parallel convergence operation $\text{pcon} \in E_{\iota^2}$ is defined by: $\text{pcon } x\, y = 1$, if $x \neq \perp$ or $y \neq \perp$, and $\text{pcon } x\, y = \perp$, otherwise. Then $h \notin R(E)$, by Lemma 4.16. It remains to show that $h \equiv^c \text{Test}$.

In the remainder of the proof, we work in the result of adding to PCF a constant PCon of sort $\iota^2$ whose interpretation is pcon. All of the results preceding Lemma 4.16 hold for the extended language, with the exception of Lemma 3.9. This lemma can be repaired, however, by adding PCon to the list of constants in its hypothesis. The logical relation defined in the proof of Lemma 3.10 is also satisfied by PCon and thus this lemma is true for the extended language. (The original proof that parallel or is not definable from parallel convergence can be found in [Abr90]. [Plo77] did not show that the continuous function model of the extended language was not inequationally fully abstract.)

It is sufficient to show $h \equiv^c \text{Test}$, and, since $h \in R(E)$, this will be a consequence of showing that $h\, x\, y \equiv^c \text{Test } x\, y$ for all $x, y \in R(E)_\iota$. If $x \neq \perp$ or $y \neq \perp$, then $h\, x\, y = \text{Test } x\, y$. But $\text{Test} \perp \perp \equiv^c \perp$ was shown in Counterexample 4.7. □

Although we were able to solve negatively the question of whether norm preserves application, the following problem is still open.

**Open Problem 4.20** *Is* norm $\sigma = \sigma$ *for all constants* $\sigma \in \Sigma$? *In particular, is* (norm K) $x\,y$ *ever strictly less than* $x$?

Now, we are able to show how the unique inequationally fully abstract, order-extensional model lives inside the continuous function model.

**Definition 4.21** We define the ordered algebra $N(\mathcal{E})$ as follows. For all $s \in S$, $N(E)_s$ consists of norm $R(E)_s$, ordered by the restriction of $\sqsubseteq_{E_s}$ to norm $R(E)_s$. For all $x \in N(E)_{s_1 \to s_2}$ and $y \in N(E)_{s_1}$, $x \cdot_{N(\mathcal{E})} y = \text{norm}(x \cdot_{\mathcal{E}} y)$. For all constants $\sigma$, $\sigma_{N(\mathcal{E})} = \text{norm}\,\sigma_{\mathcal{E}}$.

$N(E)$ is a subcpo of $R(E)$ and $N(\mathcal{E})$ is well-defined, since norm is strict and continuous.

**Theorem 4.22** $N(\mathcal{E})$ *is an order-extensional model and* norm *is a surjective morphism from* $R(\mathcal{E})$ *to* $N(\mathcal{E})$.

**Proof.** $N(\mathcal{E})$ is a complete ordered algebra by the preceding remark and the continuity of norm. Condition (i) of the definition of model holds by Lemma 4.14, and the remaining conditions can be shown using Lemma 4.12(ii) and (v) and (for condition (iv)) the continuity of norm. For the order-extensionality of $N(\mathcal{E})$, suppose that $x_1, x_2 \in N(E)_{s_1 \to s_2}$ are such that $x_1 \cdot_{N(\mathcal{E})} y \sqsubseteq x_2 \cdot_{N(\mathcal{E})} y$ for all $y \in N(E)_{s_1}$. Then, for all $y \in R(E)_{s_1}$,

$$\text{norm}(x_1 \cdot_{\mathcal{E}} y) = x_1 \cdot_{N(\mathcal{E})} \text{norm}\,y \sqsubseteq x_2 \cdot_{N(\mathcal{E})} \text{norm}\,y = \text{norm}(x_2 \cdot_{\mathcal{E}} y),$$

and thus $x_1 \cdot_{\mathcal{E}} y \leq^c x_2 \cdot_{\mathcal{E}} y$. But then $x_1 \leq^c x_2$ by Lemma 4.4(ii), so that $x_1 = \text{norm}\,x_1 \sqsubseteq \text{norm}\,x_2 = x_2$. Finally, norm is a surjective morphism from $R(\mathcal{E})$ to $N(\mathcal{E})$ because of the way $N(\mathcal{E})$ was defined. □

By Lemma 3.5, we know that $N(E)$ is a consistently complete, $\omega$-algebraic cpo.

**Lemma 4.23** *For all terms* $M$, $[\![M]\!]_{N(\mathcal{E})} = \text{norm}[\![M]\!]_{\mathcal{E}}$.

**Proof.** A consequence of norm being a morphism from $R(\mathcal{E})$ to $N(\mathcal{E})$. □

**Theorem 4.24** $N(\mathcal{E})$ *is inequationally fully abstract.*

**Proof.** Follows from Lemmas 4.2(iii) and 4.23. □

## 5 Full Abstraction and Lambda Definability

There appears to be no clear definition of what the "full abstraction problem" for PCF really is. By Milner's construction [Mil77] we know that there is a unique inequationally fully abstract, order-extensional model $\mathcal{F}$ (which we refer to below as *the fully abstract model*) that is made up out of Scott-domains of continuous (set-theoretic) functions. Why are we not satisfied? The answer to this question, as one often reads, is that Milner's model is "syntactic in nature". The same words are used against Mulmuley's description [Mul87] of the fully abstract model. What people vaguely imagine is that there ought to be a description of $\mathcal{F}$ using cpo's enriched with some additional structure (order-theoretic, topological, etc.) which allows the domains of the fully abstract model to be constructed without recourse to the syntax of PCF. Of course, nobody can specify what this additional structure will be or should be. Stated this way, there is no chance to falsify this research programme, in the sense that there is no way one can prove a result saying that there is no "semantic" presentation of $\mathcal{F}$.

We would therefore like to give a weak but precise minimal condition that a semantic solution of the full abstraction problem should satisfy. Namely, it should allow us to *effectively* construct the finite domains $F_s$ of the fully abstract model $\mathcal{F}$ of *Finitary PCF*, i.e., the variant of PCF in which the sort $\iota$ is interpreted as the booleans ($\{\bot, 0, 1\}$) rather than the natural numbers. (The results of this paper can be trivially adapted to Finitary PCF.) Clearly, neither Milner's nor Mulmuley's constructions achieve this. On the other hand, even if we can find such an algorithm for presenting $\mathcal{F}$, we may still be unsatisfied with it as a semantic description.

The results of this paper give one of the simplest descriptions of the fully abstract model to date. In order to satisfy the above condition, all one needs to find is an algorithm that decides whether an element of $E$ is denotable, since then one will be able to effectively present $R(E)$ and thus $N(E)$.

The problem of deciding which elements of a model are definable in the case of the typed lambda calculus (without constants) and the full set-theoretic type hierarchy based on a finite set is known as "Plotkin's conjecture". (It seems that the term was coined by Statman in his 1982 paper [Sta82]. We do not know whether Plotkin ever considered the question nor whether he ever conjectured anything.) The "conjecture" is that the problem is decidable. We prefer to call it the "lambda definability problem" (cf. [JT93]). This problem can be studied in all kinds of contexts, and it certainly makes sense to ask whether it is decidable which elements of $E$ are denotable. We refer to this as the lambda definability problem for Finitary PCF.

Since a positive solution to the lambda definability problem for Finitary PCF will mean that $N(\mathcal{E})$ and thus $\mathcal{F}$ are effectively presentable, it is natural to ask whether the converse is also true. We conjecture that it is.

## Acknowledgments

Part of this work was done while the second author was a Guest Researcher at the Technische Hochschule Darmstadt in September, 1991.

## References

[Abr90] S. Abramsky. The lazy lambda calculus. In D. L. Turner, editor, *Research Topics in Functional Programming*, pages 65-116. Addison-Wesley, 1990.

[BCL85] G. Berry, P.-L. Curien and J.-J. Lévy. Full abstraction for sequential languages: the state of the art. In M. Nivat and J. C. Reynolds, editors, *Al-*

*gebraic Methods in Semantics*, pages 89–132. Cambridge University Press, 1985.

[Cur86] P.-L. Curien. *Categorical Combinators, Sequential Algorithms and Functional Programming*. Research Notes in Theoretical Computer Science. Pitman/Wiley, 1986.

[JT93] A. Jung and J. Tiuryn. A new characterization of lambda definability. In this proceedings, 1993.

[Mil77] R. Milner. Fully abstract models of typed $\lambda$-calculi. *Theoretical Computer Science*, 4:1–22, 1977.

[Mul87] K. Mulmuley. *Full Abstraction and Semantic Equivalence*. MIT Press, 1987.

[Plo77] G. D. Plotkin. LCF considered as a programming language. *Theoretical Computer Science*, 5:223–256, 1977.

[Plo80] G. D. Plotkin. Lambda-definability in the full type hierarchy. In J. Seldin and J. Hindley, editors, *To H. B. Curry: Essays on Combinatory Logic, Lambda Calculus and Formalism*, pages 363–374. Academic Press, 1980.

[Sie92] K. Sieber. Reasoning about sequential functions via logical relations. In M. P. Fourman, P. T. Johnstone and A. M. Pitts, editors, *Applications of Categories in Computer Science*, volume 177 of *LMS Lecture Note Series*, pages 258–269. Cambridge University Press, 1992.

[Sta82] R. Statman. Completeness, invariance and $\lambda$-definability. *Journal of Symbolic Logic*, 47:17–26, 1982.

[Sto88] A. Stoughton. *Fully Abstract Models of Programming Languages*. Research Notes in Theoretical Computer Science. Pitman/Wiley, 1988.

[Sto90] A. Stoughton. Equationally fully abstract models of PCF. In M. Main, A. Melton, M. Mislove and D. Schmidt, editors, *Fifth International Conference on the Mathematical Foundations of Programming Semantics*, volume 442 of *Lecture Notes in Computer Science*, pages 271–283. Springer-Verlag, 1990.

# A New Characterization of Lambda Definability

Achim Jung[1] and Jerzy Tiuryn[*,2]

[1] Fachbereich Mathematik, Technische Hochschule Darmstadt, Schloßgartenstraße 7, D-6100 Darmstadt, Germany, jung@mathematik.th-darmstadt.de
[2] Instytut Informatyki, Uniwersytet Warszawski, ul. Banacha 2, PL-02-097 Warszawa, Poland, tiuryn@mimuw.edu.pl

**Abstract.** We give a new characterization of lambda definability in Henkin models using logical relations defined over ordered sets with varying arity. The advantage of this over earlier approaches by Plotkin and Statman is its simplicity and universality. Yet, decidability of lambda definability for hereditarily finite Henkin models remains an open problem. But if the variable set allowed in terms is also restricted to be finite then our techniques lead to a decision procedure.

## 1 Introduction

An *applicative structure* consists of a family $(A_\sigma)_{\sigma \in \mathbb{T}}$ of sets, one for each type $\sigma$, together with a family $(app_{\sigma,\tau})_{\sigma,\tau \in \mathbb{T}}$ of application functions, where $app_{\sigma,\tau}$ maps $A_{\sigma \to \tau} \times A_\sigma$ into $A_\tau$. For an applicative structure to be a model of the simply typed lambda calculus (in which case we call it a *Henkin model*, following [4]), one requires two more conditions to hold. It must be *extensional* which means that the elements of $A_{\sigma \to \tau}$ are uniquely determined by their behavior under $app_{\sigma,\tau}$, or, more intuitively, that $A_{\sigma \to \tau}$ can be thought of as a set of functions from $A_\sigma$ to $A_\tau$. Secondly, the applicative structure must be rich enough to interpret every lambda term. (This requirement can be formalized using either the combinatory or the environment model definition, see Sect. 2 below.) The simplest examples for Henkin models are derived if one takes a set $A_\iota$ for the base type $\iota$ (more base types could be accommodated in the same way) and then defines $A_{\sigma \to \tau}$ to be the set of all functions from $A_\sigma$ to $A_\tau$. The application functions are in this case just set-theoretic application of a function to an argument. These models are sometimes called the *full type hierarchy over $A_\iota$*.

Simple as this construction is, there remains a nagging open question. Suppose $A_\iota$ is finite (in which case every $A_\sigma$ is finite), is there an algorithm which, given an element of some $A_\sigma$, decides whether it is the denotation of a closed lambda term? We could also ask for an algorithm which works uniformly for all finite sets $A_\iota$, but the essential difficulty seems to arise with the first question. The assumption that a positive solution exists is known under the name *lambda definability conjecture*. We shall speak of the *lambda definability problem* instead. Besides this being an intriguing question in itself, there are also connections to

---

[*] Supported by Polish KBN grant No. 2 1192 91 01

other open problems, such as the *higher order matching problem* (cf. [9], and also [13]) and the *full abstraction problem for PCF* (cf. [1]).

Let us quickly review the existing literature on the question. A first attempt to characterize lambda definable elements in the full type hierarchy was made by H.Läuchli [2]. He showed that lambda definable elements are invariant under permutations of the ground set $A_\iota$ which is a not too surprising result as there are no means by which the lambda calculus could speak about particular elements of $A_\iota$. He also observed that permutation invariance was too weak a property for full characterization at all types. This line of thought was taken up by G.Plotkin in [7] (a precursor of this is [6]). He replaced permutation invariance by invariance under logical relations and proved that for infinite ground sets this characterizes lambda definability at types of rank less than three. Using more complicated logical relations defined over a quasi-ordered set he could remove the restriction on the rank. The restriction on the size of $A_\iota$, however, remained. In both cases the proof is by coding the theory of lambda terms into the ground set. The problem is also discussed in papers by R.Statman (cf. [9, 10, 11, 12]). In [12] a characterization is stated (without proof) which is applicable to all Henkin models and which employs logical relations on a free extension of the given model by infinitely many variables. More recently, K.Sieber [8] used logical relations in a novel fashion to tackle the full abstraction problem for PCF. His logical relations have large arity and are reminiscent of value tables. It was this paper from which we got the initial idea for the results presented here. By looking at logical relations which are defined over an ordered set (as in [7]) but which in addition increase their arity as we pass to later "worlds", we derive a characterization theorem which works for all ground sets $A_\iota$ and, in fact, every Henkin model, which again contrasts to the characterization in [7], which can not be generalized to arbitrary Henkin models. (A counterexample is given in [12].) Furthermore, our characterization theorem has a very straightforward proof. Indeed, the proof is so simple that it suggests a positive solution to the lambda definability problem. Even though we do not achieve this, at least we can make the obstacles very clear. These lie in the fact that higher order terms (even when they are in normal form) can contain arbitrarily many auxiliary variables. For a restricted set of variables one would expect a decidability result. This can be achieved as we show in Sect. 5, but the proof becomes somewhat technical.

Our definition of logical relation will still make sense if we replace the ordered set by a small category and, in fact, it reduces to a logical predicate on the presheaf model built from the initial Henkin model (for details, see [3] or [5]). A bit of this generality indeed simplifies our presentation of Kripke logical relations with varying arity in the next section. The characterization theorem in Sect. 3, however, works with a very simple fixed ordered set.

## 2 Kripke Logical Relations with Varying Arity

Suppose $A_\iota$ is a set and we are studying the semantics of the simply typed lambda calculus in the full type hierarchy over $A_\iota$. (We could take an arbitrary Henkin model instead but would have to write out the application functions explicitly in every instance.) Let $\mathcal{C}$ be a small category of sets. We want to build a logical relation over each object $w$ of $\mathcal{C}$, taking the cardinality of $w$ as the arity of the relation at $w$. Thus elements of the relations are tuples indexed by elements of $w$. It makes no difference whether $w$ is finite or infinite.

We start with ground relations $R_\iota^w \subseteq A_\iota^w$ which have the following compatibility property: Whenever $f: v \to w$ is a map in $\mathcal{C}$ and $(x_j)_{j \in w}$ is an element of $R_\iota^w$ then $(x_{f(i)})_{i \in v}$ is an element of $R_\iota^v$ (note the contravariance). The ground relations are extended to higher types as usual. For a function type $\sigma \to \tau$ let

$$R_{\sigma \to \tau}^w = \{(g_j)_{j \in w} \mid \forall j \in w. g_j \in A_{\sigma \to \tau} \land \forall f: v \to w \forall (x_i)_{i \in v} \in R_\sigma^v.$$
$$(g_{f(i)}(x_i))_{i \in v} \in R_\tau^v\}.$$

(A tuple of functions at $w$ must have the defining property of logical relations at all $v$ reachable - via a map in $\mathcal{C}$ - from $w$.) Relations $(R_\sigma^w)_{\sigma \in \mathbb{T}}^{w \in Obj(\mathcal{C})}$ constructed this way we shall call *Kripke logical relations with varying arity*. Ordinary logical relations are subsumed by this concept - just take a one object one morphism category $\mathcal{C}$ - as well as Plotkin's "I-relations": take a category all of whose objects have the same cardinality and all of whose morphisms are bijections such that the category is isomorphic to a quasi-ordered set.

We observe that for each type $\sigma$ we have the compatibility with morphisms of $\mathcal{C}$ we required at ground level:

**Lemma 1.** Let $(R_\sigma^w)_{\sigma \in \mathbb{T}}^{w \in Obj(\mathcal{C})}$ be a Kripke logical relation with varying arity. For all types $\sigma$, objects $v, w$ of $\mathcal{C}$, morphisms $f: v \to w$, and tuples $(x_j)_{j \in w}$ in $R_\sigma^w$, the tuple $(x_{f(i)})_{i \in v}$ is in $R_\sigma^v$.

*Proof.* By induction on types. For $\iota$ it is part of the definition. If $\sigma \to \tau$ is a function type we have to show that $(g_j)_{j \in w} \in R_{\sigma \to \tau}^w$ implies $(g_{f(i)})_{i \in v} \in R_{\sigma \to \tau}^v$. By definition we have to supply arguments $(x_l)_{l \in u} \in R_\sigma^u$, for $h: u \to v$ to the functions. The resulting tuple has the form $(g_{f(h(l))}(x_l))_{l \in u}$ which belongs to $R_\tau^u$ because $f \circ h: u \to w$ is also a map in $\mathcal{C}$ and was taken account of in the definition of $R_{\sigma \to \tau}^w$. □

Our logical relations have the usual "un-Currying" property.

**Lemma 2.** Let $(R_\sigma^w)_{\sigma \in \mathbb{T}}^{w \in Obj(\mathcal{C})}$ be a Kripke logical relation with varying arity. For any type $\sigma = \sigma_1 \to \ldots \to \sigma_n \to \iota$ and any object $w$, a tuple $(g_j)_{j \in w}$ is in $R_\sigma^w$ if and only if for every chain of maps $v_n \xrightarrow{f_n} \ldots \xrightarrow{f_2} v_1 \xrightarrow{f_1} w$ and tuples $(x_i^k)_{i \in v_k} \in R_{\sigma_k}^{v_k}$, $k = 1, \ldots, n$, the result of applying the functions coordinatewise to all $n$ arguments is in $R_\iota^{v_n}$.

*Proof.* Easy induction on the length of the unfolded types $\sigma_1 \to \ldots \to \sigma_n \to \iota$. □

In order to prove the "Fundamental Theorem of Logical Relations" (in the words of [12]) let us recall how the simply typed lambda calculus is interpreted over $A_\sigma$. Free variables are assigned values by environments $\rho: Var \to \bigcup_{\sigma \in \mathbb{T}} A_\sigma$ (where a variable $x^\sigma$ of type $\sigma$ is mapped to $A_\sigma$) and the denotation of a lambda term $M$ is then defined with respect to environments as follows:

$M \equiv x^\sigma$: $[\![x^\sigma]\!]\rho = \rho(x^\sigma)$.

$M \equiv M_1 M_2$: $[\![M_1 M_2]\!]\rho = [\![M_1]\!]\rho([\![M_2]\!]\rho)$.

$M \equiv \lambda x^\sigma.M_1$: $[\![\lambda x^\sigma.M_1]\!]\rho =$ the map which assigns to $a \in A_\sigma$ the value $[\![M_1]\!]\rho[x^\sigma \mapsto a]$. (In a general extensional applicative structure there need not be a representative in $A_{\sigma \to \tau}$ for this map. This is the "richness" of Henkin models we referred to in the Introduction.)

**Theorem 3.** *For every Kripke logical relation with varying arity $(R_\sigma^w)_{\sigma \in \mathbb{T}}^{w \in Obj(\mathcal{C})}$, object $w$ of $\mathcal{C}$, and closed term $M$ of type $\sigma$ the constant tuple $([\![M]\!])_{j \in w}$ is in $R_\sigma^w$.*

*Proof.* The proof is for all objects of $\mathcal{C}$ simultaneously by induction on the term structure. Hence we must also take open terms into account. For $w \in Obj(\mathcal{C})$ let $(\rho_j)_{j \in w}$ be a tuple of environments such that for every free variable $x^\sigma$ of $M$ the tuple $(\rho_j(x^\sigma))_{j \in w}$ is in $R_\sigma^w$. We show that under this condition the tuple $([\![M]\!]\rho_j)_{j \in w}$ is in $R_\sigma^w$. We check the three cases in the definition of $[\![\cdot]\!]$:

$M \equiv x^\sigma$: $([\![x^\sigma]\!]\rho_j)_{j \in w} = (\rho_j(x^\sigma))_{j \in w} \in R_\sigma^w$ by assumption.

$M \equiv M_1 M_2$: $([\![M_1 M_2]\!]\rho_j)_{j \in w} = ([\![M_1]\!]\rho_j([\![M_2]\!]\rho_j))_{j \in w}$. By induction hypothesis $([\![M_1]\!]\rho_j)_{j \in w}$ is in $R_{\sigma \to \tau}^w$ and $([\![M_2]\!]\rho_j)_{j \in w}$ is in $R_\sigma^w$. Because a category contains an identity morphism for every object, we get that the tuple resulting from pointwise application is in $R_\tau^w$.

$M \equiv \lambda x^\sigma.M_1$: $([\![\lambda x^\sigma.M_1]\!]\rho_j)_{j \in w}$ is a tuple of functions from $A_\sigma$ to $A_\tau$. To check that it is in relation we to apply to it a tuple $(a_i)_{i \in v}$ of arguments from $R_\sigma^v$ for an object $v$ and a morphism $f: v \to w$. We get the tuple $([\![M_1]\!]\rho_{f(i)}[x^\sigma \mapsto a_i])_{i \in v}$. From Lemma 1 we know that each of the tuples $(\rho_{f(i)}(y))_{i \in v}$, $y$ a variable, is in relation at $v$. Updating the environments at $x^\sigma$ to $(a_i)_{i \in v}$ retains this property. So we can conclude from the induction hypothesis that $([\![M_1]\!]\rho_{f(i)}[x^\sigma \mapsto a_i])_{i \in v}$ is in $R_\tau^v$. □

Let us emphasize again that the preceding theorem is neither a surprise nor a generalization over already established results. Our Kripke logical relation with varying arity is nothing more than a logical predicate over a particular Henkin model in the Cartesian closed functor category $Set^{\mathcal{C}^{op}}$. The point is that we want to look at a complicated logical relation over a simple Henkin model in order to characterize lambda definability in the latter. We included the proof of the Fundamental Theorem in order to acquaint the reader with the technical apparatus.

## 3  A Characterization of Lambda Definability

We will now characterize lambda definability in the full type hierarchy over some ground set $A_\iota$. (The proof for an arbitrary Henkin model is the same but involves more notational overhead.) From $A_\iota$ we construct a concrete category $\mathcal{A}$ as follows. Objects are finite products $A_{\sigma_1\ldots\sigma_n} = A_{\sigma_1} \times \ldots \times A_{\sigma_n}$ of our denotational domains, one for each sequence $\sigma_1\ldots\sigma_n$ of types. The empty sequence $\epsilon$ is represented by an arbitrary one-point set $A_\epsilon$. If $\sigma_1\ldots\sigma_n$ is a prefix of the sequence $\sigma_1\ldots\sigma_n\tau_1\ldots\tau_m$ then our category contains the projection morphism from $A_{\sigma_1} \times \ldots \times A_{\sigma_n} \times A_{\tau_1} \times \ldots \times A_{\tau_m}$ to $A_{\sigma_1} \times \ldots \times A_{\sigma_n}$. So $\mathcal{A}$ is really an ordered set, namely, the dual of $\mathbb{T}^*$ with the prefix ordering. The logical relation $(T_\sigma^w)_{\sigma \in \mathbb{T}}^{w \in Obj(\mathcal{A})}$ which will give us the characterization, has arity $|A_{\sigma_1} \times \ldots \times A_{\sigma_n}|$ at the object $w = A_{\sigma_1} \times \ldots \times A_{\sigma_n}$. A tuple from $T_\sigma^w$ is therefore indexed by tuples $a = (a_1, \ldots, a_n) \in A_{\sigma_1} \times \ldots \times A_{\sigma_n}$. At ground level we take those tuples $(x_a)_{a \in w}$ into $T_\iota^w$ for which there is a closed lambda term $M$ of type $\sigma_1 \to \ldots \to \sigma_n \to \iota$ such that for each $a = (a_1, \ldots, a_n)$ in $A_{\sigma_1} \times \ldots \times A_{\sigma_n}$ we have $x_a = [\![M]\!](a_1)\ldots(a_n)$. Intuitively, we take only those tuples which are "value tables" of lambda definable functions. This idea is taken directly from [8]. These relations have the compatibility property with morphisms in $\mathcal{A}$. Indeed, if $M$ defines the tuple $(x_a)_{a \in w}$ at $w = A_{\sigma_1} \times \ldots \times A_{\sigma_n}$ then $\lambda x_1^{\sigma_1}\ldots x_n^{\sigma_n} y_1^{\tau_1}\ldots y_m^{\tau_m}.M x_1^{\sigma_1}\ldots x_n^{\sigma_n}$ defines the corresponding tuple at $v = A_{\sigma_1} \times \ldots \times A_{\sigma_n} \times A_{\tau_1} \times \ldots \times A_{\tau_m}$. The following lemma asserts that the lambda definable functions can be read off at $A_\epsilon$, the one element object in $\mathcal{A}$.

**Lemma 4.** *A one-element tuple $(x)$ is in $T_\sigma^{A_\epsilon}$ if and only if $x$ is the denotation of a closed lambda term of type $\sigma$.*

*Proof.* We prove by induction on types (simultaneously for all objects $w = A_{\sigma_1} \times \ldots \times A_{\sigma_n}$ of $\mathcal{A}$) that $T_\sigma^w$ only contains tuples which are definable by closed lambda terms of type $\sigma_1 \to \ldots \to \sigma_n \to \sigma$. For $\sigma = \iota$ this is the definition of $T_\iota^w$, so let us look at a function type $\sigma \to \tau$.

If $M$ is a closed term of type $\sigma_1 \to \ldots \to \sigma_n \to \sigma \to \tau$ which defines the tuple $(f_a)_{a \in w}$ we want to assert that it is in relation at $w$. To this end we supply an argument tuple $(x_b)_{b \in v}$ for an object $v = A_{\sigma_1} \times \ldots \times A_{\sigma_n} \times A_{\tau_1} \times \ldots \times A_{\tau_m}$. By induction hypothesis, this tuple is represented by a closed term $N$ of type $\sigma_1 \to \ldots \to \sigma_n \to \tau_1 \to \ldots \to \tau_m \to \sigma$. The resulting tuple $(f_{\pi(b)}(x_b))_{b \in v}$ is represented by the term $\lambda x_1^{\sigma_1}\ldots x_n^{\sigma_n} y_1^{\tau_1}\ldots y_m^{\tau_m}.(M x_1^{\sigma_1}\ldots x_n^{\sigma_n})(N x_1^{\sigma_1}\ldots x_n^{\sigma_n} y_1^{\tau_1}\ldots y_m^{\tau_m})$ and so, by induction hypothesis, is contained in $T_\tau^v$.

Conversely, assume that the tuple $(f_a)_{a \in w}$ belongs to $T_{\sigma \to \tau}^w$. We supply it with the argument tuple over the object $v = A_{\sigma_1} \times \ldots \times A_{\sigma_n} \times A_\sigma$ which is given by the term $\lambda x_1^{\sigma_1}\ldots x_n^{\sigma_n} x^\sigma.x^\sigma$. By induction hypothesis it is contained in $T_\sigma^v$. The resulting tuple $(f_a(a))_{aa \in v}$ is in $T_\tau^v$ and, again by induction hypothesis, there is a closed term $N$ of type $\sigma_1 \to \ldots \to \sigma_n \to \sigma \to \tau$ representing it. We claim that $N$ also represents $(f_a)_{a \in w}$: Using the denotation of $N$ we get a tuple $(g_a)_{a \in w}$ of functions of type $\sigma \to \tau$ where $g_{(a_1,\ldots,a_n)} = [\![N]\!](a_1)\ldots(a_n)$. In order to see that such a function is equal to the corresponding $f_a$ we supply a generic argument $a$ from $A_\sigma$. We get $f_a(a) = [\![N]\!](a_1)\ldots(a_n)(a) = g_a(a)$, which completes our argument. □

Theorem 3 and Lemma 4 together give our main result:

***Theorem 5.*** *An element of a Henkin model is lambda definable if and only if it is invariant under all Kripke logical relations with varying arity.*

Somewhat slicker but maybe less transparent is the following description of the relations $T_\sigma^w$. We replace sequences of types by finite sets of variables. The objects of $\mathcal{A}$ remain almost the same, $\{x_1^{\sigma_1}, \ldots, x_n^{\sigma_n}\}$ corresponds to the set $Env\{x_1^{\sigma_1}, \ldots, x_n^{\sigma_n}\}$ of finite environments over this collection of variables. The tuples should now be labeled by our symbol for environments $\rho$. For $w = Env\{x_1^{\sigma_1}, \ldots, x_n^{\sigma_n}\}$ we take a tuple $(x_\rho)_{\rho \in w}$ into $T_\iota^w$ if there is a lambda term $M$ whose free variables are contained in $\{x_1^{\sigma_1}, \ldots, x_n^{\sigma_n}\}$ such that for all $\rho \in w$ we have $x_\rho = [\![M]\!]\rho$. The proof of Lemma 4 can be changed accordingly.

## 4 The Lambda Definability Problem

We return to the problem of finding an effective characterization of lambda definability for hereditarily finite Henkin models. Indeed, studying the definability lemma 4 one gets the impression that for a particular type $\phi$ only a finite piece of the category $\mathcal{A}$ is used. More formally, we can precisely define the objects from $\mathcal{A}$ that occur in the proof of Lemma 4. Fix a type $\phi$ and define two relations $\vdash_\phi$ and $\Vdash_\phi$ between strings of types and types as follows:

(i) $\epsilon \vdash_\phi \phi$,
(ii) if $s \vdash_\phi \sigma \to \tau$ then $s\sigma \vdash_\phi \tau$ and $s\sigma \Vdash_\phi \sigma$,
(iii) if $s \Vdash_\phi \sigma_1 \to \ldots \to \sigma_n \to \iota$ and if for strings $s_1 \leq \ldots \leq s_n$ there are types $\xi_1, \ldots, \xi_n$ such that for all $k = 1, \ldots, n$, $s_k \vdash_\phi \xi_k$ then for all $k = 1, \ldots, n$, $s_k \vdash_\phi \sigma_k$.

Now let $\mathcal{F}_\phi$ be the full sub-category of $\mathcal{A}$ whose objects are given by $\{A_s \in \mathcal{A} \mid \exists \xi \in \mathbb{T}. s \vdash_\phi \xi\}$. (Note that the strings occurring on the left hand side of the relation $\Vdash_\phi$ all occur on the left hand side of $\vdash_\phi$ already.) We show that the proof of Lemma 4 for a particular type $\phi$ can be based on $\mathcal{F}_\phi$. At ground type we start with the same logical relation $(T_\sigma^w)_{\sigma \in \mathbb{T}}^{w \in Obj(\mathcal{F}_\phi)}$ as before.

***Lemma 6.*** *Given a type $\phi \in \mathbb{T}$ the following is true for all $\sigma \in \mathbb{T}$ and $s \in \mathbb{T}^*$:*

(i) *If $s \vdash_\phi \sigma$ then every element of $T_\sigma^{A_s}$ is lambda definable.*
(ii) *If $s \Vdash_\phi \sigma$ then every lambda definable tuple is in $T_\sigma^{A_s}$.*

*Proof.* By induction on $\sigma$. If $\sigma$ is the ground type $\iota$ then both statements follow from the definition of $T_\iota^w$. The proof of (i) for a function type $\sigma \to \tau$ works as in Lemma 4: Assume $s = \sigma_1 \ldots \sigma_n \vdash_\phi \sigma \to \tau$ and $(f_j)_{j \in A_s} \in T_{\sigma \to \tau}^{A_s}$ (where we have identified $w$ with $A_s$). We have $s\sigma \Vdash_\phi \sigma$ and so by induction hypothesis we can apply the tuple given by the term $\lambda x_1^{\sigma_1} \ldots x_n^{\sigma_n} x^\sigma . x^\sigma$ to it. The result is in $T_\tau^{A_s \times A_\sigma}$ and since $s\sigma \vdash_\phi \tau$ it is given by a term $N$. As before we see easily that $N$ also defines $(f_j)_{j \in A_s}$.

To prove part (ii) we have to un-Curry completely: $\sigma \to \tau = \sigma_1 \to \ldots \to \sigma_n \to \iota$ (we have re-named $\sigma$ to $\sigma_1$). Assume that the tuple $(f_j)_{j \in A_*}$ is given by a term $M$. By Lemma 2 we have to apply the functions to argument tuples from $T^{A_{*k}}_{\sigma_k}$, $k = 1, \ldots, n$ for strings $s \leq s_1 \leq \ldots \leq s_n$ from $\mathcal{F}_\phi$. By our rule (iii) we have for each $k$, $s_k \vdash_\phi \sigma_k$. Hence we can use the induction hypothesis and conclude that all argument tuples are lambda definable. The application of $(f_j)_{j \in A_*}$ to these argument tuples results in a tuple which again is lambda definable and of type $\iota$. But at ground type lambda definable tuples are in relation and we are done. $\square$

*Theorem 7.* *An element of type $\phi$ of a Henkin model is lambda definable if and only if it is invariant under all logical relations based on $\mathcal{F}_\phi$.*

If we are looking at a hereditarily finite Henkin model, for example the full type hierarchy over a finite ground set, and if for some type $\phi$ the category $\mathcal{F}_\phi$ happens to have only finitely many objects then we can effectively determine the lambda definable elements of $A_\phi$ by simply checking the finitely many Kripke logical relations with varying arity over $\mathcal{F}_\phi$. Unfortunately, this approach can only succeed for types of rank less than 3:

*Lemma 8.* *For every type $\phi$ of rank at least 3 the category $\mathcal{F}_\phi$ has infinitely many objects.*

*Proof.* We illustrate the idea for the type $\phi = ((\iota \to \iota) \to \iota) \to \iota$. The general proof is exactly the same but involves a lot of indices. Using rules (i)–(iii) above, we get

(1) $\epsilon \vdash_\phi \phi$ by (i).
(2) $(\iota \to \iota) \to \iota \vdash_\phi \iota$ and $(\iota \to \iota) \to \iota \Vdash_\phi (\iota \to \iota) \to \iota$ by (1) and (ii).
(3) $(\iota \to \iota) \to \iota \vdash_\phi \iota \to \iota$ by (2) and (iii).
(4) $\langle (\iota \to \iota) \to \iota \rangle \langle \iota \rangle \vdash_\phi \iota$ by (3) and (ii).
(5) $\langle (\iota \to \iota) \to \iota \rangle \langle \iota \rangle \vdash_\phi \iota \to \iota$ by (2), (4), and (iii).

The last two steps can be repeated forever. $\square$

Behind this proof is the observation that from rank 3 on we can no longer bound the number of variables occurring in a normal form. What happens if we do impose a bound is the topic of the next section.

## 5 Lambda Definability with Fixed Sets of Variables

### 5.1 Two-layered logical relations

We proceed by further refining the notion of logical relation and we begin by studying this refinement for ordinary logical relations, letting the varying arity and the Kripke universe at the side for the moment.

Observe that the definition of the extension of a logical relation to a type $\sigma \to \tau$ falls naturally into two halves:

(1) If $f: A_\sigma \to A_\tau$ belongs to $R_{\sigma \to \tau}$ then it maps each element of $R_\sigma$ into $R_\tau$.
(2) If $f: A_\sigma \to A_\tau$ maps each element of $R_\sigma$ into $R_\tau$ then it belongs $R_{\sigma \to \tau}$.

In the proof of the Fundamental Theorem the first condition is needed in order to show that an application remains invariant if the constituents are, and the second is needed for abstraction. Of course, the power of logical relations resides in the fact that the two properties are fulfilled simultaneously. Nevertheless, we shall study these two conditions separately and thus tie up our logical relations more closely with the structure of lambda terms. To this end we define the following two-layer type system $(\mathbb{T}_0, \mathbb{T}_1)$ (over a single ground type $\iota$ and over the set Var of variables):

- $\iota \in \mathbb{T}_0$
- $\sigma, \tau \in \mathbb{T}_0 \Longrightarrow \sigma \to \tau \in \mathbb{T}_0$
- $B \in \text{Var}^*, \tau \in \mathbb{T}_0 \Longrightarrow B \to \tau \in \mathbb{T}_1$

Note that $\mathbb{T}_0$ may be viewed as a subset of $\mathbb{T}_1$ by virtue of the empty string in Var$^*$. We will also need the forgetful map $e: \mathbb{T}_1 \to \mathbb{T}_0$ which maps $x_1^{\sigma_1} \ldots x_n^{\sigma_n} \to \tau$ to $\sigma_1 \to \ldots \to \sigma_n \to \tau$.

Now let $R_\iota \subseteq A_\iota$ be an arbitrary relation (for simplicity we let the arity be 1). It is extended to the types of $\mathbb{T}_0$ and $\mathbb{T}_1$ as follows. For $\sigma \to \tau \in \mathbb{T}_0$ let $R_{\sigma \to \tau}$ be *any subset* of

$$\{f \in A_{\sigma \to \tau} \mid \forall \Sigma \in \mathbb{T}_1. (e(\Sigma) = \sigma \Longrightarrow \forall a \in R_\Sigma . f(a) \in R_\tau)\}$$

and for $n \geq 1, B = x_1^{\sigma_1} \ldots x_n^{\sigma_n}, B \to \tau \in \mathbb{T}_1$ let $R_{B \to \tau}$ be *any superset* of

$$\{f \in A_{e(B \to \tau)} \mid \forall a_1 \in R_{\sigma_1} \ldots \forall a_n \in R_{\sigma_n} . f(a_1) \ldots (a_n) \in R_\tau\} .$$

Obviously, such two-layered relations are no longer determined by their value at ground type. But starting from some $R_\iota$ we can always construct a two-layered logical relation. The Fundamental Theorem now reads as follows:

**Theorem 9.** *Let $M \equiv \lambda x_1^{\sigma_1} \ldots x_n^{\sigma_n} . N$ be a lambda term in normal form and of type $\sigma_1 \to \ldots \to \sigma_n \to \tau$ such that $N$ is not an abstraction and let $\rho$ be an environment which maps each free variable $y^\sigma$ of $M$ into $R_\sigma$. Then $[\![M]\!]\rho \in R_{x_1^{\sigma_1} \ldots x_n^{\sigma_n} \to \tau}$.*

*Proof.* We have to argue more carefully, but the proof is essentially as usual. Variables can't cause any problems. In the case that $M$ is an application $M_1 M_2$, we employ the assumption that $M$ is in normal form, hence the denotation of $M_1$ under $\rho$ is in $R_{\sigma \to \tau}$ where $\sigma \to \tau$ is an ordinary type. The denotation of $M_2$ under $\rho$ is in some $R_\Sigma$ where $e(\Sigma) = \sigma$. So the composed term is in $R_\tau$ as required.

The case that $M$ is an abstraction is characterized by the fact that $n \geq 1$. Unlike in the usual proof we have to unwind all leading lambdas at one stroke. We want $[\![M]\!]\rho$ to be in $R_{x_1^{\sigma_1} \ldots x_n^{\sigma_n} \to \tau}$ and to check this we have to apply it to arguments $a_i$ from $R_{\sigma_i}$, $i = 1, \ldots, n$, and see whether the result is in $R_\tau$.

This is indeed the case, as $[\![M]\!]\rho(a_1)\ldots(a_n) = [\![N]\!]\rho[x_1 \mapsto a_1,\ldots,x_n \mapsto a_n]$ and the induction hypothesis applies to $N$ and the updated environment. (The updating must be read from left to right. This way the lemma remains true also for sequences $x_1^{\sigma_1}\ldots x_n^{\sigma_n}$ which contain some variables more than once.) $\square$

### 5.2 Two-layered Kripke logical relations with varying arity

Let us now combine the techniques of Sect. 2 with these two-layered logical relations. We use the presentation of Kripke logical relations with varying arity via environments as briefly described at the end of Sect. 2.

So let $V \subseteq \mathrm{Var}$ be a set of variables. It is our goal to characterize all functionals which are definable by lambda terms containing variables (free or bound) only from $V$. Our base category $\mathcal{V}$ is the set of all subsets of $V$ together with inclusion morphisms. There is a contravariant equivalence between $\mathcal{V}$ and the category $\mathcal{E}$ of environments $\mathrm{Env}(F)$ over sets $F$ of variables contained in $V$ with restriction maps. It no longer helps to think of $\mathcal{E}$ as a concrete example of a general category, as we make use of its particular structure. In other words, we have so far no abstract concept for a two-layered Kripke logical relation with varying arity.

For each object in $\mathcal{V}$, that is, for each set $F$ of variables contained in $V$, we want a two-layered logical relation $(R_\Sigma^F)_\Sigma$ of arity $\left|\prod_{x^\sigma \in F} A_\sigma\right|$. (Elements from the set $\prod_{x^\sigma \in F} A_\sigma$ serve a double purpose. We use them to index elements from the relations and we use them as environments. From now on, we will always use the letter $\mu$ to denote them.) Since we have restricted the set of variables available we cannot allow arbitrary types $\Sigma$ to occur, only those $\Sigma = B \to \sigma$ for which the sequence $B = x_1^{\sigma_1}\ldots x_n^{\sigma_n}$ contains each variable at most once and all variables are contained in $V$. Let us call such sequences and types built from them *non-repeating* and let $\mathbb{T}_1(V)$ stand for the collection of all non-repeating types over $V$. We will also allow ourselves to treat $B$ as a set sometimes, just to keep the complexity of our formulas within manageable range.

**Definition 10.** Let $(R_\Sigma^F)_{\Sigma \in \mathbb{T}_1(V)}^{F \subseteq V}$ be a family of relations such that the following conditions are satisfied:

(1) $\forall \sigma \to \tau \in \mathbb{T}_0 . R_{\sigma \to \tau}^F \subseteq \{(f_\mu)_{\mu \in \mathrm{Env}(F)} \in A_{\sigma \to \tau}^{\mathrm{Env}(F)} \mid \forall \Sigma \in \mathbb{T}_1(V).(e(\Sigma) = \sigma$
$\implies \forall (x_\mu)_{\mu \in \mathrm{Env}(F)} \in R_\Sigma^F.(f_\mu(x_\mu))_{\mu \in \mathrm{Env}(F)} \in R_\tau^F)\}$,

(2) $\forall \Sigma \in \mathbb{T}_1(V)$ where $\Sigma = B \to \tau$ and $B = x_1^{\sigma_1}\ldots x_n^{\sigma_n}, n \geq 1$ it is the case that $R_\Sigma^F \supseteq \{(f_\mu)_{\mu \in \mathrm{Env}(F)} \in A_{e(\Sigma)}^{\mathrm{Env}(F)} \mid f_\mu = f_{\mu'}$ if $\mu\vert_{F \setminus B} = \mu'\vert_{F \setminus B}$ and
$\forall (a_\mu^1)_{\mu \in \mathrm{Env}(F \cup B)} \in R_{\sigma_1}^{F \cup B},\ldots,\forall (a_\mu^n)_{\mu \in \mathrm{Env}(F \cup B)} \in R_{\sigma_n}^{F \cup B}.$
$(f_{\mu\vert_F}(a_\mu^1)\ldots(a_\mu^n))_{\mu \in \mathrm{Env}(F \cup B)} \in R_\tau^{F \cup B}\}$,

(3) $\forall \sigma \in \mathbb{T}_0 \forall F \subseteq F' \subseteq V.(x_\mu)_{\mu \in \mathrm{Env}(F)} \in R_\sigma^F \implies (x_{\mu'\vert_F})_{\mu' \in \mathrm{Env}(F')} \in R_\sigma^{F'}$.

If these three conditions are satisfied then we call the family $(R_\Sigma^F)_{\Sigma \in \mathbb{T}_1(V)}^{F \subseteq V}$ a *two-layered Kripke logical relation with varying arity over $V$*.

We need to check carefully whether the Fundamental Theorem remains valid:

**Theorem 11.** *Let $M \equiv \lambda x_1^{\sigma_1} \ldots x_n^{\sigma_n}.N$ be a lambda term in normal form and of type $\sigma_1 \to \ldots \to \sigma_n \to \tau$ such that $N$ is not an abstraction, let $F \subseteq V \subseteq \text{Var}$ be sets of variables such that all variables occurring in $M$ are contained in $V$ and such that all its free variables are contained in $F$, let $(R_\Sigma^F)_{\Sigma \in \mathbb{T}_1(V)}^{F \subseteq V}$ be a two-layered Kripke logical relation with varying arity over $V$ and, finally, let $(\rho_\mu)_{\mu \in \text{Env}(F)}$ be a family of environments such that for all $x^\sigma \in FV(M)$, $(\rho_\mu(x^\sigma))_{\mu \in \text{Env}(F)}$ is in $R_\sigma^F$. Then the tuple $(\llbracket M \rrbracket \rho_\mu)_{\mu \in \text{Env}(F)}$ is in $R_{x_1^{\sigma_1} \ldots x_n^{\sigma_n} \to \tau}^F$.*

*Proof.* The proof is by induction on the complexity of $M$, simultaneously for all appropriate $F, V$, and $(\rho_\mu)_{\mu \in \text{Env}(F)}$. The situation is trivial as usual for variables. If $M \equiv M_1 M_2$ is an application then because $M$ is in normal form, $M_1$ is not an abstraction. The free variables of $M_1$ and $M_2$ are also contained in $F$. We can therefore apply the induction hypothesis and get that $(\llbracket M_1 \rrbracket \rho_\mu)_{\mu \in \text{Env}(F)}$ is in $R_{\sigma \to \tau}^F$ and $(\llbracket M_2 \rrbracket \rho_\mu)_{\mu \in \text{Env}(F)}$ is in $R_\Sigma^F$ where $e(\Sigma) = \sigma$. So $(\llbracket M \rrbracket \rho_\mu)_{\mu \in \text{Env}(F)}$ is in $R_\tau^F$ by part (1) of the definition.

Let now $M \equiv \lambda x_1^{\sigma_1} \ldots x_n^{\sigma_n}.N$ be an abstraction, that is, $n \geq 1$. We want to see that $(\llbracket M \rrbracket \rho_\mu)_{\mu \in \text{Env}(F)}$ is in $R_{B \to \tau}^F$ where we have introduced $B$ as an abbreviation for $x_1^{\sigma_1} \ldots x_n^{\sigma_n}$. Let $F'$ stand for $F \cup B$. By part (2) of our definition we have to supply the functions $\llbracket M \rrbracket \rho_{\mu|_F}$ with arguments $(a_\mu^i)_{\mu \in \text{Env}(F')}$ from $R_{\sigma_i}^{F'}$, $i = 1, \ldots, n$. But since $(\llbracket M \rrbracket \rho_{\mu|_F} (a_\mu^1) \ldots (a_\mu^n))_{\mu \in \text{Env}(F')}$ equals $(\llbracket N \rrbracket \rho_{\mu|_F} [x_1 \mapsto a_\mu^1, \ldots, x_n \mapsto a_\mu^n])_{\mu \in \text{Env}(F')}$ we may apply the induction hypothesis to $N, F'$, and $(\rho_{\mu|_F} [x_1 \mapsto a_\mu^1, \ldots, x_n \mapsto a_\mu^n])_{\mu \in \text{Env}(F')}$. That the new family of environments meets the requirements of the theorem is a consequence of the persistency part (3) of our definition. □

### 5.3 The term relation

As usual, a term construction will give us completeness of the characterization. So fix a set $V$ of variables and let $(T_\Sigma^F)_{\Sigma \in \mathbb{T}_1(V)}^{F \subseteq V}$ be the family of relations for which each $T_\Sigma^F$ is the collection of all $(f_\mu)_{\mu \in \text{Env}(F)} \in A_\Sigma^{\text{Env}(F)}$ definable by lambda terms, i.e., $(f_\mu)_{\mu \in \text{Env}(F)}$ is in $T_\Sigma^F$ if there exists a lambda term $M \equiv \lambda x_1^{\sigma_1} \ldots x_n^{\sigma_n}.N$ ($N$ not an abstraction) in normal form all of whose variables belong to $V$, all of whose free variables belong to $F$, for which $x_1^{\sigma_1} \ldots x_n^{\sigma_n} = B$ is non-repeating and $B \to \tau = \Sigma$ such that $\forall \mu \in \text{Env}(F)$ we have $f_\mu = \llbracket M \rrbracket \mu$. We have to check that we get a valid relation this way.

**Lemma 12.** $(T_\Sigma^F)_{\Sigma \in \mathbb{T}_1(V)}^{F \subseteq V}$ *is a two-layered Kripke logical relation with varying arity over $V$.*

*Proof.* (1) Let $(f_\mu)_{\mu \in \text{Env}(F)}$ be in $T^F_{\sigma \to \tau}$. Then this tuple is given by a term $M$ of type $\sigma \to \tau$ which is not an abstraction. Let further $N$ be term which defines a tuple $(x_\mu)_{\mu \in \text{Env}(F)}$ from $T^F_\Sigma$ where $e(\Sigma) = \sigma$. Then $MN$ is in normal form and defines $(f_\mu(x_\mu))_{\mu \in \text{Env}(F)}$.

(2) Let $(f_\mu)_{\mu \in \text{Env}(F)}$ be an element from the right hand side of (2) in Definition 10. We can apply it to the tuples defined by $x_1^{\sigma_1}, \ldots, x_n^{\sigma_n}$ and the result $(f_{\mu|_F}(\llbracket x_1 \rrbracket \mu) \ldots (\llbracket x_n \rrbracket \mu))_{\mu \in \text{Env}(F')}$ will be in $T^{F'}_\tau$, hence given by a lambda term $M$ which is not an abstraction. We claim that $(f_\mu)_{\mu \in \text{Env}(F)}$ is given by $\lambda x_1^{\sigma_1} \ldots x_n^{\sigma_n}.M$. Indeed, if $a_1 \in A_{\sigma_1}, \ldots, a_n \in A_{\sigma_n}$ are arguments for the function $f_\mu$ then

$$f_\mu(a_1)\ldots(a_n) = f_{\mu'|_F}(\llbracket x_1 \rrbracket \mu') \ldots (\llbracket x_n \rrbracket \mu')$$
$$= \llbracket M \rrbracket \mu'$$
$$= \llbracket \lambda x_1^{\sigma_1} \ldots x_n^{\sigma_n}.M \rrbracket \mu(a_1) \ldots (a_n)$$

where $\mu'(y) = \begin{cases} a_i & \text{if } y \equiv x_i; \\ \mu(y) & \text{otherwise.} \end{cases}$ Here we have used the fact that $f_\mu = f_{\mu'|_F}$ because $\mu$ and $\mu'|_F$ differ only at variables from $F \cap \{x_1, \ldots, x_n\}$. Also the fact that $x_1^{\sigma_1} \ldots x_n^{\sigma_n}$ is non-repeating is crucial here.

(3) Persistency is clear as the denotation of a term only depends on its free variables. □

**Theorem 13.** *A functional is definable from a fixed set $V$ of variables if and only if it is invariant under all two-layered Kripke logical relations with varying arity over $V$.*

## 5.4 Decidability

We are now ready to harvest the fruit from our hard labor in this section. Unlike for full definability, the notion of definability from a fixed set of variables becomes decidable if we restrict to finite ground sets $A_\sigma$ and finite sets $V$ of variables. The reason for this simply is that there are only finitely many relations to check. Thus we have:

**Theorem 14.** *The problem whether a given functional from a hereditarily finite Henkin model is lambda definable by a term over a fixed finite set of variables is decidable.*

Although two-layered Kripke logical relations with varying arity over some set of variables may seem complicated, there is nevertheless a fairly simple underlying idea. The relations can be thought of as value tables for functionals where the new types $\Sigma \in \mathbb{T}_1$ and the restrictions to subsets $F$ of $V$ are just a way of keeping track of free and bound variables in defining terms. (Note that a variable may be re-used several times, that is, may occur both bound and free.) Gordon Plotkin has suggested to us that one may obtain Theorem 14 by working backwards from the given value table for a functional in search for a defining term. At each stage, one determines the *set* of value tables which, applied in the right order, give a value table in the set of sought after tables. Each branch of the search stops if either a projection (which corresponds to a variable) can satisfy the requirements or if only tables occur which we are looking for already. Since the set of variables is restricted the tables are finite objects and the search must eventually end. Bookkeeping over free and bound variables is also necessary in this approach and while we haven't formally carried through this approach, we think that it will amount to a scheme with probably the same complexity as ours.

## Acknowledgement

The results reported here were obtained while the second author was holding a visiting professorship at Technische Hochschule Darmstadt. We would like to thank Allen Stoughton for directing our attention to Kurt Sieber's paper on sequential logical relations.

The results presented in Sect. 5 were obtained while the first author visited the University of Sussex at the invitation of Matthew Hennessy and Allen Stoughton.

## References

1. A. Jung and A. Stoughton. Studying the Fully Abstract Model of PCF within its Continuous Function Model. In this proceedings, 1993.
2. H. Läuchli. An Abstract Notion of Realizability for which Intuitionistic Predicate Calculus is Complete. In A. Kino, J. Myhill, and R. E. Vesley, editors, *Intuitionism and Proof Theory, Proc. summer conference at Buffalo N.Y., 1968*, pages 227–234. North-Holland, 1970.
3. J. Lambek and P. J. Scott. *Introduction to Higher Order Categorical Logic*. Cambridge Studies in Advanced Mathematics Vol. 7. Cambridge University Press, 1986.
4. J. C. Mitchell. Type Systems for Programming Languages. In J. van Leeuwen, editor, *Handbook of Theoretical Computer Science*, pages 365–458. North Holland, 1990.
5. J.C. Mitchell and E. Moggi. Kripke-style models for typed lambda calculus. *Annals of Pure and Applied Logic*, 51:99–124, 1991. Preliminary version in *Proc. IEEE Symp. on Logic in Computer Science,* 1987, pages 303–314.
6. G. D. Plotkin. Lambda-Definability and Logical Relations. Memorandum SAI-RM-4, University of Edinburgh, October 1973.
7. G. D. Plotkin. Lambda-Definability in the Full Type Hierarchy. In Jonathan P. Seldin and J. Roger Hindley, editors, *To H. B. Curry: Essays on Combinatory Logic, Lambda Calculus and Formalism*, pages 363–373. Academic Press, London, 1980.
8. K. Sieber. Reasoning about Sequential Functions via Logical Relations. In M. P. Fourman, P. T. Johnstone, and A. M. Pitts, editors, *Proc. LMS Symposium on Applications of Categories in Computer Science, Durham 1991*, volume 177 of *LMS Lecture Note Series*, pages 258–269. Cambridge University Press, 1992.

9. R. Statman. Completeness, Invariance and λ-definability. *Journal of Symbolic Logic*, 47:17–26, 1982.
10. R. Statman. Embeddings, Homomorphisms and λ-definability. Manuscript, Rutgers University, 1982.
11. R. Statman. λ-definable Functionals and βη Conversion. *Arch. Math. Logik*, 23:21–26, 1983.
12. R. Statman. Logical Relations and the Typed λ-Calculus. *Information and Control*, 65:85–97, 1985.
13. R. Statman and G. Dowek. On Statman's Finite Completeness Theorem. Technical Report CMU-CS-92-152, Carnegie Mellon University, 1992.

# Combining Recursive and Dynamic Types*

Hans Leiß
internet: leiss@cis.uni-muenchen.de
CIS, Universität München
Leopoldstraße 139
D-8000 München 40
Germany

**Abstract.** A denotational semantics of simply typed lambda calculus with a basic type Dynamic, modelling values whose type is to be inspected at runtime, has been given by Abadi e.a.[1]. We extend this interpretation to cover (formally contractive) recursive types as well. Soundness of typing rules and freeness of run-time type errors for well-typed programs hold.

The interpretation works also for implicitly polymorphic languages like *ML* with Dynamic and recursive types, and for explicitly polymorphic languages under the types-as-ideals interpretation.

## 1 Introduction

Static typing of programming languages has well known advantages like error detection at compile time and efficient object code free of type-checking at run-time. However, for programs that interact with storage media, other programs, or humans, it is often impossible to determine all relevant type information at compile time.

For example, consider a program operating on data that are modified after it has been compiled, are provided by a user interactively, or are fetched from external storage media. One would like the running program to inspect the type of the data, continue computation if this type is compatible with the type expected by the program, and terminate computaion or raise an exception, otherwise. Clearly, the use of such dynamic type checking should not compromise the soundness of the typing system.

In the last few years, several attempts have been made to make restricted use of dynamic typing in statically typed programming languages. A. Mycroft[13], building on ideas of M. Gordon to model dynamic typing by pairs $\langle v, \tau \rangle$ of values $v$ with their types $\tau$, introduced a type Dynamic, henceforth called dyn, as an infinite disjoint sum of types, each summand containing values tagged with their type. Inspection of these "dynamic values" is accomplished by a case-statement that branches according to finitely many type patterns, which exhaust the infinitely many types a dynamic value could have.

Mycroft proposed an extension of *ML*[6] by dynamic values. Actually, a version of dyn has been built into *CAML*[5]. The background of this implementation and alternative designs for adding dynamics to *ML* are treated by Leroy/Mauny[8]. Other

---

* This work has been supported by the Esprit Working Group BRA 7232, GENTZEN.

languages like L. Cardelli's[3] *AMBER* also have dynamic values. In fact, the implementation of *AMBER* took advantage of treating compiled modules as dynamic values.

M. Abadi, L. Cardelli, B. Pierce and G. Plotkin[1] investigated dynamic typing in a systematic way. They gave operational and denotational semantics of the simply typed lambda calculus with a basic type dyn. Additionally, they presented some ideas on polymorphically typed languages with dyn, focussing on difficulties in matching polymorphic types against patterns. This last aspect has been further investigated very recently by M. Abadi, L. Cardelli, B. Pierce and D. Remy[2], with extensions to languages with subtyping and abstract data types. A rather different study of dynamic type checking has been given by F. Henglein[7].

To give a denotational semantics for dyn, one has to establish

$$\langle v, \tau \rangle \in [\![\text{dyn}]\!] \quad \text{if and only if} \quad v \in [\![\tau]\!],$$

for arbitrary values $v$ and (closed) types $\tau$, including $\tau = \text{dyn}$. In the case of simply typed $\lambda$-calculus, Abadi e.a.[1] work in the ideal model of D. MacQueen, G. Plotkin and R. Sethi[9], and define $[\![\text{dyn}]\!]$ recusively by mimicking the construction of types over the set of pairs $\langle v, \tau \rangle$. However, for the combinations of dyn with polymorphic languages, suggested in [1, 2] and [8], no denotational semantics was known so far.

Our main concern is to extend the model of Abadi e.a.[1] to cover recursive types $\mu\alpha.\tau$ as well, such as the trees and lists of real programming languages. When recursive types are present, the method of defining $[\![\text{dyn}]\!]$ by mimicking the construction of types seems impossible: $v \in [\![\mu\alpha.\tau]\!]$ cannot be characterized by conditions of the form $u \in [\![\sigma]\!]$ for types $\sigma$ simpler than $\tau$.

A more general approach is needed in order to define the meaning of types when dyn and recursive types are combined. The main contribution of this paper is that while -in this situation- the recursive equations for type interpretations do *not* constitute a well-founded recursion on types, we can solve them simultaneously for all types, using Banach's fixed point theorem on an infinite product space of the metric space of ideals.

An advantage of our method is that it allows to give a denotational semantics for dynamic (and recursive) types in polymorphic languages as well. On the other hand, in this case we have neglected somewhat the semantics of terms in that only closed type-tags and first-order pattern-variables in typecase-expressions have been treated.

In Section 2.1 we present the syntax of simply typed $\lambda$-calculus with dynamic and recursive types, and in Section 2.2 review the construction of the ideal model and some preliminaries for our extension. In Section 2.3 we will construct the ideal interpretation for simply typed $\lambda$-calculus with dyn and recursive types. Section 3 extends this to an interpretation of dynamic and recursive types combined with explicit and implicit polymorphism.

## 2 Dynamic and recursive types in a simply typed language

Following ideas of M. Gordon, we view *dynamic types* as sets of pairs $\langle v, \tau \rangle$ containing a value $v$ of type $\tau$ together with its type. Programs are allowed to access the type-tag and make their actions depend on it. Compared with sum types $(\tau_1 + \tau_2)$, whose elements $\langle v, i \rangle$ contain a tag $i$ indicating that $v$ is of type $\tau_i$, dynamic types are more flexible in that their tags range over an infinite set. They can be modelled as infinite sums, with the restriction that the infinite set of tags has some algebraic structure that allows programs to use branching according to finitely many tag-*patterns*, instead of an infinite distinction on tags. A. Mycroft[12] first considered dynamic types as infinite sums of this kind.

Instead of working with different dynamic types, it is sufficient to consider one type dyn containing all these ⟨value,tag⟩-pairs, with the type as the algebra of tags.

## 2.1 Simply typed λ-calculus with dyn and recursive types

In this section we give the syntax of a simply typed λ-calculus with recursive types and a basic type dyn. It is similar to that of [1], to which we also refer for example programs. The types we consider are basic types including dyn, disjoint sums, product, function space, and recursive types.

**Definition 1.** *Types $\tau$ and terms $t$ are defined by the following grammar:*

$$\tau = \alpha \mid \text{bool} \mid \text{nat} \mid \text{dyn} \mid (\tau + \tau) \mid (\tau \times \tau) \mid (\tau \to \tau) \mid \mu\alpha.\tau$$

$$\begin{aligned}
t = {} & x \mid \textbf{wrong} \\
  & \mid \mathit{tt} \mid \mathit{ff} \mid (\text{case } t \text{ of } \mathit{tt} \text{ then } t, \mathit{ff} \text{ then } t) \\
  & \mid 0 \mid S(t) \mid (\text{case } t \text{ of } 0 \text{ then } t, S(x) \text{ then } t) \\
  & \mid (\text{dynamic } t : \tau) \mid (\text{typecase } t \text{ of } \{\alpha_1, \ldots, \alpha_n\} \, x : \tau \text{ then } t \text{ else } t) \\
  & \mid in_{1,\tau+\tau}(t) \mid in_{2,\tau+\tau}(t) \mid (\text{case } t \text{ of } \langle x, 1\rangle \text{ then } t, \langle x, 2\rangle \text{ then } t) \\
  & \mid (t, t) \mid \pi_1 t \mid \pi_2 t \\
  & \mid \lambda x : \tau.t \mid (t \cdot t)
\end{aligned}$$

The intended meaning of an expression (dynamic $t : \tau$) is to create a *dynamic value* $\langle v, \tau \rangle$, which may then be stored to external media or otherwise brought out of the control of the running program.

Conversely, the meaning of (typecase $d$ of $\{\alpha_1, \ldots, \alpha_n\}$ $x : \tau$ then $r$ else $s$) is to match the type $\sigma$ of $\langle v, \sigma \rangle$, the value of $d$, against the pattern $\tau$, and if the match succeeds, then continue with $r$ –using $v$ for $x$ and the types found by the match for the pattern-variables $\alpha_i$–, else with $s$. If the value of $d$ is not a dynamic, an error occurs. (C.f. Section 2.4 and 2.5 for details.) The variables $x$ and $\alpha_i$ are bound variables with scope $x : \tau$ and $r$.[2] To handle embedded patterns, the distinction between free and bound pattern-variables is made explicit by binding guards $\{\alpha_1, \ldots, \alpha_n\}$, as in

$$\lambda x : \text{dyn}.\lambda f : \text{dyn}. \;(\text{typecase } x \text{ of } \{\alpha\} \; y : \alpha \times \alpha$$
$$\text{then } (\text{typecase } f \text{ of } \{\beta\} \; g : \alpha \times \alpha \to \beta$$
$$\text{then } (\text{dynamic } gy : \beta)$$
$$\text{else } (\text{dynamic } \pi_1 y : \alpha))$$
$$\text{else } (\text{dynamic "not a nice pair"} : \text{string}))$$

---

[2] A more flexible syntax, (typecase $d$ of $\{x_1, \ldots, x_m, \alpha_1, \ldots, \alpha_n\}$ $p : \tau$ then $r$ else $s$) with a term-pattern $p$, would allow to inspect the value component as well. Essentially, the case (typecase $d$ of $\{x_1, \ldots, x_m\}$ $p : \tau$ then $r$ else $s$) is available in *CAML*.

To simplify notation, from now on we will only use typecase-expressions binding a single pattern-variable.

The meaning of the remaining expressions is fairly standard and omitted. (See also [1] for operational and denotational semantics of terms.) It may be sufficient to mention that the variable $x$ in the case-statements is a pattern-variable bound in the corresponding then-branch.

From the terms generated by the above grammar, a subclass of *well-typed* terms is defined using a type inference system. We only give those rules of the system in Abadi e.a.[1] that deal with dynamics[3], those of [9] that deal with recursive types, and those for function types.

A *type basis* is a finite sequence $\Gamma$ of *typing statements* $x : \tau$, assigning types $\tau$ to object variables $x$. If $\Gamma$ contains several assumptions $x : \sigma$ for the same object variable $x$, then $\Gamma(x)$, –*the type assigned to $x$ by $\Gamma$*– is the type $\sigma$ of the rightmost of these. When writing $\Gamma \triangleright t : \tau$, we always assume that $\Gamma$ a typing assumption for each free variable of $t$.

**Definition 2 (Typing Rules).**

(Var) $\quad \dfrac{}{\Gamma \triangleright x : \Gamma(x)}$

(Dyn I)$_e$ $\quad \dfrac{\Gamma \triangleright t : \tau}{\Gamma \triangleright (\text{dynamic } t : \tau) : \text{dyn}}$, if $\tau$ is closed

(Dyn E)$^-$ $\quad \dfrac{\Gamma \triangleright d : \text{dyn}, \quad \Gamma, x : \tau[\rho/\alpha] \triangleright r[\rho/\alpha] : \sigma \text{ for all closed } \rho, \quad \Gamma \triangleright s : \sigma}{\Gamma \triangleright (\text{typecase } d \text{ of } \{\alpha\}\ x : \tau \text{ then } r \text{ else } s) : \sigma}$

($\mu$ I) $\quad \dfrac{\Gamma \triangleright t : \mu\alpha.\tau}{\Gamma \triangleright t : \tau[\mu\alpha.\tau/\alpha]}$ $\qquad$ ($\mu$ E) $\quad \dfrac{\Gamma \triangleright t : \tau[\mu\alpha.\tau/\alpha]}{\Gamma \triangleright t : \mu\alpha.\tau}$

($\rightarrow$ I)$_e$ $\quad \dfrac{\Gamma, x : \sigma \triangleright t : \tau}{\Gamma \triangleright \lambda x : \sigma.t : (\sigma \rightarrow \tau)}$ $\qquad$ ($\rightarrow$ E) $\quad \dfrac{\Gamma \triangleright t : \sigma \rightarrow \tau, \quad \Gamma \triangleright s : \sigma}{\Gamma \triangleright (t \cdot s) : \tau}$

Note that rule (Dyn E)$^-$ has an infinite number of premises; these capture what is needed in the soundness Theorem 18. For practical purposes, it is more natural to use the following more restrictive rule:

(Dyn E) $\quad \dfrac{\Gamma \triangleright d : \text{dyn}, \quad \Gamma, x : \tau \triangleright r : \sigma, \quad \Gamma \triangleright s : \sigma}{\Gamma \triangleright (\text{typecase } d \text{ of } \{\alpha\}\ x : \tau \text{ then } r \text{ else } s) : \sigma}$, $\alpha$ not free in $\Gamma, \sigma$.

This demands that there is a uniform proof of $\Gamma, x : \tau[\rho/\alpha] \triangleright r[\rho/\alpha] : \sigma$ for all $\rho$.

*Example 1.* Assuming a further base type Unit with single element (), one can define the type of lists of naturals as the recusive type nat-list $:= \mu\alpha.(\text{Unit} + (\text{nat} \times \alpha))$. From assumptions

$$\Gamma_{\text{nat}} = () : \text{Unit}, \text{in}_1 : \text{Unit} \rightarrow (\text{Unit} + (\text{nat} \times \text{nat-list})),$$
$$\text{in}_2 : \text{nat} \times \text{nat-list} \rightarrow (\text{Unit} + (\text{nat} \times \text{nat-list}))$$

---

[3] The added restriction on closed types in (Dyn I)$_e$ is needed to define the semantics for types as in Theorem 13.

we can derive typings like $\Gamma_{\text{nat}} \triangleright [0,1,3] : \text{nat-list}$, using $[] := \text{in}_1()$, $[n,x] := \text{in}_2(n,x)$, $1 := S(0)$ etc. Indirectly, we can also define a type of inhomogeneous lists as $\text{dyn-list} := \mu\alpha.(\text{Unit} + (\text{dyn} \times \alpha))$. With corresponding typing basis, one can derive type dyn-list for the list

$$[(\text{dynamic } t\!t : \text{bool}), (\text{dynamic } \lambda x : \text{nat}.S(S(x)) : \text{nat} \to \text{nat})].$$

## 2.2 Preliminaries for the ideal model

A *domain* is a complete partial order $(D, \leq, \bot)$ with least element $\bot$ such that (i) every bounded subset $X \subseteq D$ has a least upper bound, $\bigsqcup X \in D$, (ii) $D$ has only countably many finite elements, and (iii) for any $d \in D$, $\{e \mid e \text{ finite}, e \leq d\}$ is directed and $d$ is its least upper bound. An element $d \in D$ is *finite*, if for all directed $X \subseteq D$ with $d \leq \bigsqcup X$ there is some $x \in X$ such that $d \leq x$.

In the following we will work in a domain $\mathcal{V}$ satisfying the recursion equation

$$V \cong D_{\text{bool}} + D_{\text{nat}} + (V+V) + (V \times V) + (V \to V) + (V * \mathit{clType}) + \{error\}_\bot, \quad (1)$$

where $D_{\text{bool}}$ and $D_{\text{nat}}$ are the flat domains of the booleans and natural numbers, $\{error\}_\bot$ the flat domain of an element *error* modelling run-time errors of programs, $+$ denotes the disjoint sum, $\times$ the cartesian product and $\to$ the space of continuous functions of two domains. Finally, $*$ constructs from a domain $D$ and a set $A$ the domain with universe

$$\{\langle v, a \rangle \mid v \in D - \{\bot\}, a \in A\} \cup \{\bot\}$$

and the natural partial ordering inherited from $D$. If $D$ is one of the summands of $V$, we write $d^V$ for the injection of $d \in D$ into $V$, and use

$$v\lceil_D = \begin{cases} d, & \text{if } v = d^V \text{ and } d \in D \\ \bot_D, & \text{else} \end{cases}$$

**Definition 3.** Let $\mathcal{V} = (V, \leq, \bot)$ be a complete partial order. $I \subseteq V$ is an *ideal* of $\mathcal{V}$, iff (i) $\bot_\mathcal{V} \in I$, (ii) the supremum of every directed subset of $I$ belongs to $I$, and (iii) $I$ is downward closed, i.e. $a \leq b \in I$ implies $a \in I$. Let $\mathcal{I}$ be the set of all ideals of $\mathcal{V}$.

**Definition 4.** Let $I$ and $J$ be ideals of $\mathcal{V}$, and $\tau$ a single and $A$ a set of closed types. Define the following subsets of $V$:

$$(I+J) := \{\langle i,1 \rangle^V \mid i \in I - \{\bot_V\}\} \cup \{\langle j,2 \rangle^V \mid j \in J - \{\bot_V\}\} \cup \{\bot_V\}$$
$$(I \times J) := \{\langle i,j \rangle^V \mid i \in I, j \in J\} \cup \{\bot_V\}$$
$$(I \to J) := \{f^V \mid f \in (V \to V), f(I) \subseteq J\} \cup \{\bot_V\}$$
$$I * A \quad := \{\langle i, \tau \rangle^V \mid i \in I, \tau \in A\} \cup \{\bot_V\}$$

These subsets are partially ordered as follows. Take $\bot_V$ as least element, compare pairs $\langle v, \tau \rangle^V$ of $I * A$ with the same type component according to their value component, compare $f^V$'s of $(I \to J)$ according to the ordering on $V \to V$, and on $(I \times J)$ let $\langle i,j \rangle^V \leq \langle k,l \rangle^V$ be true iff $i \leq k$ and $j \leq l$ on $I$ and $J$. On $I + J$, elements with the same tag are compared as their value components are on $I$ or $J$.

**Proposition 5.** *([1])* $(I+J)$, $(I \times J)$, $(I \to J)$ *and* $I * A$ *are ideals.*

**Definition 6.** From now on we assume that the domain solution $\mathcal{V}$ of equation (1) we are working in is the limit of domains

$$V_0 = \{\bot_V\}$$
$$V_{n+1} = D_{\text{bool}} + D_{\text{nat}} + (V_n + V_n) + (V_n \times V_n)$$
$$+ (V_n \to V_n) + (V_n * clType) + \{error\}_\bot.$$

The *rank* $rk(v)$ of $v \in V$ is the least $n \in \omega$ such that $v \in V_n$. The *distance* $d(I,J)$ of ideals $I, J \in \mathcal{I}$ is

$$d(I,J) = \begin{cases} 0, & \text{if } I = J \\ 2^{-\min\{rk(v)\ |\ v \in I \bowtie J\}}, & \text{if } \emptyset \neq I \bowtie J \end{cases}$$

where $I \bowtie J := \{v \mid v \text{ a finite element of } I - J \text{ or } J - I\}$.

**Lemma 7.** *([1])* $(\mathcal{I}, d)$ *is a complete metric space. In fact, $d$ is an ultrametric, i.e. satisfies* $d(I,J) \leq \max\{d(I,K), d(K,J)\}$ *rather than just the triangle inequality of a metric.*

A function $f : (X, d_X) \to (Y, d_Y)$ between metric spaces is *c-contractive* (resp. *non-expansive*), if $d_Y(f(x_1), f(x_2)) \leq c \cdot d_X(x_1, x_2)$ for all $x_1, x_2 \in X$, where $0 \leq c < 1$ (resp. $0 \leq c \leq 1$). Recall that by Banach's theorem, every c-contractive mapping $f : (X, d_X) \to (X, d_X)$ on a complete metric space has a unique fixed point, $fix\, x.\, f(x)$. Moreover, we will need the following facts:

**Lemma 8.** *Let $(X_j, d_j)_{j \in J}$ be a family of complete ultra-metric spaces, such that $d_j(x,y) \leq 1$ for all $j \in J$ and $x, y \in X_j$. Let $(X, d) := (\Pi_{j \in J} X_j, \sup_{j \in J} d_j)$ be their cartesian product, equipped with $d(< x_j >_{j \in J}, < y_j >_{j \in J}) := \sup\{d_j(x_j, y_j) \mid j \in J\} \leq 1$*

1. *$(X, d)$ is a complete ultra-metric space.*
2. *If $\{f_j : X_j \to X \mid j \in J\}$ a family of c-contractive mappings (for fixed c), then $f : X \to X$, defined by $f(< x_j >_{j \in J}) := < f_j(x_j) >_{j \in J}$, is c-contractive.*
3. *If, for some $c < 1$, $f$ is c-contractive in $x_k$ when keeping the others components $x_j$ fixed, then $\lambda x \in X. fix\, x_k. < f_j(x_j) >_{j \in J}$ is c-contractive.*
4. *$h \circ g$ is contractive if $g$ is contractive and $h$ is non-expanding, or vice versa.*

**Proposition 9.** *([1]) On $(\mathcal{I}, d)$, $+$, $\times$, $\to$ and $*\{\tau\}$ are 1/2-contractive functions.*

The intersection of an arbitrary nonempty set $\mathcal{J}$ of ideals is an ideal. Since its union in general is not, one has to consider $\bigsqcup \mathcal{J} := \bigcap \{I \in \mathcal{I} \mid \bigcup \mathcal{J} \subset I\}$.

**Proposition 10.** *Let $\{I_k \mid k \in K\}$ and $\{J_k \mid k \in K\}$ be nonempty families of ideals. Then (i) $d(\bigcap_{k \in K} I_k, \bigcap_{k \in K} J_k) \leq \sup_{k \in K} d(I_k, J_k)$ and (ii) $d(\bigsqcup_{k \in K} I_k, \bigsqcup_{k \in K} J_k) \leq \sup_{k \in K} d(I_k, J_k)$.*

*Proof.* (i) Since $Y := \bigcap\{I_k \mid k \in K\} \bowtie \bigcap\{J_k \mid k \in K\} \subseteq \bigcup_{k \in K}(I_k \bowtie J_k)$, we have $2^{\min rk(Y)} \geq 2^{\min_{k \in K} \min rk(I_k \bowtie J_k)} = \min_{k \in K} 2^{\min rk(I_k \bowtie J_k)}$, which implies the claim.
(ii) Similarly, we use that $X := \bigsqcup\{I_k \mid k \in K\} \bowtie \bigsqcup\{J_k \mid k \in K\} \subseteq \bigcup_{k \in K}(I_k \bowtie J_k)$. For this, note that a finite element of $\bigsqcup\{I_k \mid k \in K\}$ already belongs to some $I_k$.

## 2.3 Semantics of types as ideals

Following MacQueen e.a.[9], we use Banach's fixed point theorem to define the meaning $[\![\mu\alpha.\tau]\!]\eta$ of a recursive type $\mu\alpha.\tau$ as the fixed-point of $\lambda J \in \mathcal{I}.[\![\tau]\!]\eta[J/\alpha]$. Since only contractive mappings are guaranteed to have (unique) fixed points, we have to restrict ourselves to a subclass of all types.

**Definition 11.** ([9]) *Well-formed types* are the following subclass of types:

$$\tau = \alpha \mid \text{bool} \mid \text{nat} \mid \text{dyn} \mid (\tau + \tau) \mid (\tau \times \tau) \mid (\tau \to \tau) \mid$$
$$\mid \mu\alpha.\tau, \quad \text{provided } \tau \text{ is formally contractive in } \alpha.$$

A type $\tau$ is *formally contractive in* $\alpha$, if either (i) $\tau$ is atomic and $\alpha$ is not free in $\tau$, or (ii) $\tau$ is of the form $(\tau_1 + \tau_2)$, $(\tau_1 \times \tau_2)$ or $(\tau_1 \to \tau_2)$, or (iii) $\tau$ is of the form $\mu\beta.\sigma$ and $\sigma$ is contractive in $\alpha$, or $\alpha \equiv \beta$.

*From now on, "type" means "well-formed type"*. We write *clType* for the set of *closed* types, containing no free variables, and *Type* for the set of all types.

Well-formed types are those not containing a subtype of the form $\mu\alpha_1 \ldots \mu\alpha_n.\alpha_i$, $1 \leq i \leq n$. They form a rich class of type expressions which define contractive mappings on $(\mathcal{I}, d)$.

According to the informal discussion of dynamic values in Section 1, the type dyn should be interpreted by the collection of all pairs $\langle v, \tau \rangle^V$ where $v \in [\![\tau]\!]$ and $\tau \in clType$. Abadi e.a.[1] give such an interpretation by mimicking the ordinary type constructors $+$, $\times$ and $\to$ by new "dynamic" constructors $\dot{+}$, $\dot{\times}$, and $\dot{\to}$ on the universe $V * clType$ of dynamic values. These constructors combine *dynamic* types $I, J \subseteq (V * clType)^V$ as follows (but also work for arbitrary ideals):

$$(I \dot{+} J) := \{\langle\langle i, 1\rangle^V, \sigma + \tau\rangle^V \mid \langle i, \sigma\rangle^V \in I, \tau \in clType\}$$
$$\cup \{\langle\langle j, 2\rangle^V, \sigma + \tau\rangle^V \mid \sigma \in clType, \langle j, \tau\rangle^V \in J\} \cup \{\bot_V\}.$$
$$(I \dot{\times} J) := \{\langle\langle i, j\rangle^V, \sigma \times \tau\rangle^V \mid \langle i, \sigma\rangle^V \in I, \langle j, \tau\rangle^V \in J\} \cup \{\bot_V\}.$$
$$(I \dot{\to} J) := \{\langle f^V, \sigma \to \tau\rangle^V \mid f \in (V \to V),$$
$$\langle f(i), \tau\rangle^V \in J \text{ for all } i \in V \text{ with } \langle i, \sigma\rangle^V \in I\} \cup \{\bot_V\},$$

Since these fucntions are contractive on $\mathcal{I}$, $[\![\text{dyn}]\!]$ can recursively be defined by

$$[\![\text{dyn}]\!] = [\![\text{bool}]\!] * \{\text{bool}\} \cup [\![\text{nat}]\!] * \{\text{nat}\} \cup [\![\text{dyn}]\!] * \{\text{dyn}\}$$
$$\cup ([\![\text{dyn}]\!] \dot{+} [\![\text{dyn}]\!]) \cup ([\![\text{dyn}]\!] \dot{\times} [\![\text{dyn}]\!]) \cup ([\![\text{dyn}]\!] \dot{\to} [\![\text{dyn}]\!]).$$

In the presence of recursive types, however, this method seems no longer be applicable, as it relies on the fact that for non-basic $\tau$, $\langle v, \tau\rangle^V \in [\![\text{dyn}]\!]$ is characterized by elements $\langle u, \sigma\rangle^V \in [\![\text{dyn}]\!]$ with *simpler* types $\sigma$. This is not the case when choosing

$$\dot{\mu}(I) := \{\langle v, \mu\alpha.\tau\rangle^V \mid \langle v, \tau[\mu\alpha.\tau/\alpha]\rangle^V \in I\} \cup \{\bot_V\}.$$

Also, $d(\dot{\mu}([\![\text{bool}]\!] * \{\text{bool}\}), \dot{\mu}([\![\text{nat}]\!] * \{\text{nat}\})) = d([\![\text{bool}]\!] * \{\text{bool}\}, [\![\text{nat}]\!] * \{\text{nat}\})$ shows that we would not get a contractive mapping.

So we take a more general approach to define $[\![\text{dyn}]\!]$ that is fairly independent of the choice of type constructors, and formalizes the intuitive meaning directly.

**Definition 12.** Let $Env$ be the set of all assignments $\eta : TypeVar \to \mathcal{I}$. A *meaning of types* as ideals in $\mathcal{V}$ is a function $[\![\cdot]\!]\cdot$, such that the following conditions hold for all types $\tau$ and assignments $\eta \in Env$:

$$[\![\alpha]\!]\eta = \eta(\alpha) \qquad\qquad [\![(\tau_1 + \tau_2)]\!]\eta = ([\![\tau_1]\!]\eta + [\![\tau_2]\!]\eta)$$
$$[\![\text{bool}]\!]\eta = D_{\text{bool}} \cup \{\bot_\mathcal{V}\} \qquad [\![(\tau_1 \times \tau_2)]\!]\eta = ([\![\tau_1]\!]\eta \times [\![\tau_2]\!]\eta)$$
$$[\![\text{nat}]\!]\eta = D_{\text{nat}} \cup \{\bot_\mathcal{V}\} \qquad [\![(\tau_1 \to \tau_2)]\!]\eta = ([\![\tau_1]\!]\eta \to [\![\tau_2]\!]\eta)$$
$$[\![\text{dyn}]\!]\eta = \bigcup_{\tau \in clType} [\![\tau]\!]\eta * \{\tau\} \qquad [\![\mu\alpha.\tau]\!]\eta = [\![\tau]\!]\eta[[\![\mu\alpha.\tau]\!]\eta / \alpha]$$

Without the clause for dyn, it is easily seen by induction on $\tau$ that $[\![\tau]\!]\eta$ exists for all $\eta$, once it is clear that the fixed point of $\lambda J \in \mathcal{I}.[\![\tau]\!]\eta[I/\alpha]$ is an ideal of $\mathcal{V}$. However, with the clause for dyn we can no longer apply induction on $\tau$ to ensure the existence of $[\![\tau]\!]\eta$: the above equations specify $[\![\text{dyn}]\!]\eta$ using the values of arbitrarily complicated types - possibly containing the type dyn.

The main point of this paper is to show that, in spite of the apparently non-well-founded recursion in the clauses of Definition 12, $[\![\tau]\!]\eta$ makes perfect sense.[4] An infinite simultaneous recursion and the uniqueness of fixed-points in complete metric spaces is used to ensure that $[\![\cdot]\!]\cdot$ exists.

**Theorem 13.** *There is a mapping* $[\![\cdot]\!]\cdot : Type \times Env \to \mathcal{I}$ *satisfying the conditions of Definition 12, i.e.* $[\![\tau]\!]\eta$ *is well-defined for all $\tau$ and $\eta$.*

*Proof.* Let $\Pi\mathcal{I} = \Pi_{\tau \in Type} \mathcal{I}_\tau$, with $\mathcal{I}_\tau = \mathcal{I}$, be the product space of the space of ideals of $\mathcal{V}$, indexed by all types. Define $F = <F_\tau>_{\tau \in Type} : \Pi\mathcal{I} \times Env \to \mathcal{I}$ as follows, where $I \in \Pi\mathcal{I}$ is written as a function $I(\tau)$:

$$F_\alpha(I,\eta) = \eta(\alpha) \qquad\qquad F_{(\tau_1+\tau_2)}(I,\eta) = F_{\tau_1}(I,\eta) + F_{\tau_2}(I,\eta)$$
$$F_{\text{bool}}(I,\eta) = D_{\text{bool}} \cup \{\bot_\mathcal{V}\} \qquad F_{(\tau_1 \times \tau_2)}(I,\eta) = F_{\tau_1}(I,\eta) \times F_{\tau_2}(I,\eta)$$
$$F_{\text{nat}}(I,\eta) = D_{\text{nat}} \cup \{\bot_\mathcal{V}\} \qquad F_{(\tau_1 \to \tau_2)}(I,\eta) = F_{\tau_1}(I,\eta) \to F_{\tau_2}(I,\eta)$$
$$F_{\text{dyn}}(I,\eta) = \bigcup \{I(\tau) * \{\tau\} \mid \tau \in clType\} \qquad F_{\mu\alpha.\tau}(I,\eta) = fix J \in \mathcal{I}. F_\tau(I, \eta[J/\alpha]).$$

**Claim 1.** For each type $\tau$ and $\eta \in Env$, $d(F_\tau(I,\eta), F_\tau(I',\eta)) \leq 1/2 \cdot d(I,I')$.

*Proof* by induction on $\tau$:

$\tau \in \{\alpha, \text{bool}, \text{nat}\}$: Then $d(F_\tau(I,\eta), F_\tau(I',\eta)) = 0 \leq 1/2 \cdot d(I,I')$.

$\tau = \text{dyn}$ : First note that $F_{\text{dyn}}(I,\eta) \in \mathcal{I}$, because if any two ideals of an arbitrary union of ideals have incomparable non-bottom elements only, then this union is an ideal. Next, we have

$$d(F_{\text{dyn}}(I,\eta), F_{\text{dyn}}(I',\eta))$$
$$\leq sup\{d(I(\tau) * \{\tau\}, I'(\tau) * \{\tau\}) \mid \tau \in clType\} \quad \text{(by Proposition 10)}$$
$$\leq 1/2 \cdot sup\{d(I(\tau), I'(\tau)) \mid \tau \in clType\}$$
$$\leq 1/2 \cdot d(I,I').$$

---

[4] Without recursive types, we might add a clause for variables like

$$[\![\text{dyn}]\!]\eta = \bigcup \{[\![\alpha]\!]\eta * \{\alpha\} \mid \alpha \text{ a type variable}\} \cup \ldots$$

to the definition and still show that $[\![\tau]\!]\eta$ exists. But then if $\tau$ contains dyn, $[\![\tau]\!]\eta$ depends on *all* the $[\![\alpha]\!]\eta$ and hence the substitution lemma below fails for such $\tau$.

$\tau = (\tau_1 \circ \tau_2)$, where $\circ$ is one of $+$, $\times$, or $\to$: We can use Proposition 9.

$\tau = \mu\alpha.\sigma$: By induction, we have $d(F_\sigma(I,\theta), F_\sigma(I',\theta)) \leq 1/2 \cdot d(I,I')$, for each $\theta$, in particular for each $\theta = \eta[J/\alpha]$, where $J \in \mathcal{I}$. It is sufficient to show the following claim, whose proof by induction on $\sigma$ is standard.

**Claim 2.** Suppose $\sigma$ is formally contractive in $\alpha_1, \ldots, \alpha_n$. The map $(I, J_1, \ldots, J_n) \mapsto F_\sigma(I, \eta[J_1/\alpha_1, \ldots, J_n/\alpha_n])$ is 1/2-contractive.

Using Claim 2 and Lemma 8, we get that $I \mapsto fixJ.\, F_\sigma(I, \eta[J/\alpha])$ is 1/2-contractive, and hence

$$d(F_{\mu\alpha.\sigma}(I,\eta), F_{\mu\alpha.\sigma}(I',\eta)) = d(fixJ.\, F_\sigma(I, \eta[J/\alpha]), fixJ.\, F_\sigma(I', \eta[J/\alpha]))$$
$$\leq 1/2 \cdot d(I, I'),$$

which finishes the proof of Claim 1.

From Claim 1 we conclude that on the product space $\Pi\mathcal{I}$, $d(F(I,\eta), F(I',\eta)) \leq 1/2 \cdot d(I,I')$, and so $F$ is contractive, for fixed $\eta$. By Banach's theorem, for each $\eta$ there is a (unique) element $I_\eta \in \Pi\mathcal{I}$ such that $F(I_\eta, \eta) = I_\eta$. We now define

$$[\![\tau]\!]\eta := I_\eta(\tau) = F_\tau(I_\eta, \eta).$$

Note that induction on types cannot be used to show that $[\![\cdot]\!]$ satisfies the conditions of Definition 12.

**Claim 3.** $F_\tau(I, \rho)$ only depends on $\rho(\alpha)$, with $\alpha$ free in $\tau$, and on $I(\sigma)$ for closed $\sigma$.

This is easily seen by induction on $\tau$, since the case for $\tau = \text{dyn}$ is obvious. Next we use the uniqueness of fixed points to show:

**Claim 4.** If $I_\eta = F(I_\eta, \eta)$ and $I_\theta = F(I_\theta, \theta)$, then $I_\eta(\sigma) = I_\theta(\sigma)$ for all *closed* $\sigma$.

*Proof:* By Claim 3, $I_\eta(\sigma) = F_\sigma(I_\eta, \eta)$ depends only on all the $I_\eta(\sigma')$ for closed $\sigma'$. Note that this dependency is contractive, so

$$d(I_\eta(\sigma), I_\theta(\sigma)) \leq 1/2 \cdot sup\{d(I_\eta(\sigma'), I_\theta(\sigma')) \mid \sigma' \in clType\}.$$

Since this holds for all closed types, the right hand side must be 0.

**Claim 5.** $[\![\cdot]\!]$ satisfies the conditions of Definition 12.

*Proof:* This is obvious for all types except the recursive ones. For these, use

$$\begin{aligned}
[\![\mu\alpha.\tau]\!]\eta &= F_{\mu\alpha.\tau}(I_\eta, \eta) \\
&= fixJ \in \mathcal{I}.\, F_\tau(I_\eta, \eta[J/\alpha]) \\
&= fixJ \in \mathcal{I}.\, F_\tau(I_{\eta[J/\alpha]}, \eta[J/\alpha]) \quad \text{(by Claims 3 and 4)} \\
&= fixJ \in \mathcal{I}.\, [\![\tau]\!]\eta[J/\alpha] \\
&= [\![\tau]\!]\eta[[\![\mu\alpha.\tau]\!]\eta/\alpha].
\end{aligned}$$

**Corollary 14.** *(i)* $[\![\tau]\!]\eta$ *does not depend on* $\eta(\alpha)$ *for* $\alpha$ *not free in* $\tau$. *(ii) If* $\tau$ *is contractive in* $\alpha$, *then* $\lambda J \in \mathcal{I}.\, [\![\tau]\!]\eta[J/\alpha]$ *is 1/2-contractive.*

**Corollary 15.** *(Substitution Lemma)* $[\![\sigma]\!]\eta[[\![\tau]\!]\eta/\alpha] = [\![\sigma[\tau/\alpha]]\!]\eta.$

*Proof.* By induction on $\sigma$. The claim is obvious if $\sigma$ is a type variable, and immediate by Proposition 9 if $\sigma$ is $(\sigma_1 + \sigma_2)$, $(\sigma_1 \times \sigma_2)$, or $(\sigma_1 \to \sigma_2)$. Let $\theta$ be $\eta[[\tau]\eta/\alpha]$.
$\sigma \in \{\text{bool}, \text{nat}, \text{dyn}\}$: $[\sigma]\theta = I_\theta(\sigma) = I_\eta(\sigma) = [\sigma]\eta = [\sigma[\tau/\alpha]]\eta$, using Claim 4.
$\sigma = \mu\beta.\rho$: We may assume $\alpha \not\equiv \beta$ and $free(\tau) \cap bound(\sigma) = \emptyset$, and hence

$$\begin{aligned}
[\sigma]\theta = [\mu\beta.\rho]\theta &= fix J \in \mathcal{I}. [\rho]\theta[J/\beta] \\
&= fix J \in \mathcal{I}. [\rho]\eta[J/\beta][[\tau]\eta/\alpha] &&\text{(by disjointness of variables)} \\
&= fix J \in \mathcal{I}. [\rho[\tau/\alpha]]\eta[J/\beta] &&\text{(by induction and Cor. 14 (i))} \\
&= [\mu\beta.\rho[\tau/\alpha]]\eta \\
&= [(\mu\beta.\rho)[\tau/\alpha]]\eta = [\sigma[\tau/\alpha]]\eta &&\text{(by disjointness of variables)}.
\end{aligned}$$

Immediate consequences are theorems 18 and 19 of the following sections, extending those of Abadi e.a.[1] for the corresponding type system without recursive types.

### 2.4 Soundness of typing rules

The meaning of terms is defined along Milner's[11] original description of the ideal model. We only give the clauses for **wrong** and **dynamics**:

**Definition 16.** Let $match_{\{\alpha_1,\ldots,\alpha_n\}}(\sigma,\tau) = S$ say that $S : \{\alpha_1,\ldots,\alpha_n\} \to Type$ is a substitution such that $\sigma \equiv \tau S$.

$[\textbf{wrong}]\eta := error^V$

$[(\textbf{dynamic } t : \tau)]\eta := \begin{cases} error^V, & \text{if } [t]\eta = error^V \\ \langle [t]\eta, \tau \rangle^V, & \text{otherwise} \end{cases}$

$[(\textbf{typecase } d \textbf{ of } \{\alpha\} \ x : \tau \textbf{ then } r \textbf{ else } s)]\eta :=$
$\begin{cases} [r[\rho/\alpha]]\eta[v/x], & \text{if } \langle v,\sigma \rangle = [d]\eta\lceil_{(V*clType)} \text{ and } match_{\{\alpha\}}(\sigma,\tau) = [\rho/\alpha] \neq fail \\ [s]\eta, & \text{if } \langle v,\sigma \rangle = [d]\eta\lceil_{(V*clType)} \text{ and } match_{\{\alpha\}}(\sigma,\tau) = fail, \\ & \text{or } [d]\eta = \bot_V \\ error^V, & \text{else} \end{cases}$

The typing rules (including the familiar ones not given) can be shown to be sound with respect to the denotational meanings of types and terms. We have to restrict type assignments to have values in the set of *semantic types* of $\mathcal{V}$,

$$\mathcal{T} := \{I \in \mathcal{I} \mid error^V \notin I\}.$$

The typing rules are chosen such that **wrong** is untypable. It is easily seen that *no* type contains $error^V$:

**Lemma 17.** *If* $\eta : TypeVar \to \mathcal{T}$, *then* $[\tau]\eta \in \mathcal{T}$ *for each type* $\tau$.

Note that we cannot use induction on types to prove Lemma 17. Instead, observe that in the metric space $(\mathcal{I}, d)$, a Cauchy sequence contained in $\mathcal{T}$ never converges to an ideal $I \notin \mathcal{T}$.

The following theorem, which can be shown by induction on the proof of $\Gamma \rhd s : \sigma$, implies that no typable term denotes $error^V$.

**Theorem 18.** *(c.f. [1]) Suppose* $\eta(x) \in \Gamma(x) \in \mathcal{T}$ *whenever* $\Gamma(x)$ *is defined. If* $\Gamma \rhd t : \tau$ *is provable, then* $[t]\eta \in [\tau]\eta$.

## 2.5 Soundness of evaluation

There is an operational notion of evaluation, as defined in Abadi e.a.[1], which is correct with respect to the denotational one. Only closed expressions are evaluated, and the result is a term in canonical form. *Terms in canonical form*, or (operational) *values* $v$, are given by the grammar

$$
\begin{aligned}
v &= \text{wrong} \mid u & &(\textit{values}) \\
u &= \text{tt} \mid \text{ff} \mid n & &(\textit{proper values}) \\
&\mid (u, u) \\
&\mid in_{1,\tau+\sigma}\, u \mid in_{2,\tau+\sigma}\, u \\
&\mid \lambda x : \tau.t, \quad \text{if } \tau \text{ is closed and } \textit{free}(t) \subseteq \{x\} \\
&\mid (\text{dynamic } u : \tau), \quad \text{if } \tau \text{ is closed} \\
n &= 0 \mid S(n) & &(\textit{natural values}).
\end{aligned}
$$

Inductively, it is defined when *closed term $t$ reduces to canonical form $v$*, written as $t \Rightarrow v$. Again, we only give the rules for expressions dealing with **wrong** and dynamics (with $u$ and $v$ as above):

$(\Rightarrow\text{wrong})\quad \overline{\text{wrong} \Rightarrow \text{wrong}}$

$(\Rightarrow_{\text{dyn},1})\quad \dfrac{t \Rightarrow u}{(\text{dynamic } t : \tau) \Rightarrow (\text{dynamic } u : \tau)}$

$(\Rightarrow_{\text{dyn},2})\quad \dfrac{t \Rightarrow \text{wrong}}{(\text{dynamic } t : \tau) \Rightarrow \text{wrong}}$

$(\Rightarrow_{\text{tc},1})\quad \dfrac{d \Rightarrow (\text{dynamic } u : \sigma),\quad r[\rho/\alpha][u/x] \Rightarrow v}{(\text{typecase } d \text{ of } \{\alpha\}\, x : \tau \text{ then } r \text{ else } s) \Rightarrow v},\quad \textit{match}_{\{\alpha\}}(\sigma,\tau) = [\rho/\alpha]$

$(\Rightarrow_{\text{tc},2})\quad \dfrac{d \Rightarrow (\text{dynamic } u : \sigma),\quad s \Rightarrow v}{(\text{typecase } d \text{ of } \{\alpha\}\, x : \tau \text{ then } r \text{ else } s) \Rightarrow v},\quad \textit{match}_{\{\alpha\}}(\sigma,\tau) = \textit{fail}$

$(\Rightarrow_{\text{tc},3})\quad \dfrac{d \Rightarrow v \quad v \neq (\text{dynamic } u : \sigma)}{(\text{typecase } d \text{ of } \{\alpha\}\, x : \tau \text{ then } r \text{ else } s) \Rightarrow \text{wrong}}$

Next one can show that operational evaluation preserves types and denotational value. Together with the results of the previous section, this ensures that well-typed expressions $t$ "do not cause run-time errors", i.e. $t \Rightarrow$ **wrong** is impossible.

**Theorem 19.** *(c.f. [1]) Let $t$ be closed with respect to object- and typevariables. If $t \Rightarrow v$, then (a) $[\![t]\!] = [\![v]\!]$ and (b) if $\triangleright t : \sigma$ is provable, so is $\triangleright v : \sigma$.*

# 3 Dynamic and recursive types in polymorphic languages

We can extend the combination of recursive and dynamic types from simply to polymorphically typed $\lambda$-calculus.

## 3.1 Explicit polymorphism

The set of types for explicit polymorphism is given by the grammar

$$\tau = \alpha \mid \texttt{bool} \mid \texttt{nat} \mid \texttt{dyn} \mid (\tau + \tau) \mid (\tau \times \tau) \mid (\tau \to \tau) \mid \mu\alpha.\tau \mid \forall\alpha.\tau \mid \exists\alpha.\tau.$$

The intended meaning of type quantifiers in the ideal model is given by

$$[\![\forall\alpha.\tau]\!]\eta = \bigcap_{J \in \mathcal{T}} [\![\tau]\!]\eta[J/\alpha] \quad \text{and} \quad [\![\exists\alpha.\tau]\!]\eta = \bigsqcup_{J \in \mathcal{T}} [\![\tau]\!]\eta[J/\alpha], \qquad (2)$$

where $\mathcal{T}$ is the set of ideals of $\mathcal{V}$ that do not contain $error^V$. Define $\forall\beta.\tau$ and $\exists\beta.\tau$ to be formally contractive in $\alpha$ just as for $\mu\beta.\tau$ in Section 2.3, and let $\forall\beta.\tau$ and $\exists\beta.\tau$ be well-formed if $\tau$ is. Restricting to well-formed types, we obtain:

**Theorem 20.** *There is a meaning function $[\![\tau]\!]\eta$ for the ideal interpretation of polymorphic types satisfying the conditions of Definition 12 and the equations (2).*

*Proof.* We modify the function $F$ from the proof of Theorem 13 by adding component functions $F_{\forall\alpha.\tau}$ and $F_{\exists\alpha.\tau}$ defined by

$$F_{\forall\alpha.\tau}(I,\eta) := \bigcap_{J \in \mathcal{T}} F_\tau(J, \eta[J/\alpha]), \quad \text{and} \quad F_{\exists\alpha.\tau}(I,\eta) := \bigsqcup_{J \in \mathcal{T}} F_\tau(J, \eta[J/\alpha]).$$

The proof of Theorem 13 carries over, once we have shown:

**Claim 6.** $F_{\forall\alpha.\tau}$ and $F_{\exists\alpha.\tau}$ are 1/2-contractive in their first arguments.

For $F_{\exists\alpha.\tau}$, the proof is similar to the one for $F_{\forall\alpha.\tau}$:

$$\begin{aligned}
&d(F_{\forall\alpha.\tau}(I,\eta), F_{\forall\alpha.\tau}(I',\eta)) \\
&= d(\bigcap_{J \in \mathcal{T}} F_\tau(I, \eta[J/\alpha]), \bigcap_{J \in \mathcal{T}} F_\tau(I', \eta[J/\alpha])) \\
&\leq \sup_{J \in \mathcal{T}} d(F_\tau(I, \eta[J/\alpha]), F_\tau(I', \eta[J/\alpha])) \quad &\text{(by Proposition 10)} \\
&\leq \sup_{J \in \mathcal{T}} 1/2 \cdot d(I, I') = 1/2 \cdot d(I, I') \quad &\text{(by induction).}
\end{aligned}$$

The following is shown exactly as for the case $\mu\beta.\rho$ in Proposition 15.

**Proposition 21.** *(Substitution Lemma)*

$$[\![\forall\beta.\tau]\!]\eta[[\![\sigma]\!]\eta/\alpha] = [\![(\forall\beta.\tau)[\sigma/\alpha]]\!]\eta \quad \text{and} \quad [\![\exists\beta.\tau]\!]\eta[[\![\sigma]\!]\eta/\alpha] = [\![(\exists\beta.\tau)[\sigma/\alpha]]\!]\eta.$$

## 3.2 Implicit polymorphism

We now present an interpretation of dynamic and recursive types in a language with implicit polymorphism in the style of *ML*. This gives a denotational interpretation of A. Mycroft's[12] proposal to extend the functional language *ML*. He pointed out that functions one would like to have for *ML*, like

$$print\ :\ \texttt{dyn} \to \texttt{string} \quad \text{or} \quad eval\ :\ \texttt{expression} \times \texttt{environment} \to \texttt{dyn},$$

could be defined when *ML* had a type dyn.

The previous notion of *types* is modified by adding universal type quantifiers in prenex form only, according to the grammar

$$\tau = \alpha \mid \text{bool} \mid \text{nat} \mid \text{dyn}$$
$$\mid (\tau + \tau) \mid (\tau \times \tau) \mid (\tau \to \tau) \mid \mu\alpha.\tau \qquad (monotypes)$$
$$\overline{\sigma} = \tau \mid \forall \alpha.\overline{\sigma}, \qquad\qquad\qquad\qquad\qquad (polytypes).$$

Let *MType* and *PType* be the set of *well-formed* mono- and polytypes, respectively. By *clMType* and *clPType* we mean the *closed* well-formed mono- and polytypes, respectively.

*Terms* are modified in that (dynamic $t : \tau$) is replaced by (dynamic $t$), and $\lambda x : \tau.t$ by $\lambda x.t$, and so terms do not contain type information any more, except in type patterns.

**Semantics of types** According to Milner's[11] interpretation for implicit polymorphism, type quantifiers are meant to range over closed monotypes only. In the ideal model, separate a universe $\mathcal{MT}$ of monotypes from the universe $\mathcal{T}$ of all types by

$$\mathcal{MT} := \{[\![\tau]\!] \mid \tau \in clMType\} \subseteq \mathcal{T} = \{J \in \mathcal{I} \mid error^V \notin J\}.$$

The meaning of polytypes is reduced to that of monotypes by induction on the quantifier-rank, using

$$[\![\forall\alpha.\,\overline{\sigma}]\!]\eta := \bigcap\{[\![\overline{\sigma}[\tau/\alpha]]\!]\eta \mid \tau \in clMType\}. \tag{3}$$

In order to cover type-tags with quantifiers, Definition 12 is changed by

$$[\![\text{dyn}]\!]\eta = \bigcup\{[\![\overline{\sigma}]\!]\eta * \{\overline{\sigma}\} \mid \overline{\sigma} \in clPType\}. \tag{4}$$

**Theorem 22.** *There is a meaning function $[\![\tau]\!]\eta$ for implicitly polymorphic types satisfying the conditions of Definition 12 and equation (3).*

*Proof.* Again, we modify the function $F$ from the proof of Theorem 13 by adding component functions $F_{\forall\alpha.\overline{\sigma}}$ for polytypes $\overline{\sigma}$. The former *clType* has to be replaced by *clPType* everywhere. We define

$$F_{\forall\alpha.\overline{\sigma}}(I,\eta) := \bigcap\{F_{\overline{\sigma}[\tau/\alpha]}(I,\eta) \mid \tau \in clMType\}. \tag{5}$$

**Claim 7.** $F_{\forall\alpha.\overline{\sigma}}$ is 1/2-contractive in its first argument.

The proof of this is analogous to that of Claim 6, using Proposition 10. It follows that $\lambda I \in \Pi\mathcal{I}.F(I,\eta)$ is contractive, whence the meaning of types can again be defined by $[\![\overline{\sigma}]\!]\eta := I_\eta(\overline{\sigma})$, using the unique fixed point $I_\eta = F(I_\eta, \eta)$ of $F$.

The substitution lemma holds in the following form:

**Proposition 23.** *For each monotype $\tau$, $[\![\overline{\sigma}]\!]\eta[[\![\tau]\!]\eta/\alpha] = [\![\overline{\sigma}[\tau/\alpha]]\!]\eta$.*

*Proof.* By induction on the quantifier-rank of $\overline{\sigma}$. Let $\theta$ be $\eta[\![\tau]\!]\eta/\alpha]$. In the case of $\forall \beta.\overline{\sigma}$, we may assume $\alpha \not\equiv \beta$ and $free(\tau) \cap bound(\forall\beta.\overline{\sigma}) = \emptyset$, and thus

$$
\begin{aligned}
[\![\forall \beta.\overline{\sigma}]\!]\eta[\![\tau]\!]\eta/\alpha] &= F_{\forall\beta.\overline{\sigma}}(I_\theta, \theta) \\
&= \bigcap\{ F_{\overline{\sigma}[\rho/\beta]}(I_\theta, \theta) \mid \rho \in clMType \} && \text{(by definition)} \\
&= \bigcap\{ [\![\overline{\sigma}[\rho/\beta]]\!]\theta \mid \rho \in clMType \} && \text{(by definition)} \\
&= \bigcap\{ [\![\overline{\sigma}[\rho/\beta]]\!][\tau/\alpha]]\eta \mid \rho \in clMType \} && \text{(by induction)} \\
&= \bigcap\{ [\![\overline{\sigma}[\tau/\alpha][\rho/\beta]]\!]\eta \mid \rho \in clMType \} && \text{(since } \rho \text{ is closed,} \\
&= [\![\forall\beta(\overline{\sigma}[\tau/\alpha])]\!]\eta && \beta \notin free(\tau)) \\
&= [\![(\forall\beta.\overline{\sigma})[\tau/\alpha]]\!]\eta. && \text{(by variable disjointness)}
\end{aligned}
$$

**Corollary 24.** $[\![\forall \alpha.\overline{\sigma}]\!]\eta = \bigcap_{J \in \mathcal{MT}} [\![\overline{\sigma}]\!]\eta[J/\alpha]$.

*Remark.* In the absence of recursive types, one can avoid to define the meaning function for types by approximations, by using the (finite) recursive definition

$$
\begin{aligned}
[\![dyn]\!]\eta = [\![bool]\!]\eta &* \{bool\} \cup [\![nat]\!]\eta * \{nat\} \cup [\![dyn]\!]\eta * \{dyn\} \\
&\cup ([\![dyn]\!]\eta \mathbin{\dot{+}} [\![dyn]\!]\eta) \cup ([\![dyn]\!]\eta \mathbin{\dot{\times}} [\![dyn]\!]\eta) \cup ([\![dyn]\!]\eta \mathbin{\dot{\to}} [\![dyn]\!]\eta) \\
&\cup \dot{\forall}([\![dyn]\!]\eta)
\end{aligned}
$$

For an ideal $I$ over $\mathcal{V}$, we define $\dot{\forall}(I)$ to be

$$\{\langle i, \forall\alpha.\overline{\sigma}\rangle^V \mid \forall\alpha.\overline{\sigma} \in clPType, \langle i, \overline{\sigma}[\tau/\alpha]\rangle^V \in I \text{ for all } \tau \in clMType\} \cup \{\bot_V\}.$$

In contrast to $\dot{+}$, $\dot{\times}$, and $\dot{\to}$, the operation $\dot{\forall}$ is *not* a contractive mapping on the space $(\mathcal{I}, d)$ of ideals of $\mathcal{V}$ – it is just non-expanding. To ensure that the recursion equation for $[\![dyn]\!]$ has a solution, we modify the metric $d$ on $\mathcal{V}$ as follows:

For polytypes $\overline{\sigma}$, let $qrk(\overline{\sigma})$, the *quantifier-rank of* $\overline{\sigma}$, be the number of (leading) quantifiers in $\overline{\sigma}$. For $v \in \mathcal{V}$ and $I, J \in \mathcal{I}$, define a modified rank and distance by

$$\widetilde{rk}(v) := \begin{cases} rk(v) + qrk(\overline{\sigma}) & \text{if } v\lceil_{(V*clType)} = \langle i, \overline{\sigma}\rangle \\ rk(v), & \text{otherwise,} \end{cases}$$

$$\widetilde{d}(I, J) := 2^{-\min\{\widetilde{rk}(v) \mid v \in I \bowtie J\}}.$$

The reader may check that the functions $+$, $\times$, $\to$, $\dot{+}$, $\dot{\times}$, $\dot{\to}$, $*\{\tau\}$, and $\dot{\forall}$ are contractive on $(\mathcal{I}, \widetilde{d})$.

**Semantics of terms** One might wish to define the meaning of (**dynamic** $e$) in terms of the principal type of $e$, but $e$ need not have a principal type: for example, $e = \lambda x.(\mathbf{dynamic}\ x)$ is of type $\mathrm{dyn} \to \mathrm{dyn}$ and $\mathrm{nat} \to \mathrm{dyn}$, but has no principal type. For type inference, use the rules of *ML* together with the implicit version (Dyn E) of dyn-elimination and the following implicit version of dyn-introduction:

$$\text{(Dyn I)} \quad \frac{\Gamma \triangleright e : \tau}{\Gamma \triangleright (\mathbf{dynamic}\ e) : \mathrm{dyn}}, \quad \text{if } \overline{\tau}^\Gamma \text{ is closed.}$$

Since terms $t$ lack principal types, the meaning $[\![t]\!]\eta$ can only be given relative to a typing derivation $D$ for $t$. For each subterm (dynamic $e$) of $t$, $D$ uniquely fixes a closed type $\overline{\tau}^\Gamma$ that can be used for tagging the value of $e$. To do so, we assign to each subterm $r$ of $t$ a sequence $\pi$ of numbers $i \in \{1, 2, 3\}$, coding the branch leading from the root to $r$ in the tree representation of $t$. Then we can define

$$[\![(\text{dynamic } e^{\pi 1})^\pi]\!]\eta := \begin{cases} error^V, & \text{if } [\![e^{\pi 1}]\!]\eta = error^V \\ \langle [\![e^{\pi 1}]\!]\eta, \overline{\tau}^\Gamma \rangle^V, & \text{if } \Gamma \triangleright e^{\pi 1} : \tau \text{ is the inference step of } D \\ & \text{that assigns a type to } e^{\pi 1}. \end{cases}$$

To define the meaning of typecase-terms, we match polytypes as follows. For $\overline{\sigma} \equiv \forall \gamma_1 \ldots \gamma_n.\sigma(\gamma_1, \ldots, \gamma_n) \in clType$ and $\overline{\tau}(\alpha) \equiv \forall \beta_1 \ldots \beta_m.\tau(\alpha, \beta_1, \ldots, \beta_m) \in PType$, define

$$match_{\{\alpha\}}(\overline{\sigma}, \overline{\tau}) = [\rho/\alpha] : \iff \rho \text{ is closed and } \sigma[\rho_1/\gamma_1, \ldots, \rho_n/\gamma_n] \equiv \tau[\rho/\alpha],$$

where $[\rho/\alpha, \rho_1/\gamma_1, \ldots, \rho_n/\gamma_n]$ is the most general unifier of monotypes $\sigma$ and $\tau$ when considering the $\beta_i$ in $\tau$ as constants. Since every instantiation of $\tau[\rho/\alpha]$ by closed monotypes is an instance of $\sigma$ by closed monotypes as well, $[\![\overline{\sigma}]\!] \subseteq [\![\forall \beta_1 \ldots \beta_m.\tau[\rho/\alpha]]\!]$.

Leaving out the path annotation which relativize the meaning to $D$, we can now define the meaning of typecase-expressions just as in Section 2.4:

$$[\![(\text{typecase } d \text{ of } \{\alpha\} \ x : \overline{\tau} \text{ then } r \text{ else } s)]\!]\eta :=$$
$$\begin{cases} [\![r[\rho/\alpha]]\!]\eta[v/x], & \text{if } \langle v, \overline{\sigma} \rangle = [\![d]\!]\eta\lceil_{(V * clType)} \text{ and } match_{\{\alpha\}}(\overline{\sigma}, \overline{\tau}) = [\rho/\alpha] \\ [\![s]\!]\eta, & \text{if } \langle v, \overline{\sigma} \rangle = [\![d]\!]\eta\lceil_{(V * clType)} \text{ and } match_{\{\alpha\}}(\overline{\sigma}, \overline{\tau}) = fail, \\ & \text{or } [\![d]\!]\eta = \bot_V \\ error^V, & \text{else.} \end{cases}$$

Observe that pattern-variables $\alpha$ in patterns range over *closed* types $\rho$ only; if $\rho$ were allowed to contain free variables (bound by the quantifiers of the pattern or the type-tag), it would be unclear what $\alpha$ means in the then-branch of the typecase-statement. Compared with the treatment of non-closed patterns in Section 3 of [8], the above corresponds to their case of patterns with existential type quantifier prefixes.

In the presence of recursive types, it seems preferable that in the definition of the *match*-function obove we read $\equiv$ not as syntactical identity of type expressions, but rather as identity of the rational trees obtained by infinite unfolding of the $\mu$-operator (c.f. [4]). The fold and unfold-rules for recursive types should then be replaced by the stronger rule of equality for recursive types, see [4].

## 4 Open problems

By providing an interpretation in the ideal model, it has been shown that dynamic types can be combined with recursive types and explicit or implicit polymorphism in a sound way. This gives a partial answer to questions of Abadi e.a.[1], and adds semantical support to implementations integrating dynamic types into polymorphic languages like *CAML*[5].

The main drawbacks of the extension of *ML* by dynamic typing sketched in Section 3.2 and the similar ones of [2] and [8] are the failure of the principal types property and the restriction to closed type-tags. For languages where types can be passed as parameters, models for dynamics with *open* type-tags are needed.

An open point is to remove the well-formedness restriction on types. Domains with a notion of approximation were introduced by Cardone and Coppo[4] in order to give meanings to arbitrary –not just contractive– recursive types. We believe that dyn can be added to the simple and recursive types of [4]. But we do not know whether the completeness result of [4] carries over to the system of Section 2.1.

For typecase-expressions with higher-order matching as in [2], or just the systems of Section 3 above, even soundness theorems have not yet been given.

**Acknowledgement** I wish to thank the referees for some very helpful proposals to improve the results and presentation. Thanks also to Fritz Henglein for sending a copy of Mycroft's papers and for a hint to [4].

# References

1. M. Abadi, L. Cardelli, B. Pierce, and G. Plotkin. Dynamic typing in a statically-typed language. In *16th POPL*, pages 213–227, 1989.
2. M. Abadi, L. Cardelli, B. Pierce, and D. Remy. Dynamic typing in polymorphic languages. In *ACM SIGPLAN Workshop on ML and its Applications. San Francisco, California, June 20-21, 1992*, pages 92–103, 1992.
3. L. Cardelli. Amber. In G. Cousineau, P. L. Curien, and B. Robinet, editors, *Combinators and Functional Programming Languages*. Springer LNCS 242, 1986.
4. F. Cardone and M. Coppo. Type inference with recursive types: Syntax and semantics. *Information and Computation*, 92(1):48–80, May 1991.
5. G. Cousineau and G. Huet. The CAML Primer. Version 2.6. Project Formel, INRIA-ENS, April 1989.
6. R. Harper, R. Milner, and M. Tofte. The definition on Standard ML - Version 2. LFCS Report Series ECS-LFCS-88-62, Dept. of Computer Science, Univ. of Edinburgh, 1988.
7. F. Henglein. Dynamic typing. In *European Symposium on Programming (ESOP). Rennes, France*, pages 233–253. Springer LNCS, vol. 582, 1992.
8. X. Leroy and M. Mauny. Dynamics in ML. In *Conf. on Functional Programming Languages and Computer Architecture. Cambridge, Massachusetts, August 1991*, pages 406–426. Springer LNCS 523.
9. D. MacQueen, G. Plotkin, and R. Sethi. An ideal model for recursive polymorphic types. In *Proceedings of the 11th ACM Symposium on Principles of Programming Languages*, 1984.
10. D. C. J. Matthews. Static and Dynamic Type-Checking. In: Papers on Poly/ML. Technical Report 161, Computer Laboratory, University of Cambridge, February 1989.
11. R. Milner. A theory of type polymorphism in programming. *Journal of Computer and System Sciences*, 17:348–375, 1978.
12. A. Mycroft. Dynamic types in statically typed languages (preliminary draft). Unpublished typescript, December 1983.
13. A. Mycroft. Dynamic types in statically typed languages (2nd draft version). Unpublished typescript, August 1984.

# Lambda calculus characterizations of poly-time[1]

Daniel Leivant and Jean-Yves Marion
Department of Computer Science
*Indiana University, Bloomington, IN 47405*

### Abstract

We consider typed $\lambda$-calculi with pairing over the algebra **W** of words over $\{0,1\}$, with a destructor and discriminator function. We show that the poly-time functions are precisely the functions (1) $\lambda$-representable using simple types, with abstract input (represented by Church-like terms) and concrete output (represented by algebra terms); (2) $\lambda$-representable using simple types, with abstract input and output, but with the input and output representations differing slightly; (3) $\lambda$-representable using polymorphic typing with type quantification ranging over multiplicative types only; (4) $\lambda$-representable using simple and list types (akin to ML style) with abstract input and output; and (5) $\lambda$-representable over the algebra of flat lists (in place of **W**), using simple types, with abstract input and output.

## 1 Introduction

Machine-independent characterizations of computational complexity classes, such as poly-time, lend further credence to the importance of the classes considered, provide insight into their nature, relate them to issues relevant to programming and to verification, and suggest directions and tools for generalizing computational complexity classes to new settings, such as computational complexity for higher-type functionals and computational complexity in database theory. These characterizations fall, by and large, into three major classes: database queries (i.e. global methods of finite model theory), proof-theoretic characterizations, and applicative programs over free algebras.

---

[1] This preliminary report contains only a few selected proofs. The full paper is to appear in **Fundamenta Informaticae**.

Characterizations of poly-time are known within each of the three approaches. Poly-time queries are characterized by recursion equations [Saz80,Gur83], pure uninterpreted logic programs [Pap85], positive first-order fixpoints [Var82,Imm86], first-order inflationary fixpoints [GS86,Lei90a], and alternating transitive-closure [Imm87]. The poly-time functions are characterized proof theoretically by a weak system of Bounded Arithmetic [Bus86], and using weak second-order logic [Lei91,Lei93]. We are interested here in characterizations by applicative programs.

Functional schemas are the canonical definitions of classes such as the primitive recursive and the double-recursive functions [Pet66], but analogous characterizations of smaller complexity classes, more relevant to Computer Science, have faced some difficulties. Cobham's seminal work in this area [Cob65] characterizes poly-time using a rather ad hoc collection of initial functions, a modified recurrence schema over the natural numbers ("recursion on notations"), and a growth-restriction condition inspired by Grzegorczyk's classification of the primitive recursive functions [Grz53]. The defects of this characterization have been corrected in [BC91] and [Lei93a], to which we return momentarily.

An orthogonal research direction attempted to identify the functions that are definable using purely functional means, namely within typed $\lambda$-calculi. It is well-known that the (numeric) functions representable in the untyped $\lambda$-calculus are exactly the computable numeric functions (see e.g. [Bar80]). In contrast, the numeric functions representable in the simply-typed $\lambda$-Calculus **1$\lambda$** form the very small class of *extended polynomials* [Sch76,Sta79], that is, the functions defined by composition from 0,1, addition, multiplication, and the function $case(x,y,z) =_{df}$ if $x=0$ then $y$ else $z$ (see e.g. [FLO83] for an exposition and further references). In particular, the predecessor function is not representable. When the representation of input and output are allowed to differ, the predecessor is representable, but not the subtraction function [Sta79,Mai92,FLO83]. Only when the type-abstraction power of the calculus is extended to stratified polymorphism do we get a genuine complexity class, namely the super-exponential functions ($\mathcal{E}_4$ in Grzegorczyk's Hierarchy) [Lei91a]. This class, however, is too large to be of genuine computational interest.

Capturing the class of functions representable in **1$\lambda$** by a recurrence schema, in an algebraic style, presents two difficulties. First, the predecessor function is not representable, in spite of its apparent computational simplicity. This is related to the computation model implicit in $\lambda$-terms, in which there is no representation of memory, not even the single-character memory that allows a child to find that $q$ is the predecessor of $r$ in the Roman alphabet. The other difficulty is the limitation on size: the representability of multiplication arises from iterating over addition; but exponentiation, which is also definable from addition by a single recurrence ($2^{n+1} = 2^n + 2^n$), is not representable. One purely functional characterization of the functions representable in **1$\lambda$** is given in [Zai87]. An alternative characterization was given in [Lei90], using the surprisingly generic concept of "data tiering." The uses of data objects are classified into "tiers", where a use of an object $a$ is of higher tier if it is "global," i.e. $a$ is used as an iterator for functions over lower tiers. For example, each 0-1 word $w$ can be used computationally in two distinct ways: "locally," by accessing bits of $w$, and as a whole object, by using $w$ as a template for iterating function application. The functions representable in $1\lambda$ are then precisely the functions definable using two tiers and monotone recurrence (i.e. the successor case of a recurrence is of the form $f(sn, \vec{x}) = g(\vec{x}, f(n, \vec{x}))$, where $n$ is not a direct argument of $g$) [Lei90]. A similar phenomenon is implicit in the proof theoretic characterization of poly-time in [Lei91,Lei93].

The concept of "tiering" was rediscovered independently (in a slightly different guise) in [BC91] (motivated by the implicit tiering in [Lei91]). There a subrecursive characterization of poly-time is given without using Cobham's bounding condition. This method has been extended in [Blo92] to other complexity classes. In [Lei93a] similar characterizations are given, based more squarely on the approach of [Lei90], and doing away completely with initial functions other than the constructors for the free algebras in hand.

In this paper we use tiers within typed $\lambda$-calculi, in what seems to be the most appealing and transparent application of tiering to date. We refer to $1\lambda$ (with pairing) on top of the free algebra of words (with destructor and selector). In this formalism we have present simultaneously two familiar representations of data: as words (terms of base type), and as Church-like abstraction terms (expressions of higher type). These two representations play now the roles of tiers. We show that the functions representable in this formalism are precisely the poly-time functions. The tiering limits the growth of the representable functions to polynomial rate, whereas the basic computation power infused into the base type "fills in the gaps", and allows the representation of all poly-time functions. This seems to be the first characterization to date of a computational complexity class of practical interest within a $\lambda$-calculus.

First, we consider $1\lambda^p(\mathbf{W})$, the simply typed $\lambda$-calculus with pairing over the algebra $\mathbf{W}$ of words over $\{0, 1\}$, with a destructor and discriminator function. We show that the functions representable in $1\lambda^p(\mathbf{W})$ with abstract input and concrete output are precisely the poly-time functions. The same result holds for function representation that refers to abstracted output, provided the abstraction used to represent the output is allowed to differ slightly from that for the input. This difference disappears when a weak form of polymorphism is allowed.

## 2 Data and function representation

### 2.1 The typed $\lambda$-calculus over a word algebra

Let $\mathbf{W}$ be the free algebra inductively generated from two unary constructors **0** and **1** and one constant (i.e. 0-ary constructor) $\epsilon$. The intent is that $\epsilon$ denotes the empty word, so that a term such as $0(0(1(\epsilon)))$ can be identified with the corresponding word $001 \in \{0,1\}^*$. Two important functions over **W** are the destructor (predecessor) function **p**, defined by

$$\begin{aligned} \mathbf{p}(\epsilon) &= \epsilon \\ \mathbf{p}(c(x)) &= x \qquad c = 0, 1, \end{aligned}$$

and the discriminator (conditional) **d**, defined by

$$\begin{aligned} \mathbf{d}(\epsilon, a, b, c) &= a \\ \mathbf{d}(0(x), a, b, c) &= b \\ \mathbf{d}(1(x), a, b, c) &= c \end{aligned}$$

We can now define $\mathbf{1}\lambda^p(\mathbf{W})$, the simply typed $\lambda$-calculus with pairing over **W**, expanded with the destructor and discriminator functions. The *types* of $\mathbf{1}\lambda^p(\mathbf{W})$ are generated inductively by: $o$ is a type; if $\sigma$ and $\tau$ are types, then so are $\sigma \times \tau$ and $\sigma \to \tau$. The type $o$ is intended to denote a set of basic data objects. We abbreviate $o \times \cdots \times o$ with $k$ factors, i.e. $o \times (o \times \cdots (o \times o) \cdots)$, by $o^k$.

The *expressions* of $\mathbf{1}\lambda^p(\mathbf{W})$ are defined inductively, as follows.

- $\epsilon$, **0**, **1**, **p** and **d** are expressions, of types $o$, $o \to o$, $o \to o$, $o \to o$ and $o^4 \to o$, respectively.

- The pairing function for base type, $\langle -, - \rangle$, is an expression of type $o \to (o \to o \times o)$; i.e. we write $\langle E, F \rangle$ for the result of pairing $E$ and $F$.

- The projections at base type, $\pi_0$ and $\pi_1$, are expressions of type $(o \times o) \to o$.

- For each type $\tau$ there is a countable collection of variables of type $\tau$, where each variable of type $\tau$ is also an expression of type $\tau$.

- If $x$ is a variable of type $\sigma$ and $E$ is an expression of type $\tau$, then $\lambda x\, E$ is an expression of type $\sigma \to \tau$.

- If $E$ and $F$ are expressions of types $\sigma \to \tau$ and $\sigma$, respectively, then $EF$ is an expression of type $\tau$.

We use $\langle E, F, G \rangle$ as an abbreviation for $\langle E, \langle F, G \rangle \rangle$. If $E$ is of type $o^3$, then $\pi_1(E)$ and $\pi_2(E)$ abbreviate $\pi_0(\pi_1(E))$ and $\pi_1(\pi_1(E))$, respectively. Similar notational conventions apply to tuples of higher arities.

The variables *free* in an expression are defined as usual. An expression without free variables is *closed*. We write $[F/x]E$ for the result of substituting $F$ for all free occurrences of the variable $x$ in $E$. We write $E[\vec{x}]$ for $E$ if all free variables in $E$ are among $\vec{x} = x_1, \cdots, x_k$. Assuming some canonical ordering of the variables, we then write $E[F_1 \ldots F_k]$ for $[F_1/x_1] \cdots [F_k/x_k]\, E$.

## 2.2 Reductions and normal form

The computational meaning of $\lambda$-abstraction and of the basic functions is conveyed by the *reduction rules*, which for $1\lambda^p(\mathbf{W})$ take the following forms.

$$\beta: \quad (\lambda x\, E)F \;\vdash\; [F/x]E$$
$$\text{projection:} \quad \pi_i\langle E_0, E_1\rangle \;\vdash\; E_i \qquad i = 0, 1$$
$$\text{destruction:} \quad \mathbf{p}(\epsilon) \;\vdash\; \epsilon$$
$$\mathbf{p}(cE) \;\vdash\; E, \qquad c = 0, 1.$$
$$\text{selection:} \quad \mathbf{d}\langle \epsilon, E, F_0, F_1\rangle \;\vdash\; E$$
$$\mathbf{d}\langle ct, E, F_0, F_1\rangle \;\vdash\; F_c \qquad c = 0, 1.$$

If $F \vdash F'$ then we say that $F$ *reduces* to $F'$. A subexpression $F$ of $E$ that can be reduced by one of the rules above is a *redex* of $E$. If $F \vdash F'$ then we write $[F/x]G \succ [F'/x]G$. In particular, if $E \vdash E'$ then $E \succ E'$. An expression $E$ is irreducible, or **normal**, if it has no redex, i.e. if for no $E'$ do we have $E \succ E'$. We write $=_{\text{red}}$ for the symmetric and transitive closure of $\succ$.

PROPOSITION 2.1 *If $E$ is a closed normal expression of type $o$ then $E \in \mathbf{W}$.*

Call an expression *non-abstracted* if it does not start with $\lambda$. The normal expressions have the following property (related to the *subformula property* of natural deductions, cf. [Pra65]), as follows. We say that a type expression $\sigma$ is a *subtype* of $\tau$ if $\sigma$ is a subexpression of $\tau$.

PROPOSITION 2.2 *Let $E$ be a normal expression of $1\lambda^p(\mathbf{W})$ of type $\tau$, $S$ a subexpression of $E$, of type $\sigma$. Then either (1) $\sigma$ is a subtype of $\tau$, or (2) $\sigma$ is a subtype of the type of a constant of $1\lambda^p(\mathbf{W})$ occurring in $E$, or a free variable in $E$. If $S$ is non-abstracted, then (2) holds.*

## 2.3 Abstraction representation of algebra terms

Every term $w \in \mathbf{W}$ is an expression of $1\lambda^p(\mathbf{W})$. At the same time, each such $w$ has a representation as a $\lambda$-expression. Church's well-known $\lambda$-representation maps a natural number $n$ to $\bar{n} =_{\text{df}} \lambda s \lambda z.\, s^n(z)$, (where $f^k$ denotes the $k$'th iterate of the function $f$). Put differently, $\bar{n} =_{\text{df}} \lambda s \lambda z.\, n[s, z]$, where $0[s, z] =_{\text{df}} z$ and $(sn)[s, z] =_{\text{df}} s(n[s, z])$. That is, $n$ is represented by its unary notation, with zero and successor abstracted.

Similar $\lambda$-representations can be given for any free algebra [BB85]. If $\mathbf{A}$ is a free algebra with constructors $\vec{c} = c_1 \ldots c_k$, of arities $r_1 \ldots r_k$ respectively, then an element $a \in \mathbf{A}$ is $\lambda$-represented by $\bar{a}$, where $\bar{a} =_{\text{df}} \lambda v_{c_1} \cdots v_{c_k} a[\vec{v}]$, and $a[\vec{v}]$ is the "variable-form" of $a$, defined recursively by $(c_i(a_1 \ldots a_{r_i}))[\vec{v}] =_{\text{df}}$

$v_{c_i}(a_1[\vec{v}], \cdots, a_{r_i}[\vec{v}])$. For example, for $\mathbf{A}=\mathbf{W}$, and $w = 011\epsilon$, we have $w[v_0 v_1 v_\epsilon] = v_0(v_1(v_1(v_\epsilon)))$, and $\bar{w} = \lambda\, v_0\, v_1\, v_\epsilon .\, v_0(v_1(v_1(v_\epsilon)))$.

The Church-Böhm representation of algebra terms has a canonical typing in $\mathbf{1\lambda}$. Every Church numeral can be assigned all types of the form $(\tau \to \tau) \to (\tau \to \tau)$. Similarly, the expressions $\bar{w}$, for $w \in \mathbf{W}$, are naturally assigned the type $\omega[\tau] =_{\mathrm{df}} (\tau \to \tau) \to (\tau \to \tau) \to \tau \to \tau$. We write $\bar{w}^q$ for $\bar{w}$ with the type $\omega[o^q]$.

## 2.4 Function representation

Expressions of $\mathbf{1\lambda^p(W)}$ convey two forms of computation. On the one hand, there are "low level" computation, via the projection, destruction, and selection reductions. On the other hand, we have "high level" computation, via $\beta$-reduction. Computable functions can be defined using either one of these aspects. The combination of the two forms is the salient feature of our main result.

DEFINITION 2.3 A function $f : \mathbf{W}^k \to \mathbf{W}^r$ is *explicitly defined* in $\mathbf{1\lambda^p(W)}$ by an expression $E$, of type $o^k \to o^r$, if $E w_1 \cdots w_k =_{\mathrm{red}} f(\vec{w})$, for every $\vec{w} = (w_1 \ldots w_k) \in \mathbf{W}^k$. A function $f : \mathbf{W}^k \to \mathbf{W}$ is *2-tier represented* by an expression $E$, if, for some $m_1 \ldots m_k$, $E \bar{w}_1^{m_1} \cdots \bar{w}_k^{m_k} =_{\mathrm{red}} f(\vec{w})$ for all $\vec{w} = (w_1 \ldots w_k) \in \mathbf{W}^k$. Note that $E$ here is of type $\omega[o^{m_1}] \to \cdots \to \omega[o^{m_k}] \to o$.

THEOREM 2.4 *A function over $\mathbf{W}$ is poly-time iff it is 2-tier representable in $\mathbf{1\lambda^p(W)}$.*

We mention the main points of the proof in the next two sections. We then discuss alternative forms of representability, which are closed under composition.

# 3 Poly-time functions are 2-tier representable

## 3.1 Explicit definability of Turing machine transitions

In this section we prove the forward direction of Theorem 2.4:

PROPOSITION 3.1 *Every poly-time function over $\mathbf{W}$ is 2-tier representable in $\mathbf{1\lambda^p(W)}$.*

The idea of the proof is this. Given a Turing machine $M$, the configuration-change of $M$ under a single transition can be described by a function explicitly-defined in $\mathbf{1\lambda^p(W)}$. On the other hand, polynomial-size iterators are representable. Using such iterators over transitions we obtain the configuration of $M$ after a polynomial number of transitions starting with the initial configuration.

LEMMA 3.2 *Given a deterministic TM $M$, there is an explicitly definable function* **cnvt** *such that* $(s, u, v) \vdash_M (s', u', v')$ *iff* $\mathbf{cnvt}(\#s, u, v) = (\#s', u', v')$.

## 3.2 Polynomial-length iteration of TM transitions

It is well known (see e.g. [FLO83]) that Church's $\lambda$-representation for numeric addition and multiplication are in fact Church-representations in $1\lambda$. Moreover, these representations lift unchanged to $1\lambda^p(\mathbf{W})$. The primitive recursive definition of numeric addition, when applied to $\mathbf{W}$, yields concatenation:

$$\mathbf{S} =_{df} \lambda w\, y.\, \lambda v_0 v_1 v_\epsilon.\, w v_0 v_1 (y v_0 v_1 v_\epsilon).$$

Corresponding to numeric multiplication we have for $\mathbf{W}$ the function that, on input $w, y$, returns $y$ concatenated to itself $|w|$ times; this is represented by

$$\mathbf{P} =_{df} \lambda w\, y.\, \lambda v_0 v_1 v_\epsilon.\, w(y v_0 v_1)(y v_0 v_1) v_\epsilon.$$

$\mathbf{S}$ and $\mathbf{P}$ can both be assigned the type $\omega[\tau] \to \omega[\tau] \to \omega[\tau]$, for any type $\tau$. Consequently, we have:

LEMMA 3.3 *Let $k, c \geq 0$, and $Q(x) = x^k + c$. There is an expression $\mathbf{Q}$ of $1\lambda$, such that, for every $w \in \mathbf{W}$, if $|w| = n$, then*

$$\mathbf{Q}\bar{w} =_{red} \lambda v_0 v_1 v_\epsilon.\, v_0^{q(n)} v_\epsilon.$$

*Moreover, $\mathbf{Q}$ can be assigned the type $\omega[\tau] \to \omega[\tau]$ for any type $\tau$.*

LEMMA 3.4 *Let $q \geq 1$, and set $\tau =_{df} o^q$. There is an expression $\mathbf{T}_q^-$ of $1\lambda^p(\mathbf{W})$, of type $\omega[\tau] \to o$, such that, for all $w \in \mathbf{W}$, $\mathbf{T}_q^- \bar{w}^q =_{red} w$.*

**Proof of Proposition 3.1.** We give the proof for a unary function. The proof for the general case is similar. Let $M$ be a deterministic TM as above, computing a function $f : \mathbf{W} \to \mathbf{W}$, and whose run-time for input of size $\ell$ is $\leq Q(\ell) = \ell^k + c$.

Let **cnvt** be the transition function for $M$, and let $\mathbf{D}$ be an explicit definition of **cnvt**, as defined in Lemma 3.2. Let $s_0$ be the initial state of $M$. Let $\mathbf{Q}$ represent $Q$ in $1\lambda$, as in Lemma 3.3, and let $\mathbf{D}$ be as in Lemma 3.2. Let $x$ be a variable of type $\omega[o^3]$, and set

$$\begin{aligned}\mathbf{J}[x] &=_{df} \langle \#s_0, \epsilon, \mathbf{T}_3^- x \rangle, \\ \mathbf{E}[x] &=_{df} (\mathbf{Q}x)\mathbf{D}\mathbf{D}\mathbf{J}, \\ \mathbf{F} &=_{df} \lambda x^{\omega[o^3]}.\, \pi_2 \mathbf{E}[x],\end{aligned}$$

Then $\mathbf{J}[\bar{w}] =_{\text{red}} \langle \#s_0, w, \epsilon \rangle$, the initial configuration of $M$, and so

$$\mathbf{E}[\bar{w}] =_{\text{red}} \bar{k}\,\mathbf{DDJ}$$
$$=_{\text{red}} \mathbf{D}^k \langle \#s_0, w, \epsilon \rangle,$$

where $k = Q(|w|)$ and $\bar{k} =_{\text{df}} \lambda v_0 v_1 v_\epsilon.\ v_0^k\, v_\epsilon$. That is, $E[\bar{w}]$ is $=_{\text{red}}$ to the $k$'th iterate of $\mathbf{D}$ applied to the initial configuration. Therefore, $\mathbf{F}\bar{w}$ is $=_{\text{red}}$ to the portion of $M$'s tape to the right of the head, after $Q(|w|)$ transitions, i.e. $f(w)$. Thus $\mathbf{F}$ 2-tier represents $f$, and is of type $\omega[o^3] \to o$. $\square$

## 4 Computability of functions over W

PROPOSITION 4.1 *If a function over* $\mathbf{W}$ *is 2-tier representable in* $1\lambda^p(\mathbf{W})$ *then it is poly-time.*

Consider a *linked-pointer representation* of abstraction-free expressions of $1\lambda^p(\mathbf{W})$, as follows. $\epsilon$ is represented by itself; if $\mathbf{f}$ is $\mathbf{0}$, $\mathbf{1}$, $\mathbf{p}$, $\mathbf{d}$, $\pi_i$, then $\mathbf{f}(\mathbf{t}_1 \ldots \mathbf{t}_n)$ is represented by an $n+1$ tuple consisting of $\mathbf{f}$ followed by pointers to representations of $\mathbf{t}_1 \ldots \mathbf{t}_n$, respectively; and a pair $\langle a, a' \rangle$ is represented by a pair of pointers to representations of $a$ and $a'$, respectively. In particular, we say that a function $h$ is constant-time if it is constant time with respect to the representation above. For example, the function $f(v) = (v, \mathbf{p}v)$ is constant-time, because given a representation $r(v)$ of $v \in \mathbf{W}$, a representation of $f(v)$ is obtained as a pointer to $r(v)$ paired with a pointer to a representation of $\mathbf{p}(v)$. Note that $f(v)$ is not constant-time under the representation of expressions by their syntax-tree, because $f(v)$ would then require the creation of a second copy of $v$.

LEMMA 4.2 *Suppose* $E[x_1 \ldots x_p, y_1 \ldots y_q]$ *is a normal expression of type* $o^{r_1} \to o^{r_2} \to \cdots \to o^{r_n} \to o^{r_0}$, *with free variables* $x_1 \ldots x_p$, *where* $x_i$ *is of type* $\omega[o^{s_i}]$, *and free variables* $y_1 \ldots y_q$ *where* $y_j$ *is of type* $o^{t_j}$. *Then, for* $w_1 \ldots w_p \in \mathbf{W}$, $v_j \in \mathbf{W}^{t_j}$ $(j = 1 \ldots q)$, *and* $z_\ell \in \mathbf{W}^{r_\ell}$ $(\ell = 1 \ldots n)$, *the normal form of* $E^* z_1 \ldots z_n$, *where* $E^* =_{\text{df}} E[\bar{w}_1 \ldots \bar{w}_p, v_1 \ldots v_q]$, *is computable in time polynomial in* $L_1 =_{\text{df}} \max(|w_1| \ldots |w_p|)$ *and constant in* $L_2 =_{\text{df}} \max(|v_1|, \ldots |v_q|, |z_1| \ldots |z_n|)$.

**Proof of Proposition 4.1.** Suppose $f: \mathbf{W} \to \mathbf{W}$ is represented by $\mathbf{F}$, of type $\omega[o^m] \to o$, i.e. $\mathbf{F}\bar{w} =_{\text{red}} f(w)$, for every $w \in \mathbf{W}$. By the Normal Form Theorem we may assume that $\mathbf{F}$ is normal, i.e. it is of the form $\lambda x E[x]$, where $E$ is of type $o$. So $f(w) =_{\text{red}} E[\bar{w}]$. Since, by Lemma 4.2, the normal form of $E[\bar{w}]$ is computable within time polynomial in $|w|$, it follows that $f$ is poly-time. $\square$

# 5 Church-representation using changing arities

The concept of 2-tier representability differs from Church-style $\lambda$-representation of functions in that the output is a concrete rather than an abstracted expression. In particular, the definition of 2-tier representability does not imply that the functions representable are closed under composition. This difference is related to our working with the type system and constructors of $\mathbf{1}\lambda^p(\mathbf{W})$, and is not inherent to the result. In this section we describe alternatives choices, that yield Church-style $\lambda$-characterizations of the poly-time functions over $\{0,1\}^*$.

**DEFINITION 5.1** A function $f : \mathbf{W} \to \mathbf{W}$ is *Church-represented in* $\mathbf{1}\lambda^p(\mathbf{W})$ *by $E$, with arity changing from $n$ to $m$*, if $E\bar{w}^n =_{\text{red}} \overline{f(w)}^m$ for all $w \in \mathbf{W}$. The function $f$ is *Church-represented in* $\mathbf{1}\lambda^p(\mathbf{W})$ *with changing arity* if for all $m$ there is an $n$ such that $f$ is representable with arity changing from $n$ to $m$.

More generally, a function $f : \mathbf{W}^k \to \mathbf{W}$ is *Church-represented in* $\mathbf{1}\lambda^p(\mathbf{W})$ *by $E$, with arity changing from $n_1 \ldots n_k$ to $m$*, if

$$E\bar{w}_1^{n_1} \cdots \bar{w}_k^{n_k} =_{\text{red}} \overline{f(w_1 \ldots w_k)}^m$$

for all $w_1 \ldots w_k \in \mathbf{W}$. $f$ is *Church-represented in* $\mathbf{1}\lambda^p(\mathbf{W})$ *with changing arity* if for all $m$ there are $n_1 \ldots n_k$ such that $f$ is representable with arity changing from $n_1, \ldots n_k$ to $m$.

The following lemma states that $y \in \mathbf{W}$ can be mapped uniformly to an abstract form $\bar{y}^m$, provided a term $\bar{w}^{m+q}$ is given with $w$ at least as long as $y$, and $q \geq 1$.

**LEMMA 5.2** Let $m, q \geq 1$. There is an expression $\mathbf{T}^+_{mq}$, of type $\omega[o^{m+q}] \to o \to \omega[o^m]$, such that, for all $w, y \in \mathbf{W}$, if $|w| \geq |y|$, then $\mathbf{T}^+_{mq} \bar{w}^{m+q} y =_{\text{red}} \bar{y}^m$.

Using Lemma 5.2 we obtain:

**LEMMA 5.3** Let $n \geq m$. There is an expression $\mathbf{L}_{nm}$ of the simply typed $\lambda$-calculus with pairing, such that $\mathbf{L}_{nm} \bar{w}^n =_{\text{red}} \bar{w}^m$ for all $w \in \mathbf{W}$.

From Lemma 5.3 we derive:

**LEMMA 5.4** *The functions representable with changing arities are closed under composition.*

**THEOREM 5.5** *A function is representable in* $\mathbf{1}\lambda^p(\mathbf{W})$ *with changing arity iff it is poly-time.*

**Proof.** Suppose $f$ runs in time $\leq q(\ell) =_{\mathrm{df}} \ell^k + c$ for all input $w$ of length $\leq \ell$. The proof of Theorem 2.4 implies that there is an expression $\mathbf{F}$, of type $\omega[o^3] \to o$, which 2-tier represents $f$. The same proof shows that, for every $n \geq 3$ there is an expression $\mathbf{F}_n$, of type $\omega[o^n] \to o$, which 2-tier represents $f$. Let $\mathbf{Q}$ be as in Lemma 3.3. Given $m$, set $n = m+3$, and define

$$\mathbf{F}_{nm} =_{\mathrm{df}} \lambda x^{\omega[o^n]} . \mathbf{T}_{m3}^+(\mathbf{Q}x)(\mathbf{F}_n x).$$

Then $\mathbf{F}_{nm}$ represents $f$ with arity changing from $n$ to $m$.

Conversely, suppose $\mathbf{F}$ represents a function $f$ with arity changing from $n$ to $m$. Let

$$\begin{aligned}
\epsilon_m &=_{\mathrm{df}} \langle \epsilon, \ldots, \epsilon \rangle \quad (m\text{-tuple}), \\
\mathbf{0}_m &=_{\mathrm{df}} \lambda x^{o^m} . \langle \mathbf{0}(\pi_0 x), \pi_1 x, \ldots, \pi_1 x \rangle, \\
\mathbf{1}_m &=_{\mathrm{df}} \lambda x^{o^m} . \langle \mathbf{1}(\pi_0 x), \pi_1 x, \ldots, \pi_1 x \rangle.
\end{aligned}$$

Define

$$\mathbf{F}_0 =_{\mathrm{df}} \lambda x^{\omega[o^n]} \pi_0 (\mathbf{F} x \mathbf{0}_m \mathbf{1}_m \epsilon_m).$$

The expression $\mathbf{F}_0$ is of type $\omega[o^m] \to o$, and $\mathbf{F}_0(\bar{w}^n) =_{\mathrm{red}} v$ iff $\mathbf{F}(\bar{w}^n) =_{\mathrm{red}} \bar{v}^m$. Thus $\mathbf{F}_0$ 2-tier represents $f$, and so $f$ is poly-time, by Theorem 2.4. □

## 6 Representation using weak polymorphism

The slightly messy role of pairing in the $\lambda$-representation above is streamlined when all objects whose type is $o^m$ for some $m$ form one entity, in one way or another. In this section we achieve this by using a weak form of polymorphism. In the next two sections we use variants of lists, one represented algebraically, and the other in the type system.

Let $2\lambda$ be the Girard-Reynolds polymorphically typed $\lambda$-calculus [Gir72, Rey74] (see e.g. [FLO83] for an exposition). Let $2\lambda_o^p(\mathbf{W})$ be defined like $1\lambda^p(\mathbf{W})$, but with type variables and type abstraction as in $2\lambda$, and with type application only for type arguments without $\to$ or $\forall$ (i.e. for types generated from $o$ and type variables using only $\times$). That is, type quantifiers in $2\lambda_o^p(\mathbf{W})$ range over multiplicative types only. Note that the constants, $\epsilon$, $\mathbf{0}$, $\mathbf{1}$, $\mathbf{p}$, and $\mathbf{d}$, have here the same type as in $1\lambda^p(\mathbf{W})$.

Let $w \in \mathbf{W}$, and let $\bar{w}$ be the Church-Böhm abstraction term for $W$. In analogy to the Fortune-O'Donnell numerals [For79, ODo79] (see [FLO83]), we have the *polymorphic form* of $\bar{w}$,

$$\bar{w}^{FO} =_{\mathrm{df}} \Lambda t. \lambda v_0^{t \to t} v_1^{t \to t} v_\epsilon^t . w[v_0, v_1, v_\epsilon],$$

which is of type $\forall t. \omega[t]$.

DEFINITION 6.1 A function $f : \mathbf{W} \to \mathbf{W}$ is *Fortune-O'Donnell represented in* $2\lambda_o^p(\mathbf{W})$ *by* $E$ if $E\bar{w}^{FO} =_{\text{red}} \overline{f(w)}^{FO}$ for all $w \in \mathbf{W}$.

We have for $2\lambda_o^p(\mathbf{W})$ a weak analogue of Proposition 2.2. The *subtypes* of a type $\tau$ are defined inductively, as follows. Every type is a subtype of itself; every subtype of $\tau$ is a subtype of $\tau \to \sigma$, $\sigma \to \tau$, $\tau \times \sigma$, and $\sigma \times \tau$; every type $[\sigma/t]\tau$ is a subtype of $\forall t.\tau$, where $\sigma$ is a type without $\to$ or $\forall$, and $[\sigma/t]\tau$ is the result of substituting $\sigma$ for all free occurrences of $t$ in $\tau$.

PROPOSITION 6.2 *Let* $E$ *be a normal expression of* $2\lambda_o^p(\mathbf{W})$, *of type* $\tau$, $S$ *a subexpression of* $E$, *of type* $\sigma$. *Then* $\sigma$ *is either a subtype (in the sense above) of* $\tau$, *or of the type of a constant of* $1\lambda^p(\mathbf{W})$ *occurring in* $E$, *or of a free variable in* $E$.

The following lemma is analogous to Lemma 5.2.

LEMMA 6.3 *There is an expression* $\mathbf{T}^t$, *of type* $(\forall s \omega[s]) \to o \to \omega[t]$, *such that, for all* $w, y \in \mathbf{W}$, *if* $|w| \geq |y|$, *then* $\mathbf{T}^t \bar{w}^{FO}(o \times t)y =_{\text{red}} \bar{y}^t$.

THEOREM 6.4 *A function over* $\mathbf{W}$ *is poly time iff it is Fortune-O'Donnell representable in* $2\lambda_o^p(\mathbf{W})$.

**Proof.** We prove the theorem for a unary function; the general case is similar. Suppose $f : \mathbf{W} \to \mathbf{W}$ runs in time bounded by $\leq q(\ell) = \ell^k + c$ for all input $w$ of length $\leq \ell$. Then, by the proof of Theorem 3.1, there is an expression $\mathbf{F} = \lambda x^{\omega[o^3]} \mathbf{F}_0$ of $1\lambda^p(\mathbf{W})$ that represents $f$ with input of type $\omega(o^3)$ and output of type $o$. Let $\mathbf{F}_0'$ be obtained from $\mathbf{F}_0$ by replacing each occurrence of $x$ by $yo^3$, where $y$ is a fresh variable of type $\forall s \omega[s]$. Then, by a simple induction on expressions, we see that $\mathbf{F}_0$ is a legal expression of $2\lambda_o^p(\mathbf{W})$, of type $o$.

Also, it is well-known that addition and multiplication are represented in $2\lambda$ using only atomic types as arguments of type applications [FLO83]. Therefore, as in Lemma 3.3, we obtain an expression $\mathbf{Q}$ that represents $q$ over $\mathbf{W}$ in $2\lambda_o^p(\mathbf{W})$. Let $\mathbf{F}' =_{\text{df}} \lambda y^{\forall s \omega[s]}. \Lambda t. \mathbf{T}^t(\mathbf{Q}y)(\mathbf{F}_0' y)$. Then $\mathbf{F}'$ Fortune-O'Donnell represents $f$ in $2\lambda_o^p(\mathbf{W})$.

Conversely, suppose $f$ is a unary function over $\mathbf{W}$ which is Fortune-O'Donnell represented in $2\lambda_o^p(\mathbf{W})$ by a normal expression $\mathbf{F} = \lambda x^{\forall s. \omega[s]}. \Lambda t. \mathbf{F}_0$, where $\mathbf{F}_0$ is of type $\omega[t]$. By Proposition 6.2 $\mathbf{F}_0$ is without type abstraction, and every occurrence of $x$ in $\mathbf{F}_0$ is applied to a type argument. Each type argument $\sigma$ of $x$ is of the form $\alpha_1 \times \cdots \times \alpha_{k(\sigma)}$, where each $\alpha_i$ is either $t$ or $o$.

Let $\mathbf{F}_0'$ be obtained from $\mathbf{F}_0$ by: (1) replacing every $t$ by $o$; (2) replacing every subexpression $x\sigma$ by $y_\sigma$, where $y_\sigma$ is a variable of type $\omega[o^{k(\sigma)}]$, i.e. the type of $x([o/t]\sigma)$. Let $\sigma_1 \ldots \sigma_n$ be a list of all type arguments of $x$ in $\mathbf{F}_0$, and define
$$\mathbf{F}' =_{\text{df}} \lambda y_{\sigma_1} \ldots y_{\sigma_n}. \mathbf{F}_0'.$$

Then
$$\mathbf{F}'\bar{w}^{k(\sigma_1)}\ldots\bar{w}^{k(\sigma_n)} =_{\mathrm{red}} \overline{f(w)}^1.$$

$\mathbf{F}'$ represents, with changing arity, an $n$-ary function $g$, which by Theorem 5.5 is poly-time. But $f(w) = g(w,\ldots,w)$, so $f$ too is poly-time. □

## 7 Representation using algebraic lists

Consider the free algebra $\mathbf{L}$ generated from the 0-ary constructors $\mathbf{0}$ and $\mathbf{1}$ and the binary constructor $\mathbf{p}$. Let $\mathbf{1\lambda(L)}$ be the simply typed $\lambda$-calculus (without pairing) over the algebra $\mathbf{L}$, augmented with the destructor functions $\mathbf{p}_0, \mathbf{p}_1$ and the discriminator function $\mathbf{d}$, defined by

$$\mathbf{p}_i(\mathbf{0}) = \mathbf{0}$$
$$\mathbf{p}_i(\mathbf{1}) = \mathbf{1}$$
$$\mathbf{p}_i(\mathbf{p}(E_0, E_1)) = E_i$$

$$\mathbf{d}(\mathbf{0}, E_0, E_1, F) = E_0$$
$$\mathbf{d}(\mathbf{1}, E_0, E_1, F) = E_1$$
$$\mathbf{d}(\mathbf{p}(x,y), E_0, E_1, F) = F$$

Let $\mathbf{L}_W$ be the subset of $\mathbf{L}$ inductively generated by: $\mathbf{0}, \mathbf{1} \in \mathbf{L}_W$; and if $x \in \mathbf{L}_W$ then $\mathbf{p}(\mathbf{0}, x), \mathbf{p}(\mathbf{1}, x) \in \mathbf{L}_W$. Thus $\mathbf{L}_W$ consists of the list-representations of words over $\{0, 1\}$.

Terms of $\mathbf{L}$ are represented in $\mathbf{1\lambda(L)}$ both as themselves, and by abstraction expressions. For example, the term $\mathbf{p}(\mathbf{0}, \mathbf{p}(\mathbf{1}, \mathbf{p}(\mathbf{1}, \mathbf{0}))) \in \mathbf{L}_W$, which represents in $\mathbf{L}_W$ the string 0110, is represented by the abstraction expression $\lambda v_p^{o\to(o\to o)} v_0^o v_1^o . v_p v_0 (v_p v_1 (v_p v_1 v_0))$. For $w \in \{0,1\}^*$ let $\bar{w}^L$ be the representation of $w$ as above in $\mathbf{1\lambda(L)}$, i.e., of $w$ seen as an object in $\mathbf{L}_W$.

DEFINITION 7.1 *A function* $f : \{0,1\}^* \to \{0,1\}^*$ *is Church-represented in* $\mathbf{1\lambda(L)}$ *by* $E$ *if* $E\bar{w}^L =_{\mathrm{red}} \overline{f(w)}^L$ *for all* $w \in \{0,1\}^*$.

Again, the proof of Theorem 5.5 can be easily modified to yield:

THEOREM 7.2 *A function* $f : \{0,1\}^* \to \{0,1\}^*$ *is Church-representable in* $\mathbf{1\lambda(L)}$ *iff it is poly-time.*

## 8 Representation using a list type

Here we incorporate all types $o^m$ in a type $o^*$ of "flat" lists over $o$. Consider a calculus $\mathbf{1\lambda^*(W)}$ that extends $\mathbf{1\lambda^p(W)}$ as follows. The base type of $\mathbf{1\lambda^*(W)}$ is $o^*$. Modify the defining clauses for expressions in $\mathbf{1\lambda^p(W)}$ as follows.

- Every expression of type $o$ is also an expression of type $o^*$.

- **d** is a constant of type $o \times (o^*)^3 \to o^*$; other constants have the same type as before.

- Every expression of type $o \times o^*$ is also an expression of type $o^*$.

- The projection functions $\pi_0$ is of type $o^* \to o$, and $\pi_1$ is of type $o^* \to o^*$.

We are particularly interested in the assignment of the type $\omega[o^*]$ to expressions $\bar{w}$ for $w \in \mathbf{W}$, which we denote by $\bar{w}^*$.

**Definition 8.1** A function $f : \mathbf{W} \to \mathbf{W}$ is *Church-represented in* $\mathbf{1\lambda^*(W)}$ *by* $E$ if $E\bar{w}^* =_{\text{red}} \overline{f(w)}^*$ for all $w \in \mathbf{W}$.

**Theorem 8.2** *A function $f$ is Church-representable in $\mathbf{1\lambda^*(W)}$ iff it is poly-time.*

The proof is molded after the proof of Theorem 5.5.

One might wonder about the relation between the weak polymorphic function representation and the representation that uses the list-type $o^*$, as both can be regarded as using an implicit quantification over the exponent $i$ in $o^i$. The two are related, but not directly. The polymorphic type quantification over multiplicative types is a universal quantifier applied to potentially compound types, so that the type of abstracted words, $\forall t \omega[t]$, can be regarded as the conjunction of all types $(\tau \to \tau)^2 \to \tau \to \tau$, for $\tau$ multiplicative. In contrast, the type $o^*$ is, intuitively, the existential type $\exists i\, o^i$, and the type of abstracted numerals in $\mathbf{1\lambda^*(W)}$, $(o^* \to o^*)^2 \to o^* \to o^*$, has that quantifier appear six times. The proofs of the characterization of poly-time by these two forms of representations are therefore significantly different.

## References

**Bar80** Hendrick P. Barendregt, **The Lambda-Calculus: Its Syntax and Semantics**, North-Holland, 1980.

**BB85** Corrado Böhm and Allessandro Berarducci, *Automatic synthesis of typed $\lambda$-programs on term algebras*, **Theoretical Computer Science 39** (1985) 135–154.

**BC92** Stephen Bellantoni and Stephen Cook, *A new recursion-theoretic characterization of the poly-time functions,* to appear in **Computational Complexity** 1992.

**Blo92** Stephen Bloch, *Functional characterizations of uniform log-depth and polylog-depth circuit families,* to appear in the Proceedings of the 1992 IEEE Conference on Structure in Complexity.

**Bus86** Samuel Buss, **Bounded Arithmetic**, Bibliopolis, Naples, 1986.

**Cob65** A. Cobham, *The intrinsic computational difficulty of functions,* in Y. Bar-Hillel (ed.), **Proceedings of the International Conference on Logic, Methodology, and Philosophy of Science**, North-Holland, Amsterdam (1962) 24–30.

**FLO83** Steven Fortune, Daniel Leivant, and Michael O'Donnell, *The expressiveness of simple and second order type structures,* **Journal of the ACM 30** (1983), pp 151-185.

**For79** Steven Fortune, *Topics in Computational Complexity,* PhD Dissertation, Cornell University, Ithaca, NY, 1979.

**Gir72** Jean-Yves Girard, *Interprétation fonctionelle et élimination des coupures dans l'arithmétique d'ordre superieur,* Thèse de Doctorat d'Etat, 1972, Paris.

**Grz53** A. Grzegoczyk, *Some classes of recursive functions,* **Rozprawy Mate. IV**, Warsaw, 1953.

**GS86** Yuri Gurevich and Saharon Shelah, *Fixed-point extensions of first-order logic,* **Annals of Pure and Applied Logic 32** (1986) 265–280.

**Gur83** Yuri Gurevich, *Algebras of feasible functions,* **Twenty Fourth Symposium on Foundations of Computer Science**, IEEE Computer Society Press, 1983, 210–214.

**Imm86** Neil Immerman, *Relational queries computable in polynomial time,* **Information and Control 68** (1986) 86–104. Preliminary report in **Fourteenth ACM Symposium on Theory of Computing** (1982) 147–152.

**Imm87** Neil Immerman, *Languages which capture complexity classes,* **SIAM Journal of Computing 16** (1987) 760–778.

**Lei90** Daniel Leivant, *Subrecursion and lambda representation over free algebras* (Preliminary Summary), in Samuel Buss & Philip Scott (eds.), **Feasible Mathematics**, *Perspectives in Computer Science*, Birkhauser-Boston, New York (1990) 281–291.

**Lei90a** Daniel Leivant, *Inductive definitions over finite structures,* **Information and Computation 89** (1990) 95–108.

**Lei91** Daniel Leivant, *A foundational delineation of computational feasiblity*, in **Proceedings of the Sixth IEEE Conference on Logic in Computer Science** (Amsterdam), IEEE Computer Society Press, Washington, D.C., 1991.

**Lei91a** Daniel Leivant, *Finitely stratified polymorphism*, **Information and Computation**, 1991.

**Lei93** Daniel Leivant, *A foundational delineation of poly-time*, **Information and Computation**, 1993.

**Lei93a** Daniel Leivant, *Stratified functional programs and computational complexity*, **Conference Record of the Twentieth Annual ACM Symposium on Principles of Programming Languages**, 1993.

**Mai92** Harry Mairson, *A simple proof of a theorem of Statman,* to appear in **Theoretical Computer Science**, 1992.

**ODo79** Michael O'Donnell, *A programming language theorem which is independent of Peano Arithemtic*, **Proceedings of the Eleventh Annual ACM Symposium on Theory of Computing**, ACM, New York, 1979, 176–188.

**Pap85** Christos Papadimitriou, *A note on the expressive power of PROLOG*, **Bull. EATCS 26** (June 1985) 21–23.

**Pet66** Rósza Péter, **Rekursive Funktionen**, Akadémiai Kiadó, Budapest, 1966. English translation: **Recursive Functions**, Academic Press, New York, 1967.

**Pra65** Dag Prawitz, **Natural Deduction**, Almqvist and Wiskel, Uppsala, 1965.

**Rey74** John Reynolds, *Towards a theory of type structures,* in J. Loeckx (ed.), **Conference on Porgramming**, Springer-Verlag (LNCS #19), Berlin, 1974, pp. 408–425.

**Saz80** Vladimir Sazonov, *Polynomial computability and recursivity in finite domains,* **Electronische Informationsverarbeitung und Kybernetik 7** (1980) 319–323.

**Sch76** Helmut Schwichtenberg, *Definierbare Funktionen im Lambda-Kalkul mit Typen,* **Archiv f. Logik u. Grundlagenforsch. 17** (1976) 113–114.

**Sta79** Richard Statman, *The typed $\lambda$-calculus is not elementary recursive,* **Theoretical Computer Science 9** (1979) 73–81.

**Var82** Moshe Vardi, *Complexity and relational query languages,* **Fourteenth ACM Symposium on Theory of Computing** (1982) 137–146.

**Zai87** Marek Zaionc, *Word operations definable in typed $\lambda$-calculus,* **Theoretical Computer Science 52** (1987).

# Pure Type Systems Formalized *

James McKinna
jhm@dcs.ed.ac.uk

Robert Pollack
rap@dcs.ed.ac.uk

Laboratory for Foundations of Computer Science
The King's Buildings, University of Edinburgh, EH9 3JZ, U.K.

## 1  Introduction

This paper is about our hobby. For us, machine-checked mathematics is a passion, and constructive type theory (in the broadest sense) is the way to this objective. Efficient and correct type-checking programs are necessary, so a formal theory of type systems leading to verified type synthesis algorithms is a natural goal. For over a year the second author has been developing a machine-checked presentation of the elementary meta-theory of Pure Type Systems (PTS) [Bar91], (formerly called Generalized Type Systems (GTS)). This project was blocked until the first author collaborated with a fresh idea. Here we describe the state of this ongoing project, presenting a completely formal, machine checked development of this basic meta-theory, including the underlying language of (explicitly typed) lambda calculus. We discuss some of the choices involved in formalization, some of the difficulties encountered, and techniques to overcome these difficulties.

PTS, a "framework of purely functional type theories" has a beautiful meta-theory, developed informally in [Bar92, Ber90, GN91, vBJ92]. The cited papers are unusually clear and mathematical, and there is little doubt about the correctness of the results we will discuss, so why write a machine-checked development? The informal presentations leave many decisions unspecified and many facts unproved. They are far from the detail of representation needed to write a computer program for proofchecking, and the lemmas needed to prove correctness of such a program. Our long-term goal is to fill these gaps.

Another goal of the project is to develop a realistic example of formal mathematics. In mathematics and computer science we do not prove one big theorem and then throw away all the work leading up to that theorem; we want to build a body of formal knowledge that can continually be extended and developed. Representations and definitions must be suitable for the whole development, not specialized for a single theorem. The theory should be structured, like computer programs, by abstraction, providing "isolation of components" so that several parts of the theory can be worked on simultaneously, perhaps by several workers, and so that the inevitable wrong decisions in underlying representations can later be fixed without affecting too seriously a large theory that depends on them. We are not working from a completed informal theory; there are still open questions such as the Expansion Postponement problem [Pol92]. In particular, the theory of type-checking and type-synthesis has recently made much progress, and involves several related systems with similar properties [vBJMP92]. While the basic informal theory is well understood, we suggest reformulations which clarify the presentation. On the other hand, in some ways type theory is not a realistic example of formal mathematics: it is especially suitable for

---

*This work was supported by the ESPRIT Basic Research Action on Logical Frameworks, and by grants from the British Science and Engineering Research Council.

formalization because the objects are inductively constructed, their properties are proved by induction over structure, and there is little equality reasoning, one of the weak points of intentional type theory.

A novelty in our presentation is the use of named variables instead of de Bruijn "nameless variables". Most of the formalizations of type theory or lambda calculus that we know of (e.g. [Sha85, Alt93]) use de Bruijn indices to avoid formalizing the Curry-style renaming of variables to prevent capture. While de Bruijn notation is very elegant and suitable for formalization, there are reasons to formalize the theory with named variables. For one thing, implementations must use names at some level, whether internally or only for parsing and printing. In either case this use of names must be formally explained. Also interesting is the insight to be gained into what we mean by binding. It is common wisdom among researchers that de Bruijn representation "really is" what we mean by lambda terms, in the sense that there is no quotient of terms by alpha-conversion, i.e. intensional equality on de Bruijn terms corresponds with what we intend to mean by identity of terms. We feel that this perception is correct for bound variables, but not for free variables: the names of free variables matter, their order of occurrence does not. Thus, in de Bruijn representation, we must consider a quotient with respect to the order of free variables, and while this notion is formalizable (as is Curry style alpha-conversion), it is inconvenient, and shows that de Bruijn representation is not the perfect answer to the problem. We use a formulation suggested by Coquand [Coq91] in which explicit alpha-conversion of named variables is not necessary in the theory of reduction, conversion, and typing.

The LEGO Proof Development System [LP92] was used to check the work in an implementation of the Pure Calculus of Constructions extended with inductive types in the style of the Martin-Löf logical framework (or monomorphic set theory) [SNP90] (first implemented in ALF). LEGO is a refinement style proof checker, publicly available by ftp, with a User's Manual [LP92] and a large collection of examples. Interesting examples formalized in LEGO include the Tarski fixedpoint theorem (Pollack), construction of the reals from the rationals and completion of a metric space [Jon91], the chinese remainder theorem [McK92], program specification and data refinement [Luo91], and the Schröder-Bernstein theorem (M. Hofmann, proof in [LP92]). Recently Thorsten Altenkirch has used LEGO's inductive types to give a very elegant and informative proof of strong normalization for System F [Alt93].

**Acknowledgement** We especially thank Thierry Coquand for many inspiring conversations, and many ideas that appear in this work.

## 2  Some Basic Types

We use LEGO's built-in library of impredicative definitions for the usual logical connectives and their properties, although inductive definitions would do just as well. We also use LEGO's library of basic inductive types. These include a type of booleans, BB, containing tt and ff, with the usual classical boolean operators, conjunction andd, disjunction orr and conditional if, together with the lifting functions is_tt and is_ff, which convert booleans to (decidable) propositions. Also, a type of polymorphic lists, LL, with its induction principle LLrec, and many common operations such as append, member and assoc, and a type of polymorphic cartesian products, PROD, with pairing Pr, and projections Fst and Snd. An inductive type of natural numbers, NN is included in the library, and used to support induction on the length of terms.

We use an inductive equality relation, Q, although Leibniz equality is sufficient for our purposes.

## 3 The Presentation of a PTS

A PTS is a 5-tuple (PP, VV, SS, Ax, Rule) where

- PP is an infinite set of *parameters*, ranged over by $p, q$. Parameters are the global, or free, variables.
- VV is an infinite set of *variables*, ranged over by $x, v, u$. Variables are the local, or bound, variables.
- SS, a set of *sorts*, ranged over by $s$. Sorts are the constants.
- Ax $\subseteq$ SS $\times$ SS, a set of *axioms* of the form Ax$(s{:}s)$
- Rule $\subseteq$ SS $\times$ SS $\times$ SS, a set of *rules* of the form Rule$(s_1, s_2, s_3)$

The parameters, variables, and sorts are used in term construction; the axioms and rules parameterize an inductively defined typing judgement. We now formalize this presentation.

Assume there is a type of parameters, PP, that is infinite and has a decidable equivalence relation. In fact we also assume that the decidable equivalence relation on PP is the same as intensional equality; this extra assumption is not necessary but it vastly simplifies our formal development.

```
[PP : Prop];                                          (* Parameters *)
[PPeq : PP->PP->BB];
[PPeq_decides_Q : {p,q:PP}iff (is_tt (PPeq p q)) (Q p q)];
[PPs = LL|PP];                                        (* lists of parameters *)
[PPinf : {l:PPs}ex[p:PP] is_ff (member PPeq p l)];
```

PPinf is a "local gensym" operation. It says that for every list of parameters, l, there is a parameter, p, that is not a member of l. member is defined with respect to some decidable equality, in this case PPeq.

These are not mathematical principles we are assuming, but part of the presentation of the language of a PTS. Having developed some theory of PTS in LEGO, we may **Discharge** these assumptions, making the whole theory parametric in such a type of parameters. The assumption that PPeq is equivalent to Q means that we may instantiate PP with, for example the type of natural numbers and its inductively definable decidable equality, or with the type of lists of characters, but not with the type of integers defined as a quotient over pairs of naturals, because in this latter type the intended equality, definable by induction, is not equivalent to Q.

Also, assume a type of variables, VV, with similar properties, VVeq, VVeq_decides_Q, VVinf; and a type of sorts, SS, which has decidable equality, SSeq, SSeq_decides_Q, but need not be infinite.

What remains to complete a presentation is the axioms and rules that parameterize the typing judgement. They are just relations.

```
[ax : SS->SS->Prop];
[rl : SS->SS->SS->Prop];
```

We usually intend ax and rl to be decidable, but the elementary theory developed in this paper does not use such an assumption. If we are interested in decidability of typechecking or algorithms for type synthesis, even stronger assumptions about decidability are needed [Pol92, vBJMP92].

## 4 Terms

The syntax of terms of PTS (PP, VV, SS, Ax, Rule) is informally given by

```
atoms   α ::= p | x | s
terms   M ::= α | [x:M]M | {x:M}M | M M    atoms, lambda, pi, application
```

The type of terms is formalized as an inductive type.

```
[Trm:Prop];
[sort:SS->Trm];                    (* Trm constructors (introduction rules) *)
[var:VV->Trm];
[par:PP->Trm];
[pi:VV->Trm->Trm->Trm];
[lda:VV->Trm->Trm->Trm];
[app:Trm->Trm->Trm];
```

Every term is a finitely branching well-founded tree. In particular, the lda and pi constructors do *not* have type Trm->(VV->Trm)->Trm, which would create well-founded but infinitely branching terms. The induction principle for Trm is

```
[Trec:{C:Trm->Prop}                (* Trm induction (elimination) principle *)
      {TSORT:{s:SS}C (sort s)}
      {TVAR:{n:VV}C (var n)}
      {TPAR:{n:PP}C (par n)}
      {TPI:{n:VV}{A,B:Trm}(C A)->(C B)->C (pi n A B)}
      {TLDA:{n:VV}{A,B:Trm}(C A)->(C B)->C (lda n A B)}
      {TAPP:{M,N:Trm}(C M)->(C N)->C (app M N)}
      {t:Trm}C t];
```

There is a boolean-valued structural equality function, Trm_eq, inductively definable on terms. Because PPeq, VVeq, and SSeq are equivalent to Q, Trm_eq is also provably equivalent to Q.

**Substitution** For the machinery on terms, we need two kinds of substitution, both defined by primitive recursion over term structure using the induction principle Trec. Substitution of a term for a parameter is entirely textual, not preventing capture. Since parameters have no binding instances in terms (we may view them as being globally bound by the context in a judgement), there is no hiding of a parameter name by a binder.

```
[psub [M:Trm][n:PP] : Trm->Trm =
  Trec ([_:Trm]Trm)
       ([s:SS]sort s)
       ([v:VV]var v)
       ([p:PP]if (PPeq n p) M (par p))
       ([v:VV][_,_,l,r:Trm]pi v l r)
       ([v:VV][_,_,l,r:Trm]lda v l r)
       ([_,_,l,r:Trm]app l r)];
```

Substitution of a term for a variable does respect variable binders that hide their bound instances from substitution, but does not prevent capture.

```
[vsub [M:Trm][n:VV] : Trm->Trm =
  Trec ([_:Trm]Trm)
```

```
 ([s:SS]sort s)
 ([v:VV]if (VVeq n v) M (var v))
 ([p:PP]par p)
 ([v:VV][_,or,nl,nr:Trm]pi v nl (if (VVeq n v) or nr))
 ([v:VV][_,or,nl,nr:Trm]lda v nl (if (VVeq n v) or nr))
 ([_,_,l,r:Trm]app l r)];
```

Both of these will be used only in safe ways in the type theory and the theory of reduction and conversion, so as to prevent unintended capture of variables.

There is a frequently used abbreviation for substituting a parameter for a variable

```
[alpha [p:PP][v:VV] = vsub (par p) v];
```

This is not alpha conversion in the usual sense.

**Free and Bound Occurrences** The list of parameters occurring in a term is computed by primitive recursion over term structure, and the boolean judgement whether or not a given parameter occurs in a given term is computed by the **member** function on this list of parameters.

```
[params : Trm->PPs =
  Trec ([_:Trm]PPs)
       ([_:SS]NIL|PP)
       ([_:VV]NIL|PP)
       ([p:PP]unit p)
       ([_:VV][_,_:Trm][l,r:PPs]append l r)
       ([_:VV][_,_:Trm][l,r:PPs]append l r)
       ([_,_:Trm][l,r:PPs]append l r)];
[poccur [p:PP][A:Trm] : BB = member PPeq p (params A)];
```

Similarly **sorts** and **soccur** are defined.

We do not compute the list of variables occurring free in a term, but instead define inductively a notion **Vclosed** of *closed term* (Table 1). In fact, only the **Vclosed** terms are considered to be well formed for the theory of reduction in the same way that only typable terms will be considered well formed for the type theory in later sections. Thus **Vclosed** is used as an induction principle over well formed terms.

**Vclosed** is provably equivalent to having no free variables, although the proof is not as trivial as might be expected. (Thanks to Thierry Coquand for the crucial step in this proof.)

# 5  Reduction and Conversion

Table 2 shows the reduction and conversion relations. **par_red1** is the *one-step parallel reduction* used in the Tait–Martin-Löf proof of the Church-Rosser property for $\beta$. The interesting point in these rules is how reduction under binders is handled: to go under a binder, replace the free variable occurrences by a suitably fresh parameter, operate on the closed subterm, and undo the "closing up". For example, consider informally one-step $\beta$-reduction of untyped lambda calculus. In the style of our formalization the $\xi$ rule is

$$\xi \quad \frac{[q/x]M \to [q/y]N}{\lambda x.M \to \lambda y.N} \quad q \notin M, q \notin N$$

```
[Vclosed : Trm->Prop];

[Vclosed_sort : {s:SS} Vclosed (sort s)];

[Vclosed_par : {p:PP} Vclosed (par p)];

[Vclosed_pi : {n|VV}{A,B|Trm}{p|PP}
     {premA:Vclosed A}   {premB:Vclosed (alpha p n B)}
     (****************************************************)
         Vclosed (pi n A B)];

[Vclosed_lda : {n|VV}{A,B|Trm}{p|PP}
     {premA:Vclosed A}   {premB:Vclosed (alpha p n B)}
     (****************************************************)
         Vclosed (lda n A B)];

[Vclosed_app : {A,B|Trm}
     {premA:Vclosed A}   {premB:Vclosed B}
     (****************************************************)
         Vclosed (app A B)];
```

Table 1: The inductive property `Vclosed`

where $[q/x]M$ is our (alpha q x M). Here is an instance of this rule, contracting the underlined redex.
$$\frac{(\lambda v.\lambda x.v)\, q \to \lambda x.q}{\lambda x.(\underline{(\lambda v.\lambda x.v)\, x}) \to \lambda y.\lambda x.y}$$

After removing the outer $\lambda x$, replacing its bound instances by a fresh parameter, $q$, and contracting the closed weak-head redex thus obtained[1], we must re-bind the hole now occupied by $q$. According to the rule $\xi$, we require a variable, $y$, and a term, $N$, such that $[q/y]N = \lambda x.q$. Such a pair is $y$, $\lambda x.y$ (the one we have used above), as is $z$, $\lambda x.z$ for any $z \ne x$. However $x$, $\lambda x.x$ will not do, for vsub will not substitute $q$ for $x$ under the binder $\lambda x$.

Parallel n-step reduction, par_redn (Table 2), is the transitive closure of par_red1. One must take care in defining conversion, conv, or the second Church-Rosser property will fail. In particular, the diamond property only holds for Vclosed terms. While reduction preserves Vclosed, expansion does not, so the transitivity rule for conversion, transConv in Table 2, must guarantee that the intermediate term is Vclosed.

**The Church-Rosser Theorem** is now proved by the argument of Tait and Martin-Löf.

```
[CommonReduct [t|Type][R,S:t->t->Prop]
           = [b,c:t]Ex [d:t]and (S b d) (R c d)];
[DiamondProperty [t|Type][P:t->Prop][R:t->t->Prop]
           = {a,b,c|t}(P a)->(R a b)->(R a c)->CommonReduct R R b c];

Goal DiamondProperty Vclosed par_red1;
Goal DiamondProperty Vclosed par_redn;
```

[1] There is no possibility of variable capture when contracting a closed weak-head redex.

```
[par_red1 : Trm->Trm->Prop]              (* parallel 1-step reduction *)

[par_red1_refl : refl par_red1];

[par_red1_beta:
    {U:Trm}{A,A'|Trm}{premA:par_red1 A A'}
    {u,v|VV}{B,B'|Trm}{p|PP}
    {noccB:is_ff (poccur p B)}{noccB':is_ff (poccur p B')}
    {premB:par_red1 (alpha p u B) (alpha p v B')}
 (*********************************************************)
    par_red1 (app (lda u U B) A) (vsub A' v B')];

[par_red1_pi:
    {A,A'|Trm}{premA:par_red1 A A'}
    {u,v|VV}{B,B'|Trm}{p|PP}
    {noccB:is_ff (poccur p B)}{noccB':is_ff (poccur p B')}
    {premB:par_red1 (alpha p u B) (alpha p v B')}
 (*********************************************************)
    par_red1 (pi u A B) (pi v A' B')];

[par_red1_lda:
    {A,A'|Trm}{premA:par_red1 A A'}
    {u,v|VV}{B,B'|Trm}{p|PP}
    {noccB:is_ff (poccur p B)}{noccB':is_ff (poccur p B')}
    {premB:par_red1 (alpha p u B) (alpha p v B')}
 (*********************************************************)
    par_red1 (lda u A B) (lda v A' B')];

[par_red1_app:
    {A,A'|Trm}{premA:par_red1 A A'}
    {B,B'|Trm}{premB:par_red1 B B'}
 (*********************************************************)
    par_red1 (app A B) (app A' B')];


[par_redn : Trm->Trm->Prop];             (* parallel n-step reduction *)
[par_redn_red1 : {A,B|Trm}(par_red1 A B)->par_redn A B];
[par_redn_trans : trans par_redn];

[conv : Trm->Trm->Prop];                 (* conversion *)
[reflConv  : refl conv];
[symConv   : sym conv];
[transConv : {A,D,B|Trm}(conv A D)->(Vclosed D)->(conv D B)->conv A B];
[rednConv  : {A,B|Trm}(par_redn A B)->conv A B];
```

Table 2: The 1-Step Reduction, Many-step Reduction, and Conversion Relations

```
Goal {A,B|Trm}(conv A B)->(Vclosed A)->(Vclosed B)->
        Ex[C:Trm] and (par_redn A C) (par_redn B C);
```

In these proofs there is no need define or reason about any notion of α-conversion. The proofs are by structural induction, but use a technique we will describe in Section 9.2.

## 6 Contexts

The type of global bindings, [p:A], that occur in contexts, is a cartesian product of PP by Trm. We also give the two projections of global bindings, and a defined structural equality that is provably equivalent to Q.

```
[GB : Prop = PROD|PP|Trm];              (* type of global bindings .. *)
[Gb : PP->Trm->GB = Pr|PP|Trm];         (* .. its constructor .. *)
[namOf = Fst][typOf = Snd];             (* .. its destructors .. *)
[GBeq [b,c:GB]
    = andd (PPeq (namOf b) (namOf c)) (Trm_eq (typOf b) (typOf c))];
```

Contexts are lists of global bindings.

```
[Cxt = LL|GB];
[nilCxt = NIL|GB];
```

**Occurrences** Occurrence of a global binding in a context is defined using the **member** function and the defined structural equality GBeq.

```
[GBoccur [b:GB][G:Cxt] : BB = member GBeq b G];
```

Now we define the list of parameters "bound" by a context, and, using the **member** function as before, the boolean relation deciding whether or not a given parameter is bound by a given context.

```
(* names with binding instances in a context *)
[globalNames : Cxt->PPs
    = LLrec ([_:Cxt]PPs)
            (NIL|PP)                                    (* nil *)
            ([b:GB][_:Cxt][rest:PPs]CONS (namOf b) rest)];  (* cons *)
[Poccur [p:PP][G:Cxt] : BB = member PPeq p (globalNames G)];
```

**Subcontexts** The subcontext relation is defined by containment of the respective occurrence predicates.

```
[subCxt [G,H:Cxt] = {b:GB}(is_tt (GBoccur b G))->is_tt (GBoccur b H)];
```

This is exactly the definition used informally in [Bar92, GN91, vBJ92]. Imagine expressing this property in a representation using de Bruijn indices for global variables.

## 7 The Typing Judgement

The typing judgement has the shape $\Gamma \vdash M : A$, meaning that in context $\Gamma$, term $M$ has type $A$. An informal presentation of the rules inductively defining this relation is shown in Table 3. First observe that the handling of parameters and variables in the PI and LDA rules is similar to that in the rules of Table 2: to operate under a binder, locally extend the context

| | | |
|---|---|---|
| Ax | $\bullet \vdash c : s$ | Ax($c$:$s$) |
| Start | $\dfrac{\Gamma \vdash A : s}{\Gamma[p{:}A] \vdash p : A}$ | $p \notin \Gamma$ |
| vWeak | $\dfrac{\Gamma \vdash q : C \quad \Gamma \vdash A : s}{\Gamma[p{:}A] \vdash q : C}$ | $p \notin \Gamma$ |
| sWeak | $\dfrac{\Gamma \vdash s : C \quad \Gamma \vdash A : s}{\Gamma[p{:}A] \vdash s : C}$ | $p \notin \Gamma$ |
| Pi | $\dfrac{\Gamma \vdash A : s_1 \quad \Gamma[p{:}A] \vdash [p/x]B : s_2}{\Gamma \vdash \{x{:}A\}B : s_3}$ | $p \notin B$, Rule$(s_1, s_2, s_3)$ |
| Lda | $\dfrac{\Gamma[p{:}A] \vdash [p/x]M : [p/y]B \quad \Gamma \vdash \{y{:}A\}B : s}{\Gamma \vdash [x{:}A]M : \{y{:}A\}B}$ | $p \notin M$, $p \notin B$ |
| App | $\dfrac{\Gamma \vdash M : \{x{:}A\}B \quad \Gamma \vdash N : A}{\Gamma \vdash M\,N : [N/x]B}$ | |
| tConv | $\dfrac{\Gamma \vdash M : A \quad \Gamma \vdash B : s}{\Gamma \vdash M : B}$ | $A \simeq B$ |

Table 3: The Informal Typing Rules of PTS

replacing the newly freed variable by a new parameter, do the operation, then forget the parameter and bind the variable again. Natural deduction always has this aspect of locally extending and discharging the context, and implementors know that assumption/discharge is related to variable names. Huet's Constructive Engine [Hue89], for example, "weaves" back and forth between named global variables and de Bruijn indices for local binding in a manner similar to that of Table 3. Even an implementation that uses only one class of named variables may have to change names in the Pi and Lda rules because locally bound names may be reused, but global names must be unique. In this regard, notice that the Lda rule allows "alpha converting" in the conclusion. Informal presentations [Bar92, Ber90, GN91] suggest the rule

| Lda' | $\dfrac{\Gamma[p{:}A] \vdash [p/x]M : [p/x]B \quad \Gamma \vdash \{x{:}A\}B : s}{\Gamma \vdash [x{:}A]M : \{x{:}A\}B}$ | $p \notin M$, $p \notin B$ |
|---|---|---|

To see why we use the rule Lda instead of Lda', consider deriving the judgement

$$[A{:}*][P{:}A{\rightarrow}*] \vdash [x{:}A][x{:}Px]x : \{x{:}A\}\{y{:}Px\}Px$$

in the Pure Calculus of Constructions. While both systems, using Lda, and using Lda', derive this judgement, they do so with different derivations, as the system with Lda' must use the rule tConv to alpha-convert $x$ to $y$ in the type part of the judgement. In particular, the system with Lda' has different derivations than a presentation using de Bruijn indices.

Another thing to notice about Table 3 is that we have replaced the usual weakening rule

| Weak | $\dfrac{\Gamma \vdash M : C \quad \Gamma \vdash A : s}{\Gamma[p{:}A] \vdash M : C}$ | $p \notin \Gamma$ |
|---|---|---|

having an arbitrary term as subject, with vWeak and sWeak, restricting weakening to atomic subjects, and later show that Weak is derivable in our system. This presentation improves the meta-theory (see Section 8) and makes the presentation more syntax-directed for type checking [Pol92].

Finally observe that instances of $[M/v]N$ in Table 3 only occur at the top level (not under a binder) and with $M$ Vclosed: it is safe to use vsub in this manner, even though vsub is not a correct substitution operation.

**The Formal Typing Judgement** The typing rules, formalized as the constructors of an inductive relation, gts, are shown in Table 4. To understand this formalization, consider the "genericity" of the informal rules, presented in Table 3. The rule START states that for any context $\Gamma$, term $A$, and parameter $p \notin \Gamma$, to derive $\Gamma[p:A] \vdash p : A$ it suffices to derive $\Gamma \vdash A : s$. What is the quantification of $s$ in this explanation? The view that rules are constructors of derivations suggests that one must actually provide $s$ and a derivation of $\Gamma \vdash A : s$, i.e. that the correct formalization is "for any context $\Gamma$, term $A$, parameter $p$, and sort $s$,...". This question is more interesting for the PI rule: must we actually supply $p$, is it sufficient to know the existence of such a $p$, or is there another way to characterize those parameters that may be used in an instance of the rule? Here we feel that $p$ really doesn't matter, as long as it is "fresh enough", i.e. doesn't occur in $\Gamma$ or $B$. Again, we decide that the correct formalization actually requires a particular parameter $p$ to be supplied. Another way to understand the choice we have made is that the meaning of the informal rules is as constructors of well-founded and finitely branching trees. It is also possible to accept a type of well-founded but infinitely branching trees, and the issue is discussed again in Section 9.2.

## 8  Some Basic Lemmas

Most of the basic results about gts are straightfoward to prove by induction. For example

```
(* Free Variable Lemmas *)
Goal {G|Cxt}{M,A|Trm}(gts G M A)->and (Vclosed M) (Vclosed A);
Goal {G|Cxt}{M,A|Trm}(gts G M A)->
   {p|PP}(or (is_tt (poccur p M)) (is_tt (poccur p A)))->is_tt (Poccur p G);
(* Start Lemmas *)
Goal {G|Cxt}{M,A|Trm}(gts G M A)->
     {s1,s2:SS}(ax s1 s2)->(gts G (sort s1) (sort s2));
Goal {G|Cxt}{M,A|Trm}(gts G M A)->
     {b|GB}(is_tt (GBoccur b G))->gts G (par (namOf b)) (typOf b);
```

**Generation Lemmas** For each term constructor, there is one rule in Table 3 with that term constructor as its subject. In our presentation, only the rule tConv is not syntax-directed, giving a form of "uniqueness of generation" for judgements, up to conversion. This observation is formalized in the Generation Lemmas. For example, the Generation Lemma for sorts, whose types are generated by Ax, and that for parameters, whose types are generated by START are

```
Goal {G|Cxt}{s|SS}{C|Trm}{d:gts G (sort s) C}
          Ex [s1:SS] and (ax s s1) (conv (sort s1) C);
Goal {G|Cxt}{C|Trm}{p|PP}{d:gts G (par p) C}
          Ex [B:Trm]and (is_tt (GBoccur (Gb p B) G)) (conv B C);
```

In the standard presentation of PTS, say [Bar92], with the general non-syntax-directed weakening rule WEAK, uniqueness of generation holds only up to conversion *and weakening*;

```
[gts : Cxt->Trm->Trm->Prop];

[Ax : {s1,s2|SS}{sc:ax s1 s2}
         gts nilCxt (sort s1) (sort s2)];

[Start : {G|Cxt}{A|Trm}{s|SS}{p|PP}{sc:is_ff (Poccur p G)}
         {prem:gts G A (sort s)}
         (****************************************************)
         gts (CONS (Gb p A) G) (par p) A];

[vWeak : {G|Cxt}{D,A|Trm}{s|SS}{n,p|PP}{sc:is_ff (Poccur p G)}
         {l_prem:gts G (par n) D}
         {r_prem:gts G A (sort s)}
         (****************************************************)
         gts (CONS (Gb p A) G) (par n) D];

[sWeak : {G|Cxt}{D,A|Trm}{t,s|SS}{p|PP}{sc:is_ff (Poccur p G)}
         {l_prem:gts G (sort t) D}
         {r_prem:gts G A (sort s)}
         (****************************************************)
         gts (CONS (Gb p A) G) (sort t) D];

[Pi : {G|Cxt}{A,B|Trm}{s1,s2,s3|SS}{p|PP}{n|VV}
           {sc:rl s1 s2 s3}{sc':is_ff (poccur p B)}
         {l_prem:gts G A (sort s1)}
         {r_prem:gts (CONS (Gb p A) G) (alpha p n B) (sort s2)}
         (****************************************************)
         gts G (pi n A B) (sort s3)];

[Lda : {G|Cxt}{A,M,B|Trm}{s|SS}{p|PP}{n,m|VV}
           {sc:and (is_ff (poccur p M)) (is_ff (poccur p B))}
         {l_prem:gts (CONS (Gb p A) G) (alpha p n M) (alpha p m B)}
         {r_prem:gts G (pi m A B) (sort s)}
         (****************************************************)
         gts G (lda n A M) (pi m A B)];

[App : {G|Cxt}{M,A,B,L|Trm}{n|VV}
         {l_prem:gts G M (pi n A B)}
         {r_prem:gts G L A}
         (****************************************************)
         gts G (app M L) (vsub L n B)];

[tConv : {G|Cxt}{M,A,B|Trm}{s|SS}{sc:conv A B}
         {l_prem:gts G M A}
         {r_prem:gts G B (sort s)}
         (****************************************************)
         gts G M B];
```

Table 4: The PTS Judgement as an Inductive Relation

then the generation lemmas depend on the Thinning Lemma, while in our presentation with atomic weakening, they are provable before the Thinning Lemma.

## 9 The Thinning Lemma

The Thinning Lemma is important to our formulation because it shows that full weakening is derivable in our system from atomic weakening. First a definition

[Valid [G:Cxt] = Ex2[M,A:Trm] gts G M A];

(** Thinning Lemma **)
Goal  {G,G'|Cxt}{M,A|Trm}(gts G M A)->(subCxt G G')->(Valid G')->gts G' M A;

When we come to prove the Thinning Lemma, a serious difficulty arises from our use of parameters. We temporarily revert to informal notation.

**Naive attempt to prove the Thinning Lemma** By induction on the derivation of $G \vdash M : A$. Consider the case for the Pi rule: the derivation ends with

$$\frac{G \vdash A : s_1 \qquad G[p{:}A] \vdash [p/x]B : s_2}{G \vdash \{x{:}A\}B : s_3} \quad \text{Rule}(s_1, s_2, s_3)$$

**left IH** For any $K$, if $G$ is a subcontext of $K$ and $K$ is Valid, then $K \vdash A : s_1$.

**right IH** For any $K$, if $G[p{:}A]$ is a subcontext of $K$ and $K$ is Valid, then $K \vdash [p/x]B : s_2$.

**to show** For any $G'$, if $G$ is a subcontext of $G'$ and $G'$ is Valid, then $G' \vdash \{x{:}A\}B : s_3$.

So let $G'$ be a valid extension of $G$. Using the Pi rule, we need to show $G' \vdash A : s_1$ and $G'[p{:}A] \vdash [p/x]B : s_2$. The first of these is proved by the left IH applied to $G'$. In order to use the right IH applied to $G'[p{:}A]$ to prove the second, we need to show $G'[p{:}A]$ is valid. However, this may be false, e.g. if $p$ occurs in $G'$!

The left premiss of the LDA rule presents a similar problem. All other cases are straightfoward.

**Naive attempt to fix the problem** Let us change the name of $p$ in the right premiss. We prove a global renaming lemma to the effect that any injective renaming of parameters preserves judgements. (This is straightfoward given the technology of renaming we will develop below.) However, this does not help us finish the naive proof of the Thinning Lemma, for the right IH still mentions $p$, not some fresh parameter $q$. That is, although we can turn the right subderivation $G[p{:}A] \vdash [p/x]B : s_2$ into a derivation of $G[q{:}A] \vdash [q/x]B : s_2$ for some fresh $q$, this is *not* a structural subderivation of the original proof of $G \vdash \{x{:}A\}B : s_3$, and is no help in structural induction.

### 9.1 Some correct proofs of the Thinning Lemma

We present three correct proofs of the Thinning Lemma, based on different analyses of what goes wrong in the naive proof. The two proofs in this section are, we think, possible to formalize, but a little heavy. Section 9.2 gives the more beautiful formalization we have chosen.

**Changing the names of parameters.** Once we say "by induction on the derivation..." it is too late to change the names of any troublesome parameters, but one may fix the naive proof by changing the names of parameters in a derivation *before* the structural induction. First observe, by structural induction, that if $d$ is a derivation of $G \vdash M : A$, $G'$ is a Valid extension of $G$, and no parameter occurring in $d$ is in globalNames($G'$)\globalNames($G$), then $G' \vdash M : A$. Now for the Thinning Lemma, given $G$, $G'$, and a derivation $d$ of $G \vdash M : A$, change parameters in $d$ to produce a derivation $d'$ of $G \vdash M : A$ such that no parameter occurring in $d'$ is in globalNames($G'$)\globalNames($G$). The previous observation completes the proof.

This proof takes the view that it is fortunate there are many derivations of each judgement, for we can find one suitable for our purposes among them. To formalize it we need technology for renaming parameters in derivations, which is heavier than the technology for renaming parameters in *judgements* that we will present below.

**Induction on length.** A different analysis of the failure of the naive proof shows that it is the use of *structural* induction that is at fault. Induction on the height of derivations appears to work, but we must show that renaming parameters in a derivation doesn't change its height; again reasoning about derivations rather than judgements. Induction on the length of the subject term $M$ does not seem to work, as some rules have premisses whose subject is not shorter than that of the conclusion.

## 9.2 A Better Solution

We take the view that there are too many derivations of each judgement, but instead of giving up on structural induction, we present a different inductive definition of the same relation which is more suitable for the problem at hand. Consider an "alternative PTS" relation, apts:Cxt->Trm->Trm->Prop, that differs from gts only in the right premiss of the Pi rule and the left premiss of the Lda rule:

```
[aPi : {G|Cxt}{A,B|Trm}{s1,s2,s3|SS}{n|VV}{sc:rl s1 s2 s3}
        {l_prem:apts G A (sort s1)}
        {r_prem:{p|PP}{scG:is_ff (Poccur p G)}
                        apts (CONS (Gb p A) G) (alpha p n B) (sort s2)}
     (****************************************************************)
        apts G (pi n A B) (sort s3)];

[aLda : {G|Cxt}{A,M,B|Trm}{s|SS}{n,m|VV}
        {l_prem:{p|PP}{scG:is_ff (Poccur p G)}
                        apts (CONS (Gb p A) G) (alpha p n M) (alpha p m B)}
        {r_prem:apts G (pi n A B) (sort s)}
     (****************************************************************)
        apts G (lda n A M) (pi m A B)];
```

In these premisses we avoid the problem of choosing a particular parameter by requiring the premiss to hold for all parameters for which there is no reason it should not hold, that is, for all "sufficiently fresh" parameters. apts identifies all those derivations of a gts judgement that are inessentially different because of parameters occurring in the derivation but not in its conclusion. Notice that apts constructs well-founded but infinitely branching trees.

It is interesting to compare the side conditions of Pi with those of aPi. In Pi, we intuitively want to know (is_ff (Poccur p G)), but this is derivable from the right premiss so is not required as a side condition. In aPi, we cannot require the right premiss for all p, but only for those such that (is_ff (Poccur p G)), while the condition (is_ff (poccur p B)) is not required because of genericity.

Once we prove that apts and gts have the same derivable judgements, the Thinning Lemma becomes

{G,G'|Cxt}{M,A|Trm}(apts G M A)->(subCxt G G')->(Valid G')->apts G' M A;

which is straightfoward to prove by structural induction on (apts G M A).

### 9.2.1  apts is equivalent to gts

There are two implications to prove; one is straightfoward. The interesting one is

Goal {G|Cxt}{M,A|Trm}(gts G M A)->apts G M A;

This is the essence of our solution to the problem of name clash with the parameters introduced by the right premiss of Pi and the left premiss of Lda. In order to prove it we introduce a concept of "renaming" and prove an induction loaded version of the goal.

**Renamings** A *renaming* is, informally, a finite function from parameters to parameters. They are represented formally by their graphs as lists of ordered pairs

[rp = PROD|PP|PP];
[Renaming = LL|rp];

rho and sigma range over renamings. Renamings are applied to parameters by assoc, and extended compositionally to Trm, GB and Cxt.

[renPar [rho:Renaming][p:PP] : PP = assoc (PPeq p) p rho];
[renTrm [rho:Renaming] : Trm->Trm =
  Trec ([_:Trm]Trm)
       ([s:SS]sort s)
       ([v:VV]var v)
       ([p:PP]par (renPar rho p))
       ([v:VV][_,_,l,r:Trm]pi v l r)
       ([v:VV][_,_,l,r:Trm]lda v l r)
       ([_,_,l,r:Trm]app l r)];
[renGB [rho:Renaming] : GB->GB =
  GBrec ([_:GB]GB) ([p:PP][t:Trm](Gb (renPar rho p) (renTrm rho t)))];
[renCxt [rho:Renaming] : Cxt->Cxt =
  LLrec ([_:Cxt]Cxt)
        (nilCxt)
        ([b:GB][_:Cxt][rest:Cxt]CONS (renGB rho b) rest)];

This is a "tricky" representation. First, if there is no pair (p,q) in the list rho, (assoc (PPeq p) p rho) returns p (the second occurrence of p in this expression), so renPar rho is always a total function with finite support. Also, while there is no assumption that renamings are the graphs of functional or injective relations, *application* of a renaming is functional, because assoc only finds the first pair whose domain matches a given parameter. Conversely, consing a new pair to the front of a renaming will "shadow" any old pair with the same domain. Interestingly we do not have to formalize these observations.

We prove an induction-loaded form of the goal

```
Goal {G|Cxt}{M,A|Trm}(gts G M A)->
    {rho|Renaming}(Valid (renCxt rho G))->
        apts (renCxt rho G) (renTrm rho M) (renTrm rho A);
```

by induction on the derivation of gts G M A. (The desired goal is obtained using the identity renaming.) Reverting again to informal notation, consider the case where the derivation ends with the rule Pi

$$\frac{G \vdash A : s_1 \quad G[p{:}A] \vdash [p/x]B : s_2}{G \vdash \{x{:}A\}B : s_3} \quad p \notin B, \text{Rule}(s_1, s_2, s_3)$$

We have

**left IH** For any $\sigma$, if $\sigma G$ is Valid, then $\sigma G \vdash_{apts} \sigma A : s_1$.

**right IH** For any $\sigma$, if $\sigma(G[p{:}A])$ is Valid, then $\sigma(G[p{:}A]) \vdash_{apts} \sigma([p/x]B) : s_2$.

**to show** For any $\rho$, if $\rho G$ is Valid, then $\rho G \vdash_{apts} \rho(\{x{:}A\}B) : s_3$.

So choose $\rho$ with $\rho G$ Valid. Using the aPi rule we need to show

**left subgoal** $\rho G \vdash_{apts} \rho A : s_1$

**right subgoal** for all $q$ not occurring in $\rho G$, $(\rho G)[q{:}\rho A] \vdash_{apts} [q/x]\rho B : s_2$

The left subgoal follows from the left IH.

For the right subgoal, choose a parameter $q$ not occurring in $\rho G$. Let $\sigma$ = (CONS (Pr p q) rho) be the renaming that first changes $p$ to $q$, then behaves like $\rho$. Notice $\sigma p = q$, and, because $p$ does not occur in $B$, $A$, or $G$, $\sigma B = \rho B$, $\sigma A = \rho A$, and $\sigma G = \rho G$. Thus $\sigma(G[p{:}A]) = \rho(G)[q{:}\rho A]$ is Valid (using the left IH and the Start rule), and by the right IH, $(\rho G)[q{:}\rho A] \vdash_{apts} [q/x]\rho B : s_2$ as required.

The aLda case is similar, and all other cases are straightfoward, except the tConv case, which requires that renaming preserves conversion. This is proved by similarly using an alternative inductive definition of reduction. ■

Notice that the reason we don't need any assumption that renamings are injective is that the assumption Valid (renCxt rho G) guarantees the parts of rho actually used are, in fact, injective.

## 10 Other Results

With the techniques of the last section, the three main results of the elementary theory of PTS are now straightfoward.

**The Substitution Lemma,** a cut property

```
Goal {Gamma,Delta:Cxt}{N,A,M,B:Trm}{p:PP}
    [sub = psub N p][subGB [pA:GB] = Gb (namOf pA) (sub (typOf pA))]
    (gts Gamma N A)->(gts (append Delta (CONS (Gb p A) Gamma)) M B)->
    gts (append (map subGB Delta) Gamma) (sub M) (sub B);
```

**Correctness of Types**

```
{G|Cxt}{M,A|Trm}(gts G M A)->
    Ex [s:SS]or (is_tt (Trm_eq A (sort s))) (gts G A (sort s));
```

**Subject Reduction Theorem,** also called closure under reduction.

```
Goal {G|Cxt}{M,A|Trm}(gts G M A)->{M'|Trm}(par_redn M M')->gts G M' A;
```

The proof in [GN91, Bar92] goes by simultaneous induction on

i  If $G \vdash M : A$ and $M \rightarrow M'$ then $G \vdash M' : A$.
ii If $G \vdash M : A$ and $G \rightarrow G'$ then $G' \vdash M : A$.

where $\rightarrow$ is one-step-reduction, i.e. contraction of one $\beta$-redex in a term or context. (The reason for this simultaneous induction is that some of the typing rules move terms from the subject to the context, e.g. `Pi`.) This approach produces a large number of cases, all of which are trivial except for the case where the one redex contracted is the application constructed by the rule `App`. All of these case distinctions are inessential except to isolate the one non-trivial case. Thus, in place of one-step-reduction, we use a new relation of *non-overlapping reduction*, `no_red1`, that differs from `par_red1` only in the $\beta$ rule

```
[no_red1_beta:{u:VV}{U,A,B:Trm}no_red1 (app (lda u U B) A) (vsub A u B)];
```

Further, the distinction between a redex in the term and a redex in the context is also inessential, so we extend `no_red1` compositionally to contexts, and then to pairs of a context and a term. We suggest that the correct meaning of *subject* of a PTS judgement $G \vdash M : A$ is the pair $\langle G, M \rangle$, and call this non-overlapping reduction on subjects

```
[red1Subj : Cxt->Trm->Cxt->Trm->Prop];
```

Now we prove

```
Goal {G|Cxt}{M,A|Trm}(gts G M A)->
       {G'|Cxt}{M'|Trm}(red1Subj G M G' M')->gts G' M' A;
```

by straightfoward induction (with far fewer cases), and the usual Subject Reduction Theorem is an easy corollary.

## 10.1 Work in Progress

As mentioned in the Introduction, one of our long term goals is a completely verified proofchecker. The formalization reported here has been used to verify typechecking algorithms for subclasses of PTS [Pol92]. We are currently working with van Benthem Jutting on formalizing results leading to more general typechecking algorithms and the strengthening theorem [vBJMP92]. (Typechecking is not enough for real proofcheckers; e.g. LEGO uses strengthening in its `Discharge` tactic.) This project uses the basic theory of several systems similar to PTS, and by "cutting and pasting" existing proofs we have been able to effectively reuse some of the vast amount of work expended so far.

# 11 Conclusion

In doing this work of formalizing a well known body of mathematics, we spent a large amount of time solving mathematical problems, e.g. the Thinning Lemma. Another big problem was maintaining and organizing the formal knowledge, e.g. allowing two people to extend different parts of the data base at the same time, and finding the right lemma in the mass of checked material. We feel that better understanding of mathematical issues of formalization (e.g. names/namefree, intentional/extentional), and organization of formal development are the most useful areas to work on now for the long-term goal of formal mathematics.

Finally, it is not so easy to understand the relationship between some informal mathematics and a claimed formalization of it. Are you satisfied with our definition of reduction? It might be more satisfying if we also defined de Bruijn terms and their reduction, and proved a correspondence between the two representations, but this only changes the degree of the problem, not its nature. What about the choice between the typing rules Lda and Lda'? There may be no "right" answer, as we may have different ideas in mind informally. There is no such thing as certain truth, and formalization does not change this state of affairs.

# References

[Alt93] Thorsten Altenkirch. A formalization of the strong normalization proof for System F in LEGO. In *Proceedings of the International Conference on Typed Lambda Calculi and Applications, TLCA '93*, March 1993.

[Bar91] Henk Barendregt. Introduction to generalised type systems. *J. Functional Programming*, 1(2):124–154, April 1991.

[Bar92] Henk Barendregt. Lambda calculi with types. In Gabbai Abramsky and Maibaum, editors, *Handbook of Logic in Computer Science*, volume II. Oxford University Press, 1992.

[Ber90] Stefano Berardi. *Type Dependence and Constructive Mathematics*. PhD thesis, Dipartimento di Informatica, Torino, Italy, 1990.

[Coq91] Thierry Coquand. An algorithm for testing conversion in type theory. In G. Huet and G. D. Plotkin, editors, *Logical Frameworks*. Cambridge University Press, 1991.

[GN91] Herman Geuvers and Mark-Jan Nederhof. A modular proof of strong normalization for the calculus of constructions. *Journal of Functional Programming*, 1(2):155–189, April 1991.

[Hue89] Gérard Huet. The constructive engine. In *The Calculus of Constructions; Documentation and users' guide*. INRIA-Rocquencourt, Aug 1989. Technical Report 110.

[Jon91] Claire Jones. Completing the rationals and metric spaces in LEGO. In *2nd Workshop of Logical Frameworks, Edinburgh*, pages 209–222, May 1991. available by ftp.

[LP92] Zhaohui Luo and Robert Pollack. LEGO proof development system: User's manual. Technical Report ECS-LFCS-92-211, LFCS, Computer Science Dept., University of Edinburgh, The King's Buildings, Edinburgh EH9 3JZ, May 1992. Updated version.

[Luo91] Zhaohui Luo. Program specification and data refinement in type theory. In *TAPSOFT '91 (Volume 1)*, number 493 in Lecture Notes in Computer Science, pages 143–168. Springer-Verlag, 1991.

[McK92] James McKinna. *Deliverables: a Categorical Approach to Program Development in Type Theory*. PhD thesis, University of Edinburgh, 1992.

[Pol92] R. Pollack. Typechecking in pure type systems. In *1992 Workshop on Types for Proofs and Programs, Båstad, Sweden*, pages 271–288, June 1992. available by ftp.

[Sha85] N. Shankar. A mechanical proof of the church-rosser theorem. Technical Report 45, Institute for Computing Science, University of Texas at Austin, March 1985.

[SNP90] Jan Smith, Bengt Nordström, and Kent Petersson. *Programming in Martin-Löf's Type Theory. An Introduction*. Oxford University Press, 1990.

[vBJ92] L.S. van Benthem Jutting. Typing in pure type systems. *Information and Computation*, 1992. To appear.

[vBJMP92] L. van Benthem Jutting, James McKinna, and Robert Pollack. Typechecking in pure type systems. in preparation, 1992.

# Orthogonal Higher-Order Rewrite Systems are Confluent

Tobias Nipkow*
TU München[†]

## 1 Introduction

Two important aspects of computation are *reduction*, hence the current infatuation with term-rewriting systems (TRSs), and *abstraction*, hence the study of $\lambda$-calculus. In his ground-breaking thesis, Klop [10] combined both areas in the framework of *combinatory reduction systems* (CRSs) and lifted many results from the $\lambda$-calculus to the CRS-level. More recently [15], we proposed a different although related approach to rewriting with abstractions, *higher-order rewrite systems* (HRSs). HRSs use the simply-typed $\lambda$-calculus as a meta-language for the description of reductions of terms with bound variables. Furthermore we showed that the well-known *critical pairs* can be extended to this setting, thus proving that confluence of terminating HRSs is decidable.

This paper is a second step towards a general theory of higher-order rewrite systems. It considers the complementary case, when there are no critical pairs and all rules are left-linear (no free variable appears twice on the left-hand side). Such systems are usually called *orthogonal*. The main result of this paper is that all orthogonal HRSs are confluent, irrespective of their termination. To a specialist in the field this will probably not come as a surprise because the same result holds for TRSs and CRSs, and the latter are close to HRSs. There are also many partial results in this direction [18, 1]. Hence the contribution of the paper is not so much the result itself but the combination of the following points:

- The demonstration that simply-typed $\lambda$-calculus is an ideal meta-language for rewriting terms with bound variables, because of its simple definition and because of the many powerful results available, especially w.r.t. reduction and unification.

- The simple proof, the idea of which goes back to Aczel [1] (in a more restrictive framework) and is similar to the classical confluence proof due to Tait and Martin-Löf.

- The meta-theorem itself.

This consolidates an emerging theory of higher-order rewrite systems based on simply-typed $\lambda$-calculus.

In the larger scheme of things it confirms the view that simply-typed $\lambda$-calculus is not a computational mechanism but a language for representing formulae. The above results apply directly to programming languages (like $\lambda$Prolog [14]) and theorem provers (like Isabelle [17]) whose meta-language is simply-typed $\lambda$-calculus. In fact, Amy Felty has implemented some aspects of HRSs in $\lambda$Prolog [6].

---

*Research supported by ESPRIT BRA 3245, *Logical Frameworks*, and WG 6028, *CCL*.
[†]Address: Institut für Informatik, Technische Universität München, Postfach 20 24 20, W-8000 München 2, Germany. Email: Tobias.Nipkow@Informatik.TU-Muenchen.De

## 2 Preliminaries

The standard theory of the simply-typed $\lambda$-calculus is assumed (see [8, 15]). In particular we use the following terminology and notation.

$\mathcal{T}$ is the set of all **types** and the letter $\tau$ ranges over $\mathcal{T}$. $\mathcal{T}$ is generated from a set of **base types** using the function space constructor $\to$. Instead of $\tau_1 \to \cdots \to \tau_n \to \tau$ we write $\overline{\tau_n} \to \tau$. The latter form is used only if $\tau$ is a base type.

Terms are generated from a set of typed **variables** $V = \bigcup_{\tau \in \mathcal{T}} V_\tau$ and a set of typed **constants** $C = \bigcup_{\tau \in \mathcal{T}} C_\tau$, where $V_\tau \cap V_{\tau'} = C_\tau \cap C_{\tau'} = \{\}$ if $\tau \neq \tau'$. Bound variables are denoted by $x$, $y$ and $z$, free variables by upper-case letters ($F$, $G$, ...), and constants by $c$. **Atoms** are constants or variables and are denoted by $a$. Terms are denoted by $l$, $r$, $s$, $t$, and $u$. In the sequel all $\lambda$-terms are assumed to be simply typed. We write $t : \tau$ if $t$ is of type $\tau$.

Instead of $\lambda x_1 \ldots \lambda x_n.s$ we write $\lambda x_1, \ldots, x_n.s$ or just $\lambda \overline{x_n}.s$, where the $x_i$ are assumed to be distinct. Similarly we write $t(u_1, \ldots, u_n)$ or just $t(\overline{u_n})$ instead of $(\ldots(t \; u_1)\ldots)u_n$. The **free variables** occurring in a term $s$ are denoted by $fv(s)$.

The $\beta$-normal form of a term $s$ is denoted by $s{\downarrow}_\beta$. Let $t$ be in $\beta$-normal form. Then $t$ is of the form $\lambda \overline{x_n}.a(\overline{u_m})$, where $a$ is called the **head** of $t$ and the $u_i$ are in $\beta$-normal form. The $\eta$-**expanded form** of $t$ is defined by

$$t{\uparrow}^\eta = \lambda \overline{x_{n+k}}.a(\overline{u_m}{\uparrow}^\eta, x_{n+1}{\uparrow}^\eta, \ldots, x_{n+k}{\uparrow}^\eta)$$

where $t : \overline{\tau_{n+k}} \to \tau$ and $x_{n+1}, \ldots, x_{n+k} \notin fv(\overline{u_m})$. Instead of $t{\downarrow}_\beta{\uparrow}^\eta$ we also write $t{\updownarrow}^\eta_\beta$. A term $t$ is in **long $\beta\eta$-normal form** if $t{\updownarrow}^\eta_\beta = t$.

Most of the time, except at intermediate stages of a computation, we are working with long $\beta\eta$-normal forms. Hence their explicit inductive definition may be helpful:

$$\frac{a : \overline{\tau_n} \to \tau \quad t_1 : \tau_1 \quad \ldots \quad t_n : \tau_n}{a(t_1, \ldots, t_n) : \tau} \qquad \frac{x : \tau \quad t : \tau'}{\lambda x.t : \tau \to \tau'}$$

Terms can also be viewed as trees. Subterms can be numbered by so-called **positions** which are the paths from the root to the subterm in Dewey decimal notation. Details can be found in [9, 4]. We just briefly review the notation. The **positions** in a term $t$ are denoted by $\mathcal{P}os(t) \subseteq \mathbb{N}^*$. The letters $p$ and $q$ stand for positions. The root position is $\varepsilon$, the empty sequence. Two positions $p$ and $q$ are appended by juxtaposing them: $pq$. Given $p \in \mathcal{P}os(t)$, $t/p$ is the subterm of $t$ at position $p$; $t[u]_p$ is $t$ with $t/p$ replaced by $u$.

Abstractions and applications yield the following trees:

```
   λx           ·
   |          / \
   s         s   t
```

Hence positions in $\lambda$-terms are sequences over $\{1, 2\}$. Note that the bound variable in an abstraction is not a separate subterm and can therefore not be accessed by the $s/p$ notation.

In order to simplify the treatment of $\lambda$-terms we use Barendregt's [2, p.26] variable conventions: $\alpha$-equivalent terms are identified and free and bound variables in terms are kept disjoint. We cannot enforce this convention completely. For example in the congruence rule $s = t \Rightarrow \lambda x.s = \lambda x.t$ the variable $x$ occurs both free and bound. Similarly the names of bound variables matter if the notation $t/p$ is used.

**Substitutions** are finite mappings from variables to terms of the same type. Substitutions are denoted by $\theta$ and $\sigma$. If $\theta = \{x_1 \mapsto t_1, \ldots, x_n \mapsto t_n\}$ we define $\mathcal{D}om(\theta) = \{x_1, \ldots, x_n\}$ and $\mathcal{C}od(\theta) = \{t_1, \ldots, t_n\}$. A **renaming** $\rho$ is an injective substitution with $\mathcal{C}od(\rho) \subset V$ and $\mathcal{D}om(\rho) \cap \mathcal{C}od(\rho) = \{\}$. Renamings are always denoted by $\rho$.

Using the above convention about free and bound variables, $t\theta$, the application of a substitution to a term, can be defined naïvely:

$$c\theta = c$$
$$F\theta = \theta F$$
$$(s\ t)\theta = (s\theta\ t\theta)$$
$$(\lambda x.s)\theta = \lambda x.(s\theta)$$

Composition of substitutions is juxtaposition: $t(\theta_1\theta_2) = (t\theta_1)\theta_2$. Application and normalization are postfix operators which associate to the left: $s\theta_1\theta_2\downarrow_\beta = ((s\theta_1)\theta_2)\downarrow_\beta$. In particular $s\theta\downarrow_\beta$ is the $\beta$-normal form of $s\theta$.

We say that two terms $s$ and $t$ **unify** if there is a substitution $\theta$ such that $s\theta$ and $t\theta$ are $\beta\eta$-equivalent. It is well-known that this holds iff $s\theta\uparrow^\eta_\beta = t\theta\uparrow^\eta_\beta$.

A relation $\to$ has the **diamond property** if $y_0 \leftarrow x \to y_1$ implies that there is a $z$ such that $y_0 \to z \leftarrow y_1$; $\to$ is **confluent** if $\overset{*}{\to}$ has the diamond property. The following lemma is folklore:

**Lemma 2.1** *If* $\to\ \subseteq\ >\ \subseteq\ \overset{*}{\to}$ *and* $>$ *has the diamond property, then* $\to$ *is confluent.*

## 3 Orthogonal Higher-Order Rewrite Systems

Higher-Order Rewrite Systems (HRSs) were introduced in [15]. Let us briefly review their definition and some basic results.

**Definition 3.1** A term $t$ in $\beta$-normal form is called a (**higher-order**) **pattern** if every free occurrence of a variable $F$ is in a subterm $F(\overline{u_n})$ of $t$ such that $\overline{u_n}\downarrow_\eta$ is a list of distinct bound variables.

Examples of higher-order patterns are $\lambda x.c(x)$, $X$, $\lambda x.F(\lambda z.x(z))$, and $\lambda x,y.F(y,x)$, examples of non-patterns are $F(c)$, $\lambda x.F(x,x)$ and $\lambda x.F(F(x))$.

The following crucial result about unification of patterns is due to Dale Miller [12]:

**Theorem 3.2** *It is decidable whether two patterns are unifiable; if they are unifiable, a most general unifier can be computed.*

This result ensures both the computability of the rewrite relation defined by an HRS and of critical pairs. Nipkow [15] presented a simplified form of Miller's unification algorithm which is further developed towards a practical implementation in [16].

**Definition 3.3** A **rewrite rule** is a pair $l \to r$ such that $l$ is a pattern but not $\eta$-equivalent to a free variable, $l$ and $r$ are long $\beta\eta$-normal forms of the same base type, and $fv(l) \supseteq fv(r)$. A **Higher-Order Rewrite System** (for short: **HRS**) is a set of rewrite rules. The letter $R$ always denotes an HRS. An HRS $R$ induces a relation $\to_R$ on terms in long $\beta\eta$-normal form:

$$s \underset{R}{\to} t \iff \exists (l \to r) \in R, p \in \mathcal{P}os(s), \theta.\ s/p = l\theta\downarrow_\beta \land t = s[r\theta\downarrow_\beta]_p.$$

A few explanatory remarks are in order, especially since this definition differs from the one in [15]:

- In contrast to [15] we have defined $\to_R$ only on long $\beta\eta$-normal forms and require the rules to be in that form as well. This is no restriction — every term has a unique long $\beta\eta$-normal form — and simplifies some technicalities.

- The term $t = s[r\theta\!\downarrow_\beta]_p$ is again in long $\beta\eta$-normal form provided $s$, $r$ and the elements of $Cod(\theta)$ are: $t$ is in $\beta$-normal form because $r\theta\!\downarrow_\beta$ is of base type; $t$ is in $\eta$-expanded form because $r\theta\!\downarrow_\beta$ is.

- The rewriting mechanism works just as well for rules not of base type. However, rewriting and equality fail to coincide in that case (see Theorem 3.5 below) and both the proof of the critical pair lemma [15] and of confluence of orthogonal HRSs rely on this restriction. Hence it is easier to build it into the framework.

- Because all rules in $R$ are of base type and left-hand sides must not be $\eta$-equivalent to free variables, each left-hand side is of the form $c(\overline{l_n})$.

- The subterm $s/p$ may contain free variables which used to be bound in $s$ but whose binding $\lambda$s are above $p$. Felty [6] uses a slightly different but equivalent definition of $\to_R$ which makes this fact explicit.

**Example 3.4** The syntax of the pure lambda-calculus involves just the type *term* of terms and two constants for abstraction and application:

$$abs : (term \to term) \to term$$
$$app : term \to term \to term$$

The rewrite rules are:

$$beta : app(abs(\lambda x.F(x)), S) \to F(S)$$
$$eta : abs(\lambda x.app(S, x)) \to S$$

Note how the use of meta-level application and abstraction removes the need for a substitution operator (in the *beta*-rule) and side conditions (in the *eta*-rule).

In the sequel it will be convenient to have an inference-rule based formulation of rewriting:

$$\frac{(l \to r) \in R}{l\theta\!\downarrow_\beta \to_R r\theta\!\downarrow_\beta} \qquad \frac{s \to_R t}{a(\overline{s_m}, s, \overline{u_n}) \to_R a(\overline{s_m}, t, \overline{u_n})} \qquad \frac{s \to_R t}{\lambda x.s \to_R \lambda x.t}$$

where $a$ is an atom of type $\overline{\tau_{m+1+n}} \to \tau$. The two definitions of $\to_R$ are equivalent.

Every HRS $R$ also induces an equality $=_R$ obtained by combining all instances of the rules in $R$ with reflexivity, symmetry, transitivity, the two congruence rules for application and abstraction, and $\beta$- and $\eta$-conversion. Because $\to_R$ acts only on long $\beta\eta$-normal forms, Theorem 3.4 of [15] becomes

**Theorem 3.5** $s =_R t \Leftrightarrow s\!\uparrow_\beta^\eta \leftrightarrow_R^* t\!\uparrow_\beta^\eta$

This equivalence between $=_R$ and $\leftrightarrow_R^*$ only holds because of the restriction to rules of base type (which is built into the above definition of an HRS). Given the rule

$$\lambda x.c(x, F(x)) \to \lambda x.d(F(x), x),$$

the relation $\leftrightarrow_R^*$ is strictly weaker than $=_R$: $c(a, f(a)) =_R d(f(a), a)$ holds but $c(a, f(a)) \leftrightarrow_R^* d(f(a), a)$ does not hold. The reason is that rewriting acts only on terms in $\beta$-normal form. Consequently we do not consider it a problem that the main result of this paper also relies on the restriction to rules of base type.

Confluence of rewrite systems always depends on an analysis of critical pairs. Critical pairs involve the unification of one left-hand side with the subterm of another left-hand side. But taking the subterm of a $\lambda$-term may turn a bound into a free variable. For example $x$ in the subterm $abs(S,x)$ of the left-hand side of *eta* used to be bound. To cope with this phenomenon, the lost binders have to be reintroduced, thus leading to $\lambda x.abs(S,x)$. In the example this happens to be a subterm itself, which is coincidental. In order to unify $\lambda x.abs(S,x)$ with the left-hand side of *beta*, the latter needs to be lifted:

**Definition 3.6** If $p \in \mathcal{P}os(s)$ let $bv(t,s)$ be the set of all $\lambda$-abstracted variables on the path from the root of $s$ to $p$.

An $\overline{x_k}$-**lifter of a term** $t$ **away from** $W$ is a substitution $\sigma = \{F \mapsto (\rho F)(\overline{x_k}) \mid F \in fv(t)\}$ where $\rho$ is a renaming such that $\mathcal{D}om(\rho) = fv(t), \mathcal{C}od(\rho) \cap W = \{\}$ and $\rho F : \tau_1 \to \cdots \to \tau_k \to \tau$ if $x_1 : \tau_1, \ldots, x_k : \tau_k$ and $F : \tau$.

For example $\{F \mapsto G(x), S \mapsto T(x)\}$ is an $x$-lifter of $app(abs(\lambda y.F(y)), S))$ away from any $W$ not containing $G$ or $T$.

Now we can give the (w.r.t. [15] slightly corrected) definition of critical pairs for HRSs.

**Definition 3.7** Two rules $l_i \to r_i$, $i = 1, 2$, a position $p \in \mathcal{P}os(l_1)$, and a most general unifier $\theta$ of $\lambda \overline{x_k}.(l_1/p)$ and $\lambda \overline{x_k}.(l_2\sigma)$ such that

- the free and bound variables in $l_1$ are disjoint and the head of $l_1/p$ is not free in $l_1$, and
- $\{\overline{x_k}\} = bv(l_1, p)$ and $\sigma$ is an $\overline{x_k}$-lifter of $l_2$ away from $fv(l_1)$

determine a **critical pair** $\langle r_1\theta\downarrow_\beta, (l_1[r_2\sigma]_p)\theta\downarrow_\beta\rangle$.

The critical pairs of an HRS are the critical pairs between any two of its rules, with the exception of those pairs arising from overlapping a rule with itself at position $\varepsilon$.

In Example 3.4 we have two critical pairs, one of which arises by taking $l_1 = abs(\lambda x.app(S,x))$ and $l_1/p = app(S,x)$. Then $bv(l_1,p) = \{x\}$ and $\sigma = \{F \mapsto G(x), S \mapsto T(x)\}$ is an $x$-lifter of $l_2 = app(abs(\lambda y.F(y)), S)$ away from $fv(l_1)$. Because $\theta = \{S \mapsto abs(\lambda y.H(y)), T \mapsto \lambda x.x, G \mapsto \lambda x, y.H(y)\}$ is a most general unifier of $\lambda x.app(S,x)$ and $\lambda x.app(abs(\lambda y.G(x,y)), T(x))$, the actual critical pair is $\langle abs(\lambda y.H(y)), abs(\lambda x.H(x))\rangle$.

Since our main concern is not the critical pairs themselves but their absence, we can work with a more concise definition:

**Definition 3.8** $l_1$ at $p$ **overlaps with** $l_2$ if the head of $l_1/p$ is not a variable free in $l_1$ and $\lambda \overline{x_k}.(l_1/p)$ and $\lambda \overline{x_k}.(l_2\sigma)$ unify, where $\{\overline{x_k}\} = bv(l_1, p)$, and $\sigma$ is an $\overline{x_k}$-lifter of $l_2$ away from $fv(l_1)$. Two terms $l_1$ and $l_2$ **overlap** if, for some $p \in \mathcal{P}os(l_1)$, $l_1$ at $p$ overlaps with $l_2$.

It follows easily that

**Lemma 3.9** *An HRS has no critical pairs iff each left-hand side overlaps only with itself at $\varepsilon$.*

Finally we can define our object of interest:

**Definition 3.10** An HRS is **orthogonal (OHRS)** if it is left-linear (no free variable appears twice on the left-hand side of a rule) and has no critical pairs.

It is not difficult to see that the pure lambda-calculus (Example 3.4) with only *beta* or *eta* is an OHRS. However, combining *beta* and *eta* leads to two (trivial) critical pairs [15], one of which is shown above.

For technical reasons we need the following lemma:

**Lemma 3.11** *If $l_0 = \lambda \overline{x_k}.l$, the head of $l_0$ is not a free variable and $l\theta\downarrow_\beta = l'\theta'\downarrow_\beta$, then $l_0$ and $l'$ overlap.*

**Proof** We claim that $l_0$ at $p = 1^k$, the sequence of $k$ 1s, overlaps with $l'$. Let $\sigma$ be an $\overline{x_k}$-lifter of $l'$ away from $fv(l_0)$. By definition $l_0$ at $p$ and $l'$ overlap if $\lambda\overline{x_k}.(l_0/p) = l_0$ and $\lambda\overline{x_k}.(l'\sigma)$ unify. We show that $\theta'' = \theta \cup \theta_1$, where $\theta_1 = \{\rho F \mapsto \lambda\overline{x_k}.(\theta'F) \mid F \in fv(l')\}$, is such a unifier.

For any substitution $\theta$ let $\theta\downarrow_\beta := \{F \mapsto (\theta F)\downarrow_\beta \mid F \in \mathcal{D}om(\theta)\}$. Using this definition we get $(\sigma\theta_1)\downarrow_\beta = (\{F \mapsto (\lambda\overline{x_k}.(\theta'F))(\overline{x_k}) \mid F \in fv(l')\} \cup \theta_1)\downarrow_\beta = \{F \mapsto (\theta'F)\downarrow_\beta \mid F \in fv(l')\} \cup \theta_1\downarrow_\beta = \theta'\downarrow_\beta \cup \theta_1\downarrow_\beta$. Now the required unifiability follows easily: $l_0\theta''\downarrow_\beta = l_0\theta\downarrow_\beta = \lambda\overline{x_k}.(l\theta\downarrow_\beta) = \lambda\overline{x_k}.(l'\theta'\downarrow_\beta) = \lambda\overline{x_k}.(l'(\theta'\downarrow_\beta))\downarrow_\beta) = \lambda\overline{x_k}.(l'(\theta'\downarrow_\beta \cup \theta_1\downarrow_\beta)\downarrow_\beta) = \lambda\overline{x_k}.(l'((\sigma\theta_1)\downarrow_\beta)\downarrow_\beta) = (\lambda\overline{x_k}.(l'\sigma))\theta_1\downarrow_\beta = (\lambda\overline{x_k}.(l'\sigma))\theta''\downarrow_\beta$. □

Note that this lemma is only true because by our conventions the free variables in the domain and codomain of $\theta$ are disjoint from the bound variables in $l_0$.

## 4 Confluence of OHRSs

Having set the stage, we can now proceed to prove the main result of this paper. The proof is based on the combination of two classical techniques for showing confluence of orthogonal reduction systems. In the sequel let $R$ be an OHRS. Parallel bottom-up reduction w.r.t. $R$ is the smallest relation $>$ on terms which is closed under the following inference rules:

$$\frac{s_i > t_i \quad (i = 1, \ldots, n)}{a(\overline{s_n}) > a(\overline{t_n})} \text{ (A)} \qquad \frac{s > t}{\lambda x.s > \lambda x.t} \text{ (L)} \qquad \frac{\theta >^{fv(l)} \theta'}{l\theta\downarrow_\beta > r\theta'\downarrow_\beta} \text{ (R)}$$

where $a$ is an atom of type $\overline{\tau_n} \to \tau$ and $s_i : \tau_i$, $(l \to r) \in R$, and $\theta >^W \theta'$ is short for $\forall F \in W.\ \theta F > \theta' F$.

This definition is essentially a combination of proof techniques for orthogonal TRSs and $\lambda$-calculus. It is interesting to compare it with the corresponding definitions for these two classical cases:

- for orthogonal TRSs, (A) remains unchanged, (L) must be dropped, and (R) becomes $l\theta > r\theta$;
- for pure $\lambda$-calculus, (A) becomes $s_0 > t_0 \land s_1 > t_1 \Rightarrow (s_0\ s_1) > (t_0\ t_1)$, (R) becomes $s > s' \land t > t' \Rightarrow (\lambda x.s)t > s'\{x \mapsto t'\}$, and we need to add $s > s$. The latter is implicit in the above formulation of (A).

Various further combinations of these two techniques appear in the literature [18, 5].

In contrast to the above versions of $>$, the definition given by Aczel is slightly more general:

$$\frac{s_1 > t_1 \quad \ldots \quad s_n > t_n \quad a(\overline{t_n}) = l\theta\downarrow_\beta}{a(\overline{s_n}) > r\theta\downarrow_\beta}$$

This rule allows the reduction of $a(\overline{s_n})$ even if the latter became a redex only through the reduction of some of its immediate subterms $s_i$. In orthogonal systems the two versions of (R) coincide because left-hand sides do not overlap. In *weakly* orthogonal systems, however, a certain amount of overlapping is permitted. This generalization is studied by Femke van Raamsdonk [19] who uses Aczel's version of $>$ in her confluence proof.

For inductive purposes we also define $|s > t|$ as the size of the proof of $s > t$. Although there may be many proofs of $s > t$, the one in question can usually be inferred from the context.

$$|a(\overline{s_n}) > a(\overline{t_n})| = 1 + \sum_{i=1}^{n} |s_i > t_i|$$

$$|\lambda x.s > \lambda x.t| = 1 + |s > t|$$
$$|t\theta\!\downarrow_\beta > r\theta'\!\downarrow_\beta| = 1 + |\theta >^{fv(l)} \theta'|$$

where $|\theta >^W \theta'| = \sum_{F \in W} |s_F > s'_F|$ if for all $F \in W$: $\theta F = \lambda \overline{x_{k_F}}.s_F$, $\theta' F = \lambda \overline{x_{k_F}}.s'_F$ and $s_F$ and $s'_F$ are not abstractions.

## 4.1 Lemmas

Before we can embark on the proof of the diamond property of $>$, we need a number of supporting lemmas. The first few are obvious enough.

**Lemma 4.1** $s > s$ *holds for all terms $s$ in long $\beta\eta$-normal form.*
**Proof** by induction on the structure of $s$. □

**Lemma 4.2** *If $s > t$ then both $s$ and $t$ are in long $\beta\eta$-normal form.*
**Proof** by induction on the structure of $s > t$. All three inferences (A), (L) and (R) preserve long $\beta\eta$-normal forms. For (A) this is because $a : \overline{\tau_n} \to \tau$ and $\tau$ is atomic. For (R) it follows because $s\theta\!\downarrow_\beta$ is not only in $\beta$-normal form but also in long $\beta\eta$-normal form if both $s$ and the terms in the codomain of $\theta$ are. □

**Lemma 4.3** $\to_R \subseteq > \subseteq \to_R^*$
**Proof** using the inference rule based definition of $\to_R$.

- $s \to_R t \Rightarrow s > t$ is proved by induction on the structure of $s \to_R t$. The only non-trivial case is $a(\overline{s_n}) \to_R a(s_1, \ldots, s_{k-1}, t_k, s_{k+1}, \ldots, s_n)$ and $s_k \to_R t_k$. Then $s_k > t_k$ by induction hypothesis and $s_i > t_i := s_i$ by Lemma 4.1 for $i \neq k$. Hence $a(\overline{s_n}) > a(\overline{t_n})$ follows by rule (A).

- $s > t \Rightarrow s \to_R^* t$ is proved by induction on the structure of $s > t$. The only non-trivial case is $a(\overline{s_n}) > a(\overline{t_n})$ and $s_i > t_i$ for $i = 1, \ldots, n$. By induction hypothesis $s_i \to_R^* t_i$ and hence $a(\overline{s_n}) \to^* a(t_1, s_2, \ldots, s_n) \to^* \cdots \to^* a(t_1, \ldots, t_{n-1}, s_n) \to^* a(\overline{t_n})$.

□

**Lemma 4.4** *If $\lambda x.s > t'$ then $t' = \lambda x.t$ and $s > t$ for some $t$.*
**Proof** Derivation (A) is syntactically inapplicable and $R$ cannot apply because all its rules are of base type. □

The first non-trivial lemma is the following. It relies crucially on the presence of a type structure which serves as a well-founded set for inductive purposes.

**Lemma 4.5** *If $s > t$ and $\theta >^{fv(s)} \theta'$ then $s\theta\!\downarrow_\beta > t\theta'\!\downarrow_\beta$.*
**Proof** by induction on the **order** of $\theta$

$$ord(\tau) = 0 \quad \text{if } \tau \text{ is a base type}$$
$$ord(\tau_0 \to \tau_1) = max\{ord(\tau_0) + 1, ord(\tau_1)\}$$
$$ord(\theta) = max\{ord(\tau) \mid V_\tau \cap \mathcal{D}om(\theta) \neq \{\}\}$$

immediately followed by an induction on the structure of $s > t$.

1. If $s = a(\overline{s_n})$, $t = a(\overline{t_n})$ and $s_i > t_i$ for $i = 1, \ldots, n$, then by the inner induction hypothesis $s_i\theta\downarrow_\beta > t_i\theta\downarrow_\beta$ holds.

   If $a$ is a constant or not in the domain of $\theta$ (and hence $\theta'$), then $s\theta\downarrow_\beta = a(\overline{s_n\theta\downarrow_\beta}) > a(\overline{t_n\theta\downarrow_\beta}) = t\theta'\downarrow_\beta$ follows immediately.

   Otherwise $a \in \mathcal{D}om(\theta)$. From Lemmas 4.2 and 4.4 it follows that $\theta a = \lambda \overline{x_n}.u$, $\theta'a = \lambda \overline{x_n}.u'$ and $u > u'$. Thus we have $s\theta\downarrow_\beta = (\theta a)(\overline{s_n\theta\downarrow_\beta})\downarrow_\beta = u\theta_1\downarrow_\beta$ and $t\theta'\downarrow_\beta = (\theta' a)(\overline{t_n\theta'\downarrow_\beta})\downarrow_\beta = u'\theta_1'\downarrow_\beta$ where $\theta_1 = \{\overline{x_n \mapsto s_n\theta\downarrow_\beta}\}$ and $\theta_1' = \{\overline{x_n \mapsto t_n\theta'\downarrow_\beta}\}$. If $a$ is of type $\overline{\tau_n} \to \tau$, the definition of $ord$ implies $ord(\tau_i) < ord(\overline{\tau_n} \to \tau)$ and hence $ord(\theta_1) < ord(\theta)$, i.e. the outer induction hypothesis applies. Hence $u\theta_1\downarrow_\beta > u'\theta_1'\downarrow_\beta$ follows, which completes the proof of $s\theta\downarrow_\beta > t\theta'\downarrow_\beta$.

2. If $s = \lambda x.u$, $t = \lambda x.u'$ and $u > u'$ then the inner induction hypothesis yields $s\theta\downarrow_\beta = (\lambda x.u)\theta\downarrow_\beta = \lambda x.(u\theta\downarrow_\beta) > \lambda x.(u'\theta'\downarrow_\beta) = (\lambda x.u')\theta'\downarrow_\beta = t\theta'\downarrow_\beta$.

3. If $s = l\theta_1\downarrow_\beta$, $t = r\theta_1'\downarrow_\beta$ and $\theta_1 >^{fv(l)} \theta_1'$ then the inner induction hypothesis yields $F\theta_1\theta\downarrow_\beta > F\theta_1'\theta'\downarrow_\beta$ for all $F \in fv(l)$. Let $\theta_2 = \{F \mapsto F\theta_1\theta\downarrow_\beta \mid F \in fv(l)\}$ and $\theta_2' = \{F \mapsto F\theta_1'\theta'\downarrow_\beta \mid F \in fv(l)\}$. Thus we have $s\theta\downarrow_\beta = l\theta_1\downarrow_\beta\theta\downarrow_\beta = l\theta_1\theta\downarrow_\beta = l\theta_2\downarrow_\beta > r\theta_2'\downarrow_\beta = r\theta_1'\theta'\downarrow_\beta = r\theta_1'\downarrow_\beta\theta'\downarrow_\beta = t\theta'\downarrow_\beta$.

□

**Corollary 4.6** *If $s$ is in long $\beta\eta$-normal form and $\theta >^{fv(s)} \theta'$ then $s\theta\downarrow_\beta > s\theta'\downarrow_\beta$.*
**Proof** by combining Lemmas 4.1 and 4.5. □

The following lemma looks very technical but expresses a simple idea: in a reduction step $l\theta\downarrow_\beta > s$, where $l$ is linear and does not overlap with any rule, the $l$-part cannot change, i.e. all reductions must have taken place inside the terms introduced via $\theta$. Thus there is a $\theta'$ such that $\theta >^{fv(l)} \theta'$ and $l\theta'\downarrow_\beta = s$.

**Lemma 4.7** *Let $l_0 = \lambda\overline{x_n}.l$ be a linear pattern that does not overlap with any left-hand side of $R$ and let $W = fv(l_0)$. If the elements of $Cod(\theta)$ are in long $\beta\eta$-normal form and $l\theta\downarrow_\beta > s$ then there exists a substitution $\theta'$ such that $\theta >^W \theta'$, $|\theta >^W \theta'| \leq |l\theta\downarrow_\beta > s|$ and $l\theta'\downarrow_\beta = s$.*
**Proof** by induction on the structure of $l$.

1. If $l = c(\overline{l_m})$, then the last inference in $l\theta\downarrow_\beta > s$ must be (A): it cannot be (L) and Lemma 3.11 implies that it cannot be (R) either, otherwise $l_0$ would overlap with the left-hand side of a rule in $R$. Therefore $s = c(\overline{s_m})$ and $l_i\theta\downarrow_\beta > s_i$. Because each $\lambda\overline{x_n}.l_i$ is also a linear pattern that does not overlap with any left-hand side of $R$, the induction hypothesis yields substitutions $\theta_i$ such that $\theta >^{W_i} \theta_i$, $|\theta >^{W_i} \theta_i| \leq |l_i\theta\downarrow_\beta > s_i|$ and $l_i\theta_i\downarrow_\beta = s_i$, where $W_i = fv(\lambda\overline{x_n}.l_i)$. Now let $\theta' := \bigcup_{i=1}^m \theta_i$, which is a well-defined substitution because the $W_i$, and hence the domains of the $\theta_i$, are disjoint, thanks to linearity. Obviously $\theta >^W \theta'$ and $l\theta'\downarrow_\beta = c(\overline{l_m\theta_m\downarrow_\beta}) = s$ and $|\theta >^W \theta'| = \sum_{i=1}^m |\theta >^{W_i} \theta_i| \leq \sum_{i=1}^m |l_i\theta\downarrow_\beta > s_i| < 1 + \sum_{i=1}^m |l_i\theta\downarrow_\beta > s_i| = |l\theta\downarrow_\beta > s|$.

2. If $l = F(\overline{s_m})$ then $l\downarrow_\eta = F(\overline{y_m})$ because $l_0$ is a pattern. Wlog $\theta F = \lambda\overline{y_m}.r$. Hence $l\theta\downarrow_\beta = l\downarrow_\eta\theta\uparrow^\eta_\beta = (\lambda\overline{y_m}.r)(\overline{y_m})\uparrow^\eta_\beta = r$ and thus $r > s$. Let $\theta' = \{F \mapsto \lambda\overline{y_m}.s\}$. Obviously $\theta >^{\{F\}} \theta'$, $|\theta >^{\{F\}} \theta'| = |r > s| = |l\theta\downarrow_\beta > s|$ and $l\theta'\downarrow_\beta = l\downarrow_\eta\theta'\uparrow^\eta_\beta = (\lambda\overline{y_m}.s)(\overline{y_m})\uparrow^\eta_\beta = s$.

3. If $l = \lambda x_{n+1}.l'$ then the claim follows easily from the induction hypothesis.

□

## 4.2 The Proof

**Theorem 4.8** *If $R$ is orthogonal then $>$ has the diamond property.*

**Proof** We assume that $s > s^i$, $i = 0, 1$, and show by induction on $|s > s^0| + |s > s^1|$ that there is a $t$ such that $s^i > t$, $i = 0, 1$.

If $s = \lambda x.s'$, Lemma 4.4 implies $s^i = \lambda x.t^i$ and $s' > t^i$, for $i = 0, 1$. By induction hypothesis there exists a $t'$ such that $t^i > t'$ and hence $s^i > \lambda x.t' =: t$.

If $s = a(\overline{s_n})$, there are three possible combinations of (A) and (R) inferences.

1. If both $s > s^0$ and $s > s^1$ via (A) then $s = a(\overline{s_n})$ and $s^i = a(\overline{s_n^i})$ and $s_j > s_j^i$. By induction hypothesis there exist $t_j$ such that $s_j^i > t_j$ and hence $s^i > a(\overline{t_n}) =: t$.

2. If both $s > s^0$ and $s > s^1$ via (R) then $s = l_i\theta_i\downarrow_\beta > r_i\theta_i'\downarrow_\beta = s^i$ for some $(l_i \to r_i) \in R$, $i = 0, 1$. Since $R$ is orthogonal, the two rules must be the same (otherwise they would give rise to an overlap at $\varepsilon$) and hence $(l_0 \to r_0) = (l_1 \to r_1) =: (l \to r)$ and $\theta_0 = \theta_1 =: \theta$. Because $|s > s^i| = 1 + |\theta >^{fv(l)} \theta_i'|$, the induction hypothesis applies and there is a $\theta'$ such that $\theta_i' >^{fv(l)} \theta'$. Since $fv(r) \subseteq fv(l)$, Lemma 4.5 yields the desired $s^i = r\theta_i'\downarrow_\beta > r\theta'\downarrow_\beta =: t$.

3. If $s > s^0$ via (R) and $s > s^1$ via (A) then $s = l\theta\downarrow_\beta = a(\overline{s_n})$, $s^0 = r\theta'\downarrow_\beta$, $\theta >^{fv(l)} \theta'$, $(l \to r) \in R$, $s^1 = a(\overline{s_n^1})$, and $s_i > s_i^1$. By an observation above $l$ must be of the form $c(\overline{l_n})$ and hence $a = c$ and $l_i\theta\downarrow_\beta = s_i$. Because $l$ is a linear pattern that can only overlap with itself at $\varepsilon$, the $l_i$ are linear patterns which do not overlap with any left-hand side of $R$. Thus Lemma 4.7 yields the existence of substitutions $\theta_i$ such that $\theta >^{fv(l_i)} \theta_i$, $|\theta >^{fv(l_i)} \theta_i| \leq |s_i > s_i^1|$ and $l_i\theta_i\downarrow_\beta = s_i^1$. Since $l$ is linear, the free variables in the $l_i$ are disjoint and hence $\theta'' := \bigcup_{i=1}^n \theta_i$ is a well-defined substitution with $\theta >^{fv(l)} \theta''$. Now the induction hypothesis yields a substitution $\sigma$ such that $\theta' >^{fv(l)} \sigma$ and $\theta'' >^{fv(l)} \sigma$. Hence $s^1 = c(\overline{s_n^1}) = c(\overline{l_n\theta_n\downarrow_\beta}) = c(\overline{l_n\theta''\downarrow_\beta}) = l\theta''\downarrow_\beta > r\sigma\downarrow_\beta =: t$. Corollary 4.6 yields $s^0 = r\theta'\downarrow_\beta > r\sigma\downarrow_\beta = t$, thus closing the diamond. □

Combining Lemmas 2.1 and 4.3 with the above theorem immediately yields the main result of this paper:

**Corollary 4.9** *Every OHRS is confluent.*

# 5 Applications

The standard application of our result is to various lambda-calculi, in particular pure lambda-calculus itself (Example 3.4). It was already mentioned that *beta* and *eta* on their own form orthogonal systems. Thus we can conclude that *beta* and *eta* reductions are confluent. The combination of the two is treated in the next section. Since previous work on rewriting lambda-terms [1, 10, 18] has already treated many reduction systems arising in proof theory, we refrain from adding more examples to this list. However, we would like to point out that our result should help to eliminate sentences like "A proof can be given by adapting the method of Tait and Martin-Löf using a form of parallel reduction (see also Stenlund)" [7, pp. 9 and 29] in favour of a direct application of our meta-theorem.

A second application area concerns the combination of term rewriting systems with *beta*-reduction as studied by Breazu-Tannen [3]. Breazu-Tannen combines a fixed higher-order reduction, namely *beta*, with an arbitrary set of first-order reductions $R$ and is able to show that confluence of $R$ is preserved by adding simply-typed *beta*. This result was generalized significantly by Dougherty [5] who showed that it is not the type system (simple, polymorphic, etc.) but the strong normalization property which makes the result hold.

We can also obtain some results in this area by combining our object-level *beta*-rule with other reductions. The first-order term $f(s_1, \ldots, s_n)$ translates to $app(\ldots(app(f, t_1), \ldots), t_n)$, where $t_i$ is the translation of $s_i$. It is easy to see that the left-hand side of *beta* does not overlap with the left-hand side of the translation of a first-order reduction. In fact, we have the more general lemma:

**Lemma 5.1** *Let $R$ be an OHRS such that no left-hand side has abs as a head constant or a subterm of the form $app(F, t)$, where $F$ is free and $t$ is arbitrary. Then $R \cup \{beta\}$ is orthogonal and hence confluent.*

Comparing this with the result by Dougherty, we find that the above lemma admits more general rules (not just first-order reductions) and needs no (object-level) types or strong normalization property, but requires orthogonality of $R$. In fact, our lemma is much closer to a recent result by Fritz Müller [13] who proves that $R \cup \{beta\}$ is confluent if $R$ is a confluent left-linear "applicative term rewriting system" (ATRS). Again, his and our result are incomparable: his ATRSs are less general than the rules in the above lemma in that the left-hand sides may not contain *abs* at all, but he allows overlapping rules.

## 6 Extensions

Although our main result yields the confluence of *beta* and *eta* in isolation, it does not extend to the combination of *beta* and *eta* because the two rules overlap. However, their two critical pairs are trivial, i.e. of the form $\langle t, t \rangle$ [15]. Left linear TRSs with only trivial critical pairs are sometimes called **weakly orthogonal** and are confluent [11]. In fact, Huet [9, p. 815] proved confluence for the even larger class of left-linear "parallel closed" TRSs. Very recently, Femke van Raamsdonk [19] has shown that weakly orthogonal CRSs are confluent. Hence it is tempting to assume that weakly orthogonal HRSs are confluent as well. Analyzing van Raamsdonk's proof, it becomes clear why we failed to prove her result for HRSs: our definition of $>$ is too weak and Aczel's definition should be used (see Section 4).

In order to prove confluence for the combination of *beta* and *eta* we had to resort to a proof technique that is usually employed in classical confluence proofs [2]: the two reductions are separately shown to be confluent and their combined confluence follows by a certain kind of commutation. Abstracting from the particulars of the *eta*-rule we call a term in $\beta$-normal form **non-duplicating** if it is fully linear (no free or bound variable occurs twice) and has no subterm of the form $F(\ldots G \ldots)$, where $F$ and $G$ are free. A rule is called non-duplicating if its right-hand side is. The definition ensures that non-duplicating rules cannot duplicate subterms.

**Theorem 6.1** *If $R_0$ and $R_1$ are left-linear and confluent, all critical pairs between rules in $R_0$ and $R_1$ are trivial, i.e. of the form $\langle t, t \rangle$, and $R_1$ is non-duplicating, then $R := R_0 \cup R_1$ is confluent.*

**Proof** sketch. According to the lemma of Hindley/Rosen it suffices to show that $\to_{R_0}^*$ and $\to_{R_1}^*$ commute, which in turn is implied if for every fork $s \to_{R_i} s_i$, $i = 0, 1$, there is a term $t$ such that $s_0 \to_{R_1}^* t$ and $s_1 \to_{R_0}^= t$, i.e. either $s_1 \to_{R_0} t$ or $s_1 = t$ [2, p. 65].

If the two redexes overlap, they are instances of a critical pair and hence $s_0 = s_1 =: t$. If the $R_0$-redex is above the $R_1$-redex, it is easy to find a $t$ such that $s_0 \to_{R_1}^* t$ and $s_1 \to_{R_0} t$. If the $R_0$-redex $r_0$ is below the $R_1$-redex, there is at most one descendant of $r_0$ in $s_1$ ($R_1$ is non-duplicating!) and we can find a $t$ such that $s_0 \to_{R_1} t$ and $s_1 \to_{R_0}^= t$. □

Since *beta* and *eta* on their own are left-linear and confluent, the two critical pairs between *beta* and *eta* are trivial, and *eta* is non-duplicating, *beta*+*eta* is again confluent. The same argument applies to some other well-known weakly-orthogonal systems, for example parallel-or:

$$Por = \left\{ \begin{array}{ll} or(X, true) \rightarrow true & or(true, X) \rightarrow true \\ or(X, false) \rightarrow X & or(false, X) \rightarrow X \end{array} \right\}$$

*Por* is locally confluent (all critical pairs are trivial), terminating and hence confluent. Adding *beta* preserves confluence because there are no critical pairs between *Por* and *beta*, and *Por* is non-duplicating. This confluence result also follows from the work of Müller. Note that in the absence of termination, the violation of left-linearity is again fatal: adding the rules $or(X, X) \rightarrow X$ and $or(not(X), X) \rightarrow true$ leads to non-confluence just as in Breazu-Tannen's example [3].

## 7 Conclusion

The results about higher-order critical pairs and the confluence of OHRSs provide a firm foundation for the further study of higher-order rewrite systems. It should now be interesting to lift more results and techniques both from term-rewriting and $\lambda$-calculus to the level of HRSs. For example termination proof techniques are much studied for TRSs and are urgently needed for HRSs; similarly the extension of our result to weakly orthogonal HRSs or even to Huet's "parallel closed" systems is highly desirable. Conversely, a large body of $\lambda$-calculus reduction theory has been lifted to CRSs [10] already and should be easy to carry over to HRSs.

Finally there is the need to extend the notion of an HRS to more general left-hand sides. For example the *eta*-rule for the *case*-construct on disjoint unions [15]

$$case(U, \lambda x.F(inl(x)), \lambda y.G(inr(y))) \rightarrow F(U)$$

is outside our framework, whichever way it is oriented.

**Acknowledgement** I would like to thank Jan Willem Klop, Christian Prehofer and Femke van Raamsdonk for their careful reading and for the corrections and improvements they suggested.

## References

[1] P. Aczel. A general Church-Rosser theorem. Technical report, University of Manchester, 1978.

[2] H. P. Barendregt. *The Lambda Calculus, its Syntax and Semantics*. North Holland, 2nd edition, 1984.

[3] V. Breazu-Tannen. Combining algebra and higher-order types. In *Proc. 3rd IEEE Symp. Logic in Computer Science*, pages 82–90, 1988.

[4] N. Dershowitz and J.-P. Jouannaud. Rewrite systems. In J. van Leeuwen, editor, *Formal Models and Semantics, Handbook of Theoretical Computer Science, Vol. B*, pages 243–320. Elsevier - The MIT Press, 1990.

[5] D. J. Dougherty. Adding algebraic rewriting to the untyped lambda-calculus. In R. V. Book, editor, *Proc. 4th Int. Conf. Rewriting Techniques and Applications*, pages 37–48. LNCS 488, 1991.

[6] A. Felty. A logic-programming approach to implementing higher-order term rewriting. In L.-H. Eriksson, L. Hallnäs, and P. Schroeder-Heister, editors, *Extensions of Logic Programming, Proc. 2nd Int. Workshop*, pages 135–158. LNCS 596, 1992.

[7] J. Gallier. Constructive Logics. Part I. Technical Report 8, DEC Paris Research Laboratory, 1991.

[8] J. Hindley and J. P. Seldin. *Introduction to Combinators and λ-Calculus.* Cambridge University Press, 1986.

[9] G. Huet. Confluent reductions: Abstract properties and applications to term rewriting systems. *J. ACM*, 27:797–821, 1980.

[10] J. W. Klop. *Combinatory Reduction Systems.* Mathematical Centre Tracts 127. Mathematisch Centrum, Amsterdam, 1980.

[11] J. W. Klop. Term rewriting systems. Technical Report CS-R9073, CWI, Amsterdam, 1990.

[12] D. Miller. A logic programming language with lambda-abstraction, function variables, and simple unification. In P. Schroeder-Heister, editor, *Extensions of Logic Programming*, pages 253–281. LNCS 475, 1991.

[13] F. Müller. Confluence of the lambda-calculus with left-linear algebraic rewriting. *Information Processing Letters*, 41:293–299, 1992.

[14] G. Nadathur and D. Miller. An overview of λProlog. In R. A. Kowalski and K. A. Bowen, editors, *Proc. 5th Int. Logic Programming Conference*, pages 810–827. MIT Press, 1988.

[15] T. Nipkow. Higher-order critical pairs. In *Proc. 6th IEEE Symp. Logic in Computer Science*, pages 342–349, 1991.

[16] T. Nipkow. Functional unification of higher-order patterns. Technical report, TU München, Institut für Informatik, 1992.

[17] L. C. Paulson. Isabelle: The next 700 theorem provers. In P. Odifreddi, editor, *Logic and Computer Science*, pages 361–385. Academic Press, 1990.

[18] S. Stenlund. *Combinators, λ-Terms, and Proof Theory.* D. Reidel, 1972.

[19] F. van Raamsdonk. A simple proof of confluence for weakly orthogonal combinatory reduction systems. Technical Report CS-R9234, CWI, Amsterdam, 1992.

# Monotonic versus Antimonotonic Exponentiation

Daniel F. Otth

Massachusetts Institute of Technology, Laboratory for Computer Science, 545 Technology Square, Cambridge MA 02139, USA

**Abstract.** We investigate the relationship between the monotonic ($\rightarrow$) and the antimonotonic exponentiation ($\rightarrowtail$) in a type system with subtyping. We present a model in which we can develop both exponentiations at the same time. In this model the monotonic and the antimonotonic exponentiation enjoy a duality, namely $\alpha \rightarrowtail \beta = \complement(\alpha \rightarrow \complement\beta)$ where $\complement$ is the type constructor complement. We give a sound and complete system of axioms for the type system with the type constructors $\rightarrow$, $\rightarrowtail$, $\cup$, $\cap$, $\complement$, $\bot$, $\top$.

## 1 Introduction

In a type system with subtyping we have an inclusion relation between types which interacts with the type constructors of the type system according to certain rules. For some type constructors, it is quite clear how such interaction rules have to look, *e.g.* the product type constructor is monotonic in both arguments, *i.e.* if $\alpha$, $\beta$, $\gamma$ and $\delta$ are types, where $\alpha$ is a subtype of $\beta$, $\alpha \subseteq \beta$ and $\gamma$ is a subtype of $\delta$, $\gamma \subseteq \delta$ then we have the following behavior

$$\alpha \times \gamma \subseteq \beta \times \delta$$

and everybody will agree, that this is the only reasonable behavior for the product. But for other type constructors, like exponentiation, it is not evident what the adequate behavior is. Our interest lies in studying the interaction between the inclusion relation and the type exponentiation constructor. The exponentiation of two types $\alpha$, $\beta$ gives us the function space $\alpha \rightarrow \beta$. Every function $f$ of type $\alpha \rightarrow \beta$, that is $f : \alpha \rightarrow \beta$, maps objects of type $\alpha$ to objects of type $\beta$ and so we have

$$f : \alpha \rightarrow \beta \Rightarrow \forall x.(x : \alpha \Rightarrow fx : \beta).$$

This is the minimal requirement for the exponentiation. The relevant literature presents two different constructions which satisfy this requirement.

One is the antimonotonic approach where the exponentiation is antimonotonic in the first and monotonic in the second argument, *i.e.* if $\alpha \subseteq \beta$ and $\gamma \subseteq \delta$ then $\beta \rightarrow \gamma \subseteq \alpha \rightarrow \delta$. For example, the integers Int are a subtype of the real numbers Real. So, if we have a function $f : \text{Real} \rightarrow \alpha$ which maps real numbers to objects of type $\alpha$ then surely $f$ also maps integers to objects of type $\alpha$, that is $f : \text{Int} \rightarrow \alpha$. Therefore we can deduce that $\text{Real} \rightarrow \alpha \subseteq \text{Int} \rightarrow \alpha$, but not the converse. We therefore choose as a function space $\alpha \to \beta$ exactly those functions which maps objects of type $\alpha$ to objects of type $\beta$

$$f : \alpha \to \beta \iff \forall x.(x : \alpha \Rightarrow fx : \beta).$$

In other words $\alpha \to \beta$ contains all functions which satisfy the minimal requirement for the exponentiation. Therefore this antimonotonic exponentiation is the greatest possible exponentiation.

The second possibility is the monotonic approach. Here we treat the exponentiation in the same way as the product mentioned above. For the types $\alpha \subseteq \beta$ and $\gamma \subseteq \delta$ we have $\alpha \to \gamma \subseteq \beta \to \delta$. The intuition behind this exponentiation is different than in the antimonotonic case. We give the following example to motivate the monotonic approach. We can describe the quality of a wind by the wind direction and the wind force. Knowledge about the wind direction is knowledge of type D and knowledge about both the wind direction and the wind force is knowledge of type DF. Hence, D is a subtype of DF. If we have a function $f : D \to \alpha$ which maps knowledge of type D to knowledge of type $\alpha$, then $f$ also maps knowledge of type DF to knowledge of type $\alpha$, that is $f : DF \to \alpha$ is given simply by ignoring additional knowledge that may be given in DF. Therefore we obtain $D \to \alpha \subseteq DF \to \alpha$, but not the other way round. So the function space $\alpha \to \beta$ consists of those functions $f$ which map knowledge of type $\alpha$ to knowledge of type $\beta$ and ignore all knowledge outside of $\alpha$. We obtain

$$f : \alpha \to \beta \iff \forall x.fx = f(x \cap \alpha) : \beta$$

where $x \cap \alpha$ is $x$ restricted to $\alpha$.

Now we can ask whether or not there are other possibilities for the exponentiation, *e.g.* one that is antimonotonic in the second argument (rather than in the first). But this is impractical, as the following example shows. Let $\bot$ denote the least type and for a type $\alpha$ we have that $\bot \to \alpha$ denotes the constant functions of $\alpha$. Now, if $\alpha, \beta$ are two types with $\alpha \subseteq \beta$ then we would have the type of constant functions $\bot \to \beta$ as a subtype of $\bot \to \alpha$, that is, $\bot \to \beta \subseteq \bot \to \alpha$. But normally one has a 1-1 correspondence between the constant functions $\bot \to \beta$ and the objects of type $\beta$ and we obtain $\beta \simeq \bot \to \beta \subseteq \bot \to \alpha \simeq \alpha$.

There is a third possibility; an exponentiation which is neither monotonic nor antimonotonic, *i.e.* if $\alpha \subseteq \beta$ and $\gamma \subseteq \delta$ then $\alpha \to \gamma \not\subseteq \beta \to \delta$ and $\beta \to \gamma \not\subseteq \alpha \to \delta$. We take as function space $\alpha \to \beta$ those functions with the property

$$f : \alpha \to \beta \iff \forall x.(x : \alpha \iff fx : \beta).$$

It is left to the reader to show that this is indeed a valid exponentiation which is neither monotonic nor antimonotonic.

We construct exponentiations with behaviors other than the antimonotonic one by taking appropriate subsets of the function space of the antimonotonic exponentiation. For example, if we define the exponentiation of two types $\alpha, \beta$ as follows

$$f : \alpha \to \beta \iff fx : \bot \vee (x : \alpha \wedge fx : \beta)$$

then we obtain a monotonic exponentiation too, but a different one from that defined earlier. The problem with these definitions is that we formulated without any basis. If we want to realize these exponentiations in concrete models, then it may happen that our definitions of function spaces lead to empty sets or sets which contain too few functions to model a typed $\lambda$-calculus. In a concrete model normally only one of the above definitions of the exponentiation (usually the monotonic or the antimonotonic one) leads to useful function spaces. Since we want to investigate the relationship between the different kinds of exponentiations, we need a model where we can realize both of them adequately.

The principle ideal model of Otth [2] allows us to do just that. In this model we can define the monotonic and the antimonotonic exponentiation as we described above. As we will see, the monotonic and the antimonotonic exponentiation enjoy a duality in this model, namely

$$\alpha \rightarrowtail \beta = \mathsf{C}(\alpha \rightarrow \mathsf{C}\beta)$$

where we denote with $\rightarrow$ the monotonic exponentiation, with $\rightarrowtail$ the antimonotonic exponentiation and with $\mathsf{C}$ the type constructor complement. So for every property of the monotonic exponentiation, we have a dual property of the antimonotonic exponentiation and vice versa. For example we have

$$(\alpha \rightarrow \beta) \cap (\gamma \rightarrow \delta) = (\alpha \cap \gamma) \rightarrow (\beta \cap \delta)$$

and the dual property is

$$(\alpha \rightarrowtail \beta) \cup (\gamma \rightarrowtail \delta) = (\alpha \cap \gamma) \rightarrowtail (\beta \cup \delta)$$

where we used DeMorgan's law $\alpha \cup \beta = \mathsf{C}(\mathsf{C}\alpha \cap \mathsf{C}\beta)$. Basically we can express the same types with the type constructors $\rightarrow$, $\cup$, $\mathsf{C}$ and $\bot$ as with the type constructors $\rightarrow$, $\rightarrowtail$, $\cup$, $\cap$, $\bot$ and $\top$. In the rest of this paper we will develop a sound and complete system of axioms for these type systems.

## 2 Definitions

First we introduce the type system we want to model.

**Definition.** The set $T$ of types is the least set with the properties that:

$$\bot \in T$$
$$\alpha, \beta \in T \Rightarrow \alpha \rightarrow \beta, \alpha \cup \beta, \mathsf{C}\alpha \in T$$

We define the following type constructors:

$$\alpha \rightarrowtail \beta := \mathsf{C}(\alpha \rightarrow \mathsf{C}\beta)$$
$$\alpha \cap \beta := \mathsf{C}(\mathsf{C}\alpha \cup \mathsf{C}\beta)$$
$$\top := \mathsf{C}\bot$$
$$+ := \top \rightarrow \top$$
$$\div := \mathsf{C}+$$

With $T' \subseteq T$ we denote those types where the type constructor $C$ only occurs in the type constructors $\rightarrow$, $\cap$ and $\top$, i.e. $T'$ is the least set with the property that:

$$\{\bot, \top\} \subseteq T'$$
$$\alpha, \beta \in T' \Rightarrow \alpha \rightarrow \beta, \alpha \rightarrowtail \beta, \alpha \cup \beta, \alpha \cap \beta \in T'$$

The intended meaning of these type constructors are, $\rightarrow$ monotonic exponentiation, $\rightarrowtail$ antimonotonic exponentiation, $\cup$ union, $\cap$ intersection, $C$ complement, $\bot$ the least type, $\top$ the greatest type, $+$ the functions and $*$ the constants.

As usual we omit parentheses whenever possible. The type constructors $\rightarrow$ and $\rightarrowtail$ are grouped to to the right and $\cap$, $\cup$ to the left.

$$\alpha \rightarrow \beta \rightarrow \gamma = \alpha \rightarrow (\beta \rightarrow \gamma) \qquad \alpha \rightarrowtail \beta \rightarrowtail \gamma = \alpha \rightarrowtail (\beta \rightarrowtail \gamma)$$
$$\alpha \cup \beta \cup \gamma = (\alpha \cup \beta) \cup \gamma \qquad \alpha \cap \beta \cap \gamma = (\alpha \cap \beta) \cap \gamma$$

We have the convention that $\rightarrow$ and $\rightarrowtail$ bind more than $\cap$ or $\cup$.

$$\alpha \rightarrow \beta \cup \gamma = (\alpha \rightarrow \beta) \cup \gamma \qquad \alpha \rightarrowtail \beta \cap \gamma = (\alpha \rightarrowtail \beta) \cap \gamma$$

We will realize this type system in the principal ideal model, which is based on the graph model [5]. The graph model is a set theoretical model for the combinatory algebra. Typical examples of graph models are the Engeler graph model $\mathcal{D}_A$ [1] and the Scott-Plotkin model $P_\omega$ [6] [4].

**Definition.** A *graph model* consists of a set X and an embedding $(\cdot, \cdot)$ into X from the Cartesian product $X^{<\omega} \times X$ of finite subsets of X and X. For two subsets $a, b \in 2^X$ of X we define the application $\bullet$:

$$a \bullet b := \{s | \exists \sigma \subseteq b, (\sigma, s) \in a\}.$$

We write a graph model as a pair $< 2^X, \bullet >$ (This is possible because the correspondence between embeddings $X^{<\omega} \times X \hookrightarrow X$ and the application $\bullet$, defined by the former is 1-1, see [5]).

Often we will omit the operator $\bullet$ and write $xy$ instead of $x \bullet y$. We also group nested applications to the left, i.e. $xyz = (xy)z = (x \bullet y) \bullet z$.

We call elements of a graph model $< 2^X, \bullet >$ graphs. We can think of a graph $M \subseteq X$ of a graph model as a graph in the set theoretic sense of the function $2^X \rightarrow 2^X, x \mapsto Mx$. M describes which finite sets are mapped to which elements of X. A graph M does not behave exactly like a graph in set theory; e.g. there is usually more then one graph which denotes the same function, but this motivates the naming graph model, application etc.

We understand by a type, a collection of similar graphs. We choose principal ideals (in the lattice theoretic sense) as types. In this way we have a 1-1 correspondence between types and graphs, since every graph generates a principal ideal and every ideal is generated by an unique graph. Let $[\alpha]$ be the principal ideal generated by $\alpha$, that is $[\alpha] := \{x | x \subseteq \alpha\}$. The 1-1 correspondence is now realized by the bijection $x \mapsto [x]$. Moreover $\phi$ is an isomorphism of Boolean algebras, *i.e.* the infimum of two principal ideals $[\alpha]$ and $[\beta]$ is $[\alpha \cap \beta]$, the supremum is $[\alpha \cup \beta]$ and the complement of $[\alpha]$ with respect to this infimum and supremum is $[C\alpha]$. Here we should remark that if a graph $x \in [\alpha \cup \beta]$ then it is not necessarily the case that $x \in [\alpha]$ or $x \in [\beta]$. Further it is not true for all $x$ that $x \in [\alpha]$ or $x \in [C\alpha]$. This behavior is due to the fact that we cannot take $[\alpha] \cup [\beta]$ as supremum, since it is not a principal ideal.

Now, we know how to model the type constructors $\cup$, $\cap$ and $C$. The least type $\bot$ is $\emptyset$ (the empty set), the greatest type is the whole set X, the type of function is the range of the embedding function $(\cdot, \cdot)$ and the type of constants is its complement. It remains to construct the exponentiations $\to$ and $\twoheadrightarrow$. Since there is a 1-1 correspondence between graphs and principal ideals we can define the exponentiations $\to$ and $\twoheadrightarrow$ as functions on graphs, rather than as functions on principal ideals. So from now on we understand by types just graphs, and we say that a graph $x$ is of type $\alpha$ if and only if $x \subseteq \alpha$.

**Definition.** Let $\mathcal{X} := < 2^X, \bullet >$ be a graph model and $\alpha, \beta \subseteq X$ two graphs. With $\mathcal{AT}(\mathcal{X})$ we denote the complement of the range of the embedding function $(\cdot, \cdot)$, that is $\mathcal{AT}(\mathcal{X}) := \{x \in X | \forall y, \{x\}y = \emptyset\}$. The functions $\to: 2^X \times 2^X \to 2^X, (x, y) \mapsto x \to y$ and $\twoheadrightarrow: 2^X \times 2^X \to 2^X, (x, y) \mapsto x \twoheadrightarrow y$ are defined as follows:

$$\alpha \to \beta := \{(\sigma, s) | \sigma \subseteq_{\text{fin}} \alpha, s \in \beta\}$$

$$\alpha \twoheadrightarrow \beta := \{(\sigma, s) | \text{if } \sigma \subseteq_{\text{fin}} \alpha \text{ then } s \in \beta\} \cup \mathcal{AT}(\mathcal{X})$$

No confusion need arise about the multiple use of the symbols $\to$, $\twoheadrightarrow$, $\cap$, $\cup$ or $C$ as operations on graphs (sets) and as type constructors, since it will always be clear from the context what is meant.

In the introduction we presented the antimonotonic exponentiation in a different way. But the definition above and the definition in the introduction are the same, that is

$$f \subseteq \alpha \twoheadrightarrow \beta \iff \forall x, (x \subseteq \alpha \Rightarrow fx \subseteq \beta).$$

The monotonic exponentiation given in the introduction is slightly different than that given above. Since the set of atoms of a graph model $\mathcal{AT}(\mathcal{X})$ has no extension, we can enlarge every graph $f$ by $\mathcal{AT}(\mathcal{X})$ without changing the functional behavior, *i.e.* for all graphs $x$ we have $(f \cup \mathcal{AT}(\mathcal{X}))x = fx \cup \mathcal{AT}(\mathcal{X})x = fx \cup \emptyset = fx$. Therefore we obtain for the monotonic exponentiation

$$f \subseteq \alpha \to \beta \cup \mathcal{AT}(\mathcal{X}) \iff \forall x, fx = f(x \cap \alpha) \subseteq \beta.$$

In [3] it is shown that we can model a typed $\lambda$-calculus with the monotonic exponentiation as defined above. Since $\alpha \to \beta \subseteq \alpha \twoheadrightarrow \beta$ we can also model a typed $\lambda$-model with the antimonotonic exponentiation. Therefore these exponentiations are powerful enough to model type systems.

Now we are ready to give an interpretation of the type system $\mathcal{T}$. This interpretation is a mapping of types to graphs.

**Definition.** Let $\mathcal{X} := <2^X, \bullet>$ be graph model. The interpretation $[\![\cdot]\!]^{\mathcal{X}} : \mathcal{T} \to 2^X, \alpha \mapsto [\![\alpha]\!]^{\mathcal{X}}$ is defined recursively as follows:

$$[\![\bot]\!]^{\mathcal{X}} := \emptyset$$
$$[\![C\alpha]\!]^{\mathcal{X}} := C[\![\alpha]\!]^{\mathcal{X}}$$
$$[\![\alpha \cup \beta]\!]^{\mathcal{X}} := [\![\alpha]\!]^{\mathcal{X}} \cup [\![\beta]\!]^{\mathcal{X}}$$
$$[\![\alpha \to \beta]\!]^{\mathcal{X}} := [\![\alpha]\!]^{\mathcal{X}} \to [\![\beta]\!]^{\mathcal{X}}$$

We write $\models \alpha = \beta$ iff for all graph models $\mathcal{X}$, $[\![\alpha]\!]^{\mathcal{X}} = [\![\beta]\!]^{\mathcal{X}}$.

We have to verify that the defined type constructors $\twoheadrightarrow$, $\cap$, $\top$, $*$ and $+$ are interpreted as intended.

**Lemma 1.** *Let* $\mathcal{X} := <2^X, \bullet>$ *be graph model and* $\alpha, \beta \in \mathcal{T}$ *two types. We have:*

$$[\![\top]\!]^{\mathcal{X}} := X$$
$$[\![*]\!]^{\mathcal{X}} := \mathcal{AT}(\mathcal{X})$$
$$[\![+]\!]^{\mathcal{X}} := \mathcal{CAT}(\mathcal{X})$$
$$[\![\alpha \cap \beta]\!]^{\mathcal{X}} := [\![\alpha]\!]^{\mathcal{X}} \cap [\![\beta]\!]^{\mathcal{X}}$$
$$[\![\alpha \twoheadrightarrow \beta]\!]^{\mathcal{X}} := [\![\alpha]\!]^{\mathcal{X}} \twoheadrightarrow [\![\beta]\!]^{\mathcal{X}}$$

## 3 The Axiom system

We present a system of axioms for $\mathcal{T}$.

**Definition.** The *system of axioms for* $\mathcal{T}$ is

(A1) $\quad\quad\quad\quad\quad \alpha \cup \beta = \beta \cup \alpha$
(A2) $\quad\quad\quad\quad\quad (\alpha \cup \beta) \cup \gamma = \alpha \cup (\beta \cup \gamma)$
(A3) $\quad\quad\quad\quad\quad \alpha \cup \alpha = \alpha$
(A4) $\quad\quad\quad\quad\quad \alpha \cup (\beta \cap \alpha) = \alpha$
(A5) $\quad\quad\quad\quad\quad \alpha \cup (\beta \cap \gamma) = (\alpha \cup \beta) \cap (\alpha \cup \gamma)$
(A6) $\quad\quad\quad\quad\quad \alpha \cup \top = \top$
(A7) $\quad\quad\quad\quad\quad \alpha \cup \bot = \alpha$
(A8) $\quad\quad\quad\quad\quad \alpha \cup C\alpha = \top$
(A9) $\quad\quad\quad\quad\quad CC\alpha = \alpha$
(A10) $\quad\quad\quad\quad\quad \alpha \to \bot = \bot$
(A11) $\quad\quad\quad\quad\quad \alpha \to (\beta \cup \gamma) = \alpha \to \beta \cup \alpha \to \gamma$
(A12) $\quad\quad\quad\quad\quad (\alpha \cup \beta) \to \gamma = (\alpha \cup \beta) \to \gamma \cup \alpha \to \gamma$
(A13) $\quad\quad\quad\quad\quad \alpha \to \beta \cap \gamma \to \delta = (\alpha \cap \gamma) \to (\beta \cap \delta)$

where $\alpha$, $\beta$, $\gamma$ and $\delta$ are place holders for types of $\mathcal{T}$.

Let us be clear, the type constructor $\cap$ is defined as a combination of $\cup$ and $C$. For example axiom scheme A4 is actually $\alpha \cup C(C\beta \cup C\alpha) = \alpha$.
Now we provide rules to infer new equations from the axioms.

**Definition.** If $\alpha = \beta$ is an axiom then $\vdash \alpha = \beta$. Further we have the following rules:

$$(S) \frac{\vdash \alpha = \beta}{\vdash \beta = \alpha} \qquad (T) \frac{\vdash \alpha = \beta, \vdash \beta = \gamma}{\vdash \alpha = \gamma} \qquad (R) \frac{}{\vdash \alpha = \alpha}$$

$$(I\cup) \frac{\vdash \alpha = \beta, \vdash \gamma = \delta}{\vdash \alpha \cup \gamma = \beta \cup \delta} \qquad (I\to) \frac{\vdash \alpha = \beta, \vdash \gamma = \delta}{\vdash \alpha \to \gamma = \beta \to \delta} \qquad (IC) \frac{\vdash \alpha = \beta}{\vdash C\alpha = C\beta}$$

On the one hand, we have $\vdash \alpha = \beta$ if the two types can be proved equal, *i.e.* there is a proof for $\alpha = \beta$ using the axiom system and the rules. On the other hand, we have $\models \alpha = \beta$ if $\alpha, \beta$ are semantically equal, *i.e.* they have the same interpretation in all graph models. The goal of this section is to show that these two notions of equality are equivalent, or in other words, the axiom system is sound and complete.

**Theorem.** *If $\alpha, \beta \in T$ are types then*

$$\vdash \alpha = \beta \iff \models \alpha = \beta.$$

In order to prove this theorem we need some lemmas. That the axiom system and the rules are sound can be easily verified. The hard part is to show completeness.

We will prove the theorem by induction on the level of nested exponentiations.

**Definition.** The function $\ell : T \to \mathbb{N}, \alpha \mapsto \ell(\alpha)$ is defined recursively as follows:

$$\ell(\bot) := 0$$
$$\ell(C\alpha) := \ell(\alpha)$$
$$\ell(\alpha \cup \beta) := \max\{\ell(\alpha), \ell(\beta)\}$$
$$\ell(\alpha \to \beta) := \max\{\ell(\alpha), \ell(\beta)\} + 1$$

In the next lemma we state that we can hide the type constructor $C$ in the defined type constructors $\to$, $\cap$ and $\top$.

**Lemma 2.** *Let $\alpha \in T$ be a type. There exists a type $\alpha' \in T'$ with $\vdash \alpha' = \alpha$ and $\ell(\alpha') = \ell(\alpha)$.*

*Proof.* Basically we have to show that for a type $\alpha \in T'$ there exists a type $\alpha' \in T'$ with the property $\vdash \alpha' = C\alpha$ and $\ell(\alpha') = \ell(\alpha)$. This we show by induction on type $\alpha$.
If $\alpha = \bot$ or $\alpha = \top$ then $\alpha' = \top$ respectively $\alpha' = \bot$ and $\vdash \alpha' = C\alpha, \ell(\alpha') = \ell(\alpha)$.
If $\alpha = \beta \cup \gamma$ or $\alpha = \beta \cap \gamma$ then by induction hypothesis there exists types $\beta', \gamma' \in T'$ with $\vdash \beta' = C\beta, \vdash \gamma' = C\gamma$ and $\ell(\beta') = \ell(\beta), \ell\gamma' = \ell(\gamma)$. Now we have $\alpha' = \beta' \cap \gamma'$ respectively $\alpha' = \beta' \cup \gamma', \vdash \alpha' = C\alpha$ and $\ell(\alpha') = \max\{\ell(\beta'), \ell(\gamma')\} =$

$\max\{\ell(\beta), \ell(\gamma)\} = \ell(\alpha)$.
If $\alpha = \beta \to \gamma$ or $\alpha = \beta \dashrightarrow \gamma$ then by induction hypothesis there exists a type $\gamma' \in T'$ with $\vdash \gamma' = C\gamma$ and $\ell(\gamma') = \ell(\gamma)$. Now we have $\alpha' = \beta \dashrightarrow \gamma'$ respectively $\alpha = \beta \to \gamma'$, $\vdash \alpha' = C\alpha$ and $\ell(\alpha') = \max\{\ell(\beta), \ell(\gamma')\}+1 = \max\{\ell(\beta), \ell(\gamma)\}+1 = \ell(\alpha)$. □

In the next lemma we state some equalities which we will use later.

**Lemma 3.** *If $\alpha, \beta, \gamma, \delta \in T$ are types then:*

$$\vdash \alpha \cap \beta = \beta \cap \alpha, \ \vdash (\alpha \cap \beta) \cap \gamma = \alpha \cap (\beta \cap \gamma), \ \vdash \alpha \cap \alpha = \alpha$$
$$\vdash \alpha \cap (\beta \cup \alpha) = \alpha, \ \vdash \alpha \cap (\beta \cup \gamma) = (\alpha \cap \beta) \cup (\alpha \cap \gamma)$$
$$\vdash \alpha \cap \bot = \bot, \ \vdash \alpha \cap C\alpha = \bot$$
$$\vdash \alpha \dashrightarrow \beta \cup \gamma \dashrightarrow \delta = (\alpha \cap \gamma) \dashrightarrow (\beta \cup \delta)$$

By the above lemma, we see that our type system is a Boolean algebra. So we have disjunctive normal forms for our types. Furthermore we can combine in a conjunction all terms of the form $\alpha \dashrightarrow \beta$ to one term. We introduce a normal form for our types.

**Definition.** A type $\alpha \in T$ is in *normal form* iff

$$\alpha = \alpha_0 \cap \cdots \cap \alpha_n$$

where

$$\alpha_i = \beta_{i0} \dashrightarrow \gamma_{i0} \cup \beta_{i1} \to \gamma_{i1} \cup \cdots \cup \beta_{in_i} \to \gamma_{in_i}, \qquad n_i \geq 0$$

and $\beta_{ij}, \gamma_{ij}$ $0 \leq j \leq n_i$ are in normal form, or

$$\alpha_i = \beta_{i1} \to \gamma_{i1} \cup \cdots \cup \beta_{in_i} \to \gamma_{in_i}, \qquad n_i > 0$$

and $\beta_{ij}, \gamma_{ij}$ $0 < j \leq n_i$ are in normal form, or $n = 0$ and $\alpha = \alpha_0 = \bot$ or $\alpha = \alpha_0 = \top$.

In types in normal form, the type constructor $C$ is hidden in the type constructors $\dashrightarrow, \cap, \top$ and occurs nowhere else.

**Lemma 4.** *Let $\alpha \in T$ be a type. There exists a type $\alpha' \in T$ in normal form with $\vdash \alpha = \alpha'$ and $\ell(\alpha') = \ell(\alpha)$.*

*Proof.* Let $\alpha \in T$ be a type. By lemma 2 there exists a type $\beta \in T'$ with $\vdash \beta = \alpha$ and $\ell(\beta) = \ell(\alpha)$. Now all occurrences of $C$ are hidden in the type constructors $\dashrightarrow, \cap$ and $\top$. Then we take the disjunctive normal form and combine all multiple occurrences of $\dashrightarrow$ in the conjunctions and obtain $\gamma$. Since the disjunctive normal form introduces no new exponentiations we have $\ell(\gamma) = \ell(\beta) = \ell(\alpha)$. Finally, we repeat this process for every subterm of $\gamma$ and eventually obtain $\alpha'$ in normal form with $\vdash \alpha = \alpha'$ and $\ell(\alpha') = \ell(\alpha)$. □

*Proof. (of Theorem)* Let $\mathcal{X} := <2^X, \bullet>$ be a graph model with $\mathcal{AT}(\mathcal{X}) \neq \emptyset$. It suffices to show for a type $\alpha \in \mathcal{T}$ in normal form that

$$[\![\alpha]\!]^{\mathcal{X}} = X \Rightarrow \vdash \alpha = \top$$

Since, if $\models \alpha = \beta$ then $[\![\alpha \cup C\beta]\!]^{\mathcal{X}} = [\![\alpha]\!]^{\mathcal{X}} \cup C[\![\beta]\!]^{\mathcal{X}} = X$ and $[\![C\alpha \cup \beta]\!]^{\mathcal{X}} = C[\![\alpha]\!]^{\mathcal{X}} \cup [\![\beta]\!]^{\mathcal{X}} = X$. Now, there exists types $\gamma$ and $\delta$ in normal form such that $\vdash \gamma = C\alpha \cup \beta$ and $\vdash \delta = \alpha \cup C\beta$. Then it follows that $\vdash \alpha \cup C\beta = \gamma = \top$ and $\vdash C\alpha \cup \beta = \delta = \top$. Hence $\vdash \alpha = \alpha \cap \top = \alpha \cap (C\alpha \cup \beta) = \alpha \cap \beta = \beta \cap (\alpha \cup C\beta) = \beta$. So we show that for a type $\alpha \in \mathcal{T}$ in normal form $[\![\alpha]\!]^{\mathcal{X}} = X \Rightarrow \vdash \alpha = \top$ by induction on $\ell(\alpha)$.

If $\ell(\alpha) = 0$ then $\alpha = \bot$ or $\alpha = \top$ since $\alpha$ is in normal form. But $\alpha = \bot$ is not possible and if $\alpha = \top$ then surely $\vdash \alpha = \top$.

If $\ell(\alpha) > 0$ then $\alpha$ has the form $\alpha = \alpha_0 \cap \cdots \cap \alpha_m$ and for every $0 \leq i \leq m$, $[\![\alpha_i]\!]^{\mathcal{X}} = X$. There are two cases:

The case $\alpha_i = \beta_1 \to \gamma_1 \cup \cdots \cup \beta_n \to \gamma_n$ is not possible since $\emptyset \neq \mathcal{AT}(\mathcal{X}) \not\subseteq [\![\beta_i \to \gamma_i]\!]^{\mathcal{X}}$. Hence $\mathcal{AT}(\mathcal{X}) \not\subseteq [\![\alpha_i]\!]^{\mathcal{X}}$ and $[\![\alpha_i]\!]^{\mathcal{X}} \neq X$. The case $\alpha_i = \beta_0 \dashrightarrow \gamma_0 \cup \beta_1 \to \gamma_1 \cup \cdots \cup \beta_n \to \gamma_n$ remains.

If $[\![\gamma_0]\!]^{\mathcal{X}} = X$ then by induction hypothesis $\vdash \gamma_0 = \top$ since $\ell(\gamma_0) < \ell(\alpha)$ and we can derive $\vdash \beta_0 \dashrightarrow \gamma_0 = \beta_0 \dashrightarrow \top = C(\beta_0 \to \bot) = C\bot = \top$.

So we can assume that $[\![\gamma]\!]^{\mathcal{X}} \neq X$ and therefore $n \neq 0$. Since $\beta_0 \dashrightarrow \gamma_0 = C(\beta_0 \to C\gamma_0)$ we have that $[\![\beta_0 \to C\gamma_0]\!]^{\mathcal{X}} \subseteq [\![\beta_1 \to \gamma_1 \cup \cdots \cup \beta_n \to \gamma_n]\!]^{\mathcal{X}}$. Further, we can assume that the types $\beta_i \to \gamma_i$, $0 < i \leq n$ are ordered such that for a $s \leq n$, $[\![\beta_0]\!]^{\mathcal{X}} \subseteq [\![\beta_i]\!]^{\mathcal{X}}$ if $0 < i \leq s$ and $[\![\beta_0]\!]^{\mathcal{X}} \not\subseteq [\![\beta_i]\!]^{\mathcal{X}}$ if $s < i \leq n$. Hence for each $s < i \leq n$ there exists a $t_i \in [\![\beta_0]\!]^{\mathcal{X}} \setminus [\![\beta_i]\!]^{\mathcal{X}}$. Let $\tau := \{t_{s+1}, \ldots, t_n\}$. Now if $x \in [\![C\gamma_0]\!]^{\mathcal{X}}$ then $(\tau, x) \in [\![\beta_0 \to C\gamma_0]\!]^{\mathcal{X}}$. Since $\tau \not\subseteq [\![\beta_i]\!]^{\mathcal{X}}$ for all $s < i \leq n$, there exists a $0 < i_0 \leq s$ with $(\tau, x) \in [\![\beta_{i_0} \to \gamma_{i_0}]\!]^{\mathcal{X}}$ and therefore $x \in [\![\gamma_{i_0}]\!]^{\mathcal{X}}$. It follows that $[\![C\gamma_0]\!]^{\mathcal{X}} \subseteq [\![\gamma_1 \cup \cdots \cup \gamma_s]\!]^{\mathcal{X}}$ and $[\![\gamma_0 \cup \cdots \cup \gamma_s]\!]^{\mathcal{X}} = X$. Further, we have $\ell(\gamma_0 \cup \cdots \cup \gamma_s) = \max\{\ell(\gamma_0), \ldots, \ell(\gamma_s)\} < \ell(\alpha)$. There exists a type $\delta$ in normal form with $\vdash \delta = \gamma_0 \cup \cdots \cup \gamma_s$ and $\ell(\delta) = \ell(\gamma_0 \cup \cdots \cup \gamma_s) < \ell(\alpha)$. By induction hypothesis we obtain $\vdash \gamma_0 \cup \cdots \cup \gamma_s = \delta = \top$.

Further, since $[\![\beta_0]\!]^{\mathcal{X}} \subseteq [\![\beta_i]\!]^{\mathcal{X}}$ for all $0 < i \leq s$ we have $[\![C\beta_0 \cup \beta_i]\!]^{\mathcal{X}} = X$. There exists a type $\delta_i \in \mathcal{T}$ in normal form with $\vdash \delta_i = C\beta_0 \cup \beta_i$ and $\ell(\delta_i) = \max\{\ell(C\beta_0), \ell(\beta_i)\} = \max\{\ell(\beta_0), \ell(\beta_i)\} < \ell(\alpha)$. Hence we can use the induction hypothesis and obtain $\vdash C\beta_0 \cup \beta_i = \delta_i = \top$.

From $\vdash \gamma_0 \cup \cdots \cup \gamma_s = \top$ and $\vdash C\beta_0 \cup \beta_i = \top$ it follows that $\vdash C\gamma_0 \cup \gamma_1 \cup \cdots \cup \gamma_s = \gamma_1 \cup \cdots \cup \gamma_s$ and $\vdash \beta_0 \cup \beta_i = \beta_i$. Hence we obtain $\vdash \beta_i \to \gamma_i = (\beta_0 \cup \beta_i) \to \gamma_i = \beta_i \to \gamma_i \cup \beta_0 \to \gamma_i$ for all $0 < i \leq s$. It follows that $\vdash \beta_1 \to \gamma_1 \cup \cdots \cup \beta_s \to \gamma_s = \beta_1 \to \gamma_1 \cup \cdots \cup \beta_s \to \gamma_s \cup \beta_0 \to (\gamma_1 \cup \cdots \cup \gamma_s)$. But $\vdash \beta_0 \to (\gamma_1 \cup \cdots \cup \gamma_s) = \beta_0 \to (C\gamma_0 \cup \gamma_1 \cup \cdots \cup \gamma_s) = \beta_0 \to (\gamma_1 \cup \cdots \cup \gamma_s) \cup \beta_0 \to C\gamma_0$ and $\vdash \beta_0 \to C\gamma_0 = C(\beta_0 \dashrightarrow \gamma_0)$. We obtain $\vdash \beta_0 \dashrightarrow \gamma_0 \cup \beta_1 \to \gamma_1 \cup \cdots \cup \beta_n \to \gamma_n = \beta_0 \dashrightarrow \gamma_0 \cup C(\beta_0 \dashrightarrow \gamma_0) \cup \cdots = \top \cup \cdots = \top$. □

Since the proof of the theorem is constructive, we have the following corollary.

**Corollary 5.** *Let $\alpha, \beta \in \mathcal{T}$ be types. It is decidable whether or not $\vdash \alpha = \beta$ is provable.*

*Proof.* According to the proof of the theorem it suffices to have a decision procedure for equalities of the form $\vdash \alpha = \top$, where $\alpha$ is in normal form. Let *normal* be a procedure which returns for every type $\alpha$ a normal form $\alpha'$ with $\ell(\alpha) = \ell(\alpha')$. Now the procedure *topequal* is defined as follows:

```
procedure topequal(α)
begin
    if α = ⊥ then return false fi
    if α = ⊤ then return true fi
    if α = β ∩ γ then return topequal(β) & topequal(γ) fi
    if α = β₁ → γ₁ ∪ ··· ∪ βₙ → γₙ then return false fi
    if α = β₀⇢γ₀ ∪ β₁ → γ₁ ∪ ··· ∪ βₙ → γₙ then
        δ := γ₀
        for i = 1 to n
            if topequal(normal(Cβ₀ ∪ βᵢ)) then δ := δ ∪ γᵢ fi
        rof
        return topequal(δ)
    fi
end
```

We have that $topequal(\alpha)$ returns true if $\vdash \alpha = \top$ and false otherwise. Therefore we obtain that $topequal(normal(\alpha \cup C\beta))$ & $topequal(normal(C\alpha \cup \beta))$ is true if $\vdash \alpha = \beta$ and false otherwise. □

## 4 Conclusion

We have investigated the relationship between type constructors in a concrete model. Mainly we were interested in the relationship between the monotonic and the antimonotonic exponentiation. We have shown that these two exponentiations are dual notions, like union and intersection in set theory. We have developed a system of axioms and rules which are sound and complete. Furthermore, it is decidable whether an equality between types $\alpha = \beta$ is provable.

## References

1. Engeler, E.: Algebras and combinators. Algebra Universalis 13 (1981) 389–392
2. Otth, D.: Konsistente Operatoren. ETHZ Report(1990)
3. Otth, D.: Types and Consistency in Combinatory Algebras. Dissertation 9800, ETHZ (1992).
4. Plotkin, G.: A powerdomain construction. SIAM J. Comput. 5 (1976) 452–487
5. Schellinx, H.: Isomorphisms and nonisomorphisms of graph models. J. of Symbolic Logic, 56 (1991) 227–249
6. Scott, D.: Data types as lattices. SIAM J. Comput. 5 (1976) 522–587

# Inductive Definitions in the system Coq
# Rules and Properties

Christine Paulin-Mohring *

LIP-IMAG, URA CNRS 1398
Ecole Normale Supérieure de Lyon
46 Allée d'Italie, 69364 Lyon cedex 07, France
e-mail : cpaulin@lip.ens-lyon.fr

**Abstract.** In the pure Calculus of Constructions, it is possible to represent data structures and predicates using higher-order quantification. However, this representation is not satisfactory, from the point of view of both the efficiency of the underlying programs and the power of the logical system. For these reasons, the calculus was extended with a primitive notion of inductive definitions [8]. This paper describes the rules for inductive definitions in the system Coq. They are general enough to be seen as one formulation of adding inductive definitions to a typed lambda-calculus. We prove strong normalization for a subsystem of Coq corresponding to the pure Calculus of Constructions plus Inductive Definitions with only weak eliminations.

## 1 Introduction

### 1.1 Motivations

Several proof environments suitable for mechanizing mathematics and program development [10, 5] are based on the "Curry-Howard correspondence" between natural deduction proofs and typed functional programs. Such proof tools are used interactively and consequently aims at providing rules as natural as possible and avoid tedious encoding. Such a motivation was the starting point for an extension of the pure Calculus of Constructions with primitive inductive definitions.

The (Pure) Calculus of Constructions extends the powerful polymorphic programming language $F_\omega$ with dependent types and allows reasoning about programs. Its main advantage is to be a "closed system" where mathematical and computational notions can be internally represented using higher-order quantification. But this representation is not satisfactory from both the computational and the logical points of view. This leads to a proposition for an extension of the Calculus of Constructions with inductive definitions as first-class objects [8]. The rules for these definitions follow the point of view of Martin-Löf's type theory. From a specification of the introduction rules for a new inductive definition, we generate a dependent elimination with new computational rules. This elimination for natural numbers corresponds both to the construction of functions following a primitive recursive scheme

---

* This research was partly supported by ESPRIT Basic Research Action "Types for Proofs and Programs" and by Programme de Recherche Coordonnées "Programmation avancée et Outils pour l'Intelligence Artificielle".

and to proofs by induction. Our extension of the Calculus of Constructions with Inductive Definitions, unlike Martin-Löf's type theory, still preserve the property for the system to be closed.

There exists an implementation of this extension, namely the system Coq [10] developed in the Formel project at Inria-Rocquencourt and ENS Lyon. The mechanism for inductive definitions has proved to be really useful for the development of examples but its meta-theory is not yet established. The purpose of this paper is to give a precise description of the rules used in the system and state what we know about its properties. In particular we shall prove the strong normalization for a subsystem of Coq, corresponding to the pure Calculus of Constructions plus Inductive Definitions with only weak eliminations. For this, we use a similar result obtained by Ph. Audebaud [1, 2] for an extension of the Calculus of Constructions with a fixpoint operator.

The paper is organized as follow. The remaining part of this section describes the drawbacks of the impredicative coding of inductive definitions in the pure Calculus of Constructions and gives an intuitive idea, using the example of natural numbers, of the rules for elimination we shall introduce. Part 2 introduces schematic rules for adding inductive definitions to a typed lambda-calculus. Part 3 gives the rules used in the system Coq and examples of inductive definitions. In part 4 the strong normalization of a subsystem $F_\omega^{\text{Ind}}$ of Coq is established. In part 5 we shall discuss our choices and make comparisons with other systems.

## 1.2 Impredicative inductive definitions

In [4, 21, 22] a systematic way to generate a representation of an inductive definition from a description of the generative rules for it was described. But this representation is not really adequate. We list some problems :

- We can represent a type of boolean with two elements **true** and **false** but the fact that **true** and **false** are not equal is not provable in the system.
- We can represent a type of natural numbers but the induction principle is not provable.
- We can define all primitive recursive functions on natural numbers. That is given two terms $x$ and $g$, we can find $h$ such that $(h\ 0) = x$ and $(h\ n+1) = (g\ n\ (h\ n))$ holds internally for a closed natural number $n$ (and consequently are provable). But if $n$ is not closed then the proposition $(h\ n+1) = (g\ n\ (h\ n))$ (using Leibniz's equality) is provable using an induction over $n$ but is not an internal reduction rule. Moreover for $n$ closed, the reduction of $(h\ n + 1)$ into $(g\ n\ (h\ n))$ can take a time proportional to $n$. This corresponds to the well-known problem of the predecessor function (the predecessor of $n + 1$ is computed in $n$ steps) and is a problem inherent to the representation of natural numbers with what is known as Church's numerals.
- Some structures represented with an impredicative coding (like the products of two types) can contain more elements (closed normal terms) than the one built from the constructors of the type. It is explained for instance in [21].

Some points concerning the representation of inductive definitions in an impredicative type theory are reflected in our rules for primitive inductive definitions. For instance an inductive definition makes sense in any context (and not just at toplevel); the names of the type and of its constructors are not relevant for the type conversion rules (so two types with the same specification will be equal). We use a uniform rule for the definitions of disjunction, conjunction, types and relations.

### 1.3 The case of natural numbers

The problem with the inductive definitions in impredicative systems comes from the weakness of the elimination scheme. The elimination scheme is intended to make use of the assumption that the inductive type is the smallest set generated by the introduction rules.

The specification of the natural numbers is a set *nat* with two constructors for zero and the successor function : 0 of type *nat* and $S$ of type *nat* $\to$ *nat*. Each "mathematical" natural number $n$ can be represented as a term $(S^n\ 0)$. The elimination scheme internalizes the fact that the type of natural numbers only contains elements representing $(S^n\ 0)$ for some $n$.

The first problem is to be able to represent enough functions on natural numbers. For this, we need a special kind of recursive definitions. Theoretically, the representation of functions $H$ on *nat* iteratively defined by equations $H(0) = x$, $H(S\ n) = f(H(n))$ combined with a notion of product is enough to get many interesting functions (in $F_\omega$ for instance, all recursive functions provably total in higher-order arithmetic).

But this representation is not always efficient. For instance, there is no representation in $F_\omega$ of the predecessor function which computes the predecessor of $(S\ n)$ in a constant number of reductions.

The predecessor function can be efficiently obtained in a ML-like language where the basic operation for the elimination of a concrete type is the definition by pattern. In this framework, we can represent a function $H$ which satisfies the equation $H(0) = x$, $H(S\ n) = f(n)$. In ML, a general fixpoint is available to encode recursive calls.

To keep the property of strong normalization, we do not allow a general fixpoint and restrict the use of recursion. We want a direct representation of functions $H$ which satisfies the properties $H(0) = x$, $H(S\ n) = f(n, H(n))$. This can be done by the introduction of a recursor operator $R(C, x, f, n)$ such that if $C$ is a type, $x$ of type $C$, $f$ of type *nat* $\to C \to C$ and $n$ of type *nat* then $R(C, x, f, n)$ has type $C$. Furthermore $R(C, x, f, 0)$ behaves like $x$ and $R(C, x, f, (S\ n))$ behaves like $(f\ n\ (R(C, x, f, n))$. If $H(p)$ denotes $R(C, x, f, n)$ the equality $H(S\ n) = (f\ n\ H(n))$ will hold not only for closed terms $n$ but also for a variable.

For each inductive definition we shall introduce the analogous of the $R$ operator for natural numbers. The term $R(C, x, f, n)$ will be written as $\mathrm{Elim}(n, C)\{x|f\}$ in this paper, it corresponds to the notation (<C>Match n with x f) in Coq.

*Dependent elimination.* A system like Coq is intended to program functions but also to reason about programs. It is natural to want an induction principle :

$$\forall n : nat. \forall P.(P\ 0) \to (\forall u : nat.(P\ u) \to (P\ (S\ u))) \to (P\ n)$$

This proposition can be used to prove logical properties of programs. Now, in the paradigm of proofs as programs, we interpret (intuitionistic) proofs of a property $P$

as correct programs w.r.t. the specification represented by $P$. If $n$ is a natural number, if $x$ is a correct program for the specification $(P\ 0)$ and $f$ a correct program for the specification $\forall u : nat.(P\ u) \to (P\ (S\ u))$ then we can built using the induction principle a correct program w.r.t. $(P\ n)$. It is easy to see that this program should behave like the program $R(C, x, f, n)$ built from the recursor $R$. Obviously the recursor can be obtained from the induction principle just taking a constant predicate $(P\ n) \equiv C$. This suggests that we need only one operator, with the induction principle as type and which obeys the same reduction rules as the recursor operator. This is the principle behind the rules in Martin-Löf's Intuitionistic Type Theory [18]. We shall also follow the same ideas for the system Coq.

*Strong eliminations* When we have a concrete inductive structure like the natural numbers, we may want to define a property (or a type) by induction over the structure of the natural number. This possibility is called "strong elimination" because we are building a property (or equivalently a type) by computation over a program. For instance, we could define a property $P$ on natural numbers such that $(P\ 0)$ is a true proposition (for instance $(A : \text{Set})A \to A$) and $(P\ (S\ n))$ is the absurdity proposition $(A : \text{Set})A$. Then assuming $0 = (S\ n)$, because $(P\ 0)$ is true, we have a proof of $(P\ (S\ n))$ which is the absurdity. This extension corresponds to a strong modification of the underlying typed lambda-calculus.

## 2 Rules for inductive definitions

In this section we introduce schematic rules (parameterized by sorts) for inductive definitions and illustrate them on the example of the inductive definition of lists of a given length.

### 2.1 Notations

We are interested in an extension of the pure Calculus of Constructions, but actually our proposition can be defined somehow independently of the underlying Generalized Type System (GTS). The systems we are studying in that paper extends a pure type system with new constructors for inductive definitions.

**Generalized type systems** We use the now standard presentation of typed $\lambda$-calculi as functional GTS [3, 12]. The set of (pseudo) terms contains the following constructions :
$$t ::= s \mid x \mid (x{:}t)t \mid [x{:}t]t \mid (t\ t)$$
with $s$ ranging over the set $\mathcal{S}$ of sorts. For the pure Calculus of Constructions, we shall use $\mathcal{S} = \{\text{Set}, \text{Type}_\mathcal{S}\}$ corresponding respectively to the star and the square in Barendregt's $\lambda$-cube. The set of axioms is $\mathcal{A} = \{\text{Set} : \text{Type}_\mathcal{S}\}$. All rules for product are allowed : the set of rules is $\mathcal{R} = \mathcal{S} \times \mathcal{S}$.

We write $M[x \leftarrow N]$ to denote the substitution of the term $N$ to free occurrences of the variable $x$ in $M$. The notation $M \to N$ is an abbreviation for $(x : M)N$ whenever $x$ does not occur in $N$. The arrow symbol associates to the right and the application associates to the left.

We say that a term is a *type* if it is correctly typed and that its type is a sort.

**Vectorial notations** We introduce vectorial notations to write parametric rules.

*Construction.* Let $x$ and $A$ be two possibly empty sequences of resp. variables and terms with the same length. Let $M$ be a term, we define the terms $(x : A)M$, $[x : A]M$ and $(M\ A)$ for an arbitrary $M$ by induction over the structure of the sequences $x$ and $A$.

- If $x$ and consequently $A$ are empty then $(x : A)M = [x : A]M = (M\ A) = M$.
- if $x = x, x'$ and $A = A, A'$ then $(x : A)M = (x:A)(x' : A')M$, $[x : A]M = [x : A][x' : A']M$ and $(M\ A) = ((M\ A)\ A')$.

*Decomposition.* We use the same notations to describe the decomposition of a term. Let $P$ be a term, we shall write $P \equiv [x : A]M$ (resp. $P \equiv (x : A)M$, resp. $P \equiv (M\ A)$) to define $x$ and $A$ as maximal sequences of terms such that the equality $P = [x : A]M$ (resp. $P = (x : A)M$, resp. $P = (M\ A)$) holds.

For instance, a type in normal form can uniquely be written as $(x : A)s$ or $(x:A)(X\ a)$ with $X$ a variable and $s$ a sort.

## 2.2 Preliminary Definitions

**Definition 1 Arity.** An *arity of sort s* is a term generated by the following syntax : $Ar ::= s \mid (x:M)Ar$. We denote by $arity(A, s)$ the property $A$ is an arity of sort $s$.

**Definition 2 Strictly positive types.** Let $A$ be an arity and $X$ be of type $A$, the terms $P$ which are *strictly positive* w.r.t. to $X$ are generated by the syntax : $Pos ::= X \mid (Pos\ m) \mid (x:M)Pos$ with the restriction that $X$ does not occur in $M$ and $m$. We write $strict\_positive(P, X)$ the property $P$ is strictly positive w.r.t. $X$.

A strictly positive well-formed type can be written as as $(x : M)(X\ m)$ with the restriction that $X$ does not occur in any term of $M$ or $m$.

**Definition 3 Forms of constructor.** Let $A$ be an arity and $X$ be of type $A$, the terms $C$ which are a *form of constructor* w.r.t. $X$ are generated by the syntax : $Co ::= X \mid (Co\ m) \mid P \rightarrow Co \mid (x:M)Co$ with the restriction that $strict\_positive(P, X)$ and $X$ does not occur in $M$ or $m$. We say that $C$ is a *type of constructor of $X$* if furthermore, $C$ is a type.
We write $constructor(C, X)$ the property $C$ is a form of constructor w.r.t. $X$.

A well-formed form of constructor can be written as $C \equiv (z : C)(X\ a)$. A strictly positive type is a particular case of type of constructors, which we call a *non-recursive type of constructor*. A type of constructor which is not a strictly positive type is called *recursive*.

## 2.3 Inductive Operators

We introduce a new pseudo-term for inductive definitions $t ::= \mathsf{Ind}(x:t)\{t\}$ with $t$ a possibly empty list of terms, written as $t_1 \mid \ldots \mid t_n$.

In the expression $\mathsf{Ind}(x:t)\{t\}$, $x$ is a bound variable. The typing rule for this operator is, with $n$ an arbitrary number, possibly equal to 0 :

$$\frac{arity(A,s) \quad \Gamma, X:A \vdash C_i:s \quad constructor(C_i, X) \quad (\forall i = 1\ldots n)}{\Gamma \vdash \mathsf{Ind}(X:A)\{C_1|\ldots|C_n\}:A} \quad (\mathsf{Ind}_s)$$

*Example* We take as an example of inductive definition, the type of lists with a given length. It is a type scheme which associate to each natural numbers $n$ the type *(listn n)* of lists of length $n$. The arity of this definition is $nat \to \mathsf{Set}$. There is two constructors (one for *nil* and the other one for *cons*). We want *nil* and *cons* to have type respectively *(listn 0)* and $(n:nat)A \to (listn\ n) \to (listn\ (S\ n))$. We can define *listn* to be the term:

$$listn \equiv \mathsf{Ind}(X:nat \to \mathsf{Set})\{(X\ 0)|(n:nat)A \to (X\ n) \to (X\ (S\ n))\} : nat \to \mathsf{Set}$$

## 2.4 Constructors

New pseudo-terms correspond to the constructors (which can also be seen as introduction rules) of the inductive definition : $t := \mathsf{Constr}(i,t)$. with $i$ a positive integer. Let $I$ be $\mathsf{Ind}(X:A)\{C_1|\ldots|C_n\}$, the typing rule for this term is :

$$\frac{\Gamma \vdash \mathsf{Ind}(X:A)\{C_1|\ldots|C_n\}:A \quad 1 \le i \le n}{\Gamma \vdash \mathsf{Constr}(i,I):C_i[X \leftarrow I]} \quad (\mathsf{Constr})$$

*Example.* For our example of lists we can define :

$$\begin{aligned}nil &\equiv \mathsf{Constr}(1, listn) : (listn\ 0) \\ cons &\equiv \mathsf{Constr}(2, listn) : (n:nat)A \to (listn\ n) \to (listn\ (S\ n))\end{aligned}$$

## 2.5 Eliminations

The rule for the elimination is a natural (although complicated) extension of the case of natural numbers to a general inductive definition. We shall introduce two rules corresponding to dependent and non dependent elimination.

New pseudo-terms will correspond to the eliminations of the inductive definitions : $t := \mathsf{Elim}(t,t)\{t\}$ with $t$ a possibly empty list of terms, written as $t_1|\ldots|t_n$. We need to define operations on forms of constructor that are used in the types of the arguments of the elimination.

**Non dependent elimination** Let $A$ be an arity of sort $s$ ($A \equiv (x:A)s$), let $X$ be a variable of type $A$, $s'$ be a sort, $Q$ be a variable of type $(x:A)s'$ and $C$ be a type of constructor of $X$.

**Definition 4.** We define a new term $C\{X,Q\}$ by induction over the structure of the type of constructor $C$.

$$\begin{aligned}(P \to C)\{X,Q\} &= P \to P[X \leftarrow Q] \to C\{X,Q\} \text{ if } strict\_positive(P,X) \\ ((x:M)C)\{X,Q\} &= (x:M)C\{X,Q\} \qquad \text{if } X \text{ does not occur in } M \\ (X\ a)\{X,Q\} &= (Q\ a)\end{aligned}$$

*Example.* For our example, we have :
$$(X\ 0)\{X,Q\} = (Q\ 0)$$
$$(n:nat)A \to (X\ n) \to (X\ (S\ n))\{X,Q\} = (n:nat)A \to (X\ n) \to (Q\ n) \to (Q\ (S\ n))$$

The term $C\{X,Q\}$ is not always well-formed. It depends on the product rules allowed in the GTS. But in the case where $s = s'$, $C\{X,Q\}$ is a well-formed type. We can state more precisely that it is well-formed if "enough" products are allowed for $s'$.

**Lemma 5.** *If $\Gamma, X:A \vdash C:s$ with $A \equiv (x:A)s$ and $C$ is a type of constructor of $X$. Assume $s'$ is a sort which satisfies the following hypotheses :*
1. *$(x:A)s'$ is a well-formed type,*
2. *for all sorts $s''$, $(s'',s) \in \mathcal{R}$ implies $(s'',s') \in \mathcal{R}$,*
3. *if $C$ is a recursive constructor then $(s',s') \in \mathcal{R}$,*

*then the following judgment is derivable : $\Gamma, X:A, Q:(x:A)s' \vdash C\{X,Q\}:s'$.*

Let $I$ and $P$ be two terms of the appropriate type, we write $C\{I, P\}$ for the term $C\{X,Q\}[X \leftarrow I, Q \leftarrow P]$.

*Rule* Let $I$ denotes $\mathsf{Ind}(X:A)\{C_1|\ldots|C_n\}$ with $A \equiv (x:A)s$. Let $s'$ be a sort. The typing rule for non-dependent elimination $(s, s')$ is :

$$\frac{\Gamma \vdash c:(I\ a) \quad \Gamma \vdash Q:(x:A)s' \quad \Gamma \vdash f_i:C_i\{I,Q\}\ (\forall i=1\ldots n)}{\Gamma \vdash \mathsf{Elim}(c,Q)\{f_1|\ldots|f_n\}:(Q\ a)} \quad (\text{Nodep}_{s,s'})$$

*Example.* In our example, we get $\mathsf{Elim}(l, Q)\{x|f\}$ is well-typed of type $(Q\ n)$ if $c$ is of type $(listn\ n)$, $Q$ of type $nat \to s'$, $x$ of type $(Q\ 0)$ and $f$ of type $(n:nat)A \to (listn\ n) \to (Q\ n) \to (Q\ (S\ n))$.

**Dependent elimination** Let $A$ be an arity of sort $s$ ($A \equiv (x:A)s$), let $X$ be a variable of type $A$, let $s'$ be a sort, let $Q$ be a variable of type $(x:A)(X\ x) \to s'$, let $C$ be a type of constructor of $X$, and $c$ be a term of type $C$.

**Definition 6.** We define a new type $C\{X, Q, c\}$ by induction over the type of constructor $C$. Let $P$ be such that $strict\_positive(P, X)$ ($P \equiv (x:P)(X\ m)$) and $M$ be such that $X$ does not occur in $M$.
$$(P \to C)\{X,Q,c\} = (p:P)((x:P)(Q\ m\ (p\ x))) \to C\{X,Q,(c\ p)\}$$
$$((x:M)C)\{X,Q,c\} = (x:M)C\{X,Q,(c\ x)\}$$
$$(X\ a)\{X,Q,c\} = (Q\ a\ c)$$

*Example.* For our example, with $C_2(X) \equiv (n:nat)A \to (X\ n) \to (X\ (S\ n))$
$$(X\ 0)\{X,Q,c\} = (Q\ 0\ c)$$
$$C_2(X)\{X,Q,c\} = (n:nat)(a:A)(p:(X\ n))(Q\ n\ p) \to (Q\ (S\ n)\ (c\ n\ a\ p))$$

Some restrictions are necessary to ensure that this term is well-formed. Even in the case $s' = s$, we can have troubles forming a dependent arity.

**Lemma 7.** *Assume we have the same conditions as in lemma 5 the first one being replaced by $(x:A)(X\ x) \to s'$ is a well-formed type. The following judgment is derivable : $\Gamma, X:A, Q:(x:A)(X\ x) \to s', c:C \vdash C\{X,Q,c\}:s'$*

We write $C\{I, P, t\}$ for the term $C\{X, Q, c\}[X \leftarrow I, Q \leftarrow P, c \leftarrow t]$.

*Rule* Let $I$ denotes $\mathsf{Ind}(X:A)\{C_1|\ldots|C_n\}$ with $A \equiv (x:A)s$. Let $s'$ be a sort. The typing rule for dependent elimination $(s, s')$ is:

$$\frac{\Gamma \vdash c:(I\ a) \quad \Gamma \vdash Q:(x:A)(I\ x) \to s' \quad \Gamma \vdash f_i : C_i\{I, Q, \mathsf{Constr}(i, I)\}}{\Gamma \vdash \mathsf{Elim}(c, Q)\{f_1|\ldots|f_n\}:(Q\ a\ c)} \quad (\mathsf{Dep}_{s,s'})$$

$$(\forall i = 1 \ldots n)$$

*Example.* In our example, we get $\mathsf{Elim}(l, Q)\{x|f\}$ is well-typed of type $(Q\ n\ l)$ if $c$ is of type $(listn\ n)$, $Q$ of type $(n:nat)(listn\ n) \to s'$, $x$ of type $(Q\ 0\ nil)$ and $f$ of type $(n:nat)(a:A)(m:(listn\ n))(Q\ n\ m) \to (Q\ (S\ n)\ (cons\ n\ a\ m))$.

**Discussion** Obviously the non-dependent elimination can be coded using a dependent elimination over a constant predicate. We prefer to distinguish the two operations, because in some systems, the term $(x:A)(I\ x) \to s$ is not well-formed, although $(x:A)s$ is (note that the rule $\mathsf{Nodep}_{s,s}$ can be added in any GTS). We use the same syntax for both constructions. It is not problematic because first, the two eliminations behaves the same w.r.t. the reduction rule, second we can solve the ambiguity using the type information. More precisely if the system is normalizing, given an environment $\Gamma$ and a term $t$, we can find if there exists a term $M$ such that $\Gamma \vdash t : M$. If $t$ is $\mathsf{Elim}(c, Q)\{f\}$ then $t$ is well-typed if and only if :
- $c$ is well typed of type $M$.
- $M$ is reducible to $(I\ a)$ with $I \equiv \mathsf{Ind}(X:A)\{C_1|\ldots|C_n\}$ and $A \equiv (x:A)s$.
- $f$ has length $n$ (we write it as $f_1|\ldots|f_n$).
- $Q$ is well-typed of type an arity $A'$.
- Either $A'$ is convertible with $(x:A)s'$, $\mathsf{Nodep}_{s,s'}$ is an allowed rule and for each $i$, $f_i$ is well-typed in $\Gamma$ and its type is convertible to $C_i\{I, Q\}$. In that case the type of $t$ is $(Q\ a)$. Or $A'$ is convertible with $(x:A)(I\ x) \to s'$ and $\mathsf{Dep}_{s,s'}$ is an allowed rule and for each $i$, $f_i$ is well-typed in $\Gamma$ and its type is convertible to $C_i\{I, Q, \mathsf{Constr}(i, I)\}$. In that case the type of $t$ is $(Q\ a\ c)$.

## 2.6 Rules for conversion

The rules for conversion are extended w.r.t. the new operations for defining terms.

$$\frac{A \simeq A' \quad C_i \simeq C_i' \quad (\forall i = 1 \ldots n)}{\mathsf{Ind}(X:A)\{C_1|\ldots|C_n\} \simeq \mathsf{Ind}(X:A')\{C_1'|\ldots|C_n'\}}$$

$$\frac{M \simeq M' \quad i = i'}{\mathsf{Constr}(i, M) \simeq \mathsf{Constr}(i', M')} \qquad \frac{c \simeq c' \quad P \simeq P' \quad f_i \simeq f_i' \quad (\forall i = 1 \ldots n)}{\mathsf{Elim}(c, P)\{f_1|\ldots|f_n\} \simeq \mathsf{Elim}(c', P')\{f_1'|\ldots|f_n'\}}$$

A new reduction rule (called $\iota$-reduction) is added. We need some notations to state it. Let $I$ denotes $\mathsf{Ind}(X:A)\{C_1|\ldots|C_n\}$ with $A \equiv (x:A)s$. Let $F$ and $f$ be two terms, let $X$ be a variable of type $A$ and $C$ be a type of constructor of $X$.

**Definition 8.** We define a new term $C[X, F, f]$ by induction over $C$ which is a type of constructor of $X$. Let $P \equiv (x : P)(X\ m)$ be such that *strict_positive(P, X)* and $M$ such that $X$ does not occur in $M$.

$$(P \to C)[X, F, f] = [p : P]C[X, F, (f\ p\ [x : P](F\ m\ (p\ x)))]$$
$$((x : M)C)[X, F, f] = [x : M]C[X, F, (f\ x)]$$
$$(X\ a)[X, F, f] = f$$

*Example.* For our example, we have :

$$(X\ 0)[X, F, f] = (f\ 0)$$
$$(n : nat)A \to (X\ n) \to (X\ (S\ n))[X, F, f] = [n : nat][a : A][p : (X\ n)](f\ n\ a\ p\ (F\ n\ p))$$

$C[X, F, f]$ is well-typed both in the dependent and the non dependent case.

**Lemma 9.** *Let $s'$ be a sort, $C$ be a type of constructor of $X$ and $C \equiv (z : C)(X\ a)$.*

- *Let $\Delta$ be $\Gamma, X : A, Q : (x : A)s', F : (x : A)(X\ x) \to (Q\ x)$. If $C\{X, Q\}$ is a well-formed type in $\Delta$, then the judgment $\Delta, f : C\{X, Q\} \vdash C[X, F, f] : (z : C)(Q\ a)$ is provable.*
- *Let $\Delta$ be $\Gamma, X : A, Q : (x : A)(X\ x) \to s', F : (x : A)(c : (X\ x))(Q\ x\ c)$. If $C\{X, Q, c\}$ is a well-formed type in $\Delta, c : C$ then the judgment $\Delta, c : C, f : C\{X, Q, c\} \vdash C[X, F, f] : (z : C)(Q\ a\ (c\ z))$ is provable.*

As usual $C[I, G, g]$ will denote $C[X, F, f][X \leftarrow I, F \leftarrow G, f \leftarrow g]$.

*The $\iota$ reduction.* Let $s'$ be a sort, $Q$ be a term and $f$ be a sequence of $n$ terms. We define : $\mathsf{Fun\_Elim}(I, Q, f) = [x : A][c : (I\ x)]\mathsf{Elim}(c, Q)\{f\}$.
The reduction rule is, if $f_i$ denotes the $i - th$ element of the sequence $f$:

$$\mathsf{Elim}((\mathsf{Constr}(i, I)\ m), Q)\{f\} \longrightarrow_\iota (C_i[I, \mathsf{Fun\_Elim}(I, Q, f), f_i]\ m) \qquad (\iota\text{-red})$$

*Example.* For our example, we have :

$$\mathsf{Fun\_Elim}(listn, Q, x|f) = [n : nat][l : (listn\ n)]\mathsf{Elim}(l, Q)\{x|f\}$$
$$\mathsf{Elim}(nil, Q)\{x|f\} \longrightarrow_\iota x$$
$$\mathsf{Elim}((cons\ n\ a\ p), Q)\{x|f\} \longrightarrow_\iota$$
$$([n : nat][a : A][p : (listn\ n)](f\ n\ a\ p\ (\mathsf{Fun\_Elim}(listn, Q, x|f)\ n\ a\ p)$$
$$\longrightarrow_\beta^* (f\ n\ a\ p\ \mathsf{Elim}(p, Q)\{x|f\})$$

## 3 The system Coq : definition

### 3.1 The underlying GTS

The system Coq has two sorts (Set and Prop) at the impredicative level. Set and Prop distinguish between proofs that are interpreted as programs and proofs that are only a justification of some logical part.

The underlying GTS is built from $\mathcal{S} = \{\mathsf{Prop}, \mathsf{Set}, \mathsf{Type}, \mathsf{Type}_S\}$ with axioms and $\mathcal{A} = \{\mathsf{Prop} : \mathsf{Type}, \mathsf{Set} : \mathsf{Type}_S\}$ and rules $\mathcal{R} = \mathcal{S} \times \mathcal{S}$.

## 3.2 Rules for inductive definitions

Inductive definitions can be defined both for Prop and Set, so we have the rules $\mathsf{Ind}_{\mathsf{Prop}}$ and $\mathsf{Ind}_{\mathsf{Set}}$.

**Weak eliminations** In Coq, the rules for elimination are : $\mathsf{Dep}_{\mathsf{Set},\mathsf{Set}}$, $\mathsf{Dep}_{\mathsf{Set},\mathsf{Prop}}$, $\mathsf{Nodep}_{\mathsf{Prop},\mathsf{Prop}}$.
These rules make sense because we may build the same products over Set and Prop.

We do not allow $\mathsf{Nodep}_{\mathsf{Prop},\mathsf{Set}}$ because, in general, we cannot build a program by case analysis over the structure of a proof that is discarded for computation.

There is no strong reason to avoid the rule $\mathsf{Dep}_{\mathsf{Prop},\mathsf{Prop}}$. But our interpretation of $\Gamma \vdash a : A$ with $A : \mathsf{Prop}$ is that $A$ is provable, so we are not interested in $a$ as an object to reason about.

**Strong eliminations.** With the system above we shall not be able to prove for instance that 0 is not equal to 1. For this, one possibility is to allow a strong elimination, namely a rule like $\mathsf{Nodep}_{\mathsf{Set},\mathsf{Type}}$. This rule is allowed in Coq but only for a restricted class of inductive definitions that are called *small* inductive definitions.

**Definition 10.** A form of constructor of $X$ is said to be *small* if it is $(X\ a)$ or $(x:M)C$ with $C$ a small constructor of $X$ and $M$ a "small type" ie of type Prop or Set (and not Type or $\mathsf{Type}_S$).

An inductive type $\mathsf{Ind}(X:A)\{C_1|\ldots|C_n\}$ is said to be small if all the forms of constructor $C_i$ are small.

In Coq the strong elimination scheme $\mathsf{Nodep}_{\mathsf{Set},\mathsf{Type}}$ is only allowed for small inductive types. The rule can be written as :

let $I = \mathsf{Ind}(X:(x:A)\mathsf{Set})\{C_1|\ldots|C_n\}$ be a small inductive type :

$$\frac{\Gamma \vdash c:(I\ a) \qquad \Gamma \vdash Q:(x:A)\mathsf{Type} \qquad \Gamma \vdash f_i:C_i\{I,Q\}\ (\forall i=1\ldots n)}{\Gamma \vdash \mathsf{Elim}(c,Q)\{f_1|\ldots|f_n\}:(Q\ a)} \quad (\mathsf{Nodep}_{\mathsf{Set},\mathsf{Type}})$$

A problem to state this rule is to be able to write the type of $Q$ in the system. For instance in the pure Calculus of Constructions, it is only possible to write $Q$: Type with $A$ an empty list. Also we cannot express the corresponding dependent elimination. But in the Coq system, there is a hierarchy of universes, so we can build arities on Type and there is no problem in the expression of the elimination scheme.

*Other possible strong eliminations.* We cannot allow strong eliminations for non-small inductive types without getting an inconsistency. This can be shown with an adaptation of the argument developed in [6] to this system.

The point is that we can always build an inductive type $B$:Set with one constructor $\varepsilon:\mathsf{Prop}\to B$. With the rule $\mathsf{Nodep}_{\mathsf{Set},\mathsf{Type}}$ we can build a projection $E:B\to\mathsf{Prop}$ and we will have $(E\ (\varepsilon\ A))$ and $A$ convertible for each $A$ of type Prop. So we get an object of type Set which is isomorphic to Prop. Because we have impredicativity at both levels, it is not surprising we get an inconsistency.

We do not allow the rule Nodep$_{\text{Prop,Type}}$ even for small inductive definitions. Actually if such a rule was available, we could find $A:\text{Prop}, a,b:A$ and prove $a \neq b$. This goes against the intended interpretation of the Prop part of the system with a proof-irrelevance semantics (all proofs of a proposition are identified). However this rule forbids the use of other natural principle for instance the axiom for extensionality $(A,B:\text{Prop})(A\to B)\to(B\to A)\to A = B$ or the excluded middle $(A:\text{Prop})A \vee \neg A$ as shown by Berardi or Coquand [6].

We do not allow the rule Nodep$_{\text{Set,Type}_S}$ although it could have very interesting applications. But this rule destroys the interpretation of proofs in Coq as non-dependent programs of $F_\omega$ extended with inductive definitions. We could for instance build a type $T(n)$ such that $T(\tilde n)$ is the $n$-ary product of a type $A$. This is really a non-trivial extension of the system from the computational point of view.

B. Werner [23] recently proved the normalization property for a second-order polymorphic system extended with a type of natural numbers together with weak and strong eliminations. If the proof can be extended to the full system, it will be possible to introduce an extended Coq system with the rule Nodep$_{\text{Set,Type}_S}$.

**Inductive types and universes.** The system Coq is build on a GTS which extends the Calculus of Constructions with an implicit hierarchy of universes. An alternative extension of the Calculus is to introduce inductive types at the predicative level. If we want to be exact, we have to say that, in the system Coq, the rules Ind$_{\text{Type}}$, Dep$_{\text{Type,Prop}}$, and Dep$_{\text{Type,Type}}$ are possible. But they were never used in the examples developed so far. Also the interaction between the implicit hierarchy of universes and the elimination scheme was not carefully checked.

### 3.3 Examples

We give a few examples of "typical" rules to give a flavor of what we get. We shall state the introduction rules (the types of the constructor), the elimination scheme (actually the type of the "generic" elimination scheme $[Q][f]\text{Fun\_Elim}(I,Q,f)$ either in the dependent or the nondependent case) and the reduction rules.

*Absurd proposition* We can define a proposition $\bot \equiv \text{Ind}(X:\text{Prop})\{\}$ without constructors. The non-dependent elimination will have type : $(P:\text{Prop})\bot \to P$ which says that from absurdity you can prove any proposition.

*Product type* Let $A$ and $B$ be two types. The product of $A$ and $B$ will be defined as a type $A*B$ with only one constructor *pair* of type $A\to B\to A*B$.

$$A*B \equiv \text{Ind}(X:\text{Set})\{A\to B\to X\}$$

The dependent elimination has type :
$$prod\_rec : (P:A*B\to \text{Set})((a:A)(b:B)(P\ (pair\ a\ b)))\to (x:A*B)(P\ x)$$

which says that any element in the type $A*B$ is equivalent to $(pair\ a\ b)$ for some $a$ in $A$ and $b$ in $B$. The non dependent elimination of $A*B$ on a type $C$ has type $(A\to B\to C)\to A*B\to C$ which corresponds to the uncurryfication. The rule for conversion is : $(prod\_rec\ P\ f\ (pair\ a\ b)) \simeq (f\ a\ b)$

*Sum type* The disjoint union $(A+B)$ of $A$ and $B$ has two constructors *inl* of type $A \to (A+B)$ and *inr* of type $B \to (A+B)$. $A+B \equiv \mathsf{Ind}(X:\mathsf{Set})\{A \to X | B \to X\}$.
The dependent elimination has type:
$$sum\_rec : (P:A+B \to \mathsf{Set})((a:A)(P\ (inl\ a))) \to ((b:B)(P\ (inr\ b))) \to (x:A+B)(P\ x)$$
The rules for conversion are:
$$(sum\_rec\ P\ f\ g\ (inl\ a)) \simeq (f\ a) \qquad (sum\_rec\ P\ f\ g\ (inr\ b)) \simeq (g\ b)$$

Remark that the constructors and eliminations for disjoint union or product are polymorphic with respect to the types $A$ and $B$.

*Equality* An example of an inductively defined predicate is the equality. It is a predicate $eq_{A,x}$ parameterized by a set $A$ and an element $x$ of $A$. We say that the set of elements of $A$ equal to $x$ is the smallest set which contains $x$. The arity of $eq_{A,x}$ is $A \to \mathsf{Prop}$, it has one constructor *refl_equal* of type $(eq_{A,x}\ x)$.
$$eq_{A,x} \equiv \mathsf{Ind}(X:A \to \mathsf{Prop})\{(X\ x)\}$$
The non-dependent elimination has type:
$$eq_{A,x}\_ind : (P:A \to \mathsf{Prop})(P\ x) \to (y:A)(eq_{A,x}\ y) \to (P\ y)$$
This principle says that if $x$ and $y$ are equal, to prove $(P\ y)$ it is enough to prove $(P\ x)$.

*Well-founded induction* Assume we have a set $A$ and an arbitrary relation $R:A \to A \to \mathsf{Prop}$. We define the set of elements of $A$ which are accessible for the relation $R$ as the smallest set which contains $x$ if it contains all its predecessors for $R$. The accessibility set $Acc$ has arity $A \to \mathsf{Prop}$. The introduction rule *Acc_intro* has type:
$(x:A)((y:A)(R\ y\ x) \to (Acc\ y)) \to (Acc\ x)$
$$Acc \equiv \mathsf{Ind}(X:A \to \mathsf{Prop})\{(x:A)((y:A)(R\ y\ x) \to (X\ y)) \to (X\ x)\}$$
The non dependent elimination principle is:
$Acc\_ind : (P:A \to \mathsf{Prop})$
$\qquad ((x:A)((y:A)(R\ y\ x) \to (Acc\ y)) \to ((y:A)(R\ y\ x) \to (P\ y)) \to (P\ x))$
$\qquad \to (x:A)(Acc\ x) \to (P\ x)$

## 3.4 Extraction of programs

It is possible to define a realisability interpretation for the system Coq, that will extract programs in the system $F_\omega$ plus inductive definitions (namely the system $F_\omega$ plus the rules $\mathsf{Ind}_{\mathsf{Set}}$, Constr and $\mathsf{Nodep}_{\mathsf{Set},\mathsf{Set}}$).

As usual the informative contents of a term corresponds to the sort of its type. We remark that the informative contents of $\mathsf{Elim}(t,Q)\{f\}$ is the one of $Q$ and consequently $f$ and if it is informative then $t$ also is informative (this property would not be true if $\mathsf{Nodep}_{\mathsf{Prop},\mathsf{Set}}$ was a rule).

The extension of the extraction function defined in [20, 21] for informative terms involving inductive definitions is the following:
$$\mathcal{E}(\mathsf{Ind}(X:A)\{C_1|\ldots|C_n\}) = \mathsf{Ind}(X:\mathcal{E}(A))\{\mathcal{E}(C_1)|\ldots|\mathcal{E}(C_n)\}$$
$$\mathcal{E}(\mathsf{Constr}(i,ind)) \qquad = \mathsf{Constr}(i,\mathcal{E}(ind))$$
$$\mathcal{E}(\mathsf{Elim}(t,Q)\{f\}) \qquad = \mathsf{Elim}(\mathcal{E}(t),\mathcal{E}(Q))\{\mathcal{E}(f)\}$$

It is easy to check that the extraction function preserves typability.

## 4 Meta-theory

### 4.1 The Calculus of Constructions with fixpoints

In order to study the meta-theory of the system Coq, we shall use a different extension of the Calculus of Constructions introduced by Ph. Audebaud [1, 2]. His idea was to extend the Calculus of Constructions with a fixpoint in order to encode efficient inductive types and also to reason about partial objects. A fixpoint operator is introduced at the level of both types and programs. For types it is assumed that it only applies to positive operators, in a sense that will be explained after.

We do not need the full power of this calculus and in particular we are not interested in partial objects. We will only consider a subsystem of CC+ that we call $CC_\mu$. This system is the pure Calculus of Constructions with a fixpoint operator for positive types transformers. We shall also need only one sort that we also call Set.

**Definitions** The terms contain the pure terms of the Calculus of Constructions plus a term for fixpoint formation that will be written $<x:t>t'$.

The set of *positive* and *negative* terms with respect to a variable $X$ are defined by the following grammar :

$$Pos ::= M \mid X \mid (Pos\ m) \mid [x:M]Pos \mid (x:Neg)Pos$$
$$Neg ::= M \mid (Neg\ m) \mid [x:M]Neg \mid (x:Pos)Neg$$

with the restriction that $X$ does not occur in $M$ or $m$. If $M$ is strictly positive with respect to $X$ then it is also positive with respect to $X$. We shall write *positive(P, X)* to indicate that $P$ is positive with respect to $X$.

**Rules** The rules for this calculus are the ones for a GTS with axioms : $\mathcal{A} = \{$Set : Type$_S\}$ and $\mathcal{S} \times \mathcal{S}$ as the set of rules. Furthermore the conversion rules are extended with the reduction step for fixpoints :

$$<x:A>M \longrightarrow ([x:A]M\ <x:A>M) \qquad \text{(fix-red)}$$

There is one new rule for fixpoint introduction at the level of types :

$$\frac{\Gamma \vdash A : \text{Type}_S \quad \Gamma, X : A \vdash M : A \quad positive(M, X)}{\Gamma \vdash <X:A>M : A} \qquad (\text{fixintro}_{\text{Type}_S})$$

**Properties** Ph. Audebaud proved that CC+ enjoys the Church-Rosser property and also strong normalization for $\beta$-reduction (of course not for fix-red) We can embed $CC_\mu$ in CC+ just transforming Set and Type$_S$ into the $\overline{Prop}$ and $\overline{Type}$ sorts of CC+. Consequently, we have strong normalization for $\beta$-reduction in $CC_\mu$ as well.

## 4.2 The system $F_\omega^{Ind}$

We introduce a subsystem $F_\omega^{Ind}$ of Coq which is a $F_\omega$ plus inductive definitions extended with weak non dependent eliminations. We define a transformation from $F_\omega^{Ind}$ to the system $CC_\mu$ that gives the strong normalization result for $\beta$ and $\iota$ reductions in $F_\omega^{Ind}$.

The system $F_\omega^{Ind}$ is the inductive GTS with sorts $S = \{\text{Set}, \text{Type}_S\}$, axioms $\mathcal{A} = \{\text{Set} : \text{Type}_S\}$, rules $\mathcal{R} = \{(\text{Set}, \text{Set}); (\text{Type}_S, \text{Set}); (\text{Type}_S, \text{Type}_S)\}$ and with inductive definitions of sort Set and only non-dependent elimination of sort (Set,Set). The added rules are $\text{Ind}_{\text{Set}}$, Constr and $\text{Nodep}_{\text{Set,Set}}$.

## 4.3 Translation from $F_\omega^{Ind}$ to $CC_\mu$

We define a translation from terms in the Calculus of Constructions into terms in $CC_\mu$. $\overline{M}$ will denote the translation of the term $M$ and if $\Gamma$ is the environment $x_1:M_1,\ldots,x_n:M_n$ of $CC_\mu$, we write $\overline{\Gamma}$ the environment $x_1:\overline{M_1},\ldots,x_n:\overline{M_n}$ of $CC_\mu$.

We can prove that $\Gamma \vdash_{coq} M : N$, implies $\overline{\Gamma} \vdash_{CC_\mu} \overline{M} : \overline{N}$. This is done by induction over the proof of $\Gamma \vdash_{coq} M:N$. We do not give here the justifications of the correctness of this translation, full details could be find in an extended version of this paper.

**Translating the pure part** The systems $F_\omega^{Ind}$ and $CC_\mu$ have a common part (terms and rules) corresponding to $F_\omega$. The translation is defined as a structural morphism on this part.

$$\overline{\text{Set}} = \text{Set}$$
$$\overline{x} = x \quad \text{if } x \text{ is a variable}$$
$$\overline{(x:A)B} = (x:\overline{A})\overline{B}$$
$$\overline{[x:A]B} = [x:\overline{A}]\overline{B}$$
$$\overline{(A\ B)} = (\overline{A}\ \overline{B})$$

The interesting cases of the translation correspond to the terms involving inductive definitions.

**Translating the inductive definitions** In the system $CC_\mu$, using fixpoints, we can identify the definition of the type and the expected recursive scheme.

For the natural numbers, it will give us the following representation $\overline{nat} \equiv <X:*>(C:*)C\to(X\to C\to C)\to C$. The recursor is just the identity function. A similar representation was proposed in the framework of $AF_2$ by M. Parigot [19]. The main drawback of this encoding of natural numbers is that the representation of the $n$th natural number uses a space proportional to $2^n$. This is a problem for practical uses but not for our study of the theoretical properties of the system Coq.

*The general case.* Let $I$ be $\mathsf{Ind}(X:A)\{C_1|\ldots|C_n\}$ and $A \equiv (x:A)\mathsf{Set}$. We define
$$\overline{I} = <X:\overline{A}>[x:\overline{A}](Q:\overline{A})\overline{C_1}\{X,Q\}\to\cdots\to\overline{C_n}\{X,Q\}\to(Q\,x)$$

With this definition the translation of the elimination is almost trivial :
$$\overline{\mathsf{Elim}(c,Q)\{f\}} = (\overline{c}\,\overline{Q}\,\overline{f})$$
$\overline{\mathsf{Fun\_Elim}(I,Q,f)} = [x:\overline{A}][c:(\overline{I}\,x)](c\,\overline{Q}\,\overline{f})$ which has type $(x:\overline{A})(\overline{I}\,x)\to(\overline{Q}\,x)$.

The translation of constructors is as follows. We assume $I = \mathsf{Ind}(X:A)\{C_1|\ldots|C_n\}$, $i$ such that $1\leq i \leq n$ and $C_i \equiv (z:C)(X\,a)$. We expect $\overline{\mathsf{Constr}(i,I)}$ to be of type $\overline{C_i[X\leftarrow I]}$ ie $(z:\overline{C[X\leftarrow I]})(\overline{I}\,\overline{a})$. We take with $f = f_1|\ldots|f_n$ :
$$\overline{\mathsf{Constr}(i,I)} = [z:\overline{C[X\leftarrow I]}][Q:\overline{A}][f_1:\overline{C_1}\{\overline{I},Q\}]\ldots[f_n:\overline{C_n}\{\overline{I},Q\}]$$
$$(\overline{C_i}[\overline{I},\mathsf{Fun\_Elim}(I,Q,f),f_i]\,z)$$

We have to show that if $M \equiv_{coq} N$ then $\overline{M} \equiv_{CC_\mu} \overline{N}$. The $\beta$-reduction rule does not lead to any problem. For the inductive reduction, it is easy to check that
$$\overline{\mathsf{Elim}((\mathsf{Constr}(i,I)\,m),Q)\{f\}} \longrightarrow^+_\beta \overline{(C_i[I,\mathsf{Fun\_Elim}(I,Q,f),f_i]\,m)}$$

So we have if $M$ reduces to $N$ in one step of inductive reduction then $\overline{M}$ reduces to $\overline{N}$ in at least one step of $\beta$-reduction. Remark that we do not use fix-reduction at this stage, this rule is only used to make sure that the elimination and constructors are well-typed.

We can deduce from these results that there cannot be infinite sequences of $\iota$ and $\beta$ reduction in $F_\omega^{\mathsf{Ind}}$ and consequently :

**Theorem 11.** *The system $F_\omega^{\mathsf{Ind}}$ is strongly normalizing.*

The same translation shows the strong normalization of the extension of pure Calculus of Constructions with Inductive definitions and weak elimination.

### 4.4 The Calculus of Inductive Definitions

From the previous study we can deduce properties of the full Calculus of Inductive Definitions as defined in section 3.

**Weak elimination** The distinction between Prop and Set does not matter for the study of the normalization properties. We can just map Prop and Type to respectively Set and Type$_S$. First let us consider the Calculus of Constructions with only weak (possibly dependent) eliminations. We can map this calculus (called $Coq_w$) into the system $F_\omega^{\mathsf{Ind}}$ using a transformation which forgets dependencies. This is an extension of the map which transforms a term of the Calculus of Constructions into a term of $F_\omega$ as described for instance in [20, 21]. It is easy to show that as the Calculus of Constructions is conservative over $F_\omega$, the system $Coq_w$ is conservative over $F_\omega^{\mathsf{Ind}}$. The normalization for $F_\omega^{\mathsf{Ind}}$ implies its consistency (there can be no closed normal proof of $(C:\mathsf{Set})C$) and then the consistency of $Coq_w$. The translation from $Coq_w$ to $F_\omega^{\mathsf{Ind}}$ preserves the underlying pure lambda terms (without any type information). We believe (but did not check precisely the details) that the method of translation from the pure Calculus of Constructions to $F_\omega$ used by Geuvers and Nederhof [13] could also be adapted to $Coq_w$ and $F_\omega^{\mathsf{Ind}}$ in order to justify strong normalization for $Coq_w$.

**Strong elimination** To deal with strong elimination is more complicated because in a system with strong elimination we cannot anymore ignore dependencies with respect to programs. Also we know that a careless use of strong elimination leads to paradoxes.

The scheme for strong elimination in Coq is restricted to the rule Nodep$_{Set,Type}$. This rule is sufficient to prove for instance $\neg(true = false)$.

What we actually use of this rule in the examples developed so far could be obtained by just adding the axiom $\neg(true = false)$ to $Coq_w$ and not the full scheme of strong elimination. If $A$:Type in $Coq_w$ with only weak elimination, then $A \equiv (x : A)$Prop. Now with Nodep$_{Set,Type}$ we can build a function of type $nat \rightarrow A$ by giving $F$ of type $A$ and $G$ of type $nat \rightarrow A \rightarrow A$. Furthermore, we shall have $(H\ 0)$ convertible with $F$ and $(H\ (S\ n))$ convertible with $(G\ n\ (H\ n))$.

Using only $\neg(true = false)$ and weak eliminations we could internally represent a term $H$ of type $nat \rightarrow A$ such that $(H\ 0) \Leftrightarrow_A F$ and $(H\ (S\ n)) \Leftrightarrow_A (G\ n\ (H\ n))$. The equivalence $\Leftrightarrow_A$ being defined as $[p, q : A](x : A)((p\ x) \rightarrow (q\ x) \wedge (q\ x) \rightarrow (p\ x))$. Obviously the proofs are shorter to do with the strong elimination scheme than with the internal encoding.

## 5 Discussion

### 5.1 Allowing more positive types

One restriction in our proposition for an extension with inductive definition is in the shape required for a form of constructors and mainly the condition of strict positivity.

We could relax this condition in two ways. The first one is to ask only for a positivity condition (as defined in section 4.1). As it is explained in [8], with only a positivity condition and if we allow the rule Ind$_{Type}$ and the elimination Nodep$_{Type,Type}$, we get a paradox. But in the system without the rule Ind$_{Type}$, such an extension could be justified by a translation in $CC_\mu$.

We could also extend the definition of strictly positive by saying $X$ is strictly positive in $\mathsf{Ind}(Y : A)\{C_1 | \ldots | C_n\}$ if $X$ occurs strictly negatively in each $C_i$ (we say that $X$ occurs strictly negatively in $(z : C)(X\ a)$ if $X$ does not occur or occurs strictly positively in each term of $C$). This possibility follows the general habit to say that $X$ is strictly positive in $A + B$ or $A * B$ if it is strictly positive or does not occur in $A$ and $B$. This allows also to define mutually inductive types by an encoding using several levels of inductive definitions.

But the first question with a weaker notion of positivity is how to define naturally the elimination principles, namely the auxiliary functions $C\{X, P\}$ or $C\{X, P, c\}$. There are several possibilities either using a product (for the non-dependent case) or a strong sum (for the dependent case) together with projection (it works for the two extensions of the positivity condition). In the case of an extension with several levels of inductive definitions, we can use a strong elimination scheme in order to define the predicates $C\{X, P\}$ and $C\{X, P, c\}$ but it is not really satisfactory to use a so strong scheme for defining the type of even a weak elimination. Another possibility is the one proposed by Mendler [16] which can be extended to the dependent elimination in our formalism and does not involve auxiliary operations. But none of this possibilities give a really natural formulation of the elimination principle. We think that the most natural solution for mutually inductive definitions will be to introduce them as a primitive notion by a simple generalization of the rules presented in this paper.

## 5.2 Elimination for inductive predicate

As mentioned by Th. Coquand in [7], the general elimination scheme proposed in this paper is not fully adequate for the case of inductive predicate, because it does not take into account the fact that, for instance, an object in the type $(listn\ (S\ n))$ can only be built from the second constructor. He proposed an alternative presentation of the elimination scheme similar to a definition by pattern-matching in functional languages. With this method we can easily prove $\neg(true = false)$ without introducing the full power of strong elimination.

## 5.3 Conclusion

The main ideas of the formulation of the inductive definitions were already in [8] and a similar proposition was simultaneously done by [9]. The purpose of this paper was to give the precise definitions corresponding to the system Coq and also to show that strong normalization for a large useful subsystem of Coq could be obtained from previous results known about the Calculus of Constructions.

The main point about this presentation is that it corresponds to a closed system. This is different from the extension with inductive definitions provided in other systems like Lego [14] or Alf [15] also based on typed lambda-calculus. In these systems new constants and new equality rules can be added in the system. It is the user responsibility to check that adding these new objects and rules is safe.

Our proposition was intended to be of practical use and integrated to the previous implementation of the Calculus of Constructions. Indeed, with these operations, the type recognition procedure or the discharge operation were easy to extend. Also having only one kind of constants (as abbreviation) makes a lot of things simpler. It is less clear whether this presentation is well-suited for proof synthesis. For instance, defining the addition function going back to the elimination scheme is a bit delicate and surely users will prefer to define such a function using equalities or an ML-like syntax. Also assume that addition is defined for instance as :

$$add \equiv [n, m : nat]\mathsf{Elim}(n, nat)\{m, [p, addmp : nat](S\ addmp)\}$$

Then we will have $(add\ (S\ n)\ m)$ is convertible with $(S\ (add\ n\ m))$ but we do not have $(add\ (S\ n)\ m)$ reduces to $(S\ (add\ n\ m))$. Obviously we would like such an equality corresponding to the intended definition of the addition to be automatically recognized by the system. In conclusion, our opinion is that the status of names involved in the inductive definitions has still to be clearly understood.

## Acknowledgments

We thanks G. Huet, B. Werner and A. Sellink for comments on a previous version of the paper.

# References

1. Ph. Audebaud. Partial objects in the calculus of constructions. In *Proceedings of the sixth Conf. on Logic in Computer Science.* IEEE, 1991.
2. Ph. Audebaud. *Extension du Calcul des Constructions par Points fixes.* PhD thesis, Université Bordeaux I, 1992.
3. H. Barendregt. Lambda calculi with types. Technical Report 91-19, Catholic University Nijmegen, 1991. in Handbook of Logic in Computer Science, Vol II.
4. C. Böhm and A. Berarducci. Automatic synthesis of typed $\lambda$-programs on term algebras. *Theoretical Computer Science*, 39, 1985.
5. R.L. Constable et al. *Implementing Mathematics with the Nuprl Proof Development System.* Prentice-Hall, 1986.
6. Th. Coquand. Metamathematical investigations of a Calculus of Constructions. In P. Oddifredi, editor, *Logic and Computer Science.* Academic Press, 1990. Rapport de recherche INRIA 1088, also in [11].
7. Th. Coquand. Pattern matching with dependent types. In Nordström et al. [17].
8. Th. Coquand and C. Paulin-Mohring. Inductively defined types. In P. Martin-Löf and G. Mints, editors, *Proceedings of Colog'88.* Springer-Verlag, 1990. LNCS 417.
9. P. Dybjer. Comparing integrated and external logics of functional programs. *Science of Computer Programming*, 14:59–79, 1990.
10. G. Dowek et al. The Coq Proof Assistant User's Guide Version 5.6. Rapport Technique 134, INRIA, December 1991.
11. G. Huet ed. *The Calculus of Constructions, Documentation and user's guide, Version V4.10*, 1989. Rapport technique INRIA 110.
12. H. Geuvers. Type systems for Higher Order Logic. Faculty of Mathematics and Informatics, Catholic University Nijmegen, 1990.
13. H. Geuvers and M.-J. Nederhof. A modular proof of strong normalization for the Calculus of Constructions. Faculty of Mathematics and Informatics, Catholic University Nijmegen, 1989.
14. Z. Luo and R. Pollack. Lego proof development syste : User's manual. Technical Report ECS-LFCS-92-211, University of Edinburgh., 1992.
15. L. Magnusson. The new implementation of ALF. In Nordström et al. [17].
16. N. Mendler. Recursive types and type constraints in second order lambda-calculus. In *Symposium on Logic in Computer Science*, Ithaca, NY, 1987. IEEE.
17. B. Nordström, K. Petersson, and G. Plotkin, editors. *Proceedings of the 1992 Workshop on Types for Proofs and Programs*, 1992.
18. P. Martin-Löf. *Intuitionistic Type Theory.* Studies in Proof Theory. Bibliopolis, 1984.
19. M. Parigot. On the representation of data in lambda-calculus. In *CSL'89*, volume 440 of *LNCS*, Kaiserslautern, 1989. Springer-Verlag.
20. C. Paulin-Mohring. Extracting $F_\omega$'s programs from proofs in the Calculus of Constructions. In Association for Computing Machinery, editor, *Sixteenth Annual ACM Symposium on Principles of Programming Languages*, Austin, January 1989.
21. C. Paulin-Mohring. *Extraction de programmes dans le Calcul des Constructions.* PhD thesis, Université Paris 7, January 1989.
22. F. Pfenning and C. Paulin-Mohring. Inductively defined types in the Calculus of Constructions. In *Proceedings of Mathematical Foundations of Programming Semantics*, LNCS 442. Springer-Verlag, 1990. also technical report CMU-CS-89-209.
23. B. Werner. A normalization proof for an impredicative type system with large elimination over integers. In Nordström et al. [17].

# Intersection Types
# and Bounded Polymorphism

## Benjamin C. Pierce*

**Abstract:** Intersection types and bounded quantification are complementary extensions of first-order a statically typed programming language with subtyping. We define a typed $\lambda$-calculus combining these extensions, illustrate its properties, and develop proof-theoretic results leading to algorithms for subtyping and typechecking.

## 1 Introduction

Among the intriguing properties of intersection types [2, 12, 14, 29, 33] is their ability to carry detailed information about the way a function behaves when used in different contexts. For example, the function + can be given the type $Int \rightarrow Int \rightarrow Int \wedge Real \rightarrow Real \rightarrow Real$, capturing the general fact that the sum of two real numbers is a real as well as the special observation that the sum of two integers is always integral. A compiler for a language with intersection types might even provide two different object-code sequences for the different versions of +, one using a floating point addition instruction and one using integer addition. Reynolds' Forsythe language [31] illustrates this *coherent overloading* in the context of a practical type system organized around a notion of subtyping. Intersection types are also used in Forsythe to give an elegant treatment of mutable reference cells by combining the more primitive concepts of "sources for values" and "destinations for values."

Bounded quantification [11, 3, 15, 10], on the other hand, integrates parametric polymorphism [18, 30] with subtyping by allowing a quantified type to give an upper bound for its parameter; for example, an inhabitant of $\forall \alpha \leq Student. List(\alpha) \rightarrow List(\alpha)$ takes as its first parameter an arbitrary subtype of the type *Student* and returns a function on lists of values of this type. This kind of polymorphism is both broader and more rigid than the finitary polymorphism provided by intersection types, since the number of possible instantiations of a polymorphic type is infinite) but all instances must have the same basic shape. Its practical advantages are brevity and compile-time efficiency: it allows polymorphic expressions to be written, typechecked, and compiled just once.

By viewing intersection types and bounded polymorphism as extensions of a common base, a simply typed $\lambda$-calculus with subtyping, we can merge them,

---

*Author's address: INRIA, Projet Formel, BP 105, 78153 Le Chesnay Cedex, France. Electronic mail: benjamin.pierce@inria.fr. This research was sponsored in part by the Avionics Laboratory, Wright Research and Development Center, Aeronautical Systems Division (AFSC), U. S. Air Force, Wright-Patterson AFB, OH 45433-6543 under Contract F33615-90-C-1465, Arpa Order No. 7597; in part by the Office of Naval Research under Contract N00013-84-K-0415; in part by the National Science Foundation under Contract CCR-8922109; in part by Siemens; in part by the Laboratory for Foundations of Computer Science, University of Edinburgh; and in part by Esprit Basic Research Action TYPES.

yielding a compact, natural synthesis of their features in a new calculus. This calculus, called $F_\wedge$ ("F-meet"), provides a formal basis for new programming languages combining the benefits of existing languages based on intersection types [31] or bounded quantification [8, 24, 4] alone: a simple, semantically clean mechanism for overloading functions, the possibility of code optimization during typechecking, Forsythe's treatment of references, and all the well-known benefits of parametric polymorphism. $F_\wedge$ has also been used by Ma [22] as the basis for a syntactical analysis of parametricity in System F. Recent accounts of object-oriented inheritance using higher-order extensions of bounded quantification [4, 9, 23, 28] suggest that a higher-order generalization of $F_\wedge$ would be an appropriate setting for a type theoretic account of multiple inheritance.

The following section establishes some notational conventions and reviews the definitions of the pure systems of intersections and bounded quantification. Section 3 presents $F_\wedge$ in full. Section 4 gives examples illustrating its expressive power, and Sections 5 and 6 present algorithms for checking the subtype relation and synthesizing minimal types for terms and sketch their proofs of soundness and completeness. Section 7 discusses semantic issues, and Section 8 offers directions for future research. A more detailed presentation of these results can be found in [25].

## 2 Background

The metavariables $\alpha$ and $\beta$ range over type variables; $\sigma$, $\tau$, $\theta$, $\phi$, and $\psi$ range over types; $e$ and $f$ range over terms; and $x$ and $y$ range over term variables. A finite sequence with elements $x_1$ through $x_n$ is written $[x_1..x_n]$. Concatenation of finite sequences is written $X_1 * X_2$. Single elements are adjoined to the right or left of sequences with a comma: $[x_a, X_b]$ or $[X_a, x_b]$. Sequences are sometimes written using "comprehension" notation: $[x \mid ...]$; for example, if $T \equiv [\sigma \rightarrow \tau, \psi \rightarrow \psi, \sigma \rightarrow \theta]$, then the comprehension $[\zeta_2 \mid \zeta_1 \rightarrow \zeta_2 \in T \text{ and } \zeta_1 \equiv \sigma]$ stands for the sequence $[\tau, \theta]$. A *context* $\Gamma$ is a finite sequence of typing assumptions $x{:}\tau$ and subtyping assumptions $\alpha \leq \tau$, with no variable or type variable appearing twice on the left. It is convenient to view a context $\Gamma$ as a finite function and write $dom(\Gamma)$ for its domain.

A type $\tau$ is *closed* with respect to a context $\Gamma$ if its free (type) variables are all in $dom(\Gamma)$. A term $e$ is closed with respect to $\Gamma$ if its free (term and type) variables are all in $dom(\Gamma)$. A context $\Gamma$ is closed if $\Gamma \equiv \{\}$, or if $\Gamma \equiv \Gamma_1, \alpha \leq \tau$ with $\Gamma_1$ closed and $\tau$ closed with respect to $\Gamma_1$, or if $\Gamma \equiv \Gamma_1, x{:}\tau$ with $\Gamma_1$ closed and $\tau$ closed with respect to $\Gamma_1$. A subtyping statement $\Gamma \vdash \sigma \leq \tau$ is closed if $\Gamma$ is closed and $\sigma$ and $\tau$ are closed with respect to $\Gamma$; a typing statement $\Gamma \vdash e \in \tau$ is closed if $\Gamma$ is closed and $e$ and $\tau$ are closed with respect to $\Gamma$. In the following, we assume that all statements under discussion are closed; in particular, we allow only closed statements in instances of inference rules.

Types, terms, contexts, and statements that differ only in the names of bound variables are considered identical. (That is., we think of variables not as names but as pointers into the surrounding context, as suggested by deBruijn [16].)

Examples are set in a typewriter font; $\lambda$-calculus notation is transliterated as follows: $\top$ is written as `T`, $\lambda$ as `\`, $\Lambda$ as `\\`, $\forall$ as `All`, and $\leq$ as `<`. Lines of input to the typechecker are prefixed with `>` and followed by the system's response. The type constructors $\rightarrow$ and $\forall$ bind more tightly than $\wedge$. Also, $\rightarrow$ associates to the right and $\forall$ obeys the usual "dot rule" where the body $\tau$ of a quantified type $\forall \alpha \leq \sigma. \tau$ is taken to extend to the right as far as possible.

## 2.1 Simply Typed $\lambda$-Calculus with Subtyping

The first-order $\lambda$-calculus with intersection types (called $\lambda_\wedge$ here) and the second-order $\lambda$-calculus with bounded quantification (called $F_\leq$) can both be presented as extensions of the simply typed $\lambda$-calculus enriched with a subtyping relation ($\lambda_\leq$), a system proposed by Cardelli [6] as a "core calculus of subtyping" in a foundational framework for object-oriented programming languages.

The types of $\lambda_\leq$ consist of a set of *primitive types* (ranged over by the metavariable $\rho$) closed under the type constructor $\rightarrow$:

$$\tau ::= \rho \mid \tau_1 \rightarrow \tau_2$$

The terms of $\lambda_\leq$ consist of a countable set of variables together with all the phrases that can be built from these by functional abstraction and application:

$$e ::= x \mid \lambda x{:}\tau.\, e \mid e_1\, e_2$$

The typing relation of $\lambda_\leq$ is formalized as a collection of inference rules for deriving typing statements of the form $\Gamma \vdash e \in \tau$ ("under assumptions $\Gamma$, expression $e$ has type $\tau$"). The rules for variables, abstractions, and applications are exactly the same as in the ordinary simply typed $\lambda$-calculus [13]. In addition, the rule of *subsumption* states that whenever a term $e$ has a type $\sigma$ and $\sigma$ is a subtype of another type $\tau$, the type of $e$ may be promoted to $\tau$:

$$\frac{\Gamma \vdash e \in \tau_1 \quad \Gamma \vdash \tau_1 \leq \tau_2}{\Gamma \vdash e \in \tau_2} \quad \text{(SUB)}$$

Intuitively, a subtyping statement $\Gamma \vdash \sigma \leq \tau$ corresponds to the assertion that $\sigma$ is a *refinement* of $\tau$, in the sense that every element of $\sigma$ contains enough information to meaningfully be regarded as an element of $\tau$. In some models this means simply that $\sigma$ is a subset of $\tau$; more generally, it implies the existence of a distinguished *coercion function* from $\sigma$ to $\tau$.

## 2.2 Intersection Types

The first-order calculus of intersection types, $\lambda_\wedge$, is formed from $\lambda_\leq$ by adding intersections to the language of types:

$$\tau ::= \rho \mid \tau_1 \rightarrow \tau_2 \mid \bigwedge[\tau_1..\tau_n]$$

In the examples, we use the abbreviations $\sigma \wedge \tau = \bigwedge[\sigma, \tau]$ and $\top = \bigwedge[\,]$.

Two new subtyping rules capture the order-theoretic properties of the $\wedge$ operator:

$$\frac{\text{for all } i,\ \Gamma \vdash \sigma \leq \tau_i}{\Gamma \vdash \sigma \leq \bigwedge[\tau_1..\tau_n]} \quad \text{(Sub-Inter-G)}$$

$$\Gamma \vdash \bigwedge[\tau_1..\tau_n] \leq \tau_i \quad \text{(Sub-Inter-LB)}$$

One additional subtyping rule captures the relation between intersections and function spaces, allowing the two constructors to "distribute" when an intersection appears on the right-hand side of an arrow:

$$\Gamma \vdash \bigwedge[\sigma \to \tau_1 .. \sigma \to \tau_n] \leq \sigma \to \bigwedge[\tau_1..\tau_n] \quad \text{(Sub-Dist-IA)}$$

This inclusion is actually an equivalence, since the other direction may be proved from the rules for meets and arrows (listed in full in Section 3. It has a strong effect on both syntactic and semantic properties of the language; for example, it implies that $\top \leq \sigma \to \top$ for any $\sigma$.

The intersection introduction rule allows an intersection type to be derived for a term whenever each of the elements of the intersection can be derived for it separately:

$$\frac{\text{for all } i,\ \Gamma \vdash e \in \tau_i}{\Gamma \vdash e \in \bigwedge[\tau_1..\tau_n]} \quad \text{(Inter-I)}$$

The corresponding elimination rule would allow us to infer, on the basis of a derivation of a statement like $\Gamma \vdash e \in \bigwedge[\tau_1..\tau_n]$, that $e$ possesses every $\tau_i$ individually. But this follows already from the rule Sub-Inter-LB and subsumption; we need not add the elimination rule explicitly to the calculus.

The nullary case of Inter-I is worth particular notice, since it allows the type $\top$ to be derived for *every* term of the calculus, including terms whose evaluation intuitively encounters a run time error or fails to terminate.

The system as we have described it so far supports the use of intersection types in programming only to a limited degree. Suppose, for example, that the primitive subtype relation has $Int \leq Real$ and the addition function is overloaded to operate on both integers and reals:

$$+ \in Int \to Int \to Int \wedge Real \to Real \to Real.$$

Using just the constructs introduced so far, there is no way of writing our own functions that "inherit" the finitary polymorphism of $+$. For example, the doubling function $\lambda x{:}?.\ x + x$ cannot be given the type $Int \to Int \wedge Real \to Real$, since replacing the ? with either $Int$ or $Real$ (or $Int \wedge Real$, or indeed — assuming we could write it — $Int \vee Real$) gives a typing that is too restrictive:

```
> double1 = \x:Int. plus x x;
double1 : Int -> Int

> double2 = \x:Real. plus x x;
double2 : Real -> Real
```

This led Reynolds [31] to introduce a generalized form of $\lambda$-abstraction allowing explicit programmer-controlled generation of alternative typings for terms:

$$e ::= \ldots \mid \lambda x{:}\tau_1..\tau_n.\ e$$

The typing rule for this form allows the typechecker to make a choice of any of the $\sigma$'s as the type of $x$ in the body:

$$\frac{\Gamma,\ x{:}\sigma_i \vdash e \in \tau_i}{\Gamma \vdash \lambda x{:}\sigma_1..\sigma_n.\ e \in \sigma_i \to \tau_i} \qquad \text{(Arrow-I')}$$

This rule can be used together with INTER-I to generate a set of $n$ alternative typings for the body and then form their intersection as the type of the whole $\lambda$-abstraction:

```
> double = \x:Int,Real. plus x x;
double : Real->Real /\ Int->Int
```

## 2.3 Bounded Polymorphism

Like other second-order $\lambda$-calculi, the terms of $F_\leq$ include the variables, abstractions, and applications of $\lambda_\leq$ plus type abstractions and type applications. The latter are slightly refined to take account of the subtype relation: each type abstraction gives a *bound* for the type variable it introduces and each type application must satisfy the constraint that the argument type is a subtype of the bound of the polymorphic function being applied. Also, like that of $\lambda_\wedge$, the $F_\leq$ subtype relation includes a maximal element, generally called Top.

$$\tau ::= \text{Top} \mid \alpha \mid \tau_1 \to \tau_2 \mid \forall \alpha {\leq} \tau_1.\ \tau_2$$
$$e ::= x \mid \lambda x{:}\tau.\ e \mid e_1\ e_2 \mid \Lambda \alpha {\leq} \tau.\ e \mid e[\tau]$$

Type abstractions are checked by moving the bound for the type variable into the context and checking the body of the abstraction under the enriched set of assumptions (rule ALL-I in Section 3). The rule for type applications must check that the argument type is indeed a subtype of the bound of the corresponding quantifier (rule ALL-E). Like arrow types, subtyping of quantified types is contravariant in their bounds and covariant in their bodies (rule SUB-ALL).

# 3 The $F_\wedge$ Calculus

$F_\wedge$ is essentially a "least upper bound" of $\lambda_\wedge$ and $F_\leq$. To achieve a compact and symmetric calculus, however, a few modifications and extensions are needed. Since $F_\leq$ allows primitive types to be encoded as type variables, we drop the primitive types of $\lambda_\leq$ and $\lambda_\wedge$. Since $\top$ and Top both function as maximal elements of their respective subtype orderings, we drop Top and let $\top$ take over its job. Since $\forall$ behaves like a kind of function space constructor, we add a new law SUB-DIST-IQ, analogous to SUB-DIST-IA, allowing intersections to be distributed over quantifiers on the right-hand side.

The notions of type variables and type substitution inherited from $F_\leq$ can be used to define a further generalization of $\lambda_\wedge$'s $\lambda$-abstraction. We extend the syntax of terms with a new form

$$e ::= \ldots \mid \text{for } \alpha \text{ in } \sigma_1..\sigma_n.\ e$$

whose typing rule allows a choice of any of the $\sigma$'s as a replacement for $\alpha$ in the body:

$$\frac{\Gamma \vdash \{\sigma_i/\alpha\}e \in \tau_i}{\Gamma \vdash \text{for } \alpha \text{ in } \sigma_1..\sigma_n.\ e \in \tau_i} \quad \text{(FOR)}$$

This rule, like the generalized arrow introduction rule ARROW-I' of $\lambda_\wedge$, can be used together with INTER-I to generate a set of $n$ alternative typings for the body and then form their intersection as the type of the whole *for* expression:

```
> double = for A in Int,Real. \x:A. plus x x;
double : Real->Real /\ Int->Int
```

Indeed, $\lambda_\wedge$'s generalized $\lambda$-abstraction may be reintroduced as a simple syntactic abbreviation: $\lambda x{:}\sigma_1..\sigma_n.\ e \stackrel{\text{def}}{=} \text{for } \alpha \text{ in } \sigma_1..\sigma_n.\ \lambda x{:}\alpha.\ e$, where $\alpha$ is fresh.

Besides separating the mechanisms of functional abstraction and alternation, the introduction of the *for* construct extends the expressive power of the language by providing a *name* for the "current choice" being made by the type checker. For example, the explicit *for* construct may be used to improve the efficiency of typechecking even for first-order languages with intersections. The second version of poly requires that the body be checked only twice, as compared to sixteen times for the first version.

```
> poly = \w:Int,Real. \x:Int,Real. \y:Int,Real. \z:Int,Real.
>         plus (double x) (plus (plus w y) z);
poly : Real->Real->Real->Real->Real /\ Int->Int->Int->Int->Int

> poly =   for A in Int,Real.
>          \w:A. \x:A. \y:A. \z:A.
>          plus (double x) (plus (plus w y) z);
poly : Real->Real->Real->Real->Real /\ Int->Int->Int->Int->Int
```

We now define the $F_\wedge$ calculus formally. The sets of types and terms are given by the following abstract grammar:

$$\begin{aligned}
\tau ::=\ & \alpha \\
\mid\ & \tau_1 \to \tau_2 \\
\mid\ & \forall \alpha {\leq} \tau_1.\ \tau_2 \\
\mid\ & \bigwedge[\tau_1..\tau_n]
\end{aligned}$$

$$\begin{aligned}
e ::=\ & x \\
\mid\ & \lambda x{:}\tau.\ e \\
\mid\ & e_1\ e_2 \\
\mid\ & \Lambda \alpha {\leq} \tau.\ e \\
\mid\ & e[\tau] \\
\mid\ & \text{for } \alpha \text{ in } \tau_1..\tau_n.\ e
\end{aligned}$$

The three-place *subtype relation* $\Gamma \vdash \sigma \leq \tau$ is the least relation closed under the following rules:

$$\Gamma \vdash \tau \leq \tau \quad \text{(SUB-REFL)}$$

$$\frac{\Gamma \vdash \tau_1 \leq \tau_2 \quad \Gamma \vdash \tau_2 \leq \tau_3}{\Gamma \vdash \tau_1 \leq \tau_3} \quad \text{(Sub-Trans)}$$

$$\Gamma \vdash \alpha \leq \Gamma(\alpha) \quad \text{(Sub-TVar)}$$

$$\frac{\Gamma \vdash \tau_1 \leq \sigma_1 \quad \Gamma \vdash \sigma_2 \leq \tau_2}{\Gamma \vdash \sigma_1 \to \sigma_2 \leq \tau_1 \to \tau_2} \quad \text{(Sub-Arrow)}$$

$$\frac{\Gamma \vdash \tau_1 \leq \sigma_1 \quad \Gamma, \alpha \leq \tau_1 \vdash \sigma_2 \leq \tau_2}{\Gamma \vdash \forall \alpha \leq \sigma_1. \, \sigma_2 \leq \forall \alpha \leq \tau_1. \, \tau_2} \quad \text{(Sub-All)}$$

$$\frac{\text{for all } i, \, \Gamma \vdash \sigma \leq \tau_i}{\Gamma \vdash \sigma \leq \bigwedge[\tau_1..\tau_n]} \quad \text{(Sub-Inter-G)}$$

$$\Gamma \vdash \bigwedge[\tau_1..\tau_n] \leq \tau_i \quad \text{(Sub-Inter-LB)}$$

$$\Gamma \vdash \bigwedge[\sigma \to \tau_1 .. \sigma \to \tau_n] \leq \sigma \to \bigwedge[\tau_1..\tau_n] \quad \text{(Sub-Dist-IA)}$$

The three-place *typing relation* $\Gamma \vdash e \in \tau$ is the least relation closed under the following rules:

$$\Gamma \vdash x \in \Gamma(x) \quad \text{(Var)}$$

$$\frac{\Gamma, x{:}\tau_1 \vdash e \in \tau_2}{\Gamma \vdash \lambda x{:}\tau_1. \, e \in \tau_1 \to \tau_2} \quad \text{(Arrow-I)}$$

$$\frac{\Gamma \vdash e_1 \in \tau_1 \to \tau_2 \quad \Gamma \vdash e_2 \in \tau_1}{\Gamma \vdash e_1 \, e_2 \in \tau_2} \quad \text{(Arrow-E)}$$

$$\frac{\Gamma, \alpha \leq \tau_1 \vdash e \in \tau_2}{\Gamma \vdash \Lambda \alpha \leq \tau_1. \, e \in \forall \alpha \leq \tau_1. \, \tau_2} \quad \text{(All-I)}$$

$$\frac{\Gamma \vdash e \in \forall \alpha \leq \tau_1. \, \tau_2 \quad \Gamma \vdash \tau \leq \tau_1}{\Gamma \vdash e[\tau] \in \{\tau/\alpha\}\tau_2} \quad \text{(All-E)}$$

$$\frac{\Gamma \vdash \{\sigma_i/\alpha\}e \in \tau_i}{\Gamma \vdash \text{for } \alpha \text{ in } \sigma_1..\sigma_n. \, e \in \tau_i} \quad \text{(For)}$$

$$\frac{\text{for all } i, \, \Gamma \vdash e \in \tau_i}{\Gamma \vdash e \in \bigwedge[\tau_1..\tau_n]} \quad \text{(Inter-I)}$$

$$\frac{\Gamma \vdash e \in \tau_1 \quad \Gamma \vdash \tau_1 \leq \tau_2}{\Gamma \vdash e \in \tau_2} \quad \text{(Sub)}$$

The one point where $\lambda_\wedge$ and $F_\wedge$ do not fit together perfectly is the maximal types $\top$ (the empty intersection) and Top. We might hope that these would coincide in $F_\wedge$, but this, unfortunately, is not the case. The difference arises from the Inter-I rule of $\lambda_\wedge$, which, in its nullary form, states that any term whatsoever has type $\top$. $F_\leq$ has no such rule; the only way a term $e$ can be assigned type Top is by the rules Sub and Sub-Top, which require that the term already have some type $\sigma$ with $\sigma \leq \tau$. In other words, Top is the type of all *well-typed* terms, whereas $\top$ is the type of *all* terms. Order-theoretically, of course, the two types are equivalent (each is a subtype of the other), since each is maximal.

# 4 Examples

Intersection types allow very refined types to be assigned to expressions — much more refined than is possible in conventional polymorphic languages. Instead of a single description, each expression may be assigned any finite collection of descriptions, each capturing some aspect of its behavior. Since we are working in an explicitly typed calculus, this requires effort from the programmer in the form of type assumptions or annotations; in general, as more effort is expended, better typings are obtained.

Many functional languages provide a primitive type `Bool` with two elements, `true` and `false`. Here we can introduce two subtypes of `Bool`, called `True` and `False` and give more exact types for the constants `true` and `false` in terms of these refinements:

```
> Bool < T,
> True < Bool,
> False < Bool;

> true : True,
> false : False;
```

The polymorphic `if` primitive can also be given a more refined type: if we know whether the value of the test lies in the type `True` or the type `False`, we can tell in advance which of the branches will be chosen. An optimizing compiler might use this information to generate more efficient code in some cases.

```
> if : All A.    (True -> A -> T -> A)
>             /\ (False -> T -> A -> A)
>             /\ (Bool -> A -> A -> A);
```

(The third typing is needed here because $F_\wedge$'s types cannot express the idea that every element of `Bool` is an element of either `True` or `False`. This shortcoming, while not serious in practice, has motivated the investigation of a dual notion of *union types* [1, 20, 26].) The refinement in the types of `true`, `false`, and `if` can now be exploited in typing new functions:

```
> or =
>     \x:True,False,Bool. \y:True,False,Bool.
>         for R in True,False,Bool.
>             if [R] x true y;
or : Bool->(Bool->Bool/\True->True)
   /\ False->False->False
   /\ True->Bool->True
```

In fact, we can carry out the same construction in the *pure* $F_\wedge$ calculus with no assumptions about predefined types or constants, using a generalization of the familiar Church encoding of booleans in the $\lambda$-calculus.

```
> True  == All B. All TT<B. All FF<B. TT -> T  -> TT,
> False == All B. All TT<B. All FF<B. T  -> FF -> FF,
> Bool  == All B. All TT<B. All FF<B. TT -> FF -> B;
```

```
> true = \\B. \\TT<B. \\FF<B. \x:TT. \y:T. x,
> false = \\B. \\TT<B. \\FF<B. \x:T. \y:FF. y;
true : True
false : False

> or = for M in True,False,Bool.
>       for N in True,False,Bool.
>         \m:M. \n:N.
>           m [Bool] [True] [N] true n;
or : Bool->Bool->Bool
  /\ False->(False->False/\True->True)
  /\ True->Bool->True
```

## 5  Subtyping Algorithm

Next we give a semi-decision procedure for the $F_\wedge$ subtype relation. This relation is undecidable [27, 25], so a semi-decision procedure is the best we can hope for; however, the same algorithm forms a decision procedure for a large fragment of $F_\wedge$, believed to contain all but some extremely pathological terms [17]. The algorithm presented here generalizes both Reynolds' decision procedure for the subtype relation of Forsythe [personal communication, 1988] and the standard subtyping algorithm for $F_\leq$ [3, 15].

Because of the axioms SUB-DIST-IA and SUB-DIST-IQ, we cannot check whether $\Gamma \vdash \sigma \leq \tau$ just by comparing the outermost constructors of $\sigma$ and $\tau$ and making recursive calls. For example,

$$\alpha \leq \mathsf{T}, \beta \leq \mathsf{T} \vdash \alpha \to \alpha \leq \beta \to \mathsf{T},$$

is derivable (using SUB-DIST-IA and SUB-TRANS, with intermediate type $\mathsf{T}$), although

$$\alpha \leq \mathsf{T}, \beta \leq \mathsf{T} \vdash \beta \leq \alpha$$

is not. So, given $\Gamma$, $\sigma$, and $\tau$, the algorithm must first perform a complete analysis of the structure of $\tau$. Whenever $\tau$ has the form $\tau_1 \to \tau_2$ or $\forall \alpha \leq \tau_1.\ \tau_2$, it pushes the left-hand side — $\tau_1$ or $\alpha \leq \tau_1$ — onto a queue of pending left-hand sides and proceeds recursively with the analysis of $\tau_2$. When $\tau$ has the form of an intersection, it calls itself recursively on each of the elements. When $\tau$ is finally reduced to a type variable, the algorithm begins analyzing $\sigma$, matching left-hand sides of arrow and polymorphic types against the queue of pending left-hand sides from $\tau$. In the base case, when both $\sigma$ and $\tau$ have been reduced to variables, the algorithm first checks whether they are identical; if so, and if the queue of pending left-hand sides is empty, the algorithm immediately returns *true*. Otherwise, the variable $\sigma$ is replaced by its upper bound from $\Gamma$ and the analysis continues.

Formally, let $X$ be a finite sequence of elements of the set

$$\{\tau \mid \tau \text{ a type}\} \cup \{\alpha \leq \tau \mid \alpha \text{ a type variable and } \tau \text{ a type}\}.$$

Define the type $X \Rightarrow \tau$ as follows:

$$\begin{aligned}
{[\,]}\Rightarrow\tau &= \tau \\
[\sigma, X]\Rightarrow\tau &= \sigma \to (X\Rightarrow\tau) \\
[\alpha\leq\sigma, X]\Rightarrow\tau &= \forall\alpha\leq\sigma.\,(X\Rightarrow\tau).
\end{aligned}$$

Note that every type $\tau$ has either the form $X\Rightarrow\alpha$ or the form $X\Rightarrow\bigwedge[\tau_1..\tau_n]$, for a unique $X$.

The 4-place relation $\Gamma \vdash \sigma \leq X\Rightarrow\tau$ is the least relation closed under the following rules. (Note that these rules are *syntax-directed* — at most one of them can be used to establish a given conclusion.)

$$\frac{\text{for all } i,\ \Gamma \vdash \sigma \leq X \Rightarrow \tau_i}{\Gamma \vdash \sigma \leq X \Rightarrow \bigwedge[\tau_1..\tau_n]} \quad \text{(ASubR-Inter)}$$

$$\frac{\text{for some } i,\ \Gamma \vdash \sigma_i \leq X \Rightarrow \alpha}{\Gamma \vdash \bigwedge[\sigma_1..\sigma_n] \leq X \Rightarrow \alpha} \quad \text{(ASubL-Inter)}$$

$$\frac{\Gamma \vdash \tau_1 \leq [\,] \Rightarrow \sigma_1 \quad \Gamma \vdash \sigma_2 \leq X_2 \Rightarrow \alpha}{\Gamma \vdash \sigma_1\to\sigma_2 \leq [\tau_1, X_2] \Rightarrow \alpha} \quad \text{(ASubL-Arrow)}$$

$$\frac{\Gamma \vdash \tau_1 \leq [\,] \Rightarrow \sigma_1 \quad \Gamma, \beta\leq\tau_1 \vdash \sigma_2 \leq X_2 \Rightarrow \alpha}{\Gamma \vdash \forall\beta\leq\sigma_1.\,\sigma_2 \leq [\beta\leq\tau_1, X_2] \Rightarrow \alpha} \quad \text{(ASubL-All)}$$

$$\Gamma \vdash \alpha \leq [\,] \Rightarrow \alpha \quad \text{(ASubL-Refl)}$$

$$\frac{\Gamma \vdash \Gamma(\beta) \leq X \Rightarrow \alpha}{\Gamma \vdash \beta \leq X \Rightarrow \alpha} \quad \text{(ASubL-TVar)}$$

The more convenient three-place relation $\Gamma \vdash \sigma \leq \tau$ may be defined as

$$\Gamma \vdash \sigma \leq \tau \quad \text{iff} \quad \Gamma \vdash \sigma \leq X\Rightarrow\phi,$$

where $\tau \equiv X\Rightarrow\phi$ and either $\phi \equiv \bigwedge[\phi_1..\phi_n]$ or $\phi \equiv \alpha$.

**Theorem:** A statement $\Gamma \vdash \sigma \leq \tau$ is derivable from the original $F_\wedge$ subtyping rules iff it is derivable from the rules defining the algorithm.

**Proof sketch:** The soundness of the algorithm is established by a straightforward inductive argument. To show completeness, we first reformulate pure $F_\wedge$ subtyping as an equivalent relation on types in a "conjunctive normal form" where all intersections appear on the outside or on the left of arrows or quantifiers. (The advantage of this presentation is that it does not need distributivity rules.) The algorithm is then shown to be complete for this system by an extension of the method used by Curien and Ghelli [15] for showing the completeness of the subtyping algorithm for $F_\leq$: give a system of rewriting rules on derivations in this system, eliminating instances of transitivity by pushing them toward the leaves of the derivation, find a terminating reduction strategy for these rules, and observe that the resulting normal-form derivations correspond to execution traces of the algorithm.

# 6 Type Synthesis Algorithm

Given a term $e$ and a context $\Gamma$, the type synthesis algorithm constructs a *minimal type* $\sigma$ for $e$ under $\Gamma$ — that is, it finds a type $\sigma$ such that $\Gamma \vdash e \in \sigma$, and such that any other type that can be derived for $e$ is a supertype of $\sigma$. (Because it calls the subtyping algorithm, the type synthesis procedure may also fail to terminate in some pathological cases.)

The algorithm can be explained by separating the typing rules of $F_\wedge$ into two sets: the *structural* or *syntax-directed* rules VAR, ARROW-E, ALL-I, ALL-E, and FOR, whose applicability depends on the form of $e$, and the non-structural rules INTER-I and SUB, which can be applied without regard to the form of $e$. The non-structural rules are removed from the system and their possible effects accounted for by modifying the structural rules VAR, ARROW-E, ALL-E, and FOR.

The main source of difficulty is the application rules ARROW-E and ALL-E. An application $(e_1\ e_2)$ in $F_\wedge$ has *every* type $\tau_2$ such that $e_1$ can be shown to have some type $\tau_1 \to \tau_2$ and $e_2$ can be shown to inhabit $\tau_1$, where the rule SUB may be used on both sides to promote the types of $e_1$ and $e_2$ to supertypes with appropriate shapes. For example, if

$$e_1 \in (\sigma_1 \to \sigma_2) \wedge (\forall \alpha \leq \sigma_3.\ \sigma_4) \wedge (\sigma_5 \to \sigma_6) \wedge (\sigma_7 \to \sigma_8)$$
$$e_2 \in \sigma_1 \wedge (\forall \alpha \leq \sigma_3.\ \sigma_4) \wedge \sigma_5,$$

then $(e_1\ e_2)$ has both types $\sigma_2$ and $\sigma_6$, and hence (by INTER-I) also type $\sigma_2 \wedge \sigma_6$. To deal with this flexibility deterministically, we must show that the set of supertypes of $(\sigma_1 \to \sigma_2) \wedge (\forall \alpha \leq \sigma_3.\ \sigma_4) \wedge (\sigma_5 \to \sigma_6) \wedge (\sigma_7 \to \sigma_8)$ that have the appropriate shape to appear as the type of $e_1$ in an instance of ARROW-E can be characterized finitely, using an auxiliary function *arrowbasis*:

$$arrowbasis((\sigma_1 \to \sigma_2) \wedge (\forall \alpha \leq \sigma_3.\ \sigma_4) \wedge (\sigma_5 \to \sigma_6) \wedge (\sigma_7 \to \sigma_8))$$
$$= [\sigma_1 \to \sigma_2,\ \sigma_5 \to \sigma_6,\ \sigma_7 \to \sigma_8].$$

Formally, $arrowbasis_\Gamma$ and the analogous function $allbasis_\Gamma$ for dealing with type applications are defined as follows:

$$\begin{aligned}
arrowbasis_\Gamma(\alpha) &= arrowbasis_\Gamma(\Gamma(\alpha)) \\
arrowbasis_\Gamma(\tau_1 \to \tau_2) &= [\tau_1 \to \tau_2] \\
arrowbasis_\Gamma(\forall \alpha \leq \tau_1.\ \tau_2) &= [\,] \\
arrowbasis_\Gamma(\bigwedge[\tau_1 .. \tau_n]) &= arrowbasis_\Gamma(\tau_1) * \cdots * arrowbasis_\Gamma(\tau_n)
\end{aligned}$$

$$\begin{aligned}
allbasis_\Gamma(\alpha) &= allbasis_\Gamma(\Gamma(\alpha)) \\
allbasis_\Gamma(\tau_1 \to \tau_2) &= [\,] \\
allbasis_\Gamma(\forall \alpha \leq \tau_1.\ \tau_2) &= [\forall \alpha \leq \tau_1.\ \tau_2] \\
allbasis_\Gamma(\bigwedge[\tau_1 .. \tau_n]) &= allbasis_\Gamma(\tau_1) * \cdots * allbasis_\Gamma(\tau_n).
\end{aligned}$$

The fact that *arrowbasis* deserves its name — that is computes a "finite basis" for the set of all arrow-shaped supertypes of a given type — is captured by the following lemma:

**Lemma:**

1. $\Gamma \vdash \sigma \leq \bigwedge(arrowbasis_\Gamma(\sigma))$.
2. If $\Gamma \vdash \sigma \leq \tau_1 \to \tau_2$, then $\Gamma \vdash \bigwedge(arrowbasis_\Gamma(\sigma)) \leq \tau_1 \to \tau_2$.

It is then a simple matter to characterize the possible types for $(e_1\ e_2)$ by checking whether the minimal type of $e_2$ is a subtype of each of the left-hand sides in the arrow basis of the minimal type of $e_1$.

The *type synthesis* relation $\Gamma \vdash e \in \tau$ is the least relation closed under the following (syntax-directed) rules:

$$\Gamma \vdash x \in \Gamma(x) \qquad \text{(A-VAR)}$$

$$\frac{\Gamma, x{:}\tau_1 \vdash e \in \tau_2}{\Gamma \vdash \lambda x{:}\tau_1.\ e \in \tau_1 \to \tau_2} \qquad \text{(A-ARROW-I)}$$

$$\frac{\Gamma \vdash e_1 \in \sigma_1 \qquad \Gamma \vdash e_2 \in \sigma_2}{\Gamma \vdash e_1\ e_2 \in \bigwedge[\psi_i \mid (\phi_i \to \psi_i) \in arrowbasis_\Gamma(\sigma_1) \text{ and } \Gamma \vdash \sigma_2 \leq \phi_i]} \qquad \text{(A-ARROW-E)}$$

$$\frac{\Gamma, \alpha \leq \tau_1 \vdash e \in \tau_2}{\Gamma \vdash \Lambda\alpha \leq \tau_1.\ e \in \forall\alpha \leq \tau_1.\ \tau_2} \qquad \text{(A-ALL-I)}$$

$$\frac{\Gamma \vdash e \in \sigma_1}{\Gamma \vdash e[\tau] \in \bigwedge[\{\tau/\alpha\}\psi_i \mid (\forall\alpha \leq \phi_i.\ \psi_i) \in allbasis_\Gamma(\sigma_1) \text{ and } \Gamma \vdash \tau \leq \phi_i]} \qquad \text{(A-ALL-E)}$$

$$\frac{\text{for all } i,\ \Gamma \vdash \{\sigma_i/\alpha\}e \in \tau_i}{\Gamma \vdash for\ \alpha\ in\ \sigma_1..\sigma_n.\ e \in \bigwedge[\tau_1..\tau_n]} \qquad \text{(A-FOR)}$$

**Theorem:** [Soundness] If $\Gamma \vdash e \in \tau$ is derived by this algorithm, then it can also be derived from the original $F_\wedge$ typing rules.

**Proof:** Straightforward translation of derivations.

**Theorem:** [Minimal typing] If $\Gamma \vdash e \in \sigma$ is derived by the algorithm and $\Gamma \vdash e \in \tau$ can be derived from the original typing rules, then $\Gamma \vdash \sigma \leq \tau$.

**Proof:** Straightforward induction, using the properties of *allbasis* and *arrowbasis* for the application cases.

# 7 Semantics

A straightforward untyped semantics can be given for $F_\wedge$ by extending Bruce and Longo's partial equivalence relation model for $F_\leq$ [5]. The $\wedge$ type constructor is interpreted as intersection of PERs, and the bounded quantifier is interpreted as an infinite intersection.

More refined models have been given for intersections (c.f. [31]) interpreting $\wedge$ as a *limit* in the semantic category, and a typed semantics of $F_\wedge$ can be given along these lines as well, by translating $F_\wedge$ typing derivations into the pure second-order $\lambda$-calculus with surjective pairing, system $F_\times$. (This style of presentation avoids some of the subtleties involved in giving a direct denotational semantics for $F_\wedge$, since $F_\times$ itself has many well-studied models, but it still yields a useful soundness theorem relating the semantics to the $F_\wedge$ type system: valid $F_\wedge$ typing derivations are translated to well-typed (and hence well-behaved) $F_\times$ terms.) However, this construction has not been shown to be *coherent* [3], because the standard proof of this property for $\lambda_\wedge$ [32] relies on the existence of least upper bounds in the subtype relation — a property that $F_\wedge$ does not share.

An equational theory of provable equivalences between terms of pure $F_\wedge$ can be shown to be sound for both the untyped and the translation semantics.

## 8  Future Work

A primary practical concern for programming notations based on intersection types is the efficiency of typechecking for large programs. Naive implementations of the algorithms given here exhibit exponential behavior — in practice — in both type synthesis (because of the *for* construct) and subtyping (because of rules ASUBR-INTER and ASUBL-INTER). Fortunately, this behavior normally occurs as a result of explicit programmer directives — requests, in effect, for an exponential amount of analysis of the program during typechecking. Still, a serious implementation must find ways to economize; for example, by cacheing the partial results of previous analysis.

Another consideration for any language based on second-order polymorphism is the problem of verbosity. Without some means of abbreviation or partial type inference, even modest programs quickly become overburdened with type annotations. Cardelli's partial type inference method for $F_\leq$ [7] offers one promising direction of investigation. Another possibility is to use a $\omega$-order extension of $F_\wedge$, in which type operators could be used to express type information more succinctly.

Larger examples are needed to establish the practical need for intersection types and bounded quantification in their most general forms. It may be possible to obtain most of the practical power of $F_\wedge$ while remaining within a simpler, more tractable fragment.

# Acknowledgements

John Reynolds and Bob Harper supervised the thesis in which these results first appeared: their contributions can be found on every page. I am also grateful for discussions with Luca Cardelli, Tim Freeman, Nico Habermann, QingMing Ma, Frank Pfenning, and Didier Rémy.

# References

[1] Franco Barbanera and Mariangiola Dezani-Ciancaglini. Intersection and union types. In Ito and Meyer [21], pages 651–674.

[2] H. Barendregt, M. Coppo, and M. Dezani-Ciancaglini. A filter lambda model and the completeness of type assignment. *Journal of Symbolic Logic*, 48(4):931–940, 1983.

[3] Val Breazu-Tannen, Thierry Coquand, Carl Gunter, and Andre Scedrov. Inheritance as implicit coercion. *Information and Computation*, 93:172–221, 1991.

[4] Kim B. Bruce. Safe type checking in a statically typed object-oriented programming languages. In *Proceedings of the Twentieth ACM Symposium on Principles of Programming Languages*, January 1993.

[5] Kim B. Bruce and Giuseppe Longo. A modest model of records, inheritance, and bounded quantification. *Information and Computation*, 87:196–240, 1990. To appear in [19]. An earlier version appeared in the proceedings of the IEEE Symposium on Logic in Computer Science, 1988.

[6] Luca Cardelli. A semantics of multiple inheritance. *Information and Computation*, 76:138–164, 1988. An earlier version appears in *Semantics of Data Types*, Kahn, MacQueen, and Plotkin, eds., Springer-Verlag LNCS 173, 1984.

[7] Luca Cardelli. F-sub, the system. Unpublished manuscript, July 1991.

[8] Luca Cardelli. Typeful programming. In E. J. Neuhold and M. Paul, editors, *Formal Description of Programming Concepts*. Springer-Verlag, 1991. An earlier version appeared as DEC Systems Research Center Research Report #45, February 1989.

[9] Luca Cardelli. Extensible records in a pure calculus of subtyping. Research report 81, DEC Systems Research Center, January 1992. To appear in [19].

[10] Luca Cardelli, Simone Martini, John C. Mitchell, and Andre Scedrov. An extension of system F with subtyping. In Ito and Meyer [21], pages 750–770.

[11] Luca Cardelli and Peter Wegner. On understanding types, data abstraction, and polymorphism. *Computing Surveys*, 17(4), December 1985.

[12] Felice Cardone and Mario Coppo. Two extensions of Curry's type inference system. In Piergiorgio Odifreddi, editor, *Logic and Computer Science*, number 31 in APIC Studies in Data Processing, pages 19–76. Academic Press, 1990.

[13] Alonzo Church. A formulation of the simple theory of types. *Journal of Symbolic Logic*, 5:56–68, 1940.

[14] M. Coppo and M. Dezani-Ciancaglini. A new type-assignment for $\lambda$-terms. *Archiv. Math. Logik*, 19:139–156, 1978.

[15] Pierre-Louis Curien and Giorgio Ghelli. Coherence of subsumption: Minimum typng and type-checking in $f_\leq$. *Mathematical Structures in Computer Science*, 2:55–91, 1992. To appear in [19].

[16] Nicolas G. de Bruijn. Lambda-calculus notation with nameless dummies: a tool for automatic formula manipulation with application to the Church-Rosser theorem. *Indag. Math.*, 34(5):381–392, 1972.

[17] Giorgio Ghelli. A static type system for message passing. In *Conference on Object-Oriented Programming Systems, Languages, and Applications*, pages 129–143, Phoenix, Arizona, October 1991. Distributed as SIGPLAN Notices, Volume 26, Number 11, November 1991.

[18] Jean-Yves Girard. *Interprétation fonctionelle et élimination des coupures de l'arithmétique d'ordre supérieur*. PhD thesis, Université Paris VII, 1972.

[19] Carl A. Gunter and John C. Mitchell. *Theoretical Aspects of Object-Oriented Programming: Types, Semantics, and Language Design.* The MIT Press, 1993. To appear.

[20] Susumu Hayashi. Singleton, union and intersection types for program extraction. In Ito and Meyer [21], pages 701–730. To appear in Information and Computation.

[21] T. Ito and A. R. Meyer, editors. *Theoretical Aspects of Computer Software (Sendai, Japan)*, number 526 in Lecture Notes in Computer Science. Springer-Verlag, September 1991.

[22] QingMing Ma. Parametricity as subtyping. In *Proceedings of the Nineteenth ACM Symposium on Principles of Programming Languages*, Albequerque, NM, January 1992.

[23] John C. Mitchell. Toward a typed foundation for method specialization and inheritance. In *Proceedings of the 17th ACM Symposium on Principles of Programming Languages*, pages 109–124, January 1990. To appear in [19].

[24] John Mitchell, Sigurd Meldal, and Neel Madhav. An extension of Standard ML modules with subtyping and inheritance. In *Proceedings of the Eighteenth ACM Symposium on Principles of Programming Languages*, pages 270–278, Orlando, FL, January 1991.

[25] Benjamin C. Pierce. *Programming with Intersection Types and Bounded Polymorphism.* PhD thesis, Carnegie Mellon University, December 1991. Available as School of Computer Science technical report CMU-CS-91-205.

[26] Benjamin C. Pierce. Programming with intersection types, union types, and polymorphism. Technical Report CMU-CS-91-106, Carnegie Mellon University, February 1991.

[27] Benjamin C. Pierce. Bounded quantification is undecidable. *Information and Computation*, 1993. To appear; also to appear in [19]. Preliminary version in proceedings of POPL '91.

[28] Benjamin C. Pierce and David N. Turner. Object-oriented programming without recursive types. In *Proceedings of the Twentieth ACM Symposium on Principles of Programming Languages*, January 1993. Also available as University of Edinburgh technical report number ECS-LFCS-92-225.

[29] Garrell Pottinger. A type assignment for the strongly normalizable $\lambda$-terms. In *To H. B. Curry: Essays on Combinatory Logic, Lambda Calculus, and Formalism*, pages 561–577. Academic Press, New York, 1980.

[30] John Reynolds. Towards a theory of type structure. In *Proc. Colloque sur la Programmation*, pages 408–425, New York, 1974. Springer-Verlag LNCS 19.

[31] John C. Reynolds. Preliminary design of the programming language Forsythe. Technical Report CMU-CS-88-159, Carnegie Mellon University, June 1988.

[32] John C. Reynolds. The coherence of languages with intersection types. In Ito and Meyer [21], pages 675–700.

[33] P. Sallé. Une extension de la theorie des types en $\lambda$-calcul. pages 398–410. Springer-Verlag, 1982. Lecture Notes in Computer Science No. 62.

# A Logic for Parametric Polymorphism

Gordon Plotkin[*]    Martín Abadi[†]

**Abstract**

In this paper we introduce a logic for parametric polymorphism. Just as LCF is a logic for the simply-typed $\lambda$-calculus with recursion and arithmetic, our logic is a logic for System F. The logic permits the formal presentation and use of relational parametricity. Parametricity yields—for example—encodings of initial algebras, final co-algebras and abstract datatypes, with corresponding proof principles of induction, co-induction and simulation.

## 1 Introduction

In this paper we introduce a logic for parametric polymorphism, in the binary relational sense of Reynolds [Rey83]. Just as LCF is a first-order logic for the simply-typed $\lambda$-calculus, with recursion and arithmetic, so our logic is a second-order logic for System F. It is intended as a step towards a general logic of polymorphically typed programs. The terms are those of the second-order $\lambda$-calculus, and the formulae are built from equations and relations by propositional operators and quantifiers over elements of types, or over types, or over relations between types. The logic permits the formal presentation and use of relational parametricity, which is expressed by an axiom schema. Parametricity yields—for example—encodings of initial algebras, final co-algebras and abstract datatypes, with corresponding proof principles of induction, co-induction and simulation.

Our first goal is to provide a formal system for arguments that exploit relational parametricity. In all models of System F, standard constructions such as products and initial algebras are available in a weak sense (see *e.g.* [Böh85, Has90, RP90, Wra89]). If the models are relationally parametric, then these constructions become universal constructions in the usual sense of category theory. Bainbridge, Freyd, Scedrov, and Scott have given such results for the parametric Per model [BFSS90], and Hasegawa and Wadler for classes of models [Has90, Wad89]. Hasegawa [Has91] has shown that the second order minimal model—arising from the maximal consistent theory of Moggi and Statman—is parametric. By defining a formal logic, we hope to display the assumption of relational parametricity in a clear way, and to be able to obtain simple, general arguments for the results.

Our second goal is to pursue the idea of LCF. In LCF, a logic is given over a simply-typed $\lambda$-calculus with recursion and arithmetic. This $\lambda$-calculus is suitable for denotational semantics, and thus the corresponding logic acts as a rather powerful logic of programs. However, the simply-typed $\lambda$-calculus is inadequate for dealing with programming languages with abstract or polymorphic types, and an extension of the kind considered here is needed. The present work is but an intermediate step: it does not include recursion, at either

---

[*]Department of Computer Science, University of Edinburgh, King's Buildings, Edinburgh EH9 3JZ, UK. Part of this work was completed while at Digital Equipment Corporation, Systems Research Center.

[†]Digital Equipment Corporation, Systems Research Center. 130 Lytton Avenue, Palo Alto, California 94301, USA.

the level of values or the level of types. Moreover, one may also wish to consider richer type systems (*e.g.*, that of $F_\omega$ or even Constructions), richer logics (*e.g.*, that of Topos Theory), or more general notions of computation (*e.g.*, as in Moggi's suggestions for the use of monads). In yet another direction, we may consider a formal theory of subtypes, based, say, on $F_\leq$ [CG91, CG92].

There has been some work along related lines. Abadi, Cardelli, and Curien gave a system with very elementary judgments and few relations, and with a syntax as close as possible to the basic typed $\lambda$-calculus [ACC93]. Mairson used second-order logic over the untyped $\lambda$-calculus [Mai91]. He interpreted the types of System F as relations; as it happens his interpretation of polymorphic types is not parametric, but it could easily be made so. The second-order theory of subtypes of Cardelli, Martini, Mitchell and Scedrov embodies some aspects of parametricity *via* an equational rule [CMMS91].

In Section 2 of this paper we present the logic. Apart from axioms giving the equational theory of System F, the only non-logical principle is a schema for parametricity; further, only intuitionistic reasoning is employed. Other axioms may well be possible: one would expect that intuitionistic reasoning is (unsurprisingly) consistent with various choice principles. The availability of these principles would be a miniature form of Pitts' result that polymorphism can be considered as set-theoretic as long as one works constructively [Pit87]. A model can be provided as in [BFSS90]; it remains to consider a more abstract notion and other examples. We also compare other notions of polymorphism: we present a tentative formalisation of Strachey's idea; we consider the dinatural approach of Bainbridge *et al*; and, finally, we make the evident generalisation of Reynold's binary relations to relations of other degrees. We conjecture that if two terms of System F are equal in all models parametric in Strachey's sense then they are also equal in all models parametric in Reynold's sense. We prove that a schema expressing dinaturality is a consequence of the schema expressing binary relational parametricity. The relationship betweeen the various notions of relational parametricity is unknown.

In Section 3 we consider a variety of constructions showing the availability of finite products and sums, second-order existential types, initial algebras and final co-algebras. The constructions are by now standard; the main point here is that we can derive their properties within our logic. We also consider the logical properties of the constructions. For example, we give a general induction principle for initial algebras and a general bisimulation (or co-induction) principle for final co-algebras [AM89, Smy91, Pit92]. For second-order existential types, parametricity becomes a "simulation principle", namely that if there is a relation between two types respected by the two corresponding "implementations" then the corresponding elements of the existential type (the abstract types) are equal. Thus we also obtain a proof rule for abstract types, along the lines envisaged by Mitchell [Mit91]. Presumably one would obtain similar principles relative to relational parametricity of other degrees, and, again, it would be interesting to know the relationships between these. In both Sections 2 and 3, we generally omit proofs.

## 2  Basic Logic

In this section we present the basic logic. The types and terms are those of System F and they are given by the grammar:

Types: $\quad A ::= \ X \mid A \to B \mid \forall X.\, A$

Terms: $\quad t ::= \ x \mid \lambda x : A.\, t \mid u(t) \mid \Lambda X.\, t \mid t(A)$

Here $X$ ranges over type variables, $x$ over ordinary variables; we use notations such as $A[X]$ to indicate possible occurrences of variables in expressions, and then may write for example $A[B]$ to represent the result of substituting $B$ for $X$ in $A$ (avoiding capture of bound variables).

Using these terms we build up formulae from equations and binary relations.

Formulae: $\phi ::= (t =_A u) \mid R(t, u) \mid$
$\phi \supset \psi \mid \forall x : A. \phi \mid \forall X. \phi \mid \forall R {\subset} A \times B. \phi \mid$
$\bot \mid \phi \wedge \psi \mid \phi \vee \psi \mid \exists x : A. \phi \mid \exists X. \phi \mid \exists R {\subset} A \times B. \phi$

Here $R$ ranges over relation variables. The equality symbol is subscripted with a type, the type of the terms being equated. In System F, this type is unique, in a given environment, and so we often leave it implicit. The basic constructs are implication ($\supset$) and three sorts of universal quantifiers: over values; over types; and over relations between types (where $R {\subset} A \times B$ is read as "$R$ is a relation between $A$ and $B$"). The other constructs are useful but not altogether necessary. We make use of standard abbreviations when writing formulae, terms and types.

Typing judgments are used to specify which terms have which types, and which formulae are well-formed. A second-order environment $E$ is a finite sequence of type variables $X$ or typings $x : A$ in which no variable is introduced twice. The judgments $E \vdash A$ Type and $E \vdash t : A$ are defined in the usual way. To specify the well-formed formulae, we also need relation environments, which are finite sequences of relational typings $R {\subset} A \times B$ in which no variable is introduced twice. We define a judgment $E \vdash G$ REnv to assert that $G$ is a well-formed relation environment given $E$, with the obvious rule that if $R {\subset} A \times B$ appears in $G$ then $A$ and $B$ should be well-formed types given $E$. We define a judgment $E; G \vdash \phi$ Prop to assert that $\phi$ is a well-formed formula given $E; G$. The rules for atomic formulae are:

$$\frac{E \vdash t : A \quad E \vdash u : A \quad E \vdash G \text{ REnv}}{E; G \vdash t =_A u \text{ Prop}}$$

$$\frac{E \vdash t : A \quad E \vdash u : B \quad E \vdash G \text{ REnv} \quad R {\subset} A \times B \text{ in } G}{E; G \vdash R(t, u) \text{ Prop}}$$

Among the other rules we have, for example:

$$\frac{E, X; G \vdash \phi \text{ Prop}}{E; G \vdash \forall X. \phi \text{ Prop}} \qquad \frac{E; G, R {\subset} A \times B \vdash \phi \text{ Prop}}{E; G \vdash \forall R {\subset} A \times B. \phi \text{ Prop}}$$

In order to define the consequence relation of the logic we need further logical apparatus. This concerns definable relations, their substitution for relation variables in formulae (obtaining formulae) and for type variables in types (obtaining definable relations). Substitution in formulae is needed for the rules for relational quantification; substitution in types is needed to state the parametricity schema. Definable relations are given by the grammar:

Definable relations: $\rho ::= (x : A, y : B). \phi[x, y]$

We say that such a $\rho$ is a definable relation between $A$ and $B$, writing this as $\rho \subset A \times B$. We define a judgment $E; G \vdash \rho \subset A \times B$ to assert that $\rho$ is a well-formed definable relation between $A$ and $B$, given $E; G$. There is one rule for this judgment:

$$\frac{E; G, x : A, y : B \vdash \phi \text{ Prop}}{E; G \vdash (x : A, y : B). \phi \subset A \times B}$$

For example, the expression $(x : A, y : A). (x =_A y)$ defines the equality relation $eq_A$ on $A$, and $\langle f \rangle = (x : A, y : B). (y =_B f(x))$ defines the graph of a function $f$; here if $f$ has type $A \to B$ then $\langle f \rangle$ is a well-formed definable relation between $A$ and $B$ (given appropriate $E;G$). We sometimes treat a relation variable $R \subset A \times B$ as the definable relation $(y:A, z:B). R(y,z)$.

When $\rho$ is such a definable relation between $A$ and $B$, we sometimes use the abbreviations $t\rho u$ or $\rho(t,u)$ for the corresponding $\phi[t,u]$. A definable relation $\rho$ can be substituted for a relation variable $R$ in a formula $\psi[R]$, yielding $\psi[\rho]$. In particular when $\psi[R]$ is $R(t,u)$, the result of the substitution is $\rho(t,u)$.

To define the substitution of definable relations for type variables in types, we need first to be able to combine relations by exponentiation and universal quantification. If $\rho \subset A \times B$ and $\rho' \subset A' \times B'$ then

$$(\rho \to \rho') \subset (A \to A') \times (B \to B')$$

is defined by:

$$f(\rho \to \rho')g \equiv \forall x : A \forall x' : A'. (x\rho x' \supset f(x)\rho' g(x'))$$

If $\rho$ and $\rho'$ are well-formed given $E;G$ then so is $\rho \to \rho'$.

For universal quantification, if $\rho \subset A \times B$, then

$$\forall(Y, Z, R{\subset}Y \times Z)\rho \subset (\forall Y. A) \times (\forall Z. B)$$

is defined by:

$$y(\forall(Y, Z, R{\subset}Y \times Z).\rho)z \equiv \forall Y \forall Z \forall R{\subset}Y \times Z. ((yY)\rho(zZ))$$

If $\rho$ is well-formed given $E, Y, Z; G, (R{\subset}Y \times Z)$ then $\forall(Y, Z, R{\subset}Y \times Z)\rho$ is well-formed given $E;G$.

We can now define the substitution of definable relations for type variables in types. If $\vec{X} = X_1, \ldots, X_n$, $\vec{B} = B_1, \ldots, B_n$, $\vec{C} = C_1, \ldots, C_n$, and $\vec{\rho} = \rho_1, \ldots, \rho_n$ where $\rho_i \subset B_i \times C_i$, then $A[\vec{\rho}] \subset A[\vec{B}] \times A[\vec{C}]$ is the result of substituting $\vec{\rho}$ for $\vec{X}$ in $A[\vec{X}]$. The definition of $A[\vec{\rho}]$ is by cases:

- if $A$ is $X_i$ then $A[\vec{\rho}]$ is $\rho_i$;

- if $A$ is $A' \to A''$ then $A[\vec{\rho}] = A'[\vec{\rho}] \to A''[\vec{\rho}]$;

- if $A$ is $\forall Y. A'[\vec{X}, Y]$ then $A[\vec{\rho}] = \forall(Y, Z, R{\subset}Y \times Z). A'[\vec{\rho}, R]$.

If each $\rho_i$ is well-formed given $E;G$ then so is $A[\vec{\rho}]$.

It remains to give axiom schemes and rules in order to define the consequence relation of the logic. This relation is written as $\Gamma \vdash_{E;G} \phi$, where $\Gamma$ is a finite set of formulae, and all formulae involved are well-formed given $E;G$. The proof system has three parts:

- standard rules for the connectives and quantifiers;

- axioms and rules for the equational part of System F;

- a schema to express relational parametricity.

Natural deduction rules for the connectives and quantifiers are given as usual for intuitionistic logic. For example the rules for implication and universal quantification over types and relations are:

$$\frac{\Gamma, \phi \vdash_{E;G} \psi}{\Gamma \vdash_{E;G} \phi \supset \psi} \qquad \frac{\Gamma \vdash_{E;G} \phi \supset \psi \quad \Gamma \vdash_{E;G} \phi}{\Gamma \vdash_{E;G} \psi}$$

$$\frac{\Gamma \vdash_{E,X;G} \phi[X]}{\Gamma \vdash_{E;G} \forall X. \phi[X]} \qquad \frac{\Gamma \vdash_{E;G} \forall X. \phi[X] \quad E \vdash A \text{ Type}}{\Gamma \vdash_{E;G} \phi[A]}$$

$$\frac{\Gamma \vdash_{E;G,R \subset A \times B} \phi[R]}{\Gamma \vdash_{E;G} \forall R \subset A \times B. \phi[R]} \qquad \frac{\Gamma \vdash_{E;G} \forall R \subset A \times B. \phi[R] \quad E;G \vdash \rho \subset A \times B}{\Gamma \vdash_{E;G} \phi[\rho]}$$

In the above rules the usual provisions are made that the variables being bound do not appear in assumptions.

We now give that part of the logic that deals with equality. This consists of certain axioms and axioma schemes. We adopt the convention that if an axiom $\phi$ is written, what is meant is that all sequents of the form $\Gamma \vdash_{E;G} \phi$ are asserted, where $\phi$ and all formulae in $\Gamma$ are well-formed for some $E;G$. Reflexivity is the axiom

$$\forall X \forall x : X. (x =_X x)$$

Substitution is the axiom

$$\forall X \forall Y \forall R \subset X \times Y \forall x : X \forall x' : X \forall y : Y \forall y' : Y.$$
$$R(x,y) \wedge x =_X x' \wedge y =_Y y' \supset R(x',y')$$

We also use the congruence schemas

$$(\forall x : A. t =_B u) \supset (\lambda x : A. t) =_{A \to B} (\lambda x : A. u)$$

$$(\forall X. t =_B u) \supset (\Lambda X. t) =_{\forall X. B} (\Lambda X. u)$$

Lastly there are $\beta$ equalities:

$$\forall x : A. ((\lambda x : A. t)x =_B t)$$

$$\forall X. ((\Lambda X. t)X =_B t)$$

and $\eta$ equalities:

$$\forall X \forall Y \forall f : X \to Y. ((\lambda x : X. fx) =_{X \to Y} f)$$

$$\forall f : \forall X. B. ((\Lambda X. fX) =_{\forall X. B} f)$$

We can now formulate the parametricity schema:

$$\forall Y_1 \ldots \forall Y_n \forall u : (\forall X. A[X, Y_1, \ldots, Y_n]). u(\forall X. A[X, eq_{Y_1}, \ldots, eq_{Y_n}])u$$

where $A$ has free type variables $X, Y_1, \ldots, Y_n$.

To understand this, it is convenient to ignore the parameters, $Y_1, \ldots, Y_n$, and expand the definition to obtain

$$\forall u : (\forall X. A[X]). \forall Y \forall Z \forall R \subset Y \times Z. u(Y)A[R]u(Z)$$

This formula states that if one instantiates an element of a polymorphic type at two related types, then the two values obtained are themselves related; this statement expresses Reynolds' idea of binary relational parametricity.

One way to interpret this logic is to follow Bainbridge et al. [BFSS90], taking types to be pers over the natural numbers and with universal quantification being interpreted parametrically with respect to the double-negation-closed relations. Formulae are interpreted classically, with type variables ranging over pers, ordinary variables ranging over equivalence classes relative to the appropriate per, and relation variables ranging over the double-negation closed relations. There are other evident possibilities of interpretation; for example one could interpret the logic intuitionistically *via* a realisability interpretation. It would be good to have a general framework for interpreting the logic. Both Reynolds and Ma [MR92] and Hasegawa [Has90, Has91] have considered frameworks for parametric interpretations of System F. Pitts has shown that every hyperdoctrine model of System F fully embeds in a topos model [Pit87]. In this way the types of the model appear as "sets" in the topos. Perhaps something similar can be done here, in such a way that the relations of the model appear as relations in the topos, in the usual categorical sense. One would hope for completeness results for the present logic. In this connection one can ask too what is the most convenient language combining that of the present logic and the higher-order logic afforded by a topos; this question is apposite even without any consideration of parametricity.

**Lemma 1 (Identity Extension Lemma)** *For any $A[\vec{X}]$ with free variables in $\vec{X}$ it is provable in the logic that*

$$\forall \vec{X} \forall u : A, v : A. \ (u A[eq_{\vec{X}}] v \equiv (u =_A v))$$

**Lemma 2 (Logical Relations Lemma)** *Let $A_1[\vec{X}], \ldots, A_m[\vec{X}], B[\vec{X}]$ have free type variables in $\vec{X}$. Suppose $t[x_1, \ldots, x_m] : B$ given the environment $\vec{X}, x_1 : A_1, \ldots, x_m : A_m$. Then the following is provable in the logic without using the parametricity schema:*

$$\forall \vec{X} \forall \vec{Y} \forall \vec{R} \subset \vec{X} \times \vec{Y}$$
$$\forall x_1 : A_1[\vec{X}], \ldots, x_m : A_m[\vec{X}]$$
$$\forall y_1 : A_1[\vec{Y}], \ldots, y_m : A_m[\vec{Y}].$$
$$(x_1 A_1[\vec{R}] y_1 \wedge \ldots \wedge x_m A_m[\vec{R}] y_m) \supset t[x_1, \ldots, x_m] A[\vec{R}] t[y_1, \ldots, y_m]$$

## 2.1 Categorical Matters

The types form a category, in a formal sense, within our logic. The idea is to take the type $X \to X'$ as the "set" of morphisms from $X$ to $X'$. Composition is given by the combinator comp : $\forall X \forall X' \forall X''.(X' \to X'') \to (X \to X') \to (X \to X'')$, where:

$$\text{comp} = \Lambda X \Lambda X' \Lambda X'' \lambda g : X' \to X'' \lambda f : X \to X' \lambda x : X. \ g(f(x))$$

and the identity, id : $\forall X. (X \to X)$ by:

$$\text{id} = \Lambda X \lambda x : X. \ x$$

We write $t; u$ for $\text{comp}(A)(A')(A'')(u)(t)$ (where $u$ and $t$ have respective types $A' \to A''$ and $A \to A'$, given $E; G$); we write $id_A$ for $\text{id}(A)$. One can show that composition is associative, in that it is provable in the logic (without using the parametricity schema) that:

$$\forall X, X', X'', X'''. \forall f : X \to X', g : X' \to X'', h : X'' \to X'''. \ (f; g); h = f; (g; h)$$

and for the identity that:

$$\forall X, X'. \forall f : X \to X'. \ (id_X; f = f) \wedge (f; id_{X'} = f)$$

Now let $A[\vec{X}, \vec{Y}]$ be a type such that all type variables in $\vec{X}$ have only negative occurrences, and all type variables in $\vec{Y}$ have only positive occurrences. Then one may think of $A[\vec{X}, \vec{Y}]$ as acting on types $\vec{X}, \vec{Y}$ to produce a type. With an additional action on functions one obtains an associated multivariant formal functor. This action is given by a term:

$$M_A : \forall \vec{X}, \vec{Y}. \forall \vec{X'}, \vec{Y'}. (\vec{X'} \to \vec{X}) \to (\vec{Y} \to \vec{Y'}) \to (A[\vec{X}, \vec{Y}] \to A[\vec{X'}, \vec{Y'}])$$

The term is given inductively on the structure of $A$.

$$M_Z \vec{X}\vec{Y}\vec{X}'\vec{Y}'\vec{f}\vec{g} = \begin{cases} f_i & \text{(if $Z$ is the $i$th entry in $\vec{X}$)} \\ g_j & \text{(if $Z$ is the $j$th entry in $\vec{Y}$)} \\ id_Z & \text{(otherwise)} \end{cases}$$

$$M_{(B[\vec{Y},\vec{X}] \to C[\vec{X},\vec{Y}])} \vec{X}\vec{Y}\vec{X}'\vec{Y}'\vec{f}\vec{g} = \lambda h : B[\vec{Y},\vec{X}] \to C[\vec{X},\vec{Y}].$$
$$(M_{B[\vec{Y},\vec{X}]} \vec{Y}'\vec{X}'\vec{Y}\vec{X}\vec{g}\vec{f}); h; (M_{C[\vec{X},\vec{Y}]} \vec{X}\vec{Y}\vec{X}'\vec{Y}'\vec{f}\vec{g})$$

$$M_{\forall Z. B[\vec{X},\vec{Y}]} \vec{X}\vec{Y}\vec{X}'\vec{Y}'\vec{f}\vec{g} = \lambda z : (\forall Z. B[\vec{X},\vec{Y}]) \wedge Z.(B[\vec{X},\vec{Y}]\vec{X}\vec{Y}\vec{X}'\vec{Y}'\vec{f}\vec{g}(zZ))$$

Given $\vec{t},\vec{u}$ of types $\vec{B}' \to \vec{B}, \vec{C} \to \vec{C}'$, we write $A[\vec{t},\vec{u}]$ for $M_A \vec{A}\vec{B}\vec{A}'\vec{B}'\vec{t}\vec{u}$. One can show that this yields a functor. That is, it is provable in the logic (without using the parametricity schema) that:

$$\forall \vec{X},\vec{Y}.(A[id_{\vec{X}}, id_{\vec{Y}}] = id_{A[\vec{X},\vec{Y}]})$$

and also:

$$\forall \vec{X},\vec{Y},\vec{X}',\vec{Y}',\vec{X}'',\vec{Y}''. \forall \vec{f}:\vec{X}' \to \vec{X}. \forall \vec{g}:\vec{Y} \to \vec{Y}'. \forall \vec{f}':\vec{X}'' \to \vec{X}'. \forall \vec{g}':\vec{Y}' \to \vec{Y}''.$$
$$(A[(\vec{f}';\vec{f}),(\vec{g};\vec{g}')] = A[\vec{f},\vec{g}]; A[\vec{f}',\vec{g}'])$$

One can also show that the operations of applying functors and taking graphs commute. The *opposite* of a relation $\rho \subset A \times B$ is defined by:

$$\rho^{op} = (y:B, x:A).x\rho y$$

**Lemma 3 (Graph Lemma)** *Let $A[\vec{X},\vec{Y}]$ be a type all of whose free variables appear in the list $\vec{X},\vec{Y}$ and such that all type variables in $\vec{X}$ have only negative occurrences, and all type variables in $\vec{Y}$ have only positive occurrences. Then it can be proved in the logic that:*

$$\forall \vec{X},\vec{Y},\vec{X}',\vec{Y}'. \forall \vec{f}:\vec{X}' \to \vec{X}. \forall \vec{g}:\vec{Y} \to \vec{Y}'. (\langle A[\vec{f},\vec{g}]\rangle = A[\langle \vec{f}\rangle^{op}, \langle \vec{g}\rangle])$$

Here two relations are considered as equal if they coincide extensionally.

## 2.2 Other Notions of Parametricity

As well as Reynold's relational notion of parametricity, there are two others: an informal one due to Strachey [Str67] and a categorical one due to Bainbridge *et al.* [BFSS90]. Strachey's original idea was that a parametric polymorphic function behaves the same way for all types. Let us essay a formalisation of this idea. In Per models, universal types can be modelled as intersections of all their instances; so at any type a polymorphic function has the same realisor as at any other. It seems reasonable, therefore, to claim that Strachey could accept any Per model as being parametric. Now Mitchell has shown that two terms of the same type are equal in a wide class of such models iff their erasures are $\beta\eta$-equivalent [Mit90]. Let us (informally) define two terms of System F to be *Strachey equivalent* iff they are equal in all models of parametric polymorphism in Strachey's sense. Accepting Per models as parametric, we are lead to assert that Strachey equivalence implies being of the same type and having the same erasure. Conversely if two terms have the same type and erasure it seems not unreasonable that they would denote the same element of any model parametric in Strachey's sense. We therefore tentatively identify Strachey equivalence with having the same type and the same erasure (up to $\beta\eta$-equivalence).

Along similar lines, one would informally define two terms of System F to be *Reynold's equivalent* iff they are equal in all models of binary relational polymorphism in Reynold's sense. Anticipating a completeness theorem for our logic relative to models of System F which are polymorphic in the binary relational sense, we can identify Reynold's equivalence with provable equality in our logic. Since there are many examples of terms provably equal in the above theory which are not Strachey equivalent, Reynolds equivalence does not imply Strachey equivalence. We conjecture the converse does hold, (A similar conjecture appears in [ACC93].) Many positive instances encourage this conjecture. For example, $\lambda x : (\forall X. X). x$ and $\lambda x : (\forall X. X). x(\forall X. X)$ have the same erasure and the same type $(\forall X. X) \to (\forall X. X)$, and they are provably equal; if $X$ is not free in $A$ then $\lambda x : (\forall X. A). xB$ and $\lambda x : (\forall X. A). xC$ have the same erasure and the same type $(\forall X. A) \to A$, and they too are provably equal.

According to Bainbridge *et al.* parametricity is provided by the notion of a dinatural transformation: in certain models, all terms denote such transformations between types [BFSS90]. One can express dinaturality by a schema. Let $A[Y, X]$ be a type in which $Y$ occurs only negatively and $X$ occurs only positively. Then:

$$\forall X, Y \forall t : (\forall X. A[X, X] \to A[Y, Y]) \forall f : X \to Y.$$
$$A[f, id_X]; t(X); B[id_X, f] = A[id_Y, f]; t(Y); B[f, id_Y]$$

We show now that this schema follows from that for (relational) parametricity. (This seems to answer a question raised in [BFSS90].)

First, the schema can be recast in a simpler way, as:

$$\forall X, Y \forall f : X \to Y. (\cdot)_X; A[id_X, f] = (\cdot)_Y; A[f, id_Y]$$

where (again) $A[Y, X]$ is a type in which $Y$ occurs only negatively and $X$ occurs only positively and where, e.g., $(\cdot)_X$ is $\lambda u : (\forall X. A[X, X]). u(X)$. For the sake of notational simplicity let us only consider the case where there are no parameters (*i.e.* free type variables in $A$ other than $X$ or $Y$). So let $X$ and $Y$ be types and suppose that $f : X \to Y$. Choose $u$ in $(\forall X. A[X, X])$. Appying the parametricity schema to $\langle f \rangle$ we obtain:

$$u(X) A[\langle f \rangle, \langle f \rangle] u(Y)$$

We have that $A[h, g] : A[V, U] \to A[V', U']$ given the $U, V, U', V', g : U \to U', h : V' \to V$. We may therefore apply the Logical Relations Lemma, obtaining:

$$A[id_X, f](A[\langle f \rangle, \langle f \rangle] \to A[eq_X, eq_Y]) A[f, id_Y]$$

as $f(\langle f \rangle) \to eq_Y) id_Y$ and $id_X (eq_X \to \langle f \rangle) f$. But now it follows that:

$$A[id_X, f](u(X)) A[eq_X, eq_Y] A[f, id_Y](u(Y))$$

and by the Identity Extension Lemma this is just:

$$A[id_X, f](u(X)) =_{A[X,Y]} A[f, id_Y](u(Y))$$

as required.

There is an evident generalisation of relational parametricity to $n$-ary relational parametricity. One introduces $n$-ary definable relations ($n \geq 0$) by:

$$\rho = (x_1 : A_1, \ldots, x_n : A_n). \phi[x_1, \ldots, x_n]$$

We write $\rho \subset (A_1 \times \ldots \times A_n)$, and say $\rho$ is well-formed given $E; G$ provided that $\phi[x_1, \ldots, x_n]$ is well-formed given $E, x_1 : A_1, \ldots, x_n : A_n; G$. We also write $\rho(t_1, \ldots, t_n)$ for $\phi[t_1, \ldots, t_n]$. The place of the equality relation is taken by the diagonal $\Delta_A^n$ where:

$$\Delta_A^n = (x_1 : A, \ldots, x_n : A). (x_1 = x_2) \wedge \ldots \wedge (x_{n-1} = x_n)$$

Exponentiation and universal quantification of $n$-ary relations is defined analogously to the case of binary relations, and we may substitute definable $n$-ary relations in types to obtain definable $n$-ary relations. The schema of $n$-ary relational parametricity is:

$$\forall Y_1 \ldots \forall Y_m \forall u : (\forall X.\ A[X, Y_1, \ldots, Y_m]).\ (\forall X.\ A[X, \Delta^n_{Y_1}, \ldots, \Delta^n_{Y_m}])(u, \ldots, u)$$

where $A$ has free type variables $X, Y_1, \ldots, Y_n$. It is unknown what the relations between these schemas are. We do not even know if the schema for binary relational parametricity implies that for unary parametricity.

## 3 Parametricity and Datatypes

We now treat datatype constructors, comprising: finite products and sums, second-order existential quantification; and initial algebras and final co-algebras. As well as categorical properties, we derive logical properties of the constructs.

### 3.1 Products

For the empty product, let us define:

$$1 = \forall X.\ X \to X$$

The term $\star = \Lambda X \lambda x : X.\ x$ inhabits 1. It is provable, using dinaturality, that:

**Theorem 1** $\forall u : 1.\ (u =_1 \star)$

Note that we have adopted an informal style, asserting formulae. What we intend is to assert that the formulae are provable in the logic. The categorical statement

$$\forall X \exists! f : X \to 1.\ \top$$

of terminality follows at once from the theorem.

Turning to binary products, for any types $A$ and $B$, set

$$A \times B = \forall Z.\ ((A \to B \to Z) \to Z)$$

where $Z$ does not appear in $A$ or $B$. This yields a weak product in System F, with combinators $fst_{A,B} : A \times B \to A$, $snd_{A,B} : A \times B \to B$, and $pair_{A,B} : A \to B \to A \times B$, given by

$$fst_{A,B}(z) = zAK_{A,B}$$

$$snd_{A,B}(z) = zBK'_{A,B}$$

(where $K_{A,B} = \lambda x : A \lambda y : B.\ x$ and $K'_{A,B} = \lambda x : A \lambda y : B.\ y$), and:

$$pair_{A,B}(a)(b) = \Lambda Z \lambda f : A \to B \to Z.\ fab$$

Below we drop the subscripts $A, B$ and also write $\langle t, u \rangle$ for $pair(t)(u)$.

The formulae

$$\forall x : A \forall y : B.\ \bigl(fst(\langle x, y \rangle) = x\bigr)$$

and

$$\forall x : A \forall y : B.\ \bigl(snd(\langle x, y \rangle) = y\bigr)$$

are provable without using parametricity. Using dinaturality we obtain:

**Theorem 2** $\forall w : A \times B. (\langle fst(w), snd(w)\rangle =_{A \times B} w)$

The theorem yields the usual categorical characterization of binary products, namely:

$$\forall Z \forall f : Z \to A \forall g : Z \to B. \exists! h : Z \to A \times B. (f = h; fst \land g = h; snd)$$

Lastly we consider the action of binary products on relations. Given $\rho \subset A \times A'$ and $\sigma \subset B \times B'$ we take $(\rho \times \sigma) \subset (A \times B) \times (A' \times B')$ to be $\forall Z. (\rho \to (\sigma \to Z)) \to Z$. This is obtained by substituting $\rho$ and $\sigma$ for $X$ and $Y$ in $X \times Y$, which is $\forall Z. (X \to (Y \to Z)) \to Z$.

**Theorem 3**

$$\forall u : A \times A' \forall v : B \times B'. u(\rho \times \sigma)v \equiv (fst(u)\rho\, fst(v) \land snd(u)\sigma\, snd(v))$$

### 3.2 Sums

For the empty sum, define:

$$\mathbf{0} = \forall X. X$$

**Theorem 4** $\forall u : \mathbf{0}. \bot$

It follows that **0** is initial in the categorical sense.

Turning to binary sums, for types $A$ and $B$ we take

$$A + B = \forall Z. ((A \to Z) \to (B \to Z) \to Z)$$

where $Z$ does not occur in $A$ or $B$. The combinators $inl_{A,B} : A \to A+B$, $inr_{A,B} : B \to A+B$, and $cases_{A,B} : \forall Z. ((A \to Z) \to (B \to Z) \to (A+B) \to Z)$ are defined by:

$$inl_{A,B}(a) = \Lambda Z \lambda f : A \to Z \lambda g : B \to Z. f(a)$$
$$inr_{A,B}(b) = \Lambda Z \lambda f : A \to Z \lambda g : B \to Z. g(b)$$

and

$$cases_{A,B} C f g u = u C f g$$

We may drop the subscripts and may also write $[t, u]_C$ for $cases(t, u)$.

It is provable without using parametricity that

$$\forall x : A \forall Z \forall f : A \to Z \forall g : B \to Z. ([f, g]_Z(inl(x)) = f(x))$$

and

$$\forall y : B \forall Z \forall f : A \to Z \forall g : B \to Z. ([f, g]_Z(inr(y)) = g(y))$$

With dinaturality we obtain:

**Theorem 5** $\forall Z \forall h : A + B \to Z. h = [inl; h, inr; h]_Z$

from which the usual categorical characterisation of binary sums follows.

Next we consider the action of binary sums on relations. Given $\rho \subset A \times A'$ and $\sigma \subset B \times B'$ we take $(\rho + \sigma) \subset (A + B) \times (A' + B')$ to be $\forall Z. ((\rho \to Z) \to (\sigma \to Z) \to Z)$, following the same pattern as for binary products.

**Theorem 6**

$$\forall u : A + B \forall u' : A' + B'.$$
$$u(\rho + \sigma)u'$$
$$\equiv$$
$$(\exists x : A \exists x' : A'. (u = inl(x) \land u' = inl(x') \land x\rho x')$$
$$\lor$$
$$\exists y : B \exists y' : B'. (u = inr(y) \land u' = inr(y') \land y\sigma y'))$$

By the Identity Extension Lemma, $eq_A + eq_B$ and $eq_{A+B}$ are equivalent. With this and the above theorem we get that:

$$\forall u : A + B. (\exists x : A. u = inl(x)) \lor (\exists y : B. u = inr(y))$$

## 3.3 Existential Quantification

The existential quantification $\exists X. A[X]$ is taken to abbreviate

$$\forall Y. (\forall X. (A[X] \to Y) \to Y)$$

where $Y$ does not appear in $A$ and is different from $X$.

One defines the combinator $pack : \forall X.(A[X] \to \exists X. A[X])$ and the combinator $unpack : (\exists X. A[X]) \to \forall Y. (\forall X. (A[X] \to Y) \to Y)$ by:

$$pack(X)(x) = \Lambda Y \lambda f : \forall X. (A[X] \to Y). f(X)(x)$$

and

$$unpack(u)(Y)(f) = u(Y)(f)$$

and it is provable without using parametricity that

$$\forall X \forall x : A[X] \forall Y \forall f : (\forall X.(A[X] \to Y). (unpack(pack\, X\, x) Y f =_Y f X x)$$

Let us see how existential quantification operates on relations. Let $\vec{Z}$ be the type variables free in $A[X, \vec{Z}]$, other than $X$, and let $\vec{\rho} \subset \vec{B} \times \vec{C}$ be a vector of relations of the same length. Then $\exists X. A[X, \vec{\rho}]$ is taken to be $\forall Y. (\forall X. (A[X, \vec{\rho}] \to Y) \to Y)$, following the usual pattern. Using the parametricity schema, one can show:

**Theorem 7**

$$\forall u : (\exists X. A[X, \vec{B}]), v : (\exists X. A[X, \vec{C}]).$$
$$u(\exists X. A[X, \vec{\rho}])v$$
$$\equiv$$
$$(\exists X, Y \exists x : A[X, \vec{B}], y : A[Y, \vec{C}] \exists S \subset X \times Y.$$
$$u = (pack\, X\, x) \land v = (pack\, Y\, y) \land x(A[S, \vec{\rho}])y)$$

Applying the Identity Extension Lemma, we get:

$$\forall \vec{Z} \forall u, v : (\exists X. A[X, \vec{Z}]).$$
$$u = v$$
$$\equiv$$
$$(\exists X, Y \exists x : A[X, \vec{Z}], y : A[Y, \vec{Z}] \exists S \subset X \times Y.$$
$$u = (pack\, X\, x) \land v = (pack\, Y\, y) \land x(A[S, eq_{\vec{Z}}])y)$$

The implication from right to left can be viewed as stating that existentials are parametric (in a sense dual to that axiomatized for universals); it is a principle that enables one to show that two abstract datatypes are equal, if one can demonstrate a simulation relation between them. We view this as the proof principle requested by Mitchell in [Mit91].

Working from left to right, one obtains:

$$\forall \vec{Z} \forall u : (\exists X. A[X, \vec{Z}]) \exists X \exists x : A[X, \vec{Z}].(u = (pack\, X\, x))$$

and one can then obtain a categorical characterization of existential quantification in the form:

$$\forall Y \forall f : (\forall X. (A[X] \to Y)) \exists! g : (\exists X. A[X] \to Y) \forall X. (fX =_Y (pack X); g)$$

## 3.4 Initial Algebras

Let $A[Z]$ be a type where the variable $Z$ occurs only positively. The initial $A[Z]$ algebra is given by

$$\mu Z.\, A[Z] = \forall Z.\, ((A[Z] \to Z) \to Z)$$

and the combinators are

$$fold : \forall Z.\, ((A[Z] \to Z) \to ((\mu Z.\, A[Z]) \to Z))$$

and

$$in : A[\mu Z.\, A[Z]] \to \mu Z.\, A[Z]$$

where:

$$fold(Z)(f) = \lambda z : \mu Z.\, A[Z].\, z(Z)(f)$$

and

$$in(z) = \Lambda Z \lambda f : A[Z] \to Z.\, f(A[foldZf]z)$$

The structure $\langle \mu Z.\, A[Z], in \rangle$ is a weakly initial $A[Z]$ algebra:

$$\forall Z \forall f : A[Z] \to Z.\, (in; (foldZf) = A[foldZf]; f)$$

This can be shown without using parametricity (or any $\eta$-equalities). With parametricity one has the following expression of the initiality of $\langle \mu Z.\, A[Z], in \rangle$:

**Theorem 8** $\forall Z \forall f : A[Z] \to Z \exists! h : \mu Z.\, A[Z] \to Z.\, (in; h = A[h]; f)$

Turning to logical relations let $\vec{X}$ be a list of the type variables free in $A[Z, \vec{X}]$ apart from $Z$ and let $\vec{\rho} \subset \vec{B} \times \vec{C}$ be a list of relations of the same length. Then the relation $\mu Z.\, A[Z, \vec{\rho}] \subset \mu Z.\, A[Z, \vec{B}] \times \mu Z.\, A[Z, \vec{C}]$ is taken to be $\forall Z.\, ((A[Z, \vec{\rho}] \to Z) \to Z)$, following the usual pattern. Using parametricity one has:

**Theorem 9**

$$\forall x : (\mu Z.\, A[Z, \vec{B}]) \forall y : (\mu Z.\, A[Z, \vec{C}]).$$
$$x(\mu Z.\, A[Z, \vec{\rho}])y$$
$$\equiv$$
$$\forall R \subset (\mu Z.\, A[Z, \vec{B}]) \times (\mu Z.\, A[Z, \vec{C}]).$$
$$(\forall u : A[(\mu Z.\, A[Z, \vec{B}]), \vec{B}] \forall v : A[(\mu Z.\, A[Z, \vec{C}]), \vec{C}].$$
$$uA[R, \vec{\rho}]v \supset in(u)R\, in(v))$$
$$\supset$$
$$xRy$$

Put more informally, this theorem states that $\mu Z.\, A[Z, \vec{\rho}]$ is the *least* relation $R$ such that $in(A[R, \vec{\rho}]) \subset R$.

Applying the Identity Extension Lemma to this theorem we obtain the following *binary* induction principle for a relation $\rho \subset \mu Z.\, A[Z, \vec{X}] \times \mu Z.\, A[Z, \vec{X}]$:

$$\forall u, v : A[\mu Z.\, A[Z, eq_{\vec{X}}], eq_{\vec{X}}].$$
$$(uA[\rho, eq_{\vec{X}}]v \supset in(u)\rho\, in(v))$$
$$\supset$$
$$\forall x, y : \mu Z.\, A[Z, \vec{X}].\, x\rho y$$

This amounts to saying that, if $in(A[\rho, eq_{\vec{X}}]) \subset \rho$ then $x\rho y$ holds for all $x, y$ in $\mu Z.\, A[Z, \vec{X}]$. It is somewhat surprising that we do not obtain the expected *unary* induction principle, that, for any property (i.e. unary relation) $\pi$, if $in(A[\pi, \Delta^1_{\vec{X}}]) \subset \pi$ then $\pi(x)$ holds for all $x$ in $\mu Z.\, A[Z, \vec{X}]$. It would be interesting to know the relation between the two, and also with the corresponding principles of other degrees.

## 3.5 Final Co-algebras

Let $A[Z]$ be a type where the variable $Z$ occurs only positively. The final $A[Z]$ co-algebra is given by
$$\nu Z. A[Z] = \exists Z. ((Z \to A[Z]) \times Z)$$
and the combinators are
$$\textit{unfold} : \forall Z. ((Z \to A[Z]) \to (Z \to (\nu Z. A[Z])))$$
and
$$\textit{out} : \nu Z. A[Z] \to A[\nu Z. A[Z]]$$
where
$$\textit{unfold}(Z)(f)(z) = \textit{pack}(Z)(\langle f, z \rangle)$$
and
$$\begin{aligned}\textit{out}(u) &= \textit{unpack}(u)(A[\nu Z. A[Z]]) \\ &\quad (\Lambda Z \lambda w : ((Z \to A[Z]) \times Z). A[\textit{unfold}\, Z(\textit{fst}\, w)]((\textit{fst}\, w)(\textit{snd}\, w)))\end{aligned}$$

The structure $\langle \nu Z. A[Z], \textit{out} \rangle$ is a weakly final co-algebra:
$$\forall Z \forall f : Z \to A[Z]. (\textit{unfold}(Z)(f); \textit{out} = f; A[\textit{unfold}(Z)(f)])$$

This can be shown without using parametricity (or any $\eta$-equalities). With parametricity one obtains a characterization of $\langle \nu Z. A[Z], \textit{out} \rangle$ as the final $A[Z]$ co-algebra:

**Theorem 10** $\forall Z \forall f : Z \to A[Z] \exists ! h : Z \to (\nu Z. A[Z]). (h; \textit{out} = f; A[h])$

Turning to logical relations, let $\vec{X}$ be a list of the type variables free in $A[Z, \vec{X}]$ apart from $Z$ and let $\vec{\rho} \subset \vec{B} \times \vec{C}$ be a list of relations of the same length. Then the relation $\nu Z. A[Z, \vec{\rho}] \subset \nu Z. A[Z, \vec{B}] \times \nu Z. A[Z, \vec{C}]$ is taken to be $\exists Z. ((Z \to A[Z, \vec{\rho}]) \times Z)$, following the usual pattern.

**Theorem 11**
$$\begin{aligned}&\forall x : (\nu Z. A[Z, \vec{B}]) \forall y : (\nu Z. A[Z, \vec{C}]). \\ &\quad x(\nu Z. A[Z, \vec{\rho}])y \\ &\quad \equiv \\ &\exists R \subset (\nu Z. A[Z, \vec{B}]) \times (\nu Z. A[Z, \vec{C}]). \\ &\quad x R y \wedge \forall x' : (\nu Z. A[Z, \vec{B}]) \forall y' : (\nu Z. A[Z, \vec{C}]). \\ &\qquad x' R y' \supset \textit{out}(x') A[R, \vec{\rho}] \textit{out}(y')\end{aligned}$$

Put more informally, what this theorem states is that $\nu Z. A[Z, \vec{\rho}]$ is the *greatest* relation $R$ such that $\textit{out}(R) \subset A[R, \vec{\rho}]$.

¿From this theorem and the Identity Extension Lemma one obtains the following principle of co-induction (or bisimulation) [AM89, Smy91, Pit92]:

$$\begin{aligned}&\forall x, y : (\nu Z. A[Z]). \\ &\quad (\exists R \subset (\nu Z. A[Z]) \times (\nu Z. A[Z]). \\ &\qquad x R y \wedge \forall x', y' : (\nu Z. A[Z]). x' R y' \supset \textit{out}(x') A[R, eq_{\vec{X}}] \textit{out}(y')) \\ &\supset \\ &\quad x = y\end{aligned}$$

To put this in perhaps more familiar terms, say that $R \subset (\nu Z. A[Z]) \times (\nu Z. A[Z])$ is a *bisimulation* relation if $\textit{out}(R) \subset A[R, eq_{\vec{X}}]$. Then the principle of co-induction states that equality is the maximal bisimulation relation. From previous work a binary principle is expected here. There are also evident principles of other degrees. The unary principle is logically trivial. Ternary and higher degree principles seem pointless; however we do not know if they are equivalent to the usual binary one.

## Acknowledgments

We benefited from many conversations with Luca Cardelli, Pierre-Louis Curien and John Mitchell.

# References

[ACC93] Martin Abadi, Luca Cardelli and Pierre-Louis Curien. Formal parametric polymorphism. To Appear in proceedings of *Principles of Prgramming Languages '93*.

[AM89] Peter Aczel and Nax Mendler. A final co-algebra theorem. In D. H. Pitt *et al.*, editors, *Category Theory and Computer Science* Lecture Notes in Computer Science, 389:357–365 Berlin, 1989. Springer-Verlag.

[BFSS90] E. S. Bainbridge, Peter J. Freyd, Andre Scedrov, and Philip J. Scott. Functorial polymorphism. *Theoretical Computer Science*, 70(1):35–64, January 15 1990. Corrigendum in (3) 71, 10 April 1990, p. 431.

[Böh85] Corrado Böhm and A. Berarducci. Automatic synthesis of typed $\Lambda$-programs on term algebras. *Theoretical Computer Science*, 39:85–114, 1985.

[CMMS91] Luca Cardelli, Simone Martini, John Mitchell, and Andre Scedrov. An extension of system $F$ with subtyping. In T. Ito and A. R. Meyer, editors, *Theoretical Aspects of Computer Software*, volume 526 of *Lecture Notes in Computer Science*, pages 750–770, Berlin, 1991. Springer-Verlag.

[CG91] Pierre-Louis Curien and Giorgio Ghelli. Subtyping + extensionality: Confluence of $\beta\eta$top reduction in $F_\leq$. In T. Ito and A. R. Meyer, editors, *Theoretical Aspects of Computer Software*, volume 526 of *Lecture Notes in Computer Science*, pages 731–749, Berlin, 1991. Springer-Verlag.

[CG92] Pierre-Louis Curien and Giorgio Ghelli. Coherence of subsumption, minimum typing and type-checking in $F_\leq$. *Mathematical Structures in Computer Science*, 2(1):55–92, March 1992.

[Has90] Ryu Hasegawa. Categorical data types in parametric polymorphism. *Mathematical Structures in Computer Science*, 1990. To appear.

[Has91] Ryu Hasegawa. Parametricity of extensionally collapsed models of polymorphism and their categorical properties. In T. Ito and A. R. Meyer, editors, *Theoretical Aspects of Computer Software*, volume 526 of *Lecture Notes in Computer Science*, pages 495–512, Berlin, 1991. Springer-Verlag.

[MR92] QingMing Ma and John C. Reynolds. Types, abstraction, and parametric polymorphism, part 2. In S. Brookes, M. Main, A. Melton, M. Mislove, and D. A. Schmidt, editors, *Proceedings of the 1991 Mathematical Foundations of Programming Semantics Conference*, Lecture Notes in Computer Science, Berlin, 1992. Springer-Verlag. To appear.

[Mai91] Harry Mairson. Outline of a proof theory of parametricity. In *Proc. 5th International Symp. on Functional Programming Languages and Computer Architecture*, Springer-Verlag, 1991.

[MP85]  John C. Mitchell and Gordon D. Plotkin. Abstract types have existential type. In *Proceedings of the Twelfth Annual ACM Symposium on Principles of Programming Languages*, pages 37–51, 1985.

[Mit90]  John C. Mitchell. A type inference approach to reduction properties and semantics of polymorphic expressions (summary). In G. Huet, editor, *Logical Foundations of Functional Programming*, pages 195–212, Reading, 1990. Addison-Wesley.

[Mit91]  John C. Mitchell. On the equivalence of data representations. In V. Lifshitz, editor, *Artificial Intelligence and Mathematical Theory of Computation: Papers in Honor of John McCarthy*, Academic Press, pages 305–330, 1991.

[Pit87]  Andrew M. Pitts. Polymorphism is set-theoretic, constructively. In D. H. Pitt *et al.*, editors, *Category Theory and Computer Science* Lecture Notes in Computer Science, 283:12–39 Berlin, 1987. Springer-Verlag.

[Pit92]  Andrew M. Pitts. A co-induction principle for recursively defined domains. To appear in *Theoretical Computer Science*.

[Rey83]  John C. Reynolds. Types, abstraction and parametric polymorphism. In R. E. A. Mason, editor, *Information Processing 83*, pages 513–523, Amsterdam, 1983. Elsevier Science Publishers B. V. (North-Holland).

[RP90]  John C. Reynolds and Gordon D. Plotkin. On functors expressible in the polymorphic typed lambda calculus. In G. Huet, editor, *Logical Foundations of Functional Programming*, pages 127–152, Reading, 1990. Addison-Wesley.

[Smy91]  Michael B. Smyth. I-categories and duality. In M. P. Fourman, P. T. Johnstone and A. M. Pitts, *Applications of Categories in Computer Science* London Mathematical Society Lecture Note Series, 177:270–287, Cambridge, 1991 Cambridge University Press.

[Str67]  Christopher Strachey. Fundamental concepts in programming languages. Lecture Notes, International Summer School in Programming Languages, Copenhagen, Unpublished, August 1967.

[Wad89]  Philip Wadler. Recursive types in polymorphic second-order lambda-calculus Draft, University of Glasgow 1990.

[Wra89]  Gavin C. Wraith. A note on categorical datatypes. In A. M. Pitts and A. Poigné, editors, *Category Theory and Computer Science* Lecture Notes in Computer Science, 39:118–127, Berlin, 1989, Springer-Verlag.

# Call-by-Value and Nondeterminism

Kurt Sieber*

FB 14 Informatik, Universität des Saarlandes
W-6600 Saarbrücken, Germany

**Abstract.** We consider the three classical powerdomain semantics (lower, upper and convex) for $\text{PCF}_{nv}$, a *n*ondeterministic call-by-*v*alue version of Plotkin's PCF. By a 'Computational Adequacy Theorem' we describe how these three semantics are related to different notions of observable behavior of nondeterministic programs. Then we investigate full abstraction: The lower powerdomain semantics turns out to be fully abstract. For the upper powerdomain semantics, full abstraction fails irreparably, i.e. there is *no* extension of $\text{PCF}_{nv}$ by computable operators, for which it is fully abstract. Full abstraction also fails for the convex powerdomain—already at the level of the ground type powerdomain. We repair this low level failure by adding (besides parallel conditional) an **exists**-operator, which tests all paths of a nondeterministic computation in parallel. But even with this operator we do not obtain full abstraction for *all* types, and there is some evidence that adding further (reasonable) operators does not help.

## 1 Introduction

In [8] we presented a fully abstract denotational semantics for $\text{PCF}_v$, a call-by-value version of Plotkin's functional language PCF ([7]). Here we extend $\text{PCF}_v$ to a new language $\text{PCF}_{nv}$ by adding a nondeterministic choice operator **or**. Following [2] we define a generic denotational semantics for $\text{PCF}_{nv}$ in terms of an arbitrary powerdomain construction $\mathcal{P}$. Then we consider the three classical instances of $\mathcal{P}$, given by the lower (Hoare-), upper (Smyth-) and convex (Plotkin-) powerdomain.

Nondeterministic extensions of the **call-by-name** language PCF have already been considered in [1]. So what is the special thing about call-by-value? We give an example. Consider the two expressions

$$M \equiv (\lambda x.\, 1) \text{ or } (\lambda x.\, 2) \quad \text{and} \quad N \equiv \lambda x.\, (1 \text{ or } 2)$$

of type $\iota \to \iota$ (where $\iota$ is the type of integers). There is an obvious difference in the *evaluation* of both expressions, no matter whether a call-by-name or a call-by-value interpreter is used: Evaluation of $M$ leads to a nondeterministic computation which *either* returns the function $\lambda x.\, 1$ *or* the function $\lambda x.\, 2$ as a result. On the contrary, evaluation of $N$ stops immediately, because

---
* Supported by the Deutsche Forschungsgemeinschaft, SFB 124, Teilprojekt C1

$N$ is a $\lambda$-abstraction, hence there is only *one* result in this case, namely the (nondeterministic) function $N$ itself.

The crucial point is, that this difference can never be *observed* in a call-by-name language, because evaluation of a function type expression can only be enforced by applying it to some argument, and the possible results of $MP$ and $NP$ are the same for each argument $P$. Similar considerations show that for any two expressions $F$ and $G$ of function type, $(F \text{ or } G)$ can always be replaced by $\lambda x.(Fx \text{ or } Gx)$ (with a fresh variable $x$) without changing the observable behavior of the program. This transformation can be applied repeatedly, until the choice operator **or** finally occurs only between ground type expressions. In terms of a denotational semantics this means that powerdomains are only needed for the ground types, i.e. only over flat cpo's (cf. [1]).

For a **call-by-value** language the situation is totally different, because evaluation of a function type expression can also be enforced by passing it as a parameter. For example, the difference between the above two expressions $M$ and $N$ can be observed by applying the function

$$P \equiv \lambda f.(f\,0 + f\,0)$$

to them. $PM$ can only deliver the results 2 or 4, because $M$ is evaluated before the parameter passing and then *either* $\lambda x.1$ *or* $\lambda x.2$ is bound to the formal parameter $f$. On the contrary, $PN$ can also deliver 3, because the nondeterministic function $N$ which is bound to $f$ in this case, can return 1 in the first call and 2 in the second. This shows that $M$ and $N$ cannot have the same meaning in a denotational call-by-value semantics. Intuitively, $M$ must be interpreted as a *set* of two functions and $N$ as a single *set valued* function. In terms of the denotational semantics this means that powerdomains are not only needed for ground types but also for function types, i.e. over non-flat cpo's.

So much about the difference between call-by-name and call-by-value. As mentioned above, we will consider three different denotational semantics for the call-by-value language $PCF_{nv}$. Each of these semantics describes a different kind of *observable behavior* of programs. This is made precise in the Computational Adequacy Theorem (Theorem 1): Lower powerdomain semantics describes the set of possible outputs of a program; upper powerdomain semantics tells us whether a program is strongly normalizing and describes the set of possible outputs (only) in this case; convex powerdomain semantics combines both informations.

Computational adequacy is only the minimal requirement for a denotational semantics. After having established it, we address the question of full abstraction for each of the three semantics. Our results are as follows.

The lower powerdomain semantics is fully abstract for $PCF_{nv}$ (Theorem 3); *no* additional operators (like *parallel conditional* in the deterministic case, cf. [7]) are needed for this result, because the nondeterministic choice operator is already powerful enough. Full abstraction *fails* for the upper powerdomain semantics, and—even worse—it fails irreparably, i.e. there is no extension of $PCF_{nv}$ by a set of computable functions, for which this semantics is fully abstract (Theorem 6).

Full abstraction also fails for the convex powerdomain semantics, and this cannot be repaired by adding a parallel conditional or any other 'simple' operator which is definable in the lower or upper powerdomain (Theorem 7). This first failure of full abstraction already appears at a low level. The problem is that there are different sets in the ground type powerdomain which cannot be distinguished by a definable 'test function'. This can be repaired by adding an **exists**-operator, which tests all paths of a nondeterministic computation in parallel. With this operator, full abstraction is obtained for a sublanguage in which functions have only ground type results (Theorem 8) and we conjecture that a similar full abstraction proof works for a nondeterministic version of the original (call-by-name) PCF (Conjecture 9). But for the full language $PCF_{nv}$ + **pif** + **exists**, full abstraction still fails (Theorem 10). The 'nonstandard element' which causes this failure is interesting in its own. It is a nondeterministic (continuous) function which cannot be 'split' into deterministic monotone parts. The existence of this function is closely related to the fact that the underlying convex powerdomain—and hence the nondeterministic function space—is not bounded complete (see the discussion in the Conclusion). Thus a well known mathematical problem with the convex powerdomain causes a failure of full abstraction.

## 2  Syntax

$PCF_{nv}$ is a simply typed $\lambda$-value calculus over a single ground type $\iota$ of integers. The constants and their types are

$$\underline{n} : \iota \qquad \text{for each } n \in \mathbb{Z},$$
$$succ, pred : \iota \to \iota,$$
$$Y_{\sigma \to \tau} : ((\sigma \to \tau) \to (\sigma \to \tau)) \to \sigma \to \tau \qquad \text{for each pair of types } \sigma, \tau.$$

Occasionally we use superscripts to indicate the types of constants $c$ and variables $x$. With this convention, *expressions* $M, N, \ldots$ and their types are defined by:

$$\begin{array}{ll}
c^\tau : \tau & \text{(constant),} \\
x^\tau : \tau & \text{(variable),} \\
(MN) : \tau \quad \text{if } M : \sigma \to \tau \text{ and } N : \sigma & \text{(application),} \\
(\lambda x^\sigma.M) : \sigma \to \tau \quad \text{if } M : \tau & \text{(abstraction),} \\
(\text{case } M\, N) : \iota \to \tau \quad \text{if } M : \tau \text{ and } N : \tau & \text{(case expression),} \\
(M \text{ or } N) : \tau \quad \text{if } M : \tau \text{ and } N : \tau & \text{(choice expression).}
\end{array}$$

(**case** $M\, N$) denotes a function which behaves like $M$ when it is applied to 0 and like $N$ otherwise. If we interpret 0 as 'true' and all other integers as 'false', then the conditional expression (**if** $P$ **then** $M$ **else** $N$ **fi**) can be considered as syntactic sugar for (**case** $M\, N\, P$). We use **case** in the first place, because it is the more primitive operation. We will also consider an extension of $PCF_{nv}$ by parallel conditional expressions

$$(\text{pif } M\, N\, P) : \iota \quad \text{if } M, N, P : \iota.$$

as introduced in [7]. Further extensions of the language are considered when we investigate full abstraction.

We conclude this section by introducing some notation. $FV(M)$ stands for the set of free variables of an expression $M$; $M$ is called *closed* if $FV(M) = \emptyset$; a closed expression of ground type $\iota$ is called a *program*. $M[x := N]$ denotes the expression obtained from $M$ by substituting $N$ for all free occurences of $x$. $\Omega_\tau$ (or just $\Omega$) stands for some closed diverging expression of type $\tau$, say $\Omega_\tau \equiv Y_{\iota \to \tau}(\lambda x^{\iota \to \tau}. x)\underline{0}$. Constants, variables, abstractions and case expressions are called *value expressions*; we will use meta variables $U, V, \ldots$ for them. (They are often called *values*, cf. [6], but we want to reserve this word for semantic objects).

## 3 Operational Semantics

A structured operational semantics for PCF$_{\text{nv}}$ is defined in Table 1 and additional rules for the parallel conditional in Table 2. The rules define a one step transition relation '$\to$' between (closed) PCF$_{\text{nv}}$-expressions. '$\stackrel{*}{\to}$' denotes the reflexive, transitive closure of '$\to$'. $N$ is called a *normal form*, if there is no expression $P$ such that $N \to P$; it is called a *normal form of* $M$, if additionally $M \stackrel{*}{\to} N$ holds; $NF(M)$ denotes the set of normal forms of $M$. An expression $M$ is called *strongly normalizing* (notation: $SN(M)$), if there is *no* infinite sequence $M \to M_1 \to M_2 \to \ldots$. It is easy to see that the normal forms are exactly the value expressions, i.e. no computation 'gets stuck'.

**Table 1.** Structured Operational Semantics for PCF$_{\text{nv}}$

$$succ\ \underline{n} \to \underline{n+1} \qquad pred\ \underline{n} \to \underline{n-1}$$

$$case\ M\ N\ \underline{0} \to M \qquad case\ M\ N\ \underline{n} \to N \quad (n \neq 0)$$

$$\frac{M \to M'}{MN \to M'N} \qquad \frac{N \to N'}{VN \to VN'}$$

$$M\ or\ N \to M \qquad M\ or\ N \to N$$

$$Y_{\sigma \to \tau} V \to \lambda x^\sigma. V(Y_{\sigma \to \tau} V) x \quad (x \notin FV(V))$$

$$(\lambda x^\sigma. M)V \to M[x := V]$$

We consider three different notions of *observable behavior* of a program $M$, which we call the *partial correctness behavior* $beh_p(M)$, the *total correctness*

**Table 2.** Additional rules for the parallel conditional operator

$$\frac{M \to M', \ P \to P'}{\text{pif } M \, \underline{n} \, P \to \text{pif } M' \, \underline{n} \, P'} \qquad \frac{M \to M', \ N \to N'}{\text{pif } M \, N \, \underline{n} \to \text{pif } M' \, N' \, \underline{n}}$$

$$\frac{M \to M', \ N \to N', \ P \to P'}{\text{pif } M \, N \, P \to \text{pif } M' \, N' \, P'} \qquad \frac{M \to M'}{\text{pif } M \, \underline{m} \, \underline{n} \to \text{pif } M' \, \underline{m} \, \underline{n}} \ (m \ne n)$$

$$\text{pif } \underline{0} \, N \, P \to N \qquad \text{pif } \underline{n} \, N \, P \to P \ (n \ne 0) \qquad \text{pif } M \, \underline{n} \, \underline{n} \to \underline{n}$$

behavior $beh_t(M)$ and the *overall behavior* $beh_o(M)$. They are defined as follows:

$$beh_p(M) = \{n \in \mathbb{Z} \mid M \xrightarrow{*} \underline{n}\};$$
$$beh_t(M) = \begin{cases} beh_p(M) & \text{if } SN(M), \\ \mathbb{Z} \cup \{\bot\} & \text{otherwise}; \end{cases}$$
$$beh_o(M) = \begin{cases} beh_p(M) & \text{if } SN(M), \\ beh_p(M) \cup \{\bot\} & \text{otherwise}. \end{cases}$$

For each notion of observable behavior there is a corresponding notion of observational congruence, namely: Two expressions $M$ and $N$ of the same type are *observationally congruent* if $C[M]$ and $C[N]$ have the same observable behavior for every program context $C[\ ]$. Thus we obtain three different notions of observational congruence which we denote $M \approx_p N$, $M \approx_t N$ and $M \approx_o N$.

## 4 Denotational Semantics

In this section we define a *generic* denotational semantics for $\text{PCF}_{nv}$ in terms of an arbitrary powerdomain construction $\mathcal{P}$. Then we consider the three 'classical' instances of this semantics, obtained by the lower (Hoare-), upper (Smyth-) and convex (Plotkin-) powerdomain. We do not give a formal definition of the term 'powerdomain construction', the reader who is interested in such a definition may consult [4]. Instead we only make the following assumptions which are sufficient for our purposes.

Let **C** be a class of cpo's, e.g. the class of bounded complete algebraic cpo's or the class of SFP-objects. A *powerdomain construction* $\mathcal{P}$ over **C** associates with each cpo $D$ in **C** a cpo $\mathcal{P}D$ in **C** and provides the following operations for all cpo's $D, E$ in **C**:

$$\{\!|\_|\!\} : D \to \mathcal{P}D \qquad (singleton),$$
$$\_ \uplus \_ : \mathcal{P}D \times \mathcal{P}D \to \mathcal{P}D \qquad (union),$$
$$ext\_ : (D \to \mathcal{P}E) \to (\mathcal{P}D \to \mathcal{P}E) \qquad (function\ extension).$$

The notation $\{\!|\ |\!\}$ is generalized to finite sets of elements by

$$\{\!|d_1, \ldots, d_n|\!\} = \{\!|d_1|\!\} \uplus \ldots \uplus \{\!|d_n|\!\}$$

and function extension is generalized to functions of arity $n \geq 2$ by

$$ext^{(n)} : (D_1 \to \ldots \to D_n \to \mathcal{P}E) \to (\mathcal{P}D_1 \to \ldots \to \mathcal{P}D_n \to \mathcal{P}E)$$
$$ext^{(1)} = ext,$$
$$ext^{(n+1)} f \mathcal{A}_1 \ldots \mathcal{A}_{n+1} =$$
$$\quad ext(\lambda d_1 \in D_1. \, ext^{(n)}(\lambda d_2, \ldots, d_{n+1}. f d_1 d_2 \ldots d_{n+1}) \mathcal{A}_2 \ldots \mathcal{A}_{n+1}) \mathcal{A}_1.$$

For a given powerdomain construction $\mathcal{P}$ the meaning of types is then defined by[2]

$$[\![\iota]\!] = \mathbb{Z} \qquad \text{(the discrete cpo of integers),}$$
$$[\![\sigma \to \tau]\!] = ([\![\sigma]\!] \to \mathcal{P}([\![\tau]\!]_\bot)) \quad \text{(the cpo of continuous functions),}$$

and the operator $\_\bullet\_ : \mathcal{P}([\![\sigma \to \tau]\!]_\bot) \times \mathcal{P}([\![\sigma]\!]_\bot) \to \mathcal{P}([\![\tau]\!]_\bot)$ by

$$\mathcal{A} \bullet \mathcal{B} = ext^{(2)} app_\bot^{\sigma,\tau} \mathcal{A} \mathcal{B},$$

where $app_\bot^{\sigma,\tau}$ is the strict extension of the application operator, i.e.

$$app_\bot^{\sigma,\tau} : [\![\sigma \to \tau]\!]_\bot \times [\![\sigma]\!]_\bot \to \mathcal{P}([\![\tau]\!]_\bot)$$
$$app_\bot^{\sigma,\tau} d\, e = \begin{cases} \{\!|\bot|\!\} & \text{if } d = \bot \text{ or } e = \bot, \\ d\, e & \text{otherwise.} \end{cases}$$

If $\mathcal{P}([\![\tau]\!]_\bot)$ has a least element, then $[\![\sigma \to \tau]\!]$ has a least element and a least fixed point operator $fix : [\![\sigma \to \tau]\!] \to [\![\sigma \to \tau]\!]$ can be defined as usual.

Now we are ready to define the denotational semantics. An *interpretation* $\mathcal{I}(c) \in [\![\tau]\!]$ is associated with each constant $c$ of type $\tau$ by

- $\mathcal{I}(\underline{n}) = n \quad$ for each $n \in \mathbb{Z}$
- $\mathcal{I}(succ) : [\![\iota]\!] \to \mathcal{P}([\![\iota]\!]_\bot)$
  $\mathcal{I}(succ)\, n = \{\!|n+1|\!\}$
- $\mathcal{I}(pred) : [\![\iota]\!] \to \mathcal{P}([\![\iota]\!]_\bot)$
  $\mathcal{I}(pred)\, n = \{\!|n-1|\!\}$
- $\mathcal{I}(Y_{\sigma \to \tau}) : [\![(\sigma \to \tau) \to (\sigma \to \tau)]\!] \to \mathcal{P}([\![\sigma \to \tau]\!]_\bot)$
  $\mathcal{I}(Y_{\sigma \to \tau})\, f = \{\!|fix(\lambda g \in [\![\sigma \to \tau]\!]. \lambda v \in [\![\sigma]\!].(f\, g) \bullet \{\!|v|\!\})|\!\}$

Environments $\rho$ are type preserving functions from variables to values, i.e.

$$\rho : Var \to \bigcup_{\tau \in Type} [\![\tau]\!]$$

---

[2] This definition is in the spirit of Moggi's *computational $\lambda$-calculus* (cf. [5]): $[\![\tau]\!]$ is the object of values and $\mathcal{P}([\![\tau]\!]_\bot)$ the object of computations of type $\tau$; values of type $[\![\sigma \to \tau]\!]$ are functions which map values of type $\sigma$ to computations of type $\tau$.

with $\rho x \in [\![\tau]\!]$ whenever $x$ is of type $\tau$. The meaning of a $\text{PCF}_{\text{nv}}$-expression $M : \tau$ in environment $\rho$ is the element $[\![M]\!]\rho \in \mathcal{P}([\![\tau]\!]_\bot)$ defined by

$$[\![c^\tau]\!]\rho = \{\!|\mathcal{I}(c)|\!\}$$
$$[\![x^\tau]\!]\rho = \{\!|\rho(x)|\!\}$$
$$[\![M\,N]\!]\rho = ([\![M]\!]\rho) \bullet ([\![N]\!]\rho)$$
$$[\![\lambda x^\sigma.M]\!]\rho = \{\!|\lambda v \in [\![\sigma]\!].[\![M]\!]\rho[v/x]|\!\}$$
$$[\![\text{case}\,M\,N]\!]\rho = \{\!|\lambda n \in \mathbb{Z}.\text{if}\,n = 0\,\text{then}\,[\![M]\!]\rho\,\text{else}\,[\![N]\!]\rho|\!\}$$
$$[\![M\,\text{or}\,N]\!]\rho = [\![M]\!]\rho \uplus [\![N]\!]\rho$$

and the meaning of a parallel conditional expression is

$$[\![\text{pif}\,M\,N\,P]\!]\rho = ext^{(3)}\,pcond([\![M]\!]\rho)([\![N]\!]\rho)([\![P]\!]\rho),$$

where $pcond$ is the usual (deterministic) parallel conditional

$$pcond : \mathbb{Z}_\bot \to \mathbb{Z}_\bot \to \mathbb{Z}_\bot \to \mathcal{P}(\mathbb{Z}_\bot)$$
$$pcond\,b\,d\,e = \begin{cases} \{\!|d|\!\} & \text{if } b = 0 \text{ or } d = e, \\ \{\!|e|\!\} & \text{if } b \in \mathbb{Z} \setminus \{0\}, \\ \{\!|\bot|\!\} & \text{if } b = \bot \text{ and } d \neq e. \end{cases}$$

As usual, $[\![M]\!]\rho$ only depends on the values $\rho x$ for $x \in FV(M)$; hence we will write $[\![M]\!]$ instead of $[\![M]\!]\rho$ if $M$ is closed.

We conclude this section by defining the three particular powerdomain constructions which we want to investigate. A uniform presentation can be given in terms of ideal completion (cf. [2, 3]).

Let $D$ be an algebraic cpo and let $KD$ be its basis, i.e. the set of compact elements of $D$. Further let $P_f^*(KD)$ be the set of all finite *nonempty* subsets of $KD$. Then three different preorders are defined on $P_f^*(KD)$ by:

$$A \sqsubseteq_\mathcal{L} B \Leftrightarrow \forall a \in A.\,\exists b \in B.\,a \sqsubseteq b,$$
$$A \sqsubseteq_\mathcal{U} B \Leftrightarrow \forall b \in B.\,\exists a \in A.\,a \sqsubseteq b,$$
$$A \sqsubseteq_\mathcal{C} B \Leftrightarrow A \sqsubseteq_\mathcal{L} B \wedge A \sqsubseteq_\mathcal{U} B.$$

Now, for $\mathcal{P} = \mathcal{L}, \mathcal{U}$ and $\mathcal{C}$, the powerdomain $\mathcal{P}D$ is defined as the ideal completion of $P_f^*(KD)$ wrt the preorder $\sqsubseteq_\mathcal{P}$, and the powerdomain operations are

$$\{\!|d|\!\} = \{A \in P_f^*(KD) \mid \exists e \in KD.\,A \sqsubseteq_\mathcal{P} \{e\} \wedge e \sqsubseteq d\}$$
$$\mathcal{A} \uplus \mathcal{B} = \{C \in P_f^*(KD) \mid \exists A \in \mathcal{A}, B \in \mathcal{B}.\,C \sqsubseteq_\mathcal{P} A \cup B\}$$
$$ext f \mathcal{A} = \bigcup\{f\,a_1 \uplus \ldots \uplus f\,a_n \mid \{a_1,\ldots,a_n\} \in \mathcal{A}\}$$

The following results are well-known (cf. [3, 4]): If $D$ is a bounded complete algebraic cpo, then $\mathcal{L}D$ and $\mathcal{U}D$ are bounded complete and algebraic; if $D$ is an SFP-object, then $\mathcal{C}D$ is an SFP-object. Hence $[\![\tau]\!]_\mathcal{L}$ and $[\![\tau]\!]_\mathcal{U}$ are well defined bounded complete algebraic cpos and $[\![\tau]\!]_\mathcal{C}$ is a well defined SFP-object for $\tau \neq \iota$.[3]

---

[3] Whenever a distinction between different powerdomains is necessary, we add indices $\mathcal{L}, \mathcal{U}, \mathcal{C}$ to our notations.

# 5 Computational Adequacy

A denotational semantics is called *computationally adequate* if it gives us the expected information about the behavior of a program. For our three denotational semantics this is made precise in

**Theorem 1.** (**Computational Adequacy of** $\mathcal{L}, \mathcal{U}$ **and** $\mathcal{C}$)
*Let $M$ be a program in $PCF_{nv}$ + pif. Then*

(a) lower *powerdomain semantics describes the* partial correctness *behavior by*

$$M \xrightarrow{*} \underline{n} \Leftrightarrow \exists A \in [\![M]\!]_{\mathcal{L}} . n \in A;$$

(b) upper *powerdomain semantics describes the* total correctness *behavior by*

$$SN(M) \Leftrightarrow \exists A \in [\![M]\!]_{\mathcal{U}} . \bot \notin A,$$
$$SN(M) \Rightarrow (M \xrightarrow{*} \underline{n} \Leftrightarrow \forall A \in [\![M]\!]_{\mathcal{U}} . \bot \notin A \Rightarrow n \in A);$$

(c) convex *powerdomain semantics describes the* overall behavior *by*

$$M \xrightarrow{*} \underline{n} \Leftrightarrow \exists A \in [\![M]\!]_{\mathcal{C}} . n \in A,$$
$$SN(M) \Leftrightarrow \exists A \in [\![M]\!]_{\mathcal{C}} . \bot \notin A.$$

We don't present a proof of computational adequacy in this paper, because we want to concentrate on full abstraction questions. Note that by Theorem 1 there is always a one-to-one correspondence between the behavior of a program and its denotational semantics, i.e. we have

**Corollary 2.** *For any two programs $M, N$:*

$$beh_p(M) = beh_p(N) \Leftrightarrow [\![M]\!]_{\mathcal{L}} = [\![N]\!]_{\mathcal{L}}$$
$$beh_t(M) = beh_t(N) \Leftrightarrow [\![M]\!]_{\mathcal{U}} = [\![N]\!]_{\mathcal{U}}$$
$$beh_o(M) = beh_o(N) \Leftrightarrow [\![M]\!]_{\mathcal{C}} = [\![N]\!]_{\mathcal{C}}$$

# 6 Full Abstraction

Let $[\![\ ]\!]$ be a denotational semantics which is computationally adequate for a certain notion of observable behavior. Then $[\![\ ]\!]$ is called *fully abstract* if semantic equality coincides with observational congruence (wrt the given notion of behavior). We will now investigate the question of full abstraction for each of the three powerdomain semantics.

## 6.1 Lower Powerdomain

**Theorem 3.** *Lower powerdomain semantics is fully abstract for $PCF_{nv}$, i.e. for any two expressions $M_1, M_2$ of the same type*

$$[M_1]_{\mathcal{L}} = [M_2]_{\mathcal{L}} \Leftrightarrow M_1 \approx_p M_2.$$

As usual, the clue to full abstraction is the definability of compact elements.

**Definition 4.** Let $u \in [\![\tau]\!]$ and let $U$ be a closed value expression of type $\tau$. We say that $u$ is *definable* (or: *defined*) by $U$ if $[U] = \{\!|u|\!\}$.

**Theorem 5.** *Let $\tau$ be some arbitrary type and let $u \in [\![\tau]\!]_{\mathcal{C}}$ be compact. Then $u$ and the threshold function $(u \Rightarrow \{\!|0|\!\})$ are definable by closed value expressions.*

*Proof.* by induction on types:

$u \in [\![\iota]\!]$ : Obvious.

$u \in [\![\sigma \to \tau]\!]$ :

By definition, $[\![\sigma \to \tau]\!] = ([\![\sigma]\!] \to \mathcal{L}([\![\tau]\!]_\bot))$. As $[\![\sigma]\!]$ and $\mathcal{L}([\![\tau]\!]_\bot)$ are bounded complete algebraic cpo's, we know (cf. [7, 8]) that $u$ is of the form

$$u = \bigsqcup_{i=1}^{n}(u_i \Rightarrow \mathcal{A}_i) \quad (n \geq 0) \tag{1}$$

with compact elements $u_i \in [\![\sigma]\!], \mathcal{A}_i \in \mathcal{L}([\![\tau]\!]_\bot)$ (for $n = 0$ this lub is defined as $u = \underline{\lambda}v \in [\![\sigma]\!].\{\!|\bot|\!\}$, the least element of $[\![\sigma \to \tau]\!]$). Moreover, each compact element $\mathcal{A} \in \mathcal{L}([\![\tau]\!]_\bot)$ is a principal ideal, i.e. there are compact elements $v_1, \ldots, v_m \in [\![\tau]\!]_\bot$ ($m \geq 1$) such that

$$\mathcal{A} = \{\!|v_1, \ldots, v_m|\!\} = \{\!|v_1|\!\} \uplus \ldots \uplus \{\!|v_m|\!\} = \{\!|v_1|\!\} \sqcup \ldots \sqcup \{\!|v_m|\!\}.$$

Hence (1) can be expanded to

$$u = \bigsqcup_{i=1}^{n}(u_i \Rightarrow \{\!|v_i|\!\}) \quad (n \geq 0)$$

with compact elements $u_i \in [\![\sigma]\!], v_i \in [\![\tau]\!]$. The case $n = 0$ is clear; for $n > 0$ we know by induction hypothesis that $u_i, v_i, (u_i \Rightarrow \{\!|0|\!\})$ and $(v_i \Rightarrow \{\!|0|\!\})$ are definable by closed value expressions $U_i, V_i, ABOVE\_U_i$ and $ABOVE\_V_i$. Then it is easy to see that $u$ and $(u \Rightarrow \{\!|0|\!\})$ are definable by the expressions

$$\lambda x^\sigma . \, OR_{i=1}^{n} (\text{if } ABOVE\_U_i \, x \text{ then } V_i \text{ else } \Omega_\tau \text{ fi})$$

and[4]

$$\lambda f^{\sigma \to \tau} . \bigwedge_{i=1}^{n} ABOVE\_V_i \, (fU_i).$$

□

---

[4] Symbols like $\wedge, \vee, \ldots$ are used with their standard (strict) interpretation; they are of course definable in $PCF_{nv}$.

Now Theorem 3 can be derived from Theorem 5 as usual: '$\Rightarrow$' is clear from the compositionality of denotational semantics. For '$\Leftarrow$' assume that $[\![M_1]\!] \neq [\![M_2]\!]$. Let $v_i$ ($i = 1, 2$) be defined by $\lambda \overline{x}.\, M_i$, where $\overline{x}$ is the list of free variables of $M_1$ and $M_2$. Then $v_1 \neq v_2$, say $v_1 \not\sqsubseteq v_2$, hence there is some compact element $u$ such that $u \sqsubseteq v_1, u \not\sqsubseteq v_2$. By Theorem 5, $(u \Rightarrow \{\!|0|\!\})$ is definable by a closed value expression $ABOVE\_U$, hence

$$[\![ABOVE\_U(\lambda \overline{x}.\, M_1)]\!] = \{\!|0|\!\} \neq \{\!|\bot|\!\} = [\![ABOVE\_U(\lambda \overline{x}.\, M_2)]\!]$$

and this implies $M_1 \not\approx_p M_2$ by Computational Adequacy. $\square$

## 6.2 Upper Powerdomain

**Theorem 6.** *Upper powerdomain semantics is not fully abstract for* $\text{PCF}_{\text{nv}}$. *Moreover, there is no way to extend* $\text{PCF}_{\text{nv}}$ *by a set of computable functions, such that* $[\![\ ]\!]_{\mathcal{U}}$ *remains computationally adequate and becomes fully abstract.*

*Proof.* Consider the two expressions

$$M_1 \equiv \lambda f^{(\iota \to \iota) \to \iota}.\, \text{if } f(\lambda x^\iota.\, \underline{0}) \text{ then } f(\lambda x^\iota.\, \Omega_\iota) \text{ else } \Omega_\iota \text{ fi}$$
$$M_2 \equiv \lambda f^{(\iota \to \iota) \to \iota}.\, \text{if } f(\lambda x^\iota.\, \underline{0}) \wedge f(\lambda x^\iota.\, \Omega_\iota) \text{ then } \underline{0} \text{ else } \Omega_\iota \text{ fi}$$

and let $f \in [\![(\iota \to \iota) \to \iota]\!]$ be defined by

$$f g = \begin{cases} \{\!|0|\!\} & \text{if } g \sqsupseteq (0 \Rightarrow \{\!|0|\!\}), \\ \{\!|0, 1|\!\} & \text{otherwise.} \end{cases}$$

Then $[\![M_1]\!] f = \{\!|0, 1|\!\}$ and $[\![M_2]\!] f = \{\!|\bot|\!\}$, hence $[\![M_1]\!] \neq [\![M_2]\!]$.

Now assume that $M_1 \not\approx_t M_2$ in $\text{PCF}_{\text{nv}}$ or some extension of $\text{PCF}_{\text{nv}}$. Then there must be a closed value expression $U : (\iota \to \iota) \to \iota$ with $[\![M_1 U]\!] \neq [\![M_2 U]\!]$. If $[\![U(\lambda x^\iota.\, \underline{0})]\!] \neq \{\!|0|\!\}$ or $[\![U(\lambda x^\iota.\, \Omega_\iota)]\!] = \{\!|\bot|\!\}$, then $[\![M_1 U]\!] = [\![M_2 U]\!] = \{\!|\bot|\!\}$; if $[\![U(\lambda x^\iota.\, \underline{0})]\!] = [\![U(\lambda x^\iota.\, \Omega_\iota)]\!] = \{\!|0|\!\}$ then $[\![M_1 U]\!] = [\![M_2 U]\!] = \{\!|0|\!\}$; hence the only remaining possibility is $[\![U(\lambda x^\iota.\, \underline{0})]\!] = \{\!|0|\!\}$ and $[\![U(\lambda x^\iota.\, \Omega_\iota)]\!] = \{\!|0, n_1, \ldots, n_k|\!\}$ for some $k \geq 1, n_1, \ldots, n_k \in \mathbb{Z} \setminus \{0\}$. But then it follows from computational adequacy of $[\![\ ]\!]_{\mathcal{U}}$ that

$$beh_t(U(\lambda x^\iota.\, \underline{0})) = \{0\} \quad \text{and} \quad beh_t(U(\lambda x^\iota.\, \Omega_\iota)) = \{0, n_1, \ldots, n_k\},$$

hence a computable function can be constructed (not in $\text{PCF}_{\text{nv}}$, but e.g. in the language of the next section) which returns 0 for $\lambda x^\iota.\, \underline{0}$ and 1 for $\lambda x^\iota.\, \Omega_\iota$. This is a contradiction to the unsolvability of the halting problem. $\square$

## 6.3 Convex Powerdomain

**Theorem 7.** *Convex powerdomain semantics is not fully abstract for $PCF_{nv}$. Moreover, there is no way to extend the language such that either $[\ ]_C$ or $[\ ]_U$ remains computationally adequate and $[\ ]_C$ becomes fully abstract.*

*Proof.* Consider the expressions

$$M_1 \equiv \lambda f^{(\iota \to \iota) \to \iota}.\underline{0}\,\text{or}\,(\lambda x^\iota.\underline{0})(f(\lambda x^\iota.\Omega_\iota)),$$
$$M_2 \equiv \lambda f^{(\iota \to \iota) \to \iota}.\underline{0}\,\text{or}\,(\lambda x^\iota.\underline{0})(f(\lambda x^\iota.\underline{0}\,\text{or}\,\Omega_\iota)),$$
$$M_3 \equiv \lambda f^{(\iota \to \iota) \to \iota}.\underline{0}\,\text{or}\,(\lambda x^\iota.\underline{0})(f(\lambda x^\iota.\underline{0})),$$

and let $f, g \in [\![(\iota \to \iota) \to \iota]\!]$ be defined by

$$f = ((0 \Rightarrow \{\!|0, \bot|\!\}) \Rightarrow \{\!|0|\!\}),$$
$$g = ((0 \Rightarrow \{\!|0|\!\}) \Rightarrow \{\!|0|\!\}).$$

Then $[\![M_1]\!]f = \{\!|0, \bot|\!\} \neq \{\!|0|\!\} = [\![M_2]\!]f$ and $[\![M_2]\!]g = \{\!|0, \bot|\!\} \neq \{\!|0|\!\} = [\![M_3]\!]g$, hence $[\![M_1]\!] \neq [\![M_2]\!] \neq [\![M_3]\!]$.

Now assume that $M_1 \not\approx_o M_2$ in $PCF_{nv}$ or some extension of $PCF_{nv}$. Then there must be a closed value expression $U$ with $[\![M_1 U]\!] \neq [\![M_2 U]\!]$. The only possibility is $[\![M_1 U]\!] = \{\!|0, \bot|\!\}$ and $[\![M_2 U]\!] = \{\!|0|\!\}$. But then computational adequacy of $[\ ]_C$ implies $\neg SN(M_1)$ and $SN(M_2)$. This is impossible, if $[\ ]_U$ is computationally adequate for the language, because $[\![\lambda x^\iota.\Omega_\iota]\!]_U = [\![\lambda x^\iota.\underline{0}\,\text{or}\,\Omega_\iota]\!]_U$ and hence $[\![M_1]\!]_U = [\![M_2]\!]_U$.

Similarly $M_2 \not\approx_o M_3$ leads to a contradiction, if $[\ ]_C$ is computationally adequate, because $[\![\lambda x^\iota.\underline{0}\,\text{or}\,\Omega_\iota]\!]_C = [\![\lambda x^\iota.\underline{0}]\!]_C$. □

Theorem 7 implies in particular that the convex powerdomain semantics is *not* fully abstract for $PCF_{nv} + \mathbf{pif}$ (cf. Theorem 1). Moreover, its proof suggests that full abstraction can only be expected, if we add an operator to the language which can execute all paths of a nondeterministic computation in parallel. We introduce additional expressions

$$\mathbf{exists}\,M : \iota \quad \text{if } M : \iota$$

with the following properties[5]

| | | |
|---|---|---|
| $\mathbf{exists}\,M$ | evaluates to $\underline{0}$ | if $\underline{0} \in NF(M)$, |
| $\mathbf{exists}\,M$ | evaluates to $\underline{1}$ | if $SN(M)$ and $\underline{0} \notin NF(M)$, |
| $\mathbf{exists}\,M$ | diverges | otherwise. |

If the denotational semantics of these expressions is defined by

$$[\![\mathbf{exists}\,M]\!]\rho = \begin{cases} \{\!|0|\!\} & \text{if } \exists A \in [\![M]\!]\rho.0 \in A, \\ \{\!|1|\!\} & \text{if } [\![M]\!]\rho = \{\!|n_1, \ldots, n_k|\!\}, n_1, \ldots, n_k \neq 0, \\ \{\!|\bot|\!\} & \text{otherwise,} \end{cases}$$

then part (c) of Theorem 1 generalizes to $PCF_{nv} + \mathbf{pif} + \mathbf{exists}$.

---

[5] It is straightforward but tedious to define an extension of the transition relation '$\to$' such that these properties are satisfied. We refer to [1], where this has been worked out for a similar operator.

**Theorem 8.** *Let $PCF_{nv}^{\iota}$ be the subset of $PCF_{nv}$, which 'has only functions with ground type results': Its types are defined by $\tau ::= \iota \mid (\tau \to \iota)$, expressions are defined as for $PCF_{nv}$, but with the restriction that $M$ in $\lambda x^{\sigma}.M$ must always be of type $\iota$. Then the (restriction of the) convex powerdomain semantics is fully abstract for $PCF_{nv}^{\iota} + \mathbf{pif} + \mathbf{exists}$.*

The proof idea is as usual: one shows that all compact elements (in the restricted type system) are definable. The details are tricky and are not presented here, because we want to concentrate on the full language. □

**Conjecture 9.** *Let $PCF_{nn}$ be the extension of the original (call-by-name) language PCF by a nondeterministic choice operator $\mathbf{or}$. Then a fully abstract semantics for $PCF_{nn} + \mathbf{pif} + \mathbf{exists}$ can be defined with the aid of the convex powerdomain.*

This was left as an open question in [1]; we believe that it can be proved along the same lines as Theorem 8. □

These results say that the **exists**-operator is powerful enough to obtain full abstraction in a setting where the convex powerdomain is only needed for the (flat) ground type. We will see that this is no longer true, as soon as powerdomains over function types play a role.

**Theorem 10.** *The convex powerdomain semantics is not fully abstract for the full language $PCF_{nv} + \mathbf{pif} + \mathbf{exists}$.*

*Proof.* As usual, full abstraction fails because a certain 'nonstandard element' occurs in the denotational semantics. We are now going to define this element. For each finite set $B \subseteq \mathbb{Z}$, let $g_B = \bigsqcup_{n \in B}(n \Rightarrow \{|0|\})$ and for $0 \leq i \leq 3$ define $u_i \in [\![\iota \to \iota]\!]$ and $A_i \in P_f^*(K([\![\iota \to \iota]\!]_\bot))$ by

$$u_0 = g_\emptyset \qquad A_0 = \{g_{\{0\}}, g_{\{3\}}\}$$
$$u_1 = g_{\{1\}} \qquad A_1 = \{g_{\{0,1\}}, g_{\{2,3\}}\}$$
$$u_2 = g_{\{2\}} \qquad A_2 = \{g_{\{0,2\}}, g_{\{1,3\}}\}$$
$$u_3 = g_{\{1,2\}} \qquad A_3 = \{g_{\{0,1,3\}}, g_{\{0,2,3\}}\}$$

It is easy to see that there is a monotone function $f : [\![\iota \to \iota]\!] \to P_f^*(K([\![\iota \to \iota]\!]_\bot))$ with $f u_i = A_i$ for $0 \leq i \leq 3$, and such a monotone function has a (unique) continuous extension $\hat{f} \in [\![(\iota \to \iota) \to (\iota \to \iota)]\!]$ (cf. [2]). The interesting property of the sets $A_i$ is that—although $A_0 \sqsubseteq_c A_1, A_2$ and $A_1, A_2 \sqsubseteq_c A_3$—one cannot pick elements $v_i \in A_i$ with $v_0 \sqsubseteq v_1, v_2$ and $v_1, v_2 \sqsubseteq v_3$. This means that the monotone function $f$ has no deterministic monotone 'part', i.e. there is no monotone $g : [\![\iota \to \iota]\!] \to [\![\iota \to \iota]\!]_\bot$ with $g u_i \in A_i$ for $0 \leq i \leq 3$. This is (intuitively) the reason why $\hat{f}$ is not definable and full abstraction fails. In order to prove this rigorously, we use a so called *oracle semantics*, which will now be defined.

Let $S$ be the set of infinite sequences over $\{0,1\}_\bot$ with the pointwise order. A relation $(M, s) \Downarrow V$ for $M, V$ closed and $s \in S$ is defined in Table 3. (Notation:

$s^0, s^1$ and $s^2$ are disjoint subsequences of $s$, say $s^i_j = s_{(3*j+i)}$; $|s| = n$ means that $s_j \neq \bot \Leftrightarrow j < n$.) Intuitively, $(M, s) \Downarrow V$ means that $M$ evaluates to $V$ if the nondeterministic choices in $M$ are controlled (in some tricky way) by the sequence $s$. The connection between '$\to$' and '$\Downarrow$' is made precise by

**Lemma 11.** *Let $M$ be a closed expression and $V$ a closed value expression in* $PCF + \mathbf{pif} + \mathbf{exists}$. *Then*

$$M \stackrel{*}{\to} V \Leftrightarrow \exists s \in S. (M, s) \Downarrow V$$
$$SN(M) \Leftrightarrow \exists n \in \mathbb{N}. \forall s \in S. |s| = n \Rightarrow (M, s) \Downarrow$$

**Table 3.** Main clauses of the oracle semantics for $\mathrm{PCF}_{\mathrm{nv}} + \mathbf{pif} + \mathbf{exists}$

$$(V, s) \Downarrow V$$

$$\frac{(M, s^0) \Downarrow \lambda x.P,\ (N, s^1) \Downarrow U,\ (P[x := U], s^2) \Downarrow V}{(M\,N, s) \Downarrow V}$$

$$\frac{(M, s) \Downarrow V}{(M \text{ or } N, 0.s) \Downarrow V} \qquad \frac{(N, s) \Downarrow V}{(M \text{ or } N, 1.s) \Downarrow V}$$

$$\frac{(B, s^0) \Downarrow \underline{0},\ (M, s^1) \Downarrow \underline{n}}{(\mathbf{pif}\,B\,M\,N, s) \Downarrow \underline{n}} \qquad \frac{(B, s^0) \Downarrow \underline{m}\,(m \neq 0),\ (N, s^1) \Downarrow \underline{n}}{(\mathbf{pif}\,B\,M\,N, s) \Downarrow \underline{n}}$$

$$\frac{(M, s^1) \Downarrow \underline{n},\ (N, s^2) \Downarrow \underline{n}}{(\mathbf{pif}\,B\,M\,N, s) \Downarrow \underline{n}} \qquad \frac{(M, s') \Downarrow \underline{0}}{(\mathbf{exists}\,M, s) \Downarrow \underline{0}}$$

$$\frac{(M, s_1) \Downarrow \underline{n_1}, \ldots, (M, s_k) \Downarrow \underline{n_k}}{(\mathbf{exists}\,M, s) \Downarrow \underline{1}} \text{ if } n_1, \ldots, n_k \neq 0 \text{ and } \exists n \in \mathbb{N}. \{s_1, \ldots, s_k\} = \{s \mid |s| = n\}$$

A denotational counterpart to the oracle semantics can be defined along the same lines. Its semantic domains are

$$[\![\iota]\!] = \mathbb{Z} \qquad \text{(the discrete cpo),}$$
$$[\![(\sigma \to \tau)]\!] = ([\![\sigma]\!] \to (S \to [\![\tau]\!]_\bot)) \quad \text{(the cpo of continuous functions),}$$

and the interesting clauses of the meaning function $[\![M]\!]\rho \in (S \to [\![\tau]\!]_\bot)$ are

$$[\![M \text{ or } N]\!]\rho s = \begin{cases} [\![M]\!]\rho s' & \text{if } s = 0.s' \\ [\![N]\!]\rho s' & \text{if } s = 1.s' \\ \bot & \text{if } s = \bot.s' \end{cases}$$

$$[\![\mathbf{exists}\,M]\!]\rho s = \begin{cases} 0 & \text{if } \exists s' \in S.\ [\![M]\!]\rho s' = 0 \\ 1 & \text{if } \exists n \in \mathbb{N}. \forall s' \in S.\ |s'| = n \Rightarrow [\![M]\!]\rho \neq 0, \bot \\ \bot & \text{otherwise.} \end{cases}$$

We write $[\![M]\!]_{\mathcal{O}}$ if we want to distinguish the oracle semantics from the powerdomain semantics.

**Lemma 12. (Computational Adequacy of the Oracle Semantics)**
*Let M be a program in* $PCF_{nv} + \mathbf{pif} + \mathbf{exists}$. *Then*

$$(M, s) \Downarrow \underline{n} \Leftrightarrow [\![M]\!]_{\mathcal{O}}\, s = n.$$

With these results, the proof of Theorem 10 can now be continued. We must find an equation in $PCF_{nv} + \mathbf{pif} + \mathbf{exists}$ which holds for all definable functions but *not* for the function $\hat{f}$. Here is a rough sketch, how to construct such an equation:

For $0 \leq i \leq 3$ we can find closed value expressions $U_i : \iota \to \iota$ such that $[\![U_i]\!]_{\mathcal{C}} = \{|u_i|\}$ and $[\![U_0]\!]_{\mathcal{O}} \sqsubseteq [\![U_1]\!]_{\mathcal{O}}, [\![U_2]\!]_{\mathcal{O}}$ and $[\![U_1]\!]_{\mathcal{O}}, [\![U_2]\!]_{\mathcal{O}} \sqsubseteq [\![U_3]\!]_{\mathcal{O}}$. Then—by extensive use of **pif**- and **exists**-operators—it is possible to construct expressions $P1$ and $P2$ with a free variable $F : (\iota \to \iota) \to (\iota \to \iota)$, which check the following (semidecidable) properties for $F$:

$P1: \ (\exists V \in NF(FU_0).\, V\underline{0} \overset{*}{\to} \underline{0}) \wedge$
$\qquad (SN(FU_1) \wedge \forall V \in NF(FU_1).\, |\{n \mid V\underline{n} \overset{*}{\to} \underline{0}\}| \geq 2) \wedge$
$\qquad (SN(FU_2) \wedge \forall V \in NF(FU_2).\, |\{n \mid V\underline{n} \overset{*}{\to} \underline{0}\}| \geq 2)$

$P2: \ (\exists V \in NF(FU_1).\, V\underline{0} \overset{*}{\to} \underline{0} \wedge \exists n.\, n \neq 0, 1 \wedge V\underline{n} \overset{*}{\to} \underline{0}) \vee$
$\qquad (\exists V \in NF(FU_2).\, V\underline{0} \overset{*}{\to} \underline{0} \wedge \exists n.\, n \neq 0, 2 \wedge V\underline{n} \overset{*}{\to} \underline{0}) \vee$
$\qquad (\exists V \in NF(FU_3).\, V\underline{0} \overset{*}{\to} \underline{0} \wedge V\underline{1} \overset{*}{\to} \underline{0} \wedge V\underline{2} \overset{*}{\to} \underline{0})$

Then $P1$ holds for $\hat{f}$ but $P2$ doesn't, i.e. $[\![P1]\!]_{\mathcal{C}}\, \rho[\hat{f}/F] = \{|0|\} \neq \{|\bot|\} = [\![P1 \wedge P2]\!]_{\mathcal{C}}\rho[\hat{f}/F]$.

Now assume that $P1$ holds for some *definable* $F$. Then—by Lemma 11—there are some $s \in S$ and some value expression $V_0$ such that $(FU_0, s) \Downarrow V_0$ and $V_0\underline{0} \overset{*}{\to} \underline{0}$. Computational adequacy (and monotonicity) of the oracle semantics then implies that there are value expressions $V_i$ ($0 \leq i \leq 3$) with $(FU_i, s) \Downarrow V_i$, $[\![V_0]\!]_{\mathcal{O}} \leq [\![V_1]\!]_{\mathcal{O}}, [\![V_2]\!]_{\mathcal{O}}$ and $[\![V_1]\!]_{\mathcal{O}}, [\![V_2]\!]_{\mathcal{O}} \leq [\![V_3]\!]_{\mathcal{O}}$, hence in particular $V_i\, \underline{0} \overset{*}{\to} \underline{0}$ for all $i$. Again by $P1$, there must be further elements $n_1, n_2$ with $V_1\, \underline{n_1} \overset{*}{\to} \underline{0}$ and $V_2\, \underline{n_2} \overset{*}{\to} \underline{0}$. If $n_1 \neq 1$ or $n_2 \neq 2$ then $P2$ holds. If $n_1 = 1$ and $n_2 = 2$ then $V_3\, \underline{1} \overset{*}{\to} \underline{0}$ and $V_3\, \underline{2} \overset{*}{\to} \underline{0}$ by monotonicity and $P2$ holds again. Hence we have proved that $P1$ implies $P2$ for all definable $F$, i.e. $P1 \approx_o (P1 \wedge P2)$ and the proof of Theorem 10 is finished. □

## 7 Conclusion

The main result of our paper is a negative result: The convex powerdomain semantics—which is the most important of all three, because it describes the overall behavior of programs—is *not* fully abstract, even if we extend our language by powerful 'parallel features' like the **pif**- and the **exists**-operator. Of course the question arises whether this can be repaired, either by adding further operators to the language or by inventing a new denotational semantics. We believe that the search for further operators is hopeless, because some kind of oracle semantics will be possible for *every* reasonable extension of the language. Hence the question is, whether the denotational semantics can be improved in such a way, that certain nonstandard elements at function type are thrown out. More recent powerdomain constructions like the mixed or sandwich powerdomain (cf. [2, 3]) do not seem to solve this problem, because they are even 'larger' than the convex powerdomain ([3]). Hence a possible approach might be to repair the nondeterministic function space instead of the powerdomain.

By the way, the failure of full abstraction is closely related to a well known mathematical problem: The convex powerdomain—and hence also the nondeterministic function space—need not be bounded complete, even if the underlying cpo is bounded complete. It is exactly this property which allowed us to construct the nonstandard element $\hat{f}$ in the proof of Theorem 10, namely by choosing the 'wrong' minimal upper bound $A_3$ of $A_1$ and $A_2$ as a result for $fu_3$.

## Acknowledgements

I'm grateful to C. Gunter who suggested to investigate these full abstraction questions, to R. Heckmann from whom I learnt a lot about powerdomains and to H. Sprenger who draw my attention to the work of Astesiano and Costa.

# References

1. E. Astesiano and G. Costa. Nondeterminism and fully abstract models. *RAIRO*, 14(4):323–347, 1980.
2. C. A. Gunter. Relating total and partial correctness interpretations of nondeterministic programs. In $17^{th}$ *Annual ACM Symposium on Principles of Programming Languages*, pages 306–319, 1990.
3. C. A. Gunter and D. S. Scott. Semantic domains. In J. van Leeuwen, editor, *Handbook of Theoretical Computer Science*, chapter 12, pages 635–674. Elsevier Science Publishers, 1990.
4. R. Heckmann. *Power Domain Constructions*. PhD thesis, Universität des Saarlandes, Saarbrücken, 1990.
5. E. Moggi. Notions of computation and monads. *Information and Computation*, 93:55–92, 1991.
6. G. D. Plotkin. Call-by-name, call-by-value and the λ-calculus. *Theoretical Computer Science*, 1:125–159, 1975.
7. G. D. Plotkin. LCF considered as a programming language. *Theoretical Computer Science*, 5:223–256, 1977.
8. K. Sieber. Relating full abstraction results for different programming languages. In K. Nori and C. V. Madhavan, editors, $10^{th}$ *Conference on Foundations of Software Technology and Theoretical Computer Science*, Springer LNCS 472, pages 373–387, Bangalore, India, 1990.

# Lower and Upper Bounds for Reductions of Types in $\lambda\underline{\omega}$ and $\lambda P$

(Extended abstract)

Jan Springintveld [*]

*Department of Philosophy, University of Utrecht*
*P.O. Box 80126, 3508 TC Utrecht, The Netherlands*
*E-mail: Jan.Springintveld@phil.ruu.nl*

**Abstract**

For several important systems of the $\lambda$-cube we study the time-complexity of type conversion. Non-elementary lower bounds are given for the type-conversion problem for $\lambda\underline{\omega}$ and $\lambda P$ and hence for the systems that include one of these systems. For $\lambda\underline{\omega}$ and $\lambda P$ a super-exponential upper bound is given to the length of reduction sequences starting from types that are legal in these systems.

## 1 Introduction

Barendregt's $\lambda$-cube ([1]) gives a finestructure of the Calculus of Constructions (usually abbreviated as $\lambda C$) of Coquand and Huet ([2]). It consists of eight subsystems of $\lambda C$ among which the simply typed $\lambda$-calculus, Girard's second order (polymorphic) $\lambda$-calculus and $\lambda P$, a system similar to one of the Automath languages.

The systems of the $\lambda$-cube are not presented in their original form, but in the style of Pure Type Systems (PTS for short). In the PTS presentation the systems come equipped with a Conversion Rule:

$$\frac{\Gamma \vdash A : B' \quad \Gamma \vdash B : s \quad B' =_\beta B}{\Gamma \vdash A : B}.$$

Here the condition $B' =_\beta B$ is external, i.e. it is not derivable in the system. In this paper we are interested in the question how difficult it is to determine whether $B' =_\beta B$. Besides having an intrinsic interest, an answer to this question will be an indispensable step towards the determination of the complexity of type inference and type checking for the various systems of the $\lambda$-cube.

Coquand has shown that all systems of the $\lambda$-cube are strongly normalizing (for an alternative proof, see [3]); i.e. for all systems: if $\Gamma \vdash A : B$, then all reduction sequences starting from $A$ or $B$ are finite. So the problem is in any case decidable: just compare the normal forms of $B$ and $B'$. But what is the time complexity of the problem $B =_\beta B'$?

For $\lambda\to$ and $\lambda 2$ it can be shown that $B =_\beta B'$ iff $B \equiv B'$. So we only have to check that $B$ and $B'$ are syntactically equal, which needs an amount of time that is linear in the length of $B$ and $B'$.

But for the other systems it is considerably more difficult. In Section 3 we show that the type-convertibility problem for $\lambda\underline{\omega}$ and $\lambda P$ (and hence for the systems which contain one

---

[*]The investigations were supported by the Foundation for Computer Science in the Netherlands (SION) with financial support from the Netherlands Organization for Scientific Research (NWO).

of these systems as a subsystem) is not elementary recursive (see [5] for a definition of this notion).

This follows from a result of [7] which states that the problem whether two *objects* of the simply typed $\lambda$-calculus (without type variables) are $\beta$-convertible is not elementary recursive. We give translations from the *objects* of the simply typed $\lambda$-calculus to the *types* and *constructors* of $\lambda\underline{\omega}$ and $\lambda P$ such that two $\lambda\tau$-objects are $\beta$-convertible iff their translations are.

Given this intrinsic difficulty of the question whether two types are $\beta$-convertible, one may ask whether the Conversion Rule is necessary for type inference or type checking. The answer is yes; it is needed to make types match when applying the Application Rule:

$$\frac{\Gamma \vdash A_1 : \Pi x{:}B_1.B_2 \quad \Gamma \vdash A_2 : B_1}{\Gamma \vdash A_1 A_2 : B_2[x:=A_2]}.$$

For $A_1$ to be applicable to $A_2$ the type of $A_2$ must match with $B_1$ and sometimes the original type of $A_2$ may only be $\beta$-convertible with $B_1$. Then one has to use the Conversion Rule to patch up the differences. We give an example in $\lambda\underline{\omega}$: let $B_1$ be $((\lambda\alpha{:}*.\alpha{\to}\alpha)A)$ in $\Pi x{:}B_1.B_2$ and let $\Gamma \vdash_{\lambda\underline{\omega}} A_2 : A{\to}A$. By reducing $(\lambda\alpha{:}*.\alpha{\to}\alpha)A$ one step we get $A{\to}A$, which is equal to the type of $A_2$.

The question arises whether it is possible to give an upper bound for the time complexity of type conversion. In Section 4 we give for two systems of the $\lambda$-cube, $\lambda\underline{\omega}$ and $\lambda P$, an upper bound to the length of reduction sequences starting from types that are legal in those systems. To obtain our upper bound we use a result of [6] which gives a super-exponential upper bound to the number of steps needed to reduce an object from the simply typed $\lambda$-calculus to normal form. We also use existing translations (see [3]) from $\lambda\underline{\omega}$ and $\lambda P$ to $\lambda{\to}$ to transfer Schwichtenberg's upper bound to these systems. This gives us immediately an upper bound for the length of reduction sequences starting from a type $B$; this upper bound is super-exponential in the length of $B$. But we are more interested in an upper bound expressed not in terms of the length of $B$ but in terms of the length of $\Gamma$ and the length of $A$, for $\Gamma$ and $A$ such that $\Gamma \vdash A : B$. The reason for this is twofold. The first reason is that such an upper bound can be applied to type inference and type checking. The second reason is that in this way we can in some interesting cases obtain a better upper bound. We will do this for slightly simplified versions of $\lambda\underline{\omega}$ and $\lambda P$.

Suppose $\Gamma \vdash A : B$. To obtain an upper bound we construct a term $\mathcal{B}$ such that every reduction sequence starting from $B$ is a subsequence of a reduction sequence starting from $\mathcal{B}$ and such that the longest reduction sequence in $\mathcal{B}$ has an 'optimal' length in terms of the length of $\Gamma$ and $A$.

The main problem one encounters when constructing this term $\mathcal{B}$ in $\lambda P$ is that by using the Application Rule terms are substituted in types. To simulate this substitution we would have to substitute for variables bound by a $\Pi$. Our main technical tool is a replacement operation that performs this "substitution".

**Acknowledgements.** This paper has benefitted much from discussions with Marc Bezem and Jan Friso Groote.

## 2 The $\lambda$-cube

In this section we briefly present the systems of the $\lambda$-cube. For more information on these systems the reader is referred to [1], [3].

**Definition 2.1.** *Pseudo-terms* are given by the following abstract syntax:

$$\mathcal{T} ::= \mathcal{C} \mid \mathcal{V} \mid \mathcal{T}\mathcal{T} \mid \lambda\mathcal{V}{:}\mathcal{T}.\mathcal{T} \mid \Pi\mathcal{V}{:}\mathcal{T}.\mathcal{T}.$$

Here $\mathcal{C}$ is an infinite set of constants and $\mathcal{V}$ is an infinite set of variables.

Substitution on pseudo-terms, one-step $\beta$-reduction (denoted by $\to_\beta$), $\beta$-reduction (denoted by $\twoheadrightarrow_\beta$) and $\beta$-conversion (denoted by $=_\beta$) are defined as usual.

We will often use the *variable convention*: this states that the bound variables occurring in an expression are different from the free ones.

Let $A$ be a pseudo-term. Then $\mathcal{FV}(A)$ denotes the free variables of $A$. Pseudo-terms of the form $\Pi x{:}A.B$ such that $x \notin \mathcal{FV}(B)$ are sometimes written as $A \to B$.

A *statement* is of the form $A : B$, where $A$ and $B$ are pseudo-terms. $A$ is called the *subject* and $B$ is called the *predicate*. If $A \in \mathcal{V}$ then $A : B$ is called a *declaration*. A *pseudo-context* is a finite ordered sequence of declarations, where each subject is distinct. Pseudo-contexts are denoted by capital Greek letters $\Delta$, $\Gamma$, $\Gamma_0, \ldots$ The empty context is denoted by $\langle\rangle$. If $\Gamma \equiv \langle x_1{:}A_1, \ldots, x_n{:}A_n\rangle$, then $\mathcal{FV}(\Gamma) := \{x_1, \ldots, x_n\}$ and $\Gamma, x{:}B$ denotes $\langle x_1{:}A_1, \ldots, x_n{:}A_n, x{:}B\rangle$.

A *judgement* is of the form $\Gamma \vdash A : B$, where $\Gamma$ is a pseudo-context and $A$ and $B$ are pseudo-terms. When we want to indicate that such a judgement is derived in a particular system $\lambda-$, then we write $\Gamma \vdash_{\lambda-} A : B$. $\Gamma \vdash A : B : C$ means that $\Gamma \vdash A : B$ and $\Gamma \vdash B : C$.

A pseudo-term $A$ is called *legal* when there exist a pseudo-context $\Gamma$ and a pseudo-term $B$ such that $\Gamma \vdash A : B$ or $\Gamma \vdash B : A$. A pseudo-context $\Gamma$ is called legal when there exist pseudo-terms $A$ and $B$ such that $\Gamma \vdash A : B$.

Let $A$ be a pseudoterm; then $|A|$ denotes the number of symbols of $A$ in $\Pi$-notation, not counting '.' and ':'. If $\Gamma \equiv \langle x_1{:}A_1, \ldots, x_n{:}A_n\rangle$, then $|\Gamma| := n + |A_1| + \cdots + |A_n|$. Let $A$ be a pseudoterm of the form $\Pi x_1{:}A_1 \ldots \Pi x_n{:}A_n.B$, where $B$ does not start with a $\Pi$. Then $B$ is called the *body* of $A$. We write $A^\star$ for the body of $A$.

**Definition 2.2.** The *specification* of a PTS consists of a triple $S = (\mathcal{S}, \mathcal{A}, \mathcal{R})$ where

1. $\mathcal{S}$ is a subset of $\mathcal{C}$, called the *sorts*.
2. $\mathcal{A}$ is a set of *axioms* of the form $c : s$, where $c \in \mathcal{C}$ and $s \in \mathcal{S}$.
3. $\mathcal{R}$ is a set of *rules* of the form $(s_1, s_2, s_3)$, where $s_1, s_2, s_3 \in \mathcal{S}$.

A PTS $\lambda S$ is determined by the specifications $S = (\mathcal{S}, \mathcal{A}, \mathcal{R})$ and the rules given by Table 1; here $s$ ranges over $\mathcal{S}$ and $x$ ranges over $\mathcal{V}$.

**Definition 2.3.** The systems of the $\lambda$-cube are defined by:

- $\mathcal{S} = \{*, \square\}$.
- $\mathcal{A} = \{* : \square\}$.
- The rules given in Table 2, where $(s_1, s_2)$ is an abbreviation of $(s_1, s_2, s_2)$.

$\lambda P\omega$ is also called $\lambda C$.

We highlight the two systems that play an important role in the sequel: $\lambda\underline{\omega}$ and $\lambda P$. The system $\lambda\underline{\omega}$ is determined by the pairs $(*, *)$ and $(\square, \square)$. In $\lambda\underline{\omega}$ one can derive statements like $\langle\rangle \vdash_{\lambda\underline{\omega}} * \to * : \square$ and $A{:}* \vdash_{\lambda\underline{\omega}} (\lambda\alpha{:}*.\alpha \to \alpha)A : *$. Note that $(\lambda\alpha{:}*.\alpha \to \alpha)A$ is a type not in normal form.

The system $\lambda P$ is determined by the pairs $(*, *)$ and $(*, \square)$. In $\lambda P$ one can derive statements like $A{:}* \vdash_{\lambda P} A \to * : \square$ and $A{:}*, f{:}A \to A \to A, P{:}A \to * \vdash_{\lambda P} \Pi y{:}A.P((\lambda x{:}A.fxx)y) : *$. Note that $\Pi y{:}A.P((\lambda x{:}A.fxx)y)$ is a type not in normal form.

It is convenient to divide the set of variables for the systems of the $\lambda$-cube into two infinite subsets $\mathcal{V}_*$ and $\mathcal{V}_\square$. Elements of the former set are denoted by ${}^*x$, ${}^*y$, etc., elements of the latter set are denoted by ${}^\square x$, ${}^\square y$, etc. The intuition for this is the following. In a derivation

| | | |
|---|---|---|
| Axiom | $<> \vdash c : s$ | if $c : s \in \mathcal{A}$ |
| Start | $\dfrac{\Gamma \vdash B : s}{\Gamma, x{:}B \vdash x{:}B}$ | if $x \notin \mathcal{FV}(\Gamma)$ |
| Weakening | $\dfrac{\Gamma \vdash A : B \quad \Gamma \vdash B' : s}{\Gamma, x{:}B' \vdash A : B}$ | if $x \notin \mathcal{FV}(\Gamma)$ |
| Product | $\dfrac{\Gamma \vdash A_1 : s_1 \quad \Gamma, x{:}A_1 \vdash A_2 : s_2}{\Gamma \vdash \Pi x{:}A_1.A_2 : s_3}$ | if $(s_1, s_2, s_3) \in \mathcal{R}$ |
| Application | $\dfrac{\Gamma \vdash A_1 : \Pi x{:}B_1.B_2 \quad \Gamma \vdash A_2 : B_1}{\Gamma \vdash A_1 A_2 : B_2[x := A_2]}$ | |
| Abstraction | $\dfrac{\Gamma, x{:}A_1 \vdash A_2 : B_2 \quad \Gamma \vdash \Pi x{:}A_1.B_2 : s}{\Gamma \vdash \lambda x{:}A_1.A_2 : \Pi x{:}A_1.B_2}$ | |
| Conversion | $\dfrac{\Gamma \vdash A : B' \quad \Gamma \vdash B : s \quad B' =_\beta B}{\Gamma \vdash A : B}$ | |

Table 1: Rules for Pure Type Systems

| | | | | |
|---|---|---|---|---|
| $\lambda{\to}$ | $(*,*)$ | | | |
| $\lambda 2$ | $(*,*)$ | $(\Box,*)$ | | |
| $\lambda P$ | $(*,*)$ | | $(*,\Box)$ | |
| $\lambda P2$ | $(*,*)$ | $(\Box,*)$ | $(*,\Box)$ | |
| $\lambda \underline{\omega}$ | $(*,*)$ | | | $(\Box,\Box)$ |
| $\lambda \omega$ | $(*,*)$ | $(\Box,*)$ | | $(\Box,\Box)$ |
| $\lambda P \underline{\omega}$ | $(*,*)$ | | $(*,\Box)$ | $(\Box,\Box)$ |
| $\lambda P \omega$ | $(*,*)$ | $(\Box,*)$ | $(*,\Box)$ | $(\Box,\Box)$ |

Table 2: The systems of the $\lambda$-cube

declarations $x{:}A$ are added to a context $\Gamma$ by the Start Rule or the Weakening Rule. They are justified by the premiss: $\Gamma \vdash A : s$. Now if $s \equiv *$ then we write $x \equiv {}^*x$ and if $s \equiv \square$ then we write $x \equiv {}^\square x$. When irrelevant or clear from the context, superscripts are often omitted.

There exists a very useful stratification of the legal terms and types in $\lambda C$.

**Definition 2.4.** Let $\mathcal{T}$ be the set of pseudo-terms of $\lambda C$. We define $\#(\cdot) : \mathcal{T} \to \{0,1,2,3\}$ by: $\#(\square) := 3$; $\#(*) := 2$; $\#({}^\square x) := 1$; $\#({}^*x) := 0$; $\#(\Pi x{:}A_1.A_2) = \#(\lambda x{:}A_1.A_2) = \#(A_2 A_1) := \#(A_2)$.

We put $\mathcal{T}_i := \{M \in \mathcal{T} \mid \#(M) = i\}$, $\mathcal{T}_{i,j} := \mathcal{T}_i \cup \mathcal{T}_j$ and similarly for $\mathcal{T}_{i,j,k}$.

**Proposition 2.5.** *For $A$ legal in $\lambda C$:*

1. $\#(A) = 3$ *iff* $A \equiv \square$.

2. $\#(A) = 2$ *iff* $\exists \Gamma(\Gamma \vdash_{\lambda C} A : \square)$. *(Such $A$ are called kinds.)*

3. $\#(A) = 1$ *iff* $\exists \Gamma, B(\Gamma \vdash_{\lambda C} A : B : \square)$.
   *(Such $A$ are called constructors (or types if $B \equiv *$).)*

4. $\#(A) = 0$ *iff* $\exists \Gamma, B(\Gamma \vdash_{\lambda C} A : B : *)$. *(Such $A$ are called objects.)*

5. *if* $\#(x) = \#(B)$, *then* $\#(A[x{:=}B]) = \#(A)$.

6. *if* $\Gamma \vdash_{\lambda C} A : B$, *then* $\#(B) = \#(A) + 1$.

# 3 Lower Bounds for $\lambda\underline{\omega}$ and $\lambda P$

In this section we give lower bounds for the problem of $\beta$-convertibility of types for $\lambda\underline{\omega}$ and $\lambda P$ and hence for the systems that contain one of these systems as a subsystem. We derive these lower bounds from a lower bound result in [7]. This was originally formulated for the traditional version of the simply typed $\lambda$-calculus (which we call $\lambda\tau$). Since we work with the PTS versions of the systems of the $\lambda$-cube, it is convenient to define $\lambda\tau$ as a PTS and rephrase Statman's result in the formalized setting of a PTS.

**Definition 3.1.** $\lambda\tau$ denotes the PTS determined by: $\mathcal{S} = \{*\}$; $\mathcal{A} = \{O{:}*\}$; $\mathcal{R} = \{(*,*)\}$ (here $O$ is a constant).

So $\lambda\tau$ is just like $\lambda{\to}$ except that there are no legal type-variables: types are $O$, $O{\to}O$, etcetera. A typical judgement is: $y{:}O{\to}O \vdash_{\lambda\tau} (\lambda x{:}O{\to}O.x)y : O{\to}O$.

**Theorem 3.2 (Statman).** *Put $\mathcal{N} := \{A \mid \exists \Gamma\; \Gamma \vdash_{\lambda\tau} A : (O{\to}O){\to}O{\to}O\}$. Then the problem whether two arbitrary closed $\lambda\tau$-objects in $\mathcal{N}$ are $\beta$-convertible cannot be solved in elementary time.*

**Corollary 3.3.** *The problem whether two arbitrary closed constructors in $\lambda\underline{\omega}$ are $\beta$-convertible cannot be solved in elementary time.*

**Proof.** Define a translation $(\cdot)^*$ from the terms of $\lambda\tau$ to the constructors of $\lambda\underline{\omega}$ as follows. Replace in the terms and types all $O$'s by $*$'s; do the same in contexts. Then one can prove: if $\Gamma \vdash_{\lambda\tau} A : B$, then $\Gamma^* \vdash_{\lambda\underline{\omega}} A^* : B^*$. Thereafter one easily verifies that for all terms $A$ and $B$ that are legal in $\lambda\tau$, $A^*$ and $B^*$ are legal and $A =_\beta B$ iff $A^* =_\beta B^*$. ☒

**Corollary 3.4.** *The problem whether two arbitrary types in $\lambda P$ are $\beta$-convertible cannot be solved in elementary time.*

**Proof.** It is easy to see that if $\Gamma \vdash_{\lambda\tau} A : B$, then $O{:}*, \Gamma \vdash_{\lambda P} A : B$, where $O$ is now a variable of $\lambda P$. Consider a term $M$ from the set $\mathcal{N}$. Then by definition there exists a context $\Gamma$ such that $\Gamma \vdash_{\lambda\tau} M : (O{\to}O){\to}O{\to}O$ and so $O{:}*, \Gamma \vdash_{\lambda P} M : (O{\to}O){\to}O{\to}O$. By applying the Weakening Rule we get: $O{:}*, \Gamma, P{:}((O{\to}O){\to}O{\to}O){\to}* \vdash_{\lambda P} M : (O{\to}O){\to}O{\to}O$. From this we can derive: $O{:}*, \Gamma, P{:}((O{\to}O){\to}O{\to}O){\to}* \vdash_{\lambda P} PM : *$.

Now fix this variable $P$ and define: $\mathcal{P} := \{PM \mid M \in \mathcal{N}\}$. Then two terms $PM_1, PM_2$ in $\mathcal{P}$ are $\beta$-convertible iff $M_1$ and $M_2$ in $\mathcal{N}$ are $\beta$-convertible. ⊠

We immediately get the same result for the systems that contain $\lambda\underline{\omega}$ or $\lambda P$ as a subsystem i.e. $\lambda\underline{\omega}$, $\lambda P2$, $\lambda P\underline{\omega}$, $\lambda C$.

## 4 Upper Bounds for $\lambda\underline{\omega}$ and $\lambda P$

In this section we give for the systems $\lambda\underline{\omega}$ and $\lambda P$ an upperbound to the length of reduction sequences starting from types that are legal in these systems. First we present an upper bound result obtained by Schwichtenberg (see [6]) for the length of reduction sequences starting from legal objects of $\lambda\tau$. The upper bound is a function of the (maximal) level and (maximal) arity of such an object and of its height (these notions will be defined below). To apply this result, we use existing translations (see [3]) from the types and constructors of these systems to the legal objects of $\lambda\tau$ that preserve reduction. More precisely: if $A$ reduces to $B$ in one step in $\lambda\underline{\omega}$ or $\lambda P$, then the translation of $A$ reduces in at least one step to the translation of $B$ in $\lambda\tau$.

At first sight we are done. For, suppose $\Gamma \vdash A : B$. To estimate the maximum length of a reduction sequence starting from $B$, translate $B$ to a $\lambda\tau$ term $B'$, give an estimate for the level, arity and height of $B'$ and apply Schwichtenbergs result. The level, arity and height of $B'$ can be estimated as a linear function of $|B| + |\Gamma|$. However, it is more interesting to obtain an upper bound to the level, arity and height of $B'$ not in terms of the length of $B$ but in terms of the length of $\Gamma$ and the length of $A$. The reason for this is twofold. First, we are interested in this upper bound mainly from the point of view of type inference. A type inference algorithm can be viewed as an algorithm that receives as input a context $\Gamma$ and a term $A$ and returns (if successful) a type $B$ such that $\Gamma \vdash A : B$. Many of the existing type inference algorithms use reductions of types. So to obtain an estimate of the complexity of such an algorithm in terms of the input $\Gamma$ and $A$ we need a bound to the length of reduction sequences in terms of the length of the input. Secondly, in this way we can give a much better upper bound.

To make such an estimate possible at all, we have to impose a restriction. It is necessitated by the fact that types may become arbitrarily expanded and still be legal in the same context. We give an example in $\lambda\underline{\omega}$: suppose $\Gamma \vdash_{\lambda\underline{\omega}} A : \alpha$ and $\Gamma \vdash_{\lambda\underline{\omega}} \alpha : *$ where $\alpha$ is a free variable that, since it is free, occurs in $\Gamma$. Fix a natural number $n$. Let $\beta_1, \ldots, \beta_n$ be fresh variables. By using the Conversion Rule we can derive: $\Gamma \vdash_{\lambda\underline{\omega}} A : (\lambda\beta_1{:}* \ldots \lambda\beta_n{:}*.\alpha)\alpha \ldots \alpha$ ($n$ $\alpha$'s as arguments). This shows that it is in general impossible to estimate the length of a type for $A$ in terms of $|\Gamma| + |A|$.

This leads us to the restriction that all applications of the Conversion Rule are in fact applications of the Reduction Rule:

$$\frac{\Gamma \vdash A : B' \quad \Gamma \vdash B : s \quad B' \twoheadrightarrow_\beta B}{\Gamma \vdash A : B}.$$

(The converse is called the Expansion Rule: it is similar to the Reduction Rule, but with the $\twoheadrightarrow_\beta$ reversed.) We call derivations that contain no application of the Expansion Rule *expansion free* (for a discussion of the role of the Reduction Rule and the Expansion Rule, see [4]).

Suppose $\Gamma \vdash A : B$ is the conclusion of an expansion free derivation. We construct a term $\mathcal{B}$ such that every reduction sequence starting from $B$ is a subsequence of a reduction sequence starting from $\mathcal{B}$ and such that the level, arity and height of the translation of $\mathcal{B}$ can be expressed as a linear function of $|\Gamma|$ and $|A|$ (call this property $(\star)$).

In some interesting cases, this gives a better upper bound than an upper bound in terms of the length of types. Consider the Reduction Rule as given above; assume that we have a term $\mathcal{B}'$ for $B'$ that satisfies $(\star)$. $B$ can get exponentially longer than $B'$. So an upper bound in terms of the length of $B$ could become quite large. We know, however, that no reduction sequence in $B$ is longer than the longest reduction sequence in $B'$. Since $\Gamma$ and $A$ remain the same we can use $\mathcal{B}'$ for $\mathcal{B}$.

So this method is indeed useful. In $\lambda P$, however, there is one complication. Consider the Application Rule:

$$\frac{\Gamma \vdash A_1 : \Pi x{:}B_1.B_2 \quad \Gamma \vdash A_2 : B_1}{\Gamma \vdash A_1 A_2 : B_2[x{:=}A_2]}.$$

Assume that we have a term $\mathcal{B}$ for $\Pi x{:}B_1.B_2$ that satisfies $(\star)$. It is not evident how to use $\mathcal{B}$ to get such a term for $B_2[x{:=}A_2]$. (This problem does not occur in $\lambda\underline{\omega}$, because in that system we can prove that if $\Pi x{:}B_1.B_2$ is legal, then $x \notin \mathcal{FV}(B_2)$.)

The idea is that any reduction in $B_2[x{:=}A_2]$ can also be performed in $\mathcal{B}$ with $A_2$ substituted "at the right places". Then the question is, of course: what are the right places? We can't simply substitute $A_2$ for $x$ in $\mathcal{B}$, because $x$ occurs bound in $\Pi x{:}B_1.B_2$ and - as it turns out - also bound in $\mathcal{B}$.

Well, if we label all the $\Pi$'s occurring in the derivation (changing $\Pi x{:}A.B$ to $\Pi^i x{:}A.B$ for some natural number $i$) then the $\Pi$ in $\Pi x{:}B_1.B_2$ has label $l$, say. This $\Pi^l$ also occurs in $\mathcal{B}$. Now we change all occurrences of subterms in $\mathcal{B}$ of the form $\Pi^l y{:}C_1.C_2$ to $\Pi^l y{:}C_1.(C_2[y{:=}A_2])$. Of course, this is not an ordinary substitution. So we have to be careful that this operation is sound.

The proof for the $\lambda P$ case consists of four steps. First we state Schwichtenberg's result. Then we give a translation from the terms of $\lambda P$ to the terms of $\lambda \tau$. Next we present a replacement operation that performs the abovementioned "substitution". Finally we prove that with this operation we can obtain the desired upper bound for $\lambda P$. Since the proof for the $\lambda\underline{\omega}$ case is much simpler, we state the result for this system without proof.

## 4.1 Preliminaries

We rephrase a result of Schwichtenberg for our PTS-style definition of $\lambda \tau$. To do this, we redefine three notions in PTS style. The first of these notions is the well-known notion of type level. This notion is extended to the legal terms of $\lambda \tau$ in such a way that if a term is a type then the two notions coincide. Simultaneously, $\tau\text{-}maxar_\Gamma(A)$ is defined; it is the maximal arity of a variable occurring in $A$.

**Definition 4.1.** Suppose $\Gamma \vdash_{\lambda\tau} A : B$. We define $\tau\text{-}maxlevel_\Gamma(A)$ and $\tau\text{-}maxar_\Gamma(A)$ by induction on the derivation of $\Gamma \vdash_{\lambda\tau} A : B$. We distinguish cases according to the last rule used.

- Axiom. So $\Gamma \vdash_{\lambda\tau} A : B$ is $\langle\rangle \vdash_{\lambda\tau} O : *$.
  Then $\tau\text{-}maxlevel_{\langle\rangle}(O) := 0$ and $\tau\text{-}maxar_{\langle\rangle}(O) := 0$.

- Start Rule. So $\Gamma \vdash_{\lambda\tau} A : B$ is $\Gamma', x{:}B \vdash_{\lambda\tau} x : B$ and is a direct consequence of $\Gamma' \vdash_{\lambda\tau} B : *$. Put $\tau\text{-}maxlevel_{\Gamma',x:B}(x) := \tau\text{-}maxlevel_{\Gamma'}(B)$ and $\tau\text{-}maxar_{\Gamma',x:B}(x) := \tau\text{-}maxar_{\Gamma'}(B)$.

- Weakening Rule. So $\Gamma \vdash_{\lambda\tau} A : B$ is $\Gamma', x{:}C \vdash_{\lambda\tau} A : B$ and is a direct consequence of $\Gamma' \vdash_{\lambda\tau} A : B$ and $\Gamma' \vdash_{\lambda\tau} C : *$. Put $\tau\text{-}maxlevel_{\Gamma',x:C}(A) = \tau\text{-}maxlevel_{\Gamma'}(A)$ and $\tau\text{-}maxar_{\Gamma',x:C}(A) = \tau\text{-}maxar_{\Gamma'}(A)$.

- Application Rule. So $\Gamma \vdash_{\lambda\tau} A : B$ is $\Gamma \vdash_{\lambda\tau} A_1 A_2 : B_2[x{:=}A_2]$ and is a direct consequence of $\Gamma \vdash_{\lambda\tau} A_1 : \Pi x{:}B_1.B_2$ and $\Gamma \vdash_{\lambda\tau} A_2 : B_1$.
  Put $\tau\text{-}maxlevel_\Gamma(A_1 A_2) := \max(\{\tau\text{-}maxlevel_\Gamma(A_1), \tau\text{-}maxlevel_\Gamma(A_2)\})$.
  Put $\tau\text{-}maxar_\Gamma(A_1 A_2) := \max(\{\tau\text{-}maxar_\Gamma(A_1), \tau\text{-}maxar_\Gamma(A_2)\})$.

- Abstraction Rule. So $\Gamma \vdash_{\lambda\tau} A : B$ is $\Gamma \vdash_{\lambda\tau} \lambda x{:}A_1.A_2 : \Pi x{:}A_1.B_2$ and is a direct consequence of $\Gamma, x{:}A_1 \vdash_{\lambda\tau} A_2 : B_2$ and $\Gamma \vdash_{\lambda\tau} \Pi x{:}A_1.B_2 : *$.
  Put $\tau\text{-}maxlevel_\Gamma(\lambda x{:}A_1.A_2) := \max(\{\tau\text{-}maxlevel_\Gamma(A_1) + 1, \tau\text{-}maxlevel_{\Gamma,x:A_1}(A_2)\})$.
  Put $\tau\text{-}maxar_\Gamma(\lambda x{:}A_1.A_2) := \tau\text{-}maxar_{\Gamma,x:A_1}(A_2)$.

- Product Rule. So $\Gamma \vdash_{\lambda\tau} A : B$ is $\Gamma \vdash_{\lambda\tau} \Pi x{:}A_1.A_2 : *$ and is a direct consequence of $\Gamma \vdash_{\lambda\tau} A_1 : *$ and $\Gamma, x{:}A_1 \vdash_{\lambda\tau} A_2 : *$.
  Put $\tau\text{-}maxlevel_\Gamma(\Pi x{:}A_1.A_2) := \max(\{\tau\text{-}maxlevel_\Gamma(A_1) + 1, \tau\text{-}maxlevel_{\Gamma,x:A_1}(A_2)\})$.
  Put $\tau\text{-}maxar_\Gamma(\Pi x{:}A_1.A_2) := 1 + \tau\text{-}maxar_{\Gamma,x:A_1}(A_2)$.

- Conversion Rule. So $\Gamma \vdash_{\lambda\tau} A : B$ is a direct consequence of $\Gamma \vdash_{\lambda\tau} A : B'$ for some $B' =_\beta B$ and $\Gamma \vdash_{\lambda\tau} B : *$. Then $\tau\text{-}maxlevel_\Gamma(A)$ and $\tau\text{-}maxar_\Gamma(A)$ are already defined.

Moreover we define:

- $\tau\text{-}maxlevel_\Gamma(*) := 0$ and $\tau\text{-}maxar_\Gamma(*) := 0$.

If $\Gamma \vdash_{\lambda\tau} A : B$ then $\tau\text{-}maxlevel_\Gamma(A)$ and $\tau\text{-}maxar_\Gamma(A)$ are well-defined. This follows roughly from the fact that the derivation of $\Gamma \vdash_{\lambda\tau} A : B$ is unique up to applications of the Weakening Rule and the Conversion Rule and these two rules do not change $\tau\text{-}maxlevel$ or $\tau\text{-}maxar$.

We give two examples:

1. $\tau\text{-}maxlevel_{()}(\lambda f{:}(O{\to}O){\to}O{\to}O.\lambda x{:}O{\to}O.f(f(x))) = 3$.

2. $\tau\text{-}maxar_{()}(\lambda f{:}(O{\to}O){\to}O{\to}O.\lambda x{:}O{\to}O.f(f(x))) = 2$.

**Lemma 4.2.**

1. Let $\Xi \equiv \Gamma, x{:}C, \Delta$. Suppose $\Xi \vdash_{\lambda\tau} A : B$ and $\Gamma \vdash_{\lambda\tau} A' : C$. Then

$$\tau\text{-}maxlevel_{\Gamma,\Delta[x:=A']}(A[x{:=}A']) \leq \max(\{\tau\text{-}maxlevel_\Xi(A), \tau\text{-}maxlevel_\Gamma(A')\}).$$

2. Let $\Gamma \vdash_{\lambda\tau} A : B$ and $A \twoheadrightarrow_\beta A'$. Then $\tau\text{-}maxlevel_\Gamma(A) \geq \tau\text{-}maxlevel_\Gamma(A')$.

**Lemma 4.3.**

1. Let $\Xi \equiv \Gamma, x{:}C, \Delta$. Suppose $\Xi \vdash_{\lambda\tau} A : B$ and $\Gamma \vdash_{\lambda\tau} A' : C$. Then

$$\tau\text{-}maxar_{\Gamma,\Delta[x:=A']}(A[x{:=}A']) \leq \max(\{\tau\text{-}maxar_\Xi(A), \tau\text{-}maxar_\Gamma(A')\}).$$

2. Let $\Gamma \vdash_{\lambda\tau} A : B$ and $A \twoheadrightarrow_\beta A'$. Then $\tau\text{-}maxar_\Gamma(A) \geq \tau\text{-}maxar_\Gamma(A')$.

**Definition 4.4.** Let $A$ be a pseudo-term of $\lambda\tau$. Then we define $\tau\text{-}height(A)$ as the height of the parse tree of $A$ with the type-information stripped off. Types have height 0.

It is easy to see that $\tau\text{-}height$ is well-defined.

**Lemma 4.5.**

1. $\tau\text{-}height(A[x:=A']) \leq \tau\text{-}height(A) + \tau\text{-}height(A')$.

2. $\tau\text{-}height$ can increase under $\beta$-reduction.

**Proof.**

1. Induction on $A$.

2. Consider: $A \equiv (\lambda x_1{:}O.\lambda x_2{:}O.\lambda x_3{:}O.x_1)((\lambda y_1{:}O.\lambda y_2{:}O.y_1)tt)$.
   Then $t{:}O \vdash_{\lambda\tau} A : O \to O \to O$. The height of $A$ is 6.
   $A$ reduces in one step to $\lambda x_2{:}O.\lambda x_3{:}O.(\lambda y_1{:}O.\lambda y_2{:}O.y_1)tt$, which has height 7.

$\boxtimes$

We define the function $itexp$ as follows: $itexp(0,x) := x$; $itexp(m+1,x) := 2^{itexp(m,x)}$.

**Theorem 4.6** (Schwichtenberg). *Let $\Gamma \vdash_{\lambda\tau} A : B$. The number of steps to reduce $A$ to normal form is bounded by:*

$$k^{itexp(m,(m+2h+2k+2))},$$

where $k = \tau\text{-}maxar_\Gamma(A)$, $m = \tau\text{-}maxlevel_\Gamma(A)$ and $h = \tau\text{-}height(A)$.

We now turn to the system $\lambda P$. Recall that the system $\lambda P$ is formally given by the triple: $\mathcal{S} = \{*, \Box\}$; $\mathcal{A} = \{* : \Box\}$; $\mathcal{R} = \{(*,*),(*,\Box)\}$. We list two characteristic properties of legal types and constructors of $\lambda P$.

**Lemma 4.7.**

1. Suppose $\Gamma \vdash_{\lambda P} \Pi x{:}A_1.A_2 : s$. Then $\Gamma \vdash_{\lambda P} A_1 : *$ and so $\#(A_1) = 1$.

2. Suppose $\Gamma \vdash_{\lambda P} A_1 A_2 : B : \Box$. Then $\#(A_2) = 0$.

In this section we work with terms that are not necessarily legal in $\lambda P$. These terms, however, do respect the $\#$-stratification, in the following sense.

**Definition 4.8.** A pseudo-term $A$ is $\#P$-legal if:

- for all subterms of the form $A_1 A_2$ we have: $\#(A_2) = 0$;
- for all subterms of the form $\lambda x{:}A_1.A_2$ we have: $\#(x) = 0$ and $\#(A_1) = 1$;
- for all subterms of the form $\Pi x{:}A_1.A_2$ we have $\#(x) = 0$, $\#(A_1) = 1$ and $\#(A_2) > 0$;
- moreover, we require that for all subterms of the form $Qx{:}A_1.A_2$ (for $Q \in \{\lambda, \Pi\}$) we have: $x \notin \mathcal{FV}(A_1)$.

It is easy to prove that every term that is legal in $\lambda P$ is $\#P$-legal and that $\#P$-legality is preserved under $\beta$-reduction.

In [3] a two step translation from $\lambda C$ to $\lambda\omega$ is defined. We use a (slight variation of a) restricted version of the translation that maps $\lambda P$ to $\lambda\tau$.

First a function $\tau : T_{1,2,3} \to T_{1,2,3}$ is defined (and extended to pseudo-contexts) that maps types of $\lambda P$ to types of $\lambda\tau$. For our purposes, it is not required that $\tau$ preserves reduction steps. We state its properties as a lemma.

**Lemma 4.9** (Geuvers & Nederhof).

1. Suppose $\Gamma \vdash_{\lambda P} B : \Box$ or $B \equiv \Box$. Then $\Gamma \vdash_{\lambda P} A : B \Rightarrow \tau(\Gamma) \vdash_{\lambda\tau} \tau(A) : *$.

2. Let $A \in \mathcal{T}_{1,2,3}$ and $\sharp(x) = \sharp(B) = 0$. Then $\tau(A[x:=B]) \equiv \tau(A)$.

3. If $A \in \mathcal{T}_{1,2,3}$ is $\sharp P$-legal and $A \twoheadrightarrow_\beta B$, then $\tau(A) \equiv \tau(B)$.

Next a function $[\cdot] : \mathcal{T}_{0,1,2} \to \mathcal{T}_{0,1,2}$ is defined that maps terms of $\lambda P$ (types as well as objects) to objects of $\lambda \tau$. This map preserves reduction steps.

**Lemma 4.10** *(Geuvers & Nederhof).*

1. $\Gamma \vdash_{\lambda P} A : B \Rightarrow \tau(\Gamma) \vdash_{\lambda \tau} [A] : \tau(B)$.
   In particular: if $\Gamma \vdash_{\lambda P} A : *$, then $\tau(\Gamma) \vdash_{\lambda \tau} [A] : O$.

2. $[A[^*x:=B]] \equiv [A][^*x:=[B]]$, for $A, B \in \mathcal{T}_{0,1,2}$.

3. Let $A, B \in \mathcal{T}_{0,1,2}$, $\sharp P$-legal. Then $(A \twoheadrightarrow_\beta B) \Rightarrow ([A] \twoheadrightarrow_\beta^{\neq 0} [B])$. Here the superscript $\neq 0$ means that at least one reduction step is performed.

## 4.2 The Replacement Operation

In this section we define Labeled Pure Type Systems and we introduce a *replacement* operation. We work in an arbitrary PTS $\lambda S$. From $\lambda S$ we construct a labeled PTS $\lambda^l S$ as follows. We usually omit the superscript.

**Definition 4.11.**

1. Labeled pseudo-terms are given by the following syntax:
   
   $$T^l ::= \mathcal{C} \mid \mathcal{V} \mid T^l T^l \mid \lambda \mathcal{V}{:}T^l.T^l \mid \Pi^N \mathcal{V}{:}T^l.T^l.$$

2. Substitution, $\beta$-reduction, $\beta$-conversion are as before and respect the labels; derivation rules too are as before, but disregard the labels.

Note that $\beta$-reduction does not create new labels.
The next definition describes how to label an expansion free derivation $\mathcal{D}$ with conclusion $\Gamma \vdash A : B$ in such a way that each label in $A$ has a unique occurrence in $A$ and the sets of labels in $\Gamma$, $A$ and $B$ are disjoint.

**Definition 4.12.** From an expansion free derivation $\mathcal{D}$ of $\Gamma \vdash A : B$ we construct an expansion free, *labeled derivation* of $\Gamma \vdash A : B$. We describe it with reference to Table 1 (except for the Reduction Rule).

1. Axiom. No labels are assigned.

2. The Start Rule. The distribution of the labels in the context $\Gamma, x{:}B$ is the same as the distribution of the labels in the premiss. The distribution of the labels in (the predicate) $B$ is obtained by replacing every label $l_i$ in $B$ in the premiss by a fresh label $m_i$. Fresh means: distinct from all labels used sofar.

3. The Weakening Rule. The distribution of the labels of the conclusion is inherited from the left premiss, except for $B'$; the distribution of labels in $B'$ is obtained by a renaming of the labels in $B'$ in the right premiss as described in (2).

4. The Product Rule. The distribution is inherited from the right premiss. The new $\Pi$ is given a fresh label.

5. **The Application Rule.** Rename every label in $A_2$ as described in (2). The distribution of the labels in $A_1 A_2$ is the combination of the distribution of the labels in $A_1$ and the new distribution of labels in $A_2$. The distribution of the labels in $\Gamma$ is inherited from the left premiss. The distribution in $B_2[x:=A_2]$ is obtained as follows: again rename the labels in $A_2$; then the distribution of the labels in $B_2[x:=A_2]$ is a combination of the distribution of the labels in $\Pi x{:}B_1.B_2$ and the new distribution of the labels in $A_2$.

6. **The Abstraction Rule.** The distribution of labels of $\Gamma$ and $\lambda x{:}A_1.A_2$ is inherited from the left premiss. The distribution of the labels in $\Pi x{:}A_1.B_2$ is a combination of the distribution of labels in $B_2$ in the left premiss and a renaming of labels in $A_1$ in the left premiss. The label of the new $\Pi$ is chosen fresh.

7. **The Reduction Rule.** The distribution of the labels of $\Gamma$ and $A$ is inherited from the left premiss. The distribution of the labels in $B$ is the result of the existing distribution of $B'$ under the reduction to $B$.

**Lemma 4.13.** *Suppose that $\Gamma \vdash A : B$ is the conclusion of an expansion free, labeled derivation $\mathcal{D}$. Then each label in $A$ has a unique occurrence, each label in $\Gamma$ has a unique occurrence and moreover the sets of labels in $\Gamma$, $A$ and $B$ are mutually disjoint.*

Now we define our *replacement* operation $(N)\{M\}_i$ which replaces all occurrences of subterms of $N$ of the form $\Pi^i x{:}A.B$ by $\Pi^i x{:}(A[x:=M]).(B[x:=M])$.

**Definition 4.14.**

- $c\{M\}_i := c$, for $c \in \mathcal{C}$.
- $x\{M\}_i := x$, for $x \in \mathcal{V}$.
- $(A_1 A_2)\{M\}_i := (A_1\{M\}_i)(A_2\{M\}_i)$.
- $(\lambda x{:}A_1.A_2)\{M\}_i := \lambda x{:}(A_1\{M\}_i).(A_2\{M\}_i)$.
- $(\Pi^j x{:}A_1.A_2)\{M\}_i := \begin{cases} \Pi^j x{:}(A_1\{M\}_i).(A_2\{M\}_i) & \text{if } i \neq j \\ \Pi^j x{:}(A_1\{M\}_i[x:=M]).(A_2\{M\}_i[x:=M]) & \text{otherwise.} \end{cases}$

Note that in the $\Pi^j$ and the $\lambda$ case, by the variable convention, the variable $x$ does not occur free in $M$. Also note that if the label $i$ does not occur in $A$ then $A\{M\}_i \equiv A$. We give an example in $\lambda C$.

$$((\lambda \alpha{:}*.\Pi^i x{:}\alpha.\alpha)(\Pi^j y{:}A.Py))\{M\}_j \equiv (\lambda \alpha{:}*.\Pi^i x{:}\alpha.\alpha)(\Pi^j y{:}A.PM).$$

The most important property of the replacement operation is that it respects $\beta$-reduction.

**Lemma 4.15.** *Let $A$ be a labeled pseudo-term and suppose that $A \twoheadrightarrow_\beta A'$. Then $A\{M\}_i \twoheadrightarrow_\beta A'\{M\}_i$.*

Suppose that $A$ is $\sharp P$-legal and that $A \twoheadrightarrow_\beta A'$. It is possible that a $\beta$-reduction step duplicates occurrences of labels or makes them disappear. So in general it is not true that each occurrence of a label $i$ in $A$ corresponds to a unique occurrence of $i$ in $A'$ and vice versa. For our purposes it is crucial that when we use the replacement operation the label $i$ that functions as index in $A\{M\}_i$ has a unique occurrence in $A$ and a unique occurrence in $A'$. Fortunately, this requirement is satisfiable because such a label $i$ occurs at a so called *active* place. We show that reductions of $\sharp P$-legal terms do not duplicate or erase active occurrences of labels.

**Definition 4.16.** Let $A$ be a pseudo-term.

- Let $l$ be a label. By induction on the structure of $A$ we define when $l$ has an *active occurrence* in $A$:
  - $l$ has no active occurrence in $x$, for $x \in \mathcal{V}$.
  - $l$ has no active occurrence in $c$, for $c \in \mathcal{C}$.
  - $l$ has an active occurrence in $A_1 A_2$ iff $l$ has an active occurrence in $A_1$.
  - $l$ has an active occurrence in $\lambda x{:}A_1.A_2$ iff $l$ has an active occurrence in $A_2$.
  - $l$ has an active occurrence in $\Pi^i x{:}A_1.A_2$ iff $l = i$ or $l$ has an active occurrence in $A_2$.

- We call $A$ *active-unique* if for every label $l$ it holds that if $l$ has an active occurrence in $A$, then it has a unique occurrence in $A$.

**Lemma 4.17.** *Let $A$ be a $\sharp P$-legal pseudo-term and suppose $A \twoheadrightarrow_\beta A'$.*

1. *For all labels $l$ there is a bijective correspondence between active occurrences of $l$ in $A$ and active occurrences of $l$ in $A'$.*

2. *If $A$ is active-unique, then $A'$ is active-unique.*

## 4.3 The Construction

Suppose $\Gamma \vdash_{\lambda P} A : B$. In this section we use the replacement operation to construct for $B$ a $\sharp P$-legal term $\mathcal{B}$ such that $\mathcal{B}$ reduces to a term $D$ that contains all redexes that occur in $B$. The intuition is that any reduction sequence starting from $B$ is a subsequence of a reduction sequence starting from $\mathcal{B}$ and hence is at most as long as the longest reduction in $\mathcal{B}$. Having constructed this term $\mathcal{B}$, we need an upper bound to the length of reduction sequences starting from $\mathcal{B}$ in terms of $|\Gamma|$ and $|A|$. By Lemma 4.10 (3) it suffices to get an upperbound for the length of reduction sequences starting from $[\mathcal{B}]$. To apply Schwichtenberg's result to $[\mathcal{B}]$, we have to show that $[\mathcal{B}]$ is legal in $\lambda \tau$; otherwise $\tau$-*maxlevel* and $\tau$-*maxar* are not defined. In case we could prove that $\Gamma \vdash_{\lambda P} \mathcal{B} : *$ we could apply Lemma 4.10 (1) to get $\tau(\Gamma) \vdash_{\lambda \tau} [\mathcal{B}] : O$. However, because of the fact that the definition of $\mathcal{B}$ uses the replacement operation, we cannot prove that $\Gamma \vdash_{\lambda P} \mathcal{B} : *$. Namely, it is not true that if $\Gamma \vdash_{\lambda P} B : *$, then $\Gamma \vdash_{\lambda P} B\{M\}_i : *$. Although we cannot prove that $\mathcal{B}$ is legal, we can prove that $\mathcal{B}$ is $\sharp P$-legal. This turns out to be sufficient, for we can prove the following proposition.

**Proposition 4.18.** *Suppose $\Gamma \vdash_{\lambda \tau} [A] : B$, $A$ is $\sharp P$-legal and active-unique. Let $i$ be a label that has an active occurrence in $A$. Suppose that $M$ is $\sharp P$-legal, $\sharp(M) = 0$ and that for the subterm of $A$ of the form $\Pi^i x{:}A_1.A_2$ we have: $\Gamma \vdash_{\lambda \tau} [M] : \tau(A_1)$. Then $\Gamma \vdash_{\lambda \tau} [A\{M\}_i] : B$.*

We now construct the term $\mathcal{B}$.

**Definition 4.19.** *Suppose that $\Gamma \vdash_{\lambda P} A : B$ is the conclusion of an expansion free, labeled derivation $\mathcal{D}$. Then we define a term $\mathcal{B}$ by induction on the height of $\mathcal{D}$.*

- Axiom.  $\langle\rangle \vdash_{\lambda P} * : \square$.

  Put $\mathcal{B} \equiv \square$.

- Start Rule. $\dfrac{\Gamma' \vdash_{\lambda P} B : s}{\Gamma', x{:}B \vdash_{\lambda P} x{:}B}$.

  Put $\mathcal{B} \equiv B$.

- **Weakening Rule.** $\dfrac{\Gamma' \vdash_{\lambda P} A : B \quad \Gamma' \vdash_{\lambda P} B' : s}{\Gamma', x{:}B' \vdash_{\lambda P} A : B}$ .

By the induction hypothesis on the left premiss we have a term $\mathcal{B}'$. Put $\mathcal{B} \equiv \mathcal{B}'$.

- **Product Rule.** $\dfrac{\Gamma \vdash_{\lambda P} A_1 : s_1 \quad \Gamma, x{:}A_1 \vdash_{\lambda P} A_2 : s_2}{\Gamma \vdash_{\lambda P} \Pi^l x{:}A_1.A_2 : s_3}$ .

Put $\mathcal{B} \equiv s_3$.

- **Reduction Rule.** $\dfrac{\Gamma \vdash_{\lambda P} A : B' \quad \Gamma \vdash_{\lambda P} B : s \quad B' \twoheadrightarrow_\beta B}{\Gamma \vdash_{\lambda P} A : B}$ .

The induction hypothesis on the leftmost premiss gives us a term $\mathcal{B}'$. Put $\mathcal{B} \equiv \mathcal{B}'$.

- **Abstraction Rule.** $\dfrac{\Gamma, x{:}A_1 \vdash_{\lambda P} A_2 : B_2 \quad \Gamma \vdash_{\lambda P} \Pi^j x{:}A_1.B_2 : s}{\Gamma \vdash_{\lambda P} \lambda x{:}A_1.A_2 : \Pi^l x{:}A_1.B_2}$ .

The induction hypothesis on the left premiss gives us a term $\mathcal{B}'$. Put $\mathcal{B} \equiv \Pi^l x{:}A_1.\mathcal{B}'$.

- **Application Rule.** $\dfrac{\Gamma \vdash_{\lambda P} A_1 : \Pi^l x{:}B_1.B_2 \quad \Gamma \vdash_{\lambda P} A_2 : B_1}{\Gamma \vdash_{\lambda P} A_1 A_2 : B_2[x{:=}A_2]}$ .

By the induction hypothesis on the left premiss we have a term $\mathcal{B}'$. Put $\mathcal{B} \equiv \mathcal{B}'\{A_2\}_l$, as defined in section 4.2.

**Lemma 4.20.** *Suppose that $\Gamma \vdash_{\lambda P} A : B$ is the conclusion of an expansion free, labeled derivation $\mathcal{D}$. Then $\mathcal{B}$, as defined in Definition 4.19, is $\sharp P$-legal and active-unique.*

Now we prove that each reduction sequence starting from $B$ can be embedded in a reduction sequence starting from $\mathcal{B}$. Recall that $B^\star$ is the body of $B$, as defined in section 2.

**Lemma 4.21.** *Suppose that $\Gamma \vdash_{\lambda P} A : B$ is the conclusion of an expansion free, labeled derivation $\mathcal{D}$. Write $B$ as $\Pi y_1{:}E_1 \ldots \Pi y_k{:}E_k.B^\star$ (for some $k \geq 0$). Then $\mathcal{B}$, as defined in Definition 4.19, satisfies: there is a term $D \equiv \Pi^{l_1} x_1{:}C_1 \ldots \Pi^{l_n} x_n{:}C_n.B^\star$ (for some $n \geq 0$, $x_i$, $C_i$) such that:*

- $\mathcal{B} \twoheadrightarrow_\beta D$.

- *Every $\Pi y_i{:}E_i$ $(1 \leq i \leq k)$ also occurs among the $\Pi x_j{:}C_j$'s $(1 \leq j \leq n)$.*

- *If $i < j$ then $\Pi y_i{:}E_i$ occurs in $D$ in a position to the left of $\Pi y_j{:}E_j$.*

**Proof.** By induction on the structure of $\mathcal{D}$. We only treat the Application Rule: $\Gamma \vdash_{\lambda P} A : B$ is $\Gamma \vdash_{\lambda P} A_1 A_2 : B_2[x{:=}A_2]$ and is a direct consequence of (i) $\Gamma \vdash_{\lambda P} A_1 : \Pi^l x{:}B_1.B_2$ and (ii) $\Gamma \vdash_{\lambda P} A_2 : B_1$. Let $\mathcal{B}'$ be the term that corresponds via Definition 4.19 to $\Pi^l x{:}B_1.B_2$. By the induction hypothesis on (i) we have a term $D'$ such that:

$$\mathcal{B}' \twoheadrightarrow_\beta D' : \equiv \Pi^{l_1} x_1{:}C'_1 \ldots \Pi^{l_{i-1}} x_{i-1}{:}C'_{i-1}.\Pi^l x{:}B_1.\Pi^{l_{i+1}} x_{i+1}{:}C'_{i+1} \ldots \Pi^{l_n} x_n{:}C'_n.B^\star_2.$$

We have defined $\mathcal{B}$ as $\mathcal{B}'\{A_2\}_l$. By Lemma 4.20, we know that $\mathcal{B}'$ is $\sharp P$-legal and active-unique. By Lemma 4.17 (2) $D'$ is also active-unique. It is easily seen that the occurrence of the label $l$ in $D'$ is active. So, by active-uniqueness, $D'$ contains no other occurrences of the label $l$. Hence, by the definition of the replacement operation and Lemma 4.15, we have

$$\mathcal{B} \twoheadrightarrow_\beta \Pi^{l_1} x_1{:}C'_1 \ldots \Pi^{l_{i-1}} x_{i-1}{:}C'_{i-1}.\Pi^l x{:}B_1.\Pi^{l_{i+1}} x_{i+1}{:}C_{i+1} \ldots \Pi^{l_n} x_n{:}C_n.B^\star_2[x{:=}A_2],$$

where each $C_j$ $(i+1 \leq j \leq n)$ is the result of substitution of $A_2$ for $x$ in $C'_j$. This is sufficient to verify that the result holds, since $(B_2[x{:=}A_2])^\star \equiv (B^\star_2[x{:=}A_2])^\star$, as can be easily verified. ☒

**Lemma 4.22.** *Suppose that $\Gamma \vdash_{\lambda P} A : B$ is the conclusion of an expansion free, labeled derivation $\mathcal{D}$. Then $\mathcal{B}$, as defined in Definition 4.19, is such that:*

1. $\tau(\Gamma) \vdash_{\lambda\tau} [\mathcal{B}] : O$.

2. 
   - $\tau\text{-maxlevel}_{\tau(\Gamma)}([\mathcal{B}]) \leq |\Gamma| + |A|$.
   - $\tau\text{-maxar}_{\tau(\Gamma)}([\mathcal{B}]) \leq |\Gamma| + |A|$.
   - $\tau\text{-height}([\mathcal{B}]) \leq 2(|\Gamma| + |A|)$.

**Proof.** Note that (2) makes sense because, by (1), $\tau\text{-maxlevel}_{\tau(\Gamma)}([\mathcal{B}])$, $\tau\text{-maxar}_{\tau(\Gamma)}([\mathcal{B}])$ are defined. We only treat (1). The proof is by induction on the derivation of $\Gamma \vdash_{\lambda P} A : B$. We only treat the case that the last rule used is the Application Rule: $\Gamma \vdash_{\lambda P} A : B$ is $\Gamma \vdash_{\lambda P} A_1 A_2 : B_2[x := A_2]$ and is a direct consequence of (i) $\Gamma \vdash_{\lambda P} A_1 : \Pi^l x{:}B_1.B_2$ and (ii) $\Gamma \vdash_{\lambda P} A_2 : B_1$. Let $\mathcal{B}'$ be the term corresponding via Definition 4.19 to $\Pi^l x{:}B_1.B_2$. By the induction hypothesis on (i) we have $\tau(\Gamma) \vdash_{\lambda\tau} [\mathcal{B}'] : O$. We have defined $\mathcal{B}$ as $\mathcal{B}'\{A_2\}_l$. We apply Proposition 4.18 to get $\tau(\Gamma) \vdash_{\lambda\tau} [\mathcal{B}'\{A_2\}_l] : O$, as required. To be able to apply Proposition 4.18, five requirements must be satisfied. First, $\mathcal{B}'$ must be $\sharp P$-legal and active-unique. This is Lemma 4.20. Secondly, $A_2$ must be $\sharp P$-legal. But we know that $\Gamma \vdash_{\lambda P} A_2 : B_1$ and so $A_2$ is $\sharp P$-legal. Thirdly, $\sharp(A_2)$ must be 0. This follows from the fact that $\Gamma \vdash_{\lambda P} A_2 : B_1 : *$ and Proposition 2.5 (4). Fourthly, the label $l$ must have an active occurrence in $\mathcal{B}'$. By Lemma 4.21 there is a $\sharp P$-legal and active-unique term $D$ such that:

$$\mathcal{B}' \twoheadrightarrow_\beta D : \equiv \Pi^{l_1} x_1{:}C_1 \ldots \Pi^{l_{i-1}} x_{i-1}{:}C_{i-1}.\Pi^l x{:}B_1.\Pi^{l_{i+1}} x_{i+1}{:}C_{i+1} \ldots \Pi^{l_n} x_n{:}C_n.B_2^\star.$$

The occurrence of $l$ is active in $D$; so by Lemma 4.17 (1), there is a unique active occurrence of $l$ in $\mathcal{B}'$. Finally, we need to show that for the subterm in $\mathcal{B}'$ of the form $\Pi^l y{:}F_1.F_2$ we have $\tau(\Gamma) \vdash_{\lambda\tau} [A_2] : \tau(F_1)$. We prove that for this subterm it holds that $\tau(F_1) \equiv \tau(B_1)$. This is sufficient since from $\Gamma \vdash_{\lambda P} A_2 : B_1$ and Lemma 4.10 (1) it follows that $\tau(\Gamma) \vdash_{\lambda\tau} [A_2] : \tau(B_1)$. The reduction sequence from $\mathcal{B}'$ to $D$ can be written as a sequence of terms $\langle T^0, \ldots, T^p \rangle$ such that $T^0 \equiv \mathcal{B}'$, $T^p \equiv D$ and for all $i$, $0 \leq i \leq p-1$, $T^{i+1} \to_\beta T^i$.

Let $S \equiv \langle \Pi^l x_0{:}F^0.G^0, \ldots, \Pi^l x_p{:}F^p.G^p \rangle$ be the sequence of terms such that:

- For all $i$, $0 \leq i \leq p$, $\Pi^l x_i{:}F^i.G^i$ is a subterm of $T^i$;

- $\Pi^l x_0{:}F^0.G^0 \equiv \Pi^l y{:}F_1.F_2$;

- $\Pi^l x_p{:}F^p.G^p \equiv \Pi^l x{:}B_1.\Pi^{l_{i+1}} x_{i+1}{:}C_{i+1}.\cdots.\Pi^{l_n} x_n{:}C_n.B_2^\star$;

- for all $i$, $0 \leq i \leq p-1$, $\Pi^l x_{i+1}{:}F^{i+1}.G^{i+1}$ is a residual of $\Pi^l x_i{:}F^i.G^i$ in $T^{i+1}$.

Note that such a sequence $S$ always exists, because $\beta$-reduction does not create new labels.

Now we prove that $\tau(F^0) \equiv \tau(B_1)$. This is done by following the path from $\Pi^l x_0{:}F^0.G^0$ to $\Pi^l x_p{:}F^p.G^p$ *backwards* and verifying that for each $\Pi^l x_{p-j}{:}F^{p-j}.G^{p-j}$ it holds that $\tau(F^{p-j}) \equiv \tau(B_1)$. The proof of this claim is by induction on $j$. The claim is trivially true for $j = 0$, since $F^p$ is defined as $B_1$. Suppose $j > 0$. So $T^{p-(j+1)} \to_\beta T^{p-j}$. By the induction hypothesis $\tau(F^{p-j}) \equiv \tau(B_1)$. We prove that $\tau(F^{p-j}) \equiv \tau(F^{p-(j+1)})$. If the reduction step from $T^{p-(j+1)}$ to $T^{p-j}$ does not involve $F^{p-(j+1)}$, then clearly $F^{p-(j+1)} \equiv F^{p-j}$ and we are done. So assume that the reduction step involves $F^{p-(j+1)}$. By analysing the possible reduction steps from $T^{p-(j+1)}$ to $T^{p-j}$ it can be seen that there are two ways in which the reduction step could change $F^{p-(j+1)}$.

1. A substitution of $N_3$ is made into $F^{p-(j+1)}$ because $F^{p-(j+1)}$ is a subterm of $N_2$ in the subterm $(\lambda z{:}N_1.N_2)N_3$. So $F^{p-j} \equiv F^{p-(j+1)}[z := N_3]$. Since $\mathcal{B}'$ is $\sharp P$-legal and $(\lambda z{:}N_1.N_2)N_3$ is a subterm of a $\beta$-reduct of $\mathcal{B}'$, $(\lambda z{:}N_1.N_2)N_3$ is $\sharp P$-legal too. So we have $\sharp(z) = \sharp(N_3) = 0$. Thus we can apply Lemma 4.9 (2) which yields that $\tau(F^{p-(j+1)}) \equiv \tau(F^{p-(j+1)}[z := N_3]) \equiv \tau(F^{p-j})$.

2. A reduction can take place in $F^{p-(j+1)}$. By Lemma 4.9 (3) we have $\tau(F^{p-(j+1)}) \equiv \tau(F^{p-j})$.

In both cases we have $\tau(F^{p-(j+1)}) \equiv \tau(F^{p-j}) \equiv \tau(B_1)$. ☒

**Theorem 4.23.** *Suppose that* $\Gamma \vdash_{\lambda P} A : B$ *is the conclusion of an expansion free derivation* $\mathcal{D}$. *Then every reduction sequence starting from* $B$ *consists of at most* $m^{itexp(m,(7 \cdot m+2))}$ *steps, where* $m = |\Gamma| + |A|$.

**Proof.** Label $\mathcal{D}$ as in Definition 4.12. In Definition 4.19 we have constructed a $\sharp P$-legal term $\mathcal{B}$ such that by Lemma 4.21 every reduction sequence starting from $B$ is at most as long as the longest reduction in $\mathcal{B}$. Now by Lemma 4.10 (3) every reduction step in $\mathcal{B}$ induces a reduction step in $[\mathcal{B}]$. By Lemma 4.22 (2) $\tau\text{-}maxlevel_{\tau(\Gamma)}([\mathcal{B}]) \leq |\Gamma| + |A|$ and $\tau\text{-}maxar_{\tau(\Gamma)}([\mathcal{B}]) \leq |\Gamma| + |A|$ and $\tau\text{-}height([\mathcal{B}]) \leq 2(|\Gamma| + |A|)$. By Schwichtenberg's result (Theorem 4.6) the maximum length of any reduction in $[\mathcal{B}]$ is less than or equal to $f := m^{itexp(m,7 \cdot m+2)}$, where $m = |\Gamma| + |A|$. So the maximum length of any reduction sequence in $\mathcal{B}$ is less than or equal to $f$ and so the maximum length of any reduction sequence starting from $B$ will be less than or equal to $f$. ☒

In an analogous but much simpler way one can prove the following result for $\lambda \underline{\omega}$. The proof uses translations from [3].

**Theorem 4.24.** *Suppose that* $\Gamma \vdash_{\lambda \underline{\omega}} A : B$ *is the conclusion of an expansion free derivation* $\mathcal{D}$. *Then every reduction sequence starting from* $B$ *consists of at most* $m^{itexp(m,(7 \cdot m+2))}$ *steps, where* $m = |\Gamma| + |A|$.

# References

[1] Barendregt, H.P., 'Lambda calculi with types' , in: Abramski, S., Gabbay, D.M. & Maibaum, T.S.E., eds., *Handbook of Logic in Computer Science*, Oxford University Press, 1992.

[2] Coquand, Th. and Huet, G., 'The calculus of constructions', *Information and Computation*, Vol. 76, pp. 95-120, 1988.

[3] Geuvers, H. and Nederhof, M.-J., 'Modular proof of strong normalization for the calculus of constructions', in: *The Journal of Functional Programming*, Vol.1, pp. 155-189, 1991.

[4] Pollack, R., 'Typechecking in pure type systems', in: Nördstrom, B., Petersson, K. and Plotkin, G., eds., *Proceedings of the 1992 Workshop on Types for Proofs and Programs*, Båstad, June 92, pp. 289-306.

[5] Rose, H.E., *Subrecursion. Functions and Hierarchies*, Clarendon Press, Oxford, 1984.

[6] Schwichtenberg, H., 'An upperbound for reduction sequences in the typed $\lambda$-calculus', *Archive for Mathematical Logic*, Vol.30, pp. 405-408, 1991.

[7] Statman, R., 'The typed lambda calculus is not elementary recursive', *Theoretical Computer Science*, Vol. 9, pp. 73-81, 1979.

# λ-Calculi with Conditional Rules

Masako Takahashi
Department of Information Science
Tokyo Institute of Technology, Tokyo 152 Japan
e-mail: masako@is.titech.ac.jp

**Abstract**

A variety of typed/untyped λ-calculi and related reduction systems have been proposed in order to study various aspects of programs, some of which contain rules subject to side conditions. As a framework to study fundamental properties of such reduction systems, we first introduce the notion of conditional λ-calculus. Then we give a sufficient condition for them to be confluent (Church-Rosser) as well as to have a normalizing strategy à la Gross. The proof, being a generalization of Tait-Martin-Löf proof for the confluence of λβ, is inductive and simple.

## 0  Introduction

During the last 20 years, a variety of typed/untyped λ-calculi and related reduction systems have been proposed in order to study various aspects of functional programs, some of which contain rules subject to side conditions. In this paper, we introduce the notion of conditional λ-calculus with the intenion of providing a framework for such reduction systems.

We formulate the notion by using a set of rule schemes of the form

$$\{ l\theta \to r\theta \mid \theta : \{m_1, ..., m_n\} \to \Lambda, \ P(m_1\theta, ..., m_n\theta)\}$$

where $l \to r$ indicates the skeleton (or pattern) of rewriting rules, which contains 'holes' named $m_1, ..., m_n$, and each $m_i\theta$ indicates the term to replace the 'holes' named $m_i$ in the skeleton. The side condition that should be satisfied by the terms $m_1\theta, ..., m_n\theta$ is specified by the predicate $P$.

By looking at the form of our rule schemes, one may wonder why the condition part is necessary from theoretical standpoint, because it is dispensable in the sense that the set of rules specified by such a rule scheme can equally be specified by a set of condition-free rule schemes. Indeed, take for instance the set of singleton rule schemes each corresponding to a member of the original rule scheme. Nevertheless we think that side conditions in the rule schemes are of theoretical importance, because they make the proof of confluence easier.

Suppose we are given a set of rewriting rules in two ways; one as a union of rule schemes with side conditions, and the other as a union of singleton (condition-free) rule schemes. In the latter case, the skeleton parts of the redexes (i.e., the lefthand sides of the rewriting rules) will tend to be large compared with the former, and naturally they have more chances to overlap each other.

But it is such overlaps of redexes that makes the proof of confluence really difficult or complicated. Although the singleton rule schemes are extreme cases, we think that it would generally be preferable to have simpler skeletons at the expense of introducing some harmless side conditions, if it is possible.

In section 1, we define the notion of conditional $\lambda$-calculus (clc, for short) and some related concepts. Among them is the notion of semi-orthogonal clc, and later we show that being semi-orthogonal is a sufficient condition for clc's to be confluent. In proving the confluence, we exploit the notion of parallel reduction which was introduced in the well-known Tait-Martin-Löf proof of the confluence of $\lambda$-calculus. The outstanding feature of the proof by Tait and Martin-Löf is its simplicity and transparency; all definitions necessary for the proof are given in just a few lines by induction on the structure of terms, and the proof proceeds thoroughly by induction (cf. [6]). As a consequence, the proof is easy to understand and also short compared with other proofs. In sections 2 and 3, we show that the feature is essentially preserved when we extend the idea to semi-orthogonal clc's. As a byproduct of the proof, we also show a normarizing strategy à la Gross for semi-orthogonal clc's.

## 1 Basic Definitions

**Definition 1.1 (terms and meta-terms)** Suppose $V = \{x, y, ...\}$ is a countable set of variables, $F^{(n)}$ is a set of $n$-ary function (or operator) symbols, and $F = \bigcup_{n \geq 0} F^{(n)}$. We denote by $\Lambda$ the set of all *terms* constructed from $V$ and $F$ as usual. In our systems it is most likely that $F^{(1)}$ contains the abstraction operator $\lambda x$ for each variable $x$, and $F^{(2)}$ contains the application operator Ap. In addition to $\lambda x$'s there may be other operators which bind variables, and the notion of free/bound occurrences of variables is defined with respect to their totality. Then by $FV(t)$ we denote the set of variables which occur free in $t$, and by $t[x_1 := t_1, x_2 := t_2, ..., x_k := t_k]$ the term obtained from $t$ by simultaneous substitution of terms $t_1, t_2, ..., t_k$ for free occurrences of $x_1, x_2, ..., x_k$ respectively in $t$.

In order to describe our systems succinctly, it is handy to have a set $M = \{m_1, m_2, ...\}$ of additional symbols called *meta-variables*, which play the same role as 'holes' in contexts. We denote by $\Lambda(M)$ the set of *meta-terms*, which are constructed from $V \cup M$ by function symbols in $F$. The set of meta-variables in a meta-term $u$ is denoted by $MV(u)$.

As in the case of terms, we can recursively define the tree-representation, tree-domain, etc. of meta-terms. For example, the tree-domain $dom(u)$ of a meta-term $u$ (i.e., the set of 'addresses' of nodes in the tree-representation of $u$ coded by finite sequences of positive integers) is defined as follows;

1. $\mathrm{dom}(x) = \{\epsilon\}^1$ if $x \in V$,

2. $\mathrm{dom}(m) = \emptyset$ if $m \in M$,

3. $\mathrm{dom}(fu_1 u_2 ... u_n) = \{\epsilon\} \cup \{\, i \cdot a \mid a \in \mathrm{dom}(u_i), 1 \le i \le n\,\}$
   if $f \in F^{(n)}$ and $u_1, u_2, ..., u_n \in \Lambda(M)$.

We will extend other standard definitions and conventions on $\lambda$-terms (cf. [2]) to meta-terms, whenever they are appropriate. For example, $\equiv$ denotes the syntactic equality (up to change of bound variables), and the application operator Ap and parentheses in $\mathrm{Ap}(...(\mathrm{Ap}(\mathrm{Ap}u_1 u_2)u_3)...)u_k$ will be suppressed.

**Definition 1.2 (rule schemes)** Suppose $l$ is a meta-term in which each $m \in \mathrm{MV}(l)$ occurs exactly once, and $l$ itself is not a meta-variable. Suppose also that $r_0, r_1, ..., r_k$ ($k \ge 0$) are meta-terms with $\mathrm{MV}(r_j) \subseteq \mathrm{MV}(l)$ ($j = 0, ..., k$). When $\mathrm{MV}(l) = \{m_1, m_2, ..., m_n\}$ and $P$ is an $n$-ary predicate on $\Lambda$, we say

$$R = \{\, l\theta \to (r_0\theta)[x_1 := r_1\theta, ..., x_k := r_k\theta] \mid \theta : \mathrm{MV}(l) \to \Lambda,\ P(m_1\theta, ..., m_n\theta)\,\}$$

is a *rule scheme*. Here by $l\theta$ we mean the term ($\in \Lambda$) obtained from $l$ by replacing each occurrence of meta-variables $m_i$ by $m_i\theta$, and similarly for $r_0\theta, r_1\theta, ..., r_k\theta$. The replacement may cause new binding relation unlike substitution. For example, $((\lambda x.m_1)m_2)\theta \equiv (\lambda x.m_1\theta)(m_2\theta)$ for any $\theta : \{m_1, m_2\} \to \Lambda$.

The rule scheme $R$ may also be written as

$$l\theta \to r\theta \quad \text{if} \quad P(m_1\theta, ..., m_n\theta)$$

where $r = r_0[x_1 := r_1, ..., x_k := r_k]$. Each member $l\theta \to r\theta$ of the rule scheme is called a *rule*, and $l \to r$ is its *skeleton*. The lefthand side $l\theta$ of a rule $l\theta \to r\theta$ (or an occurrence of it) is called a *redex*, and the righthand side $r\theta$ its *contractum*. The subterms $m_i\theta$ of the redex $l\theta$ are called its *parameters*. When we say $l\theta \to r\theta$ is a rule, we tacitly assume that $l \to r$ is its skeleton.

The followings are well-known rule schemes or unions of them.

**Example 1.3** (1) The $\beta$-reduction rules constitute the $\beta$-rule scheme

$$R_\beta = \{\, (\lambda x.t_1)t_2 \to t_1[x := t_2] \mid t_1, t_2 \in \Lambda\,\}$$

with skeleton $(\lambda x.m_1)m_2 \to m_1[x := m_2]$, while the $\eta$-reduction rules constitute the $\eta$-rule scheme

$$R_\eta = \{\, \lambda x.tx \to t \mid x \notin \mathrm{FV}(t),\ t \in \Lambda\,\}$$

---

[1] The empty sequence is denoted by $\epsilon$.

with skeleton $\lambda x.mx \to m$.
(2) Recursor rules constitute two rule schemes;

$$Rt_1 t_2 0 \to t_1$$
$$Rt_1 t_2 (St_3) \to t_2 t_3 (Rt_1 t_2 t_3) \quad \text{if } t_3 \text{ is a numeral}$$

where we assume $R \in F^{(3)}, S \in F^{(1)}, 0 \in F^{(0)}$ (or R, S, 0 $\in F^{(0)}$).
(3) Church's $\delta$-rules (cf. [4]) are also divided into two rule schemes;

$$\delta t_1 t_2 \to T \quad \text{if } t_1, t_2 \text{ are closed terms in normal form, and } t_1 \equiv t_2,$$
$$\delta t_1 t_2 \to F \quad \text{if } t_1, t_2 \text{ are closed terms in normal form, and } t_1 \not\equiv t_2$$

where $\delta \in F^{(2)}$ (or $F^{(0)}$), T and F are certain closed terms.
(4) A generalization of Church's $\delta$-rules, called Mitschke's $\delta$-rules (cf. [4]), is a finite union of rule schemes of the form

$$\delta t_1 t_2 ... t_n \to s_1 \quad \text{if } P_1(t_1, t_2, ..., t_n)$$
$$\delta t_1 t_2 ... t_n \to s_2 \quad \text{if } P_2(t_1, t_2, ..., t_n)$$
$$...\qquad\qquad ...$$
$$\delta t_1 t_2 ... t_n \to s_k \quad \text{if } P_k(t_1, t_2, ..., t_n)$$

where $\delta \in F^{(n)}$ (or $F^{(0)}$), $s_1, s_2, ..., s_k$ are closed terms, and $P_1, P_2, ..., P_k$ are mutually disjoint predicates which are closed under substitution and reduction.

**Definition 1.4 (conditional $\lambda$-calculi)** For a union $R_\Xi = \cup_{\xi \in \Xi} R_\xi$ of rule schemes $R_\xi$, we define the notion of one-step reduction $\to_\Xi$ ($\subseteq \Lambda \times \Lambda$) recursively, as follows;
(1) $t \to_\Xi t'$, if $(t \to t') \in R_\Xi$.
(2) $f t_1 t_2 ... t_n \to_\Xi f t_1 ... t_{i-1} t'_i t_{i+1} ... t_n$, if $f \in F^{(n)}, 1 \le i \le n$ and $t_i \to_\Xi t'_i$.
In other words, for any rule $(t \to t')$ and any one-hole context $C$ we write $C[t] \to_\Xi C[t']$. We call the reduction system $(\Lambda, \to_\Xi)$ a *conditional $\lambda$-calculus* (*clc*, for short), providing

- different rule schemes share no redex,

- each $R_\xi$ is closed under substitution (i.e., $(t \to t') \in R_\xi$ implies $(t[x := s] \to t'[x := s]) \in R_\xi$ for each $x \in V$ and $s \in \Lambda$),

- each $R_\xi$ is closed under reduction of parameters (i.e., if $(l\theta \to r\theta) \in R_\xi$ and $m\theta \stackrel{*}{\to}_\Xi m\theta'$ for each $m \in MV(l)$, then $(l\theta' \to r\theta') \in R_\xi$)[2].

We may simply write $\Xi$, or $\xi\xi'...\xi''$ when $\Xi = \{\xi, \xi', ..., \xi''\}$, as a shorthand for $(\Lambda, \to_\Xi)$. Note that, because of the closure property of $R_\xi$ under substitution,

---

[2] Here and in the sequel, for a binary relation $\to$, by $\stackrel{*}{\to}$ we mean its reflexive and transitive closure.

there is no free occurrence of variables in the skeletons $l \to r$ of rules ($l\theta \to r\theta$) in a clc, and we have $(l\theta)[x := s] \equiv l(\theta[x := s])$ and $(r\theta)[x := s] \equiv r(\theta[x := s])$ for each $x \in V$ and $s \in \Lambda$. Here by $\theta[x := s]$ we mean the replacement defined by $m(\theta[x := s]) \equiv (m\theta)[x := s]$ for each $m$.

**Definition 1.5** Suppose $\Xi$ is a clc. If a redex $t$ contains another redex $t_1$ as a subterm but not as a subterm of its parameters, then we say $t$ *critically contains* $t_1$, or $t_1$ is *critical* in $t$. By using the notion, we define three classes of clc's.

- If no redex critically contains others, then $\Xi$ is said to be *orthogonal*.

- Suppose $\Xi$ satisfies the condition; if a redex $t$ critically contains $t_1$, then the result obtained from $t$ by contracting $t_1$ is identical with the contractum of $t$. In this case[3] we say $\Xi$ is *weakly orthogonal*.

- Suppose $\Xi$ satisfies the condition; if a redex $t$ critically contains redexes $t_1, ..., t_h$ where $h \geq 1$ and each $t_{i-1}$ ($i=2, ..., h$) contains $t_i$ non-critically (cf. Figure 1), then the result obtained from $t$ by contracting $t_h, ..., t_1$ in this order is identical with the contractum of $t$. In this case[4] we say $\Xi$ is *semi-orthogonal*.

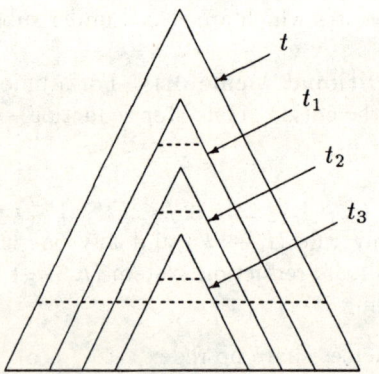

Figure 1: critical tower $(t_1, t_2, t_3)$ in $t$

When redexes $t$ and $t_1, t_2, ..., t_h$ ($h \geq 1$) satisfy the if-clause in the definition of semi-orthogonality, we say $(t_1, t_2, ..., t_h)$ is a *critical tower* (of *height* $h$) in $t$.

---

[3] i.e., if $(t \to t'), (t_1 \to t'_1) \in R_\Xi$, $t \equiv C[t_1]$ for a one-hole context $C$, and $t_1$ is critical in $t$, then $t' \equiv C[t'_1]$.

[4] a simple description of the condition is given in terms of parallel reduction in next section (just before lemma 2.3).

Clearly an orthogonal clc is semi-orthogonal, and a semi-orthogonal clc is weakly orthogonal. In particular, if there is no critical tower of height two or more in a weakly orthogonal clc, then it is semi-orthogonal.

**Example 1.6** The non-extensional $\lambda$-calculus $\beta$ is an orthogonal clc, and it remains so when we add any of rule schemes in example 1.3 (2) - (4). The rule schemes consisting of Hindley's (a)-rules for $\beta$-reduction (cf. [3]) also enjoy the property. On the other hand, the extensional $\lambda$-calculus $\beta\eta$ is a semi-orthogonal (hence weakly orthogonal) clc, but not orthogonal. Indeed, in $\beta\eta$

- The $\beta$-redex $(\lambda x.sx)t$ with $x \notin \text{FV}(s)$ critically contains $\eta$-redex $\lambda x.sx$, and it satisfies
$$(\lambda x.sx)t \xrightarrow{\beta} st, \quad (\lambda x.sx)t \xrightarrow{\eta} st.$$

- Similarly for the $\eta$-redex $\lambda x.(\lambda x.s)x$ critically containing $\beta$-redex $(\lambda x.s)x$.

- There is no other critical containment between redexes; in particular, there is no critical tower of height two or more.

When we add Hindley's (a)-rules for $\beta\eta$-reduction (cf. [3]) to $\beta\eta$, the system remains to be semi-orthogonal. However if we add the rule schemes for $\delta$-rules to $\beta\eta$, the system may no longer belong to any of the three classes. Moreover it may fail to be confluent.

We conclude this section with a simple property of weakly orthogonal clc's.

**Lemma 1.7** In a weakly orthogonal clc $\Xi$, suppose a redex $t$ critically contains two redexes $t_1$ and $t_2$ which are mutually disjoint. Then $t$ is its own contractum, and so are $t_1$ and $t_2$.
(Proof) Let $t \equiv C[t_1, t_2]$ where $C$ is a 2-hole context, and $s, s_1, s_2$ be the contractums of $t, t_1, t_2$, respectively. Then since $\Xi$ is weakly orthogonal, we have $C[s_1, t_2] \equiv s \equiv C[t_1, s_2]$. This implies $s_1 \equiv t_1$ and $s_2 \equiv t_2$, and hence $t \equiv C[t_1, t_2] \equiv C[s_1, t_2] \equiv s$. □

## 2 Parallel Reduction for clc

**Definition 2.1** For a clc $\Xi$, we define the notion of *parallel reduction* $\Rightarrow_\Xi$ ($\subseteq \Lambda \times \Lambda$) recursively, as follows;
(1) $x \Rightarrow_\Xi x$ for any $x \in V$.
(2) $ft_1t_2...t_n \Rightarrow_\Xi ft'_1t'_2...t'_n$, if $f \in F^{(n)}$ and $t_i \Rightarrow_\Xi t'_i$ ($i = 1, 2, ..., n$).
(3) $l\theta \Rightarrow_\Xi r\theta'$, if $(l\theta \to r\theta) \in R_\Xi$ and $m\theta \Rightarrow_\Xi m\theta'$ for each $m \in \text{MV}(l)$[5].

---
[5] In what follows, we simply write $\theta \Rightarrow_\Xi \theta'$ instead of '$m\theta \Rightarrow_\Xi m\theta'$ for each $m$'.

The clause (2) says that simultaneous reduction of mutually disjoint subterms by $\Rightarrow_\Xi$ amounts to a $\Rightarrow_\Xi$ reduction step of the whole term. The clause (3) says that for a redex, first reducing parameters separately by $\Rightarrow_\Xi$ and then contracting the redex itself amounts to a $\Rightarrow_\Xi$ reduction step. Note that in the case of redex, reducing the parameters alone also makes a $\Rightarrow_\Xi$ reduction step according to (2). Thus as a whole a parallel reduction step $t \Rightarrow_\Xi t'$ means to choose a number of redexes in $t$ which are mutually non-critical, and contract them all. The assumption that rule schemes be closed under reduction of parameters guarantees that the redex remains to be so after its parameters are reduced. This observation stated more precisely yeilds the following.

**Lemma 2.2** $\quad t \Rightarrow_\Xi t' \quad \Longrightarrow \quad t \xrightarrow{*}_\Xi t'$.
(Proof) By induction on the definition of $\Rightarrow_\Xi$. When $t \Rightarrow_\Xi t'$ holds by (1) or (2) of definition 2.1, it is trivial. For case (3), it suffices to show that $(l\theta \to r\theta) \in R_\Xi$ and $m\theta \xrightarrow{*}_\Xi m\theta'$ for each $m \in \mathrm{MV}(l)$ imply $l\theta \xrightarrow{*}_\Xi r\theta'$. Under the assumption, we have $l\theta \xrightarrow{*}_\Xi l\theta'$ (proof by induction on $l$), and $(l\theta' \to r\theta') \in R_\Xi$ because rule schemes are closed under reduction of parameters. Hence we get $l\theta \xrightarrow{*}_\Xi r\theta'$. □

We will denote by $\mathrm{rdx}(t \Rightarrow_\Xi t')$ the set of 'addresses' of the redexes which are contracted in the series of $\to_\Xi$ reduction steps simulating $t \Rightarrow_\Xi t'$. More precisely the set can be defined recursively depending on the cases of definition 2.1, as follows;

case (1): $\quad \mathrm{rdx}(x \Rightarrow_\Xi x) = \emptyset$,

case (2): $\quad \mathrm{rdx}(ft_1t_2...t_n \Rightarrow_\Xi ft'_1t'_2...t'_n) = \{\, i \cdot a \,|\, a \in \mathrm{rdx}(t_i \Rightarrow_\Xi t'_i), 1 \leq i \leq n\,\}$,

case (3): $\quad \mathrm{rdx}(l\theta \Rightarrow_\Xi r\theta') = \{\epsilon\} \cup \mathrm{rdx}(l\theta \Rightarrow_\Xi l\theta')$.

The notation $\mathrm{rdx}(t \Rightarrow_\Xi t')$ is ambiguous in the sense that it may not be determined by $t, t'$ alone but by specifying the way how $t \Rightarrow_\Xi t'$ is obtained. In case (3) above, by $l\theta \Rightarrow_\Xi l\theta'$ we mean the result obtained from $m\theta \Rightarrow_\Xi m\theta'$ ($\forall m \in \mathrm{MV}(l)$) by repeated application of definition 2.1 (1) and (2).

Later when it is convenient, we will write $t \Rightarrow^A_\Xi t'$ to mean that $t \Rightarrow_\Xi t'$ holds with $A = \mathrm{rdx}(t \Rightarrow_\Xi t')$. Using this notation, we can state the condition of the semi-orthogonality of a clc $\Xi$ simply as follows; if $l\theta \to r\theta$ is a rule and $l\theta \Rightarrow^A_\Xi t$ holds with a totally ordered[6] subset $A$ of $\mathrm{dom}(l)$, then $t \equiv r\theta$.

**Lemma 2.3** In a clc $\Xi$, we have $t \xrightarrow{*}_\Xi t'$ if and only if $t \stackrel{*}{\Rightarrow}_\Xi t'$.
(Proof) By induction on the structure of terms, one can easily show

$$t \to_\Xi t' \quad \Longrightarrow \quad t \Rightarrow_\Xi t',$$

---

[6]with respect to the prefix ordering $\leq$ defined by; $a \leq b$ if and only if $a \cdot c = b$ for some sequence $c$.

using the fact $t \Rightarrow_\Xi t$ (which can separately be verified by induction). Hence we get the 'only if' part. The converse is immediate from lemma 2.2. □

**Lemma 2.4** In a clc $\Xi$, $t \Rightarrow_\Xi t'$ and $s \Rightarrow_\Xi s'$ imply $t[x := s] \Rightarrow_\Xi t'[x := s']$.
(Proof) By induction on the structure of $t$.

case 1: When $t \in V$, it is trivial.

case 2: Suppose $t \equiv ft_1t_2...t_n \Rightarrow_\Xi ft'_1t'_2...t'_n \equiv t'$ with $t_i \Rightarrow_\Xi t'_i$ ($i = 1, ..., n$). When $x$ is not a bound variable of $f$ (including the case where $f$ involves no binding mechanism), we have

$$t[x := s] \equiv f(t_1[x := s])(t_2[x := s])...(t_n[x := s])$$
$$\Rightarrow_\Xi f(t'_1[x := s'])(t'_2[x := s'])...(t'_n[x := s']) \equiv t'[x := s']$$

because $t_i[x := s] \Rightarrow_\Xi t'_i[x := s']$ ($i = 1, ..., n$) by induction hypothesis. Otherwise, for example when $f \equiv \lambda x$, we have $t[x := s] \equiv t \Rightarrow_\Xi t' \equiv t'[x := s']$. Note that no rule in $\Xi$ creates free variables, because the skeletons of rule schemes in $\Xi$ contain no free occurrence of variables.

case 3: If $t \equiv l\theta \Rightarrow_\Xi r\theta' \equiv t'$ where $(l\theta \to r\theta) \in R_\xi$ for some $\xi \in \Xi$ and $\theta \Rightarrow_\Xi \theta'$, then

$$t[x := s] \equiv (l\theta)[x := s] \equiv l(\theta[x := s]),$$

$$t'[x := s'] \equiv (r\theta')[x := s'] \equiv r(\theta'[x := s']).$$

Now since $\theta[x := s] \Rightarrow_\Xi \theta'[x := s']$ by induction hypothesis (recall that $m\theta$ is a proper subterm of $l\theta$), and $(l(\theta[x := s]) \to r(\theta[x := s])) \in R_\xi$ by the closure property of $R_\xi$ under substitution, we have $l(\theta[x := s]) \Rightarrow_\Xi r(\theta'[x := s'])$ by definition 2.1(3). Thus we get $t[x := s] \Rightarrow_\Xi t'[x := s']$. □

**Corollary 2.5** In a clc $\Xi$, if $t_i \Rightarrow_\Xi t'_i$ ($i = 0, ..., k$) then $t_0[x_1 := t_1, ..., x_k := t_k] \Rightarrow_\Xi t'_0[x_1 := t'_1, ..., x_k := t'_k]$.
(Proof) By changing variables if necessary, we may assume that $x_1, ..., x_k$ do not appear in $t_i, t'_i$ ($i = 1, ..., k$). Then $t_0[x_1 := t_1, ..., x_k := t_k] \equiv t_0[x_1 := t_1][x_2 := t_2]...[x_k := t_k] \Rightarrow_\Xi t'_0[x_1 := t'_1][x_2 := t'_2]...[x_k := t'_k] \equiv t_0[x_1 := t_1, ..., x_k := t_k]$ by repeated application of lemma 2.4. □

**Corollary 2.6** For any contractum $r\theta$ of a clc $\Xi$, if $\theta \Rightarrow_\Xi \theta'$ then $r\theta \Rightarrow_\Xi r\theta'$.
(Proof) Suppose $r\theta \equiv (r_0\theta)[x_1 := r_1\theta, ..., x_k := r_k\theta]$ where $r_0, r_1, ..., r_k \in \Lambda(M)$. Since $\theta \Rightarrow_\Xi \theta'$, we have $r_i\theta \Rightarrow_\Xi r_i\theta'$ for each $i$. Then the corollary immediately follows from corollary 2.5. □

# 3 Confluence and Normalization of Semi-Orthogonal clc

The key lemma for our proof of the confluence of semi-orthogonal clc's is lemma 3.2 below, for which we need one more definition.

**Definition 3.1** For a clc $\Xi$, we define a term $t^* \in \Lambda$ for each $t \in \Lambda$ recursively as follows:
(1) $x^* \equiv x$ for each $x \in V$.
(2) $(ft_1t_2...t_n)^* \equiv ft_1^*t_2^*...t_n^*$, if $ft_1t_2...t_n$ is not a redex.
(3) $(l\theta)^* \equiv r\theta^*$, if $(l\theta \to r\theta) \in R_\Xi$.
In (3), $\theta^*$ is the replacement defined by $m\theta^* \equiv (m\theta)^*$ for each $m \in \mathrm{MV}(l)$. Note that $t^*$ is well-defined, because in (3)

- different rule schemes share no redex, and so $r$ and $\theta$ are uniquely determined by the redex $l\theta$, and

- $m\theta$ is a proper subterm of $l\theta$, and so $(m\theta)^*$ has been already defined in the recursive definition.

Note also $t \Rightarrow_\Xi t^*$ (proof by induction on $t$), and the set of redexes contracted therein is a maximal set of redexes in $t$ which are mutually non-critical.

**Lemma 3.2** In a semi-orthogonal clc $\Xi$, for any $t, t' \in \Lambda$ we have

$$t \Rightarrow_\Xi t' \implies t' \Rightarrow_\Xi t^*.$$

(Proof) By induction on $t$. We assume $t \Rightarrow_\Xi t'$.

case 1: Suppose $t$ is not a redex. If $t \equiv x \in V$, clearly $t' \equiv x \equiv t^*$. Otherwise we have $t \equiv ft_1t_2...t_n \Rightarrow_\Xi ft'_1t'_2...t'_n \equiv t'$ for some $f \in F^{(n)}$ and $t_i, t'_i$ such that $t_i \Rightarrow_\Xi t'_i$ ($i = 1, ..., n$). Then by induction hypothesis we have $t'_i \Rightarrow_\Xi t_i^*$ ($i = 1, ..., n$), and hence $t' \equiv ft'_1t'_2...t'_n \Rightarrow_\Xi ft_1^*t_2^*...t_n^* \equiv t^*$.

case 2: Suppose $t \equiv l\theta$ with $(l\theta \to r\theta) \in R_\xi$.

subcase 2.1: If $t \equiv l\theta \Rightarrow_\Xi r\theta' \equiv t'$ for some $\theta'$ such that $\theta \Rightarrow_\Xi \theta'$, then since $\theta' \Rightarrow_\Xi \theta^*$ by induction hypothesis we get $t' \equiv r\theta' \Rightarrow_\Xi r\theta^* \equiv t^*$ by corollary 2.6.

subcase 2.2: If $t \equiv ft_1t_2...t_n \Rightarrow_\Xi ft'_1t'_2...t'_n \equiv t'$ with $t_i \Rightarrow_\Xi t'_i$ ($i = 1, ..., n$), let $A = \mathrm{rdx}(t \Rightarrow_\Xi t') = \{i \cdot a \mid a \in \mathrm{rdx}(t_i \Rightarrow_\Xi t'_i), 1 \leq i \leq n\}$, and $A' = A \cap \mathrm{dom}(l)$. Among the redexes contracted in $t \Rightarrow_\Xi t'$, those specified by $A'$ are critical in $t$ and others are not. Then the parallel reduction step $t \Rightarrow_\Xi^A t'$ can be divided into two as $t \Rightarrow_\Xi^{A''} t'' \Rightarrow_\Xi^{A'} t'$ for some $t''$ and $A'' = A - A' = A - \mathrm{dom}(l)$. Note that $A'$ indeed specifies redexes in $t''$, because the redexes in $l$ specified by $A = A' \cup A''$ are mutually non-critical, and those specified by $A'$ remain to be redexes after contraction of others.

Now since the redexes contracted in the first step $t \Rightarrow_{\underline{\Xi}}^{A''} t''$ are within the parameters of $t$, we have $t'' \equiv l\theta'$ for some $\theta'$ such that $\theta \Rightarrow_\Xi \theta'$ (which is due to our assumption that meta-variables in $l$ be distinct). Here we claim

$$t' \equiv t'' (\equiv l\theta') \text{ or } t' \equiv r\theta'.$$

Once the claim is established, we easily get the result from induction hypothesis $\theta' \Rightarrow_\Xi \theta^*$. Indeed, if $t' \equiv t''$ then $t' \equiv l\theta' \Rightarrow_\Xi r\theta^* \equiv t^*$ since $(l\theta' \to r\theta') \in R_\mathcal{E}$. Otherwise $t' \equiv r\theta' \Rightarrow_\Xi r\theta^* \equiv t^*$ by corollary 2.6. Now we verify the claim in two subcases.

subcase 2.2.1: If the set $A'$ is nonempty and totally ordered with respect to the prefix ordering, then we have $t' \equiv r\theta'$ since $\Xi$ is semi-orthogonal.

subcase 2.2.2: If $A'$ contains mutually incomparable elements, say $a_1$ and $a_2$, let $B = A' - \{a_1, a_2\}$. Then we have $t'' \Rightarrow_{\underline{\Xi}}^B t'$, since the redex at $a_i (i = 1, 2)$ is its own contractum by lemma 1.7. By repeating the same argument, we eventually get a totally ordered (possibly empty) subset $B'$ of $A'$ such that $t'' \Rightarrow_{\underline{\Xi}}^{B'} t'$. If $B' = \emptyset$ then $t' \equiv t''$, and otherwise $t' \equiv r\theta'$ by subcase 2.2.1.

This completes the proof of lemma 3.2. □

**Theorem 3.3** A semi-orthogonal clc $\Xi$ is confluent; that is, $t_1 \overset{*}{\leftarrow}_\Xi \cdot \overset{*}{\to}_\Xi t_2$ implies $t_1 \overset{*}{\to}_\Xi \cdot \overset{*}{\leftarrow}_\Xi t_2$.
(Proof) Lemma 3.2 means that $\Rightarrow_\Xi$ satisfies the diamond property; i.e., $t_1 \Leftarrow_\Xi \cdot \Rightarrow_\Xi t_2$ implies $t_1 \Rightarrow_\Xi \cdot \Leftarrow_\Xi t_2$. The property can be easily generalized to that of $\overset{*}{\Rightarrow}_\Xi (= \overset{*}{\to}_\Xi)$ by induction on the number of steps. □

The lemma 3.2 also yeilds a normalizing strategy for semi-orthogonal clc's. Indeed,

$$t \Rightarrow_\Xi t^* \Rightarrow_\Xi (t^*)^* \Rightarrow_\Xi ((t^*)^*)^* \Rightarrow_\Xi \ldots$$

is a normalizing sequence in the sense that if $t$ has a $\Xi$-normal form then it appears in the sequence. More precisely, we have the following.

**Theorem 3.4** In a semi-orthogonal clc $\Xi$, let $s_0 \equiv t$ and $s_{k+1} \equiv s_k^*$ ($k = 0, 1, \ldots$). Then $t \overset{*}{\to}_\Xi t'$ implies $t' \overset{*}{\to}_\Xi s_k$ for some $k$. In particular, if $t'$ is in $\Xi$-normal form, then $t' \equiv s_k$.
(Proof) We show by induction on $k$ that if $t \overset{*}{\to}_\Xi t'$ in $k$ steps then $t' \overset{*}{\Rightarrow}_\Xi s_k$ in $k$ steps. When $k = 0$, it is trivial. Otherwise, suppose $t \equiv t_0 \to_\Xi t_1 \to_\Xi \ldots \to_\Xi t_k \to_\Xi t_{k+1}$. Then by induction hypothesis we have $t_k \equiv u_0 \Rightarrow_\Xi u_1 \Rightarrow_\Xi \ldots \Rightarrow_\Xi u_k \equiv s_k$ for some $u_0, u_1, \ldots, u_k (\in \Lambda)$, and also $t_k \Rightarrow_\Xi t_{k+1}$ by lemma 2.2. Hence we get

$$\begin{array}{ccccccccc}
u_0 & \Rightarrow_\Xi & u_1 & \Rightarrow_\Xi & u_2 & \Rightarrow_\Xi & \cdots & \Rightarrow_\Xi & u_{k-1} & \Rightarrow_\Xi & u_k \equiv s_k \\
\Downarrow & & \Downarrow & & \Downarrow & & & & \Downarrow & & \Downarrow \\
t_{k+1} \Rightarrow_\Xi & u_0^* & \Rightarrow_\Xi & u_1^* & \Rightarrow_\Xi & \cdots & \Rightarrow_\Xi & u_{k-2}^* & \Rightarrow_\Xi & u_{k-1}^* & \Rightarrow_\Xi s_k^* \equiv s_{k+1}
\end{array}$$

by repeated application of lemma 3.2. □

## 4 Concluding Remarks

In formulating the notion of clc, the idea of rule schemes with the substitution mechanism in contractums is inspired by that of combinatory reduction systems by Klop [4]. The essential difference of our rule schemes from those of combinatory reduction systems is the inclusion of side conditions. As for the side conditions, we extended the idea of the $\eta$-rule, various $\delta$-rules and the like in classical literature (cf. [2], [3], [4]) as well as numerous examples in contemporary systems.

One of the crucial points in the definition of clc is the closure property of rule schemes under substitution. One should not confuse it with the closure property of the side condition $P$ of rule schemes;

$$P(s_1, s_2, ..., s_n) \text{ implies } P(s_1[x := s], ..., s_n[x := s]) \text{ for each } s \in \Lambda.$$

The latter condition is the one stated in Mitschke $\delta$-rules, and is strictly stronger than the former. Indeed, the $\eta$-rule scheme $R_\eta = \{\lambda x.sx \rightarrow s \mid x \notin \text{FV}(s)\}$ satisfies the former, but not the latter.

Similar results as theorems 3.3 and 3.4 can be obtained by a slight modification of the method. Instead of parallel reduction, consider a stronger notion $\Rrightarrow_\Xi$ of reduction, due to Aczel [1] (cf. also [5]), which is defined recursively as follows;
(1) $x \Rrightarrow_\Xi x$ for any $x \in V$.
(2) $ft_1t_2...t_n \Rrightarrow_\Xi ft_1't_2'...t_n'$, if $f \in F^{(n)}$ and $t_i \Rrightarrow_\Xi t_i'$ $(i = 1, 2, ..., n)$.
(3) $ft_1t_2...t_n \Rrightarrow_\Xi t'$, if $f \in F^{(n)}, t_i \Rrightarrow_\Xi t_i'(i = 1, ..., n)$ and $(ft_1't_2'...t_n' \rightarrow t') \in R_\Xi$.
Informally speaking, a $\Rrightarrow_\Xi$ reduction step can be simulated by a series of $\rightarrow_\Xi$ reduction steps organized as follows: First if a term $t$ contains some redexes at nodes of depth $k$ where $k$ is the maximum depth of the nodes in the (tree-representation of) term $t$, then we contract some (possibly all or none) of them, resulting in a tree $t_k$. If $t$ has no such redexes, let $t_k$ be $t$ itself. Next, to the tree $t_k$ we do the same, but this time instead of redexes at nodes of depth $k$ we look for those at nodes of depth $k-1$, resulting in a tree $t_{k-1}$. By repeating the process, we eventually climb up to the root node (of depth 0), resulting in a tree $t_0$. Then we have $t \Rrightarrow_\Xi t_0$.

Now assume that a clc $\Xi$ satisfies the condition; for any $(l\theta \rightarrow r\theta) \in R_\Xi$ and $t \in \Lambda$, if $l\theta \Rrightarrow_\Xi^A t$ with a totally ordered subset $A$ of $\text{dom}(l)$, then $t \equiv r\theta$. Here the notation $l\theta \Rrightarrow_\Xi^A t$ means in informal terms that $A$ is the set of 'addresses' of redexes contracted in the course of $\to_\Xi$ reduction steps simulating $l\theta \Rrightarrow_\Xi t$. Under this condition, one can prove counterparts of our lemmas in section 2 and 3. In particular, one can prove

$$t \Rrightarrow_\Xi t' \implies t' \Rrightarrow_\Xi t^\circ$$

where for each $t$ the term $t^\circ$ is defined recursively as
(1) $x^\circ \equiv x$ for each $x \in V$,
(2) $(ft_1t_2...t_n)^\circ \equiv ft_1^\circ t_2^\circ...t_n^\circ$, if $ft_1^\circ t_2^\circ...t_n^\circ$ is not a redex,
(3) $(ft_1t_2...t_n)^\circ \equiv t'$, if $(ft_1^\circ t_2^\circ...t_n^\circ \to t') \in R_\Xi$.
Then based on the lemmas one can obtain the confluence and normalization result for $\Xi$ as before. The condition just stated is a little stronger than the semi-orthogonality, when it is applied to clc's. However the idea may be worth pursuing in a different but similar framework as clc.

I thank Roger Hindley and Jan Willem Klop for their valuable comments and stimuli which led me to the work.

# References

[1] P. Aczel: *A General Church-Rosser Theorem*, (Technical Report, University of Manchester, 1978).

[2] H. P. Barendregt: *The Lambda Calculus*, second edition (North-Holland, 1984).

[3] J. R. Hindley: Reduction of residuals are finite, *Trans.A.M.S.* 240 (1978) pp 345-361.

[4] J. W. Klop: *Combinatory Reduction Systems* (Mathematisch Centrum, 1980).

[5] F. van Raamsdonk: *A Simple Proof of Confluence for Weakly Orthogonal Combinatory Reduction Systems*, (Technical Report, CWI, 1992).

[6] M. Takahashi: Parallel Reductions in λ-Calculus, *J. Symbolic Computation* 7 (1989) pp 113-123. Also *Parallel Reductions in λ-Calculus*, revised version (Research Report, Tokyo Institute of Technology, 1992).

# Type reconstruction in $F_\omega$ is undecidable *

Paweł Urzyczyn
Institute of Informatics, University of Warsaw
ul. Banacha 2, 02-097 Warszawa, Poland
urzy@mimuw.edu.pl

### Abstract

We investigate the Girard's calculus $F_\omega$ as a "Curry style" type assignment system for pure lambda terms. We prove that the type-reconstruction problem for $F_\omega$ is undecidable (even with quantification restricted to constructor variables of rank 1). In addition, we show an example of a strongly normalizable pure lambda term that is untypable in $F_\omega$.

## 1 Introduction

The system $F_\omega$ was introduced by J.Y. Girard in his PhD Thesis (an English reference is [8]) as a tool to prove properties of higher-order propositional logics. The system is an extension of the second order polymorphic lambda calculus, known also as "system F". The extension allows for (quantification over) not only type variables, but also variables for functions from types to types, functions from functions to functions and so on, i.e., involves an infinite hierarchy of *type constructors* classified according to their *ranks* in a similar way that types are divided into ranks in the finitely typed lambda calculus. A general exposition of the properties of $F_\omega$ can be found in [4]. See also [3], [1], [7] and [21], the latter recommended for instructive examples.

Girard's systems F and $F_\omega$ are "Church style" calculi (in the terminology of [1]): types are parts of expressions, and there is nothing like "untypable term", since an untypable term is just not a term at all. There is however a direct connection between this presentation and the untyped lambda calculus: by a "type erasing" procedure we obtain a "Curry style" system which assigns types to pure lambda terms. This translation between typed and untyped terms creates the *type reconstruction problem*: for a given pure lambda term, determine whether it can be assigned a type or not. To be precise, terms with free variables may be typable or not, depending on types assigned to their free variables, by a *type environment*. We write $E \vdash M : \tau$ for "$M$ has type $\tau$ in the environment $E$". Thus, our problem has to be either restricted to closed terms or to be (equivalently) stated as follows:

Given $M$, does there exist $E$ and $\tau$ such that $E \vdash M : \tau$?

---

*This work is partly supported by NSF grant CCR-9113196 and by Polish KBN Grant 2 1192 91 01

The first paper we are aware of addressing the type reconstruction problem for system **F** is [17]. The decidability of this problem is still an open question, despite many attempts to solve it. There are several partial solutions, applying either to fragments of system **F**, or to certain variations of the system, see [2], [11], [6], [7]. An exponential lower bound for time complexity of type reconstruction for **F** has been obtained by Henglein [9], see also [10] for a complete exposition. This lower bound is quite an involved refinement of the following simple observation: there are terms of length $\mathcal{O}(n)$, typable in **F**, but such that their shortest types are of depth exponential in $n$ (when drawn as binary trees). Consider for example terms of the form $2(2(\ldots 2(K)\ldots))$, where 2 is the Church numeral $\lambda fx.f(fx)$, and $K$ is $\lambda xy.x$. The method of proof is to replace $K$ with a tricky representation of one step of a Turing Machine, so that the whole term can represent an exponentially long computation. This TM simulation is then put in such a context that a failing computation forces untypability.

A related issue is type reconstruction for the language ML and its modifications, see e.g. [12], [13], [14], [18] (the latter two merged into [15] for a final presentation). For instance, the above mentioned result of [9] uses a technique of [15], developed initially for ML, and the semi-unification problem of [13] is used as a tool in [7].

The type reconstruction problem for $\mathbf{F}_\omega$ has a shorter history. A modification of $\mathbf{F}_\omega$, similar to that of [2], was shown in [20] to have undecidable type reconstruction. Another variation of the original problem: the "conditional" type reconstruction is also undecidable by a result of [7]. Finally, a nonelementary lower bound has been obtained by Henglein and Mairson [10], using a similar method to that mentioned above. This time, instead of composing 2's as in the previous case, one can consider terms of the form $222\ldots 2$ and apply it to the Turing Machine simulator. A simplified version of this construction is the following exercise: assign types to terms of the form $222\ldots 2K$.

An explanation is in order here: type reconstruction for a system that types all strongly normalizable terms (e.g., involving intersection types) is immediately undecidable ([17]). Thus, one has first to ask whether the set of typable terms is a proper subset of the strongly normalizable terms. For system **F**, the first example has been given in [5], more examples can be found in [19]. The example of [5] is typable in $\mathbf{F}_\omega$, which means that the latter can type strictly more than **F**. (We conjecture that all the examples of [19] are typable too.) However, we show below (Theorem 3.1) that $\mathbf{F}_\omega$ cannot type some strongly normalizable terms. Warning: it is known that the classes of integer functions representable in our systems are all different, and are proper subclasses of the class of recursive functions. However, the fact that a certain term (function representation) cannot be assigned a particular type (integer to integer) does not mean that it cannot be assigned any type at all. In fact, all recursive functions can be represented in the untyped lambda calculus by terms in normal form and all normal forms are typable in **F**. Thus, results like that of [16] have little to do with our consideration.

The main result of this paper is that type reconstruction in $\mathbf{F}_\omega$ is undecidable (Theorem 5.1). The proof is essentially based on the following trick. We construct a term $W$, which is (roughly) of the form:

$$W \equiv \lambda x.\,\text{if } Final\ x \text{ then } \lambda w.\texttt{nil} \text{ else } \lambda w.w(Next\ x)w \text{ fi},$$

and we put it into the following context:

$$T \equiv (\lambda w.\, wC_0 w)W,$$

where $C_0$ is a term representing an initial configuration of a Turing Machine. Of course, $W$ is designed so that *Next x* and *Final x* behave as a "next configuration" function and as a test for a final state, respectively (assuming an appropriate representation of a machine ID is substituted for $x$). It is now quite easy to see that $T$ has a normal form (in fact it is strongly normalizable) iff our Turing Machine converges on the initial ID represented by $C_0$. We show more: our term is typable in $\mathbf{F}_\omega$ if the machine converges, and this means that the halting problem reduces to typability. In fact type reconstruction is also undecidable on each rank level (see below), as we need quite a restricted machinery in our $\mathbf{F}_\omega$ typings. The only way type constructors are used is based on the following trick: let $\gamma$ be a constructor variable of kind **Prop** $\Rightarrow$ **Prop** $\Rightarrow$ **Prop**, and take an object variable $x$ of type $\forall \gamma.\, \gamma\tau\sigma$ (here $\gamma$ is applied to two type arguments). Now, both $\sigma$ and $\tau$ are valid types for $x$, since $\gamma$ can be instantiated by left and right projections $\lambda\alpha_1\alpha_2.\,\alpha_1$ and $\lambda\alpha_1\alpha_2.\,\alpha_2$.

The Turing Machine coding used in this construction is taken literally from [10], which in turn follows the ideas of [14], [18], [15]. However, we cannot use directly the *typing* described in these papers — we deal with a more delicate task.

In the following Section 2 we introduce the system $\mathbf{F}_\omega$ by Curry style type inference rules for pure lambda terms. We do not consider Church style (explicit typing) variant of the system at all. The reader wishing to study the relationship between the two systems is referred to [7].

Section 3 contains a hint for a proof of Theorem 3.1. Section 4 explains the Turing Machine coding and the basic strategy used to type machine configurations. Finally, Section 5 attempts to describe the typing of the iterator term $W$, to conclude a sketch of proof of our main Theorem 5.1. Unfortunately, we are unable to provide all details in a restricted-length paper. The interested reader may want to consult [22].

## 2 The system $\mathbf{F}_\omega$

Below we describe the inference rules of our system, and state some useful technical observations. We begin with the notion of a *kind*, the set of all kinds defined as the smallest set containing a constant **Prop** and such that whenever $\nabla_1$ and $\nabla_2$ are kinds then also $(\nabla_1 \Rightarrow \nabla_2)$ is a kind. The *ranks* of kinds are defined as follows: rank(**Prop**) is 0, and rank($\nabla_1 \Rightarrow \nabla_2$) is the maximum of rank($\nabla_1$) + 1 and rank($\nabla_2$).

The next step is to introduce the notion of a *constructor*. For each kind $\nabla$, we assume an infinite set of *constructor variables of kind* $\nabla$, denoted by $Var_\nabla$ (elements of $Var_{\mathbf{Prop}}$ are called *type variables*) and we define *constructors of kind* $\nabla$ as follows:

- A constructor variable of kind $\nabla$ is a constructor of kind $\nabla$;

- If $\varphi$ is a constructor of kind $\nabla_1 \Rightarrow \nabla_2$ and $\tau$ is a constructor of kind $\nabla_1$, then $(\varphi\tau)$ is a constructor of kind $\nabla_2$.

- If $\alpha$ is a constructor variable of kind $\nabla_1$ and $\tau$ is a constructor of kind $\nabla_2$, then $(\lambda\alpha.\,\tau)$ is a constructor of kind $(\nabla_1 \Rightarrow \nabla_2)$.

- If $\alpha$ is a constructor variable of arbitrary kind and $\tau$ is a constructor of kind **Prop**, then $(\forall \alpha\, \tau)$ is a constructor of kind **Prop**.

- If $\tau$ and $\sigma$ are constructors of kind **Prop**, then $(\tau \to \sigma)$ is a constructor of kind **Prop**.

We write $\tau : \nabla$ to denote that $\tau$ is a constructor of kind $\nabla$. Constructors of kind **Prop** are called *types*. We extend to kinds and constructors the standard notations and abbreviations used in lambda calculus.

If $\tau$ is a constructor, then the symbol $FV(\tau)$ denotes the set of all constructor variables free in $\tau$. The definition of $FV(\tau)$ is by induction: $FV(\alpha) = \{\alpha\}$, $FV(\varphi \tau) = FV(\varphi) \cup FV(\tau)$, $FV(\sigma \to \tau) = FV(\sigma) \cup FV(\tau)$, $FV(\forall \alpha\, \tau) = FV(\tau) - \{\alpha\}$, and $FV(\lambda \alpha. \tau) = FV(\tau) - \{\alpha\}$. The symbol $\tau\{\rho/\alpha\}$ denotes the effect of substituting $\rho$ for all free occurrences of $\alpha$ in $\tau$ (provided of course everything is correctly *kinded*), after an appropriate renaming of bound variables, if necessary. Note that "bound" means here either $\lambda$-bound or $\forall$-bound. The notation $\{\tau_1/\alpha_1, \ldots \tau_n/\alpha_n\}$ will be used for simultaneous substitutions. From now on we identify constructors which are the same up to a renaming of bound variables.

The beta reduction on constructors is defined as in the simply typed lambda calculus: a constructor of the form $(\lambda \alpha. \tau)\rho$ is called a *redex*, and we have the usual beta rule:

$$(\beta) \qquad (\lambda \alpha. \tau)\rho \longrightarrow \tau\{\rho/\alpha\}$$

The beta rule defines the one-step reduction relation $\longrightarrow$ and the many-step reduction relation $\longrightarrow\!\!\!\rightarrow$ in the usual way. It follows from the strong normalization theorem for simply typed lambda calculus that each constructor is strongly normalizable, see e.g. [4]. From now on, we identify constructors with their normal forms: below in rule (INST) we mean the normal form of $\sigma\{\rho/\alpha\}$, rather than $\sigma\{\rho/\alpha\}$ itself.

Our type inference system $\mathbf{F}_\omega$ assigns types to terms of untyped lambda calculus. A *type environment* is a set $E$ of *type assumptions*, each of the form $(x : \sigma)$, where $x$ is a variable and $\sigma$ is a type, such that if $(x : \sigma), (x : \sigma') \in E$ then $\sigma = \sigma'$. Thus, one can think of an environment as a finite partial function from variables into types. If $E$ is an environment, then $E(x : \sigma)$ is an environment such that

$$E(x : \sigma)(y) = \begin{cases} \sigma, & \text{if } x \equiv y; \\ E(y), & \text{if } x \not\equiv y. \end{cases}$$

Our system derives *type assignments* of the form $E \vdash M : \tau$, where $E$ is an environment, $M$ is a term and $\tau$ is a type. The system consists of the following rules

(VAR) $\qquad E \vdash x : \sigma \qquad\qquad$ if $(x : \sigma)$ is in $E$

(APP) $\qquad \dfrac{E \vdash M : \tau \to \sigma,\ E \vdash N : \tau}{E \vdash (MN) : \sigma}$

(ABS) $\qquad \dfrac{E(x : \tau) \vdash M : \sigma}{E \vdash (\lambda x.M) : \tau \to \sigma}$

(GEN)  $$\frac{E \vdash M : \sigma}{E \vdash M : \forall \alpha. \sigma}$$  where $\alpha \notin FV(E)$

(INST) $$\frac{E \vdash M : \forall \alpha \sigma}{E \vdash M : \sigma\{\rho/\alpha\}}$$

In rule (INST), $\alpha$ and $\rho$ must be of the same kind. Recall once more that an application of (INST) involves the normalization process for the constructor $\sigma\{\rho/\alpha\}$.

For a given $n \geq 0$ we consider a subsystem of the above type inference system, denoted by $\mathbf{F}_\omega^{(n)}$, defined by the additional restriction that all the kinds are of rank at most $n$. That is, the the second order polymorphic lambda calculus is $\mathbf{F}_\omega^{(0)}$. (We do not use the notation $F_n$, because it is ambiguous, as different authors use it in different sense: cf. [8], [5], [10], [21], [4].)

## 3 Strongly normalizable but untypable

Here we show that the type inference rules of $\mathbf{F}_\omega$ do not suffice to type all strongly normalizable terms. Our counter-example is the following term:

$$M \equiv (\lambda x.\, z(x1)(x1'))(\lambda y.\, yyy),$$

where $1 \equiv \lambda fu.\, fu$ and $1' \equiv \lambda vg.\, gv$. Clearly, $M$ is strongly normalizable, and it is an easy exercise to see that it becomes typable in $\mathbf{F}_\omega$ after just one reduction step.

**Theorem 3.1** *The above term $M$ cannot be typed in $\mathbf{F}_\omega$, and thus the class of typable terms is a proper subclass of the class of all strongly normalizable terms.*

**Proof:** (Hint) The technique used in the proof is based on the following observation: A type (i.e., a constuctor of kind **Prop**) can be seen in the form of a binary tree with nodes labelled with arrows (we ignore quantifiers). Each arrow labelled path in a type must end with a constructor variable. We say that the variable *owns* the path. Now, if a variable $\alpha$ owns a path $X$ in a type $\forall \beta_1 \ldots \beta_k \tau$, and $\alpha$ is not among $\beta_1 \ldots \beta_k \tau$ then $\alpha$ owns the path $X$ in each type obtained from $\forall \beta_1 \ldots \beta_k \tau$ by instantiations of $\beta$'s. The proof of our theorem requires an analysis of occurrences of quantifiers in an arbitrary typing of the term $M$. It can be shown that the (discarded) type of $x$ must be of the form $\forall \alpha_1 \ldots \alpha_n \sigma \to \rho$, and that one of the $\alpha$'s must own in $\sigma$ two paths, one of which is a prefix of the other — a contradiction. ∎

## 4 Coding of a Turing Machine

As we mentioned in the Introduction, the Turing Machine coding used here is borrowed from [10]. In what follows, a tuple of terms, say $\langle M_1, \ldots, M_k \rangle$ is identified with the term $\lambda x.\, xM_1 \ldots M_k$, where $x$ is a new variable (not free in $M_1, \ldots, M_k$). Thus, we have $\langle M_1, \ldots, M_k \rangle pr_i^k \longrightarrow M_i$, for all $i = 1, \ldots, k$, where $pr_i^k = \lambda z_1, \ldots, z_k.\, z_i$ is the $i$-th projection on $k$ arguments. An empty tuple, denoted *nil*, is defined as $\lambda x.\, x$.

Lists (denoted e.g. "$[a, b, \ldots, z]$") are defined so that $cons(a, L)$ is just $\langle a, L \rangle$. Thus, for instance, $[a, b, c]$ is $\lambda x.\, xa(\lambda y.\, yb(\lambda z.\, zc\, nil))$.

Elements of finite sets are represented by projections (in particular this applies to *true* and *false* represented by $pr_1^2$ and $pr_2^2$), and functions on finite sets are tuples of values: a function $f$, defined on a $k$-element set $\{a_1, \ldots, a_k\}$ is identified with the term $\langle f(a_1), \ldots, f(a_k) \rangle$, which reduces to $f(a_i)$, when applied to $pr_i^k$. As we agreed on identifying $a_i$ with the $i$-th projection, we may assume that the application $fa_i$ is a well-formed lambda term representing the function value. This applies as well to (curried) functions on 2 arguments: each $fa_i$ is a tuple of the form $\langle f(a_i, b_1), \ldots, f(a_i, b_\ell) \rangle$, where $b_1, \ldots, b_\ell$ are the possible values for the second argument.

This machinery provides means to represent computations of Turing Machines by lambda terms. Let us fix a deterministic Turing Machine, together with an initial configuration $C_0$, and assume that the machine converges on $C_0$ in exactly $M$ steps. Following [10], we assume w/o.l.o.g., that our machine satisfies certain conditions (some of them quite peculiar, but technically useful):

- There are $m$ internal states: $s_1, \ldots, s_m$, and $s_m$ is the only accepting state;
- There are $r$ tape symbols: $d_1, \ldots, d_r$, including the blank symbol $d_r$;
- The machine never attempts to move left, when reading the leftmost tape cell;
- It never writes a blank, and it does not move if *and only if* it reads a blank;
- The transition function $D$ is total, i.e., defined also for $s_m$.

With the ordinary conventions about the input, we may assume that the tape is always of the form $a_1 a_2 \ldots a_n$ blank blank$\ldots$, where all the $a_i$'s are different than blank. If the current state is $s_j$ and the tape head is positioned at the $k$-th cell (note that $k \leq n + 1$), then the resulting configuration is identified with the term

$$\langle s_j, [a_{k-1}, a_{k-2}, \ldots, a_1, \text{blank}], [a_k, \ldots, a_n, \text{blank}] \rangle.$$

Of course states and tape symbols are identified with the appropriate projections. The extra blank at end of the first list is added to ensure that the code of the left-hand part of the tape is always a pair, not *nil*. The transition function $D$ is coded as described above, so that, for $i = 1, \ldots, m$ and $j = 1, \ldots, r$, we have $Ds_i r_j = \langle p, c, v \rangle$, where $p$, $c$ and $v$ are (projections corresponding to) the new state, the symbol to be written, and the move to be performed, respectively. (A "move" is either $\mathit{left} = pr_1^3$ or $\mathit{stay} = pr_2^3$ or $\mathit{right} = pr_3^3$. The three possible moves will also be denoted by $mv_1$, $mv_2$ and $mv_3$.)

The "next ID" function *Next* is now defined as follows:

$$
\begin{aligned}
\mathit{Next} \equiv \lambda C.\ & \text{let } \langle q, \langle a, L \rangle, \langle b, R \rangle \rangle = C \\
\text{in } & \text{let } \langle p, c, v \rangle = Dqb \\
\text{in } & \text{let } F = \langle p, L, \langle a, \langle c, R \rangle \rangle \rangle \\
\text{in } & \text{let } S = \langle p, \langle a, L \rangle, \langle c, \langle \text{blank}, \mathit{nil} \rangle \rangle \rangle \\
\text{in } & \text{let } T = \langle p, \langle c, \langle a, L \rangle \rangle, R \rangle \\
\text{in } & vFST;
\end{aligned}
$$

The let's are abbreviations. For instance, the expression "let $\langle p, c, v \rangle = Dqb$ in ..." should read "$(\lambda pcv. \ldots)(Dqb\, pr_1^3)(Dqb\, pr_2^3)(Dqb\, pr_3^3)$". It is a routine, although time-consuming, exercise to check that our function $Next$ is correct. That is, if $C$ is (a code of) a legal ID then $Next\, C$ is beta equal to the (code of) the next ID.

We assumed before that our machine converges on $C_0$ in $M$ steps. To be more specific, let $C_0, \ldots, C_M$ be all the ID's occurring in the computation, with a final configuration $C_M$, and let $C_{M+1}$ be the "after final" configuration: the ID obtained by applying a machine step to $C_M$, which is possible by the assumptions above.

The techniques of [15], [10] allow us to derive a type for the function $Next$, so that each application $Next\, C_n$ becomes well-typed. Our typing follows these techniques in part, but it has to serve additional purposes in the consideration to follow, and thus it is a lot more complicated. This means that the construction we are about to describe involves a number of additional details, or seemingly unnecessary complications, which are entirely useless and annoying at the present stage, but which will show up to be essential later on. An example is as follows: one has to be careful if universal types, like $\forall \alpha \tau$, are to be put later into a larger context, like $\sigma \to \forall \alpha \tau$, with $\alpha$ not free in $\sigma$. The problem is that the latter type cannot be instantiated to e.g. $\sigma \to \tau\{\rho/\alpha\}$, so one would prefer to have $\forall \alpha(\sigma \to \tau)$ instead. This requires leaving certain types open during construction, and keeping track of their free variables to avoid name confusion.

As we make an extensive use of tuples, it is natural to introduce product notation of the form $\tau_1 \times \cdots \times \tau_\ell$ or $\Pi_{i=1,\ldots,\ell}\tau_i$ to abbreviate $\forall \alpha((\tau_1 \to \cdots \to \cdots \tau_\ell \to \alpha) \to \alpha)$ (with $\alpha$ not free in $\tau_1, \ldots \tau_\ell$). Clearly, a "product" type will typically be assigned to a tuple $\langle M_1, \ldots, M_\ell \rangle$, where $M_i : \tau_i$ holds for all $i = 1, \ldots, \ell$. In addition, we can derive $M\, pr_i^\ell : \tau_i$ whenever $M : \tau_1 \times \cdots \times \tau_\ell$. Unfortunately, for the reasons mentioned above, we must also consider also an "open" version of a product: if $\alpha$ is not free in $\tau_1, \ldots \tau_\ell$ then $\Pi_{i=1,\ldots,\ell}^\alpha \tau_i$ denotes $(\tau_1 \to \cdots \to \cdots \tau_\ell \to \alpha) \to \alpha$. The "open product" is as good as the ordinary one as long as we use a fresh variable $\alpha$ each time we form a new product type.

A remark about types assigned to projections and lists: it is natural to expect that $pr_i^k$ is assigned a type of the form $\tau_1 \to \cdots \to \tau_k \to \tau_i$. Such a type is called a *simple projection type*. (Of course $pr_i^k$ has also types which are not simple projection types.) Similarly, the most convenient type for a $n$-element list of elements of a $k$-element set has the form $\rho_1 \times (\rho_2 \times \cdots (\rho_n \times \mathbf{OK}) \cdots)$, where $\mathbf{OK} = \forall \alpha(\alpha \to \alpha)$ is a straightforward type for *nil*, and $\rho_1, \rho_2, \cdots, \rho_n$ are simple projection types (with $k$ arguments). Such types are called *simple list types*.

In order to type the function $Next$ we must be able to type the transition function $D$, i.e., we need types for all its components. Let $i \in \{1, \ldots, m\}$ and $j \in \{1, \ldots, r\}$, and let $Ds_id_j = \langle s_n, d_k, mv_\ell \rangle$, for some $n \in \{1, \ldots, m\}$, some $k \in \{1, \ldots, r\}$, and some $\ell \in \{1, 2, 3\}$. Clearly, we have e.g. $d_k : \forall \zeta_1 \ldots \zeta_r(\zeta_1 \to \cdots \zeta_r \to \zeta_k)$, but to avoid troubles we prefer to assign to $d_k$ an open type $\zeta_{j,1} \to \cdots \zeta_{j,r} \to \zeta_{j,k}$, where $\zeta_{j,1}, \ldots, \zeta_{j,r}$ is a fresh set of type variables to be generalized (and then instantiated) in a larger scope. Similarly, we use fresh type variables $\eta_{j,1}$, $\eta_{j,2}$ and $\eta_{j,3}$ to provide an open type for the move component. Unfortunately, an instantiation of a type of the form $\alpha_1 \to \cdots \to \alpha_r \to \alpha_n$ must be a simple projection type and cannot have e.g. the form $\tau_1 \to \cdots \to \forall \alpha \tau_n \to \cdots \to \tau_r \to \tau_n\{\sigma/\alpha\}$. The latter is however a correct type for the $n$-th projection. In what follows, we will need non-simple types for the first component of $Ds_id_j$. Since such types cannot be obtained by instantiation, we must provide them somehow: now we just assume that we are given a type $P_{i,j}$ such that $s_n : P_{i,j}$ holds (in the empty environment). The family of all types $P_{i,j}$ will be denoted $\vec{P}$.

Finally, let $[Ds_i d_j]$ denote the type $P_{i,j} \times (\zeta_{j,1} \to \cdots \zeta_{j,r} \to \zeta_{j,k}) \times (\eta_{j,1} \to \eta_{j,2} \to \eta_{j,3} \to \eta_{j,\ell})$. Under the above assumptions, the type assignment $Ds_i d_j : [Ds_i d_j]$ is true for each $i,j$ in the empty environment, but we must remember that $[Ds_i d_j]$ depends on the choice of $P_{i,j}$. If we want to stress this fact, we write $[Ds_i d_j](\vec{P})$ instead of $[Ds_i d_j]$.

The next step is to define a type for $Ds_i$. Of course one is tempted to take the product of all the $[Ds_i d_j]$'s. And this is right, except we take an open product, i.e., we define $[Ds_i] := \Pi_{j=1,\ldots,r}^{\delta} [Ds_i d_j]$, where $\delta$ is a fresh variable. Again, the type $[Ds_i]$ depends on the choice of the $P_{i,j}$'s, and we easily observe that $Ds_i : [Ds_i]$ holds in the empty environment.

The function *Next* will be assigned different types in different contexts, all these types however will obey a certain pattern:

(4.1) $$\text{Next} : \mathbf{Tape}(\vec{P}) \to \mathbf{out},$$

parameterized by a choice of certain types $P_{i,j}$, for $i = 1,\ldots,m$ and $j = 1,\ldots,r$; the list of all the $P_{i,j}$'s denoted by $\vec{P}$. The type $\mathbf{Tape}(\vec{P})$ has the form $\mathbf{State}(\vec{P}) \times (\mathbf{a} \times \mathbf{L}) \times (\mathbf{b} \times \mathbf{R})$. Here, $\mathbf{a}, \mathbf{L}, \mathbf{b}, \mathbf{R}$ and $\mathbf{out}$ are fixed type variables. The type $\mathbf{State}(\vec{P})$ has the following form:

$$\mathbf{State}(\vec{P}) = \forall \beta(\forall \cdots \forall (\beta([Ds_1])([f])) \to \cdots \to \forall \cdots \forall (\beta([Ds_{m-1}])([f])) \to$$
$$\to \forall \cdots \forall (\beta([Ds_m])([t])) \to \beta(\mathbf{b} \to \mathbf{Triple})(\mathbf{Quit} \to \xi \to \gamma).$$

Types $[Ds_i]$ are as defined above, but all the rest requires explanation. First, $\xi$ and $\gamma$ are new type variables and $\beta$ is a new constructor variable of kind $\mathbf{Prop} \Rightarrow \mathbf{Prop} \Rightarrow \mathbf{Prop}$. Let us recall that instantiating it with projections $\pi_L = \lambda \alpha_1 \alpha_2.\,\alpha_1$ and $\pi_R = \lambda \alpha_1 \alpha_2.\,\alpha_2$ replaces each subexpression of the form $\beta \tau \sigma$ by $\tau$ and $\sigma$ respectively. Thus, one can think of $\mathbf{State}(\vec{P})$ as of a conjunction of two types obtained by these instantiations.

We define $\mathbf{Quit} = \forall \alpha\,(\alpha \to \mathbf{OK})$. (Recall that $nil : \mathbf{OK} = \forall \alpha\,(\alpha \to \alpha)$, and note that $\lambda w.\,nil : \mathbf{Quit}$.) Types $[f]$ and $[t]$ are meant to be assigned to *false* and *true*, and are defined as follows: $[f] = \alpha_1 \to \alpha_2 \to \alpha_2$ is the first type for *false* that comes to mind, but $[t]$ is more complex: we take $[t] = \mathbf{Quit} \to \alpha_2 \to \alpha_3 \to \mathbf{OK}$. This is still a type for *true*, since $\alpha_3 \to \mathbf{OK}$ is an instance of $\mathbf{Quit}$. Here, $\alpha_1, \alpha_2$ and $\alpha_3$ are new type variables. Another type we need is $[\text{blank}] = \varepsilon_1 \to \cdots \to \varepsilon_r \to \varepsilon_r$, with fresh $\varepsilon_1,\ldots,\varepsilon_r$, and this is of course a type for blank.

The type $\mathbf{Triple}$ has the form $\mathbf{p} \times \mathbf{c} \times ([F] \to [S] \to [T] \to \mathbf{out})$, where $\mathbf{p}$ and $\mathbf{c}$ are type variables, and $[F]$, $[S]$ and $[T]$ are defined as follows:

- $[F] = \mathbf{p} \times \mathbf{L} \times (\mathbf{a} \times (\mathbf{c} \times \mathbf{R}))$;
- $[S] = \mathbf{p} \times (\mathbf{a} \times \mathbf{L}) \times (\mathbf{c} \times ([\text{blank}] \times \mathbf{OK}))$;
- $[T] = \mathbf{p} \times (\mathbf{c} \times (\mathbf{a} \times \mathbf{L})) \times \mathbf{R}$;

One can easily guess that these types are to be assigned to the let-bound $F$, $S$ and $T$ in the definition of *Next*.

The symbol "$\forall \cdots \forall$" denotes a sequence of quantifiers binding at least all the variables $\zeta_{j,k}$, $\eta_{j,k}$, $\varepsilon_k$, $\alpha_1$, $\alpha_2$, $\alpha_3$ and $\delta$. Thus the quantifiers "$\forall \cdots \forall$" bind all type variables occurring in their scope, except possibly those free in $\vec{P}$. In fact, "$\forall \cdots \forall$" will have to bind (for technical reasons) a lot of extra variables not occurring in their scope. To avoid extra complication we leave this part of our definition unfinished, to be completed later.

In order to verify the typing (4.1), observe that if $q : \textbf{State}(\vec{P})$ then also

$$q : (\forall \cdots \forall [Ds_1]) \to \cdots \to (\forall \cdots \forall [Ds_m]) \to \textbf{b} \to \textbf{Triple}$$

after instantiating the variable $\beta$ with the left projection $\pi_L = \lambda \alpha_1 \alpha_2 . \alpha_1$. The details are left to the reader (who can forget about $\beta$ and its second argument).

The next stage in the construction is to assign some particular types $[C_n]$, for $n = 0, \ldots, M + 1$, to the configurations $C_n$. Assume that, for all $n$, we have $C_n = \langle q^n, \langle a^n, L^n \rangle, \langle b^n, R^n \rangle \rangle$. As it can be easily guessed, each $[C_n]$ is a product of the form

$$(*) \qquad [C_n] = [q^n] \times ([a^n] \times [L^n]) \times ([b^n] \times [R^n]),$$

where the three components are to be assigned to the current state and to the left and right part of tape. The type $[C_{M+1}]$ is chosen as simple as possible: we only want $C_{M+1} : [C_{M+1}]$, and thus we just assign "$\forall \alpha. \alpha \to \cdots \to \forall \alpha. \alpha \to \forall \alpha. \alpha$" to all projections occurring in the term. Types $[C_n]$, for $n \leq M$, are defined by a downward induction on $n$. The type $[q^n]$, which is the crucial part, is obtained from $\textbf{State}(\vec{P^n})$, for suitable $\vec{P^n}$, by instantiating its free variables. For each $n \leq M$ we will have (for some fresh type variable $\xi^n$):

$$(**) \qquad [q^n] = \textbf{State}(\vec{P^n})\{[a^n]/\textbf{a}, [b^n]/\textbf{b}, [c^n]/\textbf{c}, [p^n]/\textbf{p}, [L^n]/\textbf{L}, [R^n]/\textbf{R},$$
$$\textbf{Out}_n/\textbf{out}, \xi^n/\xi, \tau^n/\gamma, \textbf{E}_j^n/\varepsilon_j\}_{j=1,\ldots,r}$$

The induction hypothesis for the construction consists of the following conditions:

i) $C_n : [C_n]$;

ii) all types $[a^n]$, $[b^n]$, and $[c^n]$ are simple projection types;

iii) types $[L^n]$ and $[R^n]$ are simple list types;

iv) $\tau^M = \kappa \to \textbf{OK}$, and $\tau^n = \xi^n$, for $n < M$, where $\kappa$ is a fresh type variable;

v) the only variables that may occur free in $[C_n]$ are $\kappa$ and $\xi^k$, for $k \leq n$.

The reader will easily observe that the condition (i) implies that e.g. $L^n : [L^n]$. Together with (iii), this guarantees that the components of types $L^n$ and $R^n$ correspond to appropriate tape symbols in $C_n$.

For the induction step, i.e., for the construction of $[C_n]$ from $[C_{n+1}]$, we need to define the types $\vec{P^n}$ and the types substituted for variables as in $(**)$. In particular, we will have $\textbf{Out}_n := [C_{n+1}]$ and $[p^n] := [q^{n+1}]$. Types $[a^n]$, $[c^n]$, $[L^n]$, $[R^n]$, and $\textbf{E}_j^n$ are defined by cases depending on the value of $v^n$ (left to the reader). It remains to define $[b^n]$ and $\vec{P}^n$, i.e., a family of types $P_{i,j}^n$, such that $D(s_i, d_j) = \langle p, c, v \rangle$ implies $p : P_{i,j}^n$, for all $i,j$. For this, suppose that $q^n = s_x$ and $b^n = d_y$. We define $P_{x,y}^n := [q^{n+1}]$ (this is the same as $[p^n]$), and $P_{i,j}^n = \forall \beta (\forall \cdots \forall (\forall \alpha. \alpha) \to \cdots \to \forall \cdots \forall (\forall \alpha. \alpha) \to \forall \cdots \forall (\forall \alpha. \alpha))$, for all other pairs $i,j$.

To define $[b^n]$, note that the subtype **Triple** of **State**($\vec{P}$) does not contain free b, and thus the type **Triple**$^n$ = **Triple**$\{[a^n]/a, \ldots\}$, obtained from **Triple** by the same substitutions that make $[q^n]$ from **State**($\vec{P}^n$), is already defined. For $j \neq y$, let $\rho_j$ denote the type $[Ds_x d_j](\vec{P}^n)\{\forall \alpha. \alpha/\zeta_{j,k}, \forall \alpha. \alpha/\eta_{j,\ell}\}_{k=1,\ldots,r;\ \ell=1,2,3}$, and let $\rho_y = $ **Triple**$^n$. We define $[b^n] = \rho_1 \to \cdots \to \rho_r \to \rho_y$.

The reader is invited to complete the construction so that the induction hypothesis holds for $[C_n]$. What is still left undefined is the exact meaning of "$\forall \cdots \forall$". In the above we needed only to know that "$\forall \cdots \forall$" includes quantifiers binding all the $\zeta$'s, $\eta$'s, $\varepsilon$'s, $\alpha'$s and the variable $\delta$. We used the same set of variables for each $n$. In the sequel, we will need to distinguish them: we will assume that a fresh set of variables consisting of $\zeta_{j,k}^n$, $\eta_{j,\ell}^n$, $\varepsilon_k^n$, $\alpha_k^n$, and $\delta^n$ is used for each $n$. Since these variables are bound anyway, this does not affect our consideration above. The sequence "$\forall \cdots \forall$" is now finally defined as consisting of quantifiers binding all these variables. This guarantees that the variables corresponding to different $n$ are distinguished, and at the same time the quantifier sequence is each time exactly the same.

# 5 Typing the computation

The precise definition of our term $W$, informally announced in the Introduction, is as follows:

$$W \equiv \lambda x.\, x pr_1^3 \underbrace{false \ldots false}_{m-1}\, true(\lambda w.\, nil)(\lambda w.\, w(Next\, x)w).$$

Here, the expression $x pr_1^3 false \ldots false\, true$ stands for the final state test, in the following sense: if $C$ is a triple with an $m$-argument projection as the first coordinate (in particular when $C$ is (a representation of) a machine ID), then $WC$ evaluates either to $\lambda w.\, nil$ or to $\lambda w.\, w(Next\, C)w$. If we now inspect the behaviour of the term

$$T \equiv (\lambda w.\, wC_0 w)W,$$

we find the following sequence of reductions (we use primes and indices to mark different copies of $W$ that appear in this sequence):

$$(\lambda w.\, wC_0 w)W_0 \twoheadrightarrow W_0' C_0 W_1 \twoheadrightarrow W_1' C_1 W_2 \twoheadrightarrow \cdots$$
$$\cdots \twoheadrightarrow W_{M-1}' C_{M-1} W_M \twoheadrightarrow W_M' C_M W_{M+1} \twoheadrightarrow nil$$

There are two kinds of copies of $W$ here: the "final" (primed) copies, and the "general" (non-primed) copies $W_n$, which are still to split, according to the following scheme:

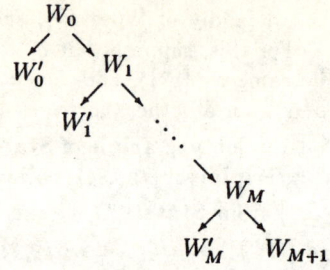

Our goal is to construct a type for $W_0$. For this, we define closed types $[W_n]$ for all the $W_n$'s ($n = 0, \ldots, M+1$), and closed types $[W'_n]$ to be assigned to the $W'_n$'s ($n = 0, \ldots, M$). The definition goes by a backward induction, starting from $n = M$, and ending with $n = 0$. (Type $[W_{M+1}]$ will be the same as $[W_M]$.)

Let $\vec{\gamma} = \gamma_0, \ldots, \gamma_{M-1}$ be a sequence of new constructor variables, each of kind **Prop** $\Rightarrow$ **Prop** $\Rightarrow$ **Prop**, and let $\gamma$ be a type variable. The overall shape of types we shall assign to the $W_n$'s will be:

$$[W_n] = \forall \gamma \vec{\gamma} \, ([x_n] \to \gamma),$$

for some $[x_n]$, to be defined below. The whole sequence $\vec{\gamma}$ is universally quantified here for (necessary) uniformity, but only $\gamma_n, \ldots \gamma_{M-1}$ and $\gamma$ will occur free in $[x_n]$. The types assigned to the "final" copies of $W$ will be of the form:

$$[W'_n] = [x'_n] \to \Delta_n, \text{ for } n < M;$$
$$[W'_M] = [x'_M] \to [W_M] \to \mathbf{OK};$$

where $[x'_n] = [C_n]\{[W_M]/\kappa\}\{\Delta_k/\xi^k\}_{k=n,\ldots,M}$, and the (closed) types $\Delta_k$ are still to be defined. Note that we have $C_n : [x'_n]$ in any case.

The above picture tells us about the structure of instantiations that one must be able to perform on our types. We need $[x_n]$ to be defined so that it can be turned into $[x'_n]$ and $[x_{n+1}]$ by appropriate instantiations of $\gamma$ and $\vec{\gamma}$. More specifically, we will have:

$$[x_n]\{\mathcal{L}_n\} = [x'_n] \quad \text{and} \quad [x_n]\{\mathcal{R}_n\} = [x_{n+1}],$$

where $\{\mathcal{L}_n\}$ (for $n < M$) abbreviates the substitution $\{\pi_L/\gamma_n, \Delta_n/\gamma\}$, and $\{\mathcal{R}_n\}$ stands for $\{\pi_R/\gamma_n\}$. An exception applies for $n = M$, when we define $\{\mathcal{L}_M\}$ to be $\{[W_M] \to \mathbf{OK}/\gamma\}$, and $\{\mathcal{R}_M\}$ to be identity. Note that we use the constructor variables here again as "switches" to choose between two cases. If we construct $[x_n]$ so that the above equations are satisfied, we will be able to obtain $[W'_n]$ as an instance of $[W_n]$, and to obtain $[W_{n+1}]$ from $[W_n]$ by instantiations and generalizations.

Each of the $[x'_n]$'s will be an instance of $\mathbf{Tape}(\vec{P}^n)$, for appropriate $\vec{P}^n$ (because $[x'_n] = [C_n]\{[W_M]/\kappa\}\{\Delta_k/\xi^k\}_{k=n,\ldots,M}$). More precisely, for $[a^n]$, $[b^n]$, etc. defined as in Section 4:

$$[x'_n] = \mathbf{Tape}(\vec{P}^n)\{[a^n]/\mathbf{a}, \, [b^n]/\mathbf{b}, \, [c^n]/\mathbf{c}, \, [p^n]/\mathbf{p}, \, [L^n]/\mathbf{L}, \, [R^n]/\mathbf{R},$$
$$\mathbf{Out}_n/\mathbf{out}, \, \xi^n/\xi, \, \tau^n/\gamma, \, \mathbf{E}^n_j/\varepsilon_j\}_{j=1,\ldots,r}\{[W_M]/\kappa\}\{\Delta_k/\xi^k\}_{k \geq n}$$

Types $[x_n]$ are obtained in a similar way as instances of some $\mathbf{Tape}(\vec{\mathbf{P}}^{*n})$:

$$[x_n] = \mathbf{Tape}(\vec{\mathbf{P}}^{*n})\{[a^n]^*/\mathbf{a},\ [b^n]^*/\mathbf{b},\ [c^n]^*/\mathbf{c},\ [p^n]^*/\mathbf{p},\ [L^n]^*/\mathbf{L},\ [R^n]^*/\mathbf{R},$$
$$\mathbf{Out}_n^*/\mathbf{out},\ \xi^n/\xi,\ \mathbf{E}_j^{*n}/\varepsilon_j\}_{j=1,\ldots,r}\{\Delta_k^*/\xi^k\}_{k\geq n}$$

Let us begin our inductive definition of the $[W_n]$'s, and $[W_n']$'s. Since we take $[W_{M+1}] = [W_M]$, we may start with $n = M$. For this put:

$$\Delta_M = \forall \gamma \vec{\vec{\gamma}} \left([C_{M+1}] \to \gamma\right) \to \mathbf{OK}.$$

To define $[x_M]$, we just ignore the stars in the general description of $[x_n]$ above: we take $\vec{\mathbf{P}}^{*M} = \vec{\mathbf{P}}^M$, $[a^M]^* = [a^M]$, etc. so that $[x_M']$ becomes equal to the type $[x_M]\{[W_M] \to \mathbf{OK}/\gamma\}$.

For the induction step we need more notation. Let $\vec{\vec{\gamma}}$ be a family of new constructor variables $\gamma_{n,k}$, for $n, k = 1, \ldots, M-1$, each of kind $\mathbf{Prop} \Rightarrow \mathbf{Prop} \Rightarrow \mathbf{Prop}$. For each $n$, by $[x_n]_n$ we denote the type $[x_n]\{\gamma_{n,k}/\gamma_k\}_{k\geq n}$ — a copy of $[x_n]$ using its own fresh variables for the "switches".

Let now $n < M$. We define:

$$\Delta_n = \forall \gamma \vec{\vec{\gamma}} \left([x_{n+1}]_{n+1} \to \gamma\right) \to \mathbf{OK},$$

which is *almost* the same as $[W_{n+1}] \to \mathbf{OK}$. Now it is easy to see that $W_n' C_n W_{n+1} : \mathbf{OK}$, provided we indeed have $W_{n+1} : [W_{n+1}]$ and $W_n' : [W_n']$.

To construct $[x_n]$, we must provide all the "starred" components $[a^n]^*$, $[b^n]^*$, etc. so that applying the substitutions $\{\mathcal{L}_n\}$ and $\{\mathcal{R}_n\}$ turns them into the appropriate components of $[x_n']$ and $[x_{n+1}]$ respectively.

We begin with the definition of $[a^n]^* := \gamma_n[a^n][a^{n+1}]^*$. It should be obvious that substituting the left projection for $\gamma_n$ provides $[a^n]$ — the appropriate part of $[C_n]$ and $[x_n']$, while the right projection chooses $[a^{n+1}]^*$ — the appropriate fragment of $[x_{n+1}]$ (type $[a^{n+1}]^*$ has no free $\gamma_n$, as a subtype of $[x_{n+1}]$). If we "unfold" our $[a^n]^*$, we perhaps can see better the basic idea:

$$[a^n]^* := \gamma_n[a^n](\gamma_{n+1}[a^{n+1}](\cdots(\gamma_{M-1}[a^{M-1}][a^M])\cdots)).$$

This type is best presented as the following "ladder", that resembles very much the relationships between all the copies of $W$:

```
            γn
           /  \
        [aⁿ]   γn+1
              /    \
          [aⁿ⁺¹]    ⋱
                     \
                    γM-1
                    /   \
                [aᴹ⁻¹] [aᴹ]
```

Types $[b^n]^*$, $[c^n]^*$, $[p^n]^*$, $[L^n]^*$, $[R^n]^*$ and $\mathbf{E}^{*n}_j$ are defined in an analogous way. Thus, such "ladders" will occur in $[x_n]$ whenever free variables occurred in **Tape**, with the exception of **out**, $\xi$ and $\gamma$ (the latter left free). Unfortunately, for **out** and $\xi$ we need a special treatment. For $\xi$, this is because we still need to have $W : [x_n] \to \gamma$, and thus the type $\Delta^*_n$ substituted for $\xi$ cannot be just a ladder: it must be a legal type for the subterm $\lambda w.\, w(Next\ x)w$. We take:

$$\Delta^*_n = \forall \gamma \vec{\gamma}\, (\sigma_n \to \gamma) \to \mathbf{OK},$$

where $\sigma$ is a ladder of $[x_{k+1}]_{k+1}$'s, i.e.,

$$\sigma_n = \gamma_n [x_{n+1}]_{n+1} (\gamma_{n+1} [x_{n+2}]_{n+2} (\cdots (\gamma_{M-1} [x_M]_M [C_{M+1}]) \cdots )).$$

Now, the reader can easily see that what we get from $\Delta^*_n$ by applying $\{\mathcal{L}_n\}$ and $\{\mathcal{R}_n\}$ is exactly $\Delta_n$ and $\Delta^*_{n+1}$, as required (remember that $\gamma$ is bound).

The definition of $\mathbf{Out}^*_n$ is even more unpleasant, and requires more notation. First, let $\{\mathcal{L}^*\}$ denote the substitution $\{\pi_L/\gamma_{k,k}, \gamma_{n+1,k}/\gamma_{n,k}\}_{0 \leq k,n \leq M-1, n \neq k}$, and let $\{\mathcal{R}^*\}$ be the substitution $\{\pi_R/\gamma_{k,k}, \gamma_{n+1,k}/\gamma_{n,k}\}_{0 \leq k,n \leq M-1, n \neq k}$. We use the abbreviations $\sigma^R_n = \sigma_n\{\mathcal{R}^*\}$ and $\sigma^L_n = \sigma_n\{\mathcal{L}^*\}$, where $\sigma_n$ is as above. We define

$$\mathbf{Out}^*_n = \sigma^L_n \{ \forall \gamma \vec{\gamma}\, (\sigma^R_n \to \gamma) \to \mathbf{OK}/\gamma \}.$$

To complete the definition of $[x_n]$, we must define a family of types $\vec{\mathbf{P}}^{*n}$, so that $\vec{\mathbf{P}}^{*n}\{\mathcal{L}_n\} = \vec{\mathbf{P}}^n\{[W_M]/\kappa\}\{\Delta^*_k/\xi^k\}_{k \geq n}$, and $\vec{\mathbf{P}}^{*n}\{\mathcal{R}_n\} = \vec{\mathbf{P}}^{*n+1}\{\Delta^*_k/\xi^k\}_{k \geq n}$. In addition, $\vec{\mathbf{P}}^{*n}$ must be types for appropriate machine states. Fortunately, if we fix any particular component, say $P^n_{i,j}$, of $\vec{\mathbf{P}}^n$, then for each $n$ it is always the same state, say $s_u$ (determined by $D(s_i, d_j)$). For $\ell \geq n$, let $\vec{\mathbf{Q}}^\ell$ abbreviate $\vec{\mathbf{P}}^\ell \{[W_M]/\kappa\}\{\Delta^*_k/\xi^k\}_{k \geq \ell}$. Then $\vec{\mathbf{Q}}^\ell$ is a family of closed types, and in addition we may assume that each $Q^\ell_{i,j}$ has the form (compare this to the definition of **State**):

$$\forall \beta (\forall \cdots \forall \tau^\ell_1 \to \cdots \to \forall \cdots \forall \tau^\ell_m \to \tau^\ell_u \{\mathcal{J}^\ell\}),$$

where $\mathcal{J}_\ell$ is some substitution. The quantifier prefix $\forall \cdots \forall$ is identical for each $\ell$, and it binds all variables occurring in its scope. The point is that the sets of variables used for each $\ell$ are disjoint, and we may define an appropriate component of $\vec{\mathbf{P}}^{*n}$ as follows:

$$P^{*n}_{i,j} := \forall \beta (\forall \cdots \forall \tau^*_1 \to \cdots \to \forall \cdots \forall \tau^*_m \to \tau^*_u \{\mathcal{J}^*\}),$$

where $\{\mathcal{J}^*\}$ is just the union of all the $\{\mathcal{J}^\ell\}$, and each $\tau^*_k$ has the form of a ladder:

$$\tau^*_k = \gamma_n \tau^n_k (\gamma_{n+1} \tau^{n+1}_k (\cdots (\gamma_{M-1} \tau^{M-1}_k \tau^M_k) \cdots )).$$

The effect of this construction is that applying the substitutions $\{\mathcal{L}_n\}$ and $\{\mathcal{R}_n\}$, we obtain $P^{*n}_{i,j}\{\mathcal{L}_n\} = Q^n_{i,j}$ and $P^{*n}_{i,j}\{\mathcal{R}_n\} = P^{*n+1}_{i,j}$, for each $i,j$. This is easily verified, using the analogous statements about $\Delta^*_n$.

The proof of the correctness of our construction is of course left to the reader. The statement to be proved for each $n = 0, \ldots, M$ is the conjunction of the following conditions:

i) Types $[x'_n]$ and $\Delta_n$ are closed, and $FV([x_n]) = \{\gamma, \gamma_n, \ldots, \gamma_{M-1}\}$;
ii) $[x_n]\{\mathcal{L}_n\} = [x'_n]$ and $[x_n]\{\mathcal{R}_n\} = [x_{n+1}]$;
iv) $W : [W_n]$ and $W : [W'_n]$;
v) The assumptions $w : [W_n]$ and $c : [x'_n]$ imply $wcw : \mathbf{OK}$.

**Theorem 5.1** *The type reconstruction problem for* $\mathbf{F}_\omega$ *is undecidable, as well as for each subsystem* $\mathbf{F}_\omega^{(n)}$, *for* $n > 0$.

**Proof:** The term $T \equiv (\lambda w. wC_0 w)W$ has been effectively obtained from a given Turing Machine and its initial configuration. If the machine converges, $T$ is typable. Indeed, we have $W : [W_0]$ and $C_0 : [x'_0]$, and we can apply the condition (vii) above to derive $T : \mathbf{OK}$. Otherwise $T$ has an infinite reduction sequence and therefore must be untypable. This means we have reduced the halting problem to the type reconstruction problem. In addition, the whole typing works in $\mathbf{F}_\omega^{(1)}$. ■

# References

[1] Barendregt, H.P., Hemerik, K., Types in lambda calculi and programming languages, *Proc. ESOP'90*, pp. 1–35.

[2] Boehm, H.J., Partial polymorphic type inference is undecidable, *Proc. 26th FOCS*, 1985, pp. 339–345.

[3] Coquand, T., Metamathematical investigations of a calculus of constructions, in: *Logic in Computer Science* (Odifreddi, ed.), Acad. Press, 1990, pp. 91–122.

[4] Gallier, J. H., On Girard's "Candidats de Reductibilité", in: *Logic in Computer Science*, (Odifreddi, ed.), Acad. Press, 1990, pp. 123–203.

[5] Giannini, P., Ronchi Della Rocca, S., Characterization of typings in polymorphic type discipline, *Proc. 3rd LICS*, 1988, pp. 61–70.

[6] Giannini, P., Ronchi Della Rocca, S., Type inference in polymorphic type discipline, *Proc. Theoretical Aspects of Computer Software* (Ito, Meyer, eds.), LNCS 526, Springer, Berlin, 1991, pp. 18–37.

[7] Giannini, P., Ronchi Della Rocca, S., Honsell, F., Type inference: some results, some problems, manuscript, 1992, to appear in *Fundamenta Informaticae*.

[8] Girard, J.-Y., The system **F** of variable types fifteen years later *Theoret. Comput. Sci.*, **45** (1986), 159–192.

[9] Henglein, F., A lower bound for full polymorphic type inference: Girard/Reynolds typability is DEXPTIME-hard, Technical Report RUU-CS-90-14, University of Utrecht, April 1990.

[10] Henglein, F., Mairson, H.G., The complexity of type inference for higher-order typed lambda calculi, *Proc. 18th POPL*, 1991, pp. 119–130.

[11] Kfoury, A.J. and Tiuryn, J., Type reconstruction in finite-rank fragments of the second-order $\lambda$-calculus, *Information and Computation*, vol. 98, No. 2, 1992, 228–257.

[12] Kfoury, A.J., Tiuryn, J. and Urzyczyn, P., An analysis of **ML** typability, to appear in Journal of the ACM (preliminary version in Proc. CAAP/ESOP'90).

[13] Kfoury, A.J., Tiuryn, J. and Urzyczyn, P., The undecidability of the semi-unification problem, to appear in Information and Computation (preliminary version in *Proc. STOC'90*).

[14] Kanellakis, P.C., Mitchell,J.C., Polymorphic unification and ML typing, *Proc. 16th POPL*, 1989, pp. 105–115.

[15] Kanellakis, P.C., Mairson H.G.; Mitchell J.C., Unification and ML type reconstruction, Chapter 13 in *Computational Logic, Essays in Honor of Alan J. Robinson*, (J-L. Lassez and G. Plotkin eds), MIT Press, 1991.

[16] Krivine, J.L., Un algorithme non typable dans le système **F**, *F.C.R. Acad. Sci. Paris*, Série I, **304** (1987), 123-126.

[17] Leivant, D., Polymorphic type inference, *Proc. 10th POPL*, 1983, pp. 88–98.

[18] Mairson, H.G., Deciding ML typability is complete for deterministic exponential time, *Proc. 17th POPL*, 1990, pp. 382–401.

[19] Malecki, S., Generic terms having no polymorphic types, *Proc. 17th ICALP*, LNCS 443, Springer, Berlin, 1990, pp. 46–59.

[20] Pfenning, F., Partial polymorphic type inference and higher-order unification, *Proc. Lisp and Functional Programming*, 1988, pp. 153–163.

[21] Pierce, B., Dietzen, S., Michaylov, S., Programming in higher-order typed lambda calculi, Research Report CMU-CS-89-111, Carnegie-Mellon University, 1989.

[22] Urzyczyn, P., Type reconstruction in $F_\omega$, Research Report BU-CS-92-014, Boston University, 1992.

# Author Index

M. Abadi ... 361
Y. Akama ... 1
Th. Altenkirch ... 13
S. van Bakel ... 29
F. Barbanera ... 45,60
N. Benton ... 75
S. Berardi ... 45
U. Berger ... 91
G. Bierman ... 75
G. Castagna ... 107
P. Di Gianantonio ... 124
G. Dowek ... 139
M. Fernández ... 60
G. Ghelli ... 107,146
Ph. de Groote ... 163
F. Honsell ... 124
M. Hyland ... 75,179
B. Jacobs ... 195,209
A. Jung ... 230,245
H. Leiß ... 258
D. Leivant ... 274
G. Longo ... 107
J.-Y. Marion ... 274
J. McKinna ... 289
T. Melham ... 209
T. Nipkow ... 306
C.-H.L. Ong ... 179
D.F. Otth ... 318
V. de Paiva ... 75
Chr. Paulin-Mohring ... 328
B.C. Pierce ... 346
G. Plotkin ... 361
R. Pollack ... 289
K. Sieber ... 376
J. Springintveld ... 391
A. Stoughton ... 230
M. Takahashi ... 406
J. Tiuryn ... 245
P. Urzyczyn ... 418

# Springer-Verlag
# and the Environment

We at Springer-Verlag firmly believe that an international science publisher has a special obligation to the environment, and our corporate policies consistently reflect this conviction.

We also expect our business partners – paper mills, printers, packaging manufacturers, etc. – to commit themselves to using environmentally friendly materials and production processes.

The paper in this book is made from low- or no-chlorine pulp and is acid free, in conformance with international standards for paper permanency.

# Lecture Notes in Computer Science

For information about Vols. 1–587
please contact your bookseller or Springer-Verlag

Vol. 588: G. Sandini (Ed.), Computer Vision – ECCV '92. Proceedings. XV, 909 pages. 1992.

Vol. 589: U. Banerjee, D. Gelernter, A. Nicolau, D. Padua (Eds.), Languages and Compilers for Parallel Computing. Proceedings, 1991. IX, 419 pages. 1992.

Vol. 590: B. Fronhöfer, G. Wrightson (Eds.), Parallelization in Inference Systems. Proceedings, 1990. VIII, 372 pages. 1992. (Subseries LNAI).

Vol. 591: H. P. Zima (Ed.), Parallel Computation. Proceedings, 1991. IX, 451 pages. 1992.

Vol. 592: A. Voronkov (Ed.), Logic Programming. Proceedings, 1991. IX, 514 pages. 1992. (Subseries LNAI).

Vol. 593: P. Loucopoulos (Ed.), Advanced Information Systems Engineering. Proceedings. XI, 650 pages. 1992.

Vol. 594: B. Monien, Th. Ottmann (Eds.), Data Structures and Efficient Algorithms. VIII, 389 pages. 1992.

Vol. 595: M. Levene, The Nested Universal Relation Database Model. X, 177 pages. 1992.

Vol. 596: L.-H. Eriksson, L. Hallnäs, P. Schroeder-Heister (Eds.), Extensions of Logic Programming. Proceedings, 1991. VII, 369 pages. 1992. (Subseries LNAI).

Vol. 597: H. W. Guesgen, J. Hertzberg, A Perspective of Constraint-Based Reasoning. VIII, 123 pages. 1992. (Subseries LNAI).

Vol. 598: S. Brookes, M. Main, A. Melton, M. Mislove, D. Schmidt (Eds.), Mathematical Foundations of Programming Semantics. Proceedings, 1991. VIII, 506 pages. 1992.

Vol. 599: Th. Wetter, K.-D. Althoff, J. Boose, B. R. Gaines, M. Linster, F. Schmalhofer (Eds.), Current Developments in Knowledge Acquisition - EKAW '92. Proceedings. XIII, 444 pages. 1992. (Subseries LNAI).

Vol. 600: J. W. de Bakker, C. Huizing, W. P. de Roever, G. Rozenberg (Eds.), Real-Time: Theory in Practice. Proceedings, 1991. VIII, 723 pages. 1992.

Vol. 601: D. Dolev, Z. Galil, M. Rodeh (Eds.), Theory of Computing and Systems. Proceedings, 1992. VIII, 220 pages. 1992.

Vol. 602: I. Tomek (Ed.), Computer Assisted Learning. Proceedings, 1992. X, 615 pages. 1992.

Vol. 603: J. van Katwijk (Ed.), Ada: Moving Towards 2000. Proceedings, 1992. VIII, 324 pages. 1992.

Vol. 604: F. Belli, F.-J. Radermacher (Eds.), Industrial and Engineering Applications of Artificial Intelligence and Expert Systems. Proceedings, 1992. XV, 702 pages. 1992. (Subseries LNAI).

Vol. 605: D. Etiemble, J.-C. Syre (Eds.), PARLE '92. Parallel Architectures and Languages Europe. Proceedings. 1992. XVII, 984 pages. 1992.

Vol. 606: D. E. Knuth, Axioms and Hulls. IX, 109 pages. 1992.

Vol. 607: D. Kapur (Ed.), Automated Deduction – CADE-11. Proceedings, 1992. XV, 793 pages. 1992. (Subseries LNAI).

Vol. 608: C. Frasson, G. Gauthier, G. I. McCalla (Eds.), Intelligent Tutoring Systems. Proceedings, 1992. XIV, 686 pages. 1992.

Vol. 609: G. Rozenberg (Ed.), Advances in Petri Nets 1992. VIII, 472 pages. 1992.

Vol. 610: F. von Martial, Coordinating Plans of Autonomous Agents. XII, 246 pages. 1992. (Subseries LNAI).

Vol. 611: M. P. Papazoglou, J. Zeleznikow (Eds.), The Next Generation of Information Systems: From Data to Knowledge. VIII, 310 pages. 1992. (Subseries LNAI).

Vol. 612: M. Tokoro, O. Nierstrasz, P. Wegner (Eds.), Object-Based Concurrent Computing. Proceedings, 1991. X, 265 pages. 1992.

Vol. 613: J. P. Myers, Jr., M. J. O'Donnell (Eds.), Constructivity in Computer Science. Proceedings, 1991. X, 247 pages. 1992.

Vol. 614: R. G. Herrtwich (Ed.), Network and Operating System Support for Digital Audio and Video. Proceedings, 1991. XII, 403 pages. 1992.

Vol. 615: O. Lehrmann Madsen (Ed.), ECOOP '92. European Conference on Object Oriented Programming. Proceedings. X, 426 pages. 1992.

Vol. 616: K. Jensen (Ed.), Application and Theory of Petri Nets 1992. Proceedings, 1992. VIII, 398 pages. 1992.

Vol. 617: V. Mařík, O. Štěpánková, R. Trappl (Eds.), Advanced Topics in Artificial Intelligence. Proceedings, 1992. IX, 484 pages. 1992. (Subseries LNAI).

Vol. 618: P. M. D. Gray, R. J. Lucas (Eds.), Advanced Database Systems. Proceedings, 1992. X, 260 pages. 1992.

Vol. 619: D. Pearce, H. Wansing (Eds.), Nonclassical Logics and Information Proceedings. Proceedings, 1990. VII, 171 pages. 1992. (Subseries LNAI).

Vol. 620: A. Nerode, M. Taitslin (Eds.), Logical Foundations of Computer Science – Tver '92. Proceedings. IX, 514 pages. 1992.

Vol. 621: O. Nurmi, E. Ukkonen (Eds.), Algorithm Theory – SWAT '92. Proceedings. VIII, 434 pages. 1992.

Vol. 622: F. Schmalhofer, G. Strube, Th. Wetter (Eds.), Contemporary Knowledge Engineering and Cognition. Proceedings, 1991. XII, 258 pages. 1992. (Subseries LNAI).

Vol. 623: W. Kuich (Ed.), Automata, Languages and Programming. Proceedings, 1992. XII, 721 pages. 1992.

Vol. 624: A. Voronkov (Ed.), Logic Programming and Automated Reasoning. Proceedings, 1992. XIV, 509 pages. 1992. (Subseries LNAI).

Vol. 625: W. Vogler, Modular Construction and Partial Order Semantics of Petri Nets. IX, 252 pages. 1992.

Vol. 626: E. Börger, G. Jäger, H. Kleine Büning, M. M. Richter (Eds.), Computer Science Logic. Proceedings, 1991. VIII, 428 pages. 1992.

Vol. 628: G. Vosselman, Relational Matching. IX, 190 pages. 1992.

Vol. 629: I. M. Havel, V. Koubek (Eds.), Mathematical Foundations of Computer Science 1992. Proceedings. IX, 521 pages. 1992.

Vol. 630: W. R. Cleaveland (Ed.), CONCUR '92. Proceedings. X, 580 pages. 1992.

Vol. 631: M. Bruynooghe, M. Wirsing (Eds.), Programming Language Implementation and Logic Programming. Proceedings, 1992. XI, 492 pages. 1992.

Vol. 632: H. Kirchner, G. Levi (Eds.), Algebraic and Logic Programming. Proceedings, 1992. IX, 457 pages. 1992.

Vol. 633: D. Pearce, G. Wagner (Eds.), Logics in AI. Proceedings. VIII, 410 pages. 1992. (Subseries LNAI).

Vol. 634: L. Bougé, M. Cosnard, Y. Robert, D. Trystram (Eds.), Parallel Processing: CONPAR 92 – VAPP V. Proceedings. XVII, 853 pages. 1992.

Vol. 635: J. C. Derniame (Ed.), Software Process Technology. Proceedings, 1992. VIII, 253 pages. 1992.

Vol. 636: G. Comyn, N. E. Fuchs, M. J. Ratcliffe (Eds.), Logic Programming in Action. Proceedings, 1992. X, 324 pages. 1992. (Subseries LNAI).

Vol. 637: Y. Bekkers, J. Cohen (Eds.), Memory Management. Proceedings, 1992. XI, 525 pages. 1992.

Vol. 639: A. U. Frank, I. Campari, U. Formentini (Eds.), Theories and Methods of Spatio-Temporal Reasoning in Geographic Space. Proceedings, 1992. XI, 431 pages. 1992.

Vol. 640: C. Sledge (Ed.), Software Engineering Education. Proceedings, 1992. X, 451 pages. 1992.

Vol. 641: U. Kastens, P. Pfahler (Eds.), Compiler Construction. Proceedings, 1992. VIII, 320 pages. 1992.

Vol. 642: K. P. Jantke (Ed.), Analogical and Inductive Inference. Proceedings, 1992. VIII, 319 pages. 1992. (Subseries LNAI).

Vol. 643: A. Habel, Hyperedge Replacement: Grammars and Languages. X, 214 pages. 1992.

Vol. 644: A. Apostolico, M. Crochemore, Z. Galil, U. Manber (Eds.), Combinatorial Pattern Matching. Proceedings, 1992. X, 287 pages. 1992.

Vol. 645: G. Pernul, A M. Tjoa (Eds.), Entity-Relationship Approach – ER '92. Proceedings, 1992. XI, 439 pages, 1992.

Vol. 646: J. Biskup, R. Hull (Eds.), Database Theory – ICDT '92. Proceedings, 1992. IX, 449 pages. 1992.

Vol. 647: A. Segall, S. Zaks (Eds.), Distributed Algorithms. X, 380 pages. 1992.

Vol. 648: Y. Deswarte, G. Eizenberg, J.-J. Quisquater (Eds.), Computer Security – ESORICS 92. Proceedings. XI, 451 pages. 1992.

Vol. 649: A. Pettorossi (Ed.), Meta-Programming in Logic. Proceedings, 1992. XII, 535 pages. 1992.

Vol. 650: T. Ibaraki, Y. Inagaki, K. Iwama, T. Nishizeki, M. Yamashita (Eds.), Algorithms and Computation. Proceedings, 1992. XI, 510 pages. 1992.

Vol. 651: R. Koymans, Specifying Message Passing and Time-Critical Systems with Temporal Logic. IX, 164 pages. 1992.

Vol. 652: R. Shyamasundar (Ed.), Foundations of Software Technology and Theoretical Computer Science. Proceedings, 1992. XIII, 405 pages. 1992.

Vol. 653: A. Bensoussan, J.-P. Verjus (Eds.), Future Tendencies in Computer Science, Control and Applied Mathematics. Proceedings, 1992. XV, 371 pages. 1992.

Vol. 654: A. Nakamura, M. Nivat, A. Saoudi, P. S. P. Wang, K. Inoue (Eds.), Prallel Image Analysis. Proceedings, 1992. VIII, 312 pages. 1992.

Vol. 655: M. Bidoit, C. Choppy (Eds.), Recent Trends in Data Type Specification. X, 344 pages. 1993.

Vol. 656: M. Rusinowitch, J. L. Rémy (Eds.), Conditional Term Rewriting Systems. Proceedings, 1992. XI, 501 pages. 1993.

Vol. 657: E. W. Mayr (Ed.), Graph-Theoretic Concepts in Computer Science. Proceedings, 1992. VIII, 350 pages. 1993.

Vol. 658: R. A. Rueppel (Ed.), Advances in Cryptology – EUROCRYPT '92. Proceedings, 1992. X, 493 pages. 1993.

Vol. 659: G. Brewka, K. P. Jantke, P. H. Schmitt (Eds.), Nonmonotonic and Inductive Logic. Proceedings, 1991. VIII, 332 pages. 1993. (Subseries LNAI).

Vol. 660: E. Lamma, P. Mello (Eds.), Extensions of Logic Programming. Proceedings, 1992. VIII, 417 pages. 1993. (Subseries LNAI).

Vol. 661: S. J. Hanson, W. Remmele, R. L. Rivest (Eds.), Machine Learning: From Theory to Applications. VIII, 271 pages. 1993.

Vol. 662: M. Nitzberg, D. Mumford, T. Shiota, Filtering, Segmentation and Depth. VIII, 143 pages. 1993.

Vol. 663: G. v. Bochmann, D. K. Probst (Eds.), Computer Aided Verification. Proceedings, 1992. IX, 422 pages. 1993.

Vol. 664: M. Bezem, J. F. Groote (Eds.), Typed Lambda Calculi and Applications. Proceedings, 1993. VIII, 433 pages. 1993.